T0418828

Metal Oxide-Carbon Hybrid Materials

The Metal Oxides Book Series Edited by Ghenadii Korotcenkov

Forthcoming titles

- Metal Oxides for Non-volatile Memory, Panagiotis Dimitrakis, Ilia Valov, Stefan Tappertzhofen, 9780128146293
- Metal Oxide Nanostructured Phosphors, H. Nagabhushana, Daruka Prasad, S.C. Sharma, 9780128118528
- Multifunctional Piezoelectric Oxide Nanostructures, Sang-Jae Kim, Nagamalleswara Rao Alluri, Yuvasree Purusothaman, 9780128193327
- Transparent Conductive Oxides, Mirela Petruta Suchea, Petronela Pascariu, Emmanouel Koudoumas, 9780128206317
- Metal Oxide-Carbon Hybrid Materials, Muhammad Akram, Rafaqat Hussain, Faheem K Butt, 9780128226940
- Metal Oxide-based heterostructures, Naveen Kumar, Bernabe Mari Soucase, 9780323852418
- Metal Oxides and Related Solids for Electrocatalytic Water Splitting, Junlei Qi, 9780323857352
- Advances in Metal Oxides and Their Composites for Emerging Applications, Sagar Delekar, 9780323857055
- Metallic Glasses and Their Oxidation, Xinyun Wang, Mao Zhang, 9780323909976
- Solution Methods for Metal Oxide Nanostructures, Rajaram S. Mane, Vijaykumar Jadhav, Abdullah M. Al-Enizi, 9780128243534
- Metal Oxide Defects, Vijay Kumar, Sudipta Som, Vishal Sharma, Hendrik Swart, 9780323855884
- Renewable Polymers and Polymer-Metal Oxide Composites, Sajjad Haider, Adnan Haider, 9780323851558
- Metal Oxides for Optoelectronics and Optics-based Medical Applications, Suresh Sagadevan, Jiban Podder, Faruq Mohammad, 9780323858243
- Graphene Oxide-Metal Oxide and Other Graphene Oxide-Based Composites in Photocatalysis and Electrocatalysis, Jiaguo Yu, Liuyang Zhang, Panyong Kuang, 9780128245262

Published titles

- Metal Oxides in Nanocomposite-Based Electrochemical Sensors for Toxic Chemicals, Alagarsamy Pandikumar, Perumal Rameshkumar, 9780128207277
- Metal Oxide-Based Nanostructured Electrocatalysts for Fuel Cells, Electrolyzers, and Metal-Air Batteries, Teko Napporn, Yaovi Holade, 9780128184967
- Titanium Dioxide (TiO_2) and Its Applications, Leonardo Palmisano, Francesco Parrino, 9780128199602
- Solution Processed Metal Oxide Thin Films for Electronic Applications, Zheng Cui, 9780128149300
- Metal Oxide Powder Technologies, Yarub Al-Douri, 9780128175057
- Colloidal Metal Oxide Nanoparticles, Sabu Thomas, Anu Tresa Sunny, Prajitha V, 9780128133576
- Cerium Oxide, Salvatore Scire, Leonardo Palmisano, 9780128156612
- Tin Oxide Materials, Marcelo Ornaghi Orlandi, 9780128159248
- Metal Oxide Glass Nanocomposites, Sanjib Bhattacharya, 9780128174586
- Gas Sensors Based on Conducting Metal Oxides, Nicolae Barsan, Klaus Schierbaum, 9780128112243
- Metal Oxides in Energy Technologies, Yuping Wu, 9780128111673
- Metal Oxide Nanostructures, Daniela Nunes, Lidia Santos, Ana Pimentel, Pedro Barquinha, Luis Pereira, Elvira Fortunato, Rodrigo Martins, 9780128115121
- Gallium Oxide, Stephen Pearton, Fan Ren, Michael Mastro, 9780128145210
- Metal Oxide-Based Photocatalysis, Adriana Zaleska-Medynska, 9780128116340
- Metal Oxides in Heterogeneous Catalysis, Jacques C. Vedrine, 9780128116319
- Magnetic, Ferroelectric, and Multiferroic Metal Oxides, Biljana Stojanovic, 9780128111802
- Iron Oxide Nanoparticles for Biomedical Applications, Sophie Laurent, Morteza Mahmoudi, 9780081019252
- The Future of Semiconductor Oxides in Next-Generation Solar Cells, Monica Lira-Cantu, 9780128111659
- Metal Oxide-Based Thin Film Structures, Nini Pryds, Vincenzo Esposito, 9780128111666
- Metal Oxides in Supercapacitors, Deepak Dubal, Pedro Gomez-Romero, 9780128111697
- Metal oxide-based nanofibers and their applications, Vincenzo Esposito, Debora Marani, 9780128206294
- Metal-oxides for Biomedical and Biosensor Applications, Kunal Mondal, 9780128230336
- Nanostructured Zinc Oxide, Kamlendra Awasthi, 9780128189009
- Transition Metal Oxide Thin Film-Based Chromogenics and Devices, Pandurang Ashrit, 9780081018996

Metal Oxides

Metal Oxide-Carbon Hybrid Materials

Synthesis, Properties and Applications

Edited by

Muhammad Akram Chaudary

Assistant Professor, Government Associate College
Raiwind (Lahore), Pakistan

Rafaqat Hussain

Professor, Department of Chemistry, COMSATS
University Islamabad, Islamabad, Pakistan

Faheem K. Butt

Associate Professor, Department of Physics,
Division of Science and Technology, University of
Education, Lahore, Pakistan

Series Editor

Ghenadii Korotcenkov

Department of Physics and Engineering, Moldova
State University, Chisinau, Republic of Moldova

ELSEVIER

Elsevier
Radarweg 29, PO Box 211, 1000 AE Amsterdam, Netherlands
The Boulevard, Langford Lane, Kidlington, Oxford OX5 1GB, United Kingdom
50 Hampshire Street, 5th Floor, Cambridge, MA 02139, United States

Notices
Knowledge and best practice in this field are constantly changing. As new research and experience broaden our understanding, changes in research methods, professional practices, or medical treatment may become necessary.

Practitioners and researchers must always rely on their own experience and knowledge in evaluating and using any information, methods, compounds, or experiments described herein. In using such information or methods they should be mindful of their own safety and the safety of others, including parties for whom they have a professional responsibility.

To the fullest extent of the law, neither the Publisher nor the authors, contributors, or editors, assume any liability for any injury and/or damage to persons or property as a matter of products liability, negligence or otherwise, or from any use or operation of any methods, products, instructions, or ideas contained in the material herein.

Library of Congress Cataloging-in-Publication Data
A catalog record for this book is available from the Library of Congress

British Library Cataloguing-in-Publication Data
A catalogue record for this book is available from the British Library

ISBN: 978-0-12-822694-0

For information on all Elsevier publications visit our website at
https://www.elsevier.com/books-and-journals

Publisher: Matthew Deans
Acquisitions Editor: Kayla Dos Santos
Editorial Project Manager: Rafael Guilherme Trombaco
Production Project Manager: Anitha Sivaraj
Cover Designer: Miles Hitchen

Typeset by TNQ Technologies

Working together
to grow libraries in
developing countries

www.elsevier.com • www.bookaid.org

Contents

List of contributors

Salem Abdulkarim Department of Physics, Faculty of Science, University of Benghazi, Al Marj City, Libya

Kiran Aftab Department of Chemistry, Government College University Faisalabad, Faisalabad, Punjab, Pakistan

Ishaq Ahmad NPU-NCP Joint International Research Center on Advanced Nanomaterials and Defects Engineering, Northwestern Polytechnical University, Xi'an, Shaanxi, China; National Centre for Physics (NCP), Islamabad, Punjab, Pakistan

Junaid Ahmad Department of Physics, Division of Science and Technology, University of Education Lahore, Lahore, Punjab, Pakistan

Muhammad Saeed Akhtar Department of Physics, Division of Science and Technology, University of Education Lahore, Lahore, Punjab, Pakistan

Deji Akinwande Microelectronics Research Center, The University of Texas at Austin, Austin, TX, United States

Ghulam Ali U.S.-Pakistan Center for Advanced Studies in Energy (USPCAS-E), National University of Sciences and Technology (NUST), Islamabad, Pakistan

Saif Ali Department of Physics, Division of Science and Technology, University of Education Lahore, Lahore, Punjab, Pakistan

Ammar Z. Alshemary Department of Biomedical Engineering, Karabük University, Karabük, Turkey; Biomedical Engineering Department, Al-Mustaqbal University College, Hillah, Babil, Iraq

Yann Aman Laboratory of Chemical Physics, University Felix Houphouet Boigny-Abidjan, Abidjan, Ivory Coast

Muhammad Arshad Nanosciences and Nanotechnology Department, National Centre for Physics, Quaid-i-Azam University Islamabad, Islamabad, Pakistan

Zeeshan Asghar Department of Physics, Division of Science and Technology, University of Education Lahore, Lahore, Punjab, Pakistan

Ebtesam E. Ateia Physics Department, Faculty of Science, Cairo University, Giza, Egypt; Academy of Scientific Research and Technology (ASRT), Cairo, Egypt

Ahmad Sher Awan Institute of Education and Research, University of the Punjab Qaid-I-Azam Campus, Lahore, Punjab, Pakistan

Sadia Zafar Bajwa Nanobiotechnology Group, Industrial Biotechnology Division, National Institute for Biotechnology and Genetic Engineering (NIBGE), Faisalabad, Punjab, Pakistan

Muhammad Bilal School of Physics, College of Physical Science and Technology and School of Environmental Science and Engineering, Yangzhou University, Yangzhou, Jiangsu, PR China

Asadullah Dawood Department of Physics, University of Wah, Wah Cantt, Punjab, Pakistan; Department of Physics, School of Sciences and Humanities, Nazarbayev University, Nur Sultan, Kazakhstan

S.I. El-Dek Materials Science and Nanotechnology Department, Faculty of Postgraduate Studies for Advanced Sciences (PSAS), Beni-Suef University, Beni-Suef, Egypt

Rengin Eltem Faculty of Engineering, Department of Bioengineering, Ege University, Izmir, Turkey

Zafer Evis Department of Engineering Sciences, Middle East Technical University, Ankara, Turkey

A.A. Farghali Materials Science and Nanotechnology Department, Faculty of Postgraduate Studies for Advanced Sciences (PSAS), Beni-Suef University, Beni-Suef, Egypt

Jameela Fatheema Physics Characterization and Simulations Lab (PCSL), Department of Physics, School of Natural Sciences (SNS), National University of Sciences and Technology (NUST), Islamabad, Pakistan; Microelectronics Research Center, The University of Texas at Austin, Austin, TX, United States

Mian Hasnain Nawaz Interdisciplinary Research Centre in Biomedical Materials (IRCBM), COMSATS University Islamabad, Islamabad, Pakistan

Akhtar Hayat Interdisciplinary Research Centre in Biomedical Materials (IRCBM), COMSATS University Islamabad, Islamabad, Pakistan

Hunza Hayat Nanobiotechnology Group, Industrial Biotechnology Division, National Institute for Biotechnology and Genetic Engineering (NIBGE), Faisalabad, Punjab, Pakistan

Jianhua Hou Jiangsu Key Laboratory of Environmental Material and Engineering, School of Environmental Science and Engineering, Yangzhou University, Yangzhou, Jiangsu, PR China; School of Physics, College of Physical Science and Technology and School of Environmental Science and Engineering, Yangzhou University, Yangzhou, Jiangsu, PR China

Asif Hussain School of Physics, College of Physical Science and Technology and School of Environmental Science and Engineering, Yangzhou University, Yangzhou, Jiangsu, PR China; Department of Physics, The University of Lahore, Lahore, Punjab, Pakistan

Rao F. Hussain Khan   Department of Radiation Oncology, Washington University School of Medicine, Missouri, WA, United States

Faryal Idrees   Department of Physics, University of the Punjab, Quaid-e-Azam Campus, Lahore, Punjab, Pakistan

Faiza Jan Iftikhar   NUTECH School of Applied Sciences and Humanities, National University of Technology, Islamabad, Pakistan

Fauzia Iqbal   Department of Physics, University of the Punjab, Quaid-e-Azam Campus, Lahore, Punjab, Pakistan

M. Zubair Iqbal   Department of Materials Engineering, College of Materials and Textiles, Zhejiang Sci-Tech University, Hangzhou, Zhejiang, PR China

Saman Iqbal   Department of Physics, University of the Punjab, Quaid-e-Azam Campus, Lahore, Punjab, Pakistan

Farwah Jameel   Nanobiotechnology Group, Industrial Biotechnology Division, National Institute for Biotechnology and Genetic Engineering (NIBGE), Faisalabad, Punjab, Pakistan

Nuzhat Jamil   Nanobiotechnology Group, Industrial Biotechnology Division, National Institute for Biotechnology and Genetic Engineering (NIBGE), Faisalabad, Punjab, Pakistan

Husnain Joan   Department of Physics, The University of Lahore, Lahore, Punjab, Pakistan

Sadia Khalid   Nanosciences & Technology Department, National Centre for Physics, Quaid-i-Azam University Campus, Islamabad, Pakistan

Waheed S. Khan   Nanobiotechnology Group, Industrial Biotechnology Division, National Institute for Biotechnology and Genetic Engineering (NIBGE), Faisalabad, Punjab, Pakistan

Mahammed Ilyas Khazi   Panzhihua Gesala Biotechnology Inc., Panzhihua, Sichuan, PR China

Muhammad Latif   Department of Physics, University of Balochistan, Quetta, Pakistan

Chuanbo Li   School of Science, Minzu University of China, Beijing, PR China; Optoelectronics Research Center, Minzu University of China, Beijing, PR China

Jian Li   School of Biological and Chemical Engineering, Panzhihua University, Panzhihua, Sichuan, PR China; Panzhihua Gesala Biotechnology Inc., Panzhihua, Sichuan, PR China

Fakhra Liaqat   Department of Biotechnology, Virtual University of Pakistan, Lahore, Punjab, Pakistan; School of Biological and Chemical Engineering, Panzhihua University, Panzhihua, Sichuan, PR China

Asif Mahmood School of Chemical and Bimolecular Engineering, The University of Sydney, Australia

Arshad Mahmood National Institute of Lasers and Optronics, Nilore, Islamabad, Pakistan

Mohammad Azad Malik School of Materials, The University of Manchester, Manchester, United Kingdom

Amira T. Mohamed Physics Department, Faculty of Science, Cairo University, Giza, Egypt

M. Morsy Building Physics and Environment Institute, Housing & Building National Research Center (HBRC), Giza, Egypt

Ali Motameni Department of Metallurgical and Materials Engineering, Middle East Technical University, Ankara, Turkey

Anam Munawar Nanobiotechnology Group, Industrial Biotechnology Division, National Institute for Biotechnology and Genetic Engineering (NIBGE), Faisalabad, Punjab, Pakistan

Saadia Mushtaq Department of Physics, Hazara University, Mansehra, KPK, Pakistan

Muhammad Naeem Ashiq Institute of Chemical Sciences, Bahauddin Zakariya University, Multan, Pakistan

Muhammad Nasir Interdisciplinary Research Centre in Biomedical Materials (IRCBM), COMSATS University Islamabad, Islamabad, Pakistan

Nazia Nasr NPU-NCP Joint International Research Center on Advanced Nanomaterials and Defects Engineering, Northwestern Polytechnical University, Xi'an, Shaanxi, China; Department of Physics, Division of Science and Technology, University of Education Lahore, Lahore, Punjab, Pakistan; School of Materials Science & Engineering, Northwestern Polytechnical University, Xi'an, Shaanxi, China; National Centre for Physics (NCP), Islamabad, Punjab, Pakistan; Faculty of Engineering Sciences, GIK Institute, Topi, Khyber Pakhtunkhwa, Pakistan

Nafeesa Nayab Nanobiotechnology Group, Industrial Biotechnology Division, National Institute for Biotechnology and Genetic Engineering (NIBGE), Faisalabad, Punjab, Pakistan

Gul Naz Institute of Physics, Faculty of Science, The Islamia University of Bahawalpur, Baghdad-ul-Jadid Campus, Bahawalpur, Pakistan

Laraib Nisar Institute of Chemical Sciences, Bahauddin Zakariya University, Multan, Pakistan

Amna Rafiq Nanobiotechnology Group, Industrial Biotechnology Division, National Institute for Biotechnology and Genetic Engineering (NIBGE), Faisalabad, Punjab, Pakistan

Muhammad Ramzan Institute of Physics, Faculty of Science, The Islamia University of Bahawalpur, Baghdad-ul-Jadid Campus, Bahawalpur, Pakistan

Asma Rehman Nanobiotechnology Group, Industrial Biotechnology Division, National Institute for Biotechnology and Genetic Engineering (NIBGE), Faisalabad, Punjab, Pakistan

Sajid Ur Rehman School of Science, Minzu University of China, Beijing, PR China; Optoelectronics Research Center, Minzu University of China, Beijing, PR China

Zia Ur Rehman Jiangsu Key Laboratory of Environmental Material and Engineering, School of Environmental Science and Engineering, Yangzhou University, Yangzhou, Jiangsu, PR China; School of Physics, College of Physical Science and Technology and School of Environmental Science and Engineering, Yangzhou University, Yangzhou, Jiangsu, PR China

Sara Riaz Department of Chemistry, COMSATS University Islamabad, Islamabad, Pakistan

Syed Rizwan Physics Characterization and Simulations Lab (PCSL), Department of Physics, School of Natural Sciences (SNS), National University of Sciences and Technology (NUST), Islamabad, Pakistan

Maira Sadaqat Institute of Chemical Sciences, Bahauddin Zakariya University, Multan, Pakistan

Ashir Saeed Nanosciences & Technology Department, National Centre for Physics, Quaid-i-Azam University Campus, Islamabad, Pakistan; Department of Physics, Khwaja Fareed University of Engineering and Information Technology, Rahim Yar Khan, Punjab, Pakistan

Muhammad Hassan Sayyad Faculty of Engineering Sciences, GIK Institute, Topi, Khyber Pakhtunkhwa, Pakistan

Amir Shehzad Shah Department of Physics, The University of Lahore, Lahore, Punjab, Pakistan

Matiullah Shah Department of Physics, University of Wah, Wah Cantt, Punjab, Pakistan

Hassina Tabassum College of Engineering, Peking University, Beijing, China; Department of Chemical and Biological Engineering, University at Buffalo, The State University of New York, Buffalo, NY, United States; Department of Materials Science and Engineering, College of Engineering, Peking University, Beijing, Haidian District, China

Muhammad Bilal Tahir Department of Physics, Khawaja Fareed University of Engineering and Information Technology, Rahim Yar Khan, Pakistan

Ayesha Taj Nanobiotechnology Group, Industrial Biotechnology Division, National Institute for Biotechnology and Genetic Engineering (NIBGE), Faisalabad, Punjab, Pakistan

Zeeshan Tariq School of Science, Minzu University of China, Beijing, PR China; Optoelectronics Research Center, Minzu University of China, Beijing, PR China; State Key Laboratory on Integrated Optoelectronics, Institute of Semiconductors, Chinese Academy of Sciences, Beijing, PR China

Qiu Tianjie College of Engineering, Peking University, Beijing, China; Department of Chemical and Biological Engineering, University at Buffalo, The State University of New York, Buffalo, NY, United States

Sami Ullah Department of Physics, Division of Science and Technology, University of Education Lahore, Lahore, Punjab, Pakistan

Xiaozhi Wang School of Physics, College of Physical Science and Technology and School of Environmental Science and Engineering, Yangzhou University, Yangzhou, Jiangsu, PR China

Sumaira Younis Nanobiotechnology Group, Industrial Biotechnology Division, National Institute for Biotechnology and Genetic Engineering (NIBGE), Faisalabad, Punjab, Pakistan; Department of Chemistry, University of Agriculture, Faisalabad, Punjab, Pakistan

Xiaoming Zhang School of Science, Minzu University of China, Beijing, PR China; Optoelectronics Research Center, Minzu University of China, Beijing, PR China

Tingkai Zhao NPU-NCP Joint International Research Center on Advanced Nano-materials and Defects Engineering, Northwestern Polytechnical University, Xi'an, Shaanxi, China; School of Materials Science & Engineering, Northwestern Polytechnical University, Xi'an, Shaanxi, China

Rabisa Zia Nanobiotechnology Group, Industrial Biotechnology Division, National Institute for Biotechnology and Genetic Engineering (NIBGE), Faisalabad, Punjab, Pakistan

Volume editor biographies

Dr. Muhammad Akram Chaudary currently serves as assistant professor of chemistry at the Government Associate College, Raiwind (Lahore, Pakistan). He holds his MSc (chemistry) from the Institute of Chemistry, University of the Punjab, Lahore (Pakistan). He taught at the University of Baluchistan, Quetta for 3 years and attended the University of the Punjab, where he completed his MPhil (chemistry). During his stay there, he worked on the spectrophotometric determination of fluoroquinolone drugs and developed a simple and cheap method for their determination. In 2011, he joined Universiti Teknologi Malaysia, Malaysia, to pursue research work in the field of nanosized ceramic materials and received his doctorate in 2014. His PhD dissertation was on the "Synthesis and Characterization of Ceramic Nanoparticles through Continuous Microwave Flow Synthesis Process." His current research interest is focused on continuous microwave flow synthesis and the microwave-assisted synthesis of nanostructured materials, including biomaterials and photocatalytic materials. His work also involves probing the properties and applications of nanostructured material in various fields especially in biomedical and electrochemical devices. He is working collaboratively with various research groups and has published more than 35 research articles/book chapters as an author/coauthor in peer-reviewed international SCI-indexed journals with various publishers, such as Elsevier, Springer, IOP, and Wiley, with an h-index of 16 and $\sim$1003 citations.

Prof. Rafaqat Hussain graduated with a BSc (Hons) in chemistry from the University College London, where he also received his MSc in chemistry and a PhD in biochemical engineering. To pursue a career in teaching, he obtained a postgraduate certificate in education (chemistry) from the Goldsmiths College, London. In 2007, he moved to Pakistan to lead an ambitious project to establish the Interdisciplinary Research Centre in Biomedical Materials (IRCBM) at the COMSATS University Islamabad (CUI),

Lahore campus. After more than 3 years at the IRCBM, he took up a teaching position at the Department of Chemistry, Universiti Teknologi Malaysia (UTM), where he established a successful materials research group. After more than 5 years teaching and supervising numerous research students at UTM, he returned to CUI to establish the Department of Chemistry at the Islamabad campus. He is actively involved in publishing research articles in leading high-impact journals, an author of over 50 research articles (impact factor >150) and several book chapters, and a winner of many research grants in the field of materials science.

Dr. Faheem K. Butt currently works as an associate professor of physics in the Division of Science and Technology at the University of Education, Lahore. Before joining the University of Education, he was an Alexander von Humboldt Fellow at the Technical University of Munich, Germany. He completed his postdoctoral research fellowship from the Centre for Sustainable Nanomaterials, Universiti Teknologi Malaysia, in 2015. He completed his PhD studies at the Beijing Institute of Technology, China. His interests is in optoelectronics devices, energy storage/conversion in nanomaterials, and theoretical studies of materials using various software. He has published more than 120 research articles/book chapters as an author/coauthor in peer-reviewed international SCI journals with various publishers, including the American Chemical Society, Royal Society of Chemistry, Elsevier, Springer, American Scientific Publishers, and Taylor & Francis, with an h-index of 28 and ~3000 citations (https://scholar.google.com/citations?user=raetpZoAAAAJ&hl=en). He has also provided his services as a reviewer/editor/editorial board member for international publishers. Dr. Faheem has won several national/international awards. He is also a Higher Education Commission (HEC)-approved supervisor for PhD awardees of HEC scholarships, as well as general PhD scholars.

Series editor biography

Ghenadii Korotcenkov received his PhD in physics and technology of semiconductor materials and devices in 1976 and his DSc degree (Doc. Hab.) in physics of semiconductors and dielectrics in 1990. He has more than 50 years' experience as a teacher and scientific researcher. For some time, he was a leader of a gas sensor group and manager of various national and international scientific and engineering projects carried out in the Laboratory of Micro-Optoelectronics, Technical University of Moldova, Chisinau, Moldova. During 2007 and 2008, he was an invited scientist at the Korea Institute of Energy Research (Daejeon). Thereafter, until 2017, Dr. Korotcenkov was a research professor in the School of Materials Science and Engineering at the Gwangju Institute of Science and Technology in Korea. Dr. Korotcenkov is currently a chief scientific researcher at Moldova State University, Chisinau, Moldova. His scientific interests since 1995 have included material sciences, focusing on metal oxide film deposition and characterization, surface science, thermoelectric conversion, and design of physical and chemical sensors, including thin film gas sensors.

Dr. Korotcenkov is the author or editor of 39 books and special issues, including the 11-volume *Chemical Sensors* series published by Momentum Press, the 15-volume *Chemical Sensors* series published by Harbin Institute of Technology Press, China, the 3-volume *Porous Silicon: From Formation to Application* issue published by CRC Press, the 2-volume *Handbook of Gas Sensor Materials* published by Springer, and the 3-volume *Handbook of Humidity Measurements* published by CRC Press. Currently, he is a series editor of the *Metal Oxides* book series published by Elsevier.

Dr. Korotcenkov is the author and coauthor of more than 650 scientific publications including 31 review papers, 38 book chapters, more than 200 peer-reviewed articles published in scientific journals (h-factor=41 (Web of Science), h = 43 (Scopus), and h = 58 (Google Scholar citation)). He is a holder of 17 patents. He has presented more than 250 reports at national and international conferences, including 17 invited talks. Dr. Korotcenkov, as a cochair or member of program, scientific, and steering committees, has participated in organizing more than 30 international scientific conferences. Dr. Korotcenkov is a member of the editorial boards of five international scientific journals. His name and activities have been listed by many biographical publications including *Who's Who*. His research activities are honored by the Honorary Diploma of the Government of the Republic of Moldova (2020); an Award of the Academy of Sciences of Moldova (2019); an Award of the Supreme Council of Science and Advanced Technology of the Republic of Moldova (2004); the Prize of the Presidents of the Ukrainian, Belarus, and Moldovan Academies of Sciences (2003); Senior Research Excellence Award of Technical University of Moldova (2001; 2003; 2005); and the National Youth Prize of the Republic of Moldova in the field of science and technology (1980); among others. Dr. Korotcenkov also received a fellowship from the International Research Exchange Board (IREX, USA, 1998), the Brain Korea 21 Program (2008−12), and the Brain Pool Program (Korea, 2015−17).

Preface to the volume

This book includes several orderly discussions on fabrications and applications of metal oxide—carbon-based nanocomposite materials. Those discussions particularly address how these materials' exclusive characteristics make them the choice in various fields like optoelectronics, energy storage, gas sensing, photonics, and catalysis.

Section 1 encompasses the historical background of hybrid materials based on metal oxide—carbon and hybridized metal oxide composites. It also includes comprehensive detail about several available, popular approaches for establishing metal oxide—carbon composites through solid-state and solution-phase reactions. This section also discusses development protocols for enabling techniques, particularly microwave-assisted methods for synthesizing these modern nanocarbon-based hybrid materials. Section 2 deals with the role of metal oxide—carbon composites in energy generation, hydrogen production, and energy storage in devices such as rechargeable batteries and supercapacitors and their extreme importance and demand in an industrialized and technology-driven society. Section 3 covers the details the use of modern metal oxide—carbon composites in water refinement and the photodegradation of industrial pollutants, and further, how biomedical applications of these synthesized materials may help ensure the availability of a green and clean environment for our modern societies. Section 4 probes problems with fabricating and developing various sensing materials. Several studies in this area have proven that the introduction of carbon-based composites significantly upgrades the operating parameters of sensors and sensing devices.

In short, this book provides a comprehensive review of science and technology related to metal oxide—carbon composites that can be used to find solutions to many energy-related and clinical field-related challenges.

We believe this book will be beneficial primarily for new researchers working in physics, chemistry, biomedical engineering, and electrochemical engineering. This book will surely catch the attention of postgraduate researchers and academicians who have a wide-ranging interest in metal oxide—carbon hybrid materials.

Dr. Muhammad Akram Chaudary

Dr. Rafaqat Hussain

Dr. Faheem K. Butt

Preface to the series

The synthesis, study, and application of metal oxides is one of the most rapidly progressing areas in science and technology. Metal oxides are groups of some of the most ubiquitous compounds on earth, with many chemical compositions, atomic structures, and crystalline shapes. In addition, they possess unique functionalities that are absent or inferior in other solid materials. In particular, metal oxides are an assorted and appealing class of materials that exhibit a full spectrum of electronic properties—insulating, semiconducting, metallic, and superconducting. Moreover, almost all the known effects, including superconductivity, thermoelectric effects, photoelectrical effects, luminescence, and magnetism, can be observed in metal oxides. Therefore, metal oxides have emerged as an important class of multifunctional materials with a rich collection of properties and great potential for numerous device applications. The specific properties of metal oxides, such as wide-ranging electrophysical, optical, and chemical characteristics, high thermal and temporal stability, and ability to function in harsh environments, make them suitable for designing transparent electrodes, high-mobility transistors, gas sensors, actuators, acoustic transducers, photovoltaic and photonic devices, photo- and heterogeneous catalysts, solid-state coolers, high-frequency and micromechanical devices, energy-harvesting and energy-storage devices, nonvolatile memories, and many others in the electronics, energy, and health sectors. In devices, metal oxides can be used successfully as sensing or active layers, substrates, electrodes, promoters, structure modifiers, membranes, and fibers—i.e., they can be used as active and passive components.

Metal oxides also have low fabrication costs and are robust in practical applications. Furthermore, they can be prepared in various forms such as ceramics, thick films, and thin films. Deposition techniques compatible with standard microelectronic technology can be used for thin film deposition. The microelectronic approach promotes low costs for mass production, offers the possibility to manufacture devices on a chip, and guarantees good reproducibility, all of which are critical for large-scale production. Various metal oxides nanostructures, including nanowires, nanotubes, nanofibers, core-shell structures, and hollow nanostructures, can also be synthesized. The field of metal oxide nanostructured morphologies (e.g., nanowires, nanorods, and nanotubes) has become one of the most active research areas within the nanoscience community.

The ability to create a variety of metal oxide-based composites and synthesize many multicomponent compounds significantly expands the range of potential properties for metal oxide-based materials, making them truly versatile and multifunctional for widespread use. Small changes in their chemical composition and atomic structure can be accompanied by spectacular variations in their properties and behaviors. Ongoing advances in synthesizing and characterizing techniques continue to reveal numerous new functions for metal oxides.

Considering the importance of metal oxides for progress in microelectronics, optoelectronics, photonics, energy conversion, sensing, and catalysis, it is not surprising that many books devoted to this class of materials have been published. However, one should note that some books on this topic are too general, some are collections of various original works without any generalizations, and still others were published many years ago. But during the past decade, great progress has been made on the synthesis and structural, physical, and chemical characterizations and applications of metal oxides in various devices, and many papers have been published on these topics. Even so, many important topics in the study and application of metal oxides had not been addressed. To remedy this situation, we decided to generalize and systematize the research results in this area and publish a series of books devoted to metal oxides.

One should note that the proposed book series, *Metal Oxides*, is the first devoted solely to metal oxides. We believe that combining books on metal oxides in a series could help readers search for specific, required information on the subject. In particular, we expect that the books in our series, with a clear specialization by content, will provide interdisciplinary discussions for various oxide materials within a wide range of topics, from material synthesis and deposition, to characterization and processing, and to device fabrications and applications. This book series is being prepared by a team of highly qualified experts, thus guaranteeing its high quality.

I hope our books will be useful as well as comfortable in use. I also hope that readers will consider this *Metal Oxides* book series an encyclopedia of metal oxides that enables one to understand the present status of metal oxides, estimate the role of multifunctional metal oxides in the design of advanced devices, and based on observed knowledge, formulate new goals for further research.

The intended audience of the present book series is scientists and researchers working in or planning to work in fields related to metal oxides, i.e., scientists and researchers whose activities are related to electronics, optoelectronics, energy, catalysis, sensors, electrical engineering, ceramics, biomedical designs, etc. I believe that practicing engineers or project managers in industries and national laboratories may also find this *Metal Oxides* book series interesting—e.g., designing metal oxide-based devices or selecting the optimal metal oxides for specific applications. With many references to the vast resource of recently published literature on the subject, this book series will serve as a significant and insightful source of valuable information that provides scientists and engineers with new insights for understanding and improving existing metal oxide-based devices and designing new metal oxide-based materials with new and unexpected properties

I believe this *Metal Oxides* book series could be helpful for university students, postdocs, and professors. The structure of these books offers a basis for courses in the materials sciences, chemical engineering, electronics, electrical engineering, optoelectronics, energy technologies, environmental control, and many others. Graduate students could also find the book series useful in their research and understanding the features of metal oxide synthesis and the study and application of this multifunctional material in various devices. We are certain that all of these readers will find some information useful in their specific activities.

Finally, I would like to thank all the contributing authors and book editors who have been involved in creating these books. I am thankful that they agreed to participate in this project and for their efforts to prepare these books for publication. Without their participation, this project would not have been possible. I also wish to express my gratitude to Elsevier for giving us the opportunity to publish this series. I especially thank the entire editorial team at Elsevier for their patience during the development of this project and for encouraging us during the various stages of preparation.

Ghenadii Korotcenkov

Section One

Metal oxide-carbon hybrid materials: Synthesis and properties

Physical and chemical aspects of metal oxide–carbon composites

A.A. Farghali and S.I. El-Dek
Materials Science and Nanotechnology Department, Faculty of Postgraduate Studies for Advanced Sciences (PSAS), Beni-Suef University, Beni-Suef, Egypt

1.1 Introduction

Nanocomposites have increasingly been studied in recent years for their ability to help in specified applications. Several types of nanocomposites have been studied and reported. The most common category is based on carbon and metal oxides in a variety of morphologies. The distinctive features of carbon nanotubes (CNTs), either single-walled or multiwalled, are interesting to prepare using versatile methods for research, industrial, medical, engineering, and commercial use. After that, graphene's extraordinary revolution makes it a sheet with tunable parameters. On the other hand, carbon spheres represent updated three-dimensional (3-D) nanostructures that have been exploited in various fields owing to their porous nature. In addition, metal oxide nanoparticles have played a major role in fine-tuning the physicochemical performance of nanocomposites.

This chapter discusses the importance of nanoscale materials and the relevance of "nanoparticles." We also report on the various types of metal oxide–carbon nanocomposites according to their dimensions and how their overall physical properties and chemical features can be tailored according to the type and ratio of metal oxide used and how it is embedded within the carbon matrix at different scales.

1.2 Materials in the nanoscale

Materials based on nanosized particles display unfamiliar chemical and physical properties that differ partially or completely from the material properties of their bulk counterparts. Their study is essential in contemporary science because they allow for the investigation of changes in the material properties of substances as they progress from the atomic or molecular scale to condensed systems.

The distinctiveness of the overall characteristics of nanoparticles is always correlated to their size and is strongly influenced by the following phenomena. The percentage of surface atoms is increased by decreasing the particle size. This results in larger reactivity due to coordination nonsaturation. Continuing here and considering that the surface-to-volume ratio increases with the transition to nanoparticles, the cause of these "unfamiliar" properties, named "dimensional effects," turns out to be direct and perfect.

Metal Oxide-Carbon Hybrid Materials. https://doi.org/10.1016/B978-0-12-822694-0.00008-9

Large surface pressure exerted on nanomaterials is the main cause of size-based influences on the physicochemical and physicomechanical behavior of nanostructures. This pressure varies inversely proportionally with particle size, increasing Gibbs free energy. Accordingly, the saturated vapor pressure above the nanoparticles increases, decreasing the liquid boiling point and the solid phase melting point. Other thermodynamical coordinates may also vary—e.g., the equilibrium constant and standard electrode potentials.[1] Additionally, when the particle size decreases, the impact of the structure on material stoichiometry is noticeably demonstrated.[2,3] Such a change in stoichiometry leads to a pronounced variation in the chemical and physical features of the material.

The emerging use of nanomaterials in different technologies is always the cause of increased publications and patents on their preparation, characterizations, properties, biological activities, applications, and sometimes commercialization. We searched using the keyword "nanoparticles" on "Scopus" to see how much interest researchers have at present in the latest updates on different types of nanoparticles.

1.3 Relevance of the term "nanoparticles"

Fig. 1.1A shows the number of documents published yearly, and the enormous increase in publications is observed to have reached 68,322 documents in 2020.

From Fig. 1.1B, we understand that the nanoparticles were repeated in nearly all subject categories from materials science, chemistry, physics, and engineering to medicine, environmental science, energy, and many other subjects.

1.4 Metal oxide-carbon nanocomposites

The production of metal-oxide nanocomposites—an extremely exciting kind of nanomaterials—endorsed by their corresponding characteristics could surpass those of its parent compound phases by several orders of magnitude. We use the carbon matrix mainly to avoid loss in metal oxide nanoparticles due to their stabilization inside the composite matrix.[4] Moreover, depending on their structure, nanocomposites have distinctive properties in a wide variety of fields and could be exploited in the following examples: assembly of novel materials in the areas concerning environmental, energy, medicine, and ecological issues.

We observed that CNT/metal oxide (CNT/MO) nanocomposite materials were reported to be synthesized using many methods. Those techniques are predominantly applied, and namely, could be mechanical mixing, sol—gel, electrospinning, chemical vapor deposition (CVD), or atomic layered deposition. Nevertheless, the uniform homogenous coating of MO on CNTs could be carried out by CVD and electrospinning ways. These methods are not simple or facile but always require special equipment. Unfortunately, the sol-gel procedure could be considered the easiest together with the low-cost approach. Even though it resulted in heterogeneity in the layer coating of CNTs by the investigated MO. The solicitation of sol—gel procedure in composite

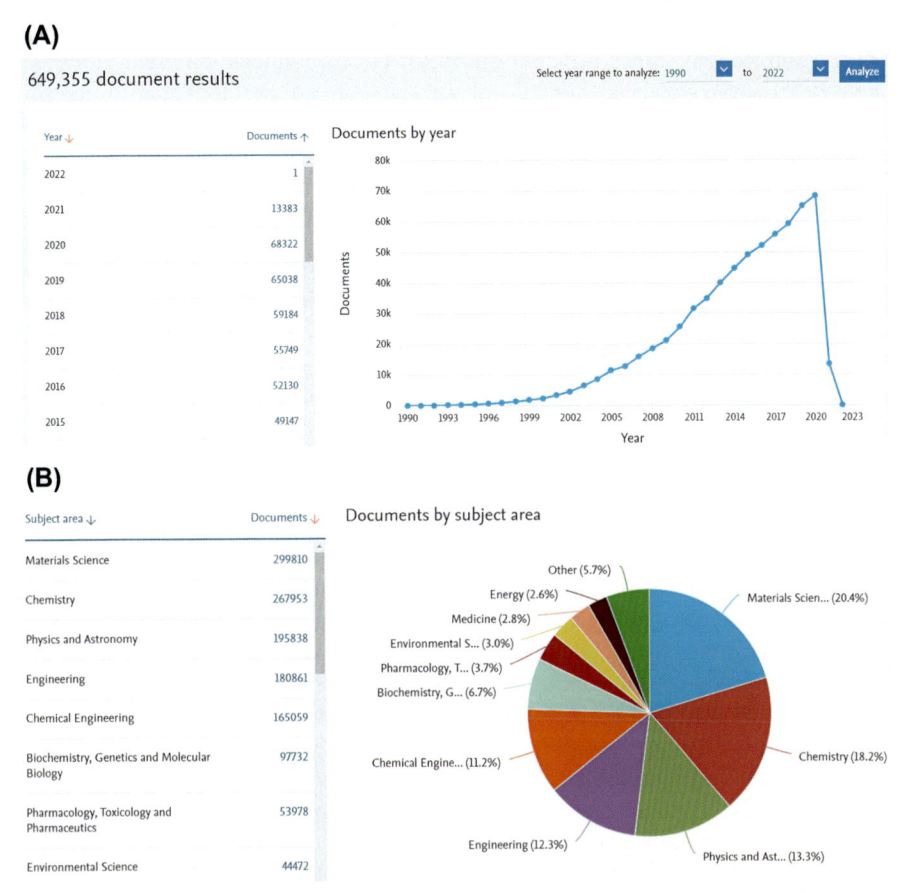

Figure 1.1 (A) The number of published articles per year using the search keyword "nanoparticles" from Scopus. (B) The number of publications in a variety of subject categories.

production has many signs of progress, e.g., its flexibility and the great purity produced materials, the stability endorsed to the chemical binding between both coating and support, and finally, the capability of controlling the coating layer thickness.[5,6]

Various investigations on carbon/metal-oxide composites were performed. Most of the researchers use the excellent mechanical strength of the CNTs and add metal oxides nanoparticles to gain some electrical properties such as large capacitance or magnetic properties such as magnetization and coercivity. This was performed by the proper choice of the metal oxide nanoparticles and the trials of change ratios.

In this part, we thoroughly report some of the features (structural, electrical, physical, and chemical properties) of these nanocomposites by classifying the carbon structure according to their dimensions.

One-dimensional (1-D), two-dimensional (2-D), and 3-D carbon structures convey plentiful benefits to carbon/metal oxide nanocomposites related to the distinctive physical property of each dimension.

Many diverse composites based on carbon and metal oxide have been employed in versatile applications. Since different dimensions of carbon have different properties owing to the structural characteristics of the material, we are motivated to describe the effects of composites with metal oxide and carbon according to the dimension of the carbon nanostructure[7]

1.5 Classification of metal oxide/carbon nanocomposites

In rapports to their dimensionality and microstructure, they are divided into three main categories: 1-D, 2-D, and 3-D nanocomposites, as shown in Fig. 1.2.

1.5.1 One-dimensional carbon—metal oxide nanocomposites

The famous CNT structure is the best example of a 1-D carbon nanostructure that affords an uninterrupted network for metal oxides possessing relatively low electrical conductivity. Herein, the large aspect in 1-D CNTs permits the creation of percolation networks in minor numbers. Furthermore, a suitable weight of carbon-based nanocomposite makes it promising to stabilize the metal oxide throughout cycling due to its great mechanical strength. Here, we report on several studies on metal oxide-decorated CNTs with various morphologies.

As an example, Fe_3O_4 decorates multiwalled CNTs in a nanocomposite with exciting features to serve versatile technologies. Ji et al.[9] investigated how MWCNT/Fe_3O_4 and MWCNTs/Fe_3O_4 functionalized by 3-aminopropyltriethoxysilane (MWCNTs/Fe_3O_4–NH_2) could affect adsorption together with the isotherms of tetrabromobisphenol A and Pb(II) removal from wastewater. After the completion of adsorption, both adsorbents were separated using an external magnetic field for a few seconds. As observed in the TEM image (Fig. 1.3), Fe_3O_4 NPs are uniformly decorating the surface of the prepared MWCNTs. The existence of MWCNT among Fe_3O_4 NPs avoids more close contact and magnetic interaction. Therefore, it develops the coalescence of pure Fe_3O_4 NPs.

The unique physical and chemical properties hand in hand with low cost, high reversible capacity, rich abundance, in addition to eco-friendliness[8,9] render iron oxide (Fe_2O_3 or Fe_3O_4) as one of the promising candidate magnetic nanocomposites with CNTs. Based on the reported literature,[10,11] embedding nano iron oxide onto the CNT surface of CNTs is achieved using different physical, chemical fabrication techniques. Superparamagnetism is the observed magnetic property of Fe_3O_4 encapsulated

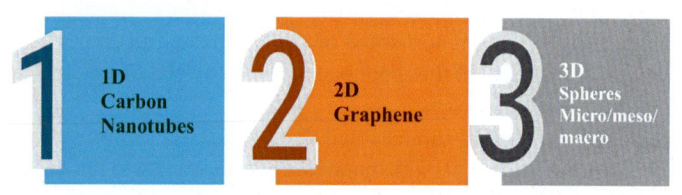

Figure 1.2 Types of Carbon structures according to their dimensions.

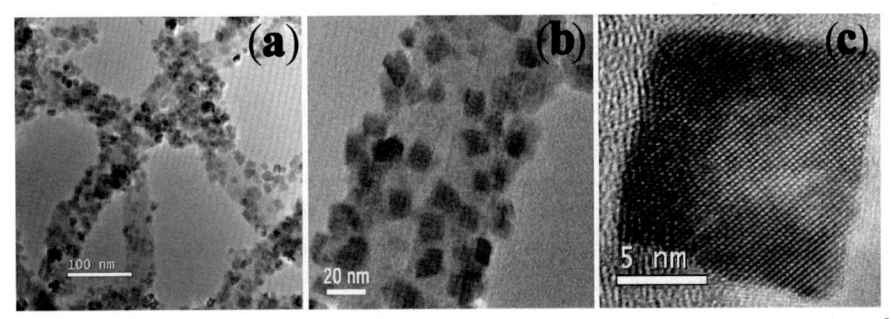

Figure 1.3 (A–C) TEM micrographs of MWCNT/Fe$_3$O$_4$ nanocomposites at different scales.[8]

on CNTs. The saturation magnetization of the pretreated MWCNTs is larger than that of primary MWCNTs, which stemmed from the existence of Fe cations.[12] Rajarao et al.[13] described a quick, solventless, simple, environmental, economical, and scalable technique for synthesizing MWCNTs/iron oxide nanocomposite using simple thermal decomposition of metal salts avoiding the addition of a solvent or any reduction.

The influence of heat treatment (annealing) temperature on both structural and magnetic properties of the MWCNTs/α-Fe$_2$O$_3$ composites was also studied.[14] The results showed that when treated at 450°C and 600°C, the nanocomposites exhibit a clear ferromagnetic character, while those annealed at 750°C are paramagnetic owing to the improved crystal growth and clustering of Fe$_3$O$_4$ in MWCNTs[15]

Explicitly, the continuous conductive grid-like networking of 1-D CNTs contributes to the charge transport enhancement due to the lowered contact resistance with neighboring nanoparticles, which is larger than that for the pure metal-oxide nanoparticle. Herein, CNTs can assist as a "facilitated electron transport path or electron highway," permitting a charge transport along the longitude direction (tube axis) rather than the cross-sectional one[16]

Usually, the mechanical macrostrain originates from a large change in the lattice volume after metal oxide nanoparticle insertion on the CNT buffer layer. In this case, additional conduction paths (electronic pathways) are initiated and help to improve the value of the composite conductivity.[17] CNTs here are used to avoid the agglomeration of the nanoparticles of metal oxides homogeneously distributed in the composites.

The benefits of the fabrication of CNT/metal-oxide nanocomposite could be mentioned as follows:

a. The metal oxide resistivity of extremely large values is easily compensated by the conductive CNTs, which assist as an uninterrupted network for the electron transport equivalent to "electronic highway".[18] CNTs can also achieve high electrical conductivity via lesser weight percent compared to nanoparticles of round shape by building up a percolation network. This is useful to be exploited for the energy storage devices as they possess high energy density.

b. The contact area between the electrolyte and electrode is improved by the large specific surface area of CNTs.

c. The extreme mechanical toughness of CNTs, compared with that of pure metal oxide, enables a carbon scaffold to form via chemical or physical bonding with metal oxide nanoparticles.[19]

When magnetic nanoparticles (MNPs) are encapsulated in the CNT cavity established on the capillary force, they generally save the adsorption sites. Accordingly, CNT adsorptivity is increased for MNPs stacked on CNT surfaces. Korneva et al. first proposed filling the hollow cavities of CNTs with an outer diameter of nearly 300 nm with Fe_3O_4-MNP nanoparticles (size = 10 nm).[20] On the other hand, alternate synthesis procedures were uninterruptedly discovered for different applications. MNP-filled MWCNTs were synthesized by Jin et al. and used to remove aromatic compounds by ultrasonic dispersion of CNTs in iron nitrate solution. After that, they reduced it using H_2.[21] Another capillary force-driven technique was implemented by Goh et al. for MNP-filled SWCNT preparation for biofuel production.[22]

Magnetic CNT nanocomposites are broadly promised to extract and precisely determine drug residuals, pesticide residues, biochemical analytes, pollutants resulting from chemical industries, food, and environmental pollutants. The various major interactions are widespread and noncovalent.

For that reason, the composites publicized excellent extractability to different types of analytes.

The mechanical properties of the CNT/MO nanocomposite are associated with numerous contributions:

(a) *The amount (yield) and dimensional consistency of CNTs.* Low quantities of CNTs are not sufficient to achieve the requested improvement for all mechanical properties of the investigated composite. Conversely, with further increasing CNT amount or weight percent, the hardness and toughness rise to extreme values and then diminish again. The larger number of CNTs in MOs seems to approve the inhibition of the main matrix's grain growth, leading to hindered densification.

(b) *The produced CNT quality.* When comparing MWCNTs with SWCNTs, we find that the former possesses more defective structures. This will lead to a decrease in CNT toughness in the metal oxide surrounding the substance. From a closer look, the outer layers of MWCNTs could relocate the load from the lattice to the inside of nanotubes. Herein, the inner layers are free and couldn't connect to the MO. Therefore, MWCNT has a touchable impact on the load transfer.

(c) The main circumstances of nanocrystals and their special virgin properties are significant for tailoring the mechanical properties of the nanocomposite.[21,22]

(d) The preparation technique will be the driving factor that controls the morphology and dispersion of the MO inside the CNT matrix (decoration on the outer surface, insertion inside the tube, or both).

We have summarized the reported physical properties of MO/CNT and listed them in Table 1.1.

1.5.2   Two-dimensional carbon–metal oxide composites

The most famous 2-D carbon nanostructure is the well-known graphene characterized by suitable mechanical strength, large values of electrical conductivity, and a sufficiently large surface area. It is considered a suitable choice to achieve the balance to compensate for the drawbacks of metal oxides nanostructures cited above. To be exact, 2-D carbon nanocomposites including metal oxides showed improved composite structural stability

Table 1.1 Physical properties of some metal oxide/carbon nanotubes.

Nanocomposite	Physical properties	Method of preparation	Ref.
CNT/Al$_2$O$_3$	0.18 WT % of MWCNT leads to low preparation limit in dc conductivity	Ultrasonic	23
CNT/Al$_2$O$_3$	High mechanical stress adsorbent for organic pollutant	Sol—gel	9
CNT/Al$_2$O$_3$	1.5 wt% CNT in increased fracture toughness 24%	Chemical vapor deposition (CVD)	24
CNT/Al$_2$O$_3$	8.4% increase in hardness and 21% increase in toughness	CVD	25
MWCNT/ZnO	Superior dielectric properties, high charge carrier of concentration, and larger relaxation polarization optimize electromagnetic properties	Precipitation	25,26
CNT/TiO$_2$	Narrowing bandgap, higher surface area, and enhanced photocatalysis invisible light	Biomimetic	27
CNT/TiO2	In a proper gas sensor largest sensitivity gas ethanol	Catalytic pyrolysis	28
CNT/TiO2	Excellent thermal and chemical stability and very good mechanical stress	Sol—gel	29
MWCNT/ZnO	Biosensor for urea with a long shelf life >16 weeks	Chemical root	30
CNT/ZnO	5% wt %CNT resulted in 35.5% higher power comparison efficiency in DSC than pure material	Ultrasonic	31
MWCNT/iron oxide	Removal of Cr^{3+} ions	Sol-gel method	32
MWCNT/iron oxide	High adsorption capacity for As^{3+} and As^{5+} by applying a magnetic field	Precipitation	33
MWCNT/iron oxide	Fast absorption with a high capacity for cation dyes in wastewater treatment	Precipitation	33
CNT/ZrO$_2$	An n-type semiconductor with a large bandgap and decreased diffraction	Coprecipitation	34,35

from tailoring various microstructures. Two-dimensional carbon with a porous microstructure enables enhanced kinetics in electronic and ionic transport as well.

The quality of graphene production depends on several factors, as shown in Fig. 1.4A. Graphene consisted of a single layer or several layers of sp^2-hybridized carbon atoms adopting a honeycomb structure. Graphene has appealed to prodigious considerations owing to its surprising properties, such as extreme surface area, outstanding stability, and extraordinary physicochemical characterizations.[37] The 2-D graphene structural model is distinguished by its large specific surface area and tremendous electrical

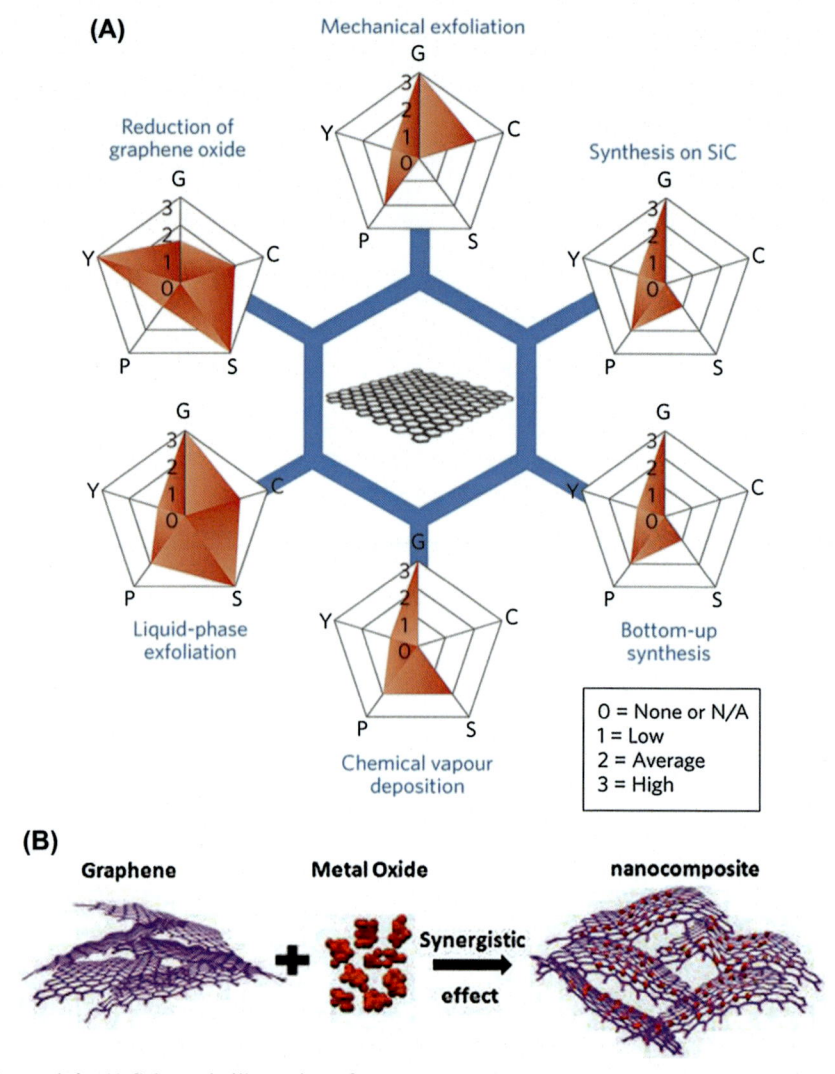

Figure 1.4 (A) Schematic illustration of most commonly used graphene production methods. Each method is evaluated according to five indicators, namely, graphene quality (G), cost aspect (C), scalability (S), purity (P), and yield (Y).[36] (B) Schematic of the preparation of graphene/metal oxide composites with synergistic effects between graphene and metal oxides.[52]

conductivity. Additionally, the hydrophilicity of both graphene oxide and reduced graphene compared to the hydrophobicity of graphene is endorsed to their large number of surface functionalization with different groups. This is profitable for combining a composite with nano metal oxide. Graphene oxide loaded with magnetic CNTs was obtained by Sun et al. to explore oxytetracycline from sewage water by implementing a quick and simple one-pot approach. The structural characterizations indicated that CNTs and MNPs served as the spacer dispensing the graphene layers, thus inhibiting efficient stacking of the sheets and increasing the graphene surface area.

Hence, the composite based on graphene oxide and magnetic CNTs has upgraded the extraction efficiency compared with two-parent carbon nanomaterials. This was magnificently employed in rapid isolation and quantification of the target analyte.[38]

The 2-D carbon nanosheet structure has been merged with a diversity of metal oxides[39] to take full advantage of the electrochemical performance of the capacitive nanomaterial. Also, some features of graphene are beneficial, as its high surface area and conductivity make it promising in the progress of multifunctional catalysts in photocatalysis, adsorbents, and antimicrobial agents. In addition, it offers an effective agent against all major water pollutants, including organic molecules, heavy metal ions, and waterborne pathogens[7] Some, such as TiO_2,[40] ZnO,[41] Cu_2O[42] $ZnFe_2O_4$,[43] $CuFe_2O_4$,[44] and Bi_2WO_6,[45] have shown impressive results in photocatalysts for the degradation of synthetic dyes. The synergy between graphene and metal oxides and the beneficial role of graphene is employed in composites for lithium-ion batteries (LIBs) like G/MoO_2,[46] G/V_2O_5,[47] G/Co_3O_4,[48] and G/Co_3O_4[49] and electrochemical capacitors like G/ZnO,[31] G/MoO_3,[50] and G/Cu_2O.[8] Significant improvements, such as high capacity values, higher rate capability, and excellent cycling stability, were also reported. Fig. 1.4B shows how graphene and metal oxide can interact together during the reaction process. Furthermore, the large value of Young's modulus (1.0 TPa) and tensile strength for the graphene greatly enhanced the device and cycle stability. For instance, a graphene/Fe_3O_4 nanocomposite showed excellent results in both high capacity and long cycle stability.[51]

Several studies on metal oxide nanoparticle-anchored rGO are listed in Table 1.2.

On the other hand, the use of graphene–metal oxide nanocomposite in gas sensors[75] is tailored by optimizing the parent compounds. While pure graphene does not easily react to sensing gases, graphene oxide and its reduced form can easily trap the analyte gas molecules, leading to a consequent variation in the conductivity value. Typically, rGO has improved conducting parameters compared with graphene oxide because of its well-known nonstoichiometry. Henceforth, the response characteristics of the investigated nanocomposite will be controlled from the two parents or the choice of one of them. Generally, oxidizing and reducing gases react differently, leading to the generation or destruction of charge carriers in the sensing layer. Consequently, an observed increase or decrease in the resistance is easily monitored as the sensor signal.

1.5.3 Three-dimensional carbon–metal oxide composites

The main characteristic of the 3-D carbon sphere is that it allows a large loading capacity of the active materials via a great storage capacity (porosity), permitting the monotonic increase in the number of active sites available for the ions throughout the electrochemical reactions. Moreover, the hierarchical structure accelerates the electrochemical reactions subject to pore physical characteristics (e.g., macro/meso/micro) and pore interconnectivity, or so-called pore connectivity.

The 3-D porous carbon physical features (wall thickness, adjustable pore size, high surface area, enormous pore volume, as well as the inner connection among the pore channel) serve them appropriate to be used as a matrix of active material.[76] In a 3-D carbon/metal oxide nanocomposite, the 3-D porous carbon contributes an essential role in building up the hierarchical structure to bring together different-sized pores

Table 1.2 Physical properties of some nanocomposites based on graphene/metal oxides.

Nanocomposite	Physical properties	Method of preparation	Ref.
$Ti_2Nb_{10}O_{29}/G$	High area capacitance	Solvothermal	53
MoO_2/G	High current density in supercapacitor	Spray pyrolysis	54
Mn_3O_4/G	High capacity	Precipitation	55
Bi_2WO_6/G	Outstanding photocatalytic performance for the degradation of dyes under visible light	Precipitation	56
WO_3/G	Lower band gap energy	Chemical precipitation	57
$ZnFe_2O_4/G$	Photoelectrochemical degradation	Coprecipitation	58
ZnO/G	Higher photocurrent	Coprecipitation	59
$CoFe_2O_4/G$	Photodegradation of methylene blue	Microwave irradiation method	60
TiO_2/G	Enhanced photocatalysis	Liquid phase deposition	61
$CoFe_2O_4/G$	Excellent recycling stability for the degradation of methylene blue	Ball-milling	62
SnO_2/G	Reducing gases at room temperature	Hydrothermal	63
ZnO/G	High-energy microsupercapacitors	Hydrothermal	64
CoO/G	Low viscosity, high diffusion, zero surface tension, and good surface wettability	Solvothermal	65
MnO_2/G	Low contact resistance, improving the charge transport efficiency of the electrolyte	Chemical vapor deposition	66
Co_3O_4/G	High-capacity energy storage	Electrochemical deposition	67
CuxOy/G	Enhance the mechanical properties	Sonication	68
MgO/G	High strength and ductility	Coprecipitation route	69
CuO/G	Enhanced thermal conductivity	Coprecipitation	70
CeO_2/G	Enhanced faradaic reactivity	Electrochemical exfoliation	71
SnO_2/G	High-performance lithium-ion batteries	Hydrothermal	72
Al_2O_3/G	Enhanced electrical conductivity	Modified hummers'	73
NiO/G	Fast charge—discharge	Hummers' method	74

(from micropores to mesopores to macropores).[77] Metal-oxide nanomaterials are well inserted in the pore structure of the carbon spherical particle to customize the nanocomposite. Herein, the interconnected pores efficiently facilitate electrolyte penetration and ion diffusion, bringing about the optimum structural stability of the overall composite.[49]

Carbon was synthesized easily from organic reactants by simple techniques, as shown in Fig. 1.5A, and then mixed with a metal oxide to be employed in various applications.

In addition to the carbonization temperature, it is taken into consideration that the other crucial factor is the treatment time that greatly affects the microstructure and then

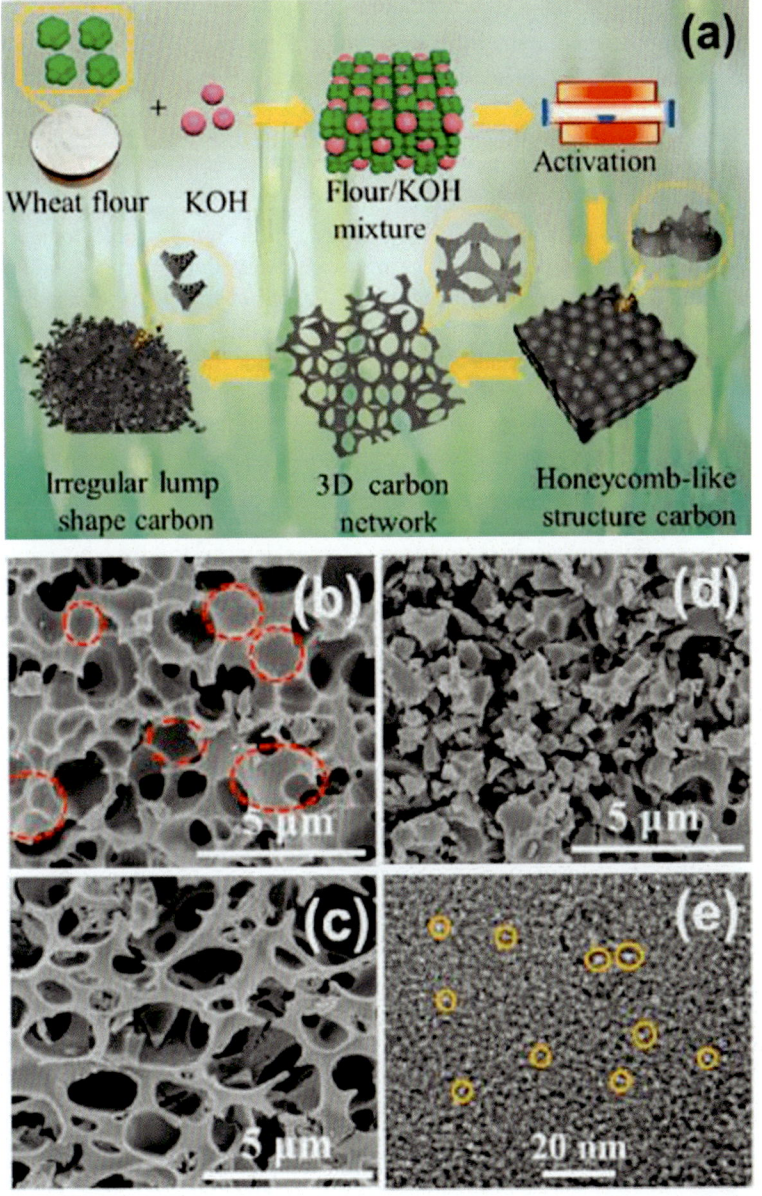

Figure 1.5 (A) The schematic illustration of nanoporous carbons prepared from wheat flour.[78] (B–E) The microstructures obtained with different carbonization times of 2 h (B: Sample S1), 3 h (C and E: Sample S2), and 4 h (D: Sample S3).[78]

dielectric and microwave absorption efficiency of the biomass-derived porous carbon (BPC) materials. Zhao et al. reported preparing nanoporous carbon structures using wheat flour as the biomass precursor via pyrolysis of wheat/KOH mixture at 700°C for 2 h, 3 and 4 h,[79] as presented in Fig. 1.5A. At a pyrolysis time of 1 h, the resultant powder revealed a honeycomb-like structure possessing some conchoidal cavities together with opened micropores (Fig. 1.5B). Meanwhile, the micropore size was much lesser than the incident wavelength. In this case, the pores could not yield multiple reflections or scatter on the incident wave.[42,43] Prolonging the calcination time to 3 h (Fig. 1.5C and E), the 3-D interconnected network was assembled, and the conduction loss formed by the carbon basic structure matrix certainly helps reduce the microwave. Extending the heating time to 4 h leads to the collapse of the 3-D framework into minuscule carbon fragments (Fig. 1.5C and D), and accordingly, its microwave absorption becomes worse to a great extent.[79]

Metal oxides, for instance, NiO, Co_3O_4, and Fe_3O_4 are valuable for accurately matching the impedance and boosting to improve polarization and magnetic loss ability.[80] Between the numerous metal oxides, Fe_3O_4 nanoparticles have acknowledged great attention owed to their exceptional chemical stability, tunable saturation magnetization value (M_S), and thermal stability over a wide range of temperatures.[45,46] Loofah sponge,[47,48] walnut shell,[81] alginates,[82] collagen,[83] cotton stalk,[84] glucose,[85] Jute fiber,[86] natural wood,[87] and wheat straw[88] were employed as biomass precursors for preparing Fe_3O_4/BPC composites for microwave absorption purposes as shown in Fig. 1.6A−F.

The most significant advantage of 3-D porous carbon/metal oxide nanostructured composite is its use as an electrode in energy-storage devices. It is the active mitigation of the mechanical contraction and expansion of the metal oxide for the duration of the electrochemical reactions. The main cause here is the hierarchical pore structure as well as pore stability.[90] Herein, Zhang et al. prepared (Fe_3O_4@C) composite where Fe_3O_4 was confined in the nanoporous carbon framework. The later composite was employed as an anode in an LIB via assembling $Fe(NO_3)_3$/resol/F127.[91]

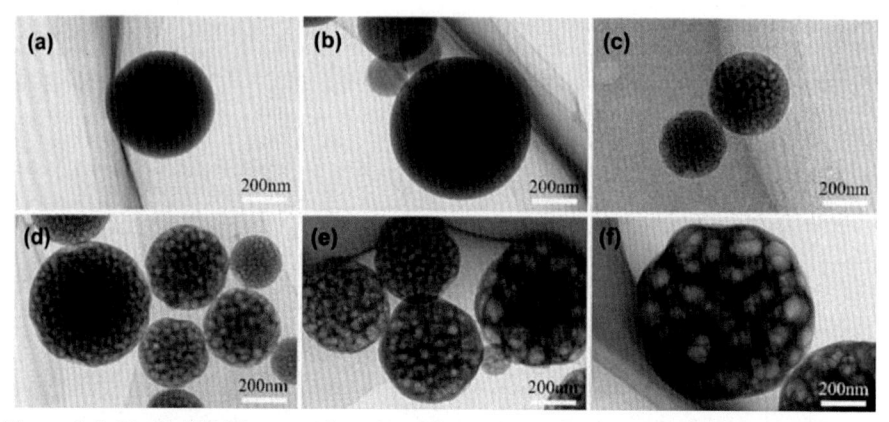

Figure 1.6 (A−F) TEM images of porous carbon spheres prepared with different F127/larch sawdust ratio: (A) 0, (B) 0.05, (C) 0.075, (D) 0.1, (E) 0.125 and (F) 0.15.[89]

(Fig. 1.6A—F) illustrated SEM and TEM micrographs for the diverse morphologies of Fe_3O_4@C-1 (Fig. 1.6A and B, Fe_3O_4@C-2 (Fig. 1.6C and D), and Fe_3O_4@C-3 (Fig. 1.6E and F) according to the ratio of Fe_3O_4 nanoparticles (see Fig. 1.7).

Among these nanocomposites based on 3-D carbon/metal oxide, the porous microstructure is fabricated to expand the surface area and enhance the electrolyte-electrode contact area to facilitate the electron transport mechanism. The 3-D carbon network

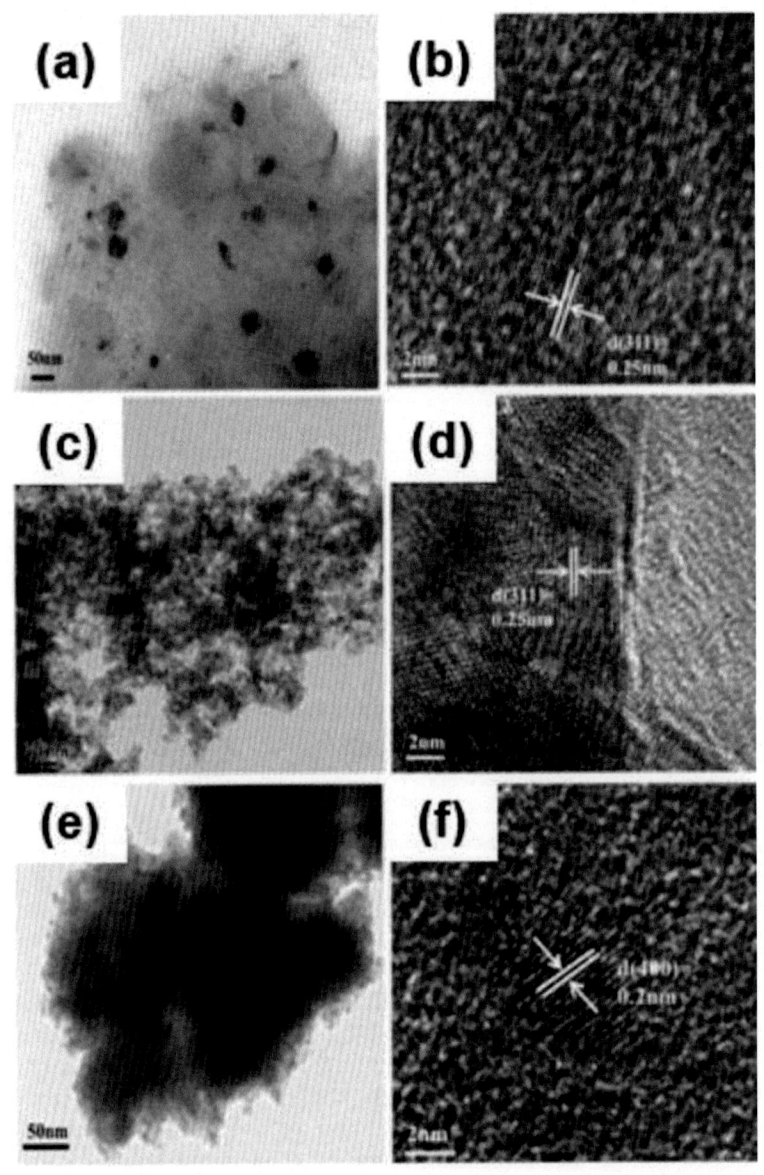

Figure 1.7 (A—F) TEM images of Fe_3O_4 confined in various nanoporous carbon frameworks: (A and B) Fe_3O_4@C-1, (C and D) Fe_3O_4@C-2, and (E and F) Fe_3O_4@C-3.[92]

Table 1.3 Preparation techniques and physical properties of three-dimensional nanocomposites formed from carbon spheres/metal oxides.

Nanocomposite	Physical properties	Method of preparation	Ref.
Carbon spheres/ Fe_3O_4	Good sorbents	Hydrothermal	94
Carbon spheres/ SiO	High capacity	Simple aldimine condensation	95
Carbon spheres/ Co_3O_4	Good mechanical properties and facility in recovering	Wet impregnation	96
Carbon spheres/ Fe_2O_3	Excellent adsorbent and removal ability for the antimony in a wide pH	Vibrated	97
Carbon spheres/ Fe_3O_4	Improved electrical conductivity	Hydrothermal	98
Carbon spheres/ CoO_x	Good conductivity	Wet penetration	99
Carbon spheres/ Co_3O_4	Current density	Hydrothermal	100
Carbon spheres/ Cu_2O	High energy density	Hydrothermal	101
Carbon spheres/ CuO	Good conductivity	Pyrolysis	102
Carbon spheres/ MnO_2	High capacity	Hydrothermal	103

works as a new connection path through which the electrons move and improve LIB performance. It also plays the buffer role for the large volumetric alteration of the metal oxide produced in battery charge $\leftrightarrow$ discharge. Wang et al. reported the investigation of how they could achieve the improved mechanical characteristics, electronic and ionic conductivities of the composites by manufacturing the composites of MnO dispersed in a 3-D porous-based carbon network. They achieved a capacity of 560.2 mA/g at a relatively high current density of 4 A/g.[93]

Herein, we list some properties of 3-D carbon sphere—metal oxide nanocomposites with their corresponding preparation methods in Table 1.3.

1.6 Conclusion and future perspectives

This chapter reported work on metal oxide/carbon nanocomposites with extensive study from the well-known CNT to nanofibers, graphene, and carbon spheres. Altogether, the work is categorized by the dimensionality of the carbon structure. The first part discussed various nanocomposites according to their preparation technique and physical and chemical properties. These composites offered great potential applications in different fields. The chapter analyzed the advantages of using CNTs to ameliorate mechanical and electrical features. For 2-D composites, graphene was the predominant composite, with metal oxide nanoparticles with versatile morphologies. For this structure, we used metal oxide to prevent stacking of the graphene layer. The third part discussed the important 3-D "sphere" structure simply prepared from BPC/metal oxides. Their most significant application nowadays is in energy conversion and storage, such as supercapacitors and various types of batteries. From the abovementioned data reported for various sections, we argued that the most important nanocomposite features could be tailored by choosing a suitable preparation method, type, and metal oxide nanoparticle ratio, as well as the morphology of both constituents. For excellent mechanical strength, CNTs are facile and cheap, and they provide high hardness and improved toughness with a tiny weight percentage of CNTs. In thinking about high electrical conductivity and chemical stability, we use rGO with metal oxides for energy storage and supercapacitors. Graphene, graphene oxide, and rGO with metal oxides in nanocomposites has harvested prospective materials with a prodigious capacity for large-scale fabrication of selective chemical sensors and biosensors for low-cost detection of environmental pollutants and clinical and pharmacological applications.

Future work will likely focus on the biomedical applications of these nanocomposites and investigate their in vitro and in vivo toxicity for different biological organs. Meanwhile, there is still much to be studied in drug delivery through porous carbon spheres and drug targeting via magnetic carbon—metal oxide nanocomposites. We must also highlight the carbon quantum dots that will be the focus of many future studies. Owing to their photothermal and photodynamic natures, the potential applications of carbon quantum dots have extended into the therapeutic fields for bioimaging, photothermal, and phototherapeutic uses. The photothermal feature could also be exploited for photothermal and photoacoustic imaging. Considering these properties, quantum dots have great prospects for future personalized theragnostic nanomedicine integrating multiple functions within one platform.

Acknowledgments

The authors thank Miss Hadeer K.El Emam for her help during the processing of this work.

References

1. Makarchuk O, Dontsova T, Perekos A, Skoblik A, Svystunov Y. Magnetic mineral nanocomposite sorbents for wastewater treatment. *J Nanomater* 2017;**2017**.
2. Zheng J, Xiao J, Zhang JG. The roles of oxygen non-stoichiometry on the electrochemical properties of oxide-based cathode materials. *Nano Today* 2016;**11**(5):678—94. https://doi.org/10.1016/j.nantod.2016.08.011.
3. Leitner J, Bartůněk V, Sedmidubský D, Jankovský O. Thermodynamic properties of nanostructured ZnO. *Appl Mater Today* 2018;**10**:1—11.

4. Santhosh C, Velmurugan V, Jacob G, Jeong SK, Grace AN, Bhatnagar A. Role of nanomaterials in water treatment applications: a review. *Chem Eng J* 2016;**306**:1116−37. https://doi.org/10.1016/j.cej.2016.08.053.

5. Leary R, Westwood A. Carbonaceous nanomaterials for the enhancement of TiO_2 photocatalysis. *Carbon N Y* 2011;**49**(3):741−72.

6. Sharma S, Islam SS. Phase transformation-dependent sensing performance of multi-walled carbon nanotube-alumina nanocomposite-based gas sensors. *Mater Sci Semicond Process* 2014;**27**(1):63−70. https://doi.org/10.1016/j.mssp.2014.06.003.

7. Seok D, Jeong Y, Han K, Yoon DY, Sohn H. Recent progress of electrochemical energy devices: metal oxide − carbon nanocomposites as materials for next-generation chemical storage for renewable energy. *Sustainability* 2019;**11**(13):3694.

8. Lin G, Ma R, Zhou Y, Hu C, Yang M, Liu Q, et al. Three-dimensional interconnected nitrogen-doped mesoporous carbons as active electrode materials for application in electrocatalytic oxygen reduction and supercapacitors. *J Colloid Interface Sci* 2018;**527**: 230−40. https://doi.org/10.1016/j.jcis.2018.05.020.

9. Ji L, Zhou L, Bai X, Shao Y, Zhao G, Qu Y, et al. Facile synthesis of multiwall carbon nanotubes/iron oxides for removal of tetrabromobisphenol A and Pb(ii). *J Mater Chem* 2012;**22**(31):15853−62.

10. Ghosh S, Remanan S, Mondal S, Ganguly S, Das P, Singha N, et al. An approach to prepare mechanically robust full IPN strengthened conductive cotton fabric for high strain tolerant electromagnetic interference shielding. *Chem Eng J* 2018;**344**:138−54. https://doi.org/10.1016/j.cej.2018.03.039.

11. Li W, Qi H, Guo F, Du Y, Song N, Liu Y, et al. Co nanoparticles supported on cotton-based carbon fibers: a novel broadband microwave absorbent. *J Alloys Compd* 2019; **772**:760−9. https://doi.org/10.1016/j.jallcom.2018.09.075.

12. Wang Y, Zhou Z, Chen M, Huang Y, Wang C, Song WL. From nanoscale to macroscale: engineering biomass derivatives with nitrogen doping for tailoring dielectric properties and electromagnetic absorption. *Appl Surf Sci* 2018;**439**:176−85. https://doi.org/10.1016/ j.apsusc.2017.12.222.

13. Rajarao R, Jayanna RP, Sahajwalla V, Bhat BR. GreenApproach to decorate multi-walled carbon nanotubes by metal/metal oxide nanoparticles. *Procedia Mater Sci* 2014;**5**:69−75. https://doi.org/10.1016/j.mspro.2014.07.243.

14. Song WL, Zhou Z, Wang LC, Cheng XD, Chen M, He R, et al. Constructing repairable meta-structures of ultra-broad-band electromagnetic absorption from three-dimensional printed patterned shells. *ACS Appl Mater Interfaces* 2017;**9**(49):43179−87.

15. CHEN C, sheng, LIU T, gui, CHEN X, hua, LIN L, wu, LIU Q, cheng, XIA Q, et al. Preparation and magnetic property of multi-walled carbon nanotube/α-Fe_2O_3 composites. *Trans Nonferrous Met Soc China* 2009;**19**(6):1567−71. https://doi.org/10.1016/S1003-6326(09)60071-6.

16. De Las Casas C, Li W. A review of application of carbon nanotubes for lithium ion battery anode material. *J Power Sources* 2012;**208**:74−85. https://doi.org/10.1016/ j.jpowsour.2012.02.013.

17. Lu B, Dong XL, Huang H, Zhang XF, Zhu XG, Lei JP, et al. Microwave absorption properties of the core/shell-type iron and nickel nanoparticles. *J Magn Magn Mater* 2008; **320**(6):1106−11.

18. Zhang De W, Xu B, Jiang LC. Functional hybrid materials based on carbon nanotubes and metal oxides. *J Mater Chem* 2010;**20**(31):6383−91.

19. Zhi M, Xiang C, Li J, Li M, Wu N. Nanostructured carbon-metal oxide composite electrodes for supercapacitors: a review. *Nanoscale* 2013;**5**(1):72−88.

20. Korneva G, Ye H, Gogotsi Y, Halverson D, Friedman G, Bradley JC, et al. Carbon nanotubes loaded with magnetic particles. *Nano Lett* 2005;**5**(5):879–84.
21. Jin J, Li R, Wang H, Chen H, Liang K, Ma J. Magnetic Fe nanoparticle functionalized water-soluble multi-walled carbon nanotubules towards the preparation of sorbent for aromatic compounds removal. *Chem Commun* 2007;(4):386–8.
22. Goh WJ, Makam VS, Hu J, Kang L, Zheng M, Yoong SL, et al. Iron oxide filled magnetic carbon nanotube-enzyme conjugates for recycling of amyloglucosidase: toward useful applications in biofuel production process. *Langmuir* 2012;**28**(49):16864–73.
23. Chem JM, Ii P. Facile synthesis of multiwall carbon nanotubes/iron oxides for removal of tetrabromobisphenol A and Pb(ii). *J Mater Chem* 2012:15853–62.
24. Zhang X, Li Y, Liu R, Rao Y, Rong H, Qin G. *High-magnetization FeCo nanochains with ultrathin interfacial gaps for broadband electromagnetic wave absorption at gigahertz.* 2016.
25. Zhong B, Sai T, Xia L, Yu Y, Wen G. *Mater Des* 2017. https://doi.org/10.1016/j.matdes.2017.02.058.
26. Cao J, Zhu C, Aoki Y, Habazaki H. *Starch-derived hierarchical porous carbon with controlled porosity for high performance supercapacitors.* 2018.
27. Jain A, Jayaraman S, Balasubramanian R, Srinivasan MP. Carbon synthesis: enhancement of chemical. *J Mater Chem A* 2014:520–8.
28. Zhang Y, Qiu M, Yu Y, Wen B, Cheng L. A novel polyaniline-coated bagasse fiber composite with core-shell heterostructure provides effective electromagnetic shielding performance. *ACS Appl Mater Interfaces* 2016;**9**.
29. Bajpai R, Rapoport L, Amsalem K, Wagner HD. *A microwave oven.* 2016. p. 230–9.
30. Chem JM, Wang J, Kaskel S. KOH activation of carbon-based materials for energy storage. *J Mater Chem* 2012:23710–25.
31. Zhao H, Cheng Y, Ma J, Zhang Y, Ji G, Du Y. A sustainable route from biomass cotton to construct lightweight and high-performance microwave absorber. *Chem Eng J* 2018;**339**: 432–41. https://doi.org/10.1016/j.cej.2018.01.151.
32. Chem JM, Liu Q, Gu J, Zhang W, Miyamoto Y, Zhang D. Biomorphic porous graphitic carbon for electromagnetic interference shielding. *J Mater Chem* 2012:21183–8.
33. Zhou X, Jia Z, Feng A, Wang X, Liu J, Zhang M, et al. Synthesis of fish skin-derived 3D carbon foams with broadened bandwidth and excellent electromagnetic wave absorption performance. *Carbon N Y* 2019;**152**:827–36. https://doi.org/10.1016/j.carbon.2019.06.080.
34. Qin F, Brosseau C. A review and analysis of microwave absorption in polymer composites filled with carbonaceous particles. *J Appl Phys* 2012:061301.
35. Liu T, Liu N, Gai L, An Q, Xiao Z, Zhai S, et al. Hierarchical carbonaceous composites with dispersed Co species prepared using the inherent nanostructural platform of biomass for enhanced microwave absorption. *Microporous Mesoporous Mater* 2020;**302**(October 2019):110210. https://doi.org/10.1016/j.micromeso.2020.110210.
36. Article P. *Energy Storage* 2015;**14**(March):271–9.
37. Li N, Jiang HL, Wang X, Wang X, Xu G, Zhang B, et al. Recent advances in graphene-based magnetic composites for magnetic solid-phase extraction. *Trends Anal Chem.* 2018; **102**:60–74.
38. Sun Y, Tian J, Wang L, Yan H, Qiao F, Qiao X. One pot synthesis of magnetic graphene/carbon nanotube composites as magnetic dispersive solid-phase extraction adsorbent for rapid determination of oxytetracycline in sewage water. *J Chromatogr A* 2015;**1422**:53–9. https://doi.org/10.1016/j.chroma.2015.10.035.

39. El Rouby WMA. Crumpled graphene: preparation and applications. *RSC Adv* 2015;**5**(82): 66767−96. https://doi.org/10.1039/C5RA10289H.

40. Meng H, Zhao X, Yu L, Jia Y, Liu H, Lv X, et al. Island-like nickel/carbon nano-composites as potential microwave absorbers - synthesis via in situ solid phase route and investigation of electromagnetic properties. *J Alloys Compd* 2015;**644**:236−41. https://doi.org/10.1016/j.jallcom.2015.04.198.

41. Gong C, Jia Y, Zhao X, Liu H, Lv X, Yu L, et al. Ni3N/Ni composites with in-situ growth heterogeneous interfaces as microwave absorbing materials. *Appl Phys Lett* 2015;**107**(15). https://doi.org/10.1063/1.4933314.

42. Pan H, Cheng X, Zhang C, Gong C, Yu L, Zhang J, et al. Preparation of Fe_2Ni_2N/SiO_2 nanocomposite via a two-step route and investigation of its electromagnetic properties. *Appl Phys Lett* 2013;**102**(1):2−6.

43. Zhang M, Jiang Z, Si H, Zhang X, Liu C, Gong C, et al. Heterogeneous iron-nickel compound/RGO composites with tunable microwave absorption frequency and ultralow filler loading. *Phys Chem Chem Phys* 2020;**22**(16):8639−46.

44. Liu S, Zhang M, Lv X, Wei Y, Shi Y, Zhang J, et al. Arousing effective attenuation mechanism of reduced graphene oxide-based composites for lightweight and high efficiency microwave absorption. *Appl Phys Lett* 2018;**113**(8).

45. Liu P, Gao S, Chen C, Zhou F, Meng Z, Huang Y, et al. Vacancies-engineered and heteroatoms-regulated N-doped porous carbon aerogel for ultrahigh microwave absorption. *Carbon N Y* 2020;**169**:276−87. https://doi.org/10.1016/j.carbon.2020.07.063.

46. Guo L, Gao SS, An QD, Xiao ZY, Zhai SR, Yang DJ, et al. Dopamine-derived cavities/Fe 3 O 4 nanoparticles-encapsulated carbonaceous composites with self-generated three-dimensional network structure as an excellent microwave absorber. *RSC Adv* 2019;**9**(2): 766−80.

47. Zhao HB, Cheng JB, Wang YZ. Biomass-derived Co@crystalline carbon@carbon aerogel composite with enhanced thermal stability and strong microwave absorption performance. *J Alloys Compd* 2018;**736**:71−9. https://doi.org/10.1016/j.jallcom.2017.11.120.

48. Liu C, Ye X, Wang X, Liao X, Huang X, Shi B. Collagen fiber membrane as an absorptive substrate to coat with carbon nanotubes-encapsulated metal nanoparticles for lightweight, wearable, and absorption-dominated shielding membrane. *Ind Eng Chem Res* 2017;**56**(30): 8553−62.

49. Guo L, An QD, Xiao ZY, Zhai SR, Cui L, Li ZC. Performance enhanced electromagnetic wave absorber from controllable modification of natural plant fiber. *RSC Adv* 2019;**9**(29): 16690−700.

50. Wang J, Kaskel S. KOH activation of carbon-based materials for energy storage. *J Mater Chem* 2012;**22**(45):23710−25.

51. Liang H, Liu J, Zhang Y, Luo L, Wu H. Ultra-thin broccoli-like $SCFs@TiO_2$ one-dimensional electromagnetic wave absorbing material. *Compos B Eng* 2019;**178**: 107507. https://doi.org/10.1016/j.compositesb.2019.107507.

52. Wu ZS, Zhou G, Yin LC, Ren W, Li F, Cheng HM. Graphene/metal oxide composite electrode materials for energy storage. *Nano Energy* 2012;**1**(1):107−31.

53. Zhang X, Deng S, Zeng Y, Yu M, Zhong Y, Xia X, et al. Oxygen defect modulated titanium niobium oxide on graphene arrays: an open-door for high-performance 1.4 V symmetric supercapacitor in acidic aqueous electrolyte. *Adv Funct Mater* 2018;**28**(44): 1−10.

54. Choi SH, Kang YC. Crumpled graphene-molybdenum oxide composite powders: preparation and application in lithium-ion batteries. *ChemSusChem* 2014;**7**(2):523−8.

55. Wang Y, Li J, Chen S, Li B, Zhu G, Wang F, et al. Facile preparation of monodisperse $NiCo_2O_4$ porous microcubes as a high capacity anode material for lithium ion batteries. *Inorg Chem Front* 2018;**5**(3):559–67.

56. Gao E, Wang W, Shang M, Xu J. Synthesis and enhanced photocatalytic performance of graphene-Bi $2WO_6$ composite. *Phys Chem Chem Phys* 2011;**13**(7):2887–93.

57. Hajishafiee H, Sangpour P, Tabrizi NS. Facile synthesis and photocatalytic performance of WO_3/rGO nanocomposite for degradation of 1-naphthol. *Nano* 2015;**10**(5):1–9.

58. Fu Y, Sun X, Wanga X. $BiVO_4$-graphene catalyst and its high photocatalytic performance under visible light irradiation. *Mater Chem Phys* 2011;**131**(1–2):325–30.

59. Luo Q, Yu X, Lei B, Chen H, Kuang D, Su C. *J Phys Chem C* 2012;**116**:8111–7.

60. Zhang D, Pu X, Gao Y, Su C, Li H, Li H, et al. One-step combustion synthesis of $CoFe_2O_4$-graphene hybrid materials for photodegradation of methylene blue. *Mater Lett* 2013;**113**:179–81. https://doi.org/10.1016/j.matlet.2013.09.088.

61. Nguyen-Phan TD, Pham VH, Shin EW, Pham HD, Kim S, Chung JS, et al. The role of graphene oxide content on the adsorption-enhanced photocatalysis of titanium dioxide/graphene oxide composites. *Chem Eng J* 2011;**170**(1):226–32. https://doi.org/10.1016/j.cej.2011.03.060.

62. He G, Ding J, Zhang J, Hao Q, Chen H. One-step ball-milling preparation of highly photocatalytic active $CoFe_2O_4$-reduced graphene oxide heterojunctions for organic dye removal. *Ind Eng Chem Res* 2015;**54**(11):2862–7.

63. Chen Y, Zhang W, Wu Q. A highly sensitive room-temperature sensing material for NH_3: SnO_2-nanorods coupled by rGO. *Sensors Actuators, B Chem* 2017;**242**:1216–26. https://doi.org/10.1016/j.snb.2016.09.096.

64. Jung J, Jeong JR, Lee J, Lee SH, Kim SY, Kim MJ, et al. In situ formation of graphene/metal oxide composites for high-energy microsupercapacitors. *NPG Asia Mater* 2020;**12**(1). https://doi.org/10.1038/s41427-020-0230-y.

65. Yuan R, Wen H, Zeng L, Li X, Liu X, Zhang C. Article supercritical CO_2 assisted solvothermal preparation of CoO/graphene nanocomposites for high performance lithium-ion batteries. *Nanomaterials* 2021;**11**(3):1–12.

66. Bai XL, Gao YL, Gao ZY, Ma JY, Tong XL, Sun HB, et al. Supercapacitor performance of 3D-graphene/MnO_2foam synthesized via the combination of chemical vapor deposition with hydrothermal method. *Appl Phys Lett* 2020;**117**(18).

67. Wu K, Geng B, Zhang C, Shen W, Yang D, Li Z, et al. Hierarchical porous arrays of mesoporous Co_3O_4 nanosheets grown on graphene skin for high-rate and high-capacity energy storage. *J Alloys Compd* 2020;**820**:153296. https://doi.org/10.1016/j.jallcom.2019.153296.

68. Hussain F, Imran M, Rasheed U, Khalil RMA, Rana AM, Kousar F, et al. A first principle study of graphene/metal-oxides as nano-composite electrode materials for supercapacitors. *J Electron Mater* 2019;**48**(4):2343–9.

69. Wang M, Zhao Y, Wang LD, Zhu YP, Wang XJ, Sheng J, et al. Achieving high strength and ductility in graphene/magnesium composite via an in-situ reaction wetting process. *Carbon N Y* 2018;**139**:954–63. https://doi.org/10.1016/j.carbon.2018.08.009.

70. Chu K, Wang X, hu, Wang F, Li Y, biao, Huang D, Jian, Liu H, et al. Largely enhanced thermal conductivity of graphene/copper composites with highly aligned graphene network. *Carbon N Y* 2018;**127**:102–12. https://doi.org/10.1016/j.carbon.2017.10.099.

71. Chen K, Xue D. In-situ electrochemical route to aerogel electrode materials of graphene and hexagonal CeO 2. *J Colloid Interface Sci* 2015;**446**:77–83. https://doi.org/10.1016/j.jcis.2015.01.013.

72. Zhu YG, Wang Y, Xie J, Cao GS, Zhu TJ, Zhao X, et al. Effects of graphene oxide function groups on SnO_2/graphene nanocomposites for lithium storage application. *Electrochim Acta* 2015;**154**:338−44. https://doi.org/10.1016/j.electacta.2014.12.065.

73. Fan Y, Kang L, Zhou W, Jiang W, Wang L, Kawasaki A. Control of doping by matrix in few-layer graphene/metal oxide composites with highly enhanced electrical conductivity. *Carbon N Y* 2015;**81**(1):83−90. https://doi.org/10.1016/j.carbon.2014.09.027.

74. Kottegoda IRM, Idris NH, Lu L, Wang JZ, Liu HK. Synthesis and characterization of graphene-nickel oxide nanostructures for fast charge-discharge application. *Electrochim Acta* 2011;**56**(16):5815−22.

75. Hazra S, Basu S. Graphene-oxide nano composites for chemical sensor applications. *Chimia* 2016;**2**(2):12.

76. Wang N, Wu F, Xie AM, Dai X, Sun M, Qiu Y, et al. One-pot synthesis of biomass-derived carbonaceous spheres for excellent microwave absorption at the Ku band. *RSC Adv* 2015;**5**(51):40531−5.

77. Liu H, Wu J, Zhuang Q, Dang A, Li T, Zhao T. Preparation and the electromagnetic interference shielding in the X-band of carbon foams with Ni-Zn ferrite additive. *J Eur Ceram Soc* 2016;**36**(16):3939−46. https://doi.org/10.1016/j.jeurceramsoc.2016.06.017.

78. Guan H, Wang Q, Wu X, Pang J, Jiang Z, Chen G, et al. Biomass derived porous carbon (BPC) and their composites as lightweight and efficient microwave absorption materials. *Compos B* 2021;**207**(August 2020):108562. https://doi.org/10.1016/j.compositesb.2020.108562.

79. Letellier M, Macutkevic J, Kuzhir P, Banys J, Fierro V, Celzard A. Electromagnetic properties of model vitreous carbon foams. *Carbon N Y* 2017;**122**:217−27. https://doi.org/10.1016/j.carbon.2017.06.080.

80. Liang C, Song P, Ma A, Shi X, Gu H, Wang L, et al. Highly oriented three-dimensional structures of Fe_3O_4 decorated CNTs/reduced graphene oxide foam/epoxy nano-composites against electromagnetic pollution. *Compos Sci Technol* 2019;**181**(March): 107683. https://doi.org/10.1016/j.compscitech.2019.107683.

81. Zhou N, An QD, Zheng W, Xiao ZY, Zhai SR. High-performance electromagnetic wave absorbing composites prepared by one-step transformation of Fe^{3+} mediated egg-box structure of seaweed. *RSC Adv* 2016;**6**(100):98128−40.

82. Gao S, An Q, Xiao Z, Zhai S, Shi Z. Significant promotion of porous architecture and magnetic Fe_3O_4 NPs inside honeycomb-like carbonaceous composites for enhanced microwave absorption. *RSC Adv* 2018;**8**(34):19011−23.

83. Wang X, Huang X, Chen Z, Liao X, Liu C, Shi B. Ferromagnetic hierarchical carbon nanofiber bundles derived from natural collagen fibers: truly lightweight and high-performance microwave absorption materials. *J Mater Chem C* 2015;**3**(39):10146−53. https://doi.org/10.1039/C5TC02689J.

84. Kong X, Li X, Wu S, Zhang X, Liu J. Efficient conversion of cotton stalks over a Fe modified HZSM-5 catalyst under microwave irradiation. *RSC Adv* 2016;**6**(34):28532−7.

85. Yang Q, Shi Y, Fang Y, Dong Y, Ni Q, Zhu Y, et al. Construction of polyaniline aligned on magnetic functionalized biomass carbon giving excellent microwave absorption properties. *Compos Sci Technol* 2019;**174**(March):176−83. https://doi.org/10.1016/j.compscitech.2019.02.031.

86. Wang L, Guan H, Hu J, Huang Q, Dong C, Qian W, et al. Jute-based porous biomass carbon composited by Fe_3O_4 nanoparticles as an excellent microwave absorber. *J Alloys Compd* 2019;**803**:1119−26. https://doi.org/10.1016/j.jallcom.2019.06.351.

87. Lou Z, Li Y, Han H, Ma H, Wang L, Cai J, et al. Synthesis of porous 3D Fe/C composites from waste wood with tunable and excellent electromagnetic wave absorption performance. *ACS Sustain Chem Eng* 2018;**6**(11):15598—607.

88. Gou G, Meng F, Wang H, Jiang M, Wei W, Zhou Z. Wheat straw-derived magnetic carbon foams: in-situ preparation and tunable high-performance microwave absorption. *Nano Res* 2019;**12**(6):1423—9.

89. Liu X, Song P, Hou J, Wang B, Xu F, Zhang X. Revealing the dynamic formation process and mechanism of hollow carbon spheres: from bowl to sphere's shape. *ACS Sustain Chem Eng* 2017;**62**.

90. He G, Duan Y, Pang H. Microwave absorption of crystalline Fe/MnO@C nanocapsules embedded in amorphous carbon. *Nano-Micro Lett* 2020;**12**(1):1—16. https://doi.org/10.1007/s40820-020-0388-4.

91. Zhang K, Wu F, Li J, Sun M, Xie A, Dong W. Networks constructed by metal organic frameworks (MOFs) and multiwall carbon nanotubes (MCNTs) for excellent electromagnetic waves absorption. *Mater Chem Phys* 2018;**208**:198—206. https://doi.org/10.1016/j.matchemphys.2018.01.008.

92. Zhang X, Hu Z, Xiao X, Sun L, Han S, Chen D, et al. Fe$_3$O$_4$@porous carbon hybrid as the anode material for a lithium-ion battery: performance optimization by composition and microstructure tailoring. *New J Chem* 2015:3435—43.

93. Quan L, Qin FX, Estevez D, Wang H, Peng HX. Magnetic graphene for microwave absorbing application: towards the lightest graphene-based absorber. *Carbon N Y* 2017;**125**:630—9. https://doi.org/10.1016/j.carbon.2017.09.101.

94. Wu H, Dong S, Huang G, Zheng Q, Huang T. The extraction of four endocrine disrupters using hollow N-doped mesoporous carbon spheres with encapsulated magnetite (Fe$_3$O$_4$) nanoparticles coupled to HPLC-DAD determination. *Microchem J* 2021;**164**(January):105984. https://doi.org/10.1016/j.microc.2021.105984.

95. Zhou X, Liu Y, Ren Y, Mu T, Yin X, Du C, et al. Engineering molecular polymerization for template-free SiOx/C hollow spheres as ultrastable anodes in lithium-ion batteries. *Adv Funct Mater* 2021;**2101145**:1—9.

96. Li W, Zhang Z, Wang J, Qiao W, Long D, Ling L. Low temperature catalytic combustion of ethylene over cobalt oxide supported mesoporous carbon spheres. *Chem Eng J* 2016;**293**:243—51. https://doi.org/10.1016/j.cej.2016.02.089.

97. Wang J, Chen Y, Zhang Z, Ai Y, Liu L, Qi L, et al. Microwell confined iron oxide nanoparticles in honeycomblike carbon spheres for the adsorption of Sb(III) and sequential utilization as a catalyst. *ACS Sustain Chem Eng* 2018;**6**(10):12925—34.

98. Xia X, Wang Y, Wang D, Zhang Y, Fan Z, Tu J, et al. Atomic-layer-deposited iron oxide on arrays of metal/carbon spheres and their application for electrocatalysis. *Nano Energy* 2016;**20**:244—53. https://doi.org/10.1016/j.nanoen.2015.12.015.

99. Liu HJ, Bo SH, Cui WJ, Li F, Wang CX, Xia YY. Nano-sized cobalt oxide/mesoporous carbon sphere composites as negative electrode material for lithium-ion batteries. *Electrochim Acta* 2008;**53**(22):6497—503.

100. Basu M. In-situ developed carbon spheres function as promising support for enhanced activity of cobalt oxide in oxygen evolution reaction. *J Colloid Interface Sci* 2018;**530**:264—73. https://doi.org/10.1016/j.jcis.2018.06.087.

101. Trukawka M, Wenelska K, Singer L, Klingeler R, Chen X, Mijowska E. Hollow carbon spheres loaded with uniform dispersion of copper oxide nanoparticles for anode in lithium-ion batteries. *J Alloys Compd* 2021;**853**:156700. https://doi.org/10.1016/j.jallcom.2020.156700.

102. Sawant SY, Somani RS, Panda AB, Bajaj HC. Utilization of plastic wastes for synthesis of carbon microspheres and their use as a template for nanocrystalline copper(II) oxide hollow spheres. *ACS Sustain Chem Eng* 2013;**1**(11):1390−7.

103. Wang N, Gao Y, Gong J, Xinyan M, Zhang X, Guo Y, et al. Synthesis of manganese oxide hollow urchins with a reactive template of carbon spheres. *Eur J Inorg Chem* 2008;**24**: 3827−32.

Further reading

1. Tian M, Du M, Qu L, Chen S, Zhu S, Han G. Electromagnetic interference shielding cotton fabrics with high electrical conductivity and electrical heating behavior: via layer-by-layer self-assembly route. *RSC Adv* 2017;**7**(68):42641−52.

2. Zhang T, Kumari L, Du GH, Li WZ, Wang QW, Balani K, et al. Mechanical properties of carbon nanotube-alumina nanocomposites synthesized by chemical vapor deposition and spark plasma sintering. *Compos A Appl Sci Manuf* 2009;**40**(1):86−93. https://doi.org/ 10.1016/j.compositesa.2008.10.003.

3. Wei Y, Shi Y, Jiang Z, Zhang X, Chen H, Zhang Y, et al. High performance and lightweight electromagnetic wave absorbers based on TiN/RGO flakes. *J Alloys Compd* 2019;**810**: 151950. https://doi.org/10.1016/j.jallcom.2019.151950.

4. Song WL, Cao MS, Fan LZ, Lu MM, Li Y, Wang CY, et al. Highly ordered porous carbon/ wax composites for effective electromagnetic attenuation and shielding. *Carbon N Y* 2014; **77**:130−42. https://doi.org/10.1016/j.carbon.2014.05.014.

5. Latham KG, Jambu G, Joseph SD, Donne SW. Nitrogen doping of hydrochars produced hydrothermal treatment of sucrose in H_2O, H_2SO_4, and NaOH. *ACS Sustain Chem Eng* 2014; **2**(4):755−64.

6. Wang L, Qiu H, Liang C, Song P, Han Y, Han Y, et al. Electromagnetic interference shielding MWCNT-Fe_3O_4@Ag/epoxy nanocomposites with satisfactory thermal conductivity and high thermal stability. *Carbon N Y* 2019;**141**:506−14.

7. Wang H, Meng F, Li J, Li T, Chen Z, Luo H, et al. Carbonized design of hierarchical porous carbon/Fe_3O_4@Fe derived from loofah sponge to achieve tunable high-performance microwave absorption. *ACS Sustain Chem Eng* 2018;**6**(9):11801−10.

8. Zhou X, Jia Z, Feng A, Qu S, Wang X, Liu X, et al. Synthesis of porous carbon embedded with NiCo/$CoNiO_2$ hybrids composites for excellent electromagnetic wave absorption performance. *J Colloid Interface Sci* 2020;**575**:130−9. https://doi.org/10.1016/ j.jcis.2020.04.099.

9. Sun H, Che R, You X, Jiang Y, Yang Z, Deng J, et al. Cross-stacking aligned carbon-nanotube films to tune microwave absorption frequencies and increase absorption intensities. *Adv Mater* 2014;**26**(48):8120−5.

Metal oxide–carbon composite: synthesis and properties by using conventional enabling technologies

2

Muhammad Bilal[1], Zia Ur Rehman[1], Jianhua Hou[1], Saif Ali[2], Sami Ullah[2] and Junaid Ahmad[2]
[1]School of Physics, College of Physical Science and Technology and School of Environmental Science and Engineering, Yangzhou University, Yangzhou, Jiangsu, PR China; [2]Department of Physics, Division of Science and Technology, University of Education Lahore, Lahore, Punjab, Pakistan

2.1 Introduction

Globally, there has been a rise in scientific progress aimed at making life on the orb sustainable. This has led to an increase in the demand for materials used in the assembling of different products. The unique physical, chemical, and mechanical properties of carbon nanostructures have triggered extensive research motivation since the discoveries by Kroto et al. in 1985,[1] followed by Iijima in 1991.[2] Many experiments and publications on carbon-based nanostructures (CNs) have been carried out since then. Numerous special properties exhibited by CNs have been driven by studies, such as elevated surface area, improved optical properties, extraordinary strength, and transport characteristics of the charge carriers (such as electrons and holes).[3,4] Such characteristics make CNs applicable for various applications, i.e., catalysis, energy storage and conversion, gas sensing, and many more.[5–8] Most of these applications demand highly efficient CNs of a specific kind. Therefore, a wide variety of CN shapes have been identified, including fibers, ribbons, tubes, and spheres.[9–12]

The different forms of CN fabrication are based on the starting materials and reaction conditions used for their synthesis. Bai et al. reported that various forms of CNs, such as carbon nanofibers (CNFs), multiwalled carbon nanotubes (MWCNTs), and single-walled carbon nanotubes (SWCNTs), have been developed using benzene and ferrocene.[13] They concluded that the product is determined by the precursor-to-catalyst ratio. Their finding was confirmed by Coville and Nyamori, where the shape and size of the CNs were also determined by the carbon-to-iron (catalyst) ratio in addition to product distribution.[14] In addition, Sevilla and coworkers documented the fabrication of graphitic carbon nanostructures via gluconate dehydrates. They stated that

Metal Oxide-Carbon Hybrid Materials. https://doi.org/10.1016/B978-0-12-822694-0.00021-1

cobalt and iron nanoparticles effectively catalyzed the reaction at 900° C/1000° C.[15] Various CNs were synthesized using silica- and gold-containing precursors, such as broken hollow and wormlike carbon nanostructures.[16]

The fabrication of CNs such as carbon nanospheres, carbon nanotubes (CNTs), and CNFs with aromatic and long-chain hydrocarbons as the carbon source in the incidence of various metal-salt catalysts is mentioned in several papers.[17−19] In general, CNs are formed by dint of pyrolysis and subsequent solvothermal action to purify the substance. Pyrolysis requires the use of heat in an inert environment to decompose a mixture having soluble metal salt and organic carbon-rich compounds into char. In certain situations, water or oxygen may be used in such a way that partial combustion takes place.[20] As a result, two steps are involved in pyrolysis: the synthesis of nanoparticles of metal and the growth of a coating substance with templates of nanoparticles.[21] Contrariwise, the solvothermal method includes the post-synthesis purification of nanostructures with solvents in an autoclave, with the term hydrothermal treatment used when water is the solvent.

The major challenges faced in the fabrication of CNs in these synthesis procedures include economic and large-scale production and the overwhelming effects of low manufacturing yields. Consequently, there is a need for groundbreaking research activities that rely on the use of readily accessible and cost-effective precursors. Renewable resources should be extensively used in the future, as they can be replenished comparatively easily. In general, biomass, that is, the whole weight of flora and fauna in a given area, may be considered and explored.[22] As curiosity drives innovation, researchers worldwide, including engineers, chemists, physicists, and materials scientists, are highly committed to discovering more techniques to achieve the ultimate potential of various technologies. For this purpose, they developed different methods to optimize the properties of materials. One of these methods is the fabrication of nanocomposites.

A composite consists of two or more diverse materials that collectively contribute to the whole system with the aim of mixing the leading properties of each component but simultaneously maintaining their own identities. The formation of metal oxide−carbon (MO−C) composites is of vital and technical importance because the composite combines their special properties and also exhibits certain new extraordinary properties induced by their interaction.[23,24] Thus, a novel class of nanocomposites can be developed with exceptional properties that satisfy a variety of applications in various arenas. For instance, due to its chemical stability, strong resistance to oxidation, and high hardness, alumina is a commonly used ceramic material. But its applications are hindered as a result of its low fracture toughness. Due to its peculiar one-dimensional structure with robust thermal and mechanical properties, CNTs have commonly been used to stabilize and improve the fracture toughness of alumina.[25]

Over the last decade, conventional enabling technologies have become appealing choices to fabricate MO−C composites owing to their remarkable energy savings and process intensification. This chapter primarily presents various conventional enabled synthesis approaches such as the hydrothermal method, electrochemical deposition, the sol−gel method, the solution mixing method, self-assembly, chemical vapor deposition, chemical bath deposition, and many others for fabricating MO−C composites. We also explore how laboratory technological developments will push the boundaries of what can be achieved using conventional enabling techniques for MO—C composites synthesis. Furthermore, we also discuss challenges and future prospects in the synthesis using conventional enabling technologies for MO—C composites.

2.2 Specific properties of metal oxide—carbon composites

MO—C composites have gained much interest among the researcher due to their specific properties such as effective surface area, high energy density, good cyclic stability, high power density, good rate capability, and specific capacitance. They are a good candidate for energy storage devices such as batteries and supercapacitors due to their abovementioned properties. MO—C composite-based electrodes for supercapacitors show excellent performance. Supercapacitors of MO—C composites provide fast charge—discharge, high power density, and high specific capacitance. Electrodes based on carbon materials have many benefits in energy storage devices due to control of porosity, high chemical stability and thermal, greater specific surface area, and cost-effectiveness that make carbon a higher desirable candidate for the electrode. However, activated carbon-based supercapacitors retain energy density at a much lower level and much lower specific capacitance than other electrical conversions/storage devices. On the other side, metal oxide (MO)-based electrodes have high specific capacitance and energy densities. But electrodes based on MOs are limited in commercially available electrodes for supercapacitors due to some drawbacks such as difficulty in porosity tailoring, poor long-term stability, and lower conductivity (due to high resistivity). The abovementioned drawbacks of MOs affect their power density and rate capability.[26,27]

Considering all the abovementioned advantages and disadvantages of carbon-based and MO-based electrode materials, scientists are working on MO—C composites to get desirable properties. Scientists have dedicated their efforts to making new symmetric and asymmetric hybrid systems in which materials based on carbon are used as substrates and MOs as active material. This approach helps to achieve higher energy and power densities because of the contribution of both pseudocapacitance and electrochemical double-layer capacitance (EDLC) by MOs and carbon, respectively.

2.2.1 Specific capacitance

The specific capacitance (C) of supercapacitors is defined as

$$C = \frac{Q}{V}$$

Here, V is the operating voltage window, and Q is the charge stored per unit mass on the electrode.

EDLC and pseudocapacitance are two ways that supercapacitors store energy.[28]

2.2.2 Electrochemical double-layer capacitance

EDLC capacitance arises from charge separation at the electrolyte/electrode interfaces because they consist of activated carbon and have a high surface area. EDLC appears from the electrical double layer that surrounds the surface of the electrode. EDLC stores energy at the electrolyte and electrode interface due to the depletion of oppositely charged species. Accumulation of electrons at an electrode is a nonfaradaic process. In this case, specific capacitance is measured by the following equation:

$$C = \frac{\varepsilon_r \varepsilon_\circ}{d} A$$

Here, $\varepsilon_\circ$ is the permittivity of free space in the electric double layer, ε_r is the permittivity of the medium, and d represents effective thickness of the electrical double-layer. The specific surface area of the electrode material is represented by A. The above relation shows that the specific capacitance can be increased by a higher specific surface area and lower electrical double-layer thickness.[26,27] A systematic illustration of EDLC is shown in Fig. 2.1.

2.2.3 Pseudocapacitance

Pseudocapacitance is initiated in the redox reaction of electrode material with electrolyte. Electrons are produced and transferred in the redox reaction across the electrolyte/

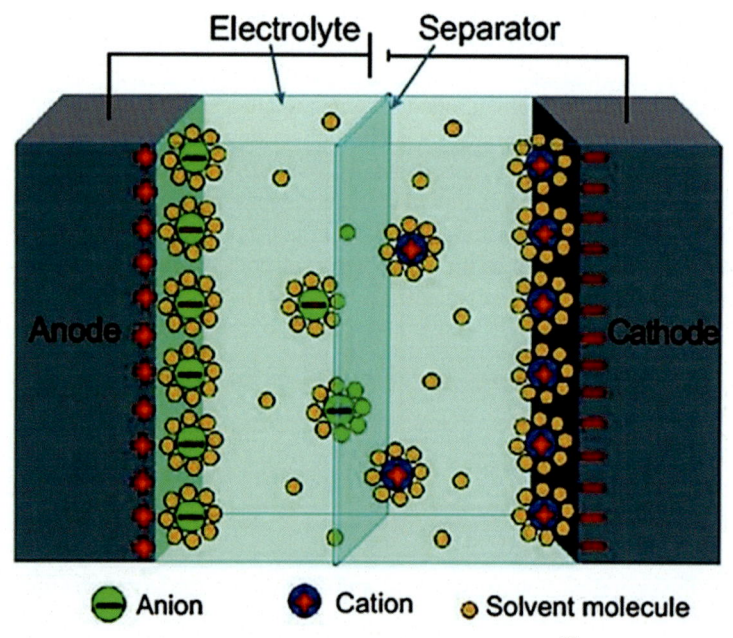

Figure 2.1 Illustration of electrochemical double-layer capacitance.[33]

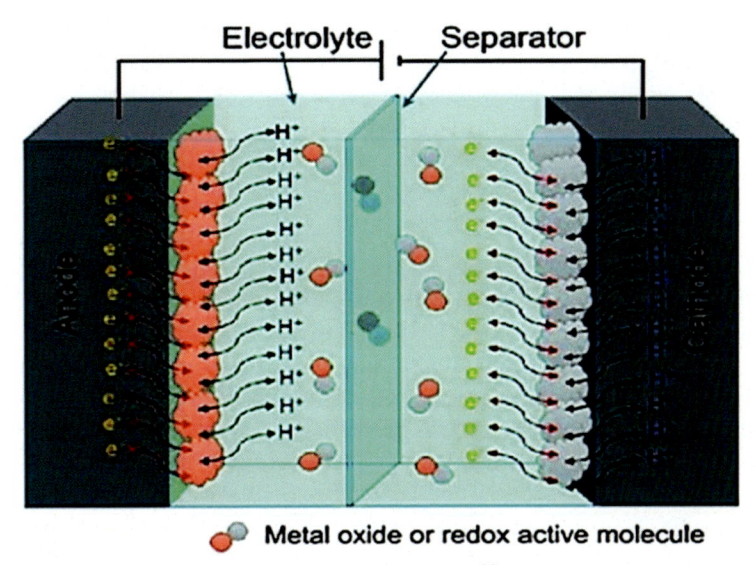

Figure 2.2 Systematic representation of pseudocapacitor.[33]

electrode interface. The process of accumulating electrons at the electrode is a faradaic process. The systematic representation of pseudocapacitors is shown in Fig. 2.2. Theoretical measurements of the specific capacitance of MOs can be derived by the following equation:

$$C = \frac{n \times F}{M \times V}$$

where F is Faraday constant, n is the mean number of electrons transferred in the redox reaction, M is the molar mass, and the operating voltage window is represented by V.[26,29,30]

2.2.4 Energy density and power density

The basic definition of energy density (E) for supercapacitors is

$$E = \frac{CV^2}{2}$$

This shows that E is proportional to the specific capacitance (C) and the square of the operating voltage window V.

The maximum power density of a supercapacitor can be measured using the following relation:

$$P_{max} = \frac{V^2}{4R}$$

P_{max} is directly proportional to the square of operating voltage and inversely proportional to the equivalent series resistance of all components in the device. To obtain a higher power density and energy density, specific capacitance and operating voltage should be increased, and the equivalent series resistance should be decreased.[26,31]

2.2.5 Hybrid supercapacitors

Lithium-ion batteries (LIBs) have the highest energy density (150−200 W h/kg) of current energy-storage devices and technologies, but their life cycle is limited. On the other hand, supercapacitors with long life cycles benefit from their high power capability (2−5 kW/kg). Recently, hybrid supercapacitors were proposed to combine the benefits of both LIBs and supercapacitors. Hybrid supercapacitors provide long life cycles, fast power capability, and the highest energy density.[32]

The storage mechanism of hybrid supercapacitors is governed by a combination of the EDLC and pseudocapacitor. Their combination removes limitations and leads to higher capacitance. A systematic illustration of hybrid supercapacitors is shown in Fig. 2.3. Without sacrificing affordability and cycling stability, hybrid supercapacitors have increased their power capacity a great deal, and their energy densities are better than EDLC and pseudocapacitors.[33,34] They utilize both faradaic and nonfaradaic processes to store charges. Researchers are focusing on three different types of hybrid supercapacitors:[34,35]

- asymmetric hybrids
- battery-type hybrids
- composite hybrids

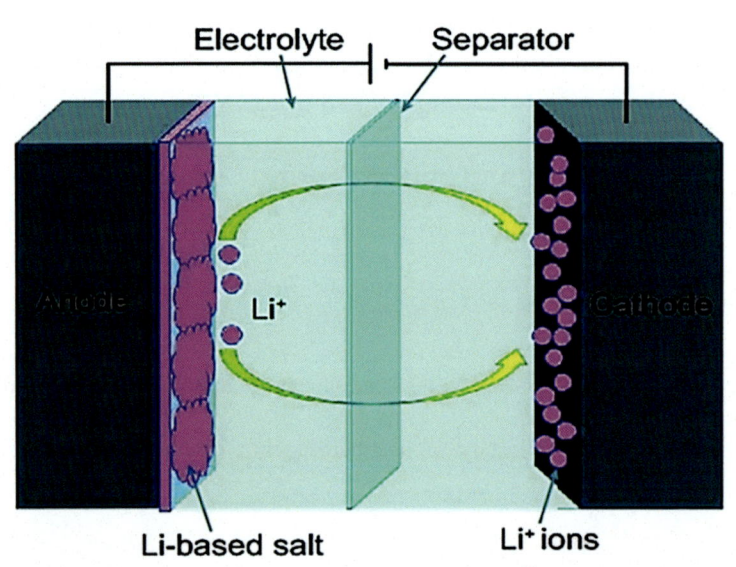

Figure 2.3 Systematic representation of hybrid supercapacitor.[33]

Asymmetric hybrids couple the EDLC electrode with a pseudocapacitive electrode that combines the faradaic and nonfaradaic processes. Similar to asymmetric hybrids, battery-type hybrids combine two different electrodes—that is, the combination of a battery-type electrode and a supercapacitor electrode. In Composite hybrids, the single electrode combines the chemical and physical charge storage mechanisms. The composite electrodes integrate carbon-based materials with either MOs or conducting polymers. Carbon-based materials with high surface areas facilitate a capacitive double layer of charge that increases the contact between electrolytic and pseudocapacitive materials. Pseudocapacitive materials then further enhance capacitance through a faradaic reaction at the composite electrode.[35–38] In Fig. 2.4 illustrates all possible types/combinations of hybrids. Table 2.1 compares the properties of various types of supercapacitors.

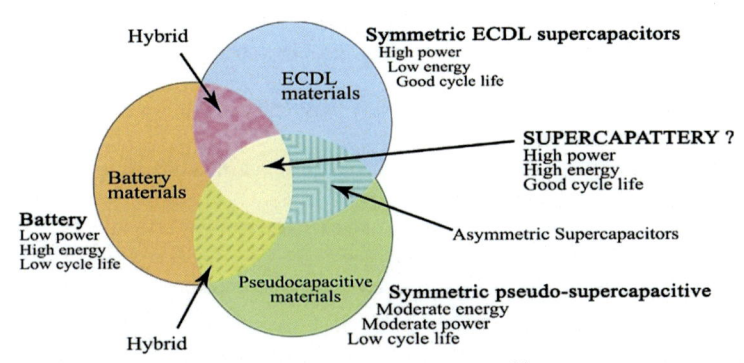

Figure 2.4 Coupling of properties of hybrid supercapacitors.[34]

Table 2.1 Comparison of various supercapacitor and lithium-ion battery properties.[39]

Properties	Supercapacitor (SC)			Lithium-ion battery
	EDLC SC	Pseudo SC	Hybrid SC	
Charge time [s]	1–10	1–10	100	600
Cell voltage [V]	2.7	2.3–2.8	2.3–2.8	3.6
Cost per kWh [USD]	$\sim$10,000	$\sim$10,000	A	$\sim$140
Cycle life	1,000,000	100,000	500,000	500
Specific energy [Wh/kg]	3–5	10	180	250
Type of electrolyte	Aprotic or protic	Protic	Aprotic	Aprotic
Self-discharge per month [%]	60	60	A	4
Operating temperature [°C]	$\sim$40–65	$\sim$40–65	$\sim$40–65	$\sim$20–60

a, Data not available.

The specific capacitance is calculated using the following equation:

$$C_s = \frac{I}{\frac{dV}{dt}(w)}$$

where C_s is specific capacitance, dV/dt is the voltage scanning rate, w is the total mass, and I is the average current in amperes.[33]

2.2.6   Properties of ideal metal oxide—carbon composites

The ideal composite for the electrode is required to possess the following properties:

- **Specific surface area:** To achieve high specific capacitance, the specific surface area should be high.
- **Porosity:** The porosity of the ideal composite should be controllable, which affects the rate capability and specific capacitance.
- **Electronic conductivity:** Electronic conductivity should be high, which is good for power density and rate capability.
- **Electroactive sites:** For a good MO—C composite electroactive sites should be desirable as they enable pseudocapacitance.
- **Thermal and chemical stability:** Thermal and chemical stability should be high because it affects cyclic stability.
- **Cost:** The cost of the composite should be low, and the manufacturing technique should be easy.

Supercapacitors based on carbon materials are operated under EDLC. Their maximum capacitance is controlled by the pore size distribution of the active electrode and surface area (typically $\sim$ 150F/g or 0.15—0.4 F/m^2 for carbon). Typically, carbon-based EDLC supercapacitors with 3—5 W h/kg energy densities are commercially available. This energy density is much lower than required to fulfill the energy needs for energy-storage devices, such as vehicles, solar power plants, etc.[40] Table 2.2 compares the specific capacitances of various carbon-based materials.

MOs have improved energy density and specific capacitance but have low conductivity. The invested MOs include NiO, MnO_2, Co_3O_4, TiO_2, ZnO, V_2O_5, CuO, WO_3,

Table 2.2 Comparison of specific capacitance of various carbon-based materials.[40]

Material	Specific capacitance (F/g)
Carbon spheres	132
Hallow mesoporous carbon spheres	120
Hollow carbon spheres	180
Porous carbon spheres	150
Carbon nanospheres	219
Mesoporous carbon spheres	218

Table 2.3 Comparison of specific capacitance and conductivity of various metal oxides.[26]

Materials	Theoretical capacitance (F/g)	Conductivity (S/cm)
NiO	2584 (0.5 V0[36])	0.1–0.32
MnO_2	1380 (0.9)[34]	10^{-8}–10^{-6}
V_2O_5	2120 (1 V)	10^{-4}–10^{-2}
Co_3O_4	3560 (0.45)[38,39]	10^{-4}–10^{-2}
RuO_2 xH_2O	1200–2200 (1.23 V)[42]	10^3 for poly crystalline

and Fe_2O_3. Some MOs show excellent pseudocapacitance. But MOs alone may not be prominent candidate materials for supercapacitors, as they have the following disadvantages:

1. Except for RuO_2, conductivity is very low for MOs, as indicated in Table 2.3. This is due to the high resistivity of MO because it increases both the charge transfer and sheet resistance of the electrode. At high current, it causes a large IR loss. Thus, poor rate capability and power density hinder the practical application of MOs.
2. The poor distribution, porosity, and surface area are difficult to modify in MOs.
3. During the process of charging and discharging, strain is developed in MO materials that may cause cracking of electrodes, and long-term stability may become poor.[26]

Scientists have developed various MO—C composites and undertaken the merits and demerits of carbon and MOs. The composite-making technique combines the merits of both carbon and MOs and removes the shortcomings of both. In this way, MO—C composites with desirable or tunable properties are obtained. In MO—C composites, the carbon provides the proper channels for charge transport and acts as physical support for MOs. In MO—C composites, the metal-oxides are the key component that stores energy and charge. MOs contribute to the high energy density and high specific capacitance of MO—C composites due to their electro-activities. The microstructures, physical properties, and compositional constituents of MO—C composites direct the performance of supercapacitors. Similarly, the electronic conductivity, specific surface area, electrode porosity, and pore size distribution affect the performance of the cell. The main objective of the scientist behind the development of MO—C composites is to grow electrode materials that have good cyclic stability, high energy density as well as high power density, and rate capability. Table 2.4 briefly compares carbon-based materials, MOs, and MO—C composites.[26,31,41,42]

2.3 General routes for making metal oxide—carbon composites

Transition MOs have proved to be the most promising material for supercapacitors and as electrode material for energy-storage devices. They are the main component that stores the charge and energy. In carbon-based MO electrodes, MOs provide high

Table 2.4 Comparison of metal oxides, carbon, and metal oxide—carbon composites.[26]

Material type	Specific capacitance	Surface area	Conductivity	Cost	Rate capability	Stability
Carbon	Low	High	High	Low	High	Good
Metal oxides	High	Normally low	Low	High	Low	Poor
MO—C composites	High	Controlled by the support of carbon	Tunable with the help of carbon support	Medium	Good (controllable)	Good

specific capacitance and high energy density. The performance of a supercapacitor depends on microstructures, physical properties, and constituents of the composite. The electrode porosity, electronic conductivity, pore size, specific surface area all govern the performance of supercapacitors. Nanostructures of carbon material have been synthesized and coupled with MOs to form composites for supercapacitor electrodes. The purpose of making MO–C composites is to develop electrodes that possess high power density, energy density, good rate capability, and cyclic ability simultaneously. Table 2.5 compares carbon, MO, and MO–C composite electrodes.[26]

Hybridized nanostructure of MOs with carbon material has been considered the most promising material to improve the performance of supercapacitors. To improve the efficiency of the cell, different techniques have been used to synthesize the MO–C composite. Some of those techniques are illustrated in Fig. 2.5.

The purpose of developing an MO–C composite is to utilize the full potential of carbon materials and MOs for supercapacitors and LIBs. Combining the effects of carbon and MOs maximizes their potential application to improve the current electrode problem and can enhance energy-storage capacity. For example, in a composite of graphene nanostructure with MOs, both materials played a vital role during the chemical process. Graphene provides chemical functionality and compatibility to easy processing for MOs, and MOs provide high capacity depending on their structure size and crystallinity. MOs dispersed on graphene sheets prevent conglomeration and restacking of graphene sheets and provide a large surface area that enables high electrochemical activity. Graphene nanostructures provide MOs and environments in which they can stimulate nucleation and the formation of fine MO nano/microstructures with uniform dispersion and controlled morphology on the surface of graphene with high chemical functionality. Significant synergistic effects often occur in graphene/MO composites because of size effects and interfacial interactions, as shown in Fig. 2.6.[43]

To synthesize a composite of graphene and MO, general wet-chemistry techniques, such as in situ deposition, sol–gel, and hydrothermal process, are widely used. First, both GO and reduced graphene oxide (rGO) are dispersed in aqueous or organic solvents by electrostatic stabilization and chemical fictionalizations. Such dispersions are a good template suspension for chemical reactions with metal ions from the precursors of inorganic and organic metal salts, which undergo hydrolysis or in situ redox reactions to anchor them on the surface of graphene with rich functionalities, followed by annealing. Special emphasis is given to the important role of graphene that can suppress the agglomeration of MO nanoparticles. For example, research under the same experimental conditions have shown that small RuO_2 nanoparticles with a size of 5–20 nm are homogeneously anchored on the surface of graphene, as shown in Fig. 2.7.[43]

Guanglin Xia et al. reported and synthesized transition MOs into ultrafine nanoparticles implanted in hierarchically porous CNFs by electro spinning with metal azides serving a pore generator, followed by a simple calcination process.[44] The template method coupled with the solvothermal and hydrothermal method is often used to prepare the nanostructure of mesoporous material and MOs. Template-directed methods can be divided into two types, hard-template and soft-template. Hard template and soft template strategies can be used for the synthesis of ordered mesoporous carbon structures (OMCs). The mesostructural stability of mesoporous carbon depends on their synthesis process.[45]

Table 2.5 The pros and cons of graphene, metal oxides, and graphene/metal oxide composites in lithium-ion batteries and electrochemical capacitors.[43]

Pros of graphene	Cons of graphene	Pros of metal oxides (MOs)	Cons of MOs	Pros of graphene \MO composites
Superior electrical conductivity	Serious agglomeration	Very large capacity\capacitance	Poor electrical conductivity	Synergistic effect
Abundant surface functional groups	Restacking	High packing density	Large volume change	Suppressing the volume change of MO
Thermal and chemical stability	Large irreversible capacity	High energy density	Severe aggregation or agglomeration	Suppressing agglomeration of MO and restacking of graphene
Large surface area	Low initial coulombic efficiency	Rich source	Large irreversible capacity	Uniform dispersion of MO
High surface-to-volume ratio	Fast capacity fading		Low initial coulombic efficiency	Highly conducting and flexible network
Ultrathin thickness	No clear lithium storage mechanisms		Poor rate capability	High capacity, good rate capability
Structural flexibility	No obvious voltage plateau		Poor cycling stability	Improved cycling stability
Broad electrochemical window	Large voltage hysteresis			Improved energy

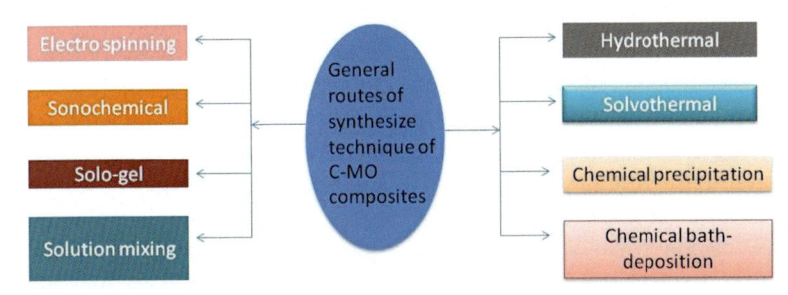

Figure 2.5 Different General routes of making C-MO composites.

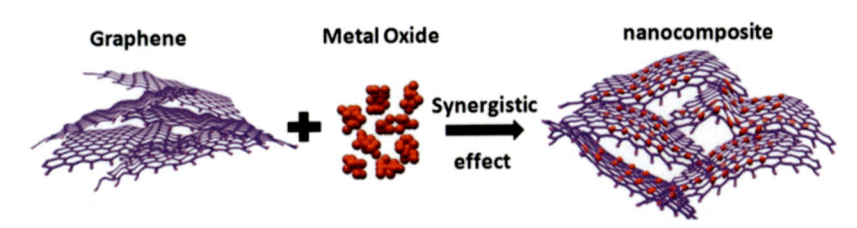

Figure 2.6 Schematic of the preparation of graphene/metal oxide composites with synergistic effects between graphene and metal oxides.[43]

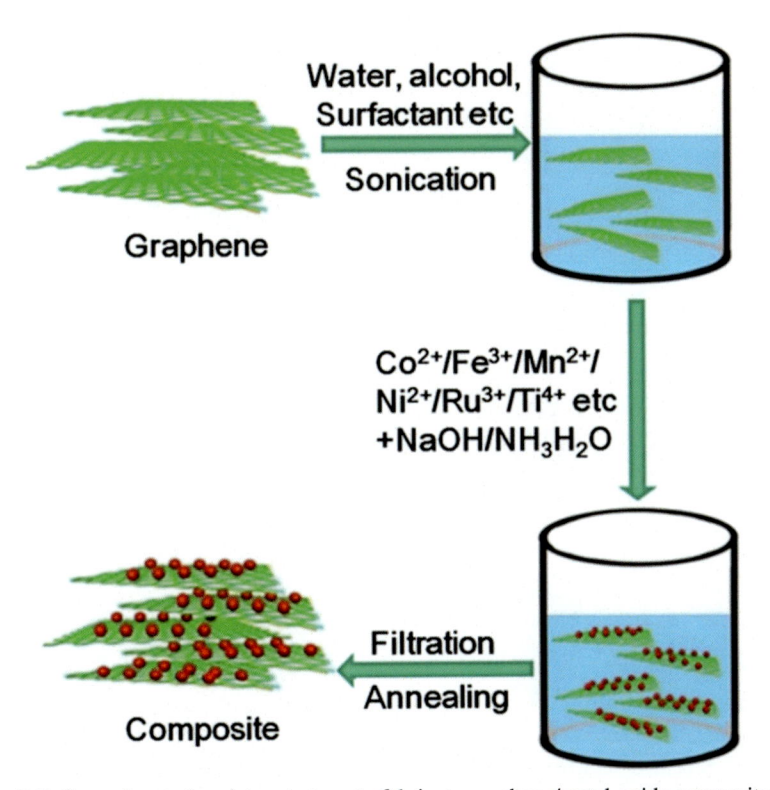

Figure 2.7 General wet-chemistry strategy to fabricate graphene/metal oxide composites.

The hard-templated mesoporous carbons have large pore volumes,[46] which are the potential for high MO loading content. Most of the nanostructure of carbon and MOs is connected by nanobridges derived from micropores. These nanobridges are easily broken and collapse during the loading of MOs or due to volume expansion during the charging and discharging process. As a result, many hard-templated MO−C anodes lose 30%−50% capacity over 100 cycles.[47,48] But soft-templated carbon nanostructures have an archlike carbon framework. They are more stable than hard-templated carbon when they are loaded with guest nanoparticles. Soft-templated carbon nanostructure has its limits due to their relatively low pore volume, which limits their MO loading amount.[49−51] As mentioned, the drawbacks of the hard and soft-templated techniques, pore volume and structural stability, need to be balanced.

Junkai Hu et al. reported the synthesis of ordered mesoporous carbon with Fe_2O_3 nanowires by dual-template technique and used it for an LIB anode.[52] This dual-template strategy presents synergetic effects combining the advantages of both soft and hard single-template methods. The resulting OMCNW/Fe_2O_3 composite enables a high pore volume, high structural stability, enhanced electrical conductivity, and Li^+ accessibility. These features collectively enable excellent electrochemical cyclability (1200 cycles) and a reversible Li^+ storage capacity as high as 819 mA h/g at a current density of 0.5 A/g (see Fig. 2.8).

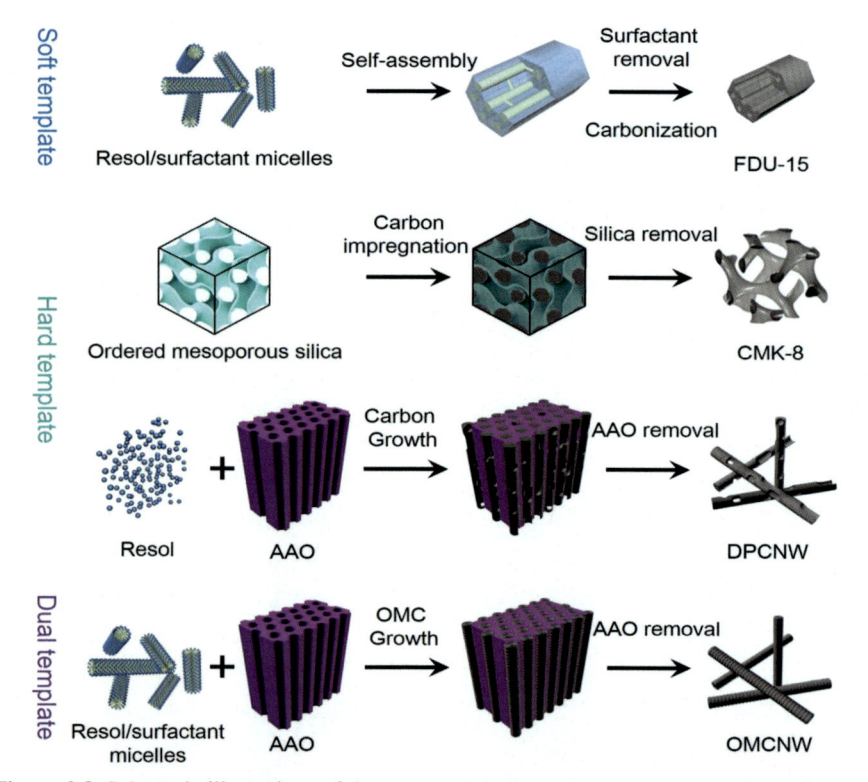

Figure 2.8 Schematic illustrations of the proposed dual-template method in comparison with soft- or hard-template methods for the controlled synthesis of porous carbon frameworks.[52]

2.4 Synthesis methods of carbon-based metal oxide composites for supercapacitors

The fabrication of MO—C composites can be achieved by utilizing various techniques such as the hydrothermal method, electrochemical deposition method, sol—gel method, chemical precipitation method, and so on.

2.4.1 Hydrothermal method

Chaitra et al.[53] synthesized the nanoparticles of hydrous ruthenium oxide (h-RuO$_2$) and the composite of h-RuO$_2$ with MWCNTs utilizing the facile hydrothermal method. They showed that this synthesized composite has potential for the supercapacitor electrode material. The synthesized nanoparticles of h-RuO$_2$ and nanocomposites of h-RuO$_2$/MWCNT were characterized by using various techniques such as BET surface area, TEM, Raman, PXRD, and SEM-EDS. Electrochemical performance of the synthesized materials was evaluated by CV, and specific capacitance for h-RuO$_2$ and h-RuO$_2$/MWCNT were observed to be 604 and 1585 F/g respectively at the scan rate of 2 mV/s with the potential window of 0—1.2 V. A device based on asymmetric supercapacitor was developed by utilizing the h-RuO$_2$/MWCNT as the positive electrode and activated carbon as the negative electrode. The device revealed a specific capacitance of 61.8 F/g at the scan rate of 2 mV/s. Furthermore, this device showed remarkable long-term stability for 20,000 cycles and 88% capacitance retention at a large current density of 25 A/g.[53] (see Fig. 2.9).

Cao et al.[54] synthesized the composites of mesoporous NiO and rGO using a simple hydrothermal technique. The synthesized materials were characterized by utilizing various techniques such as TEM, XRD, SEM, FTIR, BET, and Raman spectroscopy. The electrochemical performances of the as-synthesized materials were measured by using CV, EIS, and galvanostatic charge and discharge tests. Experimental results revealed that the synthesized composites showed excellent specific capacitance of

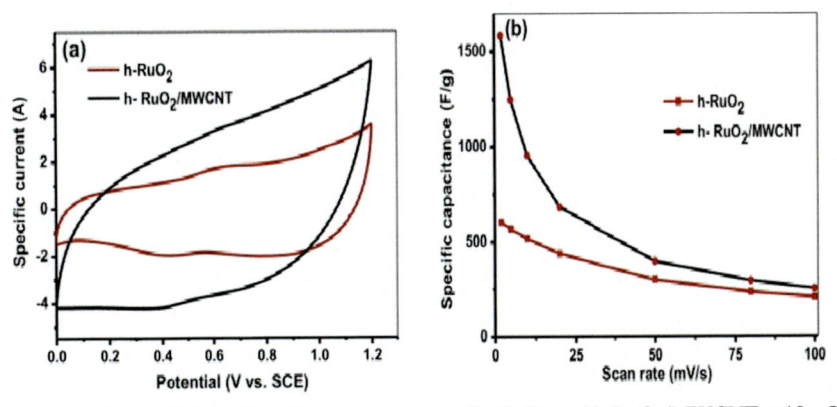

Figure 2.9 (A) Overlaid cyclic voltammetry curves of h-RuO$_2$ and h-RuO$_2$/MWCNT at 10 mV/s and (B) change in specific capacitance with scan rate.[53]

1016.6 F/g and remarkable cycling stability of 94.9% capacitance retention after 5000 cycles. These results suggested that the synthesized composites of NiO/rGO can be considered as the promising electrode material for the supercapacitors.[54]

Chen et al.[55] successfully developed the supercapacitor electrode by fabricating the nanocomposite CNT and vanadium pentoxide (V_2O_5) by utilizing the hydrothermal technique. The synthesized nanocomposites had a hierarchically porous structure, high surface area, and excellent electrical conductivity and were considered potential electrode material for the supercapacitor. The electrochemical performance of the nanocomposite was measured with the help of CV and galvanostatic charge/discharge technique. The synthesized material exhibited a specific capacitance from 200 to 440 F/g at a scan rate of 10−0.5 A/g. Furthermore, the synthesized material exhibited excellent power density and energy density.[55]

Wang et al.[56] synthesized the nanocomposite of MnO_2 with onion-like carbon by using the hydrothermal technique for the supercapacitor's electrode material. The synthesized nanocomposites were characterized with the help of XRD, SEM, TEM, FESEM, and Raman spectroscopy. The electrochemical performances were evaluated with the help of CV curves. The synthesized nanocomposites revealed the improved specific capacitance of 177.5 F/g at a high current density of 2 A/g. Furthermore, the synthesized material exhibits excellent cycling stability with retention of 99%−101% after 1000 cycles.[56]

Tang et al.[57] successfully fabricated the hybrid nanocomposite of MnO_2-nanowires and MWCNTs by using the hydrothermal technique. The presence of MWCNTs provides a pathway for quick electron movement, and MnO_2-nanowires exhibit a quick redox response. The synthesized material exhibited an excellent rate capability and provided an energy density of 17.8 Wh/kg at 400 W/kg, which is sustained even at 3340 W/kg with 0.5 M Li_2SO_4 electrolyte solution.[57]

2.4.2 Electrochemical deposition

Kim et al.[58] successfully synthesized the nanocomposites of RuO_2 and CNTs to enhance the specific capacitance and excellent cycling stability of RuO_2. They synthesized nanocomposites by using the electrochemical deposition technique. A uniform and thin deposition of RuO_2 on the substrate of CNT was attained using the promising cycling technique. The electrochemical performances of the synthesized materials were characterized using cyclic voltammetry (CV). The experimental values revealed that the synthesized material showed remarkable specific capacitance of 1170 F/g and enhanced rate capability[58](see Fig. 2.10).

Gosh et al.[59] deposited the ultrathin layer of the V_2O_5 on the CNF paper using the electrochemical deposition technique. The synthesized material exhibits the excellent specific capacitance of 1308 F/g with a 2 M KCL electrolyte solution. This improved performance was due to the enhanced external surface area of CNF and the maximum active spaces for the redox reaction of the V_2O_5 layer.[59]

Zhao et al.[60] fabricated the MnO_2/graphene/nickel foam (NF) nanocomposite for supercapacitor application by using the simple electrochemical deposition technique for the very first time. The specific capacitance of the synthesized material was

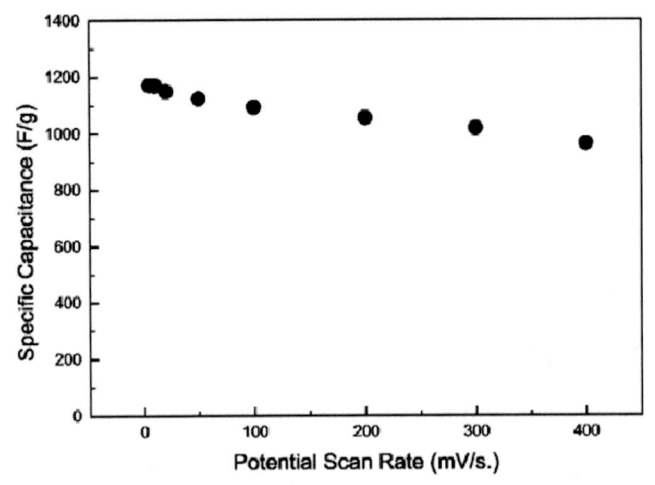

Figure 2.10 The specific capacitance of RuO$_2$ as a function of potential scan rate.[58]

observed by CV curves and was 476 F/g at a high current density of 1 A/g with 0.5 M Na$_2$SO$_4$ electrolyte solution. Furthermore, this material showed a high rate capacity (216 F/g) at 10 A/g.[60]

2.4.3 Sol−gel method

X. Liu and P.G. Pickup[61] synthesized the h-RuO$_2$ using the sol−gel technique, and annealing was performed at 110°C. They used the carbon fiber as the support of h-RuO$_2$ and measured the electrochemical performance with CV, impedance spectroscopy, and uniform current discharging. They reported that the material showed a specific capacitance of 977 F/g using 1M of H$_2$SO$_4$ as an electrolyte. During full discharge, power densities exceeded 30 W/g, and energy densities exceeded 30 Wh/Kg[61] (see Fig. 2.11).

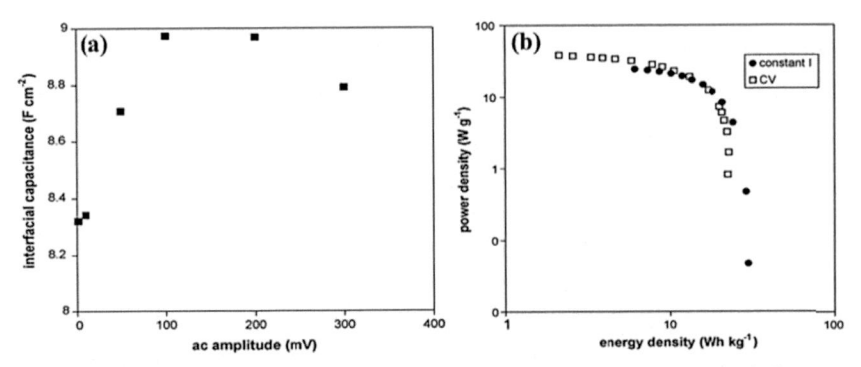

Figure 2.11 (A) Plot of Interfacial capacitance of supercapacitor versus amplitude from impedance. (B) Ragone plots between power density and energy density by different measurement methods.[61]

Rani et al.[62] successfully synthesized the nanocomposites of GO and TiO_2 by using the sol−gel method. The synthesized nanocomposite showed excellent areal-specific capacitance of 73.43 mF/cm^2 at a current density of 0.5 mA/cm^2. The developed device exhibited excellent power density of 3.5 mW/cm^2, an energy density of 0.007 mWh/cm^2, excellent flexibility, and cycling stability with 92% retention after 10,000 cycles.[62]

2.4.4   Chemical precipitation method

Lota et al.[63] successfully synthesized the nanocomposite of NiO and carbon by utilizing the chemical precipitation method. They synthesized the various composites with different concentrations of NiO and carbon. The synthesized materials were characterized by using the XRD and SEM. The electrochemical performance of the synthesized material was characterized by using CV, galvanostatic charge/discharge, and impedance spectroscopy. These electrochemical performances were observed in two- and three-electrode cells with 6 M KOH aqueous electrolytic solution. They observed the values of the capacitance for NiO, activated carbon, and a composite of NiO/activated carbon that varied from 11 to 90 F/g depending on the different concentrations of two components.[63]

Lee et al.[64] successfully synthesized the nanocomposite of NiO with CNT by using the facile chemical precipitation method. The availability of CNT in NiO remarkably enhanced the electrical conductivity of the NiO and active spaces for the redox reaction owing to its improved specific surface area, which enhanced the specific capacitance of the synthesized nanocomposites by 34%. Furthermore, they also observed that the power density and cycling stability of the nanocomposites were also improved. The specific capacitance as a function of discharge current density is illustrated in Fig. 2.12A, and its cyclic test is shown in Fig. 2.12B.[64]

2.4.5   Other methods

A. Yuan and Q. Zhang[65] successfully fabricated a supercapacitor based on nanostructured MnO_2/activated carbon and used lithium hydroxide as an electrolyte. They

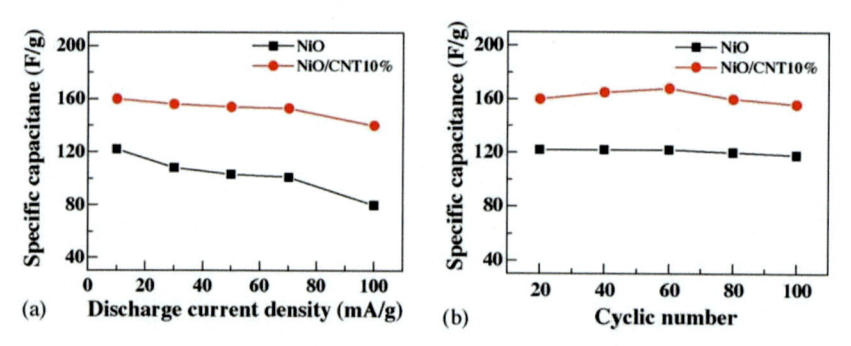

Figure 2.12 (A) Specific capacitance as a function of discharge current density for bare NiO and NiO/CNT (10%) composite. (B) Cyclic test of NiO (bare) and NiO/CNT (10%) composite for specific capacitance.[64]

observed the electrochemical performance of the electrodes by using the CV and EIS techniques. Charge and discharge analysis revealed that the discharge rate of the MnO_2/activated carbon supercapacitor having 1 M KOH electrolyte is outstanding, and the cyclic stability is better with 1 M LIOH electrolyte than with the 1 M KOH electrolyte.[65]

Wang et al.[66] prepared the nanocomposites of CNFs and MnO_2 using the in situ redox deposition and electrospinning technique. The synthesized nanocomposites exhibited a unique and conformal coaxial morphology. The electrochemical performance was measured with the help of CV, galvanostatic charging and discharging, and impedance techniques. The experimental results revealed that the synthesized material exhibits a specific capacitance of 557 F/g at a current density of 1 A/g with 0.1 M Na_2SO_4 electrolyte solution. Furthermore, this material showed excellent rate capability and long-term cycling stability of 94% retention after 1500 cycles.[66]

Lee et al.[67] synthesized the nanocomposite of MWCNT and MnO_2 using redox deposition of MnO_2. Structural analysis was carried out with the help of SEM and TEM, which revealed that the synthesized material is a potential structure for the applications of electrochemical capacitors. The synthesized material showed remarkably higher volumetric capacitance (246 F/cm^3) with excellent capacity retention up to 1000 mV/s owing to the quick movement of electrons and ions inside the electrodes.[67] Various synthesis methods are summarized in Table 2.6.

Table 2.6 Summary of different metal oxide—carbon composites prepared by various synthesis methods.

Composite	Synthesis method	Application	Reference
α-Fe$_2$O$_3$/graphene	Hydrothermal method	Li-ion batteries	68
A-Fe$_2$O$_3$/γFe$_2$O$_3$/ GO	Thermal decomposition	Li-ion storage	69
Fe$_3$O$_4$/MWCNTs	Hydrothermal method	Anodes in lithium-ion battery	70
Nickel oxide/ carbon nanotubes	Simple chemical precipitation method	Supercapacitor; electrochemical capacitance	64
RuO$_2$/MWCNTs	Microwave-polyol synthesis	Electrochemical capacitors	71
SnO$_2$/MWCNTs	Simple one-step chemical solution method	Lithium-ion batteries	72
TiO$_2$/SWCNTs	Solution mixing	Photocatalytic	73
CNF@NiCo$_2$O$_4$	Hydrothermal and annealing	High-performance supercapacitors	74
MoO$_2$/CNTs	Facile hydrothermal	Lithium-ion battery anode	75
MWNT—ZnO	Modified sol—gel	Nanoenvironmental application	76

2.5 Synthesis methods of graphene–metal oxide composites for photocatalysis

In C-based materials, graphene has gained great attention from researchers because of its unique and attractive properties, such as its strong structure, good mechanical strength, efficient oxidation and reduction reaction capability, and excellent charge transfer ability during photocatalytic activity. Consequently, the combination of graphene with MOs will increase the performance of different photocatalytic and energy conversion applications.[77,78] The graphene sheets attached with MO nanoparticles gave good results for photocatalytic activity due to the collective effects of these materials. To increase the efficiency of MO nanomaterials in different areas, the synthesis with G-based composites has attained more importance. This is due to the useful, outstanding, and extraordinary properties of these nanomaterials. Many sustainable efforts have been reported to synthesize the G-MO nanoparticles.[79] Additionally, several kinds of MOs have been fabricated with graphenes, such as TiO_2, ZnO, SnO_2, MnO_2, Co_3O_4, Fe_3O_4, Fe_2O_3, NiO, and Cu_2O. Various research articles have been written about excellent performances in the progress of developing graphene-based composites.[80,81] In this section overviewed the different synthesis techniques and effective strategies of carbon base material such as graphene with MOs and their applications in photocatalysis.

2.5.1 Different synthesis techniques of graphene–metal oxide composites

2.5.1.1 Solution mixing method

It is an effective and direct approach for preparing graphene-MO composites. Many research articles have been written on these MO–C composites. Peak et al. have synthesized G-SnO$_2$ composite by this solution maxing method.[82] In this method, SnO_2 solution is first fabricated by the hydrolysis of $SnCl_4$ with NaOH and graphene obtained by the chemical reduction of GO. To obtain the required composites, the dispersion of graphene is mixed with the prepared solution in the ethylene glycol.[83] Another technique used to make solar cells is TiO_2 nanoparticles (P25) mixed with nafion-coated graphene, and in this combination, nafion worked as glue to strongly bind the G-sheets and P25.[84] Another combination of graphene oxide collides and TiO_2 nanoparticles (P25) was accomplished ultrasonically, which is useful for UV-supported photocatalytic reduction.[85] Akhaven et al. used the same technique to obtain the thin film of G-TiO$_2$ composite.[86] So the solution mixing method is an effective technique that has been used for the fabrication of graphene-MO composites.

2.5.1.2 Sol–gel method

The sol–gel technique is the prevalent and popular method for the synthesis of MO materials and for the coating of films with the help of chlorides or M-alkoxides as ingredients that process through the different series of reactions like hydrolysis and polycondensation. TiO_2 is used as precursors for making different nanoparticles, nanorods,

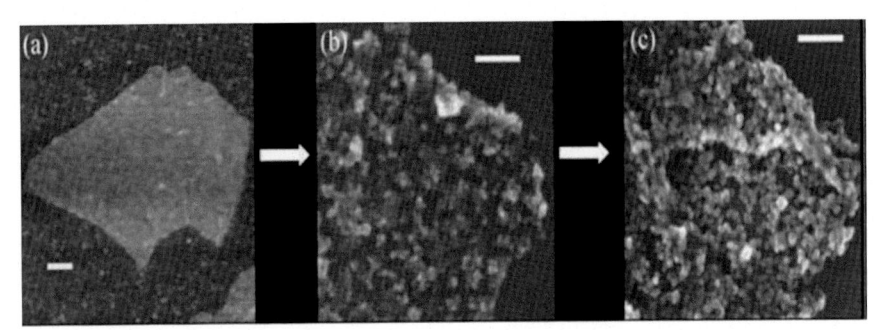

Figure 2.13 (A) AFM image of a starting graphene oxide (GO) sheet; (B), (C) SEM images of TiO$_2$ nanocrystals on GO in first and second step, respectively, at 100 nm scale.[87]

etc., with the help of various experimental conditions. In the past, with the help of two steps, the direct growth of TiO$_2$ nanocrystals on the surface of GO sheets has been attained. In this procedure, first TiO$_2$ was layered with the hydrolysis process on the graphene oxide sheets, and in the second step, with the assist of the hydrothermal technique, anatase nanocrystals were obtained. This method is helpful in the approach to nanocrystals because of the homogeneous coating and solid interaction between graphene oxide sheets and TiO$_2$ shown in Fig. 2.13. This coupling is useful in different applications as well as photocatalysis.[87] Recently sol–gel method is used in the construction of 2-D sandwich-like G-mesoporous silica nanosheets, and in these nanosheets, every graphene is separately attached with a mesoporous silica shell.[88] Moreover, with the help of the noncasting technique, G-mesoporous Co$_3$O$_4$ sheets have been made using G-mesoporous silica sheets. The benefits of this sol–gel process are that the functional groups on graphene oxide or rGO give responsive and fastening sites for the development of nanoparticles. The obtained MO structures are chemically attached to the surface of GO or rGO. So the sol–gel method has become an excellent technique for the synthesis of MO–C composites.

2.5.1.3 Hydrothermal/solvothermal method

To fabricate the inorganic nanocrystals, the hydrothermal/solvothermal method is a useful technique that is applied at a specific temperature in closed volume to increase the pressure. This process is helpful to increase the nanostructures with crystallinity without using annealing and at the same time reduce GO to rGO. In the past, many G-TiO$_2$ composites have been fabricated, as well as TiO$_2$–rGO composites with the help of the hydrothermal technique.[67] This process is easy, simple, accessible, and compatible on an industrial level. Through this hydrothermal technique, we can gain G-TiO$_2$ composites by adjusting the parameters such as the number of ingredients in making the solution and the reaction time to make the product. Huang et al. synthesized G@TiO$_2$ composites with precise visible crystal sides through the hydrothermal method.[89] In a one-step solvothermal technique, we obtained different G-TiO$_2$ composites with controlled TiO$_2$ nanostructures by adjusting various conditions. TEM images of different composites are shown in Fig. 2.14.[90]

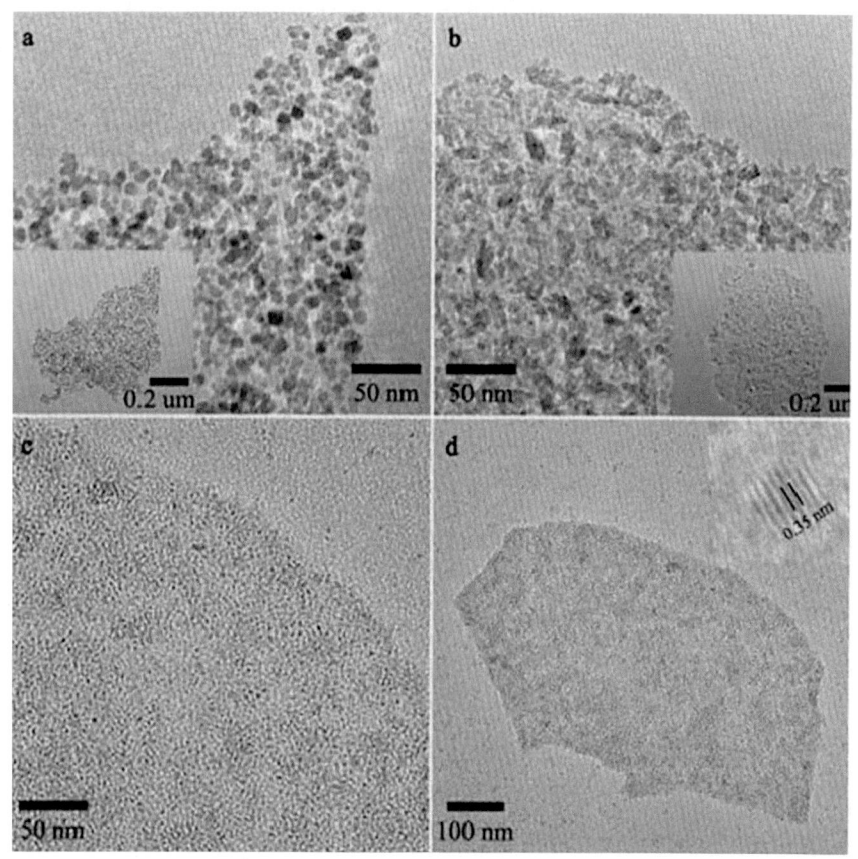

Figure 2.14 TEM images of (A) spherical TiO_2−graphene nanosheets (STGs), (B) TiO_2 nanorod−graphene nanosheets (NRTGs), (C) ultrasmall TiO_2−graphene nanosheets (USTGs) at high magnification, (D) USTGs at low magnification.[90]

These solvothermal/hydrothermal techniques have been used to fabricate various MOs like Fe_3O_4, SnO_2, and Co_3O_4 on G-nanosheets.[91,92] Co_3O_4 nanowall arrays on rGO sheets with porous structures have been synthesized by a simple and easy hydrothermal method.[92] Different properties of this composite, such as porosity, composite surface area, grain size, and its composition, were tunable by applying different annealing processes. These rGO−Co composites showed good performance for the anode material of LIBs. Huang et al. also synthesized a G-SnO_2 composite through an easy and scalable solvothermal technique.[93] The resultant composites of SnO_2 nanocrystals and G-nanosheets in which nanocrystals of SnO_2 were homogeneously distributed on sheets may be applied as the anode (+ve) materials for LIBs. So we can say these hydrothermal or solvothermal methods are very helpful to synthesize different carbon-based MOs on an economic scale and have more importance for the fabrication of materials that are useful for energy storage and conversion devices.

2.5.1.4 Self-assembly

This method is effective and often used with the preference to form an organized and macroscopic structure by assembling micro- and nanolevel objects.[94] This method is helpful for making those materials that are used as functional materials like composites and DNA structures. Wang et al. described obtaining the composites with an irregular layered structure; a new technique of self-assembly was established to make the regular G-MO hybrids by surface-assisted ternary.[95] The self-assembly process plays a vital role in making the layered composites. Some composites of reduced graphene oxide-MO with alternating layer structures such as SnO_2, MnO_2, and NiO have been synthesized by this technique. This is another easy, simple, and economical assembly procedure based on the (−ve) and (+ve) electrostatic attraction, which has become an attractive method used on a large scale to synthesize graphene-based composite materials. Kim et al. described the fabrication of firmly joined composites of layered-titanate-rGO through the self-assembly method.[96] This composite was prepared with the combination of TiO_2 nano-sol and a solution of rGO nanosheets at 60°C in the presence of reflux condition. If flocculation appears between the two types of species of materials when mixing the precursors, it can be removed by applying the zeta potential on both reactant materials, and it is the proof of electrostatic attraction between the two types of nanospecies.

This technique of electrostatic attraction is not only efficient for the layered structure G-based composite but also good for core-shell type materials. Yang et al. synthesized the core-shell structure of G-encapsulated Co_3O_4 composite.[97] This method is operated between the two different species that electrostatically interact and are obeyed by the chemical reduction process. Some other methods are also applied to manufacture G-TiO_2 composites. Liu et al. synthesized a composite of GO-TiO_2 with the help of TiO_2 nanorods and GO nanosheets under the interface self-assembly technique and gained good results for photocatalytic applications.[98] Li et al. prepared a composite of graphene and TiO_2 under the procedure of free self-assembly, and in this composite, they deposited TiO_2 nanospheres uniformly on the sheets of graphene.[99] So self-assembly is an attractive technique for the fabrication of layered stricture material and also for core-shell type materials, which may be helpful in photocatalytic activity.

2.5.1.5 Other methods

There are many other methods to synthesize the materials into a useful product. Microwave irradiation is a fast, easy and important technique to deliver energy for the reactions, and this method is very useful for the preparation of G-MO hybrids like G-MnO_2[100] and G-Co_3O_4.[101] For example, G-MnO_2 was prepared with the suspension of rGO and the powder of $KMnO_4$ through ultrasonically, and this suspension is heated in the microwave oven for 5 min to gain the final product.[102] Direct electrochemical deposition, a technique for making the composite from the inorganic crystal on graphene, has become an attractive approach for obtaining graphene-based thin film. In this aspect, some MOs such as ZnO and Cu_2O have been deposited on rGO or graphene films.[101,103] There is another simple, easy, and economical process in which graphene oxide can be reduced with the help of titanium trichloride ($TiCl_3$) at room temperature. A new method has been described for water-phase synthesis in which G/TiO_2 composite and titanium trichloride ($TiCl_3$) are used as reducing agents and as a precursor.[104] In the same way, G/SnO_2 was fabricated with the help of rGO and $SnCl_2$.[105] In this process, during oxidation and reduction reactions, GO was changed to rGO, whereas Ti^{+3} and Sn^{+2} oxidized in titanium dioxide (TiO_2) and SnO_2 respectively due to the depositing on the rGO surface.[106] So there some other methods are effective which have been published for the fabrication of carbon-based materials-MO composites and give good performances for the supercapacitors and photocatalytic applications, as summarized in Table 2.7.

2.5.2 Application of graphene–metal oxide composites in photocatalysis

2.5.2.1 Water splitting for hydrogen production

H_2 production from water through photocatalytic activity has become a favorable and emerging method to overcome increasing energy demand and reduce environmental impacts. This is a facile, sustainable, and economical technique on which many research articles have been written with good performances.[111] But their use on a large scale is limited by the rapid recombination of electrons and holes in the photocatalyst. Graphene is an efficient material with unique properties and is helpful to enhance the photocatalytic activity for H_2 production.[112] For the production of H_2 through water splitting, many graphene–MO composites have been synthesized. Zhang et al. described the photocatalytic efficiency of the G-TiO_2 composites with 5 wt% graphene.[113] H_2 evaluation rate was calculated, which is 8.6 mmol/h, and this composite has approximately double the efficiency of P25. Fan et al. synthesized the G-P25 composites with various methods and checked the photocatalytic performance for the H_2 evaluation.[88] In Fig. 2.15A, we can see that different composites gave better efficiency for the photocatalytic activity of H_2 production. In these composites, G-P25 prepared by the hydrothermal technique showed the best efficiency. It is considered that in G-P25 composites, P25 and rGO may play a significant role in the transformation of the electrons and reduce the recombination rate of charge carriers (e^- and h^+). There is a need to improve the synthesis procedures to get better performance from G-TiO_2 for the splitting of water. Another important thing is the ratio of the precursors, which is also affected by the photocatalytic efficiency. The concentration ratio 1/0.2 of P25-G gives good results for H_2 production, which is 10 times more than that of pure P25, as shown in Fig. 2.15B. This composite has more importance because it is recyclable and has good stability for the production of H_2 from the water in the absence of any other cocatalyst.[114]

Moreover, TiO_2@G sheets have been used in the water splitting for the production of H_2. Additionally, photocatalytic activity under UV radiation ($\lambda > 320$ nm) gives better performance for the H_2 evaluation due to the better connection between TiO_2 and graphene.[115] The core-shell structure could be played a vital role in the G-TiO_2 composites for the conversion of solar energy application. Some other composites

Table 2.7 Comparison of synthesis techniques for making metal oxide composites.

Synthesis techniques	Advantages	Disadvantages	Commercialization	Ref.
Sol—gel method	Producing high-quality materials with homogeneity and purity at lower temperatures can provide a thick coating for protection against corrosion	Long processing time, use of organic solutions that can be toxic, contraction during processing	Possible	107
Chemical precipitating	Easy and efficient synthesis of electrode materials	Long chemical reaction	Easy, but there is more time required to use on a commercial scale	108
Hydrothermal method	Ability to synthesize large crystals of high quality, ability to synthesize substances that are unstable near the melting point	High-cost equipment	Suitable for the growth of large production	107
Solvothermal method	Cheap and easily available precursors	Popcorn effect arises due to nucleation of sheets	Scalable	109
Electrochemical deposition	Low cost, precise control over the thickness of film and deposition rate	Large energy utilization	Can be performed but requires more energy for this process	110
Chemical bath deposition	Low temperature and inexpensive, large area substrate without voltage/current.	Low material yields, inflexible	Not feasible	108
Chemical vapor deposition	Better uniformity of film and yield than CBD	Expensive	Possible	108

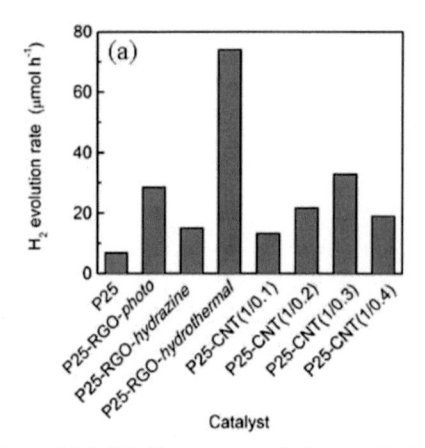
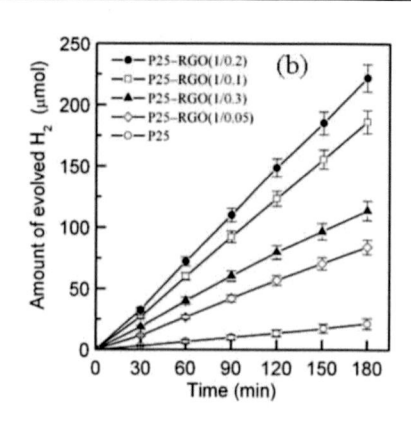

Figure 2.15 (A) Comparison of photocatalytic efficiencies for H$_2$ production rate of P25—reduced graphene oxide (rGO) composites with various methods and P25—carbon nanotube (CNT) composites with different mass ratios of P25 and CNT (B) photocatalytic performances of P25—rGO—hydrothermal series of composites with different mass ratios of P25 and rGO for the evolution of H$_2$.[114]

of graphene-MO have been synthesized and also used for H$_2$ production through photocatalyst activity like rGO-WO$_3$, rGO-BiVO$_4$. The photochemical properties of these composites and their applications have been studied for water splitting.[116,117] GO can also be used for the H$_2$ generation when we use this as a photocatalyst. These GO photocatalyst materials usually take from a bandgap range of 2.4—4.3 eV. Yeh et al. described that in the presence of visible light radiation with no other cocatalyst, GO semiconductor photocatalyst can be used for the H$_2$ evaluation from the water or aqueous solution.[118,119] Thus, many graphene—MO composites have been fabricated and used for the photocatalytic activity for H$_2$ production.

2.5.2.2 Photodegradation of pollutants

The photodegradation of the pollutants also has become an attractive field to resolving the energy and pollution problems. In this aspect, graphene is a promising and attractive material for the fabrication of graphene-MO composites for photodegradation. The G-P25 was synthesized and applied for the photodegradation for methylene blue (MB), and it was compared with the bulk-P25 and CNT-P25 composites, as shown in Fig. 2.16.[120] The results of synthesized G-P25 composite show excellent properties for photodegradation activity such as good absorption ability, long range of light absorbance, and increased charge separation ability. The authors showed the performance of photocatalytic degradation of different composites in the existence of visible light and also for the UV light for the MB in Fig. 2.16.

Liang et al. described that the G-TiO$_2$ hybrid nanocrystals showed their best performance under UV light for photodegradation of rhodamine B compared with P25, a mixture of GO and P25 fabricated by the hydrothermal method, and bare TiO$_2$, as

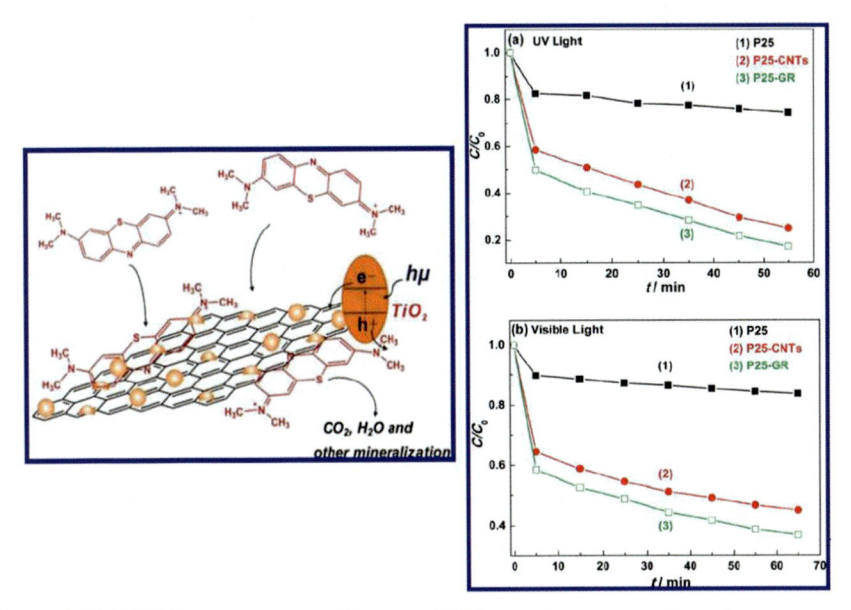

Figure 2.16 (A) Schematic structure diagram of P25–graphene composite and photodegradation for methylene blue (MB); comparison of photodegradation of MB under (B) UV light and (C) visible light for different photocatalysts like P25, P25–CNTs, and P25–GR.[120]

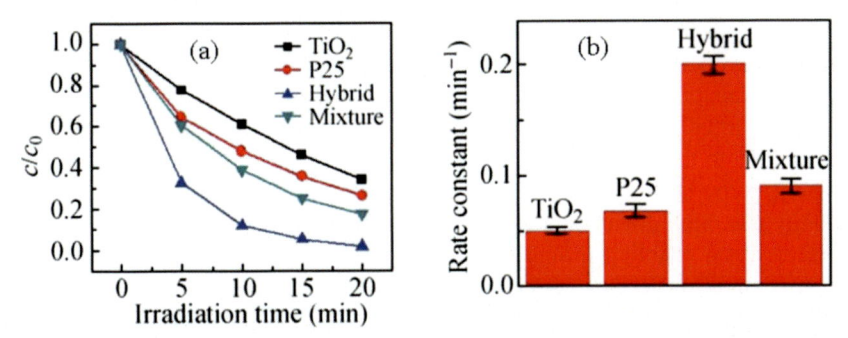

Figure 2.17 (A) Photocatalytic degradation of rhodamine B for prepared materials. (B) Average reaction rate constant (min^{-1}) of as synthesized materials for photocatalytic degradation.[87]

shown in Fig. 2.17.[87] This hybrid G-TiO$_2$ composite showed better performance for photocatalytic activity because of the robust coupling between graphene and TiO$_2$, charge transfer ability, and decrement in the e$^-$ and H$^+$ recombination.

Lee et al. described increasing the photoactivity by wrapping the TiO$_2$ nanoparticles with graphene for the photodegradation of MB.[121] The change in the concentration of graphene oxide also is important in photodegradation because graphene oxide might act as a sensitizer and can give better efficiency for photodegradation. Chen et al.

fabricated a GO-TiO$_2$ composite with a p-type or n-type heterojunction structure to increase the light absorbance for photodegradation.[122] If the bandgap of p-type SC of graphene oxide is less than 2.34 eV, it is good for photocatalytic degradation with the help of GO-TiO$_2$ composites. Moreover, some graphene-MO composites like G-SnO$_2$[123] and G-ZnO[124,125] have been synthesized and used for the photodegradation of pollutants from the water. It is observed that many G-MO composites presented outstanding efficiencies for photocatalytic degradation, but some basic and important issues still remain. The mechanism to increase the photocatalytic activity for G–MO composites is not completely understood. Some other questions include describing whether G-MO composites really differ from other C-based materials such as CNTs and activated carbons. Zhang et al. described the G-TiO$_2$ composite fundamentally the same as other carbon-based material-TiO$_2$ composites to increase the photocatalytic performance of TiO$_2$.[126] So photodegradation of pollutants has become an attractive field in photocatalysis, and many research articles have been written on this technique. Many composites of C-based materials with MOs have been prepared and give good results for photocatalytic applications.

2.6 Conclusion

The discussion in this chapter concluded that during the past few years, MOs have received much attention for various energy storage and photocatalytic applications. The C-based materials are also important for this because of their characteristics. But some deficiencies have remained while using MOs and carbon-based materials separately, so many researchers have been introduced various methods to improve the applications of materials. One of these methods is the fabrication of nanocomposites. The formation of MO–C composites has gained much interest among researchers due to their specific properties such as large specific capacitance, effective SA, good cyclic stability, tunable electrical conductivity with carbon support, good capability rate, high power density, and economical price.

Conventional enabling technologies have become appealing choices to fabricate MO–C composites because of their amazing results. This chapter primarily presents various conventional enabled synthesis approaches such as the hydrothermal method, electrochemical deposition, the solvothermal technique, the solution mixing method, self-assembly, chemical vapor deposition, the electrospinning method, chemical bath deposition, the sol–gel method, and others for the fabrication of MO–C composites. The microstructures, physical properties, and compositional constituents of MO–C composites govern the performance of supercapacitors. We also discussed the different synthesis methods of C-based MOs composites for supercapacitors. The MO–C composites are helpful to develop such electrodes materials that have a high energy density, good storage capability, and high cyclic stability for the supercapacitors. So the MO–C composites have become promising candidates for the energy storage devices such as supercapacitors.

In this chapter, we also overviewed different synthesis techniques and effective strategies of C base material such as graphene with MOs composites their properties and applications for photocatalysis. Graphene is an efficient material with unique properties for photocatalytic activity in H_2 production. This section also discussed several graphene—MO composites that showed good results for hydrogen generation through the photocatalytic activity of water splitting. Another photocatalytic method is the photodegradation of pollutants, which has become an attractive field for resolving energy and pollution problems. In this aspect, we described various promising and attractive graphene-MO composites for photodegradation. We also explore how laboratory technological developments will push the boundaries of what can be achieved using conventional enabling techniques for MO—C composite synthesis. Furthermore, we also discuss challenges and future prospects in the synthesis using conventional enabling technologies for MO—C composites.

2.7 Challenges and synthesis advancement in using conventional enabling technologies for metal oxide—carbon composites

The impetus for the fabrication of MO—C composites with distinguished features and outstanding characteristics is attracting great interest from researchers because of their outstanding applications in various fields of study, including energy storage, drug delivery, composite designing, carbon dioxide capture, optical application, bio-sensing, nanoelectronics, hydrogen storage, environmental remediation, and catalytic activities. Synthesis techniques play a crucial role not only in controlling the morphologies and sizes of these composites but also in improving their potential properties. By applying different synthesis techniques, the different potential properties such as surface area, porosity, and surface states can be improved. The other important factors are the cost of production and the commercialization of these products. The use of conventional enabling technologies to synthesize exclusive MO—C composite nanostructures has made the process cost-effective, more convenient, and with rapid kinetics that was not previously evident. In addition, it also opens new prospects for the production of novel carbon-based nanocomposites with higher specific properties that could be used in several new areas in the future.

However, there are some barriers and challenges that need to be overcome for their successful execution. The perfect feedstock, which is inexpensive, versatile, and environmentally sustainable in nature, is still debated. One fascinating solution may be the use of environmental waste gases for the production of sustainable products. One problem that could be resolved is using the catalyst, providing the highest possible recovery of the expensive catalyst, or maybe looking for a more affordable and high-performance catalyst to increase overall efficacy.

At present, these carbon-based nanocomposites are solely restricted to the batch scale or laboratory scale. So, there is a higher requirement for the establishment of industrial-scale production at the moment, and many problems may arise in this case, including scale-up complications, efficiency maintenance, cost-effectiveness, and the introduction of new products to the market. In this way, industrialists will face both managerial and technological challenges to promote these new products by replacing traditional materials. On the other hand, for the fabrication of exclusive carbon-based nanocomposites, explicit improvement in proficiency in the field of application may be responsible for its implementation and acceptability by consumers. Another problem related to nanolevel products is their proper handling and safety, and the level of awareness of this particular issue is very low, and more effort may be needed in this area. Especially, the environmental effects of carbon nanomaterials disposal should be addressed in detail. Synthesis of exclusive MO−C composites with unique morphology, high stability, controlled structure, multifunctional characteristics, and higher surface area can play an essential role and prove to be an innovation for various fields.

Acknowledgments

The financial support from the National Natural Science Foundation of China (51602281), Yangzhou University High-end Talent Support Program and the "Qinglan Project" of Jiangsu Universities.

References

1. Kroto HW, et al. C60: Buckminsterfullerene. *Nature* 1985;**318**(6042):162−3.
2. Iijima S. Helical microtubules of graphitic carbon. *Nature* 1991;**354**(6348):56−8.
3. Hwang J-Y, et al. Carbon nanotube nanocomposites with highly enhanced strength and conductivity for flexible electric circuits. *Langmuir* 2015;**31**(28):7844−51.
4. Cha C, et al. Carbon-based nanomaterials: multifunctional materials for biomedical engineering. *ACS Nano* 2013;**7**(4):2891−7.
5. Lee WJ, et al. Nitrogen-doped carbon nanotubes and graphene composite structures for energy and catalytic applications. *Chem Commun* 2014;**50**(52):6818−30.
6. Guldi DM, Sgobba V. Carbon nanostructures for solar energy conversion schemes. *Chem Commun* 2011;**47**(2):606−10.
7. Llobet E. Gas sensors using carbon nanomaterials: a review. *Sensor Actuator B Chem* 2013;**179**:32−45.
8. Yang L, Zhang L, Webster TJ. Carbon nanostructures for orthopedic medical applications. *Nanomedicine* 2011;**6**(7):1231−44.
9. Liu Y, Kumar S. Recent progress in fabrication, structure, and properties of carbon fibers. *Polym Rev* 2012;**52**(3):234−58.
10. Stamatin I, et al. Highly oriented carbon ribbons for advanced multifunctional material engineering. *Fullerenes, Nanotubes, Carbon Nanostruct* 2005;**13**(S1):543−51.
11. Mugadza K, et al. Low temperature synthesis of multiwalled carbon nanotubes and incorporation into an organic solar cell. *J Exp Nanosci* 2017;**12**(1):363−83.
12. Mhlanga S, et al. Controlled syntheses of carbon spheres in a swirled floating catalytic chemical vapour deposition vertical reactor. *J Exp Nanosci* 2010;**5**(1):40−51.

13. Bai S, et al. Influence of ferrocene/benzene mole ratio on the synthesis of carbon nano-structures. *Chem Phys Lett* 2003;**376**(1—2):83—9.
14. Nyamori VO, Coville NJ. Effect of ferrocene/carbon ratio on the size and shape of carbon nanotubes and microspheres. *Organometallics* 2007;**26**(16):4083—5.
15. Sevilla M, et al. Solid-phase synthesis of graphitic carbon nanostructures from iron and cobalt gluconates and their utilization as electrocatalyst supports. *Phys Chem Chem Phys* 2008;**10**(10):1433—42.
16. Mutuma BK, et al. Generation of open-ended, worm-like and graphene-like structures from layered spherical carbon materials. *RSC Adv* 2016;**6**(24):20399—408.
17. Yadav MD, et al. High performance fibers from carbon nanotubes: synthesis, characterization, and applications in composites a review. *Ind Eng Chem Res* 2017;**56**(44):12407—37.
18. Shah KA, Tali BA. Synthesis of carbon nanotubes by catalytic chemical vapour deposition: a review on carbon sources, catalysts and substrates. *Mater Sci Semicond Process* 2016;**41**:67—82.
19. Shoukat R, Khan MI. Growth of nanotubes using IC-PECVD as benzene carbon carrier. *Microsyst Technol* 2017;**23**(12):5447—53.
20. Bridgwater A, Bridge S. A review of biomass pyrolysis and pyrolysis technologies. In: *Biomass pyrolysis liquids upgrading and utilization*. Springer; 1991. p. 11—92.
21. Zhu M, Diao G. Review on the progress in synthesis and application of magnetic carbon nanocomposites. *Nanoscale* 2011;**3**(7):2748—67.
22. Brunori G. Biomass, biovalue and sustainability: some thoughts on the definition of the bioeconomy. *EuroChoices* 2013;**12**(1):48—52.
23. Chu H, et al. Carbon nanotubes combined with inorganic nanomaterials: preparations and applications. *Coord Chem Rev* 2010;**254**(9—10):1117—34.
24. Eder D. Carbon nanotube— inorganic hybrids. *Chem Rev* 2010;**110**(3):1348—85.
25. Mo CB, et al. Fabrication of carbon nanotube reinforced alumina matrix nanocomposite by sol—gel process. *Mater Sci Eng, A* 2005;**395**(1—2):124—8.
26. Zhi M, et al. Nanostructured carbon—metal oxide composite electrodes for supercapacitors: a review. *Nanoscale* 2013;**5**(1):72—88.
27. Abbas Q, et al. *Carbon/metal oxide composites as electrode materials for supercapacitors applications*. 2018.
28. Zhang LL, Zhao XJCSR. Carbon-based materials as supercapacitor electrodes. *Chem Soc Rev* 2009;**38**(9):2520—31.
29. Jiang H, Ma J, Li CJAm. Mesoporous carbon incorporated metal oxide nanomaterials as supercapacitor electrodes. *Adv Mater* 2012;**24**(30):4197—202.
30. Conway BE. *Electrochemical supercapacitors: scientific fundamentals and technological applications*. Springer Science & Business Media; 2013.
31. Lokhande C, Dubal D, Joo O-SJCAP. Metal oxide thin film based supercapacitors. *Curr Appl Phys* 2011;**11**(3):255—70.
32. Wang Q, Wen Z, Li JJAFM. A hybrid supercapacitor fabricated with a carbon nanotube cathode and a TiO_2—B nanowire anode. *Adv Funct Mater* 2006;**16**(16):2141—6.
33. Muzaffar A, et al. A review on recent advances in hybrid supercapacitors: design, fabrication and applications. *Renew Sustain Enery Rev* 2019;**101**:123—45.
34. Afif A, et al. Advanced materials and technologies for hybrid supercapacitors for energy storage—a review. *J Energy Storage* 2019;**25**:100852.
35. Lim E, Jo C, Lee JJN. A mini review of designed mesoporous materials for energy-storage applications: from electric double-layer capacitors to hybrid supercapacitors. *Nanoscale* 2016;**8**(15):7827—33.

36. Li B, et al. Nitrogen-doped activated carbon for a high energy hybrid supercapacitor. *Energy Environ Sci* 2016;**9**(1):102–6.

37. Gonçalves JM, et al. Trimetallic oxides/hydroxides as hybrid supercapacitor electrode materials: a review. *J Mater Chem A* 2020;**8**(21):10534–70.

38. Najib S, Erdem EJNA. Current progress achieved in novel materials for supercapacitor electrodes: mini review. *Nanoscale Adv* 2019;**1**(8):2817–27.

39. Libich J, et al. *Supercapacitors: properties and applications*, vol. 17; 2018. p. 224–7.

40. Karthikeyan GG, Boopathi G, Pandurangan AJAo. Facile synthesis of mesoporous carbon spheres using 3D cubic Fe-KIT-6 by CVD technique for the application of active electrode materials in supercapacitors. *ACS Omega* 2018;**3**(12):16658–71.

41. Frackowiak E, Beguin FJC. Carbon materials for the electrochemical storage of energy in capacitors. *Carbon* 2001;**39**(6):937–50.

42. Stoller MD, et al. Interfacial capacitance of single layer graphene. *Energy Environ Sci* 2011;**4**(11):4685–9.

43. Wu Z-S, et al. Graphene/metal oxide composite electrode materials for energy storage. *Nano Energy* 2012;**1**(1):107–31.

44. Xia G, et al. General synthesis of transition metal oxide ultrafine nanoparticles embedded in hierarchically porous carbon nanofibers as advanced electrodes for lithium storage. *Adv Funct Mater* 2016;**26**(34):6188–96.

45. Hu J, et al. Capacitance behavior of ordered mesoporous carbon/Fe_2O_3 composites: comparison between 1D cylindrical, 2D hexagonal, and 3D bicontinuous mesostructures. *Carbon* 2015;**93**:903–14.

46. Zhou H, et al. Lithium storage in ordered mesoporous carbon (CMK-3) with high reversible specific energy capacity and good cycling performance. *Adv Mater* 2003;**15**(24): 2107–11.

47. Wang X, Li Z, Yin L. Nanocomposites of SnO 2@ ordered mesoporous carbon (OMC) as anode materials for lithium-ion batteries with improved electrochemical performance. *CrystEngComm* 2013;**15**(37):7589–97.

48. Han F, et al. Nanoengineered polypyrrole-coated Fe_2O_3@ C multifunctional composites with an improved cycle stability as lithium-ion anodes. *Adv Funct Mater* 2013;**23**(13): 1692–700.

49. Patel MN, et al. Hybrid MnO_2–disordered mesoporous carbon nanocomposites: synthesis and characterization as electrochemical pseudocapacitor electrodes. *J Mater Chem* 2010; **20**(2):390–8.

50. Zhou Y, Lee CW, Yoon S. Development of an ordered mesoporous carbon/MoO_2 nanocomposite for high performance supercapacitor electrode. *Electrochem Solid State Lett* 2011;**14**(10):A157.

51. Yuan D, et al. Bi_2O_3 deposited on highly ordered mesoporous carbon for supercapacitors. *Electrochem Commun* 2009;**11**(2):313–7.

52. Hu J, et al. Dual-template ordered mesoporous carbon/Fe 2 O 3 nanowires as lithium-ion battery anodes. *Nanoscale* 2016;**8**(26):12958–69.

53. Chaitra K, et al. High energy density performance of hydrothermally produced hydrous ruthenium oxide/multiwalled carbon nanotubes composite: design of an asymmetric supercapacitor with excellent cycle life. *J Energy Chem* 2016;**25**(4):627–35.

54. Cao P, et al. Facile hydrothermal synthesis of mesoporous nickel oxide/reduced graphene oxide composites for high performance electrochemical supercapacitor. *Electrochim Acta* 2015;**157**:359–68.

55. Chen Z, et al. Design and synthesis of hierarchical nanowire composites for electrochemical energy storage. *Adv Funct Mater* 2009;**19**(21):3420–6.

56. Wang Y, et al. MnO 2/onion-like carbon nanocomposites for pseudocapacitors. *J Mater Chem* 2012;**22**(34):17584–8.
57. Tang W, et al. A hybrid of MnO_2 nanowires and MWCNTs as cathode of excellent rate capability for supercapacitors. *J Power Sources* 2012;**197**:330–3.
58. Kim I-H, Kim J-H, Kim K-B. Electrochemical characterization of electrochemically prepared ruthenium oxide/carbon nanotube electrode for supercapacitor application. *Electrochem Solid State Lett* 2005;**8**(7):A369.
59. Ghosh A, et al. High pseudocapacitance from ultrathin V_2O_5 films electrodeposited on self-standing carbon-nanofiber paper. *Adv Funct Mater* 2011;**21**(13):2541–7.
60. Zhao Y-Q, et al. MnO_2/graphene/nickel foam composite as high performance supercapacitor electrode via a facile electrochemical deposition strategy. *Mater Lett* 2012;**76**: 127–30.
61. Liu X, Pickup PG. Ru oxide supercapacitors with high loadings and high power and energy densities. *J Power Sources* 2008;**176**(1):410–6.
62. Rani P, et al. Mesoporous GO-TiO_2 nanocomposites for flexible solid-state supercapacitor applications. *Mater Res Express* 2020;**6**(12):125546.
63. Lota K, Sierczynska A, Lota G. Supercapacitors based on nickel oxide/carbon materials composites. *Int J Electrochem* 2011:2011.
64. Lee JY, et al. Nickel oxide/carbon nanotubes nanocomposite for electrochemical capacitance. *Synth Met* 2005;**150**(2):153–7.
65. Yuan A, Zhang Q. A novel hybrid manganese dioxide/activated carbon supercapacitor using lithium hydroxide electrolyte. *Electrochem Commun* 2006;**8**(7):1173–8.
66. Wang J-G, et al. Coaxial carbon nanofibers/MnO_2 nanocomposites as freestanding electrodes for high-performance electrochemical capacitors. *Electrochim Acta* 2011;**56**(25): 9240–7.
67. Lee SW, et al. Carbon nanotube/manganese oxide ultrathin film electrodes for electrochemical capacitors. *ACS Nano* 2010;**4**(7):3889–96.
68. Zhang M, et al. A green and fast strategy for the scalable synthesis of Fe_2O_3/graphene with significantly enhanced Li-ion storage properties. *J Mater Chem* 2012;**22**(9):3868–74.
69. Singh V, et al. In situ synthesis of graphene oxide and its composites with iron oxide. *New Carbon Mater* 2009;**24**(2):147–52.
70. Li X, et al. Multi-walled carbon nanotubes composited with nanomagnetite for anodes in lithium ion batteries. *RSC Adv* 2015;**5**(10):7237–44.
71. Kim J-Y, et al. Microwave-polyol synthesis of nanocrystalline ruthenium oxide nanoparticles on carbon nanotubes for electrochemical capacitors. *Electochim Acta* 2010; **55**(27):8056–61.
72. Xu C, Sun J, Gao LJTJoPCC. Synthesis of multiwalled carbon nanotubes that are both filled and coated by SnO_2 nanoparticles and their high performance in lithium-ion batteries. *J Phys Chem C* 2009;**113**(47):20509–13.
73. Dechakiatkrai C, et al. Photocatalytic oxidation of methanol using titanium dioxide/single-walled carbon nanotube composite. *J Electrochem Soc* 2007;**154**(5):A407.
74. Zhang G, Lou XWDJSr. Controlled growth of NiCo 2 O 4 nanorods and ultrathin nanosheets on carbon nanofibers for high-performance supercapacitors. *Sci Rep* 2013;**3**(1):1–6.
75. Qiu S, et al. Enhanced electrochemical performances of MoO 2 nanoparticles composited with carbon nanotubes for lithium-ion battery anodes. *RSC Adv* 2015;**5**(106):87286–94.
76. Das D, et al. An elegant method for large scale synthesis of metal oxide–carbon nanotube nanohybrids for nano-environmental application and implication studies. *Environ Sci Nano* 2017;**4**(1):60–8.

77. Wu Q, et al. Supercapacitors based on flexible graphene/polyaniline nanofiber composite films. *ACS Nano* 2010;**4**(4):1963−70.
78. Sun Y, et al. Chemically converted graphene as substrate for immobilizing and enhancing the activity of a polymeric catalyst. *Chem Commun* 2010;**46**(26):4740−2.
79. Koo HY, et al. Graphene-based multifunctional iron oxide nanosheets with tunable properties. *Chem - Eur J* 2011;**17**(4):1214−9.
80. Xiang Q, Yu J, Jaroniec M. Graphene-based semiconductor photocatalysts. *Chem Soc Rev* 2012;**41**(2):782−96.
81. Zhang N, Zhang Y, Xu Y-J. Recent progress on graphene-based photocatalysts: current status and future perspectives. *Nanoscale* 2012;**4**(19):5792−813.
82. Paek S-M, Yoo E, Honma I. Enhanced cyclic performance and lithium storage capacity of SnO_2/graphene nanoporous electrodes with three-dimensional delaminated flexible structure. *Nano Lett* 2009;**9**(1):72−5.
83. Williams G, Seger B, Kamat PV. TiO_2-graphene nanocomposites. UV-assisted photocatalytic reduction of graphene oxide. *ACS Nano* 2008;**2**(7):1487−91.
84. Sun S, Gao L, Liu Y. Enhanced dye-sensitized solar cell using graphene-TiO 2 photoanode prepared by heterogeneous coagulation. *Appl Phys Lett* 2010;**96**(8):083113.
85. Bell NJ, et al. Understanding the enhancement in photoelectrochemical properties of photocatalytically prepared TiO_2-reduced graphene oxide composite. *J Phys Chem C* 2011;**115**(13):6004−9.
86. Akhavan O, Ghaderi E. Photocatalytic reduction of graphene oxide nanosheets on TiO_2 thin film for photoinactivation of bacteria in solar light irradiation. *J Phys Chem C* 2009;**113**(47):20214−20.
87. Liang Y, et al. TiO 2 nanocrystals grown on graphene as advanced photocatalytic hybrid materials. *Nano Res* 2010;**3**(10):701−5.
88. Yang S, et al. Graphene-based nanosheets with a sandwich structure. *Angew Chem Int Ed* 2010;**49**(28):4795−9.
89. Wang Z, et al. Crystal facets controlled synthesis of graphene@ TiO 2 nanocomposites by a one-pot hydrothermal process. *CrystEngComm* 2012;**14**(5):1687−92.
90. He Z, et al. Nanostructure control of graphene-composited TiO 2 by a one-step solvothermal approach for high performance dye-sensitized solar cells. *Nanoscale* 2011;**3**(11):4613−6.
91. Wang JZ, et al. Graphene-encapsulated Fe_3O_4 nanoparticles with 3D laminated structure as superior anode in lithium ion batteries. *Chem - Eur J* 2011;**17**(2):661−7.
92. Zhu J, et al. Cobalt oxide nanowall arrays on reduced graphene oxide sheets with controlled phase, grain size, and porosity for Li-ion battery electrodes. *J Phys Chem C* 2011;**115**(16):8400−6.
93. Huang X, et al. A facile one-step solvothermal synthesis of SnO_2/graphene nanocomposite and its application as an anode material for lithium-ion batteries. *ChemPhysChem* 2011;**12**(2):278−81.
94. Du J, et al. Hierarchically ordered macro− mesoporous TiO_2− graphene composite films: improved mass transfer, reduced charge recombination, and their enhanced photocatalytic activities. *ACS Nano* 2011;**5**(1):590−6.
95. Wang D, et al. Ternary self-assembly of ordered metal oxide− graphene nanocomposites for electrochemical energy storage. *ACS Nano* 2010;**4**(3):1587−95.
96. Kim IY, et al. A strong electronic coupling between graphene nanosheets and layered titanate nanoplates: a soft-chemical route to highly porous nanocomposites with improved photocatalytic activity. *Small* 2012;**8**(7):1038−48.

97. Yang S, et al. Fabrication of graphene-encapsulated oxide nanoparticles: towards high-performance anode materials for lithium storage. *Angew Chem Int Ed* 2010;**49**(45): 8408—11.

98. Liu J, et al. Self-assembling TiO_2 nanorods on large graphene oxide sheets at a two-phase interface and their anti-recombination in photocatalytic applications. *Adv Funct Mater* 2010;**20**(23):4175—81.

99. Li N, et al. Battery performance and photocatalytic activity of mesoporous anatase TiO_2 nanospheres/graphene composites by template-free self-assembly. *Adv Funct Mater* 2011; **21**(9):1717—22.

100. Yan J, et al. Rapid microwave-assisted synthesis of graphene nanosheet/Co_3O_4 composite for supercapacitors. *Electrochim Acta* 2010;**55**(23):6973—8.

101. Wu S, et al. Electrochemical deposition of semiconductor oxides on reduced graphene oxide-based flexible, transparent, and conductive electrodes. *J Phys Chem C* 2010; **114**(27):11816—21.

102. Yan J, et al. Fast and reversible surface redox reaction of graphene—MnO_2 composites as supercapacitor electrodes. *Carbon* 2010;**48**(13):3825—33.

103. Yin Z, et al. Electrochemical deposition of ZnO nanorods on transparent reduced graphene oxide electrodes for hybrid solar cells. *Small* 2010;**6**(2):307—12.

104. Zhu C, et al. One-pot, water-phase approach to high-quality graphene/TiO 2 composite nanosheets. *Chem Commun* 2010;**46**(38):7148—50.

105. Ji F, et al. Electrochemical performance of graphene nanosheets and ceramic composites as anodes for lithium batteries. *J Mater Chem* 2009;**19**(47):9063—7.

106. Zhang J, Xiong Z, Zhao X. Graphene—metal—oxide composites for the degradation of dyes under visible light irradiation. *J Mater Chem* 2011;**21**(11):3634—40.

107. Modan EM, Plăiaşu AG. Advantages and disadvantages of chemical methods in the elaboration of nanomaterials. *The Annals of "Dunarea de Jos" University of Galati. Fascicle IX, Metallurgy and Materials Science* 2020;**43**(1):53—60.

108. Vangari M, Pryor T, Jiang L. Supercapacitors: review of materials and fabrication methods. *J Energy Eng* 2013;**139**(2):72—9.

109. Choucair M, Thordarson P, Stride JA. Gram-scale production of graphene based on solvothermal synthesis and sonication. *Nat Nanotechnol* 2009;**4**(1):30.

110. Asri R, et al. A review of hydroxyapatite-based coating techniques: sol—gel and electrochemical depositions on biocompatible metals. *J Mech Behav Biomed Mater* 2016;**57**: 95—108.

111. Sun Y, et al. Freestanding tin disulfide single-layers realizing efficient visible-light water splitting. *Angew Chem Int Ed* 2012;**51**(35):8727—31.

112. Ng YH, et al. To what extent do graphene scaffolds improve the photovoltaic and photocatalytic response of TiO_2 nanostructured films? *J Phys Chem Lett* 2010;**1**(15):2222—7.

113. Zhang X-Y, et al. Graphene/TiO 2 nanocomposites: synthesis, characterization and application in hydrogen evolution from water photocatalytic splitting. *J Mater Chem* 2010; **20**(14):2801—6.

114. Fan W, et al. Nanocomposites of TiO_2 and reduced graphene oxide as efficient photocatalysts for hydrogen evolution. *J Phys Chem C* 2011;**115**(21):10694—701.

115. Kim H-i, et al. Solar photoconversion using graphene/TiO_2 composites: nanographene shell on TiO_2 core versus TiO_2 nanoparticles on graphene sheet. *J Phys Chem C* 2012; **116**(1):1535—43.

116. Iwase A, et al. Reduced graphene oxide as a solid-state electron mediator in Z-scheme photocatalytic water splitting under visible light. *J Am Chem Soc* 2011;**133**(29):11054—7.

117. Ng YH, et al. Semiconductor/reduced graphene oxide nanocomposites derived from photocatalytic reactions. *Catal Today* 2011;**164**(1):353–7.
118. Yeh TF, et al. Graphite oxide as a photocatalyst for hydrogen production from water. *Adv Funct Mater* 2010;**20**(14):2255–62.
119. Yeh T-F, et al. Graphite oxide with different oxygenated levels for hydrogen and oxygen production from water under illumination: the band positions of graphite oxide. *J Phys Chem C* 2011;**115**(45):22587–97.
120. Zhang H, et al. P25-graphene composite as a high performance photocatalyst. *ACS Nano* 2010;**4**:380–6.
121. Lee JS, You KH, Park CB. Highly photoactive, low bandgap TiO_2 nanoparticles wrapped by graphene. *Adv Mater* 2012;**24**(8):1084–8.
122. Chen C, et al. Synthesis of visible-light responsive graphene oxide/TiO_2 composites with p/n heterojunction. *ACS Nano* 2010;**4**(11):6425–32.
123. Lightcap IV, Kosel TH, Kamat PV. Anchoring semiconductor and metal nanoparticles on a two-dimensional catalyst mat. storing and shuttling electrons with reduced graphene oxide. *Nano Lett* 2010;**10**(2):577–83.
124. Xu T, et al. Significantly enhanced photocatalytic performance of ZnO via graphene hybridization and the mechanism study. *Appl Catal B Environ* 2011;**101**(3–4):382–7.
125. Li B, Cao H. ZnO@ graphene composite with enhanced performance for the removal of dye from water. *J Mater Chem* 2011;**21**(10):3346–9.
126. Zhang Y, et al. TiO_2– graphene nanocomposites for gas-phase photocatalytic degradation of volatile aromatic pollutant: is TiO_2– graphene truly different from other TiO_2– carbon composite materials? *ACS Nano* 2010;**4**(12):7303–14.

Electrical conductivity of metal oxide–carbon composites

Sara Riaz[1], Yann Aman[2], Muhammad Nasir[3], Akhtar Hayat[3] and Mian Hasnain Nawaz[3]

[1]Department of Chemistry, COMSATS University Islamabad, Islamabad, Pakistan; [2]Laboratory of Chemical Physics, University Felix Houphouet Boigny-Abidjan, Abidjan, Ivory Coast; [3]Interdisciplinary Research Centre in Biomedical Materials (IRCBM), COMSATS University Islamabad, Islamabad, Pakistan

3.1 Nature of metal oxide–carbon substrate bindings

Recently, bulk metallic and metal oxide (MO) nanoparticles have enjoyed enormous fame owing to their unique surface and catalytic properties. Their intrinsic electronic and morphological properties have been widely explored. However, unwanted leaching and thermodynamic instability arising from their infinite size place limits on their practical application.[1] These limitations have led to investigations into attaching such nanoparticles to high-surface-area materials. Furthermore, increased knowledge of adhesion science has clarified the importance of the interfacial interactions between anchoring nanoparticles and substrate materials to eventually guarantee the stability of composite architectures.[2] Substrate-supported nanoparticles could demonstrate enhanced surface/volume and stability and eventually lead to numerous applications in emerging research fields. Particularly where increased conductive properties are desired, researchers prefer metal nanoparticle-decorated carbon interfaces.[3] Thus far, such nanocomposites have been synthesized via several tools, including physical mixing, template-assisted nanocomposites, physical/chemical adsorption, no-covalent and covalent interactions, and in situ growth of metal nanoparticles on the surfaces of carbon-based zero-dimensional (0-D), one-dimensional (1-D), two-dimensional (2-D), and three-dimensional (3-D) substrates. Fig. 3.1 is a schematic representation of various characteristics of metal oxide–carbon-based composites.

3.2 Carbon interfaces for conductive composites with metal oxides

After an overview of metal oxide linkages with carbon-based interfaces as substrate materials, we discuss carbon interfaces suitable for anchoring metallic and metal oxide nanoparticles. Carbon substrate selection depends on several parameters, including the nature, size, and functionality of anchoring bodies. It is also a function of their ability to functionalize and the availability of corresponding functional groups at the metal

Metal Oxide-Carbon Hybrid Materials. https://doi.org/10.1016/B978-0-12-822694-0.00018-1

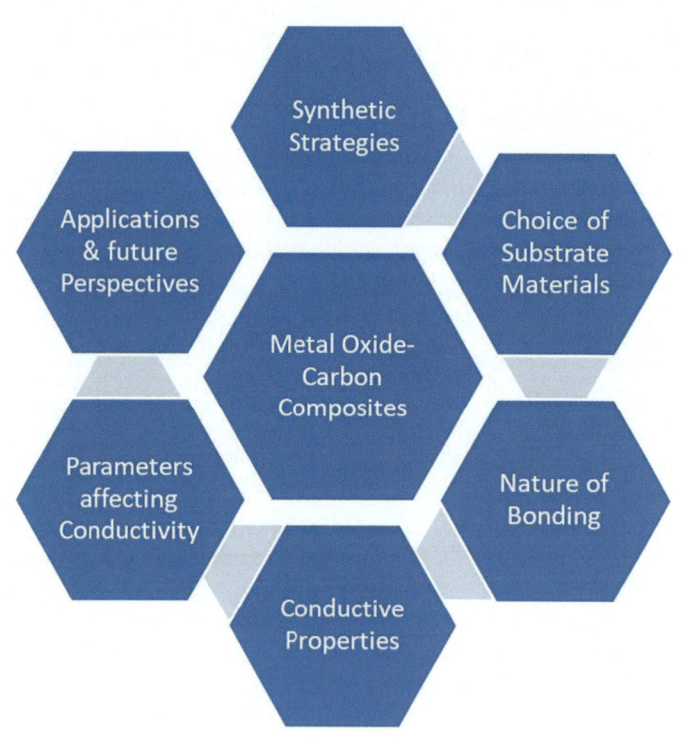

Figure 3.1 Schematic representation of choice of materials, synthetic avenues, and applications of metal oxide−carbon-based composites.

oxide nanoparticle surface. We then discuss the decoration of various carbon-based interfaces with metal oxide nanoparticles. Fig. 3.2 shows various carbon substrate material interactions with metal oxide nanoparticles.

3.2.1   Carbon fibers as a conductive interface

Carbon fibers (CFs), also known as graphite fibers, have been widely employed as carbon-based conductive interfaces, and their composites with heterogeneous materials have received considerable interest. In this regard, CFs can be applied as promising materials for conductive composites using many metal oxides, such as $NiCo_2O_4$,[4] $NiFe_2O_4$,[5] $KMnO_4$, KOH, SiO_2, Al_2O_3,[6] MnO_2, and V_2O_5,[7] owing to their remarkable electrical properties.[8] Apart from electrical traits, their mechanical reinforcement, open architecture, and unique surface states facilitate surface modifications and functionalization methodologies to engineer carbon nanofibers (CNFs) to more feasibly host metal oxides.[9−11] However, the hydrophobic surface of CFs produces difficulty in dispersing CFs into hydrophilic resins, resulting in weaker interactions with ceramics. The problem of low dispersity in hydrophilic resins has been tackled by a group of scientists via interfacial chemical bonds that have led to the production of high-strength composite fibers.[12] For instance, metal oxide decorated carbon nanofibers (CNFs)

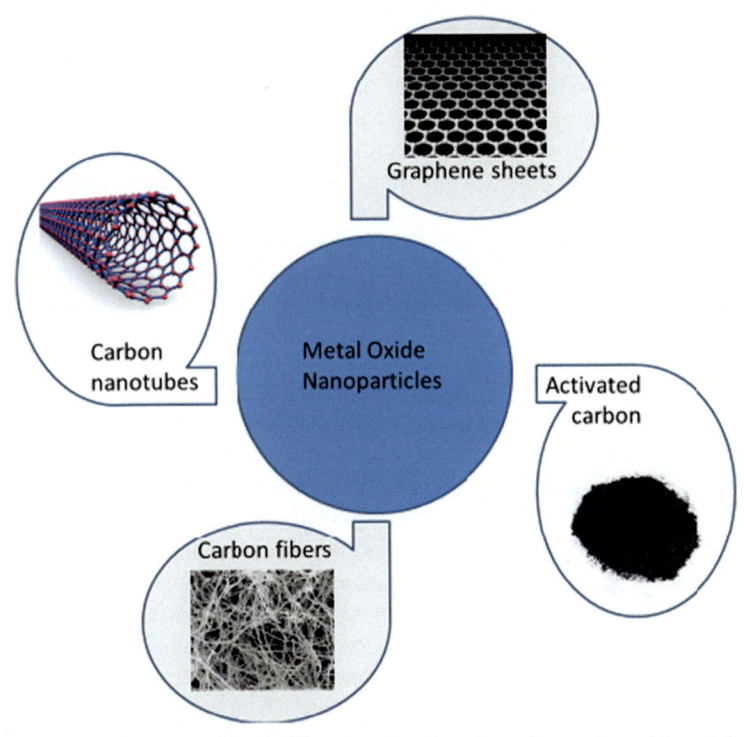

Figure 3.2 Illustrative examples of different carbon interfaces interacting with metal oxide nanoparticles to yield corresponding conducting nanohybrids.

have been synthesized, exploiting single-nozzle electrospinning owing to incompatibility of the shell (PVP) and core (PAN). The ultrafine CNF hybrid was synthesized by decomposing the PVP phase, oxidizing metal precursors to metal oxide nanocubes, and heat-assisted forming of CNFs from PAN. Furthermore, by changing the PVP concentration, the surface morphology of metal oxide nanocubes could be altered in this case. Such synthetic methodology can be exploited for future sensing technologies owing to the enhanced active surface area of synthesized CNF-based composites.[13]

3.2.2 Carbon nanotubes as a conductive interface

Carbon nanotubes (CNTs) were first reported in 1991 by Sumio Iijima in Japan. They have distinctive tubular structures with nanometer diameter along with a larger length-to-diameter ratio. The incredible electrical and mechanical properties of CNTs rely on the graphitic arrangement of C atoms and quasi-1-D structures.[14] Therefore, CNTs have high tensile strength and Young's modulus, making them a potential interface for metal oxides and desirable components for composite materials with enhanced mechanical and electrical properties. CNTs are metallic or semiconducting, depending on their chirality and diameter.[15−17] This opens new avenues for applying CNTs as a central element along with various metal oxides. Novel nanosized manganese oxide–decorated CNTs have been synthesized by electrophoretic deposition of CNTs on rigid substrates like stainless steel (SS) followed by the reduction of MnO^{4-} ions on a multi-scaled SS–CNT interface to form MnO_2. The same methodology used on metal oxide and CNTs can be employed in the future for electrochemical power sources.[18] Similarly, Murugesan et al. reported a two-stage procedure for synthesizing multiwalled CNTs (MWCNTs) with unzipped morphology and titanium dioxide-based nanocomposite (UzMWCNT/TiO_2). MWCNTs have been oxidized to O-MWCNT by modified Hummers' method followed by ultrasonication generate UzMWCNT/TiO2. The integrated conductive composite has been employed for supercapacitor applications.[19]

3.2.3   Graphene as a conductive interface

Two-dimensional (2-D) carbon sheets, named graphene owing to their exceptional properties, including superior mechanical strength, conductivity, pronounced optical and electrical properties, are one of the most explored and studied materials.[20–24] Compositing graphene with other materials like layered transition metal oxides has resulted in enhanced optoelectronic, conducting, and mechanical properties. Moreover, owing to their atomically thin layers and massive electrical, optical, and conducting properties, transition metal chalcogenides have attracted substantial interest. The layered materials provide a thin 2-D plane to the charge carrier with higher conducting properties in such composites. The favorable bandgap, atomic-level thickness, and intrinsic optoelectronic properties of such dichalcogenides make them promising entities in energy harvesting and many other applications.[25,26] Spark plasma sintering methodology has been employed to fabricate fully dense graphene nanosheet (GNS)/Al_2O_3 composites from ball-milled expanded Al_2O_3 and graphite. The integrated GNS/Al_2O_3 composites have a percolation threshold of about 3 vol.%, exceeding most reported CNT/Al_2O_3-based composites in electrical conductivity. The developed composite material behaved as a semimetal, as revealed by the temperature dependence of electrical conductivity in a temperature bound of 2–300 K. The described work overlays new opportunities for graphene-supported metal oxide composites with enhanced electrical properties.[27] Similarly, nitrogen-doped reduced graphene oxide to support spinel $ZnCo_2O_4$ porous particles ($ZnCo_2O_4$@N-RGO) via refluxing and hydrothermal treatment has been obtained. Such integrated composites can perform as superior cathode materials to overcome the shuttling of polysulfides via the chemical confinement of N-RGO and $ZnCo_2O_4$. The RGO nanosheets provide structural stability and a good conductive network along with a high specific surface area. Lithium polysulfide transfer between the cathode and anode could be inhibited by introducing doping (N) atoms and copious $ZnCo_2O_4$ porous nanoparticles. Due to their inimitable compositional and structural properties, these composites with higher sulfur loading (71%–82%) still supply increased specific capacity and cycling stability along with exceptionally high initial coulombic efficiency.[28] It can be concluded that the layered structure of graphene provides pathways to intercalate metal oxide to generate superior hybrid materials for various biomedical and energy applications.

3.2.4 Activated carbon as a conductive interface

Activated carbon (AC) has received increasing interest in recent years owing to its exceptional textural properties, including microporous structure, surface area, tunable porosity, and lower cost, along with their superior mechanical attributes to be imitated into stable interfaces.[29−31] However, there is an increasing trend of exploring composites based on activated carbon-supported semiconductor oxides due to multiple applications, such as photocatalytic dye degradation of many organic pollutants in wastewaters using AC and metal oxide composite,[32,33] gas sensors,[34] catalysis,[35] and adsorption of inorganic ions. These oxides include TiO_2, ZnO, Fe2O3, SnO_2, Al_2O_3, and WO_3.[36] Therefore, many researchers have directed particular attention to exploiting AC as a conductive interface for metal oxide immobilization to obtain desired applications.

A hybrid electrochemical capacitor was developed by Thierry et al. using AC and MnO_2 as negative and positive electrodes, respectively. The electrodes were tested in a mild aqueous solution of 0.65 M K_2SO_4 separately to define the adequate balance of active material in the capacitor along with working voltage. The developed capacitor was cycled between 0 and 2.2 V potential for almost 10,000 charge/discharge current cycles. The 10 Wh/kg of real energy density was reproduced with a real power density of 3600 W/kg. The hydrogen and oxygen evolution reactions in 0.65 M K_2SO_4 were examined for MnO_2 and AC electrodes. Consequently, a 1.5 V capacitor was verified for more than 23,000 cycles and produced interesting electrochemical activity with negligible gas evolution.[37]

A nanocomposite based on heteroatom-entrapped AC-nickel oxide (HAC-NiO) was synthesized by Yue et al. using chemical activation and hydrothermal methodologies. The as-synthesized composite was characterized by X-ray diffraction, scanning electron microscopy, and N_2 adsorption-desorption isotherms and exhibited enhanced surface area as well as improved electric conductivity. Moreover, the integrated NiO−HAC nanocomposite was employed to fabricate glass carbon electrodes (NiO−HAC/GCE) to develop a novel glucose sensor. The unique surface architectures with higher electron transfer kinetics of NiO and large active surface area of HAC resulted in exceptional electrocatalytic activity with wide linear bounds (10 µM−3.3 mM) and a very low detection limit (1 µM). Its desirable stability, high reproducibility, and anti-interference properties make it a potential material for nonenzymatic glucose detection in real samples.[38]

3.3 Synthetic strategies for conductive metal oxide-carbon composites

In this section, we discuss the various synthetic methodologies for metal oxide−carbon composites. We also discuss the effect of synthetic methods on the surfaces and properties of developed complexes.

Tugrul et al. synthesized the composite based on MnO_2/cAC and NiO/cAC by utilizing hydrothermal and precipitation methodologies to explore the effect of preparatory methods on the surface, porous structure, and electrochemical properties of integrated electrodes. The synthesized composites' surface morphology, chemistry, pore characteristics, and chemical composition have been further investigated via thermogravimetric analysis, Raman spectroscopy, X-ray photoelectron spectroscopy, and scanning electron microscopy techniques. The synthesized composites have been exploited as electrode materials in button cell supercapacitors. Additional investigation has indicated that nano-oxide deposition by hydrothermal treatment leads to enhanced active surface area and oxygen-containing surface groups that enhance electrochemical activity. However, the NiO/cAC samples acquired by the precipitation method displayed a higher specific capacitance than hydrothermally synthesized NiO/cAC. In short, the metal oxide loading methodology has a great impact on the surface chemistry and surface area along with the electrochemical performance of developed AC-based supercapacitors.[39]

Similarly, the reduction-based methodology can be employed to synthesize MnO_2@BP nanocomposites by simply reducing $KMnO_4$ with highly porous BLACK PEARLS 2000 and $Mn(CH_3COO)_2.4H_2O$ at room temperature. The porosity, conductivity, specific surface area, and crystallinity of integrated MnO_2@BP nanocomposites were analyzed via X-ray diffraction, four-point probe measurements, surface area measurements, and scanning electron microscopy, respectively. Thermogravimetric analysis was employed to determine the content in the composite. Raman spectroscopy was employed to determine the distribution of BP and MnO_2 in a composite electrode film synthesized with polytetrafluoroethylene (PTFE) as a binder. The composite electrode developed by this synthetic methodology exhibited more homogeneously dispersed MnO_2 particles than an electrode prepared by conventional physical mixing of MnO_2, BP, and PTFE (MnO_2/BP-PTFE). The electrochemical properties were investigated by cyclic voltammetry in an aqueous solution of 0.65 M K_2SO_4. The specific capacitance of the developed composite electrode (MnO_2@BP-PTFE) was calculated as 122 ± 5 F/g, which is statistically comparable to the capacitance of MnO_2/BP-PTFE-based composite electrode (129 ± 6 F/g).[40]

Heon et al. employed a liquid phase plasma process to synthesize Iron oxide/carbon composite for its application as an electrode in supercapacitors. Spherically shaped iron oxide nanoparticles with an average size of $5-10$ nm were homogeneously dispersed on a powder surface. A consecutive increase in specific capacitance of composite was observed with increasing concentration of iron oxide precipitates. However, after the threshold, there was a prominent decrease in specific capacitance even with increasing amounts of iron oxide. The iron oxide/carbon-based composite showed the smallest resistance and the largest initial resistance slope with an optimum quantity (0.33 atomic %) of iron oxide precipitate.[41]

Yanhua et al. employed an ionic-liquid-assisted synthetic strategy to facilitate the in situ growth of TiO_2 nanocrystals with controlled size on graphene and CNTs. This approach decreased the modification in carbon support to retain the graphitic infrastructure. The as-synthesized nanocomposite possessed a highly porous structure with strong interaction between carbon interface and TiO_2 nanocrystals, which offers a facile electron and ion transport pathway and enhanced mechanical stability. The nanocomposite-based electrode material manifests a long cycling lifetime and higher specific capacity in high-performance lithium-ion batteries.[42]

Chemical vapor deposition was also employed to uniformly disperse CNTs onto aluminum oxide (Al_2O_3) powder. CNT reinforced Al_2O_3 coating onto steel substrate was achieved by plasma spraying. Additionally, CNTs corroborated the higher fracture toughness of the composite coating. The enhancement in the fracture toughness is accredited to homogenous dispersion of CNTs and toughening processes such as CNT bridging, crack deflection, and enhanced interaction between CNT/Al2O3 interfaces. Tribological findings on plasma-sprayed Al_2O_3—CNT coatings depicted an improvement in wear resistance at a high load of 50 N. Meanwhile, wear volume loss and coefficient of fraction enhanced with increasing normal load from the bounds of 10—50 N.[43] Similarly, the hydrothermal crystallization process could be utilized to synthesize Zirconium oxide (ZrO_2)/CNT composites from zirconium hydroxide [($Zr(OH)_4nH_2O$; n = 8—16] and CNTs at elevated temperature of 200°C for 8. Hydrothermal crystallization of zirconium oxide occurred on the walls of the tubes resulting in a homogeneous composite powder comprising microcrystalline multiwalled nanotubes and ZrO_2. The discussed strategy is an alternative method for the production of ceramic—nanotube composites. Different electron microscopy techniques have been employed for microstructural and qualitative characterizations.[44]

You et al. synthesized a novel structured material called "ant-cave microball" using continuous ultrasonic spray pyrolysis. The decomposition of polystyrene nanobeads produces nanovoids that result in the formation of nanochannels. This ant-cave-microstructure electrochemical activity has been investigated for its application in lithium-ion batteries. The carbon component was homogenously distributed in the MoO_3-containing microballs. Furthermore, the initial charge and discharge capacities of microballs at a current density of 2 A/g were 841 and 1212 mA/g, respectively. Microballs delivered a high discharge capacity of 733 mA h/g even after 300 cycles.[45] In this regard, a series of alumina species were prepared with and without CNTs and characterized via Raman spectra, X-ray diffraction, and transmission electron microscopy. An adsorption—desorption system was employed to scrutinize the surface properties of the alumina—CNT nanocomposites. The specific surface area of the synthesized alumina—CNT nanocomposites was improved by the growth of nanoparticles and nanotubes that formed micropores inside nanocomposites. In addition, an increase in total pore volume and surface area was observed. The explosion technique was exploited for the first time to control and improve the porous structure of alumina by dual growth of alumina and CNTs at low temperatures to fulfill the special needs of catalysis and the water purification market.[46] In short, this section has detailed the various techniques for synthesizing carbon/metal oxide composites and influencing the synthetic approach for surface properties and applications of integrated nanocomposites.

3.4 Parameters affecting the conductive properties of metal oxide—carbon composites

Science and technology strive for the evolution of novel and more proficient devices for comfortable human life. To store electrical energy, an electrochemical battery is replaced with a supercapacitor to complement the battery. The performance of a supercapacitor largely depends on the used electrode material, and therefore different kinds of materials are used to make electrodes of supercapacitors for its better performance.[47] Energy crisis and environmental pollution problems are becoming increasingly dangerous worldwide.[1–3] Solar energy is stable and clean and is considered a promising method to solve these issues.[48] However, most photocatalysts experience comparatively high recombination of photo-induced charge carriers and less solar-energy usage.[49] The alternative energy storage device is a supercapacitor that provides a higher energy and power density than a standard capacitor and battery. Supercapacitors have emerged because of the enormous amount of energy stored in less time, excellent cyclic stability, internal safety, and high round trip efficiency.[50] The supercapacitor was widely studied utilizing transition metal oxides. The electrode based on MoO_3 was extensively utilized for supercapacitor applications because of unique electrocatalytic properties, including their changing oxidation states from $+2$ to $+7$, low resistivity (8.8×10^{-5}), and n-type semiconductor and variable capacitance values (2700 F/g).[51] Metal oxides are a significant class of inorganic materials with varied compositions and structures. They are broadly utilized in catalysts, sensors, contaminants adsorbents, drug delivery agents, solar cells, and skincare product additives. Over the last few years, major efforts have been focused on forming metal oxide nanostructures with controlled structures having new remarkable optical, magnetic, thermal, and catalytic properties and resulting in novel applications.[52] Sensors are used in various commercial applications, e.g., household safety measures and emission control for various industries, particularly biomedical, agricultural, and automotive. Gas sensors made of metal oxide (MO) semiconductors have high sensitivity and stability, are less expensive, and have a short time response, so they are much more focused. For detecting oxidizing, reducing, and combustible gases, numerous MOs are known.[53]

Transition metal carbides (TMCs) show remarkable catalytic properties, so they can be used as an alternate for noble metals in different reactions involving catalysis.[54] TMCs have promising physical properties and electronic characteristics. These characteristics provide them with diverse applications in optical, electrical, magnetic, and catalytic fields. Molybdenum carbide catalysts are important transition metal carbides and show interesting catalytic properties in some reactions. These properties have been investigated intensively over the last few years.[48,55]

Carbon materials, including ACs, CNTs, and graphene, are the most extensively employed electrode materials due to their beneficial properties, including high surface area and pore volume, lower costs, diverse forms, nonreacting electrochemistry, and chemical stability in acidic and basic solutions.[56,57] Moreover, electrical conductivity is significant in determining supercapacitor performance.[58] As reported earlier, the electrical conductivity of carbon materials is strongly related to their structures. Conductivity goes down with higher surface area because of smaller particle sizes.[59]

For improved electrical conductivity of carbon, Ag nanoparticles, graphene/Ag nanowires, and graphene-Ag nanoparticle–polypyrrole nanocomposites are commonly used because of their high conductivity.[60] Particularly for AC, its usually high surface area can provide a large electrode/electrolyte interface for storing supercapacitor charges, and hence, electric double-layer capacitances are produced.[60] TMCs have excellent characteristics—for instance, small resistivity, elevated melting temperature, and high electrochemical activities—for catalysis and energy storage purposes.[61] The supercapacitor is an efficient energy-storing device used to overcome the need for worldwide energy use. The carbon materials, transition metal oxides, and electronically conducting polymers are single electroactive materials and experience several shortcomings when used as electrodes in supercapacitors. Therefore, various hybrid materials are employed to attain electrodes of high specific capacitance, energy and power density, rate potential, and cycle life with lower costs and greater eco-friendliness.[62]

With the increasing growth of nanotechnology, 2-D nanomaterials as flame retardants have attracted more attention—for example, graphene, MXene, hydroxides, and MoS_2—for making flame-retardant polymer composites.[12] MoS_2 has especially shown a broad capacity to strengthen polymer composites due to its excellent mechanical properties, attractive electronic properties, low thermal conductivity, and excellent flame-retardant characteristics with a high MP of $1185°C$.[63,64]

3.5 Applications and future perspectives of conductive metal oxide−carbon nanocomposites

Overall, such conductive materials could serve both as ideal substrate materials and electroactive architectures that enhance their potential for several applications. The main thrust of research has been carried out with these materials in energy storage, exploring their catalytic properties and the sensing of various analytes. The choice of appealing morphological features and nature of interactive relations between these materials make them versatile and widen the spectrum of their applications. Given the rising demand for newer materials in emerging and novel energy-related and other analytical performance-based applications, these conductive nanohybrids have achieved much attention and have been studied extensively worldwide. Table 3.1 reports various metal oxide−carbon nanomaterials to better demonstrate the scope of such conductive materials in applied research.

Based on their improved electrical properties, MO−C nanomaterials are widely used in several energy-related applications, including fuel cells, supercapacitors, and different types of batteries. This also assures suitable and trustworthy devices that allow for sufficient energy storage for later use in electronic gadgets, electric-powered movers, transport, and other discrete purposes. Based on the established protocols and intrinsic electric properties of these materials, several future applications can be envisioned. For instance, renewable resources can clearly satisfy the world's growing need for energy, where substitutes for petroleum substances require carbon fragment roots. The innovation of compact and handy electronic devices has drawn attention to efficient energy storage.[76] To ensure maximum surface area and develop efficient electrode materials for such applications, MO−C nanohybrids could be ideal candidates. Further insight into the structural and morphological relationships between MO and various carbon shapes could yield useful data for their application. Such

Table 3.1 Metal oxide—carbon nanomaterials via different synthetic approaches for their illustrated applications.

Sr #	Nanomaterial	Application	Ref
1	ZIF-67/rGO	High performance supercapacitor	[65]
2	Titanium dioxide—carbon nanocomposite	Supercapacitors	[66]
3	Transition metal oxide/activated carbon-based composites	Supercapacitors	[67]
4	CNT/Al$_2$O$_3$	Lead ion removal from aqueous solution	[68]
5	CNT/TiO$_2$	Photocatalysis	[69]
6	CNT/TiO$_2$/graphite plate	Methyl orange adsorption	[70]
7	ZnO—MWCNT—ITO	Urea detection	[71]
8	Graphene—CNT—ZnO	Glucose detection	[72]
9	SWCNT—ZnO	Dye-sensitized solar cells	[73]
10	MWCNT—Fe$_3$O$_4$	BPA detection	[74]
11	CNT—Fe$_3$O$_4$	Superparamagnetic performance for metal ion removal	[75]

hybrid materials in the form of metal organic and covalent organic frameworks could provide interesting properties and applications. Promising areas for future electrochemical conversion technologies and energy storage, transition metal/metal oxides (TMMOs) decorated on porous carbons (PCs) have been concentrated intensively in the design of electrode materials. Incorporating TMMOs with a PC structure could provide large open surface areas, rapid reversible surface redox reaction, suitable distribution of pore size, and increased electrical conductivity. For electrochemical energy storage, transition metal oxides are favorable electrode materials because of their reversible and fast surface redox reactions, as they display higher specific capacitance.[77] Moreover, several newer research domains, including all-weather energy production devices, active packaging, and biomedical applications, could potentially provide interesting avenues for these hybrid materials. Eventually, these next-generation MO—C hybrid materials have potential in a wide range of applications, including photocatalysis, supercapacitors, photovoltaics, LIBs, and sensors, with noticeable influences on society.

3.6 Conclusion

In conclusion, we have summarized different aspects of conductive nanocomposites of metal oxide and carbon interfaces. Different strategies and types of bindings between both the intrinsic materials, as well as various carbon interfaces, have been discussed in detail. Depending on the morphology of carbon interfaces ranging from 0-D to 3-D architecture, several MO—carbon nanocomposites have been synthesized and applied. All of these carbon interfaces, including carbon fibers, CNTs, graphene sheets, and AC, possess plus points for their corresponding applications. We have also detailed a few applications of these conducting nanocomposites with possible future perspectives for their future application. The current chapter provides brief insights into the nature and types of different MO—C conductive composites and their potential applications.

References

1. Park H-Y, et al. Green synthesis of carbon-supported nanoparticle catalysts by physical vapor deposition on soluble powder substrates. *Sci Rep* 2015;**5**(1):14245.
2. Pletincx S, et al. Probing the formation and degradation of chemical interactions from model molecule/metal oxide to buried polymer/metal oxide interfaces. *NPJ Mater Degrad* 2019;**3**(1):23.
3. Sun M, et al. Graphene-based transition metal oxide nanocomposites for the oxygen reduction reaction. *Nanoscale* 2015;**7**(4):1250—69.
4. Wu L, et al. $CoFe_2O_4$/C composite fibers as anode materials for lithium-ion batteries with stable and high electrochemical performance. *Solid State Ion* 2012;**215**:24—8.
5. Santhoshkumar P, et al. Incorporation of binary metal oxide and one dimensional carbon fiber hybrid nanocomposites for electrochemical energy storage applications. *J Alloys Compd* 2020:155649.
6. Santos C, et al. Interconnected metal oxide CNT fibre hybrid networks for current collector-free asymmetric capacitive deionization. *J Mater Chem* 2018;**6**(23):10898—908.
7. Liu W, et al. A wire-shaped flexible asymmetric supercapacitor based on carbon fiber coated with a metal oxide and a polymer. *J Mater Chem* 2015;**3**(25):13461—7.
8. Qin W, et al. Mechanical and electrical properties of carbon fiber composites with incorporation of graphene nanoplatelets at the fiber—matrix interphase. *Compos B Eng* 2015;**69**:335—41.
9. Newcomb BA. Processing, structure, and properties of carbon fibers. *Compos Appl Sci Manuf* 2016;**91**:262—82.
10. Ciambella J, Stanier DC, Rahatekar SS. Magnetic alignment of short carbon fibres in curing composites. *Compos B Eng* 2017;**109**:129—37.
11. Feng L, Xie N, Zhong J. Carbon nanofibers and their composites: a review of synthesizing, properties and applications. *Materials* 2014;**7**(5):3919—45.
12. Shiba K, et al. Effective surface functionalization of carbon fibers for fiber/polymer composites with Tailor-made interfaces. *ChemPlusChem* 2014;**79**(2):197—210.
13. Lee JS, et al. Fabrication of ultrafine metal-oxide-decorated carbon nanofibers for DMMP sensor application. *ACS Nano* 2011;**5**(10):7992—8001.
14. Popov VN. Carbon nanotubes: properties and application. *Mater Sci Eng R Rep* 2004;**43**(3):61—102.
15. Xu X, et al. Bamboo-like amorphous carbon nanotubes clad in ultrathin nickel oxide nanosheets for lithium-ion battery electrodes with long cycle life. *Carbon* 2015;**84**:491—9.
16. Andersen NI, Serov A, Atanassov P. Metal oxides/CNT nano-composite catalysts for oxygen reduction/oxygen evolution in alkaline media. *Appl Catal B Environ* 2015;**163**:623—7.
17. Singh A, Chandra A. Significant performance enhancement in asymmetric supercapacitors based on metal oxides, carbon nanotubes and neutral aqueous electrolyte. *Sci Rep* 2015;**5**:15551.

18. Bordjiba T, Bélanger D. Development of new nanocomposite based on nanosized-manganese oxide and carbon nanotubes for high performance electrochemical capacitors. *Electrochim Acta* 2010;**55**(9):3428–33.

19. Krishnaveni M, Asiri AM, Anandan S. Ultrasound-assisted synthesis of unzipped multi-walled carbon nanotubes/titanium dioxide nanocomposite as a promising next-generation energy storage material. *Ultrason Sonochem* 2020:105105.

20. Papageorgiou DG, Kinloch IA, Young RJ. Mechanical properties of graphene and graphene-based nanocomposites. *Prog Mater Sci* 2017;**90**:75–127.

21. Balandin AA. Thermal properties of graphene and nanostructured carbon materials. *Nat Mater* 2011;**10**(8):569–81.

22. Baitimirova M, et al. Tuning of structural and optical properties of graphene/ZnO nano-laminates. *J Phys Chem C* 2016;**120**(41):23716–25.

23. Schöche S, et al. Optical properties of graphene oxide and reduced graphene oxide determined by spectroscopic ellipsometry. *Appl Surf Sci* 2017;**421**:778–82.

24. Kazmi SA, et al. Electrical and optical properties of graphene-TiO_2 nanocomposite and its applications in dye sensitized solar cells (DSSC). *J Alloys Compd* 2017;**691**:659–65.

25. Wang H, Feng H, Li J. Graphene and graphene-like layered transition metal dichalcogenides in energy conversion and storage. *Small* 2014;**10**(11):2165–81.

26. Hussain SZ, et al. A review on graphene based transition metal oxide composites and its application towards supercapacitor electrodes. *SN Appl Sci* 2020;**2**(4):1–23.

27. Fan Y, et al. Preparation and electrical properties of graphene nanosheet/Al_2O_3 composites. *Carbon* 2010;**48**(6):1743–9.

28. Sun Q, et al. Nitrogen-doped graphene-supported mixed transition-metal oxide porous particles to confine polysulfides for lithium–sulfur batteries. *Adv Ener Mater* 2018;**8**(22): 1800595.

29. Gomes Ferreira de Paula F, et al. Structural flexibility in activated carbon materials prepared under Harsh activation conditions. *Materials* 2019;**12**(12):1988.

30. Naeem S, et al. Development of porous and electrically conductive activated carbon web for effective EMI shielding applications. *Carbon* 2017;**111**:439–47.

31. Li X, et al. Self-supporting activated carbon/carbon nanotube/reduced graphene oxide flexible electrode for high performance supercapacitor. *Carbon* 2018;**129**:236–44.

32. Mahmoodi NM, et al. Activated carbon/metal-organic framework nanocomposite: preparation and photocatalytic dye degradation mathematical modeling from wastewater by least squares support vector machine. *J Environ Manag* 2019;**233**:660–72.

33. Andriantsiferana C, Mohamed EF, Delmas H. Photocatalytic degradation of an azo-dye on TiO_2/activated carbon composite material. *Environ Technol* 2014;**35**(3):355–63.

34. Llobet E. Gas sensors using carbon nanomaterials: a review. *Sensor Actuator B Chem* 2013; **179**:32–45.

35. Malkhandi S, et al. Electrocatalytic activity of transition metal oxide-carbon composites for oxygen reduction in alkaline batteries and fuel cells. *J Electrochem Soc* 2013;**160**(9):F943.

36. Barroso-Bogeat A, et al. *Activated carbon as a metal oxide support: a review. Activated carbon: classifications, properties and applications.* 2011. p. 297–318.

37. Brousse T, Toupin M, Belanger D. A hybrid activated carbon-manganese dioxide capacitor using a mild aqueous electrolyte. *J Electrochem Soc* 2004;**151**(4):A614.

38. Ni Y, et al. Enzyme-free glucose sensor based on heteroatom-enriched activated carbon (HAC) decorated with hedgehog-like NiO nanostructures. *Sensor Actuator B Chem* 2017; **250**:491–8.

39. Yumak T, Bragg D, Sabolsky EM. Effect of synthesis methods on the surface and electrochemical characteristics of metal oxide/activated carbon composites for supercapacitor applications. *Appl Surf Sci* 2019;**469**:983–93.

40. Gambou-Bosca A, Bélanger D. Chemical mapping and electrochemical performance of manganese dioxide/activated carbon based composite electrode for asymmetric electrochemical capacitor. *J Electrochem Soc* 2015;**162**(5):A5115.

41. Lee H, et al. Liquid phase plasma synthesis of iron oxide/carbon composite as dielectric material for capacitor. *J Nanomater* 2014;**2014**.

42. Cheng Y, et al. Ionic liquid-assisted synthesis of TiO_2—carbon hybrid nanostructures for lithium-ion batteries. *Adv Funct Mater* 2016;**26**(9):1338—46.

43. Keshri AK, et al. Synthesis of aluminum oxide coating with carbon nanotube reinforcement produced by chemical vapor deposition for improved fracture and wear resistance. *Carbon* 2010;**48**(2):431—42.

44. Lupo F, et al. Microstructural investigations on zirconium oxide—carbon nanotube composites synthesized by hydrothermal crystallization. *Carbon* 2004;**42**(10):1995—9.

45. Ko YN, et al. One-pot facile synthesis of ant-cave-structured metal oxide—carbon microballs by continuous process for use as anode materials in Li-ion batteries. *Nano Lett* 2013;**13**(11):5462—6.

46. Saber O, et al. A novel route for controlling and improving the texture of porous structures through dual growth of alumina nanoparticles and carbon nanotubes using explosion process of solid fuel. *J Mater Res Technol* 2020;**9**(1):67—75.

47. Choudhary RB, Ansari S, Purty B. Robust electrochemical performance of polypyrrole (PPy) and polyindole (PIn) based hybrid electrode materials for supercapacitor application: a review. *J Ener Stor* 2020;**29**:101302.

48. Zhou Y, et al. Sustainable hydrogen production by molybdenum carbide-based efficient photocatalysts: from properties to mechanism. *Adv Colloid Interface Sci* 2020:102144.

49. Zhou C, et al. Highly porous carbon nitride by supramolecular preassembly of monomers for photocatalytic removal of sulfamethazine under visible light driven. *Appl Catal B Environ* 2018;**220**:202—10.

50. Elshahawy AM, et al. Controllable $MnCo_2S_4$ nanostructures for high performance hybrid supercapacitors. *J Mater Chem* 2017;**5**(16):7494—506.

51. Gurusamy L, et al. Fabrication of molybdenum oxycarbide nanoparticles dispersed on nitrogen-doped carbon hollow nanotubes through anion exchange mechanism for enhanced performance in supercapacitor. *J Ener Stor* 2020;**27**:101122.

52. Hwang J, et al. Controlling the morphology of metal—organic frameworks and porous carbon materials: metal oxides as primary architecture-directing agents. *Chem Soc Rev* 2020;**49**(11):3348—422.

53. Goldoni A, et al. Advanced promising routes of carbon/metal oxides hybrids in sensors: a review. *Electrochim Acta* 2018;**266**:139—50.

54. Schlatter JC, et al. Catalytic behavior of selected transition metal carbides, nitrides, and borides in the hydrodenitrogenation of quinoline. *Ind Eng Chem Res* 1988;**27**(9):1648—53.

55. Miao M, et al. Molybdenum carbide-based electrocatalysts for hydrogen evolution reaction. *Chem A Eur J* 2017;**23**(46):10947—61.

56. Zhang LL, Zhao X. Carbon-based materials as supercapacitor electrodes. *Chem Soc Rev* 2009;**38**(9):2520—31.

57. Zang Z, et al. Tunable photoluminescence of water-soluble AgInZnS—graphene oxide (GO) nanocomposites and their application in-vivo bioimaging. *Sensor Actuator B Chem* 2017;**252**:1179—86.

58. El-Kady MF, et al. Laser scribing of high-performance and flexible graphene-based electrochemical capacitors. *Science* 2012;**335**(6074):1326—30.

59. Frackowiak E, Beguin F. Carbon materials for the electrochemical storage of energy in capacitors. *Carbon* 2001;**39**(6):937—50.

60. Jiao C, et al. Design and synthesis of phosphomolybdic acid/silver dual-modified microporous carbon composite for high performance supercapacitors. *J Alloys Compd* 2019;**791**: 1005–14.

61. Zang X, et al. Laser-sculptured ultrathin transition metal carbide layers for energy storage and energy harvesting applications. *Nat Commun* 2019;**10**(1):1–8.

62. De B, et al. Transition metal oxide-/carbon-/electronically conducting polymer-based ternary composites as electrode materials for supercapacitors. In: *Handbook of nanocomposite supercapacitor materials II*. Springer; 2020. p. 387–434.

63. Peng H, et al. Interfacial growth of 2D bimetallic metal-organic frameworks on MoS2 nanosheet for reinforcements of polyacrylonitrile fiber: from efficient flame-retardant fiber to recyclable photothermal materials. *Chem Eng J* 2020:125410.

64. Yang L, et al. Ultralight, highly thermally insulating and fire resistant aerogel by encapsulating cellulose nanofibers with two-dimensional MoS 2. *Nanoscale* 2017;**9**(32): 11452–62.

65. Sundriyal S, et al. High-performance symmetrical supercapacitor with a combination of a ZIF-67/rGO composite electrode and a redox additive electrolyte. *ACS Omega* 2018;**3**(12): 17348–58.

66. Shrivastav V, et al. Metal-organic frameworks-derived titanium dioxide–carbon nanocomposite for supercapacitor applications. *Int J Energy Res* 2020;**44**(8):6269–84. https://doi.org/10.1002/er.5328.

67. Sinha P, Banerjee S, Kar KK. Transition metal oxide/activated carbon-based composites as electrode materials for supercapacitors. In: *Handbook of nanocomposite supercapacitor materials II*. Springer; 2020. p. 145–78.

68. Gupta VK, Agarwal S, Saleh TA. Synthesis and characterization of alumina-coated carbon nanotubes and their application for lead removal. *J Hazard Mater* 2011;**185**(1):17–23.

69. Lee WJ, et al. Biomineralized N-doped CNT/TiO$_2$ core/shell nanowires for visible light photocatalysis. *ACS Nano* 2012;**6**(1):935–43.

70. Dong Y, Tang D, Li C. Photocatalytic oxidation of methyl orange in water phase by immobilized TiO$_2$-carbon nanotube nanocomposite photocatalyst. *Appl Surf Sci* 2014;**296**: 1–7.

71. Tak M, Gupta V, Tomar M. Zinc oxide–multiwalled carbon nanotubes hybrid nanocomposite based urea biosensor. *J Mater Chem B* 2013;**1**(46):6392–401.

72. Hwa K-Y, Subramani B. Synthesis of zinc oxide nanoparticles on graphene–carbon nanotube hybrid for glucose biosensor applications. *Biosens Bioelectron* 2014;**62**:127–33.

73. Omar A, et al. Characterization of zinc oxide dye-sensitized solar cell incorporation with single-walled carbon nanotubes. *J Mater Res* 2013;**28**(13):1753–60.

74. Jiao Y, et al. Determination of bisphenol A, bisphenol F and their diglycidyl ethers in environmental water by solid phase extraction using magnetic multiwalled carbon nanotubes followed by GC-MS/MS. *Anal Meth* 2012;**4**(1):291–8.

75. Zhang C, et al. Efficient removal of heavy metal ions by thiol-functionalized superparamagnetic carbon nanotubes. *Chem Eng J* 2012;**210**:45–52.

76. Siwal SS, et al. Carbon-based polymer nanocomposite for high-performance energy storage applications. *Polymers* 2020;**12**(3):505.

77. De B, et al. Transition metal oxides as electrode materials for supercapacitors. In: *Handbook of nanocomposite supercapacitor materials II*. Springer; 2020. p. 89–111.

Photoelectrochemical properties for metal oxide–carbon hybrid materials

Faryal Idrees[1], Fauzia Iqbal[1], Saman Iqbal[1], Amir Shehzad Shah[2] and Husnain Joan[2]
[1]Department of Physics, University of the Punjab, Quaid-e-Azam Campus, Lahore, Punjab, Pakistan; [2]Department of Physics, The University of Lahore, Lahore, Punjab, Pakistan

4.1 Introduction

Due to global warming, depleted fossil resources, and growing energy demands, significant advancements have been made over the years to move toward renewable energy sources. Photocatalysis is emerging as an energy conversion and storage technology, converting solar fuel into H_2 or CO_2 reduction through water treatment. Thus, photocatalysis is thought of as an efficient and economical way to convert widely available solar energy into chemical fuel and develop a carbon-neutral society in the near future.[1,2] The photocatalytic process based on semiconductors can be achieved with a system of photoelectrochemical cells (PECs) containing a direct semiconductor/liquid interface. However, attaining high-performance renewable energy systems and high stability at the industrial level in the next few years will require substantial effort to overcome fluctuating growth in demand for power generation. Thus, quick rechargeable electrochemical energy storage systems (EESs) at low cost and with high energy and power density are indispensable to meet future energy demand. EESs are categorized as (1) supercapacitors/electrochemical double-layered capacitors (EDLCs) that store energy by physical adsorption of ions at electrode surfaces and (2) batteries using faradaic reactions in the bulk volume of electrodes.[3]

This chapter discusses next-generation hybrid materials based on metal oxides composited with carbon-based materials. This hybridization has shown promise in applications for energy conversion and storage devices, i.e., PECs, EDLCs, and batteries. Moreover, the selection features required for improved efficiency of proposed energy storage and conversion devices are discussed. The chapter ends by providing the electrochemical properties of the studied hybridized materials.

4.2 Photoelectrochemical hybrid materials

The advancement of complementary, commercially feasible technologies for the production of solar fuels, i.e., molecular hydrogen production (H_2), carbon dioxide (CO_2)

Metal Oxide-Carbon Hybrid Materials. https://doi.org/10.1016/B978-0-12-822694-0.00009-0

reduction, and water-splitting (H_2 and O_2), is therefore of great significance.[4] Consequently, cost-effective, durable, and effective semiconductor materials are required to allow realistic PEC solar fuel production. The production of promising semiconductor materials has seen a great rise in the last 5 years, advancing to the appropriate engineering of superior semiconducting materials.[5,6]

Currently, TiO_2 is the benchmarked photocatalytic material for PEC applications. However, the wide bandgap of 3.2 eV limits its absorption to ultraviolet (UV) light only, and fast charge recombination has also affected its stability. Therefore, advances in photocatalytic materials are focusing on their fabrication to develop a suitable photoelectrode satisfying the requirements like (a) visible light active bandgap in the range of 1.8−2.2 eV; (b) suitable conduction-band and valence-band positions close to the redox and oxidation potential of water; (c) favorable and fast-transportation of photogenerated electron-hole (e^--h^+) pairs to slow down the recombination rate; (d) noncorrosive nature and high chemical-stability of the used electrolyte; most importantly (e) the low cost.[7−9] Considering these properties, WO_3,[10] $SrTiO_3$,[11] $BaTiO3$,[12] SnO_2,[13] ZnO,[14] etc. are favorable in aqueous electrolytes, but their bandgaps lie in the UV region, which is just 4% of the total solar spectrum. While $GaAs$,[15] NB_2O_5/g-C_3N_4,[16] $CdTe$,[17] Cu_2O,[18] etc. have a bandgap in the visible region, electrodes have a high corrosion rate toward electrolytes. Other studied materials have been limited by the fast charge recombination rate. Thus, developing stable, inexpensive, and highly efficient photoelectrode materials is the main approach to commercialize PECs for solar fuel production. Significant efforts have been considered to design such materials satisfying the five vital properties discussed above.[19]

Owing to the varied and compatible nature of the properties found in these various materials, hybrid materials/hybrids containing inorganic and organic components are emerging as an effective and exciting category of materials. Among these, metal oxide−carbon hybrid electrodes have shown potential in applications requiring efficient photoelectrochemical properties.[20] The schematic illustration of metal-hybrid material following a Z-scheme and Type-II is provided in Fig. 4.1.

4.3 Selection features for photoelectrochemical energy conversion

4.3.1 Electrode and electrolytes

For effective energy conversion using the PEC method, choosing photoelectrode and electrolyte materials is an important consideration. Other than durability, cost-effectiveness, and low maintenance, PEC electrodes should have the following characteristics: (1) effective visible light absorption, (2) effective electron-hole pairs separation, (3) fast transportation of segregated charges to prevent recombination, (4) favorable band edge positions for valence and conduction bands, (5) high electrolytic stability, and (6) noncorrosivity for high stability.[8,21,22]

The electrolyte is a significant component in PECs and serves as an electrically conductive medium in transferring holes from a photoelectrode to a counter electrode.

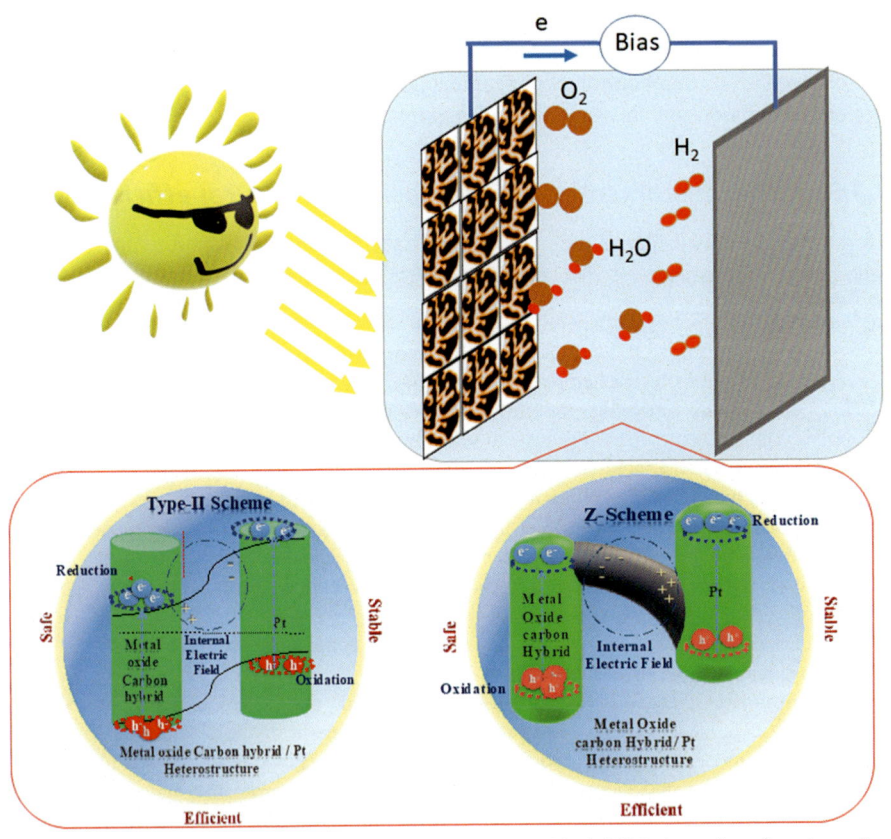

Figure 4.1 Photogeneration of electron-hole-pairs at metal hybrid/platinum interface through a direct Z-scheme and Type-II mechanism (regenerated through[16]).

An effective and efficient electrolyte (1) exhibits a fast charge transfer rate for increased redox reaction process, (2) has the lowest optical absorption, (3) shows optical and thermal stability in the solar spectrum and temperature zone during electrochemical reactions, (4) has a voltage window dependent on the electrolyte, which is important to achieve the required current densities, (5) has a concentration of oxidizing and reducing solvent materials sufficient for the supported electrodes, (6) exhibits negligible ohmic losses for ionic electrolyte conductance, (7) does not react with electrode semiconductors and is less corrosive toward electrodes, and (8) is less environmentally toxic.[21,22]

4.3.2 Favorable bandgap tuning

Since the discovery of water-splitting by Fujishima using TiO_2, significant work has been done to improve PEC water-splitting and photocatalytic degradation of organic pollutants, in which the bandgap, band edge position, and charge carrier mobility affect PEC efficiency. These finely tuned electronic band structures can help control light absorption, charge transportation, and thermodynamic driving forces for PEC and photocatalytic degradation chemical reactions.[23,24] Thus, a suitable semiconductor choice depends on the abovementioned features.

4.3.3 Other factors affecting photoelectrochemical efficiency

Morphology, crystallinity, dimensionality, surface area, porosity, and temperature are other interesting aspects of PEC energy conversion, as follows:

(a) Crystallinity: Photoelectrochemical efficiency depends on the material's crystallinity, morphology, dimension, and thickness. For amorphous materials, small photocurrent density and less photoelectrochemical performance are observed and vice versa. Monocrystalline has greater charging mobility and lower recombination rates relative to its polycrystalline counterparts because grain boundaries prohibit charge transport in polycrystalline materials and serve as places for charge recombination.[24] Therefore, in contrast to amorphous materials, crystalline materials exhibit higher PEC efficiency.[25,26]

(b) Temperature plays a critical role in improving PEC efficiency; most experiments are performed at high temperatures. High-temperature experiments decrease impurities and increase crystallinity, thus showing higher photoelectrochemical efficiency.[27] Moreover, a redshift of the bandgap is observed.

(c) Dimensionality: Based on dimensionality, nanomaterials can be categorized as zero-dimensional (0-D), one-dimensional (1-D), two-dimensional (2-D), or three-dimensional (3-D) structures that are investigated as photoelectrode materials. Quantum dot semiconductor nanocrystals have (1) favorable bandgap tuning from the UV region to visible/infrared regions, (2) high absorption ability, (3) photobleaching resistance, and (4) multiexcitation generation and energy transfer of charge carriers. Therefore, they have been used for PEC water-splitting, especially as a photoanode. But because of the confined space in the nanostructure, the charge separation in 0-D nanostructures may not be as effective as that in bulk.[28,29]

The high aspect ratio axial measurements of 1-D nanomaterials encourage the optical path lengths and transport of directional electrons along the long axis. Therefore, for PEC applications, 1-D architecture, such as vertically growing the active material on the substrate, increases charge transportation toward the current collector and slows down the recombination rate of the charge carriers. However, the specific surface area is not suitable for carrier transportation at the electrode—electrolyte interface.[30−32] Such advantages encourage efforts to promote photoanodes based on nanostructures of 1-D metal oxides. In literature, the 1-D nanostructures studied for solar energy conversion applications are nanorods (ZnO nanorods,[30,31] TiO_2 nanorods,[20,32] WO_3 nanorods[33]), nanowires (ZnO nanowires,[31] α-Fe_2O_3 nanowires,[34] SiC nanowires[35]), and nanotubes—TiO_2 nanotubes[33] and carbon nanotubes (CNTs).[26]

Two-dimensional nanostructures, such as TiO_2 thin films,[36] MoS_2 nanosheets,[37] and g−C_3N_4 nanosheets[38] have been used for water oxidation in PECs. These can accumulate a significant portion of UV light effectively because of their small thickness and high surface area. This allows the easy transportation of charges to the surfaces and increases the efficiency of hydrogen generation. Compared with thin films, nanowires and nanorods have been found more photoactive and effective for charge transportation.

Three-dimensional materials can be used as building blocks in a specific and ordered system by combining 0-D, 1-D, or 2-D nanomaterials. Three-dimensional nanostructures—dendritic α-Fe$_2$O$_3$[34] have shown high photocatalytic activity that facilitates effective light absorption by minimizing the distance to the electrolyte/electrode interface for the photogenerated holes to propagate.[36–38]

d) Size: Semiconductor efficiency is closely related to size—i.e., for 0-D, 1-D, and 2-D materials, the size of the nanostructure corresponds to diameter, transverse dimension, and thickness. If the material size falls within the nanoscale, various properties can be expressed by the materials. When the size is decreased to the nanometer scale, the number of visible atoms or ions on the surface massively increases, thus increasing the surface-to-volume ratio. For catalytic reactions, this increases the surface area and hence the number of active sites. Therefore, nanoscale materials can demonstrate high surface activity that does not occur in bulk. A cascade of electronic material properties often moves significantly because of size reduction. It is possible to increase PEC output by carefully controlling material size.[39,40]

4.4 Electrical double-layered capacitor and battery hybrid materials

Electrical double-layered capacitors (EDLCs) are emerging as fast and highly stable energy storage devices by separating electrical charges at the interface of electrodes and electrolytes. That is, electrolyte ions diffuse on the oppositely charged electrode surface, forming an electrical double layer. The electro-sorption process follows a fast charge and discharge. Therefore, materials with high electrical conductivity and surface area for better ion adsorption are required, e.g., graphene, CNTs, activated carbon (AC), carbon nanofibers (CNFs), and other carbon-based materials. Despite the fast charge—discharge process, EDLC specific energy is a few orders less than traditional lithium-ion batteries because of the small capacitance of carbon-based materials.[41,42]

Batteries, meanwhile, store energy by faradaic reactions that follow conversion or insertion, or by employing chemical reactions between electrolyte ions and electrodes. On a commercial scale, Li-ion batteries are in use, but growing environmental concerns and a scarcity of lithium (Li) have motivated the world to find alternatives such as sodium (Na) and potassium (K). Unlike EDLCs, cathodes and anodes are constructed of different materials. Thus, Li intercalation and deintercalation under faradaic charge transfer require a different electrochemical potential developed between electrodes. Therefore, layered structures like graphite and metal oxide-based materials are considered favorable for smooth intercalation and deintercalation. Metal oxides have greater capacitance than carbon-based compounds, such as NiO_x, RuO_x, CoO_x, MnO_x, and FeO_x, but their low electrical conductivity, extreme agglomeration, and short metal oxide cycle life reduce their use in high-power storage systems. Moreover, battery cycling lives are limited by the continuous corrosion of electrodes due to chemical reactions between electrolytic ions and electrodes.[43–45]

EDLCs and batteries have certain advantages and disadvantages. Therefore, hybridization of electrodes has become popular for obtaining high-energy and high-power-density electrochemical energy storage devices. An innovative hybridization uses composite active materials composed of metal oxide with a carbon host. These hybrids include metal oxides composited with AC, graphene, CNTs, carbon aerogels, and mesoporous ordered carbon and CNFs.[39,40] The schematic illustration of metal oxide–carbon hybrids have been provided in Fig. 4.2.

4.5 Metal oxide–carbon hybrid materials for energy conversion and storage

In recent years, many efficient synthesis methods have been established for metal oxide–carbon-based materials. Here their application for photocatalysis, PECs, supercapacitors, and batteries have been discussed for different combinations.

4.6 Materials studied for photocatalysis and photoelectrochemical applications

Metal oxides attract much attention for their low cost, favorable band edges, excellent chemical stability, and tunable bandgaps. This chapter revisited the recent progress of using nanostructured metal oxides as cocatalysts and as photoelectrodes for PECs.

Yufei Ma et al.[46] performed SRM (stream reforming of methanol) of different transition metals (Pt, Ni, Co, and Fe) doped molybdenum carbide via in-situ carburization. The temperature-programmed reaction maintained a final temperature of $\sim 700°C$ of reaction gas mixture containing 20% methane (CH_4) and molecular hydrogen (H_2). Fig. 4.3A and B shows XRD-patterns of undoped/metal-doped molybdenum carbide

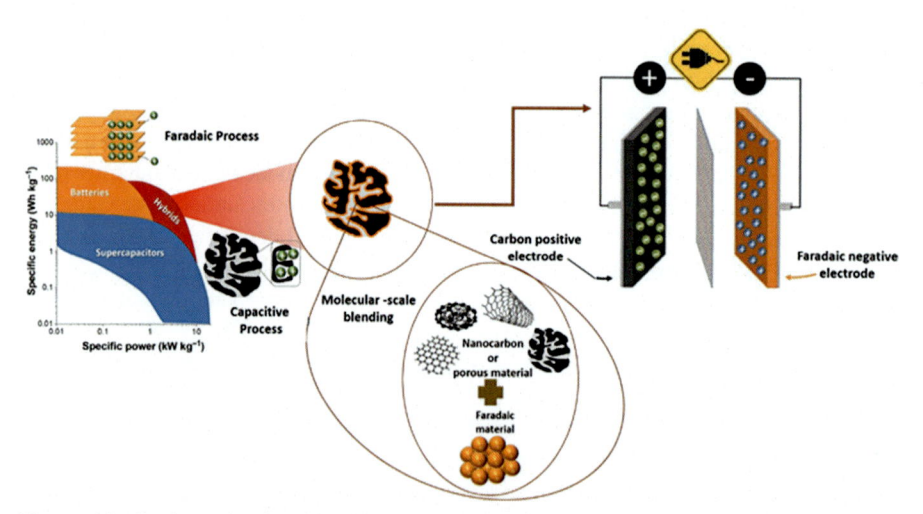

Figure 4.2 A schematic of metal oxide–carbon hybrid supercapacitor device using a capacitive positive electrode and a faradaic negative electrode, regenerated through.[3]

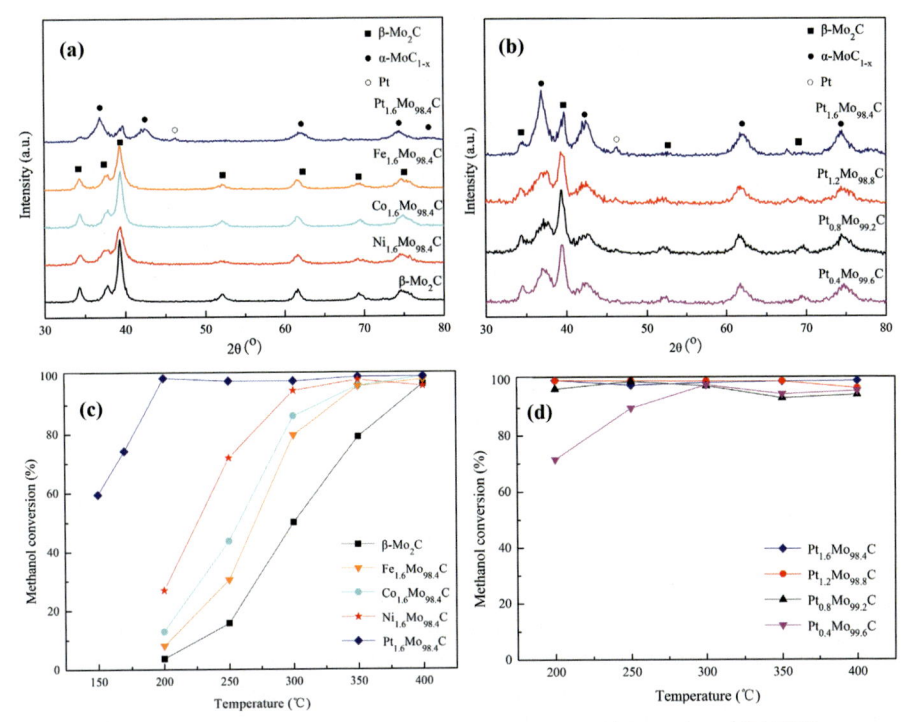

Figure 4.3 (A) XRD patterns of metal-doped molybdenum carbide catalysts; (B) XRD patterns of Pt doped molybdenum carbide with different doping amount; (C) Catalytic performances of metal-doped molybdenum carbide for SRM reaction at different temperatures; (D) Catalytic performances of Pt doped molybdenum carbide with different doping amount.

catalysts. For transition metals (Fe, Co, Ni), a hexagonal structure β-Mo$_2$C (HCP structure) was obtained. Metal and metal oxide peaks are not identified. However, Pt doping resulted in FCC structure α- MoC$_{1-x}$ along with β- Mo$_2$C. An increased amount of Pt resulted in a higher concentration of α- MoC$_{1-x}$.

A summarized result of BET surface analysis of unmodified and modified metal molybdenum carbide catalysts is shown in Table 4.1. The highest BET surface area

Table 4.1 BET surface area of β- Mo$_2$C, Fe$_{1.6}$ Mo$_{98.4}$C; Co$_{1.6}$ Mo$_{98.4}$C; Ni $_{1.6}$ Mo$_{98.4}$C; Pt$_{1.6}$ Mo$_{98.4}$C.

Catalysts	BET (m^2/g)
Mo$_2$C	10.2
Fe$_{1.6}$ Mo$_{98.4}$C	18.3
Co$_{1.6}$ Mo$_{98.4}$C	20.6
Ni$_{1.6}$ Mo$_{98.4}$C	19.3
Pt$_{1.6}$ Mo$_{98.4}$C	40

$(40\ m^2/g)$ is reported for Pt doped molybdenum carbide. Metal doping decreased the reduction temperature of MoO_3 and limited the mobility of the metal atom. As a result, particle size is reduced, and the surface area is enhanced compared with undoped.

Fig. 4.3C shows the influence of reaction temperature on methanol conversion for undoped and metal-doped molybdenum carbide. By increasing the reaction temperature, the methanol conversion reaction is increased. The strong interaction of a metal particle with a carbide substrate enhanced the methanol conversion compared with undoped Mo_2C. A higher methanol conversion and hydrogen yield were obtained for metal doping. Among all dopants (Fe, Ni, Co), Pt doping experienced the highest catalytic activity and hydrogen selectivity. A 100% methanol conversion efficiency is obtained at low temperature ($\sim 200°C$) and persists for a long-time with stable methanol conversion. Fig. 4.3D depicts the impact of the different doping concentrations of Pt on catalytic performance. A low concentration of $Pt_{0.4}\ Mo_{98.4}C$ lowered the catalytic activity below $300°C$. The higher Pt concentration (0.8, 1.2, 1.6) increased the catalytic activity for the entire range $200°C-400°C$.

Mn_3O_4/CNT supported catalysts were fabricated by Yuwei Feng et al.[47] utilizing the solvothermal method. Fig. 4.4A and B shows the SEM images of CNTs and $Mn_3O_4/CNTs$-3. An increased diameter of CNT is obtained after loading of Mn_3O_4. TEM images (Fig. 4.4C) confirmed the uniform distribution of Mn_3O_4 on CNTs. Approximate diameters of $3-5$ nm have been obtained for Mn_3O_4 nanoparticles. The crystal plane (101) of octahedral Mn_3O_4 is revealed by fringe spacing (~ 0.48 nm) as shown in HR-TEM (Fig. 4.4D). The obtained result suggested the reduction of Mn^{+7} and its conversion into Mn^{+2} and Mn^{+3} by ethylene glycol during the hydrothermal process. The octahedral and tetrahedral vacancies can be filled with Mn^{+3} and Mn^{+2} ions, respectively. However, (101) direction acted as an easy axis for Mn^{+3} ions over Mn^{+2}. Thus, the growth of the lattice plane in (101) resulted in octahedral Mn_3O_4. The exposed (101) crystal plane has active sits of Mn−O bonds which are conductive to adsorption, activate the reactants, and promote the reaction.

BET analysis (Table 4.2.) indicated a reduction in the specific surface area of CNTs after Mn_3O_4 loading, which is in close agreement with TEM results. Pore diameter and pore volume can be increased by increasing the number of CNTs. Large defect sites on CNT surfaces enhanced the interaction between CNTs and metal oxide and prevented metal oxide agglomeration. However, the number of CNTs must be optimized. The large amount of CNTs restricted the homogeneous nucleation/growth of Mn_3O_4 and resulted in lower catalytic activity.

The catalytic performance of selective oxidation of toluene to benzyl alcohol and benzaldehyde is summarized in Table 4.3. Mn_3O_4-loaded CNTs exhibited better performance than pure CNTS and Mn_3O_4. Toluene conversion rate enhanced from 7% to 24% by increasing the CNTs content in Mn_3O_4. The homogeneous distribution of Mn_3O_4 on a large CNT surface area enhances catalytic activity. Moreover, toluene activation can be improved by selective adsorption of benzene rings through π-electron of CNTs. A decreased catalytic activity is reported by increasing the CNT amount (~ 100 mg) due to acidic groups (carboxylic) at CNT surfaces that inhibit active toluene adsorption. Various parameters such as reaction temperature, reaction time, catalyst quality, oxygen flow rate, and initiator dosage must be optimized for improved results.

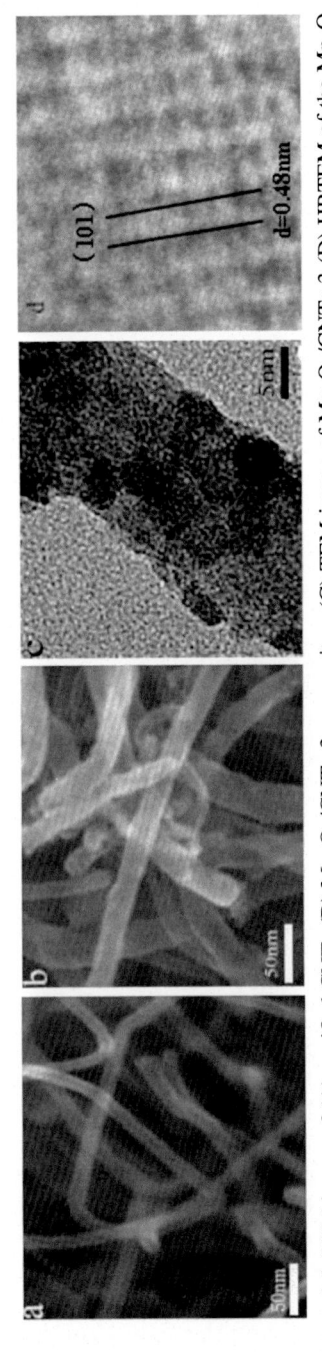

Figure 4.4 SEM image of (A) purified CNTs (B) Mn_3O_4/CNTs-3 composites; (C) TEM image of Mn_3O_4/CNTs-3 (D) HRTEM of the Mn_3O_4 supported CNTs.

Table 4.2 BET surface area of CNTs and Mn_3O_4/CNTs catalysts.

Catalysts composites[a]	Notations	CNT[b]	BET surface area (m^2/g)	Pore volume $(cm^3/g)^c$	Pore diameter(nm)[d]
CNTs	CNTs	–	341	1.2321	14.5
Mn_3O_4/CNTs (1:0.8)	Mn_3O_4/CNTs-1	0.4	251	0.7842	12.0
Mn_3O_4/CNTs (1:1.2)	Mn_3O_4/CNTs-2	0.6	268	0.8136	12.1
Mn_3O_4/CNTs (1:1.6)	Mn_3O_4/CNTs-3	0.8	269	0.8421	12.5
Mn_3O_4/CNTs (1:2.1)	Mn_3O_4/CNTs-4	1	272	0.9031	13.2
Mn_3O_4/CNTsrun	Mn_3O_4/CNTs-3-run4	–	270	0.8532	12.6

[a]The mass ratio of Mn3O4/CNTs estimated from reactants.
[b]The mass percentage of added CNTs in preparation process.
[c]BJH (Barret-Joyner-Halenda) adsorption average pore volume.
[d]BJH adsorption average pore diameter.
Regenerated from ref [47].

Table 4.3 Catalytic activity Mn_3O_4/CNTs catalyst on toluene oxidation reaction.

Catalysis	Conversion%	Selectivity%		
		Benzyl alcohol	Benzaldehyde	Other products
Mn_3O_4	3.54	40.47	58.51	1.02
Mn_3O_4[a]	1.67	38.57	61.19	0.24
CNTs	0.65	41.56	58.44	0
Mn_3O_4/CNTs-1	7.00	58.14	40.59	1.27
Mn_3O_4/CNTs-2	15.10	53.06	43.58	3.36
Mn_3O_4/CNTs-3	24.63	43.51	46.98	9.51
Mn_3O_4/CNTs-4	18.56	50.52	44.03	5.45

[a]Commercial catalyst. Reaction conditions: 10 mL toluene, 100 mg catalyst. 0.5 mL (t-butylhydroperoxide), reaction temperature 90°C, oxygen flow rate 15 mL/min, reaction time 12 h.
Regenerated from ref [47]

A combustion method was selected to prepare graphene oxide (GO)-TiO$_2$ hybrid material in a single step by Yanyan Gao et al.[48] Urea is selected as fuel, and titanyl nitrate is taken as an oxidizer, respectively. GO precursors, fuel, and oxidizer were maintained at different combustion temperatures (300°C−450°C) to start the reaction. Fig. 4.5A and B shows the TG-DSC curves of GO and GO-TiO$_2$. Weight losses in GO are observed at 100°C (water evaporation), 200°C (decomposition of partial oxygen-containing functional groups), and 500°C (complete oxidation of the carbon skeleton). A sharp exothermic peak at the range of 450°C−500°C is obtained. GO-TiO$_2$ hybrid has all three peaks of GO. A significant weight loss at 241°C produced an extrasharp exothermic peak due to a violent combustion redox reaction between the oxidizer and fuel in the precursor. The weight loss range shifts toward lower temperatures (395°C−530°C).

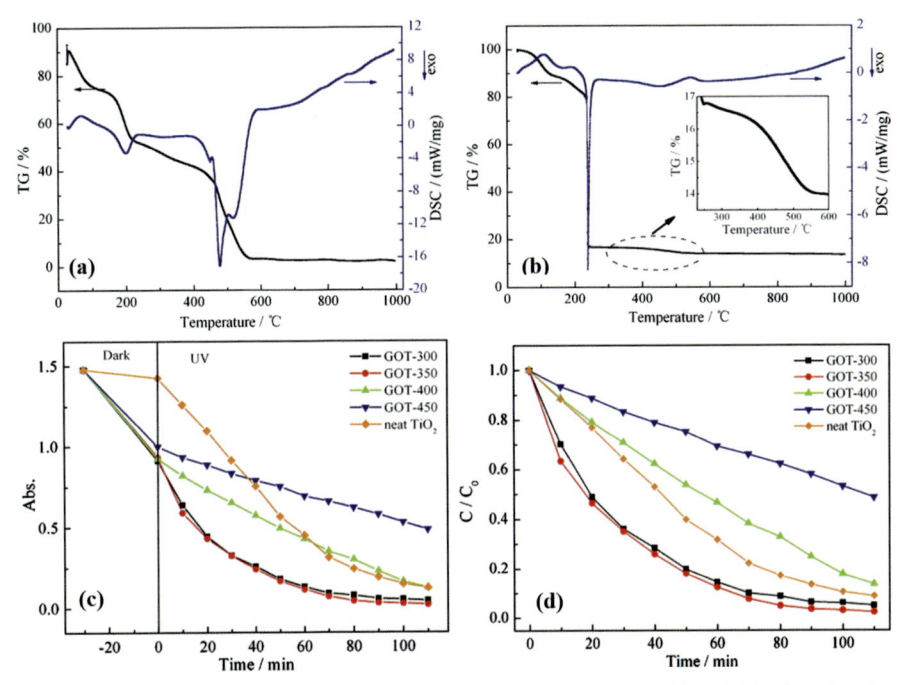

Figure 4.5 TG–DSC curves of (A) GO and (B) Precursor of GO–TiO$_2$ hybrid with a heating rate of 10°C/min using oxygen as the purge gas. The inset shows the enlargement of the TG curve of the precursor; The variations of (C) Absorbance and (D) Normalized C/C$_O$ MO concentration with different time under UV light irradiation.

Different combustion temperatures give different oxidation/reduction rates in GO. As a result, structural distortion and different chemical properties of GO lead to broad energy levels. Finally, different photodegradation performances were obtained because of the different efficiencies of photogenerated electron transfer from TiO$_2$ to graphene/GO. Fig. 4.5C and D represents the photodegradation of GO-TiO$_2$ hybrid material and pure TiO$_2$ at different combustion temperatures. GO-TiO$_2$ hybrid material shows higher adsorption ability compared with pure TiO$_2$. A large specific surface area of GO enhanced the adsorption of TiO$_2$ ($\sim$55%) at the beginning of the photodegradation, as shown in Fig. 4.5C. After 80 min of UV irradiation, hybrid material photodegraded 96% of the initial dye compared with TiO2 ($\sim$83%). The influence of combustion temperature on photodegradation of GO-TiO$_2$ hybrids is seen in Fig. 4.5D. GOT-350 reported the best photocatalytic activity. Thermal reduction of GO improved the photogenerated electron transfer. NAG further enhances the conductivity of graphene. Further increment in combustion temperature decreased the photodegradation performance due to oxidation of partially reduced O$_2$ hybrid material showed a remarkable PL quenching and enhanced photocatalytic activity due to the effective transfer of a photogenerated electron from TiO$_2$ to partially reduced GO. Adsorption of TiO$_2$ on GO sheets increased photodegradation performance by preventing agglomeration.

Agata Lamacz et al.[49] prepared CNT-supported catalysts with a uniform nanosized CeZrO2, Ni−CeZrO2, and Ni dispersion. The catalysts were tested in dry reforming methane (DRM) reaction in the presence of an excess of CO_2 ($CO_2/CH_4 = 2.5$). The hybrid material experienced a higher catalytic activity compared with Ni/CNT catalyst. XRD analysis of CNTs, $CeZro_2$/CNTs (CZ/CNT), NiCe ZrO_2/CNT(NiCZ/CNT), and Ni/CNT are shown in Fig. 4.6A. A diffraction peak confirms the presence of CNT in all samples at $\sim 26°$. Loading of CNTs with Ni, CZ, and NiCZ decreased the diffraction peak intensity. A large number of crystalline planes such as (111), (200), (311), and (222) verified the $CeZrO_2$ in NiCZ/CNT and CZ/CNT samples. Diffraction peaks found at 44.6 and 51.9° ensured the existence of Ni in NiCZ/CNT and Ni/CNT. Ni-CZ sample also contained an additional phase of NiO. The annealing of NiCZ/CNT transformed NiO to Ni (no NiO phase is detected in XRD) due to the strong interaction of Ni and $CeZro_2$. Annealing of NiCZ/CNT at $\sim 500°C$ produced oxygen vacancies due to its desorption from the $CeZrO_2$ crystal lattice.

To find the temperature range of the reaction, the catalytic activity in the DRM reaction was tested via the temperature-programmed (TP) condition (Fig. 4.6B−D). The results showed that a significant reduction in the temperature of DRM is observed for Ni. The starting temperature required for the production of carbon mono oxide and hydrogen occurred in the following order Ni/CNT (400°C)< NiCZ/CNT (630°C)< CZ/CNT (750°C). The reaction mechanism is different for all the above catalysts, as shown in the figure (right side of the TP test). Adsorption of CO_2 on $CeZrO_2$ is accompanied by its decomposition into CO and reoxidation of ceria-zirconia. Lattice oxygen of $CeZro_2$ facilities partial oxidation of methane to CO and H_2 in CZ/CNT catalyst. As a result, oxygen vacancies are produced, which are filled by oxygen created from CO_2 decomposition. Adsorption of methane on the reduced Ni sites experienced dehydrogenation in NiCZ/CNT catalyst. There is a possibility of oxidation of carbon. Oxygen came from oxygen vacancies in $CeZrO_2$ or CO_2 decomposition on Ni^0. For Ni/CNTs catalyst, Methane and carbon dioxide were reduced on active sites of Ni. Methane dehydrogenation occurs on Ni along with oxidation of carbon with CO_2. If the dehydrogenation rate of methane is greater than the carbon suppression rate with CO_2, there is a chance of coke formation or CNTs. Carbon nanofibers can be produced on Ni sites. Overall, the deposition of carbon on Ni sites is temperature-dependent. As an excellent support material, CNTs provide significant surface area for the deposition of active phases. Catalyst deactivation due to carbon deposition has the main disadvantage of the DRM process. It occurred due to insufficient suppression of carbon species created during methane dehydrogenation, as Ce^{+3} produced an active site for CO_2 activation along with oxidation of carbon species covering the catalyst's surface during DRM reaction. Thus, the combination of $CeZrO_2$ and Ni phases on CNT appeared to be a promising solution for fabricating an active catalyst that is resistant to carbon deposition. A modified fabrication method of Ni−$CeZrO_2$/CNT is suggested to achieve high catalytic activity.

Carbon materials, such as CNTs and graphene, appeared as promising supporting candidates for enhancing the photocatalytic activity of metal oxide, Pt, Mn_3O_4, Tio_2, and Ni. Surface groups at carbon provide nucleation sites for the growth of metal nanoparticles. The large surface area of carbon efficiently adsorbed and inhibited the

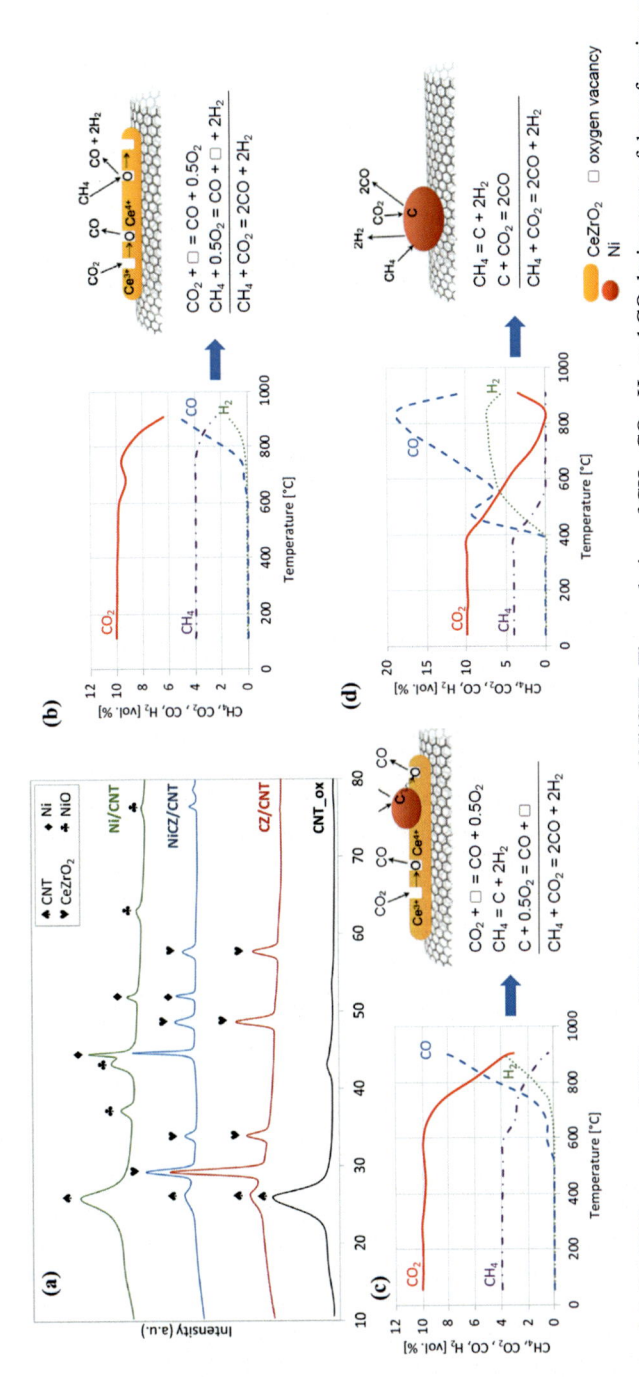

Figure 4.6 (A) XRD patterns of CNT-ox, CZ/CNT, NiCZ/CNT and Ni/CNT; The evolution of CH_4, CO_2, H_2 and CO during tests of dry reforming of methane (DRM)) carried out in temperature-programmed (TP) conditions over CZ/CNT (B), NiCZ/CNT (C) and Ni/CNT (D) and corresponding mechanisms of DRM reaction.

aggregation of metal nanoparticles. Carbon has a smaller Fermi level than the metal oxide conduction band. But it is comparatively larger than the reducing potential of H^+-H_2. Thus, many active sites, localized photothermal effects, and a high mass transfer rate effectively control electron transfer at the heterointerface during the hydrogen evolution reaction. Interference engineering between metal oxide−carbon hybrid materials is an effective way to control hydrogen production during photocatalysis.[50]

4.7 Materials studied for electrical double-layered capacitors and batteries

Different dimensions of carbon have different structural characteristics. 1-D carbon nanostructures provide a continuous network to metal oxide and stabilize metal oxide during cycling. The composites consisting of metal oxide/2-D carbon nanostructures further improve the kinetics in electron-ion transport and structural stability of the composite. 3-D carbon nanostructure, due to high storage capacity, provides a high loading of the active materials by increasing the active site for ions during electrochemical reactions. Hence the effects of composites with metal oxide and carbon improve the performance of EDLCs and Batteries. This section reviews the role of metal oxide−carbon hybrids as electrodes for EDLCs and Batteries.

Babasaheb R. et al.[51] synthesized Co_3O_4/MWNTs using the "dipping and drying method" followed by successive ionic layer adsorption and reaction method for supercapacitor application. The TEM image (Fig. 4.7A−D) shows uniform distribution of Co_3O_4 nanoparticles with a size of less than 15 nm on MWNT surfaces without any aggregation. The selected area electron diffraction also showed that Co_3O_4/MWNT thin film has crystalline characteristics (inset of Fig. 4.7D). Cyclic voltammetry (CV) was carried out to analyze the supercapacitor performance of Co_3O_4/MWNT electrodes, and the results are shown in Fig. 4.8A−D. The maximum specific capacitance (685 Fg^{-1}) of Co_3O_4/MWNT electrodes was observed at 2 M KOH electrolyte concentration within a potential window of 0.2 to +0.5 V. The performance of the electrode was decreased with an increase in further concentration. The stability of the electrode was also checked over 5000 cycles and was found to be 73% which suggested that the combined effect of Co_3O_4 and MWNTs increases the electrical and mechanical stability of the electrode. Moreover, Co_3O_4/MWNT electrodes have shown the maximum specific energy of 16.41 Wh/kg at a specific power of 300 W/kg, which was retained to 12.2 Wh/kg at a specific power of 1000 W/kg after increasing current density to 5 A/g. It was suggested that the performance of the supercapacitor was increased due to the synergetic effect of nanocomposite and fast redox reaction. Also, the diffusion path of ions and electrons was shortened due to composite. Moreover, the lower electrochemical equivalent series resistance (11.25 mF) also contributes to increasing its performance. Results revealed that Co_3O_4/MWNT composite electrodes are of potential use for supercapacitor application.

Tin and tin oxide seem to be good replacements for graphite negative electrodes in lithium-ion batteries due to their low cost, high safety, and other technical benefits. The

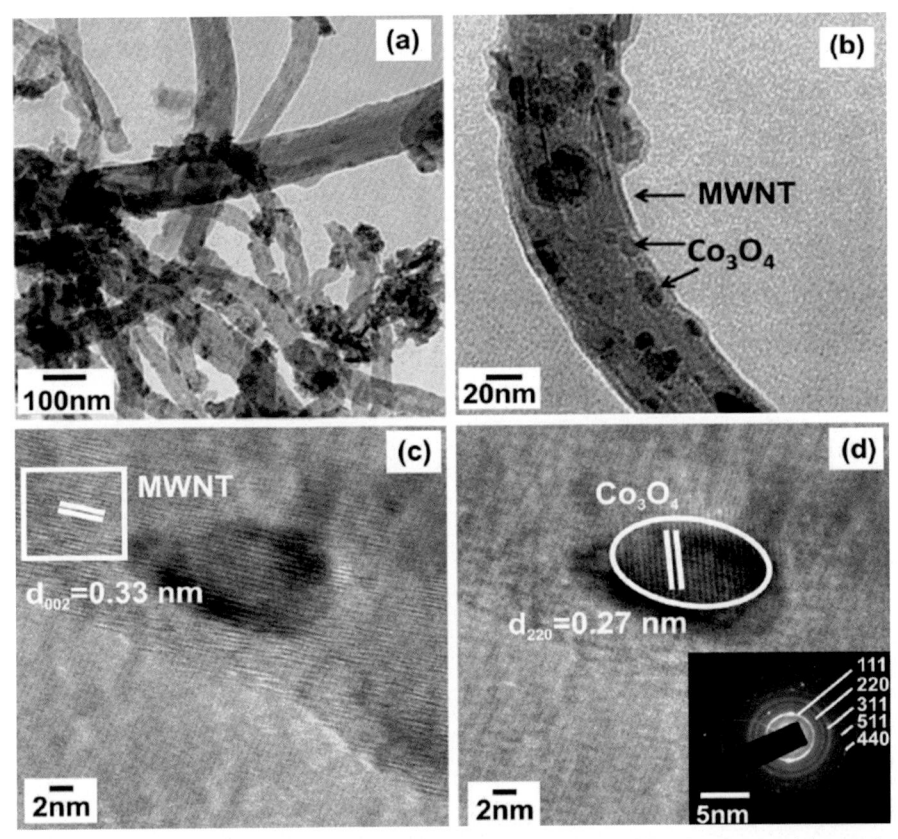

Figure 4.7 (A—D) TEM image showing the uniform distribution of Co_3O_4 nanoparticles on the surface of MWNTs. The selected area electron diffraction (SAED) of Co_3O_4/MWNT thin film (inset).

problem with using tin oxide, and other metallic materials, is that they cause great changes in volume during alloying and dealloying and hence degradation of the anode. Suresh Sagadevan et al. have synthesized SnO_2/graphene (Gr) nanocomposites using the hydrothermal method.[52] High-resolution transmission electron microscopy (HR-TEM) images (Fig. 4.9A—G) show that SnO_2 nanoparticles were uniformly deposited onto the surface of wrinkled graphene sheets. Elemental mapping of the SnO_2/graphene (Gr) nanocomposites was also done, and it showed uniform atomic distribution for O, C, and Sn, as shown in Fig. 4.9A—F. The electrochemical performance of the synthesized nanocomposites-based supercapacitor was also analyzed using cyclic voltammetry and is illustrated in Fig. 4.10A. All electrode CV curves were symmetrical and showed excellent capacitive behavior, including low contact resistance of the electrodes. This specific capacitance was due to the pseudocapacitance behavior of synthesized composite. The nanocomposite SnO_2/G also showed higher conductivity than pure SnO_2, as shown in Nyquist plots of pure SnO_2 and nanocomposite (Fig. 4.10B). The electrochemical performance of the nanocomposite was enhanced

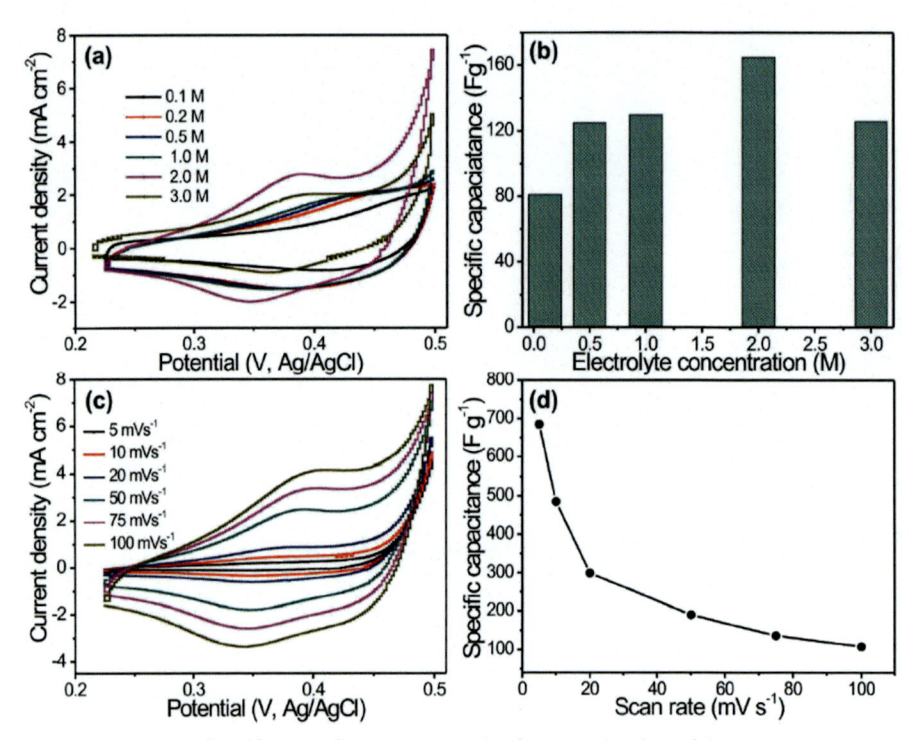

Figure 4.8 (A—D) Specific capacitance, as a result of current density, of Co_3O_4/MWNTs thin-film electrodes.

due to the incorporation of graphene into the SnO_2 sample. Cyclic stability of SnO_2, graphene, and SnO_2/G nanocomposite is shown in Fig. 4.10C, illustrating the highest reversible capacities for SnO_2/G nanocomposite. Nanocomposite electrode exhibits outstanding capacity retention by delivering discharge capacity only in the first cycle and maintained stable capacities from second to 100 cycles. High irreversible capacity and low discharge capacities for 100 cycles are reported for graphene. SnO_2 nanoparticles showed the lowest capacity among all. SnO_2 nanoparticles and reduced graphene were combined in ethylene glycol solution to form a 3-D flexible structure, as shown in Fig. 4.10D. The reduced surface area of the graphene sheets with the incorporation of SnO_2 increases the prepared electrode's stability, specific capacitance, and wettability.

Ottmann A. et al.[53] synthesized a hybrid nanomaterial Mn_3O_4@CNT for high-capacity electrode by encapsulating control-sized Mn_3O_4 nanoparticles inside multi-walled carbon nanotubes (CNT). These nanoparticles were electrochemically active inside CNT. Due to electrochemical lithiation, the filled Mn_3O_4 nanoparticles switched their behavior from ferrimagnetic to antiferromagnetic MnO.

Fig. 4.11A—D shows TEM images of uncycled (a), galvanostatic-lithiated (b), and delithiated (c) Mn_3O_4@CNT. The comparison clearly shows the benefits of the CNT encapsulation upon cycling. Single nanoparticles with limited maximum diameter can be seen inside CNT, whereas encapsulated material is deposited as patches on cycled composite. This result agrees with the volume expansion of Mn_3O_4 during the

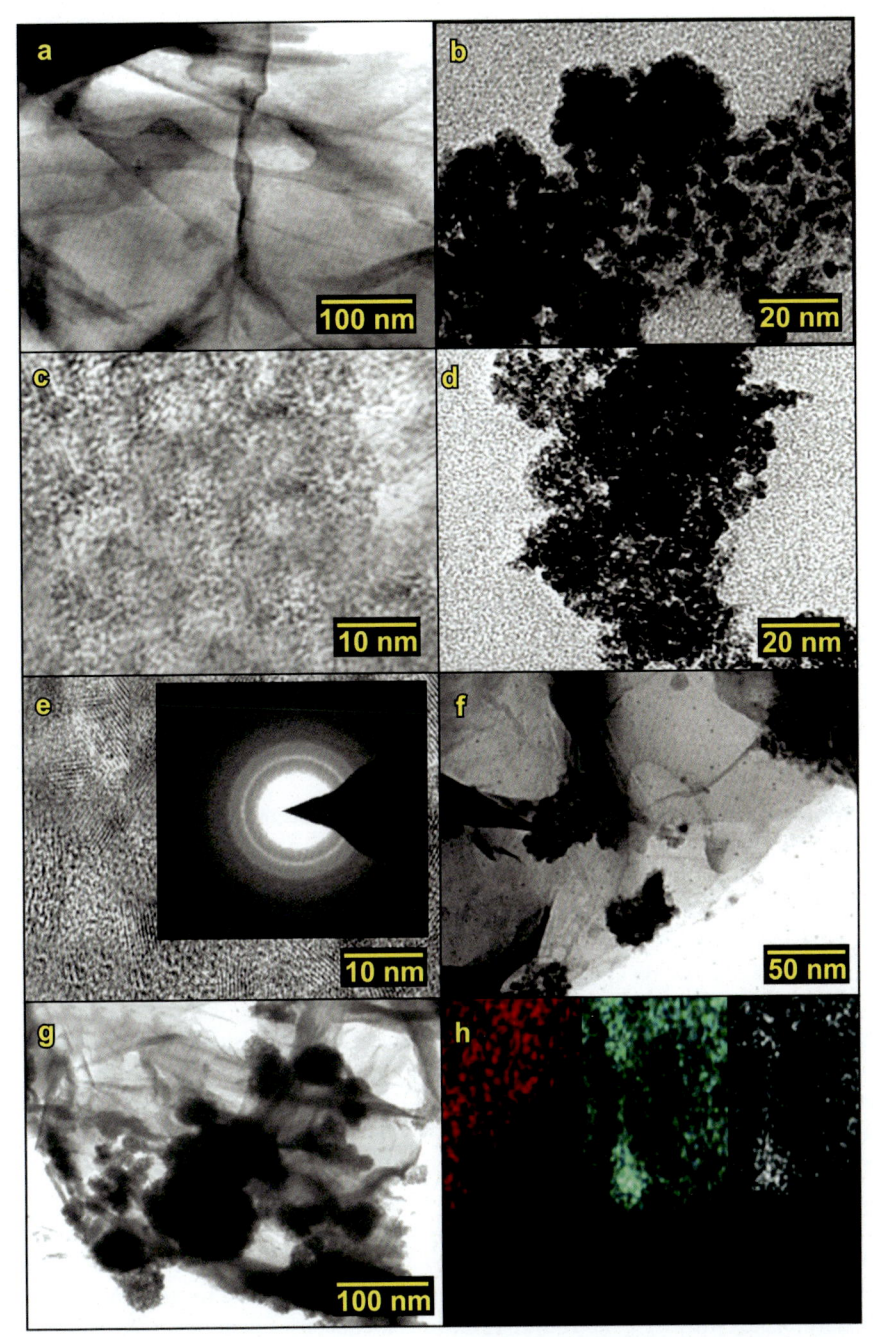

Figure 4.9 (A, B) TEM image of Graphene and SnO_2 (C–G) TEM and HR-TEM images of SnO_2/G nanocomposite and (F) Elemental mapping of SnO_2/G nanocomposite.

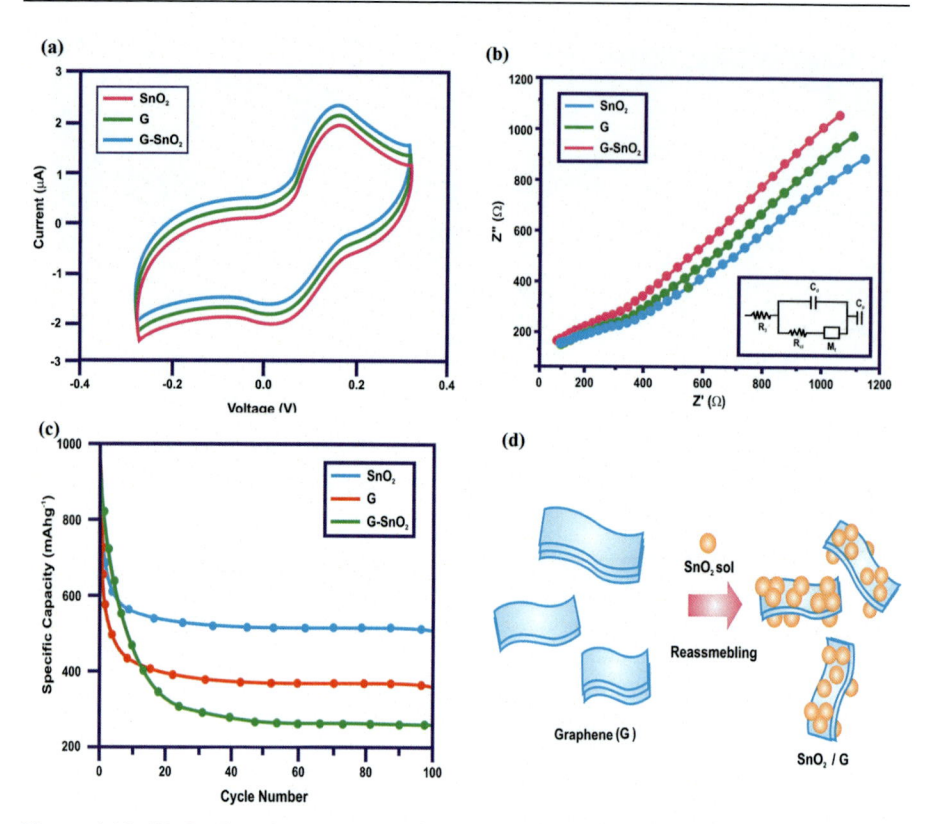

Figure 4.10 (A) Cyclic voltammetry response of SnO_2/G nanocomposites (B) Nyquist Plots of SnO_2/G-Nanocomposite (C) Cycling performances of SnO_2/G nanocomposite (D) Schematic illustration of synthesis process and structure of SnO_2/G.

electrochemical reaction, probably due to the agglomeration of many nanoparticles inside CNT. There was no significant difference between the lithiated (Fig. 4.11B) and the delithiated (Fig. 4.11C) material. Fig. 4.11D shows a high-resolution image of a delithiated CNT shell after 13 charge/discharge cycles, showing characteristic graphitic layers of multiwalled-carbon-nanotubes. Hence, during electrochemical cycling, the volume expansion of the encapsulate does not damage the structure of the carbon nanotubes. From TEM analysis, CNTs provide a stable environment and consistent electrical contact to Mn_3O_4 and can withstand the strain produced due to the volume expansion of manganese oxide during electrochemical cycling.

The electrochemical properties of composite Mn_3O_4@CNT were also analyzed. The theoretical reversible capacity of the Mn_3O_4 in the nanocomposite was fully accessible upon cycling. The Mn_3O_4@CNT nanocomposite attained a maximum discharge capacity of 461 mAh g^{-1} at 100 mAg^{-1} with a capacity retention of 90% after 50 cycles.

Fig. 4.11E compares the charge/discharge capacities of pristine CNT and Mn_3O_4@ CNT at 100 mA g^{-1}. Both the pristine CNT and the composite show a significant

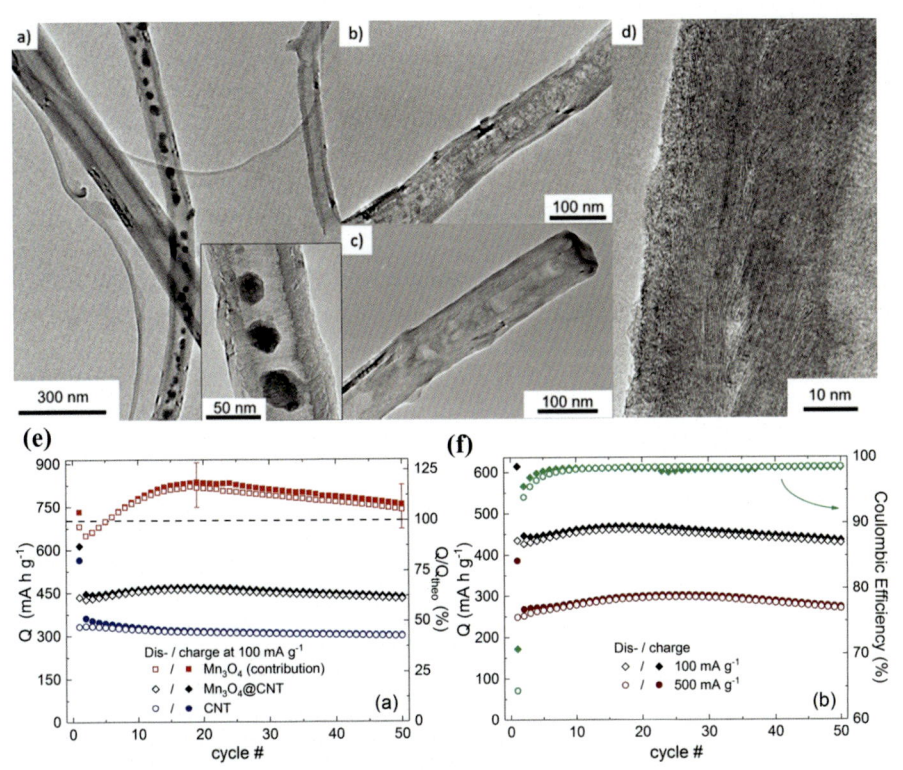

Figure 4.11 TEM images of (A) Uncycled, (B) Galavanostatically lithiated, and (C) Delithiated Mn_3O_4@CNT. (D) High-resolution TEM image of a CNT shell of delithiated material after 13 cycles; (E) Specific charge/discharge capacities of CNT (*circles*) and Mn_3O_4@CNT (*diamonds*) at 100 mA g^{-1}, and the calculated contribution of the incorporated Mn_3O_4 (*squares*), based on a filling of 29.5 wt%. (F) Charge/discharge capacities and the corresponding coulombic efficiencies of Mn_3O_4@CNT at 100 mA g^{-1} (*diamonds*) and 500 mA g^{-1} (*circles*).

irreversible contribution due to the formation of solid electrolyte interphase in the first half-cycle with initial charge/discharge capacities of 566/332 and 614/435 mAh g^{-1}, respectively. The Mn_3O_4@CNT composite shows capacity increases for approximately 15 cycles before reaching a maximum discharge capacity of 463 mAh g^{-1} in the 18th cycle. This capacitance is then almost maintained after 50 cycles (429 mAh g^{-1}). The CNT reached a maximum discharge capacity of 334 mAh g^{-1} in cycle 2 and then moderately decreased to 299 mAh g^{-1} in cycle 50. The Mn_3O_4@CNT composite performed well even at an elevated current of 500 mA g^{-1}, as shown in Fig. 4.11F compared with the specific capacities at 100 mA g^{-1} (Fig. 4.11E). Respective coulombic efficiencies were included in Fig. 4.11F. The general trends for increasing capacities after the initial cycle were similar.

Thus, results indicate that incorporating Mn_3O_4 in CNTs enhanced specific capacities on average by 40% compared with the unfilled carbon nanotube. The resulting transition-metal oxide@CNT nanocomposites can be exploited for lithium-ion batteries.

Zhong W. et al.[54] reported the synthesis of hierarchical porous titanium oxide/ carbide-derived carbon (TiO_2/CDC) composites using a solvothermal method. High-temperature etching of SiOC was carried out to produce CDC materials, and spherical TiO_2 was uniformly grown on the surface of the CDC. Fig. 4.12 shows the EDX elemental mapping and EDS images of synthesized TiO_2/CDC. EDS image confirms the existence of only Ti, C, and O in the composite. It can be seen from mapping graphs (Fig. 4.12B−F) that O and Ti elements are uniformly distributed on the surface of the composite and are spherical, and the C element of the composite was evenly distributed. This uniform distribution of TiO_2 micro-spheres onto the CDC surface causes the active sites to be highly utilized and increases charge transfer during the high rate discharging and charging process.

Moreover, the asymmetric supercapacitors were designed using synthesized TiO_2/ CDC and CDC as positive and negative electrodes, respectively. Electrochemical properties of ASC were also investigated and shown in Fig. 4.13 (schematic is shown in

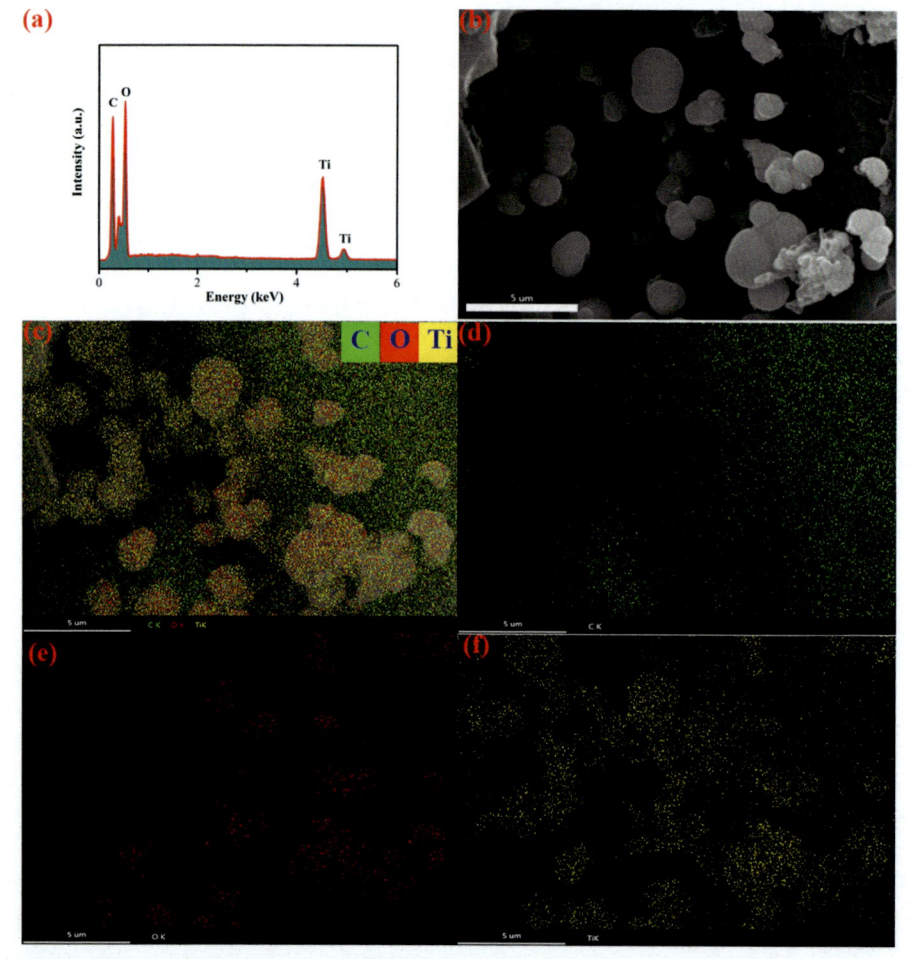

Figure 4.12 (A) EDS spectrum and (B−F) element mapping images of TiO_2/CDC.

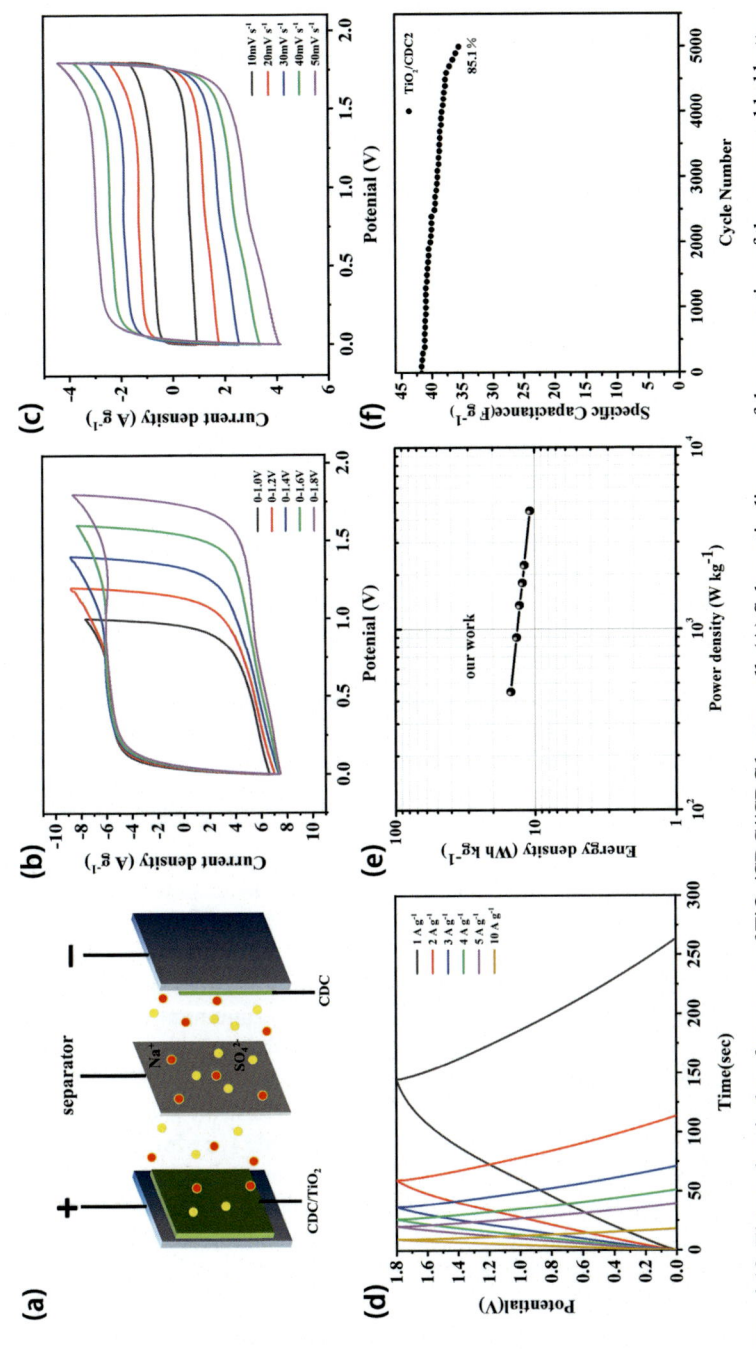

Figure 4.13 Electrochemical performances of TiO₂/CDC//CDC button cell: (A) Schematic diagram of the preparation of the as-assembled button cells; (B) CV curves at various voltage windows; (C) CV curves at 10–50 mV s⁻¹; (D) GCD curves at 1–10 A g⁻¹; (E) Ragone graph related to energy and power densities and (F) cycle performance.

Fig. 4.13A). They proposed the charge-storage mechanism for CDC//TiO_2/CDC asymmetric supercapacitors in Na_2SO_4 electrolytes. Na^+ ions were desorbed/adsorbed in the CDC electrode combined with electrons during the charge/discharge process. The cyclic voltammetry curve of the composed asymmetric supercapacitors has shown a quasi-rectangular shape in various voltage windows (Fig. 4.13B), showing that asymmetric supercapacitors can enhance the voltage window range properly. Fig. 4.13C represents the CV curves of the asymmetric supercapacitors under a large potential window of 0–1.6 V for various scanning rates; all curves showed a quasi-rectangular shape. This trend indicated the device had a very good rate of performance. The matching between two electrodes was considered to significantly affect the performance of a fabricated device, such as power density, energy density, cycle stability, and capacitance retention rate.

It is noted that different compatible electrode materials have a strong impact on the performance of a device, such as power density, energy density, capacitance retention rate, and cycle stability. GCD curves from 0.0 to 1.8 V were also investigated to check the performances of CDC//TiO_2/CDC asymmetric supercapacitors, as given in Fig. 4.13D. The capacitance was maintained to 72.4% (from 33.0 to 23.9 F g^{-1}) from 1 to 10 A g^{-1}. The energy density and power density of the asymmetric supercapacitors are demonstrated in Fig. 4.13E. Due to large specific capacitance and wide voltage region, the asymmetric supercapacitors showed an exceptional energy density of 14.85 Wh kg^{-1} and power-density of 0.45 kW kg^{-1} with a maximum reachable ideal energy density of 10.76 Wh kg^{-1} and a power-density of 4.50 kW kg^{-1}. The cycling performance of the coin cells is shown in Fig. 4.13F. The capacitance of the coin cell was decreased during the first 500 cycles, and then it was able to keep 85.1% original capacity for 5000 cycles, representing outstanding cycle stability. Hence, due to simple synthesis and outstanding performances, the TiO_2/CDC seems to be an attractive candidate for electrode material applied in commercial supercapacitors.

Syamsai R. et al.[55] has designed a supercapacitor electrode using Vanadium carbide (V_4C_3) MXene and has presented a detailed analysis of its electrochemical performance. HF etching was performed on the "A-Aluminum" layer from their parental Vanadium-aluminum carbide MAX phase. Electrolytic ions can easily penetrate the gaps between the Vanadium carbide MXenes (V4C3) layers after etching. It also facilitated the penetration of electrolytic ions into electrodes, hence increasing the electrochemical sites. The Vanadium carbide MXenes (V_4C_3) facilitates the electrolytic ions to penetrate the gaps between the layers formed due to aluminum etching, thereby allowing free movement of electrolytic ions into the electrode and increasing the number of electrochemically active sites.

The HR-TEM of both MXene samples and the MAX phase are shown in Fig. 4.14. The TEM images show complete exfoliation on the sample. The layered structure of the MXene sample with increased d-spacing can be seen in comparison to the pure MAX phase. Intragranular defects and dislocations and were also observed in the high-resolution TEM images. It was noticed that the MXene samples were transparent electronically, and the transparency was changed from lighter to a darker shade with an increase in material thickness.

The galvanostatic charge-discharge analysis is conducted to evaluate the charge-discharge behavior of prepared MXene (Fig. 4.15A). The test was processed at various

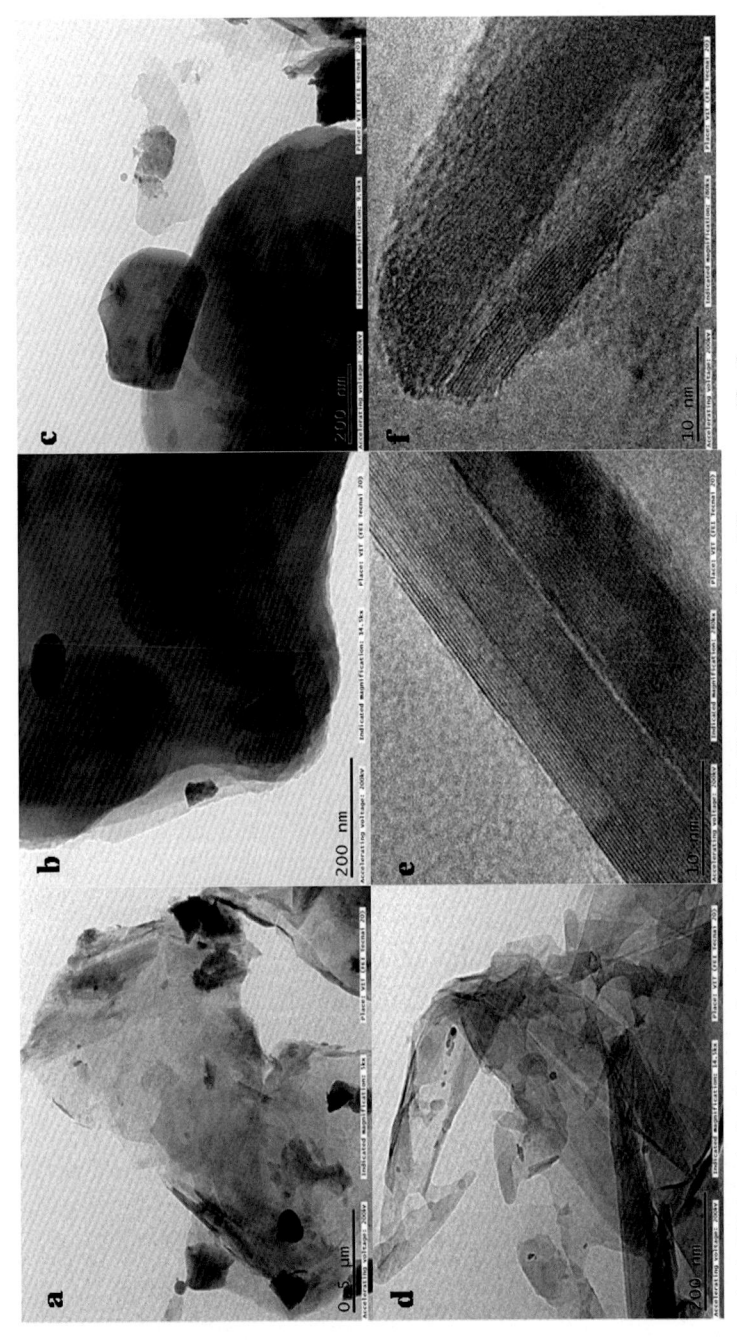

Figure 4.14 TEM analysis of the synthesized vanadium carbideV$_4$C$_3$ (MXene a-f) at different magnifications.

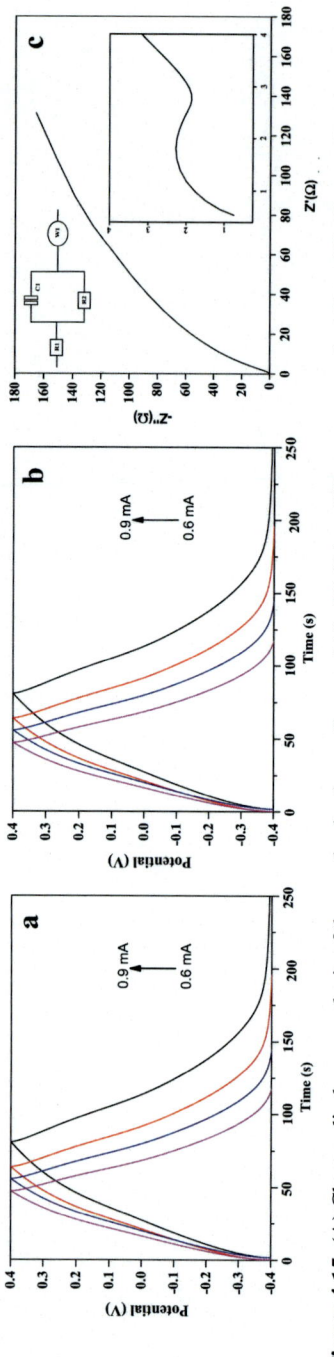

Figure 4.15 (A) Charge–discharge analysis of the synthesized vanadium carbide (V_4C_3) at different current densities; (B) Charge–discharge analysis of the synthesized vanadium carbide MXene (V_4C_3) at different current densities; (C) Nyquist plots of vanadium carbide (V_4C_3) with a varying frequency between 0.01 and 100,000 Hz (inset figure shows the Nyquist plots (zoomed) in the high-frequency region and its equivalent circuit).

current densities and voltages ranging from -0.4 V to $+0.4$ V; a triangular-shaped curve is reported for higher current densities. The extra charges stored at the electrode/electrolyte interface and faradaic reactions in the inner electrode surface reduced the discharge process and resulted in sloppy discharge at lower current densities (Fig. 4.15B). A soluble redox shuttle is produced due to the decomposition of surface groups on top layers of prepared MXene. Electrochemical Impedance Spectroscopy (EIS) was conducted to study the kinetics of fabricated MXene electrodes. A frequency range of 0.01 Hz—100 kHz is selected for electrochemical impedance measurement. An impedance $\sim 140\ \Omega$ is measured from the Nyquist plot (Fig. 4.15C); the fitted equivalent circuit model, along with the EIS spectrum, has components in the form of internal resistance, electrolytic resistance (R_1), and impedance (C/R_2). Impedance is calculated between MXene active material and Carbon paper (current collector). The electrochemical behavior of MXene electrodes for supercapacitor proved it a potential electrode material for electrochemical energy storage applications.

This section of the chapter described the role of carbon/metal oxide-based hybrid materials to increase the performance EDLCs and batteries. The synergetic effect of carbon/metal oxide composite exhibited improved properties and solved the issue related to metal oxide-based electrochemical energy devices such as low electrical conductivity and cracking during electrochemical reactions.

4.8 Conclusions

The carbon-based materials help metal oxides settle on their surface for better interactions required for future electrochemical energy conversion/storage devices. Therefore, the hybridization of metal oxides and carbon provides materials with high specific energy, high specific power, and improved and stable cycling life. Moreover, their applications as PEC have been addressed with essential selection features for improved efficiency. Literature review of studied materials for EDLCs, batteries, and PECs is also discussed to understand the ongoing research.

References

1. Shen X, Liu H, Cheng XB, Yan C, Huang JQ. Beyond lithium ion batteries: higher energy density battery systems based on lithium metal anodes. *Energy Stor Mat* 2018;**12**:161—75.
2. Chen L, Shaw LL. Recent advances in lithium—sulfur batteries. *J Power Sourc* 2014;**267**: 770—83.
3. Fleischmann S, Tolosa A, Presser V. Design of carbon/metal oxide hybrids for electrochemical energy storage. *Chem Eur J* 2018;**24**(47):12143—53.
4. Dusastre V, editor. *Materials for sustainable energy: a collection of peer-reviewed research and review articles from Nature Publishing Group.* World Scientific; 2010.
5. Acar C, Dincer I. A review and evaluation of photoelectrode coating materials and methods for photoelectrochemical hydrogen production. *Int J Hydr Energy* 2016;**41**(19):7950—9.

6. Choudhary S, Upadhyay S, Kumar P, Singh N, Satsangi VR, Shrivastav R, Dass S. Nanostructured bilayered thin films in photoelectrochemical water splitting—A review. *Int J Hydr Energy* 2012;**37**(24):18713−30.

7. Bak T, Nowotny J, Rekas M, Sorrell CC. Photo-electrochemical hydrogen generation from water using solar energy. Materials-related aspects. *Int J Hydr Energy* 2002;**27**(10): 991−1022.

8. Currao A. Photoelectrochemical water splitting. *CHIMIA Int J Chem* 2007;**61**(12):815−9.

9. Radecka M, Rekas M, Trenczek-Zajac A, Zakrzewska K. Importance of the band gap energy and flat band potential for application of modified TiO_2 photoanodes in water photolysis. *J Power Sources* 2008;**181**(1):46−55.

10. Shabdan Y, Markhabayeva A, Bakranov N, Nuraje N. Photoactive Tungsten-oxide nanomaterials for water-splitting. *Nanomaterials* 2020;**10**(9):1871.

11. Sharma D, Verma A, Satsangi VR, Shrivastav R, Dass S. Nanostructured $SrTiO_3$ thin films sensitized by Cu_2O for photoelectrochemical hydrogen generation. *Int J Hydr Energy* 2014; **39**(9):4189−97.

12. Upadhyay RK, Sharma D. Fe doped $BaTiO_3$ sensitized by Fe_3O_4 nanoparticles for improved photoelectrochemical response. *Mater Res Express* 2018;**5**(1):015913.

13. Basu K, Zhang H, Zhao H, Bhattacharya S, Navarro-Pardo F, Datta PK, Jin L, Sun S, Vetrone F, Rosei F. Highly stable photoelectrochemical cells for hydrogen production using a SnO_2−TiO_2/quantum dot heterostructured photoanode. *Nanoscale* 2018;**10**(32): 15273−84.

14. Dom R, Govindarajan S, Joshi SV, Borse PH. A solar-responsive zinc oxide photoanode for solar-photon-harvester photoelectrochemical (PEC) cells. *Nanoscale Adv* 2020;**2**(8): 3350−7.

15. Varadhan P, Fu HC, Kao YC, Horng RH, He JH. An efficient and stable photoelectrochemical system with 9% solar-to-hydrogen conversion efficiency via InGaP/GaAs double junction. *Nat Commun* 2019;**10**(1):1−9.

16. Idrees F, Dillert R, Bahnemann D, Butt FK, Tahir M. In-situ synthesis of Nb_2O_5/g-C_3N_4 heterostructures as highly efficient photocatalysts for molecular H_2 evolution under solar illumination. *Catalysts* 2019;**9**(2):169.

17. Chander S. Advancement in $CdIn_2Se_4$/CdTe based photoelectrochemical solar cells. In: *Advances in energy materials*. Cham: Springer; 2020. p. 29−47.

18. Yang Y, Xu D, Wu Q, Diao P. Cu_2O/CuO bilayered composite as a high-efficiency photocathode for photoelectrochemical hydrogen evolution reaction. *Sci Rep* 2016;**6**(1): 1−13.

19. Yang Y, Niu S, Han D, Liu T, Wang G, Li Y. Progress in developing metal oxide nanomaterials for photoelectrochemical water splitting. *Adv Energy Mat* 2017;**7**(19):1700555.

20. Saveleva MS, Eftekhari K, Abalymov A, Douglas TE, Volodkin D, Parakhonskiy BV, Skirtach AG. Hierarchy of hybrid materials—the place of inorganics-in-organics in it, their composition and applications. *Front Chem* 2019;**7**:179.

21. Acar C, Dincer I, Zamfirescu C. A review on selected heterogeneous photocatalysts for hydrogen production. *Int J Energy Res* 2014;**38**(15):1903−20.

22. Nowotny J, Dufour LC. Surface and near-surface chemistry of oxide materials. *Mater Sci Monogr* 1988;**47**.

23. Joy J, Mathew J, George SC. Nanomaterials for photoelectrochemical water splitting—review. *Int J Hydr Energy* 2018;**43**(10):4804−17.

24. Li J, Wu N. Semiconductor-based photocatalysts and photoelectrochemical cells for solar fuel generation: a review. *Catal Sci Technol* 2015;**5**(3):1360−84.

25. Gong J, Lai Y, Lin C. Electrochemically multi-anodized TiO$_2$ nanotube arrays for enhancing hydrogen generation by photoelectrocatalytic water splitting. *Electrochim Acta* 2010;**55**(16):4776—82.

26. Rai S, Ikram A, Sahai S, Dass S, Shrivastav R, Satsangi VR. CNT based photoelectrodes for PEC generation of hydrogen: a review. *Int J Hydr Energy* 2017;**42**(7):3994—4006.

27. Hankare PP, Chate PA, Sathe DJ, Jadhav BV. Effect of temperature on various properties of photoelectrochemical cell. *J Alloys Compd* 2010;**490**(1—2):350—2.

28. Cho S, Jang JW, Lee KH, Lee JS. Research update: strategies for efficient photo-electrochemical water splitting using metal oxide photoanodes. *Apl Mat* 2014;**2**(1):010703.

29. Eftekhari A, Babu VJ, Ramakrishna S. Photoelectrode nanomaterials for photo-electrochemical water splitting. *Int J Hydr Energy* 2017;**42**(16):11078—109.

30. Hu Y, Yan X, Gu Y, Chen X, Bai Z, Kang Z, Long F, Zhang Y. Large-scale patterned ZnO nanorod arrays for efficient photoelectrochemical water splitting. *Appl Surf Sci* 2015;**339**: 122—7.

31. Govatsi K, Seferlis A, Neophytides SG, Yannopoulos SN. Influence of the morphology of ZnO nanowires on the photoelectrochemical water splitting efficiency. *Int J Hydr Energy* 2018;**43**(10):4866—79.

32. Cho IS, Chen Z, Forman AJ, Kim DR, Rao PM, Jaramillo TF, Zheng X. Branched TiO$_2$ nanorods for photoelectrochemical hydrogen production. *Nano Letters* 2011;**11**(11): 4978—84.

33. Li Y, Yu H, Song W, Li G, Yi B, Shao Z. A novel photoelectrochemical cell with self-organized TiO$_2$ nanotubes as photoanodes for hydrogen generation. *Int J Hydr Energy* 2011;**36**(22):14374—80.

34. Li L, Yu Y, Meng F, Tan Y, Hamers RJ, Jin S. Facile solution synthesis of α-FeF$_3\cdot$ 3H$_2$O nanowires and their conversion to α-Fe$_2$O$_3$ nanowires for photoelectrochemical application. *Nano Letters* 2012;**12**(2):724—31.

35. Liu H, She G, Mu L, Shi W. Porous SiC nanowire arrays as stable photocatalyst for water splitting under UV irradiation. *Mat Res Bull* 2012;**47**(3):917—20.

36. Kite SV, Sathe DJ, Patil SS, Bhosale PN, Garadkar KM. Nanostructured TiO$_2$ thin films by chemical bath deposition method for high photoelectrochemical performance. *Mat Res Express* 2018;**6**(2):026411.

37. Yin Z, Chen B, Bosman M, Cao X, Chen J, Zheng B, Zhang H. Au nanoparticle-modified MoS2 nanosheet-based photoelectrochemical cells for water splitting. *Small* 2014;**10**(17): 3537—43.

38. Miao H, Zhang G, Hu X, Mu J, Han T, Fan J, Zhu C, Song L, Bai J, Hou X. A novel strategy to prepare 2D g-C3N4 nanosheets and their photoelectrochemical properties. *J Alloys Comp* 2017;**690**:669—76.

39. Yang D, Ionescu MI. Metal oxide—carbon hybrid materials for application in super-capacitors. In: *Metal oxides in supercapacitors*. Elsevier; 2017. p. 193—218.

40. Liang Y, Schwab MG, Zhi L, Mugnaioli E, Kolb U, Feng X, Müllen K. Direct access to metal or metal oxide nanocrystals integrated with one-dimensional nanoporous carbons for electrochemical energy storage. *J Am Chem Soc* 2010;**132**(42):15030—7.

41. Simon P, Gogotsi Y. Materials for electrochemical capacitors. In: *Nanoscience and technology: a collection of reviews from Nature journals*; 2010. p. 320—9.

42. Béguin F, Presser V, Balducci A, Frackowiak E. Carbons and electrolytes for advanced supercapacitors. *Adv Mat* 2014;**26**(14):2219—51.

43. Dubal DP, Ayyad O, Ruiz V, Gomez-Romero P. Hybrid energy storage: the merging of battery and supercapacitor chemistries. *Chem Soc Rev* 2015;**44**(7):1777—90.

44. Lim E, Jo C, Lee J. A mini review of designed mesoporous materials for energy-storage applications: from electric double-layer capacitors to hybrid supercapacitors. *Nanoscale* 2016;**8**(15):7827–33.

45. George SM. Atomic layer deposition: an overview. *Chem Rev* 2010;**110**(1):111–31.

46. Ma Y, Guan G, Shi C, Zhu A, Hao X, Wang Z, Kusakabe K, Abudula A. Low-temperature steam reforming of methanol to produce hydrogen over various metal-doped molybdenum carbide catalysts. *Int J Hydr Energy* 2014;**39**(1):258–66.

47. Feng Y, Zeng A. Selective liquid-phase oxidation of toluene with molecular oxygen catalyzed by Mn_3O_4 nanoparticles immobilized on CNTs under solvent-free conditions. *Catalysts* 2020;**10**(6):623.

48. Gao Y, Pu X, Zhang D, Ding G, Shao X, Ma J. Combustion synthesis of graphene oxide–TiO_2 hybrid materials for photodegradation of methyl orange. *Carbon* 2012;**50**(11):4093–101.

49. Lamacz A, Jagodka P, Stawowy M, Matus K. Dry reforming of methane over CNT-supported $CeZrO_2$, Ni and Ni-$CeZrO_2$ catalysts. *Catalysts* 2020;**10**(7):741.

50. Cao S, Yu J. Carbon-based H2-production photocatalytic materials. *J Photochem Photobiol C Photochem Rev* 2016;**27**:72–99.

51. Sankapal BR, Gajare HB, Karade SS, Dubal DP. Anchoring cobalt oxide nanoparticles on to the surface multiwalled carbon nanotubes for improved supercapacitive performances. *RSC Adv* 2015;**5**(60):48426–32.

52. Sagadevan S, Chowdhury ZZ, Johan MRB, Khan AA, Aziz FA, Rafique RF, Hoque ME. A facile hydrothermal approach for catalytic and optical behavior of tin oxide-graphene (SnO_2/G) nanocomposite. *PLoS One* 2018;**13**(10):e0202694.

53. Ottmann A, Scholz M, Haft M, Thauer E, Schneider P, Gellesch M, Nowka C, Wurmehl S, Hampel S, Klingeler R. Electrochemical magnetization switching and energy storage in manganese oxide filled carbon nanotubes. *Scientific Rep* 2017;**7**(1):1–8.

54. Zhong W, Sun H, Pan J, Zhang Y, Yan X, Guan Y, Shen W, Cheng X. Hierarchical porous TiO_2/carbide-derived carbon for asymmetric supercapacitor with enhanced electrochemical performance. *Mat Sci Semicond Process* 2021;**127**:105715.

55. Syamsai R, Grace AN. Synthesis, properties and performance evaluation of vanadium carbide MXene as supercapacitor electrodes. *Ceramics Int* 2020;**46**(4):5323–30.

Functionalized multimetal oxide−carbon nanotube-based nanocomposites and their properties

Ebtesam E. Ateia[1,2], Amira T. Mohamed[1] and M. Morsy[3]
[1]Physics Department, Faculty of Science, Cairo University, Giza, Egypt; [2]Academy of Scientific Research and Technology (ASRT), Cairo, Egypt; [3]Building Physics and Environment Institute, Housing & Building National Research Center (HBRC), Giza, Egypt

5.1 Introduction

Recently, multiferroic materials with the coexistence of anti/ferromagnetism and anti/ferroelectricity have attracted increasing attention because of their significant technological promise in novel multifunctional devices.[1−4] Currently, multifunctional perovskite oxide $LaFeO_3$ has become a focal point of research on synthesis, structural, and physical properties because of the motivating properties and prospective applications.[5−9] Fig. 5.1 shows the graph of the perovskite structure. ABO_3 is the ideal cubic structure of the perovskite containing two cations (A&B) and one anion. The A and B cations have a coordination number of 12 and 6 oxygen anions, respectively.

Double perovskites $AA'BB'O_6$ can be obtained via the existence of more than two categories of cations in other A or B sites. 1:1 A and/or B sites[10,11] can be detected in numerous arrangements such as $CaFeTi_2O_6$,[12] La_2FeNiO_6,[13] and La_2FeCoO_6.[13]

Among the numerous double perovskite structures, $A_2BB'O_6$ possesses a modified property. Fig. 5.2 illustrates A2BB'O6, where A and B sites are rare earth and transition metal ions, respectively.

The configuration of B B′ ions, the charge, and the A/B size ratio can be considered crucial parameters to control the properties of the double perovskite.[14,15] The ability of perovskite configuration to combine with the cations in the periodic table creates a large variety of properties.[16,17] Consequently, double perovskites may have more amelioration than single perovskites, and this greater modification promotes more versatile applications.[18]

Double perovskite transition metal oxides have enormous differences in properties like high magnetoresistance,[17] high transition temperature,[19] and metal-insulator transition.[20] More complexity occurs when strongly correlated 3d ions with almost completely orbital moment coexist with strongly spin-orbit coupled 5d ions in the B-site of a double perovskite. The La_2MeIrO_6 double perovskite, Me = Ni, Co, Fe was reported to be noncollinear magnetism,[21,22] Me = Mn is ferromagnetism,[23] and

Metal Oxide-Carbon Hybrid Materials. https://doi.org/10.1016/B978-0-12-822694-0.00010-7

Figure 5.1 Ideal perovskite structure ABO_3, where A, B, and oxygen are signified by silver, violet, and red spheres, respectively.

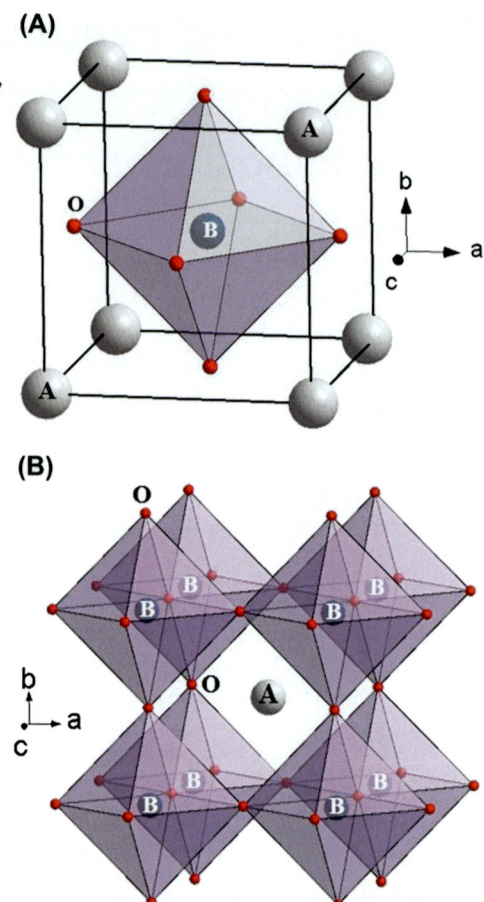

Figure 5.2 Schematic diagram for $A_2BB'O_6$ structure, where A, B, B', and oxygen are signified by silver, violet, and red spheres, respectively.

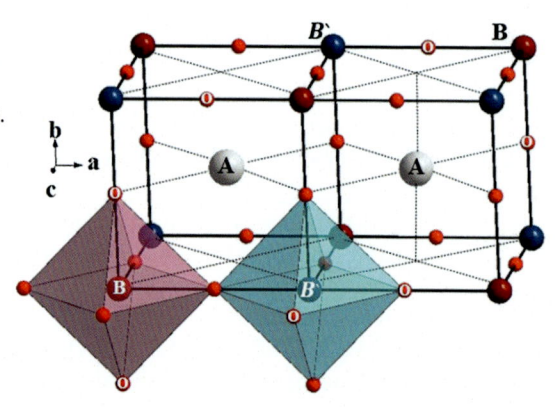

Me = Mg, Zn displays canted antiferromagnetism and diamagnetism, respectively.[24] Therefore, double perovskite composites promote new horizons for enhancing functional magnetic constituents for innovative applications.

La_2NiMnO_6 was prepared with different sizes ranging from 16 to 66 nm. The ratio of disordered phase for the prepared samples increases with decreasing the particle size, as mentioned by Zhao et al.[25] The variation of transition temperature and the exhibition of glasslike properties are the main issues for the detected behavior.

However, the price of carbon nanotubes (CNTs) has been dropping, while their performance has been enhancing with the progress of new technologies in synthetizing carbon nanomaterials. Consequently, CNTs will soon be found in most practical applications.

CNTs have gained wide fame because of their prominent properties. It can be depicted as a hollow cylinder of carbon atoms, as shown in Fig. 5.3.

One of the most important features of the CNTs is the greatness of the value of length to width, or what is known as the aspect ratio. On the other hand, compared with familiar materials such as steel, CNTs are stronger. For example, the hardness of a CNT is about 200 times greater than that of tough steel. The unusual properties of CNTs have helped them become utilized in various significant applications like catalysis, sensor, drug delivery, and adsorbents, and when the CNTs are combined with some other materials, they increase their performance and enhance their properties. CNT use has enhanced the properties of the prepared materials, increased scale, and generated many unique properties. CNTs have been subjected to many studies that deal with different preparation procedures, such as arc discharge, laser ablation, and chemical vapor deposition (CVD). CVD is reported as the most successful technique currently utilized in the production of CNTs, owing to its low cost, flexibility, ease of scale-up, and ability to control the shape and properties of the produced CNTs. The CVD method consists of synthesizing CNTs from volatile carbon-based compounds at a growing temperature from 350°C to 1150°C using a catalyst of nanoparticles.[26–30]

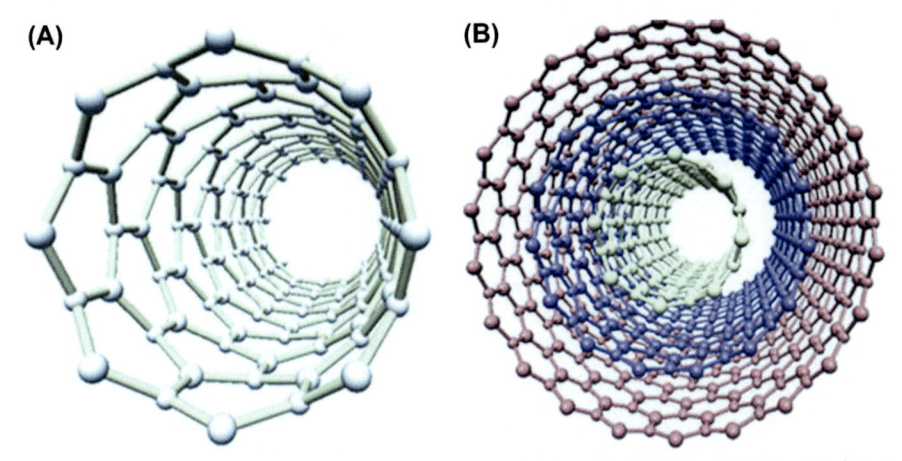

Figure 5.3 Types of carbon nanotubes (CNTs): (A) single-walled CNTs and (B) multiwalled CNTs.

Many researchers[31-36] have studied the magnetic features of double perovskite nanoparticles and the characteristics of MWCNTs, while there is no report about MWCNT-coated double perovskite.

The design of core-shell nanocomposites is a chemical or physical amalgamation of hard/soft and MWCNTs at the microscopic level. It is worth noting that the structure, nano size, and morphology of core-shell have numerous impacts on the behavior of the prepared composites.

Consequently, the present data summarize the mechanism of core-shell particles in the hope of achieving reference for the design and enhancement of new materials to be applicable in an industrial context.

Water pollution can have many causes, including heavy metals, dyes, surfactants, and radiological substances. The consumption of polluted water is recognized to be harmful to the health of living organisms. Many developing countries have detected that around 80% of problems in women and children are associated with the consumption of polluted water.[37]

However, suitable sorbent material for eliminating heavy metals from water is proving elusive. Accordingly, innovative materials with small particle sizes and surface-active sites are required. Various reports have shown considerable effort in enhancing adsorbents that have high adsorption capacity and cost-wise efficiency.[38-41]

To the best of my knowledge, no such work was directed to the double perovskite's/CNTs application as adsorbents to eliminate Cr III from wastewater.

The focus of the present work is to combine all such physical properties of $La_2Fe_2O_6$/CNT to discover its appropriate applications in the novel arena of material science research fields. Additionally, the preparation of high-quality and appropriate yield CNTs is a challenge of the present work. The present chapter is of great importance because it not only leads to the realization of controllable synthesis of CNTs directly on double perovskite throughout the careful selection of oxide faceted surface but also provides insights into the usage of such nanocomposites in wastewater treatment.

5.2 Methodology

5.2.1 Preparation of $La_2Fe_2O_6$ nanoparticles

Perovskite $La_2FeB_2O_6$ was prepared by the citrate autocombustion technique.[13,42] The metal nitrates $La(NO_3)_3 \cdot 6H_2O$, $Fe(NO_3)_3 \cdot 9H_2O$, and citric acid were utilized as starting reactants. The utilized metal nitrates having high purity (99.999%) were purchased from ACROS. Stirring and heating were sustained until the required samples were obtained, as shown in Fig. 5.4A.

5.2.2 Preparation of carbon nanotubes

MWCNs have been synthesized using atmospheric pressure CVD. Bimetallic- iron-cobalt loaded over magnesium oxide (MgO) was utilized as a catalyst. C_2H_2 and

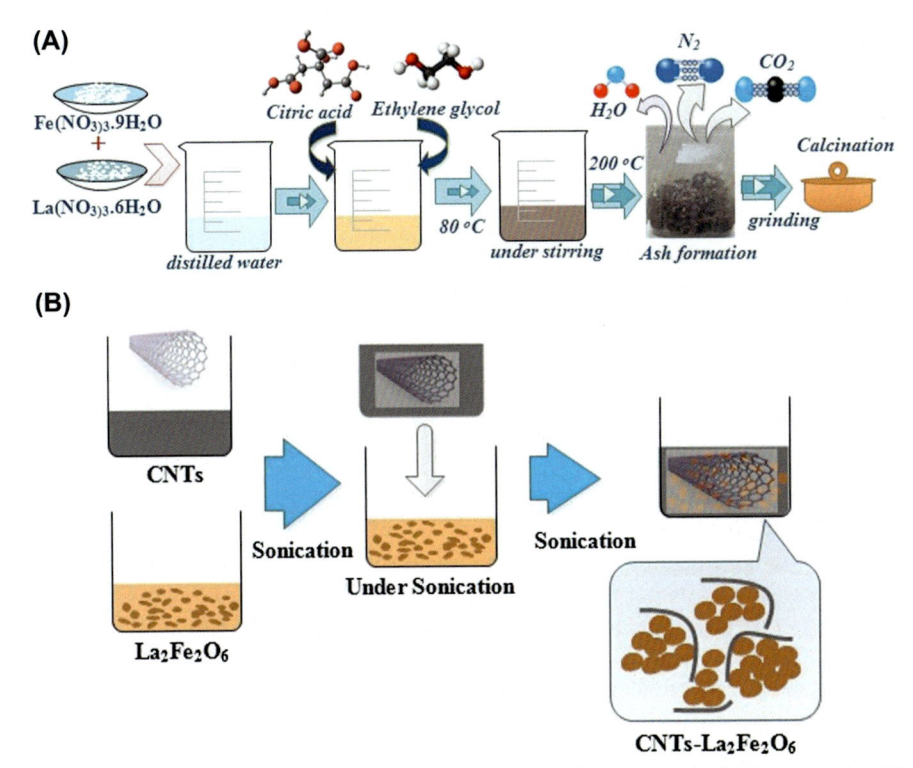

Figure 5.4 Schematic representation for (A) preparation of $La_2Fe_2O_6$, and (B) preparation of CNT$-La_2Fe_2O_6$.

dry N_2 were used for synthesizing CNTs. The deposition temperature was chosen to be 750°C: this temperature was selected based on our reported work.[36,43] In brief, a proper amount of fresh catalyst was loaded on a quartz boat and then inserted inside the middle of the reaction zone of CVD. In the beginning, the nitrogen gas was allowed to flow until reaching the deposition temperature, and then the acetylene gas was allowed to flow for 30 min. Finally, the furnace was chilled down to room temperature under N_2 gas protection. The deposited MWCNTs were collected and characterized by different techniques.

5.2.3 Preparation of carbon nanotube$-La_2Fe_2O_6$ nanocomposites

First, CNTs were washed with NH_3 acid and H_2O to eliminate the impurity; after that, they were dried at 60°C. Then, 30 wt.% of MWCNTs and 70 wt.% of $La_2Fe_2O_6$ were mixed with distilled water and dispersed in an ultrasonic bath, as shown in Fig. 5.4B. The drying process was performed for 12 h at 60°C to attain the powdered core-shell of MWCNTs coated La_2Fe2O_6.

5.2.4 Characterization techniques

The X-ray diffraction (XRD) instrument model Diano Corporation target Cu-Kα ($\lambda = 1.5418$ A$^\circ$) has been utilized to determine the crystal structure and crystallite sizes of the prepared samples. The morphology and nanostructure of the sample were examined by high-resolution transmission electron microscopy (HRTEM), and field-emission scanning electron microscopy (FESEM) attached to an energy-dispersive X-ray analysis (EDAX) unit. The surface area and pore volume were studied using adsorption/desorption isotherms of N_2 at 77K acquired with a NOVA 2200. The vibrating sample magnetometer Model Lake Shore 7410, as well as a homemade Sawyer–Tower circuit, were used to utilize the magnetic and ferroelectric properties of the studied composites.

5.2.5 Heavy metal removal experiment

The heavy metal removal (Cr^{3+}) from an aqueous solution was observed using $La_2Fe_2O_6$ and $CNT-La_2Fe_2O_6$ as adsorbents through batch experiment. The addition of 0.1 g of an adsorbent for 100 mL of a typical solution containing a chromium ion with contraction of 20 ppm and stirred on electric shaker after adjusting pH at certain values for 1 h at RT. The concentration of heavy metal ions was measured by inductively coupled plaza (aka ICP) for chromium at pH extended from 2 to 8. The removal efficiency and the sorption capacity were calculated according to Eqs. (5.1) and (5.2):

$$\text{Removal efficiency } \% = \frac{C_0 - C_f}{C_0} \cdot 100\% \tag{5.1}$$

$$\text{Sorption capacity} = \frac{C_0 - C_e}{m} \cdot V \tag{5.2}$$

where C_0 and C_f are the initial and final concentrations (mg/L) of chromium ion solution, respectively. V is the volume of the aqueous phase (L), and m (g) is the adsorbent's weight.

5.3 Results and discussion

5.3.1 Structural analyses

5.3.1.1 X-ray diffraction analysis

The phase structure of $La_2Fe_2O_6$ and $La_2Fe_2O_6$@CNT composites is studied from XRD, as shown in Fig. 5.5. The observed pattern with the (002), (112), (022), (113), (004), (114), (223), (041), and (241) lattice plans emphasize the formation of the orthorhombic structure with space group Pbnm for $La_2Fe_2O_6$ (JCDD no. 01-074-2203). Similarly, the orthorhombic structure is obtained for La_2FeCoO_6 and La_2FeNiO_6 with space group Pnma.[44] While La_2FeCoO_6 prepared by sol-gel method

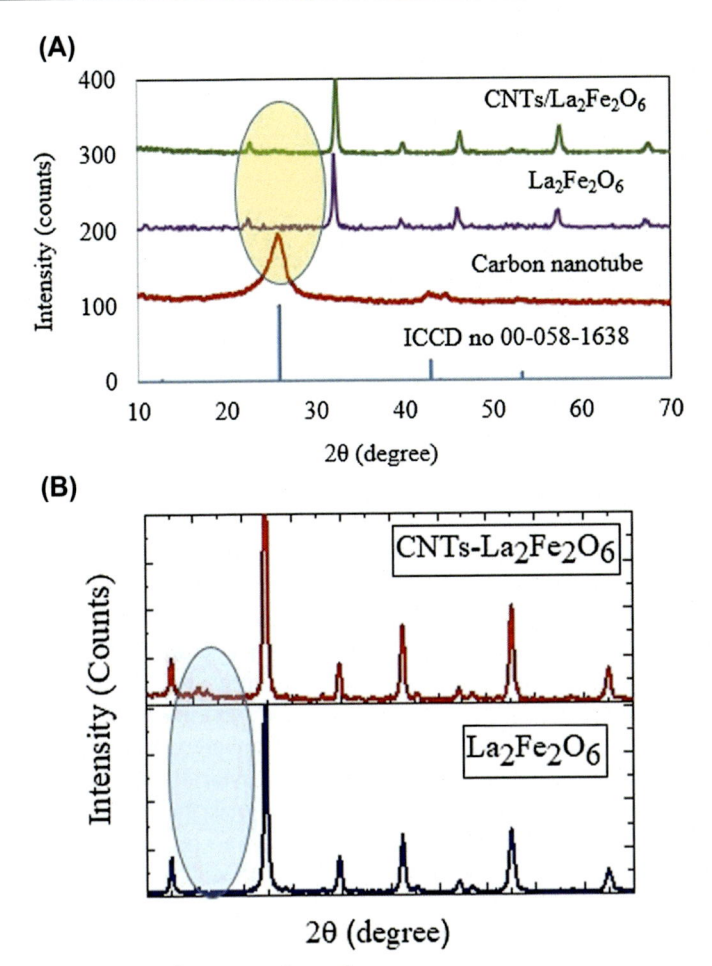

Figure 5.5 X-ray patterns for prepared samples.

is crystallized in rhombohedral structure with space group R3c.[45] For a solid-state method, a cubic structure with Pm3m space group was obtained for La_2FeMnO_6[32].

The indistinct diffraction peak at 26.4 degrees corresponding to the (002) lattice plane is detected. This weak diffraction peak is attributed to CNTs in $La_2Fe_2O_6$@ CNT composites (JCPDS Card No. 00-058-1638).

The existence of a carbon lattice plane in the XRD pattern indicates that the nanocomposites are successfully synthesized by incorporating carbon nanotubes (CNTs) into the solution of the $La_2Fe_2O_6$ sample. However, XRD does not give a good picture of the MWCNTs.[46] This can be attributed to the thin thickness of the CNT wall.[47] Consequently, the MWCNT data are further clarified by Raman spectroscopy analysis.

The lattice constants of $La_2Fe_2O_6$ and $La_2Fe_2O_6$@MWCNTs are calculated and tabulated in Table 5.1. The diffraction peaks become broader after coating with carbon nanotubes (CNTs), as shown in Fig. 5.5B, signifying a reduction of the crystallite size.

Table 5.1 Parameters obtained from X-ray diffraction analysis.

Parameters	$La_2Fe_2O_6$	$CNT-La_2Fe_2O_6$
Crystal size (nm)	19.00	17.86
Lattice parameter	a = 5.540	a = 5.547
	b = 5.558	b = 5.548
	c = 7.846	c = 7.835

The crystallite size calculation has established this via the Scherrer equation at the main peak (112).[48] Similar results was detected for $LaFeO_3-CNTs$,[49] $CoFe_2O_4-CNT$ composite,[50,51] and $ZnFe_2O_4/MWCNTs$.[52] The obtained data agree well with the Brunauer–Emmett–Teller (BET) surface area, as will be clarified later.

5.3.1.2   Raman spectroscopy analysis

Raman spectral analysis is achieved for the $La_2Fe_2O6@CNTs$ as shown in Fig. 5.6. The detected peaks for MWCNTs appear at 1336, 1558, and 2665 cm^{-1}, corresponding to D, G, and G$'$ bands, respectively.[53] The D-band is related to graphitic structure disorders resulting from broken sp^2 bonds and carbonlike impurities in sp^3 bonding. While the G band is related to the stretching mode of sp^2 bonded carbon, which is a typical graphite feature.[54] The G band matches with the hexagonal lattice structure of carbon atoms in graphitic walls. The G$'$ band can be considered an indicator for the crystallinity of the CNTs and can be explained physically as in the first-order D-band.[55] The ratio of D/G is generally utilized to evaluate the nature of carbon-based materials.[56] In the case of the study, the D/G band ratio raises from 0.75 for CNTs to 1.49 for $La_2Fe_2O_6@CNTs$, signifying that the outer wall of the CNTs deteriorates in a specified extension.[57] The Raman spectral analysis of $La_2Fe_2O_6$ samples matches well with that of E. Ateia et al.[58] The eight bands of the Raman spectrum are observed at 145, 211, 247, 283, 366, 424, 495, and 634 cm^{-1}.

5.3.1.3   High-resolution electron microscopy

Fig. 5.7 elucidates HRTEM descriptions for the $La_2Fe_2O_6$ and $La_2Fe_2O_6/CNT$ samples. The $La_2Fe_2O_6$ nanoparticle images are presented in high density with noticeable agglomeration, as shown in Fig. 5.7C. The $La_2Fe_2O_6$ nanoparticles are anchored firmly to the sidewalls of MWCNTs with an inner-external diameter of 8 nm, 18 nm, and the wall thickness of 4 nm, as revealing in Fig. 5.7D and E. Fig. 5.7F represents the particle size distribution for $MWCNTs/La_2Fe_2O_6$. The CNTs are nearly coated with the $La_2Fe_2O_6$ in a smaller ratio, as detected in Fig. 5.7E. Consequently, the mass relationship between the CNTs shell and $La_2Fe_2O_6$ core is not augmented in such a case.

Furthermore, the MWCNTs are not distributed homogeneously under the synthesis conditions. Consequently, a nonhomogeneous coating and isolated $La_2Fe_2O_6$ nanoparticles have occurred. The morphology of the samples ratifies that the MWCNTs structure is not devastated during the preparation process.

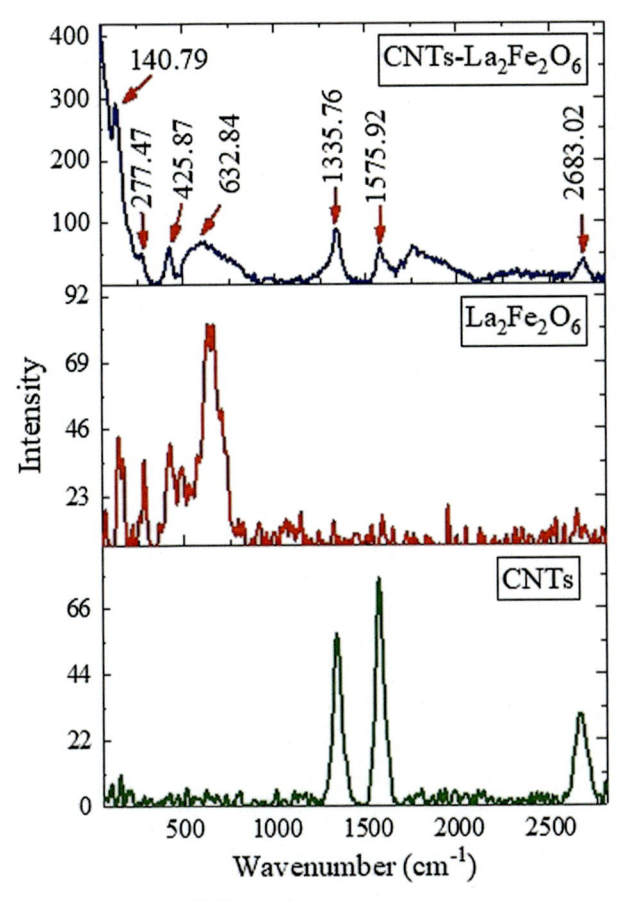

Figure 5.6 Raman spectra for studied samples.

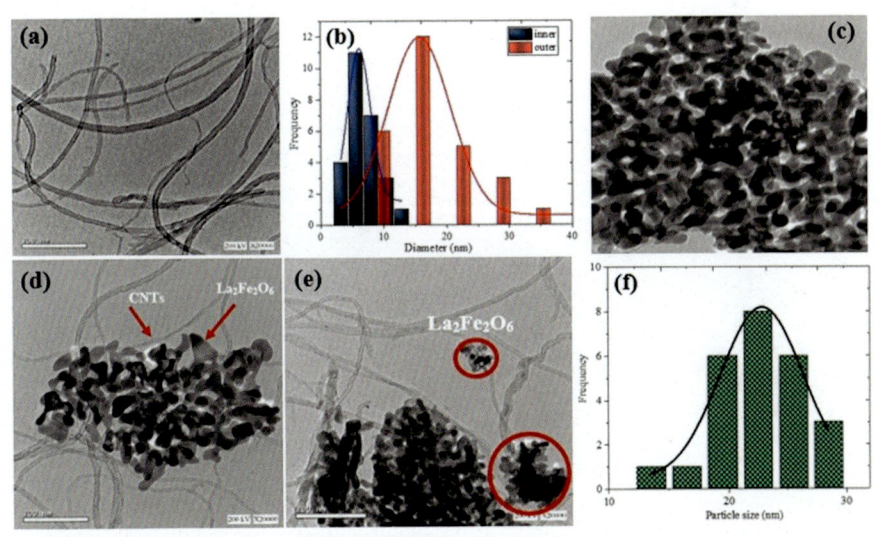

Figure 5.7 High-resolution transmission electron microscopy images for (A) Carbon nanotubes (CNTs) (C) $La_2Fe_2O_6$, (D and E) CNT–$La_2Fe_2O_6$ with the histogram for (B) CNT diameter and (F) Particle size of CNT–$La_2Fe_2O_6$.

5.3.1.4 Field-emission scanning electron microscopy

The morphologies of $La_2Fe_2O_6$ and $CNT-La_2Fe_2O_6$ nanocomposites are detected by FESEM. Fig. 5.8 reveals the image of $La_2Fe_2O_6$ and the 3D network structure of CNTs with perovskite distributed uniformly. $La_2Fe_2O_6$ exhibits a roughly orthorhombic structure with some aggregation, as presented in Fig. 5.8A. While Fig. 5.8B shows CNTs anchored on the surface of $La_2Fe_2O_6$ and rounded along $La_2Fe_2O_6$ nanoparticles to get $CNTs@La_2Fe_2O_6$ nanocomposites. The images conclude that the $CNT-La_2Fe_2O_6$ has less aggregation with a porous surface.

(A)

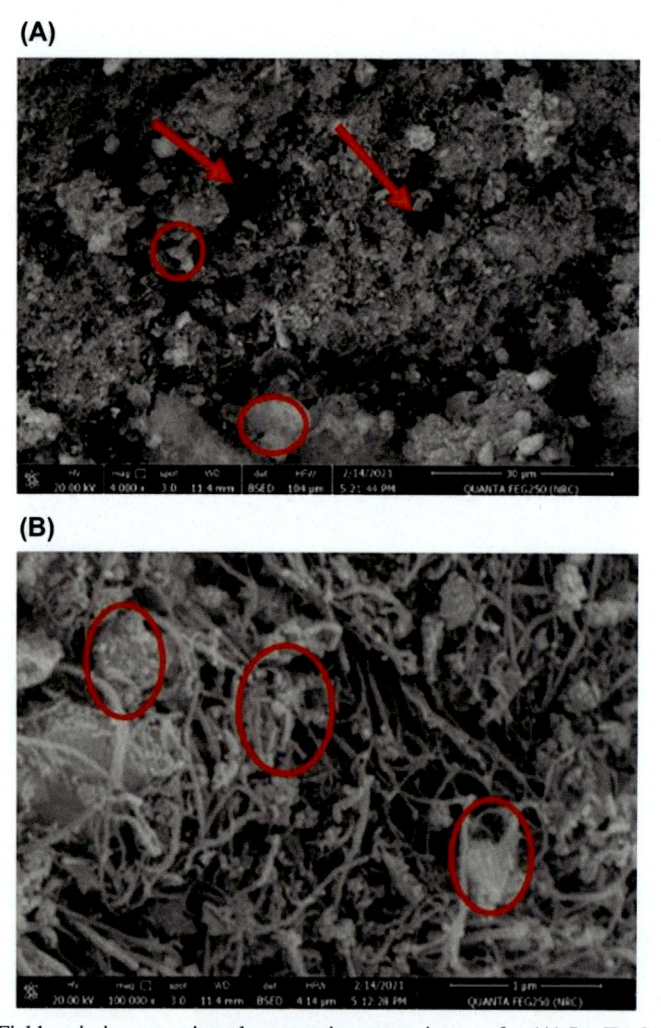

(B)

Figure 5.8 Field-emission scanning electron microscopy images for (A) $La_2Fe_2O_6$ and (B) $CNT-La_2Fe_2O_6$, the red arrows pointing to the presence of porosity, and the red circle indicates the orthorhombic structure of $La_2Fe_2O_6$.

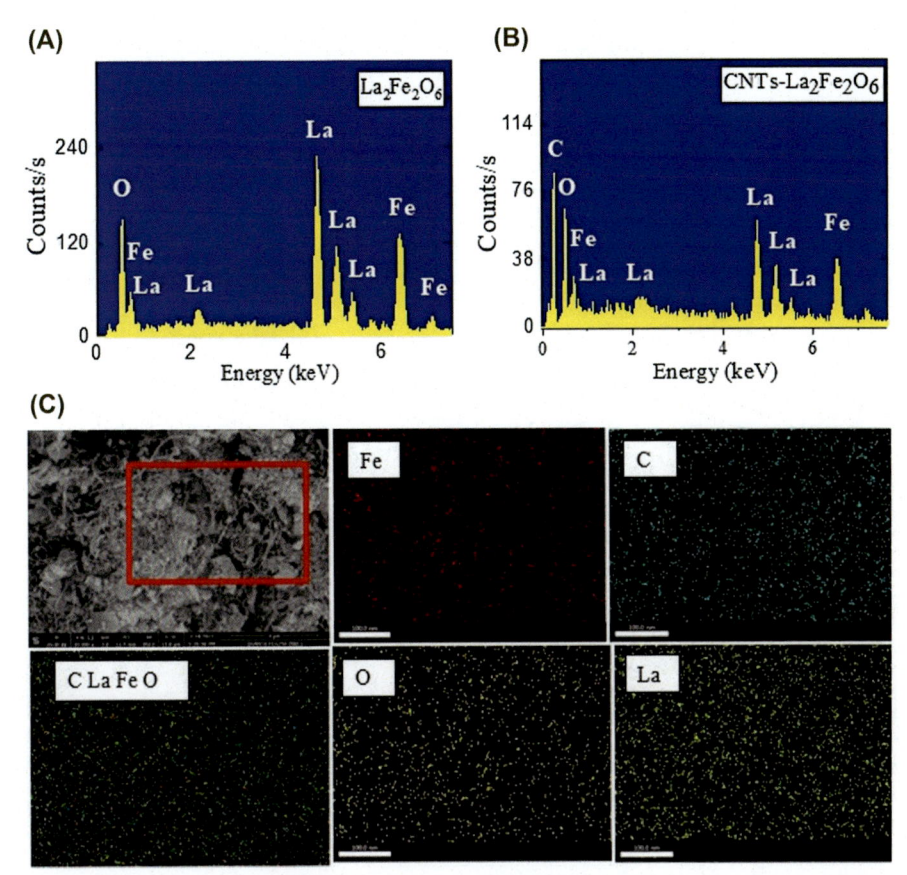

Figure 5.9 Energy-dispersive X-ray analysis spectra for (A) $La_2Fe_2O_6$, (B) $CNT-La_2Fe_2O_6$, and element mapping of (C) $CNT-La_2Fe_2O_6$.

5.3.1.5 Energy-dispersive X-ray analysis and element mapping

The chemical composition of the prepared samples is scrutinized with EDAX. Fig. 5.9A designates the presence of numerous elements (La, Fe, and O). However, the EDAX spectra of $CNT-La_2Fe_2O_6$ specify the presence of additional peaks matching to carbon, which is principally originated from the presence of MWCNTs, as shown in Fig. 5.9B. The examined composition of the nanocomposites agrees well with the nominal chemical stoichiometry of the starting powders.

Fig. 5.9D illustrates the distribution of the iron, lanthanum, carbon, and oxygen with dissimilar colors into the $CNT-La_2Fe_2O_6$ nanocomposites. It designates an identical and uniform distribution of the elements.

5.3.1.6 Brunauer–Emmett–Teller analysis

Brunauer–Emmett–Teller (BET) analysis is typically employed to describe the definite surface area. Fig. 5.10 demonstrates the N_2 adsorption-desorption isotherms for $La_2Fe_2O_6$ and $CNT-La_2Fe_2O_6$ nanocomposites. A prominent hysteresis loop between

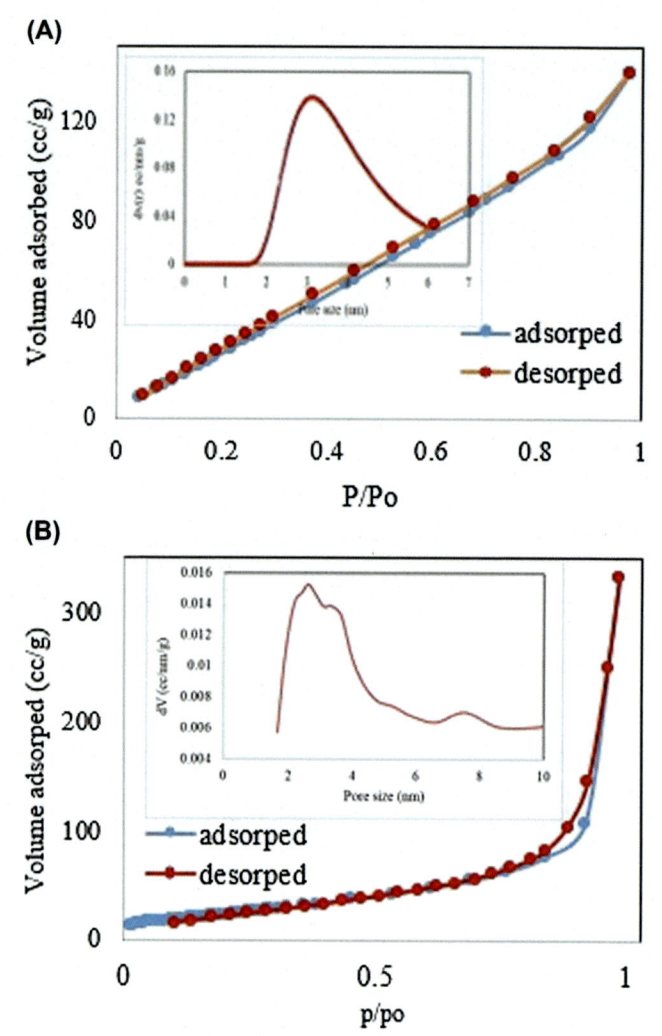

Figure 5.10 Isotherm plots for (A) La$_2$Fe$_2$O$_6$ and (B) CNT$-$La$_2$Fe$_2$O$_6$ nanocomposites. The inset shows the corresponding pore diameter distribution curve.

the desorption-adsorption for the investigated composites is detected. According to the International Union on Pure and Applied Chemistry (IUPAC), the La$_2$Fe$_2$O$_6$ is categorized as model II. This means that the La$_2$Fe$_2$O$_6$ are mesoporous particles with a size of pore ranging from 2 to 50 nm.

Furthermore, the adsorption process for the samples will occur in numerous layers. On the other hand, CNT$-$La$_2$Fe$_2$O$_6$ nanocomposite can be considered type III. This type indicates that the adsorbate and the adsorbed layer interaction is stronger than the interaction between the solid surface and the adsorbed materials.

Table 5.2 Comparison of $La_2Fe_2O_6$ and $La_2Fe_2O_6$ @ CNTs with other composites from the literature.

Sample	Surface area (m^2/g)	Pore volume (cm^3/g)	Average pore diameter (nm)	References
$CoFe_2O_4$−CNT composite	152.9	0.592	−	59
$CoFe_2O_4$	112.5	0.265	−	59
$ZnFe_2O_4/$ MWCNTs	72.9	−	−	52
$ZnFe_2O_4$	56.9	−	−	52
$La_2Fe_2O_6$	116.65	0.217	2.469	The present study
CNT−$La_2Fe_2O_6$	176.564	0.535	2.604	The present study

Generally, the pore geometry of a mesoporous sample greatly affects the shape of the hysteresis loop in Fig. 5.10 (II and III). The adsorption hysteresis in the figure can be classified according to IUPAC as H3. The inset of Fig. 5.10 displays the corresponding pore diameter distribution plot based on the Barrett, Joyner, and Halenda (aka BJH) method. CNT−$La_2Fe_2O_6$ exhibits a wide pore diameter distribution ranging from 1.65 to 10 nm.

The BET data are tabulated in Table 5.2. The specific surface area of CNT−$La_2Fe_2O_6$ greatly exceeds that of pure $La_2Fe_2O_6$. The total pore volume reaches 0.535 cc/g for CNT−$La_2Fe_2O_6$. It is clear that porosity is increased for CNT−$La_2Fe_2O_6$ owing to the lightweight, substantially porous CNTs, increasingly forming interfaces with the nanocrystals.

From the table, it can be seen that the BET parameters of the studied composites are higher compared with others from the literature. It is obvious from the tabulated data that the prepared composites are more effective in heavy metal removal.

5.3.1.7 Atomic force microscopy

Atomic force microscopy (AFM) is a relative tool utilized to characterize the system's pores and pore features down to the atomic scale. This method cannot only be used to attain topographic images of surfaces, but it also can simultaneously recognize different materials on surfaces at high resolution. The obtained data show that the investigated samples exhibit waviness surface texture. The $La_2Fe_2O_6$ and MWCNTs/$La_2Fe_2O_6$ images are shown in Fig. 5.11.

Fig. 5.11B and D show that the "valley" region of the prepared samples is comparatively smooth. While the "hill" region is composed of crystal-like structures with specific orientations. The surface roughness of the prepared samples is important because of their significant influence on device performance.[60] It is characterized by calculating the surface profile parameters. These parameters include root mean square roughness

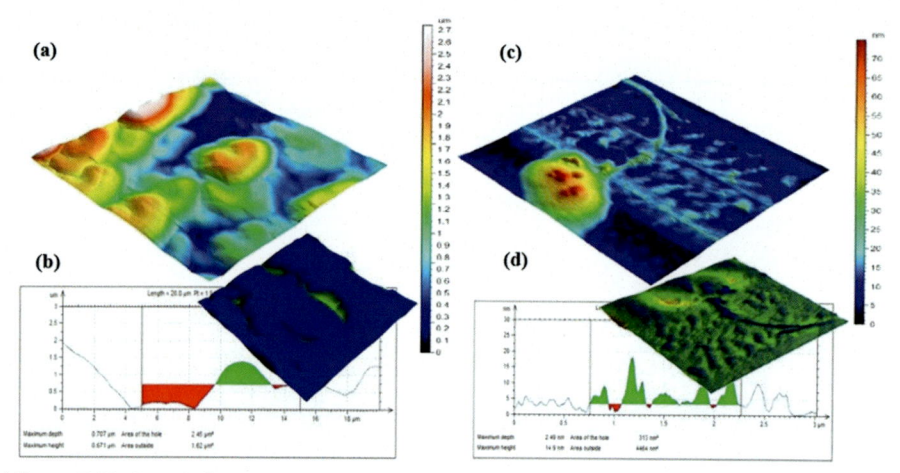

Figure 5.11 Atomic force micrographs for $La_2Fe_2O_6$ (A and B) and $CNT-La_2Fe_2O_6$ (C and D).

Table 5.3 Obtained roughness parameters.

Parameters	$La_2Fe_2O_6$	$CNT-La_2Fe_2O_6$
Rq (nm)	7.73	2.84
Ra (nm)	5.2	2.20
Rt (nm)	58.5	15.8
Rz (nm)	58.5	15.8
Rsk	-1.59	0.898
Rku	7.23	4.49

(Rq), average roughness (Ra), ten-point average roughness (Rz), maximum peak to valley height (Rt), kurtosis of the line (Rku), skewness of the line (Rsk) are tabulated in Table 5.3.

The distributed spikes are measured by roughness kurtosis (Rku). The value of roughness kurtosis is greater than 3 for spiky surfaces and less than 3 for bumpy surfaces. While ideally, random surfaces have Rku equals 3. Consequently, the prepared samples are classified as spiky surfaces, as shown in the table. Rt can also be regarded as a very significant parameter. It gives a good characterization of the total roughness of the surface. The dependence of Rt and Rz on the peak height-valley depth causes an increase in their values, as illustrated in Fig. 5.11B.

5.3.2 Magnetic properties

In double perovskite ($A_2BB'O_6$) nanoparticles, the superexchange interaction is induced between the same transition metal ions according to the Kanamori-Goodenough rules.[61,62] In the present case, the occupancy of the B and B′ sites by the same ions (Fe^{3+}) as well as their arrangement in an adequate way will produce a nonconducting antiferromagnetic material with the following electronic configurations, $t_{2g}^3 e_g^2$ (Fe^{3+}) and $t_{2g}^3 e_g^2$ (Fe^{3+}), respectively.

Comparing with another double perovskite, the antiferromagnetic coupling intensity in La_2FeCoO_6 and La_2FeNiO_6 samples are stronger than that in the $La_2Fe_2O_6$ sample.[44] The significant increase in antiferromagnetic coupling is due to the antiferromagnetic antiphase boundaries in the La_2FeNiO_6 and La_2FeCoO_6. P. M. Tirmali et al.[63] studied the exchange interactions between Fe^{3+} and Co^{3+} cations in the presence of antisite defects and antiphase boundary.

On the other side, the distribution of antisite disorders in each grain of $La_2Fe_2O_6$ reduces the magnetic coupling. Accordingly, there are antiferromagnetic and ferromagnetic moments in the nanoparticle samples with high antisite disorder degrees.

The M−H hysteresis loop of $La_2Fe_2O_6$ and its composite with CNTs are shown in Fig. 5.12. The clarified loop is characterized by a small energy loss as well as small magnetization. According to this description and the previous argument, the prepared samples can be identified as antiferromagnetic with a weak ferromagnetic component.

The calculated magnetic parameters are presented in Table 5.4. As presented in the table, the saturation magnetization (Ms) of the CNT−$La_2Fe_2O_6$ nanocomposite is less than $La_2Fe_2O_6$ nanoparticles. This drop-in magnetization can be attributed to nonmagnetic CNTs that act as voids under the effect of the magnetic field and break the magnetic circuits causing the observed drop.

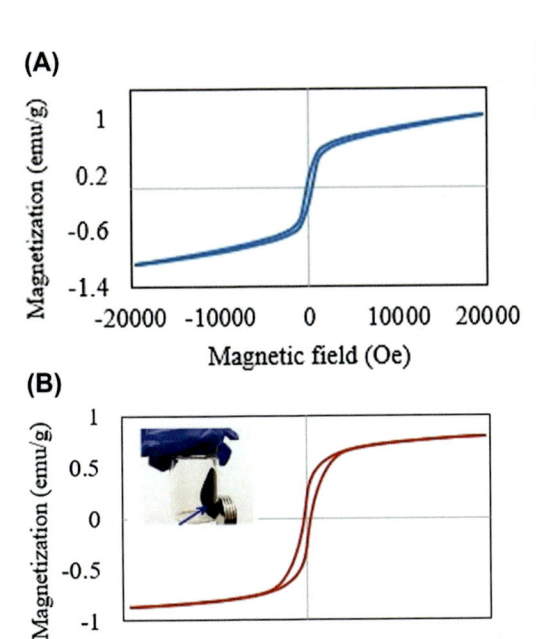

Figure 5.12 Magnetic hysteresis loop for (A) $La_2Fe_2O_6$ and (B) CNT−$La_2Fe_2O_6$.

Table 5.4 Saturation magnetization (M_s), remanent magnetization (M_r), coercivity (H_c), energy loss, squareness ratio (M_r/M_s), and EB for $La_2Fe_2O_6$ and $CNT-La_2Fe_2O_6$.

Sample	M_s (emu/g)	M_r (emu/g)	H_c (Oe)	Energy loss (erg/g)	M_r/M_s	EB
$La_2Fe_2O_6$	1.062	0.182	304.63	2018	0.171	40
$CNT-La_2Fe_2O_6$	0.832	0.233	350.61	1361.7	0.280	72

The achieved Ms is 0.832 and 1.062 emu/g for $MWCNT@La_2Fe_2O_6$ and $La_2Fe_2O_6$ composites, respectively. Therefore, the magnetic content of $MWCNT@La_2Fe_2O_6$ is about 78%. The inset shows the magnet attraction of $MWCNT@ La_2Fe_2O_6$.

The observed shift in the magnetic loop is due to the exchange bias, as detected in Fig. 5.12. Exchange bias (EB) is determined from the following equation:

$$EB = \frac{-(H_1 + H_2)}{2} \tag{5.3}$$

where H_1 and H_2 are the intercepts of the magnetization on the negative and positive magnetic field axis.[64] The obtained EB values are tabulated in the Table. The antiferromagnetic in the core and the weak ferromagnetic close to the surface create the EB in the nanocomposites. This trend is elucidated in a core/shell model and can be employed in several applications.

5.3.3 Electrical properties

The electrical hysteresis loop of nanocrystalline $La_2Fe_2O_6$ and $CNT-La_2Fe_2O_6$ samples is illustrated in Fig. 5.13. The obtained loop confirms the ferroelectric nature of the prepared composites. The electric displacement or polarization increases with an increasing electric field, as detected from the figure. A higher electric field means a rise in the ferroelectric domain or the number of dipole moments in crystals participating in polarization. Comparing $La_2Fe_2O_6$ and $CNT/La_2Fe_2O_6$ nanocomposites reveals that $CNT/La_2Fe_2O_6$ has an increase in electric polarization.

The existence of CNTs increases the polarization due to the subsequent issues. (1) every two neighboring CNTs serve as two electrodes between ceramic nanoparticles. (2) Both CNTs and $La_2Fe_2O_6$ tend to get entangled into agglomeration. However, during the preparation, CNTs, and $La_2Fe_2O_6$ interact with each other and decrease the aggregation. Consequently, the prepared composites have appropriate capacitance characteristics.

5.3.4 Adsorption and desorption performance

The pH of a solution has a great effect on the surface properties of the samples.[65] The adsorption of Cr III ions for $La_2Fe_2O_6$ and $CNT-La_2Fe_2O_6$ as a function of pH is

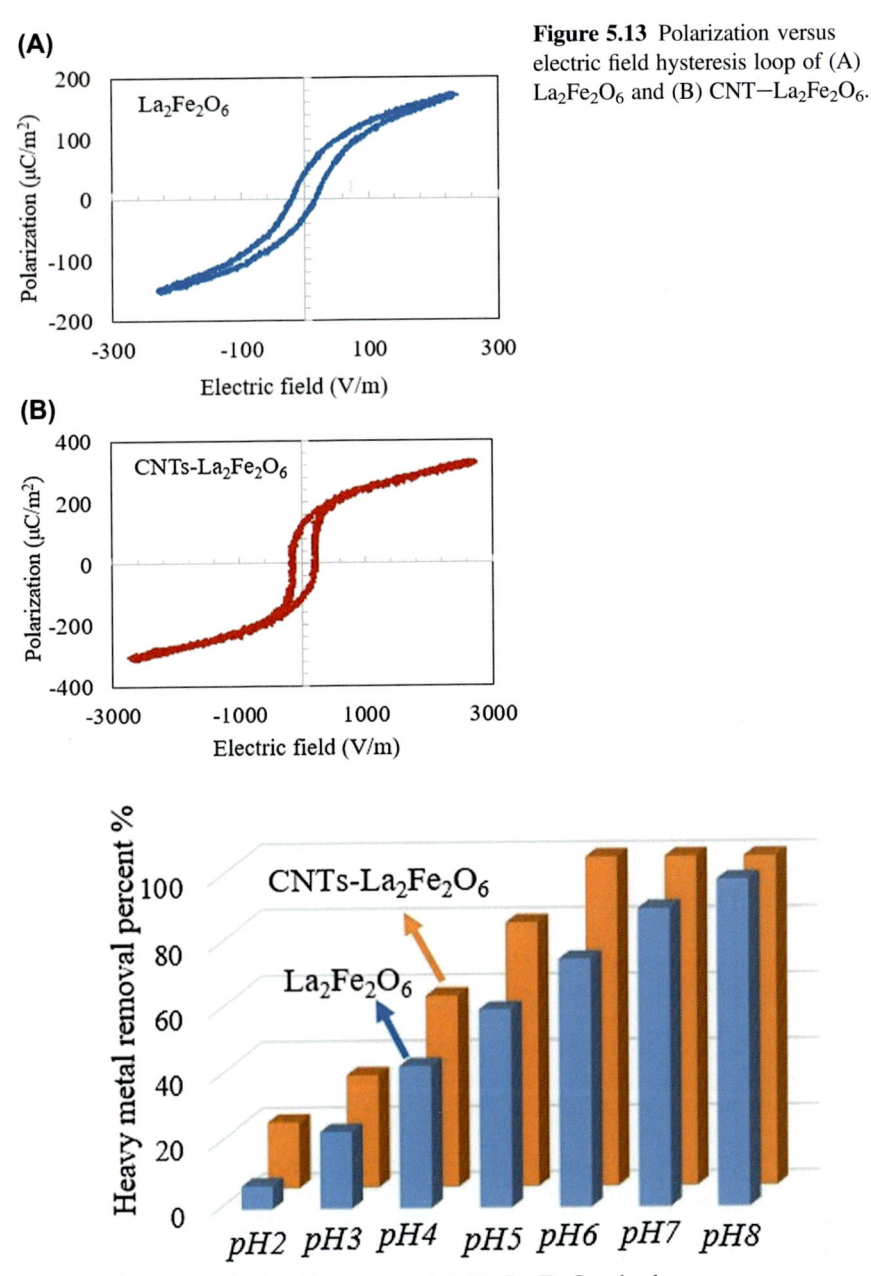

Figure 5.13 Polarization versus electric field hysteresis loop of (A) $La_2Fe_2O_6$ and (B) $CNT-La_2Fe_2O_6$.

Figure 5.14 Effect of pH on $La_2Fe_2O_6$ and $CNT-La_2Fe_2O_6$ adsorbents.

illustrated in Fig. 5.14. As shown in the figure, the experimental data indicate that Cr III removal increases to the maximum with increasing pH from 2 to 8 at 30°C. The maximum removal % of Cr (III) at pH 7 for $La_2Fe_2O_6$ and $CNT-La_2Fe_2O_6$ is nearly 90.3% and 99.89%, respectively.

The minimum adsorption of ions is detected at pH equals 2. This can be attributed to the fact that mobility and the high content of hydrogen ions existent at lower pH prefer the adsorption of H^+ rather than Cr III ions. Consequently, with increasing pH, the available proton number decreases. This can be attributed to the active sites becoming more negatively or neutrally charged, although the sorption of metal ions with a positive charge is increased through the electrostatic attraction force.[66]

At pH > 7, the formation of hydroxyl complexes increases the adsorption activity, as testified by Ali et al.[67] Subsequently, the increase of participate metal ions in the adsorption process is detected, which is not favorable for the efficiency of the process.

Many authors have scrutinized the role of CNTs in the adsorption process.[68–72] It has been emphasized that under the same circumstances, the external sites of MWCNTs have been occupied earlier than the internal sites. This can be clarified, as the external sites of CNTs are available for the adsorbing materials.[73,74] Additionally, the adsorption process is more efficient with unblocked MWCNTs. The caps of MWCNTs enhance the number of sorption sites and adsorption capacity, as mentioned by many authors.[75–79]

In the present study, the activity of MWCNTs @$La_2Fe_2O_6$ is related to the enhanced specific surface area and excessive porosity of the samples besides the hollow structure of MWCNTs. The walls of the CNTs are hydrophobically ascribable to the π electron density. Therefore, the hydrophobic interaction between the metal ions and the external wall of the MWCNTs will occur.

The isotherm models can be designated the adsorption trend of heavy metals by functionalized carbon-based nanomaterials. The Langmuir (I), Freundlich (II), and Temkin (III) models are the greatest commonly utilized models to define the obtained data presented in Fig. 5.15A−C, respectively. The previous models elucidate the adsorption process of Cr (III) on $La_2Fe_2O_6$ and MWCNT @$La_2Fe_2O_6$ at numerous pH parameters.

The I model is applied at particular binding sites on the homogenous adsorbent surface where a single layer of adsorbed molecules is covered the whole surface of the adsorbent. In addition, no interaction occurs between the adsorbed molecules on the adsorbent surface:[80,81]

$$\frac{C_e}{q_e} = \frac{1}{K_L q_m} + \frac{1}{q_m} C_e \tag{5.4}$$

where C_e is the equilibrium metal ion concentration, and q_m and K_L are the Langmuir constants associated with maximum adsorption capacity (mg/g) and the relative energy of adsorption, respectively. This model describes the adsorption of heavy ions as designated by the high correlation coefficient (R^2) of 0.97 and 0.95 for $La_2Fe_2O_6$ and CNT−$La_2Fe_2O_6$, respectively.

The separation factor R_L is a unitless constant that can describe the I model:[82]

$$R_L = \frac{1}{1 + K_L C_O} \tag{5.5}$$

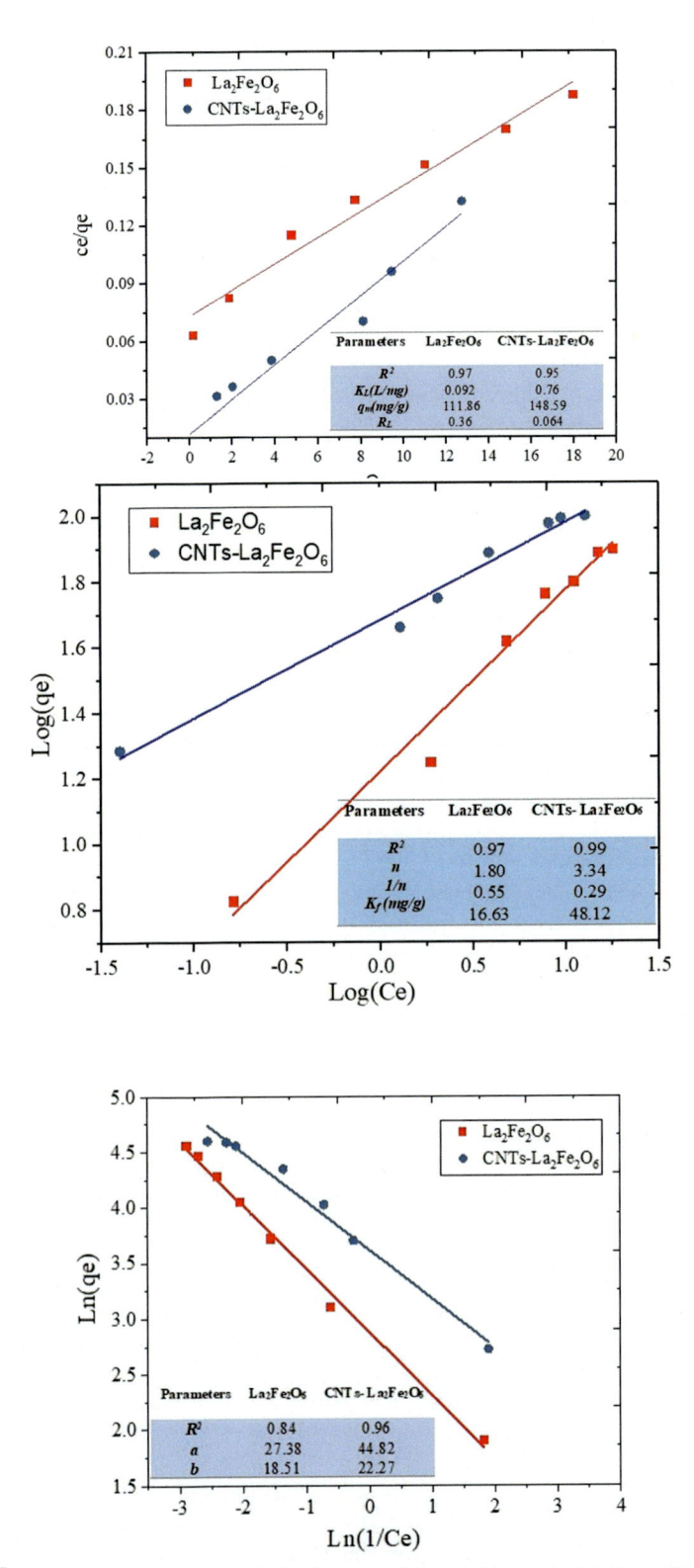

Figure 5.15 Fitting obtained data with isotherm models (A) Langmuir (B) Freundlich, and (C) Temkin. The inset table contains the obtained isotherm parameters.

The free energy variation G^0 is estimated using the following relation:[83]

$$\Delta G^0 = -RTLnK_L \tag{5.6}$$

where K_L, T, and R are the Langmuir constant, absolute temperature (°K), and gas constant.

The calculated $\Delta G°$ for the CNTs/La$_2$Fe$_2$O$_6$ and La$_2$Fe$_2$O$_6$ are -5847.794 J/mol and -689.0616 J/mol, indicating that the adsorption process of CNTs/La$_2$Fe$_2$O$_6$ is more spontaneous than the La$_2$Fe$_2$O$_6$. The $0.0 \leq R_L \leq 1$ indicates the favorability of the adsorption process under the studied conditions.[84,85] The high Langmuir monolayer adsorption capacity (qm) signifying high adsorption capacity. Additionally, the high value of K_L implying high surface energy in the process.

The II model is designated the process of adsorption on the heterogonous surfaces.[86,87] Its equation can be written as

$$\text{Log } q_e = \text{Log}K_f + \frac{1}{n}\text{Log}C_e \tag{5.7}$$

This model's correlation factors are 0.97 and 0.99 for La$_2$Fe$_2$O$_6$ and CNT–La$_2$Fe$_2$O$_6$, respectively.

The preferability of adsorption and heterogeneity degree on the surface of the adsorbent can be identified by the magnitude of $1/n$. The increase of the sorption capacity and the formation of new adsorption sites is ratified by a value of $1/n$.[88]

The III model considers the interaction between the adsorbate and adsorbent. Temkin's equation can be written as follows:[89]

$$q_e = \frac{RT}{b_T}\text{In}K_T + \frac{RT}{b_T}\text{In}C_e \tag{5.8}$$

where q_e is the quantity of the adsorbed metal ion per specific quantity of adsorbent (mg/g), a is equilibrium binding constant (g^{-1}), and b is related to the heat of the process (J/mol). The correlation factors for CNT–La$_2$Fe$_2$O$_6$ equals 0.96. It perfectly fits the data. This is because the heat of adsorption of ions decreases with increasing surface coverage, while the correlation factor for La$_2$Fe$_2$O$_6$ is 0.84. Consequently, this model does not agree well with the obtained data.

The obtained results show an excellent fit of La$_2$Fe$_2$O$_6$ and CNT–La$_2$Fe$_2$O$_6$ with the isotherm models. The correlation coefficients of the samples indicate that all of the isotherm models agree well with the results. The Freundlich isotherm model is a more convenient one for CNT–La$_2$Fe$_2$O$_6$. This indicates that the sorption of Cr (III) is a multilayer physisorption in the heterogeneous surface. While a combination of homogenous monolayer and multilayer is more convenient for La$_2$Fe$_2$O$_6$, using just 30% of CNTs with low-price double perovskite composites has been utilized for 99.89% removal of Cr(III) from wastewater. This is the prime novelty of the investigation. The removal of heavy metals using CNTs magnetic materials has been studied by many authors, as illustrated in Table 5.5. The removal efficiency of the La$_2$Fe$_2$O$_6$/

Table 5.5 Comparison of adsorption capacity of Cr ions by $La_2Fe_2O_6$/CNTs and numerous adsorbents.

Adsorbent	Adsorbates	Adsorption capacity	References
Magnetic nanoparticle-MWCNTs	Cr (VI)	17.09	90
Chitin/magnetite/MWCNTs	Cr (VI)	11.30	91
MWCNTs	Cr (VI)/Cr III	13.20	92
IL-functionalized Oxi-MWCNTs	Cr (VI)/Cr III	85.83	92
MnO_2/MWCNTs	Cr (III)	99.01	93
Magnetic MWCNTs	Cr VI	12.53 to 16.23	94
$CoFe_2O_4$/SM-MWCNTs	Cr VI	100.00	95
Cobalt ferrite-supported activated carbon	Cr VI	23.6	65
$La_2Fe_2O_6$	Cr (III)	112.00	The present study
$La_2Fe_2O_6$−CNTs	Cr (III)	149.00	The present study

CNTs is superior to the utmost of the other adsorbents as detected from the Table. The $La_2Fe_2O_6$/CNTs nanocomposites can be recommended as a practical adsorbent for Cr III metal ions removal. Finally, one can conclude that $La_2Fe_2O_6$−CNTs are more effective, environmentally friendly, economical, and sustainable than other adsorbents.

5.4 Conclusion

The citrate autocombustion and CVD techniques are utilized to prepare $La_2Fe_2O_6$/MWCNTs core-shell.

The innovative MWCNTs/$La_2Fe_2O_6$ with small particle sizes and surface-active sites is achieved.

AFM analysis of MWCNTs/$La_2Fe_2O_6$ can help tailor the deposition parameters according to surface morphology requirements for numerous applications.

The existence of nonmagnetic CNTs decreases the magnetization of the MWCNT/$La_2Fe_2O_6$ core-shell.

The appropriateness of a novel MWCNT-$La_2Fe_2O_6$ for the removal of Cr III from the wastewater is achieved.

The adsorption information is fitted well to numerous isotherm models.

The best model for Cr III with an R^2 value of 0.99 is the Freundlich model. The adsorption capacity of the nanocomposite for Cr III is greater than that of numerous other adsorbents reported in the literature.

5.5 Future prospects

The obtained data indicate that the $La_2Fe_2O_6$/CNTs are potentially applicable for heavy metals removal from an aqueous solution. However, future research should focus on lowering the mobility of different contaminants in soil environments by investigating the prepared nanocomposites as adsorbents in pilot-scale operations and soil amendments.

In addition, CNTs have outstanding physical properties that qualify them to be distinctive materials in fabricating humidity sensors. Consequently, a humidity sensor based on the $La_2Fe_2O_6-CNT$ nanocomposite is a challenge for future work.

Finally, we are looking for tools to advance photovoltaic devices using double perovskite/CNT interfaces.

Funding

This research was funded by the Academy of Scientific Research and Technology (ASRT) (No. 6621).

ASRT Acknowledgment

This chapter is supported financially by the Academy of Scientific Research and Technology (ASRT), Egypt, under initiatives of the Science Up Faculty of Science (Grant No. 6621).

References

1. Khomskii DI. Multiferroics: different ways to combine magnetism and ferroelectricity. *J Magn Magn Mater* 2006;**306**:1−8. https://doi.org/10.1016/j.jmmm.2006.01.238.
2. Cheong S-W, Mostovoy M. Multiferroics: a magnetic twist for ferroelectricity. *Nat Mater* 2007;**6**:13−20. https://doi.org/10.1038/nmat1804.
3. Thankachan RM, Balakrishnan R. Chapter 8 - synthesis strategies of single-phase and composite multiferroic nanostructures. In: Mohan Bhagyaraj S, Oluwafemi OS, Kalarikkal N, Thomas S, editors. *Micro Nano Technol.* Woodhead Publishing; 2018. p. 185−211. https://doi.org/10.1016/B978-0-08-101975-7.00008-7.
4. Ratcliff WD, Lynn JW. Chapter 5 - multiferroics. In: Fernandez-Alonso F, Price DL, editors. *Neutron Scatt - Magn Quantum Phenom.* Academic Press; 2015. p. 291−338. https://doi.org/10.1016/B978-0-12-802049-4.00005-1.
5. Mihai O, Chen D, Holmen A. Catalytic consequence of oxygen of lanthanum ferrite perovskite in chemical looping reforming of methane. *Ind Eng Chem Res* 2011;**50**:2613−21. https://doi.org/10.1021/ie100651d.
6. Zhao K, He F, Huang Z, Wei G, Zheng A, Li H, Zhao Z. CaO/MgO modified perovskite type oxides for chemical-looping steam reforming of methane. *J Fuel Chem Technol* 2016;**44**:680−8. https://doi.org/10.1016/S1872-5813(16)30032-9.
7. Shen Y, Zhao K, He F, Li H. Synthesis of three-dimensionally ordered macroporous $LaFe_{0.7}Co_{0.3}O_3$ perovskites and their performance for chemical-looping steam reforming of methane. *J Fuel Chem Technol* 2016;**44**:1168−76. https://doi.org/10.1016/S1872-5813(16)30051-2.
8. Mihai O, Chen D, Holmen A. Chemical looping methane partial oxidation: the effect of the crystal size and O content of $LaFeO_3$. *J Catal* 2012;**293**:175−85. https://doi.org/10.1016/j.jcat.2012.06.022.

9. Sun R, Yan J, Shen L, Bai H. Performance and mechanism study of LaFeO$_3$ for biomass chemical looping gasification. *J Mater Sci* 2020;**55**:11151−66. https://doi.org/10.1007/s10853-020-04890-2.

10. Vasala S, Karppinen M. A$_2$B′B″O$_6$ perovskites: a review. *Prog Solid State Chem* 2015;**43**: 1−36. https://doi.org/10.1016/j.progsolidstchem.2014.08.001.

11. Shankar U, Agarwal PK, Pandey R, Singh AK. Investigation of crystal structure of SrLa(FeTi)O$_6$ and BaLa(FeTi)O$_6$ perovskites by rietveld refinement. *Solid State Sci* 2016; **52**:78−82. https://doi.org/10.1016/j.solidstatesciences.2015.12.017.

12. Leinenweber K, Parise J. High-pressure synthesis and crystal structure of CaFeTi$_2$O$_6$, a new perovskite structure type. *J Solid State Chem* 1995;**114**:277−81. https://doi.org/10.1006/jssc.1995.1040.

13. E.E. Ateia, A.T. Mohamed, The impact of antisite disorder on the physical properties of La$_2$FeB″O$_6$ (B″ = Fe, Ni and Co) double perovskites, 6 (n.d.). https://doi.org/10.1007/s13204-020-01356-4.

14. Anderson MT, Greenwood KB, Taylor GA, Poeppelmeier KR. B-cation arrangements in double perovskites. *Prog Solid State Chem* 1993;**22**:197−233. https://doi.org/10.1016/0079-6786(93)90004-B.

15. Galasso F, Darby W. Ordering of the octahedrally coördinated cation position in the perovskite structure. *J Phys Chem* 1962;**66**:131−2. https://doi.org/10.1021/j100807a028.

16. Raveau B. The perovskite history: more than 60 years of research from the discovery of ferroelectricity to colossal magnetoresistance via high TC superconductivity. *Prog Solid State Chem* 2007;**35**:171−3. https://doi.org/10.1016/j.progsolidstchem.2007.04.001.

17. Kimura T, Sawada H, Terakura K. Room-temperature magnetoresistance in an oxide material with an ordered double-perovskite structure 1998;**395**:677−80.

18. Mahato DK, Dutta A, Sinha TP. Impedance spectroscopy analysis of double perovskite Ho$_2$NiTiO$_6$. *J Mater Sci* 2010;**45**:6757−62. https://doi.org/10.1007/s10853-010-4771-2.

19. Krockenberger Y, Mogare K, Reehuis M, Tovar M, Jansen M, Vaitheeswaran G, Kanchana V, Bultmark F, Delin A, Wilhelm F, Rogalev A, Winkler A, Alff L. Sr$_2$ CrOs O$_6$: end point of a spin-polarized metal-insulator transition by 5d band filling. *Phys Rev B - Condens Matter Mater Phys* 2007;**75**:4−7. https://doi.org/10.1103/PhysRevB.75.020404.

20. Kato H, Okuda T, Okimoto Y, Tomioka Y, Oikawa K, Kamiyama T, Tokura Y. Metal-insulator transition of ferromagnetic ordered double perovskites: (formula presented). *Phys Rev B - Condens Matter Mater Phys* 2002;**65**:1−5. https://doi.org/10.1103/PhysRevB.65.144404.

21. Currie RC, Vente JF, Frikkee E, IJdo DJW. The structure and magnetic properties of La$_2$MLrO$_6$ with M = Mg, Co, Ni, and Zn. *J Solid State Chem* 1995;**116**:199−204. https://doi.org/10.1006/jssc.1995.1202.

22. Uhl M, Matar SF, Siberchicot B. Calculated magnetic and electronic properties of the double perovskites La$_2$TlrO$_6$ (T=Mn, Fe, Co). *J Magn Magn Mater* 1998;**187**:201−9. https://doi.org/10.1016/S0304-8853(98)00122-X.

23. Demazeau G, Siberchicot B, Matar S, Gayet C, Largeteau A. A new ferromagnetic oxide La$_2$MnIrO$_6$: synthesis, characterization, and calculation of its electronic structure. *J Appl Phys* 1994;**75**:4617−20. https://doi.org/10.1063/1.355909.

24. Cao G, Subedi A, Calder S, Yan JQ, Yi J, Gai Z, Poudel L, Singh DJ, Lumsden MD, Christianson AD, Sales BC, Mandrus D. Magnetism and electronic structure of La$_2$ZnIrO$_6$ and La$_2$MgIrO$_6$: candidate Jeff=12 Mott insulators. *Phys Rev B - Condens Matter Mater Phys* 2013;**87**:1−9. https://doi.org/10.1103/PhysRevB.87.155136.

25. Zhao S, Shi L, Zhou S, Zhao J, Yang H, Guo Y. Size-dependent magnetic properties and Raman spectra of La2NiMnO6 nanoparticles. *J Appl Phys* 2009;**106**:123901. https://doi.org/10.1063/1.3269707.

26. Prasek RKJ, Drbohlavova J, Chomoucka J, Hubalek J, Jasek O. Chemical vapor depositions for carbon nanotubes synthesis. *Annu Rev Mater Res* 2013;**32**:297−319. https://doi.org/10.1146/annurev.matsci.32.012102.110247.

27. Teo K, Singh C. Catalytic synthesis of carbon nanotubes and nanofibers. *Encycl Od Nanosci Nanotechnol* 2003;**X**:1−22. https://doi.org/10.1002/ece3.515.

28. Firouzi A, Sobri S, Yasin F. Synthesis of carbon nanotubes by chemical vapor deposition and their application for CO$_2$ and CH$_4$ detection. *Ipcbee.Com.* 2011;**2**:169−72.

29. Zaytseva O, Neumann G. Carbon nanomaterials : production , impact on plant development , agricultural and environmental applications. *Chem Biol Technol Agric* 2016:1−26. https://doi.org/10.1186/s40538-016-0070-8.

30. Hersam MC. Carbon nanomaterials for electronics, optoelectronics, photovoltaics, and sensing. *Chem Soc Rev* 2013:2824−60. https://doi.org/10.1039/c2cs35335k.

31. Aguilar B, Soto TE, Medina J de la T, Navarro O. Curie temperature enhancement in the double perovskite Sr$_2$$-$xLaxFeMoO$_6$ system: an experimental study. *Phys B Condens Matter* 2019;**556**:108−13. https://doi.org/10.1016/j.physb.2018.12.016.

32. Dhilip M, Devi NA, Punitha JS, Anbarasu V, Kumar KS. Conventional synthesis and characterization of cubically ordered La$_2$ FeMnO $_6$ double perovskite compound. *Vacuum* 2019;**167**:16−20. https://doi.org/10.1016/j.vacuum.2019.05.028.

33. Sahnoun O, Bouhani-Benziane H, Sahnoun M, Driz M. Magnetic and thermoelectric properties of ordered double perovskite Ba$_2$FeMoO$_6$. *J Alloys Compd* 2017;**714**:704−8. https://doi.org/10.1016/j.jallcom.2017.04.180.

34. Mishra AK, Ramaprabhu S. Magnetite decorated multiwalled carbon nanotube based supercapacitor for arsenic removal and desalination of seawater. *J Phys Chem C* 2010;**114**:2583−90. https://doi.org/10.1021/jp911631w.

35. Mauter MS, Elimelech M. Environmental applications of carbon-based nanomaterials. *Environ Sci Technol* 2008;**42**:5843−59. https://doi.org/10.1021/es8006904.

36. Morsy M, Helal M, El-Okr M, Ibrahim M. Preparation and characterization of multiwall carbon nanotubes decorated with zinc oxide. *Der Pharma Chem* 2015;**7**:139−44.

37. Daud MK, Nafees M, Ali S, Rizwan M, Bajwa RA, Shakoor MB, Arshad MU, Chatha SAS, Deeba F, Murad W, Malook I, Zhu SJ. Drinking water quality status and contamination in Pakistan. *BioMed Res Int* 2017;**2017**:7908183. https://doi.org/10.1155/2017/7908183.

38. Wang X, Yu S, Jin J, Wang H, Alharbi NS, Alsaedi A, Hayat T, Wang X. Application of graphene oxides and graphene oxide-based nanomaterials in radionuclide removal from aqueous solutions. *Sci Bull* 2016;**61**:1583−93. https://doi.org/10.1007/s11434-016-1168-x.

39. Wang X, Fan Q, Chen Z, Wang Q, Li J, Hobiny A, Alsaedi A, Wang X. Surface modification of graphene oxides by plasma techniques and their application for environmental pollution cleanup. *Chem Rec* 2016;**16**:295−318. https://doi.org/10.1002/tcr.201500223.

40. Vikrant K, Kim K-H. Nanomaterials for the adsorptive treatment of Hg(II) ions from water. *Chem Eng J* 2019;**358**:264−82. https://doi.org/10.1016/j.cej.2018.10.022.

41. Maitlo HA, Kim JH, Kim K-H, Park JY, Khan A. Metal-air fuel cell electrocoagulation techniques for the treatment of arsenic in water. *J Clean Prod* 2019;**207**:67−84. https://doi.org/10.1016/j.jclepro.2018.09.232.

42. Ateia EE, Mohamed AT, Maged M, Abdelazim A. Crystal structures and magnetic properties of polyethylene glycol/polyacrylamide encapsulated CoCuFe$_4$O$_8$ ferrite nanoparticles. *Appl Phys A* 2020;**126**:669. https://doi.org/10.1007/s00339-020-03841-7.

43. Morsy M, Helal M, El-Okr M, Ibrahim M. Preparation, purification and characterization of high purity multi-wall carbon nanotube, Spectrochim. *Acta Part A Mol Biomol Spectrosc* 2014;**132**:594−8. https://doi.org/10.1016/j.saa.2014.04.122.

44. Ateia EE, Mohamed AT, Elshimy H. The impact of antisite disorder on the physical properties of $La_2FeB''O_6$ ($B'' = Fe$, Ni and Co) double perovskites. *Appl Nanosci* 2020;**10**:1489−99. https://doi.org/10.1007/s13204-020-01356-4.

45. Singh J, Kumar A. Facile wet chemical synthesis and electrochemical behavior of La_2Fe-CoO_6 nano-crystallites. *Mater Sci Semicond Process* 2019;**99**:8−13. https://doi.org/10.1016/j.mssp.2019.04.007.

46. Wang L, Xing H, Liu Z, Shen Z, Sun X, Xu G. Facile synthesis of net-like Fe_3O_4/MWCNTs decorated by SnO_2 nanoparticles as a highly efficient microwave absorber. *RSC Adv* 2016;**6**:97142−51. https://doi.org/10.1039/C6RA21092A.

47. Tang Y, Yin P, Zhang L, Wang J, Feng X, Wang K, Dai J. Novel carbon encapsulated zinc ferrite/MWCNTs composite: preparation and low-frequency microwave absorption investigation. *Ceram Int* 2020;**46**:28250−61. https://doi.org/10.1016/j.ceramint.2020.07.326.

48. Ingham B, Toney MF. 1 - X-ray diffraction for characterizing metallic films. In: Barmak K, Coffey KBT, editors. *Optical and Magnetic Applications*. Woodhead Publishing; 2014. p. 3−38. https://doi.org/10.1533/9780857096296.1.3.

49. Mitra A, Mahapatra AS, Mallick A, Chakrabarti PK. Enhanced microwave absorption and magnetic phase transitions of nanoparticles of multiferroic $LaFeO_3$ incorporated in multi-walled carbon nanotubes (MWCNTs). *J Magn Magn Mater* 2017;**435**:117−25. https://doi.org/10.1016/j.jmmm.2017.03.066.

50. Jelokhani F, Sheibani S, Ataie A. Adsorption and photocatalytic characteristics of cobalt ferrite-reduced graphene oxide and cobalt ferrite-carbon nanotube nanocomposites. *J Photochem Photobiol A Chem* 2020;**403**:112867. https://doi.org/10.1016/j.jphotochem.2020.112867.

51. Mubasher, Mumtaz M, Hassan M, Ullah S, Ahmad Z. Nanohybrids of multi-walled carbon nanotubes and cobalt ferrite nanoparticles: high performance anode material for lithium-ion batteries. *Carbon N Y* 2021;**171**:179−87. https://doi.org/10.1016/j.carbon.2020.08.080.

52. Chen C-H, Liang Y-H, Zhang W-D. $ZnFe_2O_4$/MWCNTs composite with enhanced photocatalytic activity under visible-light irradiation. *J Alloys Compd* 2010;**501**:168−72. https://doi.org/10.1016/j.jallcom.2010.04.072.

53. Hiura H, Ebbesen TW, Tanigaki K, Takahashi H. Raman studies of carbon nanotubes. *Chem Phys Lett* 1993;**202**:509−12. https://doi.org/10.1016/0009-2614(93)90040-8.

54. Fang J, Liu T, Chen Z, Wang Y, Wei W, Yue X, Jiang Z. A wormhole-like porous carbon/magnetic particles composite as an efficient broadband electromagnetic wave absorber. *Nanoscale* 2016;**8**:8899−909. https://doi.org/10.1039/C6NR01863G.

55. Heise HM, Kuckuk R, Ojha AK, Srivastava A, Srivastava V, Asthana BP. Characterisation of carbonaceous materials using Raman spectroscopy: a comparison of carbon nanotube filters, single- and multi-walled nanotubes, graphitised porous carbon and graphite. *J Raman Spectrosc* 2009;**40**:344−53. https://doi.org/10.1002/jrs.2120.

56. Liu T, Xie X, Pang Y, Kobayashi S. Co/C nanoparticles with low graphitization degree: a high performance microwave-absorbing material. *J Mater Chem C* 2016;**4**:1727−35. https://doi.org/10.1039/C5TC03874J.

57. Feng J, Hou Y, Wang Y, Li L. Synthesis of hierarchical $ZnFe_2O_4@SiO_2@RGO$ core−shell microspheres for enhanced electromagnetic wave absorption. *ACS Appl Mater Interfaces* 2017;**9**:14103−11. https://doi.org/10.1021/acsami.7b03330.

58. Ateia EE, Mohamed AT. Core−shell nanomaterials based on $La_2Fe_2O_6$ particles coated with polyvinylpyrrolidone for biomedical applications. *J Mater Sci Mater Electron* 2020; **31**:19355−65. https://doi.org/10.1007/s10854-020-04469-2.

59. Yue L, Zhang S, Zhao H, Feng Y, Wang M, An L, Zhang X, Mi J. One-pot synthesis $CoFe_2O_4$/CNTs composite for asymmetric supercapacitor electrode. *Solid State Ionics* 2019;**329**:15−24. https://doi.org/10.1016/j.ssi.2018.11.006.

60. Sanpo N, Berndt CC, Wang J. Microstructural and antibacterial properties of zinc-substituted cobalt ferrite nanopowders synthesized by sol-gel methods. *J Appl Phys* 2012;**112**:84333. https://doi.org/10.1063/1.4761987.

61. Morrow R, Freeland JW, Woodward PM. Probing the links between structure and magnetism in $Sr_2−xCaxFeOsO_6$ double perovskites. *Inorg Chem* 2014;**53**:7983−92. https://doi.org/10.1021/ic5006715.

62. Ahmed T, Chen A, Yarotski DA, Trugman SA, Jia Q, Zhu J-X. Magnetic, electronic, and optical properties of double perovskite Bi_2FeMnO_6. *Apl Mater* 2016;**5**:35601. https://doi.org/10.1063/1.4964676.

63. Tirmali PM, Mishra DK, Benglorkar BP, Mane SM, Kadam SL, Kulkarni SB. Structural, magnetic and dielectric relaxation behaviour study of La_2MnCoO_6 and fully substituted B-site La_2FeCoO_6. *J Chin Adv Mater Soc* 2018;**6**:207−21. https://doi.org/10.1080/22243682.2018.1446846.

64. Mumtaz A, Khan M, Janjua BH, Hasanain K. Exchange bias and vertical shift in $CoFe_2O_4$ nanoparticles. *J Mag Magn Mater* 2007. https://doi.org/10.1016/j.jmmm.2007.01.007.

65. Yahya MD, Obayomi KS, Abdulkadir MB, Iyaka YA, Olugbenga AG. Characterization of cobalt ferrite-supported activated carbon for removal of chromium and lead ions from tannery wastewater via adsorption equilibrium. *Water Sci Eng* 2020;**13**:202−13. https://doi.org/10.1016/j.wse.2020.09.007.

66. Akpomie KG, Dawodu FA, Adebowale KO. Mechanism on the sorption of heavy metals from binary-solution by a low cost montmorillonite and its desorption potential. *Alexandria Eng J* 2015;**54**:757−67. https://doi.org/10.1016/j.aej.2015.03.025.

67. Ali IH, Al Mesfer MK, Khan MI, Danish M, Alghamdi MM. Exploring adsorption process of lead (II) and chromium (VI) ions from aqueous solutions on acid activated carbon prepared from *Juniperus procera* leaves. *Process* 2019;**7**. https://doi.org/10.3390/pr7040217.

68. Zare K, Gupta VK, Moradi O, Makhlouf ASH, Sillanpää M, Nadagouda MN, Sadegh H, Shahryari-ghoshekandi R, Pal A, Wang Z, Tyagi I, Kazemi M. A comparative study on the basis of adsorption capacity between CNTs and activated carbon as adsorbents for removal of noxious synthetic dyes: a review. *J Nanostruct Chem* 2015;**5**:227−36. https://doi.org/10.1007/s40097-015-0158-x.

69. Mahmoodian H, Moradi O, Shariatzadeha B, Salehf TA, Tyagi I, Maity A, Asif M, Gupta VK. Enhanced removal of methyl orange from aqueous solutions by poly HEMA−chitosan-MWCNT nano-composite. *J Mol Liq* 2015;**202**:189−98. https://doi.org/10.1016/j.molliq.2014.10.040.

70. Yao Y, Bing H, Feifei X, Xiaofeng C. Equilibrium and kinetic studies of methyl orange adsorption on multiwalled carbon nanotubes. *Chem Eng J* 2011;**170**:82−9. https://doi.org/10.1016/j.cej.2011.03.031.

71. Ma J, Yu F, Zhou L, Jin L, Yang M, Luan J, Tang Y, Fan H, Yuan Z, Chen J. Enhanced adsorptive removal of methyl orange and methylene blue from aqueous solution by alkali-activated multiwalled carbon nanotubes. *ACS Appl Mater Interfaces* 2012;**4**:5749−60. https://doi.org/10.1021/am301053m.

72. Wang S, Ng CW, Wang W, Li Q, Hao Z. Synergistic and competitive adsorption of organic dyes on multiwalled carbon nanotubes. *Chem Eng J* 2012;**197**:34−40. https://doi.org/10.1016/j.cej.2012.05.008.

73. Zhang X, Cui H, Gui Y, Tang J. Mechanism and application of carbon nanotube sensors in SF6 decomposed production detection: a review. *Nanoscale Res Lett* 2017;**12**:177. https://doi.org/10.1186/s11671-017-1945-8.

74. Arora B, Attri P. Carbon nanotubes (CNTs): a potential nanomaterial for water purification. *J Compos Sci* 2020;**4**. https://doi.org/10.3390/jcs4030135.

75. Rawat DS, Calbi MM, Migone AD. Equilibration time: kinetics of gas adsorption on closed- and open-ended single-walled carbon nanotubes. *J Phys Chem C* 2007;**111**: 12980−6. https://doi.org/10.1021/jp072786u.

76. Burde JT, Calbi MM. Physisorption kinetics in carbon nanotube bundles. *J Phys Chem C* 2007;**111**:5057−63. https://doi.org/10.1021/jp065428k.

77. Kondratyuk P, Yates JT. Desorption kinetic detection of different adsorption sites on opened carbon single walled nanotubes: the adsorption of n-nonane and CCl4. *Chem Phys Lett* 2005;**410**:324−9. https://doi.org/10.1016/j.cplett.2005.05.073.

78. Atieh MA, Bakather OY, Tawabini BS, Bukhari AA, Khaled M, Alharthi M, Fettouhi M, Abuilaiwi FA. Removal of chromium (III) from water by using modified and nonmodified carbon nanotubes. *J Nanomater* 2010;**2010**:232378. https://doi.org/10.1155/2010/232378.

79. Atieh MA, Bakather OY, Al-Tawbini B, Bukhari AA, Abuilaiwi FA, Fettouhi MB. Effect of carboxylic functional group functionalized on carbon nanotubes surface on the removal of lead from water. *Bioinorg Chem Appl* 2010;**2010**:603978. https://doi.org/10.1155/2010/603978.

80. Maji SK, Wang S-W, Liu C-W. Arsenate removal from aqueous media on iron-oxide-coated natural rock (IOCNR): a comprehensive batch study. *Desalin Water Treat* 2013;**51**: 7775−90. https://doi.org/10.1080/19443994.2013.794551.

81. Haghseresht F, Lu GQ. Adsorption characteristics of phenolic compounds onto coal-reject-derived adsorbents. *Energy Fuels* 1998;**12**:1100−7. https://doi.org/10.1021/ef9801165.

82. Hall KR, Eagleton LC, Acrivos A, Vermeulen T. Pore- and solid-diffusion kinetics in fixed-bed adsorption under constant-pattern conditions. *Ind Eng Chem Fundam* 1966;**5**:212−23. https://doi.org/10.1021/i160018a011.

83. Hosseini M, Mertens SFL, Ghorbani M, Arshadi MR. Asymmetrical Schiff bases as inhibitors of mild steel corrosion in sulphuric acid media. *Mater Chem Phys* 2003;**78**:800−8. https://doi.org/10.1016/S0254-0584(02)00390-5.

84. Vidhyadevi T, Murugesan A, Kalaivani SS, Premkumar MP, Vinoth kumar V, Ravikumar L, Sivanesan S. Evaluation of equilibrium, kinetic, and thermodynamic parameters for adsorption of Cd^{2+} ion and methyl red dye onto amorphous poly(-azomethinethioamide) resin. *Desalin Water Treat* 2014;**52**:3477−88. https://doi.org/10.1080/19443994.2013.801323.

85. Togue Kamga F. Modeling adsorption mechanism of paraquat onto Ayous (*Triplochiton scleroxylon*) wood sawdust. *Appl Water Sci* 2018;**9**:1. https://doi.org/10.1007/s13201-018-0879-3.

86. Ayawei N, Ebelegi AN, Wankasi D. Modelling and interpretation of adsorption isotherms. *J Chem* 2017;**2017**:3039817. https://doi.org/10.1155/2017/3039817.

87. Boparai HK, Joseph M, O'Carroll DM. Kinetics and thermodynamics of cadmium ion removal by adsorption onto nano zerovalent iron particles. *J Hazard Mater* 2011;**186**: 458−65. https://doi.org/10.1016/j.jhazmat.2010.11.029.

88. Jiang JQ, Cooper C, Ouki S. Comparison of modified montmorillonite adsorbents. Part I: preparation, characterization and phenol adsorption. *Chemosphere* 2002;**47**(7):711−6. https://doi.org/10.1016/s0045-6535(02)00011-5.

89. Boudrahem F, Aissani-Benissad F, Soualah A. Adsorption of lead(II) from aqueous solution by using leaves of date trees as an adsorbent. *J Chem Eng Data* 2011;**56**:1804−12. https://doi.org/10.1021/je100770j.

90. Lu W, Li J, Sheng Y, Zhang X, You J, Chen L. One-pot synthesis of magnetic iron oxide nanoparticle-multiwalled carbon nanotube composites for enhanced removal of Cr(VI) from aqueous solution. *J Colloid Interface Sci* 2017;**505**:1134−46. https://doi.org/10.1016/j.jcis.2017.07.013.

91. Salam MA. Preparation and characterization of chitin/magnetite/multiwalled carbon nanotubes magnetic nanocomposite for toxic hexavalent chromium removal from solution. *J Mol Liq* 2017;**233**:197−202. https://doi.org/10.1016/j.molliq.2017.03.023.

92. Krishna Kumar AS, Jiang S-J, Tseng W-L. Effective adsorption of chromium(vi)/Cr(iii) from aqueous solution using ionic liquid functionalized multiwalled carbon nanotubes as a super sorbent. *J Mater Chem A* 2015;**3**:7044−57. https://doi.org/10.1039/C4TA06948J.

93. Mohammadkhani S, Gholami MR, Aghaie M. Thermodynamic study of Cr+3 ions removal by "MnO₂/MWCNT" nanocomposite, Orient. *J Chem* 2016;**32**:591−9. https://doi.org/10.13005/ojc/320167.

94. Huang Z, Wang X, Yang D. Adsorption of Cr(VI) in wastewater using magnetic multi-wall carbon nanotubes. *Water Sci Eng* 2015;**8**:226−32. https://doi.org/10.1016/j.wse.2015.01.009.

95. Verma B, Balomajumder C. Fabrication of magnetic cobalt ferrite nanocomposites: an advanced method of removal of toxic dichromate ions from electroplating wastewater. *Korean J Chem Eng* 2020;**37**:1157−65. https://doi.org/10.1007/s11814-020-0516-3.

Section Two

Metal oxide-carbon composites in energy technologies

Metal oxide–carbon composites for supercapacitor applications

Zia Ur Rehman[1], Muhammad Bilal[1], Jianhua Hou[1], Junaid Ahmad[2], Sami Ullah[2], Xiaozhi Wang[1] and Asif Hussain[1,3]
[1]School of Physics, College of Physical Science and Technology and School of Environmental Science and Engineering, Yangzhou University, Yangzhou, Jiangsu, PR China; [2]Department of Physics, Division of Science and Technology, University of Education Lahore, Lahore, Punjab, Pakistan; [3]Department of Physics, The University of Lahore, Lahore, Punjab, Pakistan

6.1 Introduction

In 1957, porous carbon-based electrodes were first used in supercapacitors (SCs) by H. Becker.[1] He developed a "low-voltage electrolytic capacitor with porous carbon electrodes." In the early 1960s, work started on fuel cell development, and good results were obtained. After attaining the licensing to NEC, the first double-layer capacitors, dubbed "supercapacitors," were successfully used on a commercial scale in 1971 as backup power for computer memory. Brian Evans Conway presented good work on electrochemical capacitors (ECs) using Ru oxide between 1975 and 1980.[2]

The market expanded slowly that changed in 1978, when Panasonic marketed its Gold capacitors. http://en.wikipedia.org/wiki/Supercapacitor - cite_note-14 Like those produced by NEC, this product was designed as an energy source for memory backup applications. At the end of the 1980s, researchers focused on synthesizing an electrode material to provide high capacitance value and better conductivity of electrolyte material. In 1982, the first SC was developed with low internal resistance by the Pinnacle Research Institute.

In 1988, the ELIT capacitor, with a capacitance of 30–40 kJ and a potential difference of 24 V, was developed and used to boost engine systems. Don Boos, a Standard Oil of Ohio (SOHIO) chemist on the project, gave a complete presentation of this work at a conference in 1991, but SOHIO did not commercialize its invention because of cuts it took in R&D spending to improve its balance sheet. In 1991, Brian Evans Conway also described the performance difference in electrochemical storage ability between the SC and the battery. In 1999 he also defined the term "supercapacitor" to make reference to the increase in observed capacitance by surface redox reactions. From about 2005, aerospace company Diehl Luftfahrt Elektronik used Maxwell Technologies' BOOSTCAP ultracapacitors to power its emergency ejection systems for doors and evacuation slides in passenger aircraft. In 2006, Joel Schindall and his team at MIT began work on a "super battery." using nanotube technology to improve upon capacitors.

Metal Oxide-Carbon Hybrid Materials. **https://doi.org/10.1016/B978-0-12-822694-0.00003-X**

After many years and struggles, developers succeeded in developing an asymmetric EC. This asymmetric EC had a nickel oxyhydroxide-based positive electrode, potassium hydroxide electrolyte, and activated carbon (AC) negative electrode. The Joint Stock Company ESMA and its business partner American Electric Power reported a low-cost lead oxide/AC-based asymmetric EC in 2007. They developed this technology to store an enormous amount of energy at night and provide it on demand when needed.[3] In modern society, advancements in technology have provided portable devices for every human hand, laptop, smartphone, and camera. Moreover, the environmental pollution produced by motorized vehicles argues that researchers should think about environmentally friendly energy sources as alternatives to fossil fuels. The need for increased power performance is driving the emergence of SCs as an alternative to conventional electric energy-storage devices.

SCs have become efficient and emerging energy-storage devices because of their fast charge and discharge rates. These promising SCs have about 100 times greater capacitance than conventional capacitors and up to 100 times greater power density than commonly used batteries. SCs have many unique properties that allow the use of carbon nanotubes (CNTs) in electrode construction, and these CNTs can raise the power density and capacitance of SCs compared with conventional capacitors. SCs have other properties, such as long cyclic ability, good working efficiency, good reversibility, and excellent power density, thus making them suitable candidates for energy storage. SCs have many potential applications: storing energy for battery-powered electric vehicles, powering devices such as mobile phones, PCs, navigation devices, cameras, medical equipment, digital-assistant devices, and flash storage cards, and providing energy and storage for many other electronic devices.[4]

Many companies are working on EC technology for fuel cells in commercial settings to improve charge/discharge and cycle performance in hybrid vehicle applications. The most prominent application of SCs is for future energy-storage and rapid-charging systems that improve on those already available in the market. Researchers have worked to further develop SC batteries for commercial purposes. Researchers at the University of Central Florida have successfully created a prototype SC battery that takes up a fraction of the space of lithium-ion cells, charges more quickly, and can recharge 30,000 times while still working like new. Many multinational companies are interested in designing electric capacitors with graphene that will be lightweight and have energy-storage capacities between 150 Fand 550 F/g. Thus, metal oxide−carbon (MO−C) composites have tremendous potential for SC applications.

6.2 Types of supercapacitors

6.2.1 Electrochemical double-layer supercapacitors

Electrochemical double-layer capacitors (EDLCs) are assembled as two carbon-based electrodes in an electrolyte with a separator between the two electrodes. In EDLCs, charge is stored on the electrodes electrostatically with the help of charges present in the electrochemical material. This type of SC has a long cycle life and good characteristics for storage and many applications.[5]

In an EDLC, when the voltage is given, electrical charges collect on the electrode surface. The ions produced in electrolyte across separator and follow the natural law of attraction of different polarity charges. These different charge carriers diffuse through the separator and attach to the pores of the opposite electrode. But in this process, no recombination occurs on the electrode surface, resulting in a double layer of charges built up at both electrodes. The surface area (SA) increases because of the double layer of electrical charges and the high energy densities attained in EDLC compared with common capacitors.[6,7]

6.2.1.1 Diagram

The ELDC schematic provided in Fig. 6.1 shows the double-layer structure of the charges.

6.2.1.2 Main parts of electrochemical double-layer capacitors

Three basic parts play vital roles in this process, with each having specific properties that are important to device performance. Many research articles have already been written on this topic, and much advancement has occurred with EDLCs. Some detail on each part is provided below.[8]

6.2.1.2.1 Electrode material

The electrode material is a critical component in EDLCs. We can check the electrical properties by determining the electrode material. In this process, the surface of the electrode material also has important impacts on SC capacitance. Many materials have been used for this purpose, such as C, metal oxides (MOs), and polymer materials. Carbon has become an attractive material for EDLCs because of its large SA, low cost, and wide availability. It has a long history and is already used in different forms such as CNTs, nanorods, fibers, and foams.

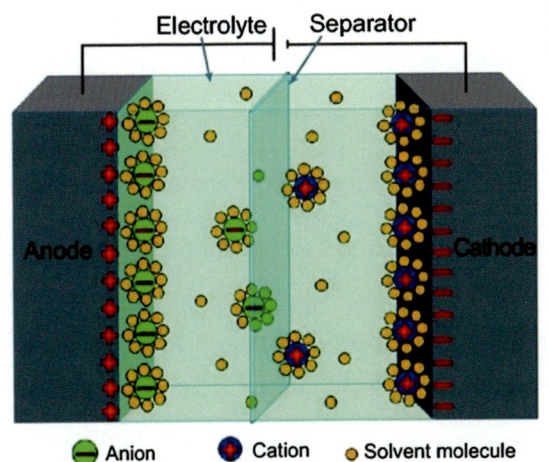

Figure 6.1 Schematic diagram of double-layered structure.
Reprinted with permission Ref. 40.

When we use carbon-based materials, the porosity on the electrode surface is increased. The movement of ions within the pores of the electrode material is different compared with the electrolyte. The electrode material pore size must be appropriate for the ions of electrolytic material. If we use CNTs as electrode material, better results are achieved, but more research is required for good capacitance.[9] MOs have also been used for electrode material due to their high capacitance, low resistance, and many other properties. These properties are useful for making efficient high-power EDLCs with ruthenium-oxides and manganese oxides as the electrodes.[6]

Electrodes also use polymer materials for charge storage. In the oxidation process of conducting polymers, charges are transferred to the polymer electrode, and during reduction, charges are escaped back into the electrolyte solution. High capacitance from polymers has been achieved and reported.[10]

6.2.1.2.2   Electrolyte

The electrolyte is a critical electrode material for EDLCs. The energy density depends on the applied voltage, and the power density of the SC depends on the electrolyte conductivity. We commonly use two kinds of electrolytes in EDLCs, organic and aqueous. The organic electrolyte is often used for commercial purposes. In organic electrolytes, cell power is limited, but resistivity is comparatively high, and the voltage range is 2−2.5 V. In the aqueous electrolyte solutions, conductivity is good compared with organic electrolytes, but the breakdown voltage is low at 1 V.

So the capacitance is dependent on the selection of electrolyte, and charge storage depends on porous SA and ion availability in EDLC. The pore size and electrolyte ions must be suitable for each other, so the electrode material and electrolyte solution must be selected with that in mind.[11]

6.2.1.2.3   Separator

In EDLCs, the separator stops the electrical connection between the electrodes, allowing the transfer of ionic charges. Organic electrolytes use mostly paper and polymer separators, whereas aqueous electrolytes commonly use glass fiber and ceramic separators. If we want the best EDLC performance, we should use separators with low thickness, good electrical resistance, and better ionic conductance.[11]

6.2.2   Pseudosupercapacitor

Transfer charges in pseudocapacitors occur between electrodes and electrolytes, and charges are stored faradaically. Compared with EDLCs, capacitances and energy densities may be achieved through the rapid redox reactions that are repeatable in pseudocapacitors. These redox reactions occur between the electrode material and electrolyte solution.[5,12]

In pseudocapacitors, faradaic redox reactions involved electrode materials that mostly made MO, conducting polymers, or metal-doped carbons play a vital role in storing charges and higher energy density.[13] Because of these electrode materials, pseudocapacitors provide high energy density but shorter cycle lives than EDLCs. In pseudosupercapacitors, electrical conductivity is low and faradaic reactions are irreversible on the surface of the electrode, but with the help of nanostructured electrodes, we can increase the capacitance and stability of pseudosupercapacitors.[14] Reduction and oxidation reactions are involved in this kind of SCs, and these processes produce charges when they occur on the electrode. In pseudosupercapacitors, faradaic systems are slow, so the potential of pseudosupercapacitor has a low power density compared with EDLCs.[15] But because of faradaic reactions, it has good capacitance and excellent energy density compared with EDLCs.[16]

Pseudosupercapacitor materials not only increase energy density but also store it in electrode materials and on their surfaces. The charge route to pass the double layer due to this faradaic current passes through the SC cell.[17] These faradaic processes are important to surge energy density and specific capacitance, but it lacks cycling stability compared with EDLCs.[18] Electrode materials are also essential in increasing SC efficiency.

Conducting polymers and MOs are important materials for the electrode material of pseudosupercapacitors. Many conducting polymers have been studied in various electrolytes.[16] The other electrode materials such as MOs, transition metal oxides (TMOs) are important and can change the oxidation and reduction states. So MOs are the more suitable candidate for the electrode materials of pseudosupercapacitors. Ruthenium oxide has relevant capacitance characteristics and unique properties but is not economical, and because of this, its use is limited.[19] Many other MOs such as Ni, Co, Au, Fe, and Al have been studied in aqueous electrolytes, and MnO_2 has attained great attention because of the high capacitance value.[20] So the composites of carbon-based materials and MO have become promising candidates for the electrode material of pseudosupercapacitors.

6.2.2.1 Diagram

The schematic diagram of the pseudosupercapacitor is given in Fig. 6.2.

6.2.3 Hybrid supercapacitor

Hybrid SCs have both pseudocapacitor and EDLC characteristics. Charges are stored in hybrid SC through both faradaic and nonfaradaic methods. Higher energy densities

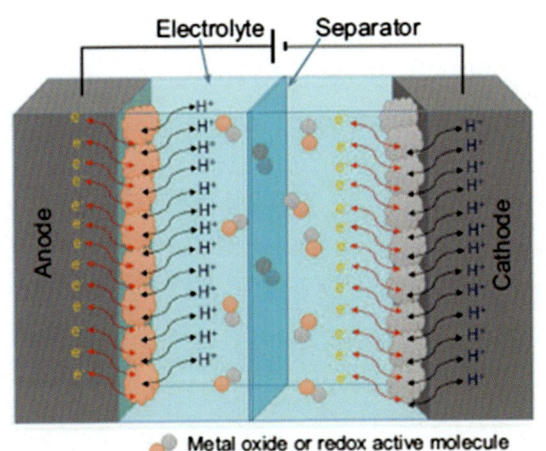

Figure 6.2 Schematic diagram of pseudosupercapacitor.
Reused with permission Ref. 40.

are attained with hybrid SCs than with EDLCs without losing the cyclic stability and affordability that define the best performance of pseudosupercapacitor. Researchers are working on two types of hybrid SCs, symmetric and asymmetric.[21]

In hybrid SCs, the mechanism is the charge stored on + ve electrode by nonfaradaic process of positive ions occurs on the surface of carbon material. The faradaic process happens on the −ve electrode of hybrid SC. Both reactions increase the voltage and energy density of SC. For −ve electrode material, low intercalation potential is necessary to attain the high energy density. In this aspect, carbon, graphite, and metal alloys are important, and many research articles have been written on these materials. To increase the cycle life, −ve electrodes made with such materials have low contraction and expansion in the extraction and insertion processes. For this purpose, some suitable materials with good electrovalent bonds between + ve and −ve ions include TMOs, sulfides, and nitrides.[22]

6.2.3.1 Diagram

A schematic diagram of a hybrid SC is given in Fig. 6.3.

Two types of hybrid SCs are used in charge storage, symmetric and asymmetric. Some description of these types of hybrid SCs is given below.

6.2.3.2 Symmetric supercapacitor

In symmetric SCs, the same electrode material is used for the + ve and −ve electrodes.[23] Compared with common SCs, it has a large energy density and good stability for the potential window. In symmetric SC, because the same electrode material has benefited but has attained high energy density, the proper selection of electrolyte and electrode materials is critical.[24] Many materials such as metals, MO, graphene, AC, CNTs, and carbon nanoparticles are already used as electrode materials in SCs because of their extraordinary and unique properties.[25] The condition in symmetric SC is that electrodes are made with the same materials. A schematic diagram of symmetric SCs is given in Fig. 6.4.

Figure 6.3 Schematic diagram of hybrid supercapacitor.
Reprinted with permission Ref. 40.

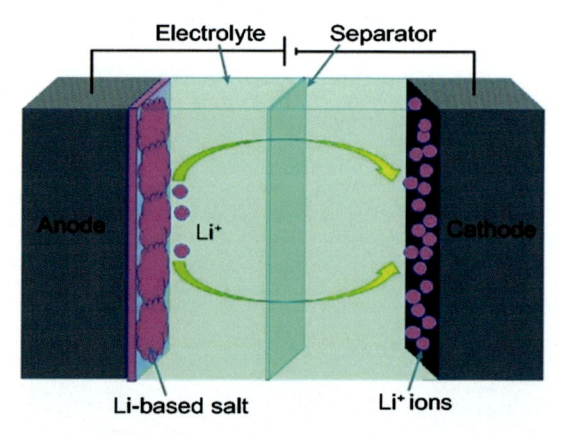

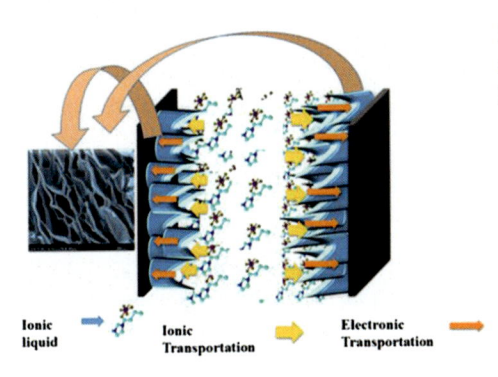

Figure 6.4 Schematic diagram of symmetric supercapacitors. Reprinted with permission Ref. 196.

6.2.3.3 Asymmetric supercapacitor

In asymmetric SCs, the two electrodes are made with different materials, so we say they are asymmetric hybrid SCs. In this hybrid SC, one electrode is made with a carbon-based material while the other is made with a conducting polymer, which has led to increased research interest.[26,27] The conducting polymer material for the negatively charged electrode is not good because it limits the efficiency in pseudocapacitors, but if it is made with AC, this problem is reduced.

The advantages of conducting polymer are capacitance is high, resistance is low compared with AC, cycling ability, and maximum voltage is low. Due to these properties of asymmetric hybrid electrodes, they attained high energy density and good cycling ability compared with EDLCs and pseudocapacitors, respectively.[10] The schematic diagram of asymmetric SCs is given in Fig. 6.5.

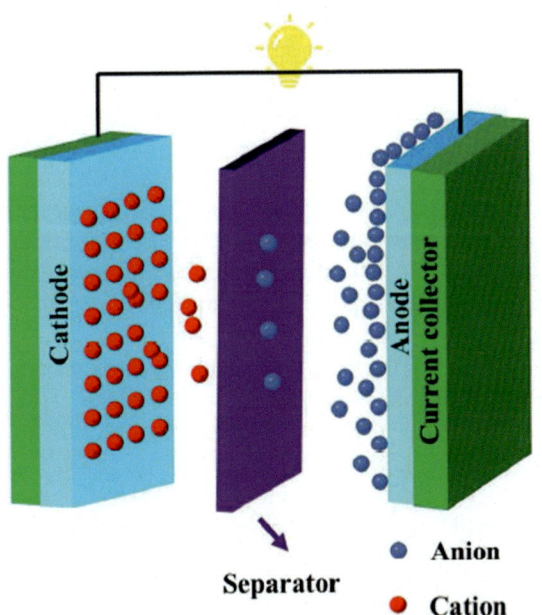

Figure 6.5 Schematic diagram of asymmetric supercapacitors. Reused with permission Ref. 197.

SCs have unique properties and various applications in the industrial field and commercial level. Because of this SCs become attractive devices for energy storage. Table 6.1 outlines different types of SCs with respect to electrode materials, storage mechanisms, and abilities.

6.2.4 Supercapattery

In the past few decades, much work has been done on electrochemical energy-storage (EES) devices because of attractive properties such as high power capabilities, good cyclic ability, and economical price. Batteries have also become a vast field of research because of rechargeable ability and high energy capacity, but each technology has its drawbacks and cannot fulfill all the necessities of commercialization. Many hybrid devices have been made using EES devices and rechargeable batteries such as Li-ion capacitors, hybrid EC, Na ion capacitors, and battery SC hybrids.[28,29] As a result, supercapattery has become a general term representing these devices, which are neither SCs nor batteries in terms of basic principles. In 2007, the term supercapattery was used for the first time in an industrial project.[30,31]

In previous years, considerable research has been published on rechargeable batteries and SCs, but they alone cannot fulfill commercialization requirements. To solve this problem, scientists found new technologies and made hybrid electrochemical energy devices. These devices have the characteristics of SCs and rechargeable batteries. This new device is called a supercapattery and has high energy capacity compared with SCs and is rechargeable like batteries but with greater power capability.[32,33]

Supercapatteries are based on both capacitive and noncapacitive faradaic reactions involved in storing charge mechanisms on electrode material. In Li-ion capacitors on one electrode, because of the noncapacitive faradaic reactions and other electrode capacitive charge storage involved due to this behavior, we cannot say it is only a capacitor or a battery. This process also involves other supercapatteries, so we can say those devices fulfill the requirements of this charge storage mechanism and are called supercapatteries. Many research articles have been written on electrode material and other parts of supercapatteries during the past few years, and now it has become an attractive area of research for storage devices.[34,35]

In supercapatteries, many combinations such as capacitive and faradaic mechanisms for charge storage are involved in the electrode material. The purpose of these combinations is to attain high power density and good storage capacity for supercapatteries.[35] The electrode materials are porous, have nanosized structures, and are redox-active for capacitive and noncapacitive use in supercapatteries. The combination of capacitor-type and battery-type electrodes has become an essential part of making supercapattery devices. In a supercapattery, one electrode, as in a capacitor, gives capacitive reactions, and another electrode, as in a battery, participates with redox or faradaic process or noncapacitive storage, as shown in Fig 6.6.

The energy capacity in a supercapattery with a negative electrode made by Li metal and a 400 F/g positive electrode, as in an SC, gives good performance compared with a Li-ion battery.[36] The reason is that the charge capacity of lithium is greater than in electrode of SCs, and the mass of Li is less in a supercapattery. For capacitor-like

Table 6.1 Various electrode materials, storage mechanisms, and shortcomings.

Type of supercapacitor		Electrode material	Charge storage mechanism	Merits/shortcomings
Electrochemical double-layer capacitor (EDLC)		Carbon	EDLC, nonfaradaic process	Good cycling stability, good rate capability, low specific capacitance, low energy density
Pseudocapacitor		Redox metal oxide or redox polymer	Redox reaction, faradaic process	High specific capacitance, relatively high energy density, relatively high power density, relatively low rate capability
Hybrid capacitor	Asymmetric hybrid	Anode: Pseudocapacitance materials, cathode: carbon	Anode: redox reaction, cathode: EDLC	High energy density, high power density and good cyclability
	Symmetric composite hybrid	Redox metal oxide/ carbon or redox polymer/carbon	Redox reaction plus EDLC	High energy density, moderate cost, and moderate stability
	Battery-like hybrid	Anode: Li-insertion material, cathode: carbon	Anode: Lithiation/ delithiation cathode: EDLC	High energy density, high cost and requires electrode material capacity match

Reused with permission Ref. 72.

Figure 6.6 Schematic diagram of
supercapattery.
Reprinted with permission Ref. 198.

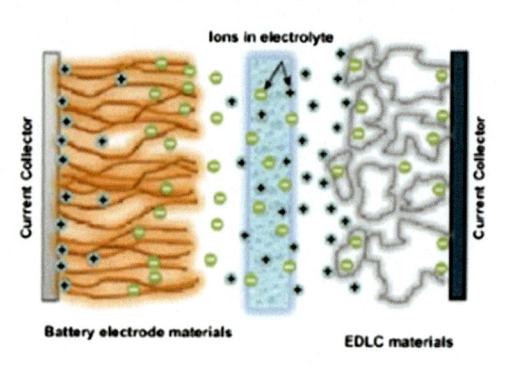

electrodes that are polarizable, nanostructured carbon materials such as CNTs, AC, and graphene are essential for large power capabilities. These carbon-based materials have exclusive properties such as high SA, porous structures, low electrical resistance, and stability for the large potential gap. Some other materials like MO (MnO_2, RuO_2) may be used for this type of electrode. For cell voltage calculations, the potential window of electrodes and the electrolyte solution in a supercapattery are critical elements.

There are wider materials options such as MOs (SnO_2, $LiFePO_4$, MnO_2, etc.), conducting polymers (PPy, PAn, PEDOT), and redox-active polymers for the batterylike electrode in a supercapattery. The redox reactions on these electrode materials are good for charge storage ability in supercapattery. The properties of electrode materials play a vital role in supercapattery efficiency. The combination of C-based and redox materials in supercapatteries has received considerable attention for charge storage. Many composites of MO and conducting polymers with graphene and CNTs have been fabricated for this in the past few years. Despite their capacitor-like behavior, the redox electrode material can come from pseudocapacitive materials such as manganese oxide (MnO2) and polypyrrole (PPy). If electrolyte is organic and the electrode is made with redox materials, supercapatteries can give higher power capacity and good energy-storage ability compared with LIBs.[37]

6.2.4.1   Future perspectives of supercapatteries

Supercapatteries have properties of both rechargeable batteries and SCs and define a large group of EES devices. Supercapatteries use various mechanisms, such as faradaic and nonfaradaic capacitive storage and noncapacitive faradaic storage, to describe the efficiency of electrode material. Redox electrolytes also play an important role in supercapatteries. Materials choices for electrodes and their synthesis methods have become more important for device efficiency, especially in several cells of supercapatteries, and can achieve high energy densities. In the past few years, supercapatteries have attained a great deal of attention for EES devices because they can have high energy capacity and great power capability within the same device. The advancements in the materials used in SCs and batteries can confer benefits to improve supercapatteries. New supercapattery designs, such as bipolar stacking, are needed in the industrial sector. In addition, more advancement is needed in making supercapatteries for commercial purposes using present equipment.[38]

6.3 Carbon-based supercapacitors

Currently, SCs are used in particular applications where high power is needed for a very short time. Device SCs followed batteries with a slow discharge rate but showed extraordinary stored energy densities. Therefore, various applications use the combination of SCs and lithium-ion batteries for the efficient working of the application. Along with the latest quick progress of flexible electronic equipment, there is a serious requirement for unified power modes with malleable and stretchable electrodes. Hence, SCs having malleable and extensible fiber structures have gained much attention.[39] In this respect, CNTs and graphene having a remarkable mechanical solidity and outstanding foldable durability have been suggested as perfect electrode materials for malleable and extensible SCs. Therefore, carbon nanostructures have been explored at a large scale to evolve the latest electrode materials in various SCs for effective energy storage.[40]

Generally, carbon structures fall into three categories: (1) diamond, (2) graphite, and (3) amorphous carbon.[41] They have different characteristics according to the order of carbon atoms. The latest progress in the field of nanotechnology has unveiled new ground in carbon materials research by developing graphitic carbon nanomaterials with a variety of dimensions, such as zero-dimensional (e.g., fullerene), one-dimensional (e.g., CNTs)[42–44], and two-dimensional (e.g., graphene).[45–48] Fullerene C_{60} exhibits geometry like a soccer ball, with 20 carbon hexagons and 12 carbon pentagons as an ideal electron acceptor with wide applications in solar cells for charge segregation. Its use is restricted for energy-storage applications owing to its contrariness, low conduction of electricity, and reduced SA. So far, graphene,[40,49–51] CNTs,[40,52,53] mesoporous carbon[54–56], and their hybrids[57–59] have been largely explored as SC electrodes owing to their unique properties such as magnificent conduction of electricity, large SA, excellent electrochemical ability, and easy transformation into multifunctional and multidimensional structures with magnificent electrical and mechanical features.

Pean et al.[60] carried out molecular dynamic simulations of carbon-based nanoporous electrodes and showed that these electrodes exhibited remarkable capacitances compared with other morphologies. For this purpose, they used nanoporous electrodes and liquid-type electrolytes (1-butyl-3-methylimidazolium hexafluorophosphate, $BMI-PF_6$) and kept them in an uncharged state, as shown in Fig. 6.7. By applying an electric potential difference between electrodes, the system abruptly changes, and electrodes begin to charge due to the movement of ions between the nanopores and electrolyte zone. Furthermore, they used these simulation results in a practical circuit to evaluate the charging rate for various macroscopic gadgets. The practical circuit used to model this simulation consists of different parameters, such as R_{bulk} as the resistance of electrolyte solution in the bulk zone, R_1 as the resistance of electrolyte adsorbed in an electrode portion, and C_1 and C_2 as the capacitances of two wafers, as illustrated in Fig. 6.7.

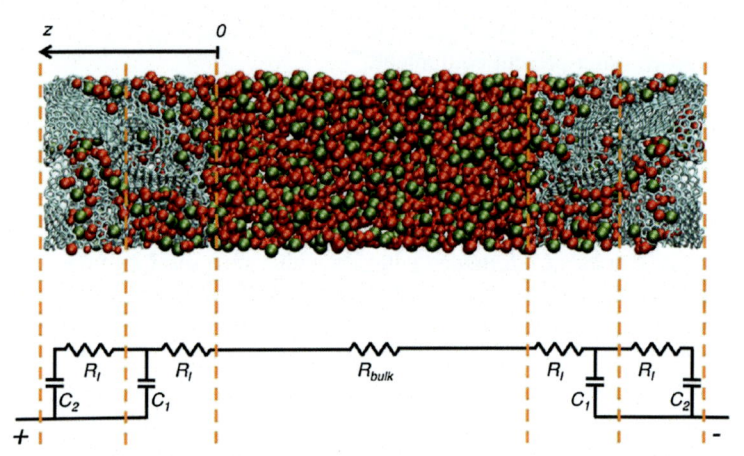

Figure 6.7 Schematic demonstration of simulation circuit.
Reprinted with permission Ref. 60.

Noked et al.[61] fabricated the composites of CNTs and AC by diffusing CNTs in a solution of vinyl dichloride monomer. The synthesized composite contained a large number of CNTs and therefore showed improved specific SA after the process of activation. SCs having CNTs/AC electrodes displayed magnificent cycling activity of 50,000 cycles with only a minor decline in capacitance in aqueous electrolyte liquid, whereas the same material in the absence of CNTs exhibited a decline in capacitance after just 30,000 cycles.

Mazurenko et al.[62] synthesized macroporous CNT/carbon-based electrodes with uniform pore size, and their characterization was carried out. This was done in four steps: (1) electrophoretic codecomposition of CNT and polystyrene beads as the template, (2) polypyrrole electropolymerization in the vacant places defined by the CNT morphology, (3) template removal, and (4) pyrolysis—as illustrated in Fig. 6.8A. The SA of the synthesized macroporous nanocomposites was five times more than the non-macroporous layer of the CNT. Furthermore, the synthesized CNT/carbon nanocomposites with uniform pore size exhibited improved specific capacitances from 6 mF/cm^2 to 200 mF/cm^2, as shown in Fig. 6.8B. The enhanced results obtained during this research are excellent evidence for the essential need for a vigorous carbonaceous grid to reinforce the nanocarbonaceous operating constituents, in contrast to the working of electrodes with nanoparticles that do not have a vigorous reinforcing platform.[62]

Physiochemical and electrochemical features of carbonaceous materials can be altered by doping with nitrogen. Furthermore, doping with nitrogen can also alter the crystalline and electronic anatomy of the carbons, surface polarity, improving their chemical durability, conduction of electricity, and electron donor characteristics. Carbons having nitrogen as an impurity are very interesting for various applications such as SCs, separation, catalysis, and adsorption.[63]

A new approach for developing nitrogen-doped substances that incorporate carbonization, nitrogen functionalization, and functioning into a single phenomenon was

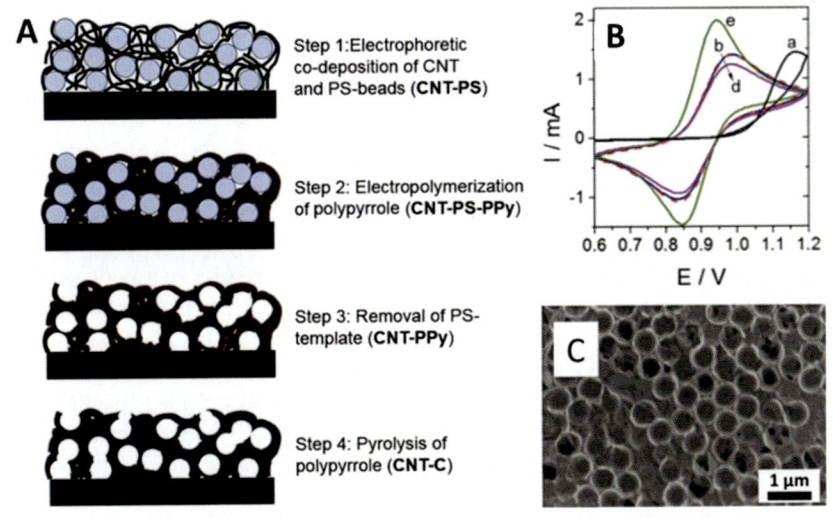

Figure 6.8 (A) Steps for the synthesis of CNT/Carbon nanocomposites, (B) Cyclic voltammograms measured in 0.15 M H$_2$SO$_4$ solution using (A) pure GCE, (B, E, D) CNT deposits on GCE and (E) a macroporous CNT layer deposited on GCE and (C) SEM image of the macroporous CNT assembly.
Reprinted with permission from Ref. 62.

reported by Wang et al.[64] The synthesized nitrogen-doped carbon materials showed a large SA (1400 m^2/g) and high nitrogen content. Using these synthesized materials as electrodes for EDLC improves the capacitance (6.8 μF/cm^2 vs. 3.2 μF/cm^2) and minimizes the ion diffusion resistance compared with carbon elements that do not incorporate nitrogen.[64]

The working of EDLC can be improved and charge transmission minimized by developing uniformly smaller-sized particles. Fabrication of nitrogen-doped carbon elements with microporous carbon spheres gives the larger capacitance, owing to the availability of the electrolyte suspension to the pores within the particles. Wu et al.[65] synthesized the microporous carbon with spherical mesopores using a unique method known as atom-transfer radical polymerization. In this technique, the precursors contain nanoparticles with organic cores overlapped by a nitrogen-containing carbon source. This synthesis method removes the complex assembly techniques and permits consistent packing with the heteroatom source. As an electrode element for SCs, these materials demonstrated a capacitance of 260 F/g with a current density of 20 mA/g.[65]

6.4 Metal oxide-based supercapacitors

Rechargeable battery lifetimes are significantly restricted by the chemical reactions involved in the charge/discharge mechanism.[66] In contrast with batteries, the energy-saving mechanism of the SCs mainly depends on the electrostatic depository in the electric double layer of the capacitor and reversible faradaic redox mechanisms with the help of electron charge movement on the facet layer of the electrode. Therefore, SCs are expected to have the ability to quickly charge/discharge under high current and longer cycle lives than batteries because no or few chemical reactions are involved.[7] Quite a bit of development has been made in conceptual and experimental research in the field of SCs; a few drawbacks of SCs, such as less energy density and expensive production, have been pointed out as well.[67]

One of the most intensive ways to overcome the barrier of less energy density is to fabricate the latest stuff for SC electrodes. Carbon-based materials have attracted much attention in this regard by exhibiting a large SA for charge storage.[68] But despite these large SAs, the charges stored in the porous electrode layers at the carbon particles limit their electrochemical characteristics. Such SCs with restricted specific capacitance and relatively low energy density are known as EDLCs.[69] SCs with electrochemically active electrode materials, such as polymers and MOs involving quick and reversible faradaic reactions, are known as faradaic supercapacitors (FSCs). FSCs can provide much greater capacitance and energy density. Thus, from the point of view of modern SCs, MOs are considered the ideal materials for the next generation of SCs.[70]

SCs based on MOs exhibit higher specific capacitance and larger energy densities than carbon-based materials.[71] A class of MOs with high conceptual performances, such as MnO_2, V_2O_5, RuO_2, Fe_3O_4, Co_3O_4, NiO, and CuO, has been investigated. Tables 6.2 and 6.3[72] summarize specific characteristics of various SCs and illustrate the values of conceptual capacitances of some particular MOs and their charge storage reactions.

MO geometry is an important component that strictly connects with the specific area of surface, surface-to-volume ratios, diffusion pathway, and supercapacitive performance. Thus, many attempts have been made to synthesize the different MO morphologies of nanowires, hollow spheres, nanorods, nanopillar formation, nanoflowers, nanotubes, and spongy thin films.

Gao et al.[73] successfully fabricated Co_3O_4 nanowires on nickel foam using a template-free technique, as illustrated in Fig. 6.9A. These synthesized nanowires with diameters of 250 nm and lengths of up to 15 μm exhibited enhanced specific capacitance of 746 F/g at a current density of 5 mA/cm^2. Yan et al.[74] successfully synthesized the hierarchically porous structure of NiO with a hollow-spherelike morphology made up of nanoflakes, as illustrated in Fig. 6.9B. They used the chemical bath deposition (CBD) method, and the synthesized structures had a thickness of ~10 nm. The synthesized nanostructures displayed remarkable supercapacitive performance (346 F/g at 1 A/g) after 2000 charge and discharge cycles.

Moosavifard et al.[75] described a facile and effective technique for developing 3-D pseudocapacitive CuO structures with strongly arranged and interlinked bimodal nanopores having walls (~4 nm) and high specific SA (149 m^2/g), as shown in Fig. 6.9C. The as-obtained nanostructures provide ease of ion movement, low ion and e$^-$ distribution tracks, and additional efficient spaces for electrochemical kinematics. As an electrode material, this material exhibits improved electrochemical behavior having a specific capacitance of 431 F/g at the rate of 3.5 mA/cm^2 and

Table 6.2 Comparison of synthesis methods and properties of various carbon-based supercapacitors.

Electrode material	Method	Specific capacitance (F/g)	Current density 7(A/g)	Stability Cycle	Stability %	Ref.
Layered NOMC	Chemical vapor deposition	810	1.0	52,000	82	[199]
Hierarchically porous NOMC	Supramolecular assembly	537	0.5	10,000	98.8	[200]
NOMC from aqueous assembly	Aqueous cooperative assembly	186	0.25	—	75	[201]
NOMC	One-pot soft templating and one-step pyrolysis	288	0.1	25,000	—	[202]
Ordered NOMC	One-pot	262	0.5	—	—	[203]
NOMC from phenol-urea-formaldehyde	—	225	0.5	1000	99	[204]
N-doped micro-mesoporous carbon	—	226	1.0	2000	75.5	[205]
rGO/PANI	In situ electrodeposition	970	2.5	1700	90	[206]
GO/PANI	In situ polymerization	746	0.2	500	73	[207]
Graphene/PANI	In situ polymerization	480	0.1	1000	70	[208]
PANI/Graphene	In situ electrochemical polymerization	878	1.0	1000	—	[209]
PEDOT/MWCNT	In situ chemical polymerization	79	1.0	1000	85	[52]
SWNT	Water-assisted chemical vapor deposition	160	1.0	1000	96.4	[210]
Exfoliated rGO	Chemical activation	166	5.7	10,000	97	[211]
N-doped graphene	Fluorination followed by thermal annealing	280	20	230,000	99.8	[212]
3-D N-doped graphene	Hydrothermal	484	1	1000	100	[213]
B-doped rGO	Solution process	200	0.1	4500	95	[214]
Graphene aerogel	Electrochemical exfoliation	325	1	5000	~98	[215]

Table 6.3 Theoretical capacitances and conductivities of particular metal oxides.

Oxide	Electrolyte	Charge storage reaction	Theoretical capacitance (F/g)	Conductivity (S/cm)
MnO_2	Na_2SO_4	$MnO_2 + M^+ + e^- = MMnO_2$ (M could be H^+, Li^+, Na^+, K^+)	1380	$10^{-5}-10^{-6}$
NiO	KOH, NaOH	$NiO + OH^- = NiOOH + e^-$	2584	$0.01-0.32$
Co_3O_4	KOH, NaOH	$Co_3O_4 + OH^-$ $+ H_2O = 3CoOOH + e^-$ $CoOOH + OH^-$ $= CoO_2 + H_2O + e^-$	3560	$10^{-4}-10^{-2}$
V_2O_5	NaCl, Na_2SO_4	$V_2O_5 + 4M^+ + 4e^-$ $= M_2V_2O_5$ (M could be H^+, Li^+, Na^+, K^+)	2120	$10^{-4}-10^{-2}$

Figure 6.9 (A) Nanowires of Co_3O_4, (B) Hollow spheres of NiO. (C) Three-dimensional nanoporous CuO and (D) Three-dimensiional nanonet hollow structured Co_3O_4.
(A) Reprinted with permission from Ref. 73, (B) Reprinted with permission from Ref. 74, (C) Reprinted with permission from Ref. 75, (D) Reprinted with permission from Ref. 76.

maintains over 70% of this capacitance by operating at the rate of 70 mA/cm^2. Wang et al.[76] successfully fabricated the 3-D-nanonet hollow structures of Co_3O_4 using a simple, cheap, and environmentally friendly technique under moderate temperature and pressure, as shown in Fig. 6.9D. The pseudocapacitive characteristics of the synthesized nanostructures were evaluated by cyclic voltammetry (CV), galvanostatic charge—discharge measurement (GCD), and electrochemical impedance spectroscopy (EIS) in a 6.0 M KOH solution. The specific capacitances of the synthesized nanostructures were 820, 755, 693, and 656 F/g at different scan rates of 5, 10, 20, and 30 mV/s, respectively. The as-obtained material exhibited excellent charge—discharge durability and sustained 902% of its original capacitance after 1000 successive charge/discharge revolutions at the current density of 5 A/g.

Furthermore, Lu et al.[77] successfully synthesized the slim (<20 nm) nanorod structure of NiO. The synthesized nanorods of NiO exhibited a maximum specific capacitance of 2018 F/g with a current density of 2.27 A/g and a maximum power density of 1536 F/g at the rate of 22.7 A/g. In general, the diameter has a vital role in one-dimensional nanomaterial. Larger specific areas and other efficient places can be obtained by the smaller diameter material, which enables the material to give improved specific capacitance. Furthermore, it is described that MnO_2 with spongy nanotube-like geometry not only improves specific capacitance but also enhances electrode durability during charge/discharge cycle.[78]

Du et al.[79] successfully prepared the hollow spheres of Co_3O_4 using the one-pot hydrothermal and calcination technique. The as-synthesized hollow spheres showed a larger SA of 60 m^2/g due to mesoporous morphology. The synthesized hollow spheres of Co_3O_4 as electrodes display extraordinary cycling performance and excellent rate capacity. This enhanced performance is due to the tininess of Co_3O_4 and the adequate room accessible to link with the electrolytes.

Xing et al.[80] fabricated the mesoporous nanostructure of NiO by the calcination of $Ni(OH)_2$ at various temperatures. NiO calcined at 250 exhibited the specific SA of 477.7 m^2/g. Mesoporous morphology of NiO exhibited a specific capacitance of 124 F/g at a current density of 100 mA/g. CV analysis revealed that the NiO exhibited good capacitive performance owing to its versatile mesoporous morphology. Yu et al.[81] fabricated the nanoporous manganese and nickel oxides by a facile method. They obtained polyhedron-like morphologies, larger SAs, and confined pore distribution with the help of guided thermal decay of the oxalate precursors. Morphological and structural characterizations were done with the help of XRD, SEM, TG-DTA, XPS, TEM, wet chemical titration, and N_2 sorption isotherm technique. The synthesized structures exhibited larger SA (for MnO_2 = 283 m^2/g and NiO_2 = 179 m^2/g). The electrochemical activity of the synthesized materials was observed by CV in a mild liquid electrolyte. Manganese oxide reveals specific capacitance of 309 F/g and nickel oxide shows 165 F/g due to their nanoporous morphology.

Purushothaman et al.[82] successfully synthesized the nanosheets of NiO with the help of hydrothermal technique. They used different temperatures (120, 140, 160 and 180) to observe the morphological, structural and electrochemical characteristics of NiO related to SC applications. Fig. 6.10 illustrates the CV curves for the different samples synthesized at different temperatures. Electrochemical results revealed that

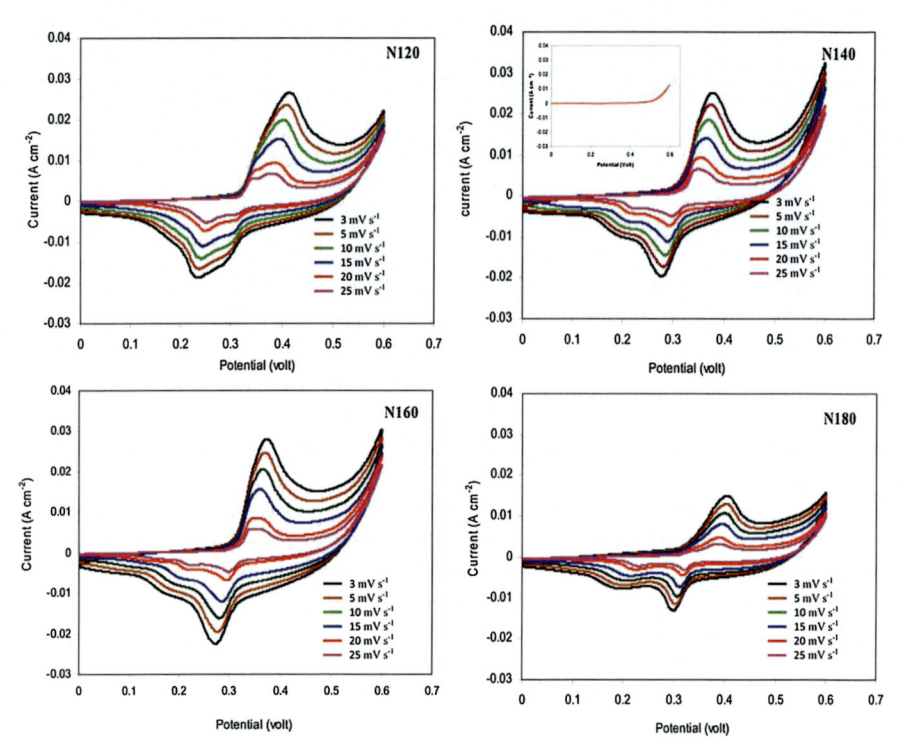

Figure 6.10 Cyclic voltammetry (CV) curves for 120, 140, 160, and 180 samples at multiple scan rates. Inset in the portion of N140 sample illustrates the CV curve of Ni-foam at 3 mV/s. Reprinted with permission from Ref. 82.

nanosheets of NiO fabricated at 160 showed maximum specific capacitance of 989 F/g at a scan rate of 3 mV/s for the potential window of 0–0.6 V, whereas the sample synthesized at 140 exhibited a specific capacitance of 925 F/g. The specimen synthesized at 180 exhibited a specific capacitance of 496 F/g, while the value of the specimen fabricated at 120 was 871 F/g. The synthesized nanosheets display enhanced capacity retention of 97% after 1000 successive charge–discharge revolutions.

Xia et al.[83] developed the self-supported hollow nanowire arrays of Co_3O_4 on different conductive substrates using a facile hydrothermal method. The synthesized nanowires of Co_3O_4 exhibited a diameter of about 200 nm, whereas the hollow cores showed a diameter of about 25 nm. Furthermore, a hierarchically spongy structure could be available in nanowires that make the perforation of electrolytes easier. Nanowires developed on nickel foam substrate are characterized as a cathode element for SC using CV measurements and GCD analysis. An electrochemical analysis reported that the synthesized nanowires exhibited excellent specific capacitance (599 F/g and 439 F/g at a current density of 2 A/g and 40 A/g) owing to their novel structure with a hollow center and porous walls that have more sites for ions diffusion at high current density. Nanowire array images are shown in Fig. 6.11.

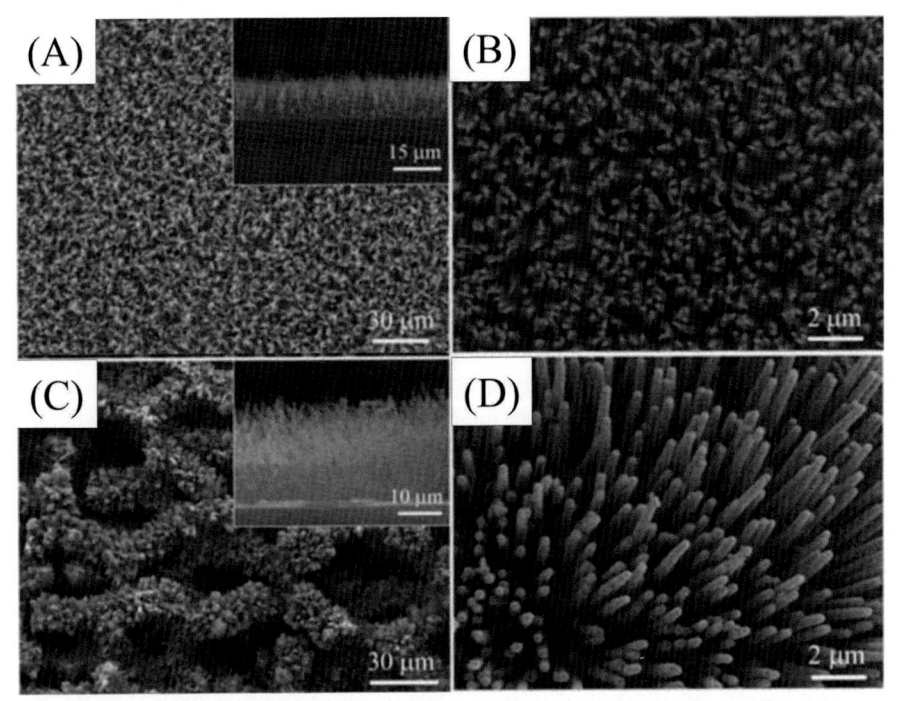

Figure 6.11 Scanning electron microscopy figures of Co_3O_4 nanowire arrays on (A and B) Silicon wafer and (C and D) nickel foam substrate. The insets in (A) and (C) correspond to the side views of Co_3O_4 nanowire arrays.
Reprinted with permission from Ref. 83.

6.5 Transition metal-based supercapacitors

TMO materials, such as MnO_2, RuO_2, V_2O_5, PbO_2, Fe_2O_3, and several other conducting polymers, including polypyrrole, polyaniline, polythiophene, and their composites or derivatives, have been studied extensively as an electrode material for energy-storage devices. The strong redox reaction abilities at the surface and adjacent to the surface and extensive diversity of compositions stabilities make promising candidates in many other fields. Therefore, MO-based electrodes have significantly higher specific capacity compared with EDLC electrodes. These electrochemical devices provide a higher energy density value but have lower power density and insignificant cyclic stability. Consequently, the electrical and ionic conductivity of TMOs is increased by integration with the synergic effect of different EDLC materials components.[84] The larger specific energy of EDLC electrodes is enhanced by faradaic reactions, not considering cyclic stability.[85]

The pseudocapacitance is practiced with larger specific energy density by TMOs with their significant cycle life. However, several other TMOs exhibited very low performance for pseudocapacitance applicants.[86] The CNTs and AC deal as encouraging candidates for SC electrode material, increasing the stored energy, power capability, and cyclic stability. CNT and AC can alter structural and electrical properties, leading to the exploration of new opportunities for diverse energy-storage device applications. The energy density of AC materials can be enhanced using various mechanisms, such as surface modification, tuning morphology, organic electrolytes, and the introduction of oxygen vacancies into the TMO is thus considered an effective strategy to increase the electrochemical kinetics for rapid pseudocapacitive charge storage.[87]

6.5.1 MnO_2-activated carbon for electrode material

EES devices are used for hybrid electric vehicles and transferable electronics instruments. Among EES devices, SCs are the only devices with large power density and long cycling stability, and they have been used in different applications.[88] Normally, SCs are of two kinds, (1) pseudocapacitors and (2) EDLCs.[89,90] EDLC capacitance is influenced by the absorption of cations and anions, which is associated with electrode material SA. Resulting, porosity was increased to enhance the specific SA of the electrode materials.

The porous carbon, like CNTs,[91,92] AC,[93,94] graphene[95,96] and carbon nanofibers[97,98] was frequently used as electrode materials for EDLCs. The aforementioned materials have excellent conductivity, large specific SA. and ability to store energy, and consequently, higher capacitance, as CNTs and graphene show higher SA and conductivity compared with other carbon materials. But the AC has been addressed as an ideal material for EDLCs electrode, which is used for commercial SCs, owning to its low cast, adaptable specific SA, porosity, inert chemical properties, and a large number of active sites for chemical reactions. Carbon-based materials have low volume energy density, which needs to be intensively enhanced[99,100]

In general, the performance of AC material can be increase by controlling pore characteristics and specific SA to enhance adsorption capacity,[101,102] furthermore, the capacitance can also be enhance by combining AC with MOs. Recently, Govindan et al. reported AC decorated with MnO_2, larger specific SA 753 m^2/g, and capacitance (388 F/g).[103] The MXene can also be incorporated with AC/MnO_2 to increase the capacitance of SC.[104] The emerging capacitive deionization technique for water desalination was reported recently, where polypyrrole (PPY)-loaded AC coupled with MnO_2 electrode has shown excellent performance with increases in PPY.

6.5.2 RuO_2-carbon nanotubes for electrode material

Like other TMOs, ruthenium oxide (RuO_2) is an encouraging electrode candidate for SC because it has a larger specific capacitance of about 1450 F/g; however, the synthesis process of RuO_2 is very complicated.[105] The Trasatti and Buzzanca reported the charge transfer reaction of RuO_2.[106] The several carbon forms were loaded with TMOs to make nanocomposites, which showed enhanced charge storage and conductivity efficiency.[107] The RuO_2 nanocomposites were acknowledged for their potential SC applications,[108] like RuO_2/GE/CNT with specific capacitance (502 F/g),[109] RuO_2/GE with capacitance 570 F/g[110] and RuO_2/PETDOT:PSS/GE with capacitance (820 F/g).[111]

6.5.3 V_2O_5-carbon nanotubes for electrode material

Vanadium pentoxide (V_2O_5) shows prevailing properties across different TMOs. It can act as a promising material for SCs owing to its elevated theoretical capacitance, low cost, multiple oxidation valences (V^{5+}, V^{4+}, V^{3+}, and V^{2+}), low toxicity, layered composition, and simple synthesis.[112–114] The charge storage mechanism of V_2O_5 is expressed by the following equation:

$$V_2O_5 + xA+ + xe- \leftrightarrow AxV_2O_5 \quad (A = Na, Li, K, etc.)$$

However, some constraints such as structural instability during the charge—discharge cycle, low electronic conductivity, and slow electrochemical kinetics result in poor cycling performance and slow the rate stability of V_2O_5.[115] So to enhance the overall electrochemical performance of VO-based electrodes, there is a need to fabricate and explore nanocomposites based on V_2O_5.

CNTs are highly desirable for SCs because of their greater specific SA, higher charge—discharge speeds, improved operating protection, and longer cycle lives.[68,116] Much work is underway to develop nanocomposite materials with CNTs and other materials to achieve desired performance characteristics. For example, Saravanakumar and coworkers[117] documented the fabrication of V_2O_5/f-MWCNT nanocomposites by means of a simple solution approach, as presented in Fig. 6.12A.

The fabricated nanocomposites showed excellent rate stability (i.e., retaining 68.3% as the current density develops from 0.5 to 10 A/g, as shown in Fig. 6.12B and C, good cycling capability (i.e., 86% retained after as many as 600 cycles), and elevated specific capacitance (410 F/g at 0.5 A/g).[117]

Saravana Kumar et al.[117] assembled a symmetrical SC using as-fabricated V_2O_5/f-MWCNT as-displayed in Fig. 6.13. The results obtained from CV and GCD curves (Fig. 6.13B and C) reveal that the fabricated nanocomposite attained the energy density of 8.9 W h k/g with a power density of 121 W k/g and specific capacitance of 64 F/g at current densities of 0.5 A/g.[117]

Qingyong Wang et al.[118] documented the fabrication of V_2O_5/CNT—superactivated carbon (SAC) ternary nanocomposite using facile hydrothermal approach, as illustrated in Fig. 6.14. They tested the fabricated hybrid composite in SCs as electrode material. It can be clearly observed in Fig. 6.14C and D that V_2O_5/CNT—SAC hybrid composite showed improved specific capacitance of 357.5 F/g at a current density of 10 A/g. The fabricated composite exhibited boosted electrochemical efficiency compared with pristine V_2O_5, binary CNT—SAC, and V_2O_5/CNTs. The elevated performance was ascribed to synergistic effect generated by MO, CNTs, and SAC.[118]

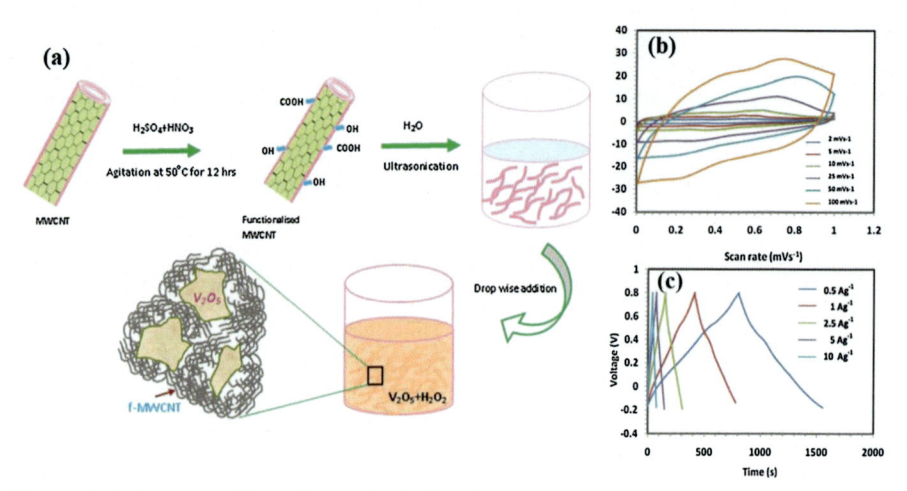

Figure 6.12 Schematic illustration for V_2O_5/f-MWCNT nanocomposite preparation (A), (B) its cyclic voltammetry plots at various sweep rates, and (C) its charge−discharge profiles at various current densities.
Reprinted with permission Ref. 117.

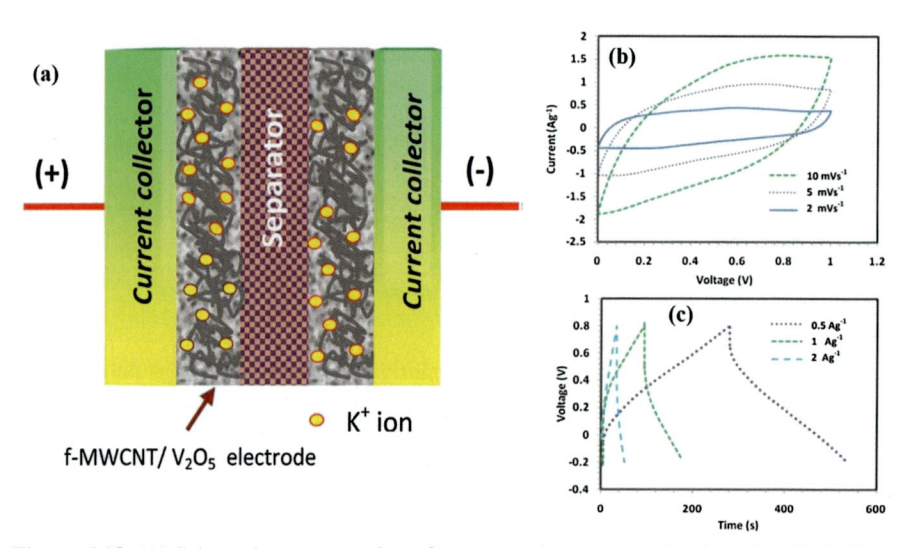

Figure 6.13 (A) Schematic representation of a symmetric supercapacitor based on V_2O_5/f-MWCNT. (B) Cyclic voltammetry plots at scan rate from 2 to 10 mV/s presented by as-manufactured symmetric supercapacitor. (C) Galvanostatic charge−discharge curves at a current density of 0.5−2 A/g.
Reused with permission Ref. 117.

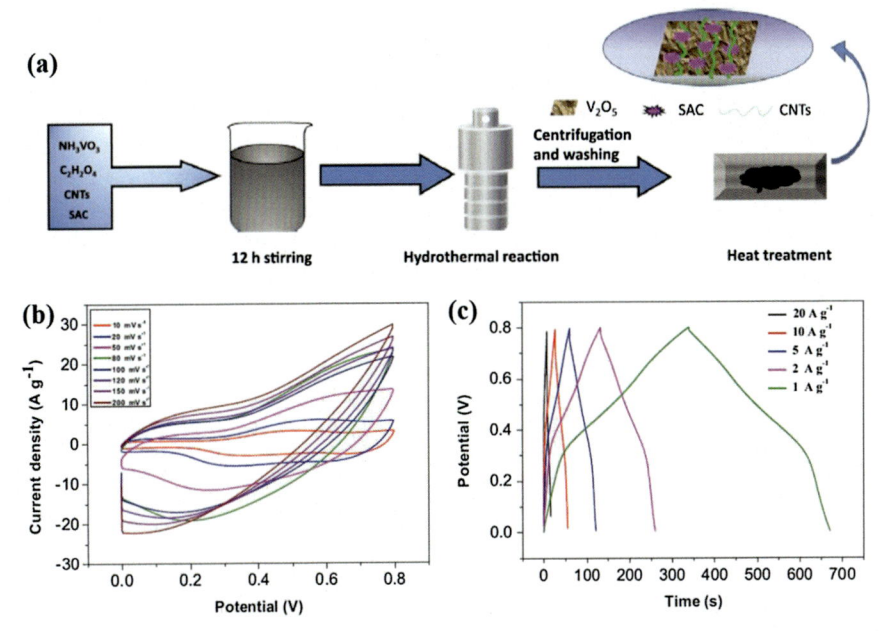

Figure 6.14 (A) Synthesis mechanism of V_2O_5/carbon nanotube—superactivated carbon (SAC) hybrid composite. (B) Cyclic voltammetry plots at various scan rates displayed by V_2O_5/CNTs—SAC. (C) Charge/discharge plots at current densities of 1—20 A/g by V_2O_5/CNTs—SAC electrode.
Reused with permission Ref. 118.

6.6 Rare-earth metal oxide-based supercapacitors

Technological development has increased the use of portable devices, further increasing the application of portable energy resources. SCs have gained significance due to their high storage capacity, high power density, fast charge—discharge rates, long cycle lives, and low weight. A significant volume of research has been dedicated to developing efficient SCs with high power density and storage capacity, fast charge—discharge rates, long cycle lives, and environmental friendliness.[15,119] Carbon-based materials have proven efficient for SC cycle lives but have failed to provide a significant value of specific capacitance.[56] In pseudocapacitors using MOs and their composites with conducting polymer as an electrode material, researchers have obtained high specific capacitance but with the drawbacks of shorter cycle life reduced mechanical stability.[120,121] However, TMOs like the oxides of ruthenium, titanium, copper, and manganese have exhibited appropriate potential as electrode materials for SCs because of their high pseudocapacitance value of 0.7.[122,123] Copper oxide (CuO) has attained worldwide attention due to its modest price rate, nontoxicity, and easy synthesis in different nanosized dimensions. CuO with a low bandgap (1.2 ev) shows a p-type semiconductor nature and proves itself to be the best candidate for SCs.[124]

Krishnamoorthy et al. reported the electrochemical properties of CuO for pseudo-supercapacitor and reported the specific capacitance 94.83 F/g at 0.005 mV/s.[125] However, it suffers from the weakness of the CuO crystal structure, which is a major drawback that reduces its efficiency. This problem can be solved by controlling the material's nanostructure and synthesizing different composites of Cu-based polymeric material.[126] Rare-earth (RE) MOs are a group of four elements that have received attention because of their unique optical, catalytical, and electrochemical properties. RE elements are rich in the earth's crust and have many applications in industrial and military fields. Researchers in the storage field were unaware of the importance of RE elements, and only a few researchers have reported on the material in energy-storage systems.[127,128] Besides traditional energy-storage devices, researchers need to focus on advanced energy-storage devices using RE-based electrodes, RE-doped electrodes, and RE nanocomposite electrodes.

RE elements consist of 15 lanthanide elements with unique 4f electron configurations and scandium (Sc) and yttrium (Y).[129,130] They have unique chemical properties because of unpaired 4f electrons because they do not participate in the chemical reaction.[131] In the last decade, RE elements have come into greater view with researchers because of their physical and chemical properties and controllable structure. The composites of RE-based nanomaterials have multiple electrochemical properties for use in energy-storage devices. They are environmentally friendly and economical and can be synthesized by the hydrothermal method, sol-gel, and coprecipitation. Compared with TMOs, they have low specific capacitance.[132,133] RE-based nanomaterial exhibits low conductivity, low SA, and poor cycle stability, limiting application in SCs. The conductivity of RE-based nanomaterials can be improved by modifying their shape and size. In addition, nanostructuring RE-based nanomaterials increase their surface-to-volume ratio to a large amount that makes more active sites during the chemical reaction. Moreover, by making a composite of TMOs with RE-based nanomaterials, the pseudocapacitance and cyclic performance of SCs can be enhanced.

For example, Wei et al. successfully prepared a porous $Co_3O_4-CeO_2$ hollow polyhedron, which displayed a high specific capacitance of 1288.3 F/g at 2.5 A/g with a capacitance loss of less than 3.3% after 6000 cycles.[134] The major drawback of RE-based nanomaterials is their poor conductivity, limiting their application in energy-storage devices.[135] Composites made with TMOs, doping, and porous materials can be improved.[136,137] Luo et al. prepared CeO_2/CNTs composites. CNTs can greatly improve the electrical conductivity of the composite so that the composite exhibits excellent electrochemical performance (818 F/g at 1 mV/s)[138] (see Fig. 6.15).

6.6.1 Graphene oxide decorated with Eu_2O_3 nanoparticles

Of RE metals recently studied for application in SCs, europium oxide (Eu_2O_3) has been studied as an electrode material for SCs. Europium oxide (Eu_2O_3) has exceptional optical properties as an electron acceptor and donor.[139] Europium oxide is known for its light-emissive and electrically insulating properties.[140] W. Sun and Z. Mo reported that PPY/graphene nanosheet/RE shows a specific capacitance of 238 F/g at a current density of 1 Ag^{-1}.[141] The polymeric nanocomposite of Eu_2O_3 has been reported to

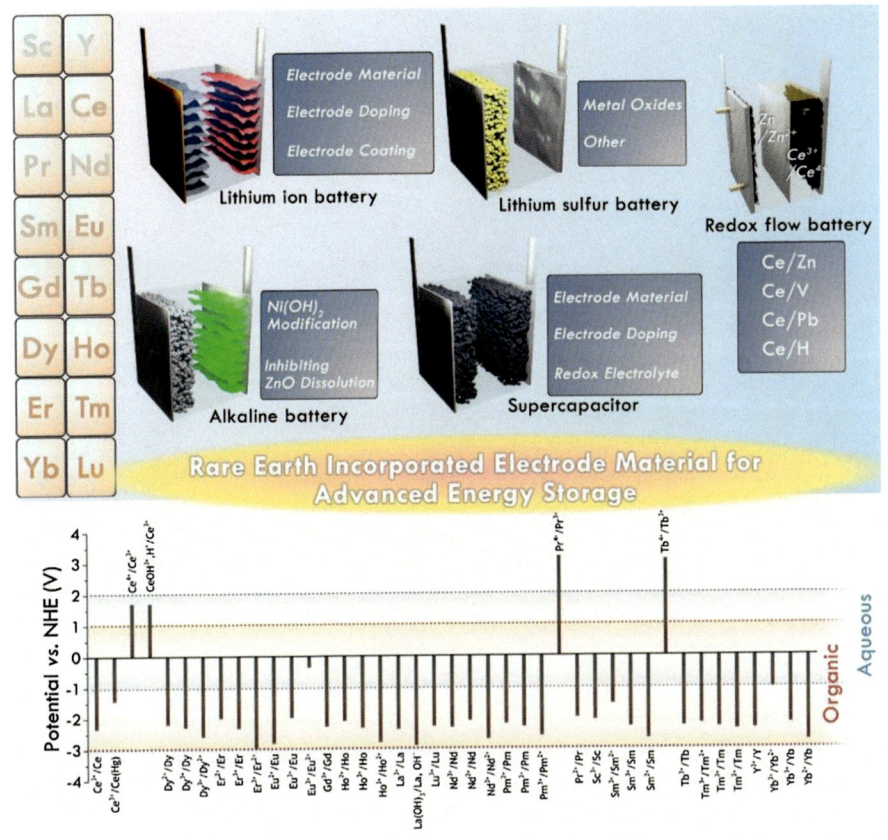

Figure 6.15 Schematic scheme for energy-storage devices in which rare-earth element-based electrodes used in supercapacitor and different energy-storage batteries. The orange range indicates the potential range of common organic electrolytes, and blue range indicates the common potential range of aqueous electrolytes.
Reused with permission Ref. 133.

promote electrochemical behavior. Europium hydroxide CNTs and europium oxide composites and their nanostructures have attracted considerable attention from researchers.[142,143] The composite of Eu_2O_3 and Cu is quite effective for electrochemical performance; Cu and Eu_2O_3 provide a large SA suitable for electron acceptors. Mandira Majumder et al.[144] reported a nanocomposite of PPY/Cu/Eu_2O_3 via in situ oxidative polymerizations as an active electrode material (see Fig. 6.16).

The pure form of polypyrrole has a spherical structure and incorporating of CuO in PPY results a sheet like morphology. Cu is uniformly distributed on the PPY matrix, enhancing porosity and forming a conductive network for charge transfer. This mesoporous conducting network enhances the probability of absorbing more ions from the electrolyte and increasing the specific capacitance value.[145] The Eu_2O_3 particles

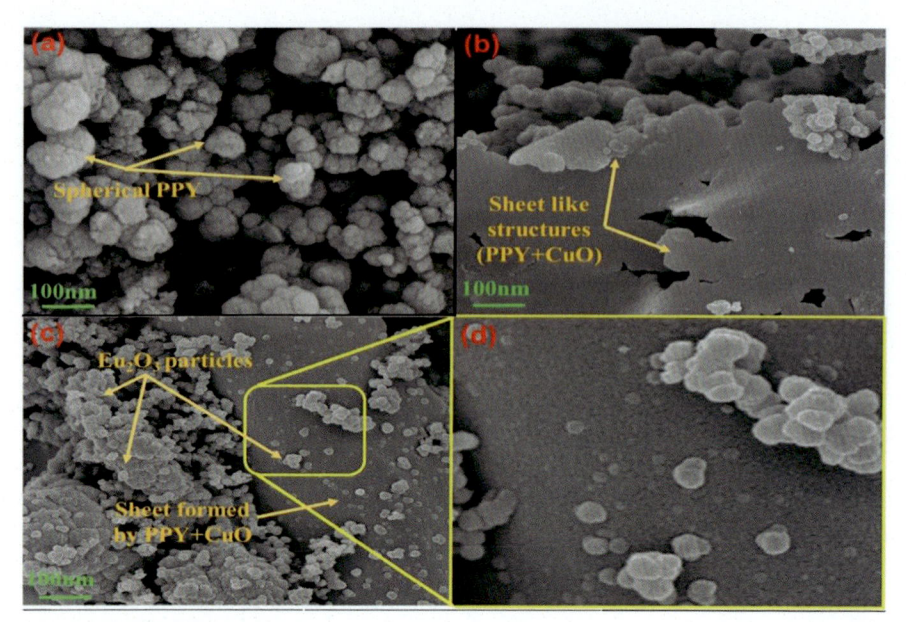

Figure 6.16 FE-SEM images of (A) pure PPY, (B) PPY/CuO, (C and D) PPY/CuO/Eu$_2$O$_3$-2 ternary nanocomposite.
Reused with permission Ref. 144.

distributed on the sheetlike PPY-CuO, spreading Eu$_2$O$_3$ nanoparticles amplified the specific SA of the composite.

6.6.2 Graphene/La$_2$O$_3$ nanocomposite as electrode material

SCs are classified as EDLC, pseudocapacitors, and hybrid SCs. In EDLCs, porous carbon material is used in electrodes, chemically electroactive material[146,147] is used in pseudocapacitors, and hybrid SC anode and cathode material is used for charge storage.[148−152] Graphene-based MOs and conducting polymers have gained considerable attention from researchers to develop high-performance electrode materials for SC.[153−156] In 2008, Stoller et al.[13] determined that graphene is a better candidate than CNT as an SC electrode material. They showed graphene-based electrode SC has higher specific capacitance compared with CNTs. They reported the specific capacitance of graphene-based SC was 135 F/g (in water electrolytes) and 99 F/g (in organic electrolytes), respectively. Some downsides of SCs include low capacitance of carbon material,[157] instability of conducting polymers, and high electrical resistance of TMOs.[158] In the past few years, graphene-based MO composites have appeared to be the best candidates for electrode materials with good electrochemical properties.[159,160]

RE MOs have many configurations and unique electrical, optical, and magnetic properties. Among them, lanthanum oxide is the most abundant product; it has good

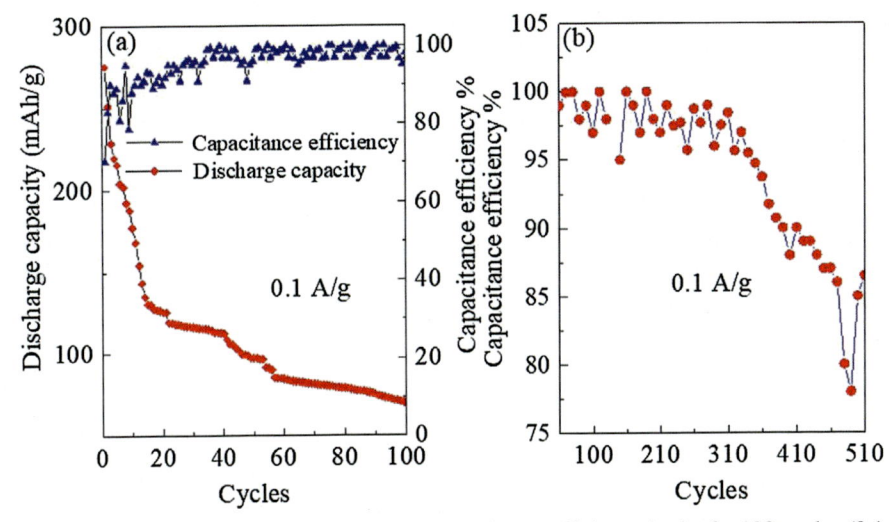

Figure 6.17 (A) The discharge capacity and capacitance efficiency in the frst100 cycles (0.1 A/g), and (B) the cycle stability of RGO/La$_2$O$_3$ at a current density of 0.1 A/g (500 cycles). Reprinted with permission Ref. 170.

physical and chemical properties. Lanthanum oxide (La$_2$O$_3$) has been extensively used in applications such as photocatalysis, protective coating layer, alloy, and gas sensor.[161–167] La$_2$O$_3$ has two different oxidation states, La$_2$+ and La$_3$+; both states can exist simultaneously during the charge–discharge process.[168,169]

A nanocomposite of graphene oxide and lanthanum oxide is used as electrode material in SCs. Jiaoxia Zhang et al.[170] synthesized a nanocomposite of reduced graphene/oxide and La$_2$O$_3$ for use as an SC electrode material and reported a specific capacitance of 156.25 F/g at a current density of 0.1 A/g. This exceptional electrochemical performance is due to lanthanum oxide nanoparticles spread on the surface of reduced graphene oxides (see Fig. 6.17).

Milon Miah et al.[136] analyzed the synthesis of La$_2$O$_3$ nanosheets decorated with graphene oxides to apply electrode material in SCs and reported a high capacitance value with good capacitance preservation. They reported the highest specific capacitance value as 751 F/g at 70°C at a current density of 1 A/g.

6.6.3 Gd$_2$O$_3$ nanostructure and its composite with electroactive polymer

Conducting polymers are used extensively in SCs on a large scale and are promising materials. The composite of conducting polymers with carbon nanostructures, graphene, and MO improved its properties.[171] The composite materials have better conductivity and cyclic ability, mechanical stability, specific capacitance, and processability.[172–174] RE-based MOs have received much interest from researchers owing to their fair electronic configuration, coordination chemistry, magnetic properties, and energy-storage characteristics. Majumder[175] and his research team synthesized an RE MO-incorporated polyindole nanocomposite for energy-storage applications. Especially, gadolinia (Gd_2O_3) is a versatile RE material/faradaic electrode material because of its most accessible layered structure, redox activity, good electrochemical properties, high temperature resistance, high photocatalytic activity/stability, and self-regeneration.[176−178]

Moreover, gadolinia can exist in two forms (monoclinic and base-centered cubic structures) with different interplanar spacing, favoring the electrochemical process. Xu et al.[179] studied the gadolinium-doped ternary composite for enhanced energy storage. Moreover, the gadolinia/polymer nanocomposite prepared by Shiri and Ehsani for SC application delivers a specific capacitance of 300 F/g.[180] A comparison table of RE-based metal composites, which shows the performance for SCs, shows that an RE-based composite material with cobalt has high capacitance compared with composites with carbon-based materials.

6.7 Synthesis methods and characteristics of metal oxide–carbon composites for supercapacitors

6.7.1 Synthesis methods of metal oxide–carbon composites for supercapacitors

Many methods are used to fabricate materials for SCs with various procedures, advantages, and disadvantages. Scientists are attempting to discover easy, green, environmentally friendly, economical, and facile methods with small routes for materials synthesis. Some methods used to synthesize SC materials in recent years are illustrated in Fig. 6.18.

There are some basics in every synthesis method which play a vital role in their efficiency. ECD is like an electrodeposition method in which a material coating is present on the electrode surface with the help of an electric current. The benefits of the ECD method include its mass production scalability, economical cost structure, use of earth-abundant material, appropriate control of features such as thickness, desirable surface uniformity of materials, and suitable deposition rate, all of which have been noticed by researchers.[181] In this ECD method, there is a need for a process setup and the required current or voltage for the synthesis process.

There is another technique for the fabrication of SC materials is CBD. In this method, the temperature is low, allowing comparatively nonexpensive materials for deposition with high SA substrate.[182] In the CBD technique, no current or voltage is needed for deposition material in the SC solution. Its low investment, ease of processing, ability to process without high temperatures, and high substrate area are some benefits of this method. Some disadvantages of CBD are its limited flexibility and low material yield for film construction on electrode material.[183]

In the chemical vapor deposition (CVD) technique, we keep the materials in suitable gas, thermal energy, and a reaction furnace tube. After the decomposition process,

Figure 6.18 Various synthesis methods for making metal oxide—carbon composites for supercapacitors.

we get the desired material on powder or thin film. This process is helpful to fabricate the types of materials that help in using SC devices.[184] Some advantages of CVD methods are their suitable and desired film uniformity and excellent material yield compared with the CBD method. But the problem is the expensive equipment used in this process, which indirectly bears the cost of the preparation material.

Sol-gel is another technique in which we obtained solid thin films. In this method, we obtained material by dip-coating surfaces in solution. The benefits of this technique are good materials composition and homogeneous surface with good structural properties and economics, but the process is not an easy one.[185]

In synthesis methods for SC materials, chemical participation is important for SCs. In this technique, materials are obtained from the solution after a chemical reaction. The two process methods are transforming the substance into an unsolvable form or making solvents with low solubility.[186] This method is efficient, easy to implement, and suitable for fabricating composite electrode materials. But in this technique, some waste products remained.

These synthesis methods discussed above are suitable, appropriate, and commonly used for materials fabrication. So when considering MO—C composites for SCs, these methods have obtained widespread attention from researchers. After using these techniques to fabricate MO—C composites, we achieved desirable properties. Tables 6.4 and 6.5 summarize various electrode materials-fabrication techniques in terms of materials, their benefits, and their drawbacks.

Table 6.4 Comparison of the performance of synthesized rare-earth-based metal oxide composites.

Rare-Earth-based metal oxide composite	Specific capacitance (F/g)	Current density (A/g)
CeO_2/Co_3O_4	1288.3	2.5
$CeO_2/CNTs$	818	—
La_2O_3/graphene oxides	751	1
Gd_2O_3/NiS_2 microsphere	354	0.5
PPY/graphene nanosheets	238	1

Table 6.5 Various electrode synthesis techniques in terms of materials, benefits, and drawbacks.[216]

Synthesis methods	Materials	Advantages	Disadvantages
Electrochemical deposition Method	• Metal oxides • Conducting polymers • Ruthenium oxide—carbon composites (electroless deposition)	Mass production; low costs; precise control on film thickness and uniformity	Process set up; current or voltage required
Chemical bath deposition (CBD)	• PANI • Ruthenium oxide	Simplicity; low temperature; inexpensive; large-area substrates.	Limited flexibility; low material yield
Chemical vapor deposition	• Carbon materials (CNTs, Nanofibers, graphene type, etc.)	High material yield compared with CBD; good film uniformity	Expensive equipment and relatively high costs
Sol-gel	• Carbon aerogels • Carbon-ruthenium aerogels • SnO_2, MnO_2 • Ruthenium oxide/ active carbon	Low costs; allow for control of the film texture, composition, homogeneity, and structural properties	Complicated process
Chemical precipitation	• MnO_2/nanofibers • MnO_2/carbon nanotubes • Nickel oxide/carbon	Good for composite electrode materials synthesis; efficient; easily implemented	May generate a waste product

Reused with permission Ref. 216.

6.7.2 Characteristics of metal oxide–carbon composites for supercapacitors

Tao et al.[187] successfully fabricated the nanocomposites of carbon with MoO_3 with a 1:1 weight ratio with the help of the ball-milling technique. After fabrication, the nanocomposites were examined by various characterization techniques such as TEM, SEM, XPS, EDS, and CV. The fabricated nanocomposite used as electrode material for SCs showed enhanced capacitance of an approximately 179 F/g at 50 mA/g charge–discharge rate, having remarkable cycling capability over 1000 cycles. Compared with bulk graphite and MoO_3, the synthesized nanocomposite showed improved electrochemical efficiency due to the unique morphology of nanocomposites. Furthermore, it is noted that there is no need for carbon black for the formation of electrodes (see Fig. 6.19).

Zhang et al.[188] synthesized the $NiMoO_4$/MWCNTs nanocomposites for improving the efficiency of SCs. They fabricated the $NiMoO_4$ electrode material using a facile solvent-thermal technique. The structure, geometry, and characteristics of synthesized material were controlled by the addition of rGO and MWCNTs. Characterizations of the synthesized composite were done with the help of SEM, CV, XRD, and EIS. The results revealed that adding rGO and MWCNTs remarkably affects the capacitance performance and morphology of $NiMoO_4$. $NiMoO4$/MWCNT composites exhibited improved supercapacitance activity due to their uniform particle size division, more active spaces, reduced charge movement, and higher ionic diffusion coefficient. This composite showed individual discharge capacitances of 702, 548, 805, and 643 F/g under current densities of 4, 10, 1, and 7 Ag^{-1}. Furthermore, capacity retention of the composite remained at 66.7% after 1000 cycles at a current of 1 Ag^{-1}, revealing enhanced charge–discharge activity for SC applications.

Mao et al. synthesized TMO–carbon-based SC material with the help of a hydrothermal treatment method for energy storage and reported poor conductivity and high stability. They reported that TMO stability could be improved by nanostructuring. They prepared Co–Mo–O–S nanospheres for SCs and characterized the material using XRD, SEM, TEM, XPS, FTIR, and EDS. This prepared material showed the high

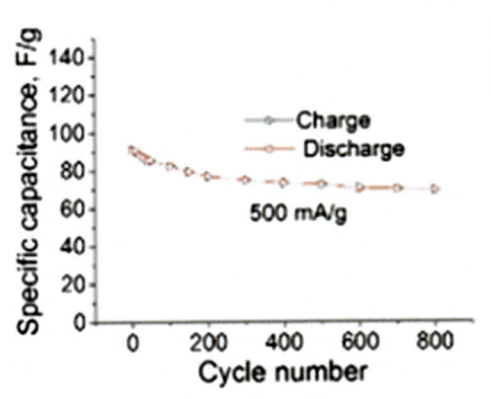

Figure 6.19 Charge–discharge cycle test of C/MoO_3 nanocomposite.
Reprinted with permission Ref. 187.

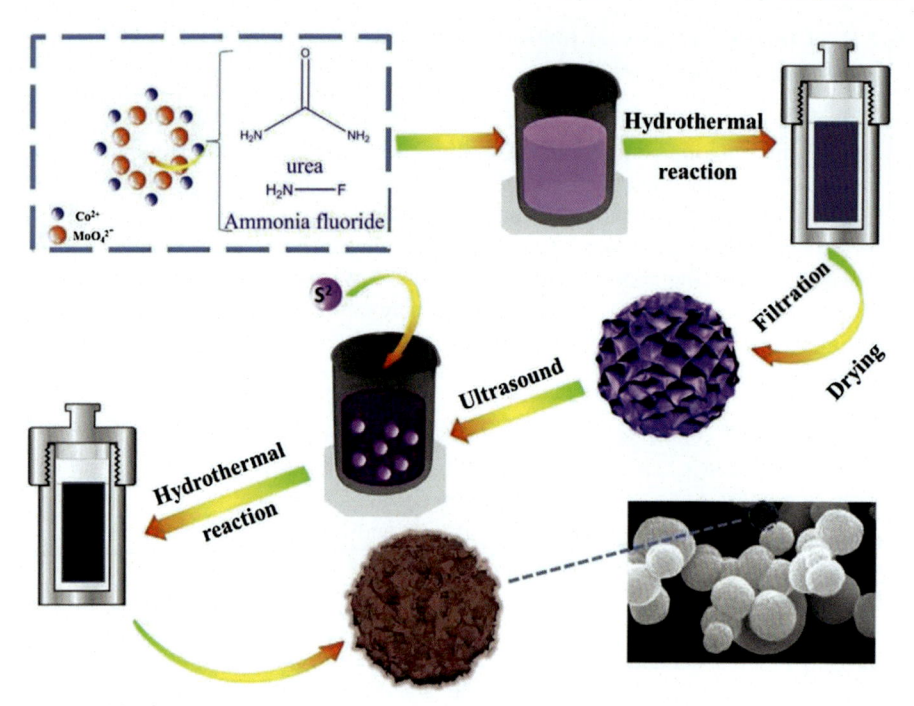

Figure 6.20 Preparation process of Co—Mo—O—S.
Reused with permission Ref. 189.

capacitance of 1134 F/g at A/g with a large capability rate. This prepared hybrid SCs showed a high energy density of 67.6 Wh/kg than reduced graphene oxide material. Thus, this material is widely used for energy-storage devices.[189] (see Fig. 6.20).

6.8 Challenges and future perspectives of metal oxide—carbon composites

6.8.1 Limitations and challenges of metal oxide—carbon composites

SC performance depends on capacitance, power and energy density, and stability and capability rates.[190] To attain high SC energy density, escalated capacitance, a larger voltage window, and reduced resistance (R) are needed. In SCs, especially EDLCs, the max operating V window (V_m) relies on an electrolytic solution that is restricted by the electrolytic solution's stability. So there is a need to develop such types of electrolytes with high V_m.[191] To achieve better SC results, it is also necessary to reduce the problems of electrode materials. In carbon-based commercial SCs, the energy density is typically 3—5 W h/Kg, which is not enough and is lower than that of EC batteries (for example, 10—250 W h/Kg for lithium-ion batteries).[192] This deficiency is not good for the commercial use of SCs.

TMOs have been investigated in recent years as substitutes for SC electrodes to increase energy density and capacitance. Some have shown excellent performance for SC capacitance. For defects, functional groups can act as redox centers on electrode surfaces and are also important for storage. Several papers have been published on MOs in recent years and have indicated good capacitance and energy density results compared with carbon electrodes. But there are some problems when using MO alone; that's why there is a need to overcome these drawbacks.

The electronic conductivity of mostly MO is low except RuO_2 and resistivity of MO create problem in charge transformation and becomes a reason for high IR loss at large current densities. The capability rate and power density are not so good, which can create problems in practical applications. In MO electrodes, the charging and discharging procedures are the reasons to produce strain and may cause the breaking of electrodes. This thing is not good for the long stability of electrodes. The porosity, large SA, and distribution of pores are difficult to adapt in MO. So there is a need to fabricate the MO—C composites for SCs electrodes and overcome hinders in both components. In MO—C composites, carbon not only provides physical support to MO but also supports in-charge transformation and the conductivity of charges, porosity, SA, and pore distribution effects on the efficiency of electrode performance.[72]

The SC is an energy-storage device with many applications in the industrial and commercial sectors but still has challenges related to its cycle life, energy density, and energy-storage capacity.[68] It faces other problems such as restricted faradaic reactions of functional groups during longer SC cycle lives. For this, carbon-based materials, e.g., CNTs, ACs, and hybrids, are helpful for better SC results. So there are some challenges and limitations while using C—Mo composite, and there is still a need to work on it. These problems can be resolved by taking some steps to improve the abilities of SCs.

6.8.2 Future perspectives

In MO—C composites for SC electrodes, MO provides high capacitance and large energy density, while carbon ensures attractive power capability at high current levels. On various dimensions, carbon nanostructures have unique properties that attract additional research for more advances and the attainment of desirable MO—C composite electrode properties. In 2010, the global market for SCs was $470 million.[193] Demand for SCs has continued to increase, with market sales expanding to $1.2 billion in 2015. The sales market for SCs was expected to reach $3.5 billion in 2020.[194]

The increasing market demand for SCs motivates researchers to further investigate and improve SCs. It is necessary to make advances in SC properties such as power capability, capacitance ability, energy density, cyclic stability, efficiency, and cost to meet the commercial requirements for SCs. It is still challenging to overcome the hurdles of a limited capability rate in MO-C composite electrodes. In the future, some important factors should be researched to improve the efficiency of MO−C composite for the electrode of SCs, which are given below.

1. Improving carbon nanostructures properties such as shape, alignment, and porosity: The carbon nanostructure helps MO physically, and its structure provides an idea about the overall structure of the MO−C composite. A large carbon SA helps improve MO loading and attain a high SA for the carbon−metal interface. Material porosity and small dimensions also play important roles in decreasing the ionic diffusion distance and the electrode surface.
2. Increase ability in carbon nanostructures for electronic conductivity: The conductivity of electrodes is necessary for the power capability rate in SCs. Generically MOs have less conductivity for charges, so carbon nanostructure is a good option to increase conductivity. The resistance in the flow of charges can be reduced by raising the connection between the carbon nanostructure and the current collector.
3. Extending the SA of C-oxide and managing the thickness of MO: The MO−C SA plays an important part in transforming charges toward the electrode. The outer layer of MO for oxidation and reduction reactions in carbon nanostructures should be thin for the maximum consumption of materials used in electronic conductivity. For MO, the distance of electron diffusion must be analogous with its thickness, in which case we can attain maximum electron transport from redox reactions to the collector.
4. Making hybrid SCs: Commonly used SCs have less energy density, so there is a need to develop hybrid SCs with greater energy density than common SCs. The voltage window and nonaqueous electrolyte are also important for energy density, so MO−C composite electrode materials may be used to develop new hybrid SCs with higher voltage windows.
5. Discovering novel and suitable materials for SCs: It is necessary to give attention to recording the demand and use of such raw materials to reduce their environmental effect.
   So we must find out such materials which are green, environmentally friendly, renewable, present in large amounts on earth and nontoxic.[72]
6. Launching testing principles: There is a need to launch the standard rules for testing the SCs and compare efficiency data from various laboratories.[195] In previously published research, capacitance data were recorded only for the MO component, but there is a need for complete performance data about composite electrode materials in SCs.

So if we take these steps, we can improve MO−C composite performance and achieve desirable SC properties. Further research is needed to find new routes of fabrication and green, renewable, environmentally friendly materials. So in future SC research, the MO−C composite is a good option as an electrode material to increase specific capacitance and provide good power capability.

6.9 Conclusion

SCs have been highly researched as next-generation energy-storage devices that can meet the demand for high energy and power density. High specific capacitance, good stability, and low internal resistance are important parameters to consider when selecting high-performance electrode materials for SCs. This chapter reviewed the fabrication and use of various electrode materials for SCs, i.e., carbon-based electrode materials, TMO−carbon composites, and RE MO−C composites. We discussed the characterization and preparation techniques of MO—C composites for SCs and explored the many characteristics exhibited by MO—C composites: intertwined porous structure, excellent wettability against the electrolyte, elevated effective high SA, and the inclusion of electrochemically active surface functionalities of carbon to boost electrochemical double-layer capacitance. MOs also increase electrochemical efficiency with the help of a faradaic charge/discharge mechanism. Our analysis indicated that MO—C composites mostly present high stability, increased mechanical flexibility, extraordinary energy density, and high specific capacitance. Finally, we discussed the challenges we faced in using MO—C composites as SC electrodes and presented some future perspectives in this research area.

Acknowledgments

Financial support was provided by the National Natural Science Foundation of China (51602281), Yangzhou University High-end Talent Support Program, and the "Qinglan Project" of Jiangsu Universities.

References

1. Becker HI, Vischers Ferry NY. *Low voltage electrolytic capacitor, assignor to General Electric Company, a corporation of New York Application April 14, 1954, Serial No. 423,042.* Google patents; 1957.
2. Andrieu X. *Ultracapacitors for portable electronics. New trends in electrochemical technology,* 1; 2000. p. 521—47.
3. Jayalakshmi M, Balasubramanian K. Simple capacitors to supercapacitors-an overview. *Int J Electrochem Sci* 2008;**3**(11):1196—217.
4. González A, et al. Review on supercapacitors: technologies and materials. *Renewable and sustainable energy reviews* 2016;**58**:1189—206.
5. Conway BE. *Electrochemical supercapacitors: scientific fundamentals and technological applications.* Springer Science & Business MEDIA; 2013. p. 698. ISBN: 1475730586.
6. Burke A. Ultracapacitors: why, how, and where is the technology. *J Power Sources* 2000;**91**:37—50.
7. Kötz R, Carlen M. Principles and applications of electrochemical capacitors. *Electrochim Acta* 2000;**45**:2483—98.
8. Allen JB, Larry RF. *Department of Chemistry and Biochemistry, University of Texas at Austin, Electrochemical methods fundamentals and applications.* John Wiley & Sons Inc.; 2001. p. 864. ISBN: 978-0-471-04372-0.
9. Frackowiak E, Beguin F. Carbon materials for the electrochemical storage of energy in capacitors. *Carbon* 2001;**39**:937—50.
10. Mastragostino M, Arbizzani C, Soavi F. Conducting polymers as electrode materials in supercapacitors. *Solid State Ionics* 2002;**148**:493—8.
11. Schneuwly A, Gallay R. *Properties and applications of supercapacitors from the state-of-the-art to future trends.* 2000.
12. Birss VI, Conway B, Wojtowicz J. *The role and utilization of pseudocapacitance for energy storage by supercapacitors.* 1997.

13. Stoller MD, et al. Graphene-based ultracapacitors. *Nano Lett* 2008;**8**(10):3498−502.
14. Lamberti A, et al. Facile fabrication of cuprous oxide nanocomposite anode films for flexible Li-ion batteries via thermal oxidation. *Electrochim Acta* 2012;**86**:323−9.
15. Wang G, Zhang L, Zhang J. A review of electrode materials for electrochemical supercapacitors. *Chem Soc Rev* 2012;**41**(2):797−828.
16. Simon P, Gogotsi Y. Materials for electrochemical capacitors. *Nanosci Technol* 2010:320−9.
17. Yu G, et al. Hybrid nanostructured materials for high-performance electrochemical capacitors. *Nano Energy* 2013;**2**(2):213−34.
18. Chuang C-M, et al. Effects of carbon nanotube grafting on the performance of electric double layer capacitors. *Energy Fuels* 2010;**24**(12):6476−82.
19. Long JW, et al. Asymmetric electrochemical capacitors—stretching the limits of aqueous electrolytes. *MRS Bull* 2011;**36**(7):513−22.
20. Wei W, et al. Manganese oxide-based materials as electrochemical supercapacitor electrodes. *Chem Soc Rev* 2011;**40**(3):1697−721.
21. Halper MS, Ellenbogen JC. *Supercapacitors: a brief overview. The MITRE Corporation.* Virginia, USA: McLean; 2006. p. 1−34.
22. Li H, Cheng L, Xia Y. A hybrid electrochemical supercapacitor based on a 5 V Li-ion battery cathode and active carbon. *Electrochem Solid State Lett* 2005;**8**(9):A433.
23. Chen PC, et al. Preparation and characterization of flexible asymmetric supercapacitors based on transition-metal-oxide nanowire/single-walled carbon nanotube hybrid thin-film electrodes. *ACS Nano* 2010;**4**(8):4403−11.
24. Patil SS, et al. Ag: BiVO 4 dendritic hybrid-architecture for high energy density symmetric supercapacitors. *J Mater Chem A* 2016;**4**(20):7580−4.
25. Simon P, Gogotsi Y. Materials for electrochemical capacitors. *Nat Mater* 2008;**7**(11):845−54.
26. Arbizzani C, Mastragostino M, Soavi F. New trends in electrochemical supercapacitors. *J Power Sources* 2001;**100**(1−2):164−70.
27. Laforgue A, et al. Activated carbon/conducting polymer hybrid supercapacitors. *Journal of the Electrochemical Society* 2003;**150**(5):A645.
28. Shabangoli Y, et al. Thionine functionalized 3D graphene aerogel: combining simplicity and efficiency in fabrication of a metal-free redox supercapacitor. *Adv Energy Mater* 2018;**8**(34):1802869.
29. Wu C, et al. Two-dimensional vanadyl phosphate ultrathin nanosheets for high energy density and flexible pseudocapacitors. *Nat Commun* 2013;**4**(1):1−7.
30. Huang Z-H, et al. Ordered polypyrrole nanowire arrays grown on a carbon cloth substrate for a high-performance pseudocapacitor electrode. *ACS Appl Mater Interfaces* 2015;**7**(45):25506−13.
31. Akinwolemiwa B, et al. Optimal utilization of combined double layer and Nernstian charging of activated carbon electrodes in aqueous halide supercapattery through capacitance unequalization. *J Electrochem Soc* 2018;**165**(16):A4067.
32. Akinwolemiwa B, Peng C, Chen GZ. Redox electrolytes in supercapacitors. *Journal of the electrochemical society* 2015;**162**(5):A5054.
33. Shimizu W, et al. Development of a 4.2 V aqueous hybrid electrochemical capacitor based on MnO2 positive and protected Li negative electrodes. *J Power Sources* 2013;**241**:572−7.
34. Akinwolemiwa B, Chen GZ. Fundamental consideration for electrochemical engineering of supercapattery. *Journal of the Brazilian Chemical Society* 2018;**29**(5):960−72.

35. Guan L, Yu L, Chen GZ. Capacitive and non-capacitive faradaic charge storage. *Electrochim Acta* 2016;**206**:464—78.
36. Yu L, Chen GZ. High energy supercapattery with an ionic liquid solution of LiClO4. *Faraday Discuss* 2016;**190**(0):231—40.
37. Chae J, Ng K, Chen G. Nanostructured materials for the construction of asymmetrical supercapacitors. *Proc IME J Power Energy* 2010;**224**(4):479—503.
38. Yu L, Chen GZ. Supercapatteries as high-performance electrochemical energy-storage devices. *Electrochem Energy Rev* 2020:1—15.
39. Cherusseri J, Sharma R, Kar KK. Helically coiled carbon nanotube electrodes for flexible supercapacitors. *Carbon* 2016;**105**:113—25.
40. Chen X, Paul R, Dai L. Carbon-based supercapacitors for efficient energy storage. *Natl Sci Rev* 2017;**4**(3):453—89.
41. Dai L, et al. Carbon nanomaterials for advanced energy conversion and storage. *Small* 2012;**8**(8):1130—66.
42. Ebbesen T, et al. Electrical conductivity of individual carbon nanotubes. *Nature* 1996;**382**(6586):54—6.
43. Graham A, et al. How do carbon nanotubes fit into the semiconductor roadmap? *Appl Phys A* 2005;**80**(6):1141—51.
44. Yu M-F, et al. Tensile loading of ropes of single wall carbon nanotubes and their mechanical properties. *Phys Rev Lett* 2000;**84**(24):5552.
45. Balandin AA. Thermal properties of graphene and nanostructured carbon materials. *Nat Mater* 2011;**10**(8):569—81.
46. Geim AK, Novoselov KS. The rise of graphene. In: *Nanoscience and technology: a collection of reviews from nature journals*. World Scientific; 2010. p. 11—9.
47. Novoselov KS, et al. Electric field effect in atomically thin carbon films. *Science* 2004;**306**(5696):666—9.
48. Park S, Ruoff RS. Chemical methods for the production of graphenes. *Nat Nanotechnol* 2009;**4**(4):217.
49. Chen Y, et al. High performance supercapacitors based on reduced graphene oxide in aqueous and ionic liquid electrolytes. *Carbon* 2011;**49**(2):573—80.
50. Hao L, et al. Structural evolution of 2D microporous covalent triazine-based framework toward the study of high-performance supercapacitors. *J Am Chem Soc* 2015;**137**(1):219—25.
51. Zhang L, Shi G. Preparation of highly conductive graphene hydrogels for fabricating supercapacitors with high rate capability. *J Phys Chem C* 2011;**115**(34):17206—12.
52. Bai X, et al. In situ polymerization and characterization of grafted poly (3, 4-ethylenedioxythiophene)/multiwalled carbon nanotubes composite with high electrochemical performances. *Electrochim Acta* 2013;**87**:394—400.
53. Wang G, et al. Improving the specific capacitance of carbon nanotubes-based supercapacitors by combining introducing functional groups on carbon nanotubes with using redox-active electrolyte. *Electrochim Acta* 2014;**115**:183—8.
54. Chmiola J, et al. Anomalous increase in carbon capacitance at pore sizes less than 1 nanometer. *Science* 2006;**313**(5794):1760—3.
55. Fernández J, et al. Performance of mesoporous carbons derived from poly (vinyl alcohol) in electrochemical capacitors. *J Power Sources* 2008;**175**(1):675—9.
56. Inagaki M, Konno H, Tanaike O. Carbon materials for electrochemical capacitors. *J Power Sources* 2010;**195**(24):7880—903.
57. Yan J, et al. Electrochemical properties of graphene nanosheet/carbon black composites as electrodes for supercapacitors. *Carbon* 2010;**48**(6):1731—7.

58. Yang Z, et al. Carbon nanotubes bridged with graphene nanoribbons and their use in high-efficiency dye-sensitized solar cells. *Angew Chem Int Ed* 2013;**52**(14):3996–9.

59. Zhou W, et al. One-step synthesis of Ni 3 S 2 nanorod@ Ni (OH) 2 nanosheet core–shell nanostructures on a three-dimensional graphene network for high-performance super-capacitors. *Energy Environ Sci* 2013;**6**(7):2216–21.

60. Péan C, et al. On the dynamics of charging in nanoporous carbon-based supercapacitors. *ACS Nano* 2014;**8**(2):1576–83.

61. Noked M, et al. Composite carbon nanotube/carbon electrodes for electrical double-layer super capacitors. *Angew Chem Int Ed* 2012;**51**(7):1568–71.

62. Mazurenko I, et al. Macroporous carbon nanotube-carbon composite electrodes. *Carbon* 2016;**109**:106–16.

63. Chen H, et al. Nitrogen doping effects on the physical and chemical properties of meso-porous carbons. *J Phys Chem C* 2013;**117**(16):8318–28.

64. Wang X, et al. Nitrogen-enriched ordered mesoporous carbons through direct pyrolysis in ammonia with enhanced capacitive performance. *J Mater Chem A* 2013;**1**(27):7920–6.

65. Wu D, et al. Templated synthesis of nitrogen-enriched nanoporous carbon materials from porogenic organic precursors prepared by ATRP. *Angew Chem Int Ed* 2014;**53**(15): 3957–60.

66. Padhi AK, Nanjundaswamy KS, Goodenough JB. Phospho-olivines as positive-electrode materials for rechargeable lithium batteries. *J Electrochem Soc* 1997;**144**(4):1188–94.

67. Zhang S, Chen GZ. Manganese oxide based materials for supercapacitors. *Energy Mater* 2008;**3**(3):186–200.

68. Zhang LL, Zhao X. Carbon-based materials as supercapacitor electrodes. *Chem Soc Rev* 2009;**38**(9):2520–31.

69. Sharma P, Bhatti T. A review on electrochemical double-layer capacitors. *Energy conversion and management* 2010;**51**(12):2901–12.

70. Babakhani B, Ivey DG. Improved capacitive behavior of electrochemically synthesized Mn oxide/PEDOT electrodes utilized as electrochemical capacitors. *Electrochim Acta* 2010;**55**(12):4014–24.

71. Conway B. Electrochemical capacitors based on pseudocapacitance. In: *Electrochemical supercapacitors*. Springer; 1999. p. 221–57.

72. Zhi M, et al. Nanostructured carbon–metal oxide composite electrodes for super-capacitors: a review. *Nanoscale* 2013;**5**(1):72–88.

73. Gao Y, et al. Electrochemical capacitance of Co_3O_4 nanowire arrays supported on nickel foam. *J Power Sources* 2010;**195**(6):1757–60.

74. Yan X, et al. Rational synthesis of hierarchically porous NiO hollow spheres and their supercapacitor application. *Mater Lett* 2013;**95**:1–4.

75. Moosavifard SE, et al. Designing 3D highly ordered nanoporous CuO electrodes for high-performance asymmetric supercapacitors. *ACS Appl Mater Interfaces* 2015;**7**(8):4851–60.

76. Wang Y, et al. Synthesis of 3D-nanonet hollow structured Co_3O_4 for high capacity supercapacitor. *ACS Appl Mater Interfaces* 2014;**6**(9):6739–47.

77. Lu Z, et al. Stable ultrahigh specific capacitance of NiO nanorod arrays. *Nano Res* 2011; **4**(7):658.

78. Huang M, et al. Self-assembly of mesoporous nanotubes assembled from interwoven ul-trathin birnessite-type MnO 2 nanosheets for asymmetric supercapacitors. *Sci Rep* 2014;**4**: 3878.

79. Du H, et al. Facile carbonaceous microsphere templated synthesis of Co 3 O 4 hollow spheres and their electrochemical performance in supercapacitors. *Nano Res* 2013;**6**(2): 87–98.

80. Xing W, et al. Synthesis and electrochemical properties of mesoporous nickel oxide. *J Power Sources* 2004;**134**(2):324−30.
81. Yu C, et al. A simple template-free strategy to synthesize nanoporous manganese and nickel oxides with narrow pore size distribution, and their electrochemical properties. *Adv Funct Mater* 2008;**18**(10):1544−54.
82. Purushothaman KK, et al. Nanosheet-assembled NiO microstructures for high-performance supercapacitors. *ACS Appl Mater Interfaces* 2013;**5**(21):10767−73.
83. Xia X-h, et al. Self-supported hydrothermal synthesized hollow Co_3O_4 nanowire arrays with high supercapacitor capacitance. *J Mater Chem* 2011;**21**(25):9319−25.
84. Fan Z, et al. Thin metal nanostructures: synthesis, properties and applications. *Chem Sci* 2015;**6**(1):95−111.
85. Simon P, Gogotsi Y, Dunn BJS. Where do batteries end and supercapacitors begin? *Science* 2014;**343**(6176):1210−1.
86. Wang T, et al. Boosting the cycling stability of transition metal compounds-based supercapacitors. *Energy Storage Mater* 2019;**16**:545−73.
87. Liu H, et al. *Transition metal based battery-type electrodes in hybrid supercapacitors: a review.* 2020.
88. Yang G, Park S-JJEA. Facile hydrothermal synthesis of $NiCo_2O_4$-decorated filter carbon as electrodes for high performance asymmetric supercapacitors. *Electrochim Acta* 2018;**285**:405−14.
89. Ahmad N, et al. Structural, morphological, and electrochemical performance of CeO_2/NiO nanocomposite for supercapacitor applications. *Appl Sci* 2021;**11**(1):411.
90. Zhan C, et al. Dual-ion hybrid supercapacitor: integration of Li-ion hybrid supercapacitor and dual-ion battery realized by porous graphitic carbon. *J Energy Chem* 2020;**42**:180−4.
91. Kshetri T, et al. Ternary graphene-carbon nanofibers-carbon nanotubes structure for hybrid supercapacitor. *Chem Eng J* 2020;**380**:122543.
92. Fan H, et al. Highly conductive $KNiF_3$@ carbon nanotubes composite materials with cross-linked structure for high performance supercapacitor. *J Power Sources* 2020;**474**:228603.
93. Cheng F, et al. Boosting the supercapacitor performances of activated carbon with carbon nanomaterials. *J Power Sources* 2020;**450**:227678.
94. Wei H, et al. Advanced porous hierarchical activated carbon derived from agricultural wastes toward high performance supercapacitors. *J Alloys Compd* 2020;**820**:153111.
95. Xiao J, et al. Electrolyte gating in graphene-based supercapacitors and its use for probing nanoconfined charging dynamics. *Nat Nanotech* 2020;**15**(8):683−9.
96. Fu M, et al. In situ growth of manganese ferrite nanorods on graphene for supercapacitors. *Ceramics Int* 2020;**46**(18):28200−5.
97. Liu Y, et al. Advanced supercapacitors based on porous hollow carbon nanofiber electrodes with high specific capacitance and large energy density. *ACS Appl Mater Interfaces* 2020;**12**(4):4777−86. https://doi.org/10.1021/acsami.9b19977. PMID: 31898452.
98. Zhang X, et al. Conjugated polyimide-coated carbon nanofiber aerogels in a redox electrolyte for binder-free supercapacitors. *Chem Eng J* 2020;**401**:126031.
99. Misnon II, et al. Conversion of oil palm kernel shell biomass to activated carbon for supercapacitor electrode application. *Waste Biomass Valorization* 2019;**10**(6):1731−40.
100. Prataap RV, et al. Effect of electrodeposition modes on ruthenium oxide electrodes for supercapacitors. *Curr Appl Phys* 2018;**18**(10):1143−8.
101. Liu N-L, et al. Enhanced desalination of electrospun activated carbon fibers with controlled pore structures in the electrosorption process. *Environ Sci* 2020;**6**(2):312−20.

102. Prauchner MJ, Rodríguez-Reinoso FJM. Chemical versus physical activation of coconut shell: a comparative study. *Microporous Mesoporous Mater* 2012;**152**:163−71.

103. Govindan B, et al. Activated carbon derived from phoenix dactylifera (palm tree) and decorated with MnO_2 nanoparticles for enhanced hybrid capacitive deionization electrodes. *Chemistry* 2020;**5**(11):3248−56.

104. Zhang C, et al. Stamping of flexible, coplanar micro-supercapacitors using MXene inks. *Adv Funct Mater* 2018;**28**(9):1705506.

105. Tale B, et al. *Graphene based nano-composites for efficient energy conversion and storage in solar cells and supercapacitors: a review*. 2021. p. 1−14.

106. Trasatti S, Buzzanca GJJoec. Ruthenium dioxide: a new interesting electrode material. Solid state structure and electrochemical behaviour. *J Electroanal Chem Interfacial Electrochem* 1971;**29**(2):A1−5.

107. Zhang C, et al. Highly flexible and transparent solid-state supercapacitors based on RuO2/PEDOT:PSS conductive ultrathin films. *Nano Energy* 2016;**28**:495−505.

108. Zhu D, et al. *Two-step preparation of carbon nanotubes/RuO 2/polyindole ternary nanocomposites and their application as high-performance supercapacitors*. 2020. p. 1−10.

109. Wang W, et al. Hydrous ruthenium oxide nanoparticles anchored to graphene and carbon nanotube hybrid foam for supercapacitors. *Sci Rep* 2014;**4**:4452.

110. Wu ZS, et al. Anchoring hydrous RuO_2 on graphene sheets for high-performance electrochemical capacitors. *Adv Funct Mater* 2010;**20**(20):3595−602.

111. Messaoudi B, et al. Anodic behaviour of manganese in alkaline medium. *Electrochim Acta* 2001;**46**(16):2487−98.

112. Rauda IE, et al. Nanostructured pseudocapacitors based on atomic layer deposition of V_2O_5 onto conductive nanocrystal-based mesoporous ITO scaffolds. *Adv Funct Mater* 2014;**24**(42):6717−28.

113. Wang G, et al. LiCl/PVA gel electrolyte stabilizes vanadium oxide nanowire electrodes for pseudocapacitors. *ACS Nano* 2012;**6**(11):10296−302.

114. Badot J, et al. Atomic layer epitaxy of vanadium oxide thin films and electrochemical behavior in presence of lithium ions. *Electrochem Solid State Lett* 2000;**3**(10):485.

115. Qu Q, et al. Core−shell structure of polypyrrole grown on V_2O_5 nanoribbon as high performance anode material for supercapacitors. *Adv Energy Mater* 2012;**2**(8):950−5.

116. Hu L, Cui Y. Energy and environmental nanotechnology in conductive paper and textiles. *Energy Environ Sci* 2012;**5**(4):6423−35.

117. Saravanakumar B, Purushothaman KK, Muralidharan G. *V2O5/functionalized MWCNT hybrid nanocomposite: the fabrication and its enhanced supercapacitive performance*. 2014.

118. Wang Q, et al. High-performance supercapacitor based on V_2O_5/carbon nanotubes-super activated carbon ternary composite. *Ceram Int* 2016;**42**(10):12129−35.

119. Wang Y-G, Wang Z-D, Xia Y-Y. An asymmetric supercapacitor using RuO2/TiO2 nanotube composite and activated carbon electrodes. *Electrochim Acta* 2005;**50**(28):5641−6.

120. Cottineau T, et al. Nanostructured transition metal oxides for aqueous hybrid electrochemical supercapacitors. *Appl Phys A* 2006;**82**(4):599−606.

121. Thakur AK, Choudhary RB. High-performance supercapacitors based on polymeric binary composites of polythiophene (PTP)−titanium dioxide (TiO_2). *Synth Met* 2016;**220**:25−33.

122. Kim Y-T, Tadai K, Mitani T. Highly dispersed ruthenium oxide nanoparticles on carboxylated carbon nanotubes for supercapacitor electrode materials. *J Mater Chem* 2005;**15**(46):4914−21.

123. Shaikh J, et al. Supercapacitor behavior of CuO−PAA hybrid films: effect of PAA concentration. *J Alloys Compd* 2011;**509**(25):7168−74.

124. Fischer AE, et al. Incorporation of homogeneous, nanoscale MnO_2 within ultraporous carbon structures via self-limiting electroless deposition: implications for electrochemical capacitors. *Nano Lett* 2007;**7**(2):281−6.

125. Krishnamoorthy K, Kim S-J. Growth, characterization and electrochemical properties of hierarchical CuO nanostructures for supercapacitor applications. *Mater Res Bull* 2013; **48**(9):3136−9.

126. Peng S, et al. Bacterial cellulose membranes coated by polypyrrole/copper oxide as flexible supercapacitor electrodes. *J Mater Sci* 2017;**52**(4):1930−42.

127. Chen K, Xue D. Colloidal supercapacitor electrode materials. *Mater Res Bull* 2016;**83**: 201−6.

128. Henao J, Martinez-Gomez L. On rare-earth perovskite-type negative electrodes in nickel−hydride (Ni/H) secondary batteries. *Mater Renew Sustain Energy* 2017;**6**(2):7.

129. Xu J, et al. Ultrathin 2D rare-earth nanomaterials: compositions, syntheses, and applications. *Adv Mater* 2020;**32**(3):1806461.

130. Li X, et al. Preparation and electrochemical properties of RuO_2/polyaniline electrodes for supercapacitors. *Synth Met* 2012;**162**(11−12):953−7.

131. Gao W, et al. Incorporation of rare earth elements with transition metal−based materials for electrocatalysis: a review for recent progress. *Mater Today Chem* 2019;**12**:266−81.

132. Chen K, Xue D. Formation of electroactive colloids via in situ coprecipitation under electric field: erbium chloride alkaline aqueous pseudocapacitor. *J Colloid Interface Sci* 2014;**430**:265−71.

133. Zhao H, et al. Rare earth incorporated electrode materials for advanced energy storage. *Coord Chem Rev* 2019;**390**:32−49.

134. Wei C, et al. Self-template synthesis of hybrid porous Co_3O_4−CeO_2 hollow polyhedrons for high-performance Supercapacitors. *Chem Asian J* 2018;**13**(1):111−7.

135. Wen H, et al. Synthesis and electrochemical properties of CeO_2 nanoparticle modified TiO_2 nanotube arrays. *Electrochim Acta* 2011;**56**(7):2914−8.

136. Miah M, et al. Temperature dependent supercapacitive performance in La_2O_3 nano sheet decorated reduce graphene oxide. *Electrochim Acta* 2018;**260**:449−58.

137. Nallappan M, Gopalan M. Fabrication of CeO_2/PANI composites for high energy density supercapacitors. *Mater Res Bull* 2018;**106**:357−64.

138. Luo Y, et al. CeO_2/CNTs hybrid with high performance as electrode materials for supercapacitor. *J Alloys Compd* 2017;**729**:64−70.

139. Shiri HM, Ehsani A. A simple and innovative route to electrosynthesis of Eu_2O_3 nanoparticles and its nanocomposite with p-type conductive polymer: characterisation and electrochemical properties. *J Colloid Interface Sci* 2016;**473**:126−31.

140. Singh M, et al. Microstructure, crystallinity, and properties of low-pressure MOCVD-grown europium oxide films. *Mater Chem Phys* 2008;**110**(2−3):337−43.

141. Sun W, Mo Z. PPy/graphene nanosheets/rare earth ions: a new composite electrode material for supercapacitor. *Mater Sci Eng, B* 2013;**178**(8):527−32.

142. Chen C, Wang X, Nagatsu M. Europium adsorption on multiwall carbon nanotube/iron oxide magnetic composite in the presence of polyacrylic acid. *Environ Sci Technol* 2009; **43**(7):2362−7.

143. Yang H, et al. Synthesis and strong red photoluminescence of europium oxide nanotubes and nanowires using carbon nanotubes as templates. *Acta Mater* 2008;**56**(5):955–67.

144. Majumder M, et al. Impact of rare-earth metal oxide (Eu 2 O 3) on the electrochemical properties of a polypyrrole/CuO polymeric composite for supercapacitor applications. *RSC Adv* 2017;**7**(32):20037–48.

145. Ng LY, et al. Polymeric membranes incorporated with metal/metal oxide nanoparticles: a comprehensive review. *Desalination* 2013;**308**:15–33.

146. Mao S, Lu G, Chen J. Three-dimensional graphene-based composites for energy applications. *Nanoscale* 2015;**7**(16):6924–43.

147. Zhang J, et al. One-step carbonization synthesis of hollow carbon nanococoons with multimodal pores and their enhanced electrochemical performance for supercapacitors. *ACS Appl Mater Interfaces* 2014;**6**(3):2192–8.

148. Lee J-W, et al. Fouling-tolerant nanofibrous polymer membranes for water treatment. *ACS Appl Mater Interfaces* 2014;**6**(16):14600–7.

149. Liu Q, Jin J, Zhang J. NiCo$_2$S$_4$@ graphene as a bifunctional electrocatalyst for oxygen reduction and evolution reactions. *ACS Appl Mater Interfaces* 2013;**5**(11):5002–8.

150. Xia X, et al. A new type of porous graphite foams and their integrated composites with oxide/polymer core/shell nanowires for supercapacitors: structural design, fabrication, and full supercapacitor demonstrations. *Nano Lett* 2014;**14**(3):1651–8.

151. Huang M, et al. NiO nanoflakes grown on porous graphene frameworks as advanced electrochemical pseudocapacitor materials. *J Power Sources* 2014;**259**:98–105.

152. Ma W, et al. Flexible all-solid-state asymmetric supercapacitor based on transition metal oxide nanorods/reduced graphene oxide hybrid fibers with high energy density. *Carbon* 2017;**113**:151–8.

153. Chen S, et al. Graphene oxide– MnO$_2$ nanocomposites for supercapacitors. *ACS Nano* 2010;**4**(5):2822–30.

154. Yang J, et al. 3D architecture materials made of NiCoAl-LDH nanoplates coupled with NiCo-carbonate hydroxide nanowires grown on flexible graphite paper for asymmetric supercapacitors. *Adv Energy Mater* 2014;**4**(18):1400761.

155. Liu T, et al. Polyaniline and polypyrrole pseudocapacitor electrodes with excellent cycling stability. *Nano Lett* 2014;**14**(5):2522–7.

156. Yin Z, et al. Hierarchical nanosheet-based CoMoO 4–NiMoO 4 nanotubes for applications in asymmetric supercapacitors and the oxygen evolution reaction. *J Mater Chem A* 2015;**3**(45):22750–8.

157. Wen L, F.L, Cheng H. *Adv Mater* 2016:4306–37.

158. Wang X, et al. Flexible energy-storage devices: design consideration and recent progress. *Adv Mater* 2014;**26**(28):4763–82.

159. Guan Q, et al. Needle-like Co$_3$O$_4$ anchored on the graphene with enhanced electrochemical performance for aqueous supercapacitors. *ACS Appl Mater Interfaces* 2014;**6**(10):7626–32.

160. Chen K, Xue D. Rare earth and transitional metal colloidal supercapacitors. *Sci China Technol Sci* 2015;**58**(11):1768–78.

161. Mu Q, Wang Y. Synthesis, characterization, shape-preserved transformation, and optical properties of La(OH)$_3$, La$_2$O$_2$CO$_3$, and La$_2$O$_3$ nanorods. *J Alloys Compd* 2011;**509**(2):396–401.

162. Chen W, Kang Z, Ding B. Nanostructured W–La$_2$O$_3$ electrode materials with high content La$_2$O$_3$ doping. *Mater Lett* 2005;**59**(10):1138–41.

163. He J-Q, et al. Enhanced acetone gas-sensing performance of La2O3-doped flowerlike ZnO structure composed of nanorods. *Sensor Actuator B Chem* 2013;**182**:170–5.

164. Michel CR, Martinez-Preciado AH. CO sensing properties of novel nanostructured La2O3 microspheres. *Sensor Actuator B Chem* 2015;**208**:355—62.

165. Cheng J-B, et al. Growth and characteristics of La_2O_3 gate dielectric prepared by low pressure metalorganic chemical vapor deposition. *Appl Surf Sci* 2004;**233**(1—4):91—8.

166. Jing F, et al. Hemocompatibility of lanthanum oxide films fabricated by dual plasma deposition. *Thin Solid Films* 2006;**515**(3):1219—22.

167. De Asha A, Nix R. Oxidation of lanthanum overlay ers on Cu (111). *Surf Sci* 1995; **322**(1—3):41—50.

168. Patil SJ, et al. Electrochemical performance of a portable asymmetric supercapacitor device based on cinnamon-like La 2 Te 3 prepared by a chemical synthesis route. *RSC Adv* 2014;**4**(99):56332—41.

169. Patil S, et al. Chemical synthesis of α-La_2S_3 thin film as an advanced electrode material for supercapacitor application. *J Alloys Compd* 2014;**611**:191—6.

170. Zhang J, et al. The graphene/lanthanum oxide nanocomposites as electrode materials of supercapacitors. *J Power Sources* 2019;**419**:99—105.

171. Chen Y, et al. Polypyrrole—polyoxometalate/reduced graphene oxide ternary nanohybrids for flexible, all-solid-state supercapacitors. *Chem Commun* 2015;**51**(62):12377—80.

172. Shiri HM, Ehsani A, Shayeh JS. Synthesis and highly efficient supercapacitor behavior of a novel poly pyrrole/ceramic oxide nanocomposite film. *RSC Adv* 2015;**5**(110):91062—8.

173. Shayeh JS, Norouzi P, Ganjali MR. Studying the supercapacitive behavior of a polyaniline/nano-structural manganese dioxide composite using fast fourier transform continuous cyclic voltammetry. *RSC Adv* 2015;**5**(26):20446—52.

174. Shayeh JS, et al. Physioelectrochemical investigation of the supercapacitive performance of a ternary nanocomposite by common electrochemical methods and fast fourier transform voltammetry. *New J Chem* 2015;**39**(12):9454—60.

175. Majumder M, et al. Rare earth metal oxide (RE 2 O 3; RE= Nd, Gd, and Yb) incorporated polyindole composites: gravimetric and volumetric capacitive performance for supercapacitor applications. *New J Chem* 2018;**42**(7):5295—308.

176. Awin EW, et al. Structural, functional and mechanical properties of spark plasma sintered gadolinia (Gd_2O_3). *Ceramics Int* 2016;**42**(1):1384—91.

177. Zhang J, et al. Enhanced photocatalytic activity for the degradation of rhodamine B by TiO_2 modified with Gd_2O_3 calcined at high temperature. *Appl Surf Sci* 2015;**344**:249—56.

178. Zou H, et al. Adsorption study of a macro-RAFT agent onto SiO_2-coated Gd_2O_3: eu3+ nanorods: requirements and limitations. *Appl Surf Sci* 2017;**394**:519—27.

179. Xu G, et al. Highly-crystalline ultrathin $Li_4Ti_5O_{12}$ nanosheets decorated with silver nanocrystals as a high-performance anode material for lithium ion batteries. *J Power Sources* 2015;**276**:247—54.

180. Shiri HM, Ehsani A. Pulse electrosynthesis of novel wormlike gadolinium oxide nanostructure and its nanocomposite with conjugated electroactive polymer as a hybrid and high efficient electrode material for energy storage device. *J Colloid Interface Sci* 2016; **484**:70—6.

181. Wei J. *The development of manganese oxide electrodes for electrochemical supercapacitors*. 2007.

182. Khallaf H, et al. Investigation of chemical bath deposition of ZnO thin films using six different complexing agents. *J Phys Appl Phys* 2009;**42**(13):135304.

183. McPeak KM, et al. In situ X-ray absorption near-edge structure spectroscopy of ZnO nanowire growth during chemical bath deposition. *Chem Mater* 2010;**22**(22):6162—70.

184. Reddy ALM, et al. Asymmetric flexible supercapacitor stack. *Nanoscale Res Lett* 2008; **3**(4):145—51.

185. Brinker C, Scherer G. *Sol-gel science: the physics and chemistry and materials applications*. San Diego: Academic Press; 1990.
186. Lota K, Sierczynska A, Lota G. Supercapacitors based on nickel oxide/carbon materials composites. *Int J Electrochem* 2011:2011.
187. Tao T, et al. MoO_3 nanoparticles distributed uniformly in carbon matrix for supercapacitor applications. *Mater Lett* 2012;**66**(1):102−5.
188. Zhang Y, et al. Morphology-dependent $NiMoO_4$/carbon composites for high performance supercapacitors. *Inorg Chem Commun* 2020;**111**:107631.
189. Mao X, et al. Facile synthesis of hierarchical Co−Mo−O−S porous microspheres for high-performance supercapacitors. *Ceram Int* 2020;**46**(2):1448−56.
190. Stoller MD, Ruoff RS. Best practice methods for determining an electrode material's performance for ultracapacitors. *Energy Environ Sci* 2010;**3**(9):1294−301.
191. Balducci A, et al. High temperature carbon−carbon supercapacitor using ionic liquid as electrolyte. *J Power Sources* 2007;**165**(2):922−7.
192. mpoweruk.Com. Electropaedia, battery performance characteristics.
193. *Gigaom, energy and environment*. 2014.
194. Ahmad. *Supercapacitors: technology developments and global markets*. 2019.
195. Gogotsi Y, Simon P. True performance metrics in electrochemical energy storage. *Science* 2011;**334**(6058):917−8.
196. Shaikh JS, et al. Symmetric supercapacitor: sulphurized graphene and ionic liquid. *J Colloid Interface Sci* 2018;**527**:40−8.
197. Gao M, et al. Porous ZnO-coated Co3O4 nanorod as a high-energy-density supercapacitor material. *ACS Appl Mater Interfaces* 2018;**10**(27):23163−73.
198. Balasubramaniam S, et al. Comprehensive Insight into the mechanism, material Selection and performance Evaluation of Supercapatteries. *Nano-Micro Lett* 2020;**12**(1):85.
199. Lin T, et al. Nitrogen-doped mesoporous carbon of extraordinary capacitance for electrochemical energy storage. *Science* 2015;**350**(6267):1508−13.
200. Han L-N, et al. Nitrogen-doped carbon nets with micro/mesoporous structures as electrodes for high-performance supercapacitors. *J Mater Chem A* 2016;**4**(42):16698−705.
201. Liu D, et al. Highly efficient synthesis of ordered nitrogen-doped mesoporous carbons with tunable properties and its application in high performance supercapacitors. *J Power Sources* 2016;**321**:143−54.
202. Wang J-G, et al. One-pot synthesis of nitrogen-doped ordered mesoporous carbon spheres for high-rate and long-cycle life supercapacitors. *Carbon* 2018;**127**:85−92.
203. Wei J, et al. A controllable synthesis of rich nitrogen-doped ordered mesoporous carbon for CO_2 capture and supercapacitors. *Adv Funct Mater* 2013;**23**(18):2322−8.
204. Xie M, et al. Ordered nitrogen doped mesoporous carbon assembled under aqueous acidic conditions and its electrochemical capacitive properties. *Microporous Mesoporous Mater* 2014;**197**:237−43.
205. Chen A, et al. Synthesis of nitrogen-doped micro-mesoporous carbon for supercapacitors. *J Electrochem Soc* 2016;**163**(9):A1959.
206. Xue M, et al. Structure-based enhanced capacitance: in situ growth of highly ordered polyaniline nanorods on reduced graphene oxide patterns. *Adv Funct Mater* 2012;**22**(6):1284−90.
207. Wang H, et al. Effect of graphene oxide on the properties of its composite with polyaniline. *ACS Appl Mater Interfaces* 2010;**2**(3):821−8.
208. Kai Z, et al. Graphene/polyaniline nanofiber composites as supercapacitor electrodes. *Chem Mater* 2010;**22**(4):1392−401.

209. Hu L, et al. In situ electrochemical polymerization of a nanorod-PANI—Graphene composite in a reverse micelle electrolyte and its application in a supercapacitor. *Phys Chem Chem Phys* 2012;**14**(45):15652—6.
210. Izadi-Najafabadi A, et al. Extracting the full potential of single-walled carbon nanotubes as durable supercapacitor electrodes operable at 4 V with high power and energy density. *Adv Mater* 2010;**22**(35):E235—41.
211. Zhu Y, et al. Carbon-based supercapacitors produced by activation of graphene. *science* 2011;**332**(6037):1537—41.
212. Liu Y, et al. Elemental superdoping of graphene and carbon nanotubes. *Nat Commun* 2016;**7**(1):1—9.
213. Zhao Y, et al. A versatile, ultralight, nitrogen-doped graphene framework. *Angew Chem Int Ed* 2012;**51**(45):11371—5.
214. Han J, et al. Generation of B-doped graphene nanoplatelets using a solution process and their supercapacitor applications. *ACS Nano* 2013;**7**(1):19—26.
215. Jung SM, et al. Controlled porous structures of graphene aerogels and their effect on supercapacitor performance. *Nanoscale* 2015;**7**(10):4386—93.
216. Vangari M, Pryor T, Jiang L. Supercapacitors: review of materials and fabrication methods. *J Energy Eng* 2013;**139**(2):72—9.

Hierarchical porous carbon-incorporated metal-based nanocomposites for secondary metal-ion batteries

7

Maira Sadaqat[1], Hassina Tabassum[2,3], Qiu Tianjie[2,3], Asif Mahmood[4], Laraib Nisar[1] and Muhammad Naeem Ashiq[1]

[1]Institute of Chemical Sciences, Bahauddin Zakariya University, Multan, Pakistan; [2]College of Engineering, Peking University, Beijing, China; [3]Department of Chemical and Biological Engineering, University at Buffalo, The State University of New York, Buffalo, NY, United States; [4]School of Chemical and Bimolecular Engineering, The University of Sydney, Sydney, Australia

7.1 Introduction

Global energy requirements are rising with the increasing population worldwide. The increment in world energy consumption is expected to be about 56% from 2010 to 2040. So the cumulative energy consumption of 524 quadrillion Btu noted in 2010 has recently increased to about 630 quadrillion Btu and will reach 820 quadrillion Btu by 2040.[1] Recently, 80% of world energy consumption was satisfied by fossil fuels, with this increased consumption causing a rapid rise in global warming. These conditions have forced the scientific community to search for secondary energy resources. Among the many energy resources, renewable energy-storage technologies are electrochemical systems, such as wind energy, solar energy, and fuel cells. Lithium-ion batteries (LIBs) have played an important role in developing smart electric appliances over the last few decades because of their long cyclic life and high storage capacities.[2] However, large-scale application of LIBs still faces issues with the low earth abundance of lithium (Li) metal and higher costs. The solution to these challenges is to replace LIBs with secondary metal-ion batteries. Well-known examples of secondary metal-ion batteries are Sodium ion batteries (SIBs), potassium ion batteries (KIBs), and metal−air batteries. The overall cost of secondary metal-ion batteries will be further reduced by the anticipated use of aluminum instead of copper for current collectors. A variety of electrode materials have been investigated for LIBs, including carbon-based electrodes, metal oxides, phosphides, and nanohybrids.[3−5] However, efficient and high storage capacity electrode materials for SIBs/KIBs are still needed. Despite the charge/discharge mechanism of these secondary metal-ion batteries being the same as that of LIBs, their storage capacity is quite low. For example, graphitic carbon has a very low storage capacity of 35 $mAhg^{-1}$ for

Metal Oxide-Carbon Hybrid Materials. https://doi.org/10.1016/B978-0-12-822694-0.00005-3

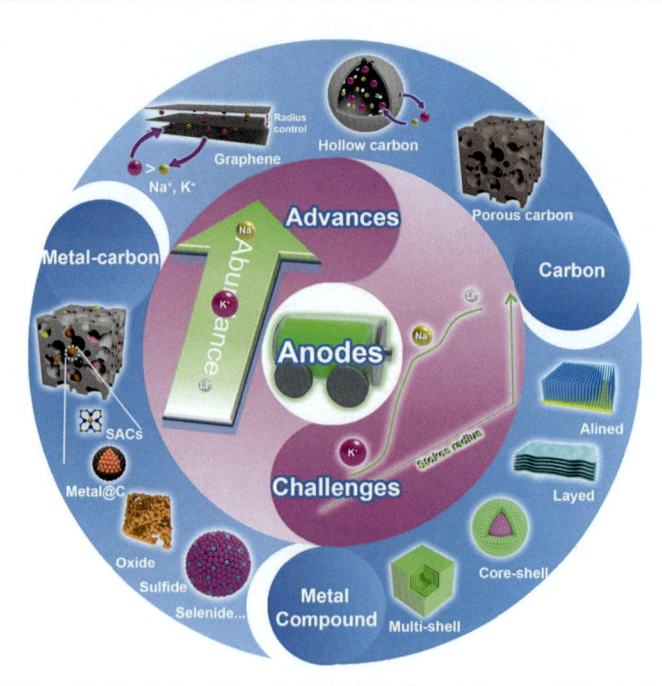

Figure 7.1 Schematic illustration of anode materials of secondary metal-ion batteries.

SIBs. Thus, suitable anode materials are needed for SIBs/KIBs that satisfy the current requirements of energy-storage technology.

In this chapter, we introduce secondary storage devices and the fundamental requirements for the development of efficient electrode material to identify the key challenges. Carbon-based electrodes, metal@C anode, noncarbon anodes, and their fabrication technologies are also discussed in detail, as shown in the schematic of Fig. 7.1. Based on these properties, we discuss the need for secondary energy-storage devices other than LIBs as well as bottleneck reactions for the LIBs. We then introduce the state-of-the-art materials and electrode design strategies used for high-performance energy-storage devices. The properties and various electrode materials of SIBs, KIBs, and metal−air batteries are surveyed. In the end, future perspectives and a summary of secondary metal-ion batteries are presented.

7.1.1   Need for secondary metal-ion batteries

LIBs have been regarded as the most advanced technology for the storage of electric energy in terms of chemical energy, and continuous extension into electric vehicles and the mobile market has increased the demand for high energy and power density Li-ion batteries. Consequently, several studies have been conducted to develop highly advanced LIBs by enhancing the storage capacity, leading to high energy density and power density technology. However, the energy storage capacity obtainable so far is inadequate to meet the rapidly growing demand for sustainable batteries with respect to high-performing energy density together with being economical and safe.[6] LIBs have dominated the global energy market for more than half of a century, which is ascribed to their high energy density and long life cyclability, and as stated, they have acted as the next-generation battery for electric vehicles and grid energy-storage systems. It is noteworthy that the efficiency of LIBs is mainly dependent on the availability of Li-metal production at a low cost. Furthermore, frequent exploitation of Li-ion technology for more than a century is responsible for diminishing Li resources. Therefore, insufficient availability of global Li resources and other safety issues will considerably hinder its use from a large-scale perspective.[7] To store a large amount of the power produced via intermittent solar and wind energy, the development of ceaseless high-capacity energy density and power supply for the next-generation battery is a prerequisite. However, to enhance energy and power density, stacks of LIBs have been added, which ultimately increases storage capacity. But this does not solve the problem long term and also augments the cost of the battery.[8] To minimize the unavailability of reliable energy-storage-and-conversion devices and to meet the needs of transportation and electricity-grid stations, new electrochemical storage devices based on nanomaterials beyond LIBs are immensely needed.[9] To fulfill daily energy demand and overtake the problems regarding clean energy resources, alternative electrochemical energy-storage systems are in a constant state of development. In this regard, KIBs and SIBs are the most promising and advanced alternatives to replace LIBs. The prime incentive behind the development of secondary metal batteries (KIBs/SIBs) comes from the high energy densities of potassium[10] and sodium ions,[11] which can be credited to the comparable reduction potentials of potassium and sodium[12] ions to Li ions (Table 7.1). Both alkali metals K/Na have lower reduction potential than Li ions. In addition, the price of raw materials for both alkali metal ions such as carbonates have low cost as well as the abundance of potassium and sodium metals have pushed the researchers to tackle the major challenges regarding state-of-the-art secondary metal-ion batteries (Table 7.1).[13] Owing to the availability of low-cost sodium and potassium elements plus similarity in chemistry to LIBs, they offer a more feasible

Table 7.1 Comparison of physical properties of Li, Na, and K ions.

Elements	Na	Li	K
Atomic radius[pm]	190	167	243
Ionic radius[pm]	102	76	138
Atomic weight[g/mol]	22.989	6.94	39.098
E_0 versus SHE[V]	−2.71	−3.04	−2.94
Melting point[°C]	97.79	180.54	63.5
Boiling point[°C]	882.94	1347	759
Crystal structure	Cubic	Cubic	Cubic
Density at 293 K[g/cm^3]	0.97	0.53	0.826
Classification	Alkali metal	Alkali metal	Alkali metal
Abundance in earth's crust	2.8	Negligible	2.6
Distribution	Everywhere	70% in South America	Everywhere
Prices/metric ton, carbonates	200−250	2000−2200	500−1500

E_0, Reduction potential, *[pm]*, Pico meter.

solution related to LIBs,[14,15] although large-scale application of these new technologies is hampered by low energy density and exploitation of highly flammable electrolytes. Such alternative technologies are required to operate with simple battery chemistries, address safety issues, and have long life cycles.[16] Here, we discuss the properties and mechanisms of batteries (potassium/sodium ion) beyond LIBs and have a look at electrode materials. Later in this section, properties of various metal–air batteries will be surveyed, and recent developments in electrode materials will be presented.

7.1.2 Energy-storage materials concept

Energy-storage technology harnesses wind, solar, and mechanical energy and channels it as either electrical or chemical energy. Therefore, a device that stores energy and mobilizes it for later use is known as an accumulator or battery. Various forms of energy are present on earth, such as gravitational, elevated temperatures, and electrical, kinetic energy.[18,19] Energy-storage devices can employ multifarious energies forms and deposit them into expedient or fuel-efficient forms. Urgent utilization of renewable and continual energy resources, such as wind and solar energy, has been swift by global warming and other severe weather perturbations related to the unceasing use of nonrenewable sources as well as the incorporation of advanced power distribution infrastructure.[20] Efficiently harnessing renewable energy sources is a decisive concern pertaining to intermittent energy and power production.[21] Hence, a popular plan of action is to devise advanced energy-storage devices capable of providing an uninterrupted source of power on demand. Currently, four major types of energy-storage systems have been presented for commercial-scale applications that include mechanical, chemical, electrical, and electrochemical. Pumped hydroelectricity is a widely available method for mechanical energy storage. Recently, electrochemical energy storage has been a swiftly leading discipline constructing on the constant surge of productive ideas.[22] As renewable energy becomes the most promising sought-after energy source to meet the increasing energy demand worldwide, providing high energy density and power density storage along with a perpetual cycle life. The development of appropriate materials for these accumulators starts with a full comprehension of the complex chemical reactions that steer the interconversion and storage of chemicals traversing long timescales.[23] Moreover, the electrochemical energy-storage batteries can store a large amount of energy to empower portable electronics and electrify the transportation sector attributed to the high round-trip efficiency, long life cycle, low cost, and sustainable material. Generally, electric energy is kept up in energy devices via a faradaic process. Typical examples of faradaic systems inculcate pseudocapacitors and numerous batteries.[24] The electric energy can be captured by transforming to chemical energy through various redox reactions of metals, and while discharging, energy is released via reversible reactions of active reagents when power is required.[21] The Li mass abundance will be lower on the earth because of its rapid use for electric appliances, proceeding to high costs of LIBs. However, the sodium and potassium metals have large reservoirs in earth crusts, which will be enough for upcoming requirements for energy-storage technology (Fig. 7.2).

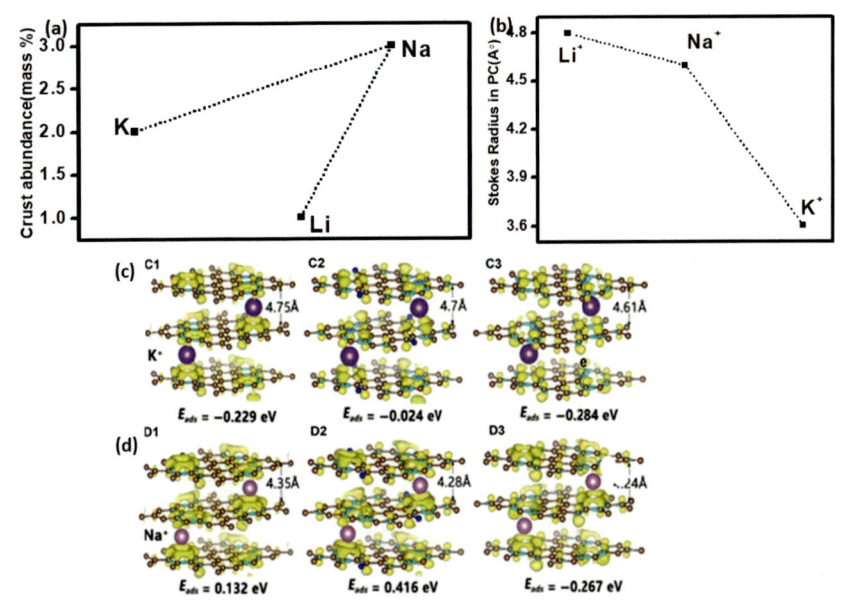

Figure 7.2 (A) Comparison of LIB, SIB, and PIB with respect to (A) Abundance of Li, sodium, and potassium metal in Earth's crust (wt%), (B) Stokes radii of Li^+, Na^+, and K^+ in PC, (C,D) Simulations of the K^+ and Na^+ adsorption capabilities and DOS of various carbon structures. (C1 to C3) Electronic density differences of K^+ ions adsorbed on pristine carbon, GN-doped carbon, and PN-doped carbon, respectively. Yellow areas represent $+0.005$ e per $Å^3$ isosurface; (D1 to D3) the electronic density differences of Na^+ ions adsorbed on pristine carbon, GN-doped carbon, and PN-doped carbon, respectively. (E) adsorption energies of K^+ and Na^+ ions on various sites; (F) DOS of different carbon structures, Copyright Elsevier.[17]

Furthermore, the storage capacity of Na^+/K^+ ions is quite low compared with Li^+ ions in graphite anode materials for LIBs (~ 372 $mAhg^{-1}$).[25] This low performance of Na^+/K^+ ion batteries is because of the larger size of Na^+/K^+ ions (55%−65%) than Li^+ ions in the interlayer spacing of graphitic layers (Fig. 7.2A−B). Mechanistically, Na^+ ions first adsorb on the surface of electrode materials and intercalate into the layers of the graphite, called the adsorption-intercalation process in batteries. The layers of the commercial graphite are quite narrow for the intercalation of the sodium ions. Thus, there is quite a need for the expansion of interlayer spacing of graphitic materials for the successful intercalation/deintercalation processes during charge storage in SIBs. Theoretically, the interlayer spacing in graphitic materials should be in the range of 0.37−0.40 nm for the sodium-ion intercalation/deintercalation in SIBs. However, the reported literature-based electrodes provide some room for the storing of Na^+/K^+ ions into carbon structures.[25] Therefore, it is necessary for producing graphitic-based materials with large interlayer spacing and tuning atomic structure through heteroatom doping (i.e., N, S), promote the storage capability and cyclic stability (Fig. 7.2C−D). Moreover, the nanohybrids electrodes materials consisting of metal selenides@C, metal phosphides@C, metal oxides@C, and metal sulfides@C materials have also been explored as electrodes for SIBs/KIBs.

7.1.3 Potassium-ion batteries

Recently, KIBs have attracted the attention of researchers due to their high energy density obtained by the low standard reduction potential of potassium (-2.93 V vs. Eo), close to Li (-3.04 V vs. Eo), and the high abundance of potassium metal. In addition, aluminum foil can be used as a current collector, which generally reduces the price and weight of the battery.[8] Further, potassium has a smaller Stokes radius than both Li and Na ions in organic solvents, which is responsible for the highest ion mobility and conductivity.[49] Efforts have been made in order to comprehend the electrochemistry of KIBs, and numerous strategies have been employed based on electrode design and material such as heteroatom doping, nanostructure design, tuning electronic structures, and preventing side reactions.[25]

7.1.3.1 Potassium-ion battery mechanism

The storage mechanism of the PIB is similar to the LIB and SIB. The electrochemical reactions depend on the intercalation/deintercalation of potassium ions at the respective electrodes (cathode and anode). The anode and cathode are separated but connected ionically by a separator soaked in an electrolyte. The potassium metal at the cathode supplies electrons by conversion into potassium ions that travel across the external circuit toward the anode, and potassium ions diffused at the anode react with electrons and store the energy in the form of chemical energy. During the discharging process, the above-mentioned process is reversed and provides energy in the form of electrical energy. In order to store energy, the KIB technology should fulfill some fundamental requirements. The cathode material is able to sustain its nanostructure morphology when operating under high and wide potentials. The cathodic nanomaterial should carry sustainable active sites as well as large lattice sites to encase and liberate the K ion freely in a reversible redox reaction. Similarly, the anodic crystal structure must be flexible enough to buffer volume changes charging and discharging process of potassium ion simultaneously; it should be prone to operate at low potentials to reap high voltage and energy density during discharging. For cycling performance, the stable anodic material should be able to elongate the reversible K ion intercalation/de-intercalation mechanism. Both materials must be chemically stable under high-current and voltage applications.[49–51]

7.1.4 Sodium ion batteries

SIBs have been regarded as the most propitious and economical technology in the placement of Li batteries owing to their abundant resources found on earth. Because sodium metal has a larger radius size than Li metal, the diffusion kinetics of sodium ions are notably reduced, and expansion in the volume of crystal structures hinders its practical potential for the high power density technology required for large-scale grid stations as electrochemical energy-storage devices.[52] Therefore, considerable efforts have been made to improve the SIB, particularly in terms of electrode material. To improve the energy density of SIBs, a pragmatic configuration of the electrode material is pivotal, facilitating the fruition of high capacities and suitable reduction-oxidation potentials. Thus far, huge attempts have been made in material engineering, achieving fortunate results largely at the cathode facet, substantially associated with the work of layered transition metal oxides[53,54] and polyanionic compounds.[11,55,56] These layered transition metal oxides and polyanion exhibit considerable thermal stability as well as the existence of covalent bonds. For instance, P and O act as support in charge states. Prussian blue and its relevant composites are regarded as an excellent host material for the intercalation/deintercalation chemistry of potassium ions. Currently, Wessells discloses low-strain nickel hexacyanoferrate[57] electrode material exhibits improved electrochemical performance with 66% retained capacity after 1200 cycles at low current density. Wessells and coworkers also prepared copper hexacyanoferrate,[58] a kind of Prussian blue analog, to undergo an extraction and insertion mechanism by a solid ion diffusion process. The definite open framework has shown excellent cycle stability.

7.2 Electrode material design for secondary metal-ion batteries

Nanostructured material has emerged as nanotechnology, and various nanomaterial designs have been fabricated to improve the electrochemical performance of the metal-ion battery. In the light of the large size of potassium ions, nanostructures have proven the most effective strategy to shorten the potassium ion diffusivity length, consequently improving potassium ion kinetics and enhancing rate capability and cycling performance.[6] Various nanostructures have been utilized as anode and cathode materials, such as carbon-based material (hollow carbon,[59] graphite, and graphene[60]) and alloy-based material (oxides, selenides,[34] and sulfides[33]) (Table 7.2). Several carbon nanostructures have been studied, such as the nanocube.[61] Of all 3D interconnected nanostructures, reduced graphene oxide provides the best channels for ion transportation and high conductivity that improve the fast transfer of ions across electrodes. In addition, the cycling stability of electrodes can be considerably enhanced by employing double-walled carbon nanotubes[62] and bamboo-shaped carbon nanotubes[63] and nanofibers.[64] The porous structures can enhance the contact of electrolyte and electrode, which ultimately improves coulombic efficiency. Lately, carbon nanofibers[42] have shown excellent rate capability and favorable long cycling stability. Moreover, the synthesis of selenium and sulfide-based alloy can be used as an electrode material owing to their high gravimetric and volumetric high specific capacities.[65] Various nanodesigns, such as nanosheets and nanorods, have been fabricated for SnP,[35] GeS,[66] SnS_2[34] and $MoSe$,[28] CuS[67] for the synthesis process of cobalt selenide quantum dots encapsulated in a metal−organic framework (MOF)-derived carbon material. Hard carbon (HC) is a promising anode material for low-cost and high-energy-density SIBs. The anode as yet described for SIBs is hard carbon, demonstrating average specific capacities, but its efficiency is decreased by low initial coulombic efficiency. Liu et al. showed that hard carbon sodiated with sodium biphenyl improved

Table 7.2 Comparison of rate capacity at particular current density after cycling stability test for various electrode designs.

Sample	Application	Rate capacity(mAh/g)	Current density(mA/g)	No. of cycles	
Hollow carbon nanospheres (HCNs)	Potassium anode	200	28	100	26
P-doped hard carbon	Potassium anode	302	20	460	27
MoSe2/N-Doped carbon	Potassium anode	258.02	100	300	28
Nitrogen-doped porous carbon (NPC)	Potassium anode	384	100	500	29
Functional phosphorus and oxygen dual-doped graphene (PODG)	Potassium anode	160	2000	600	30
Sn_4P_3 in N-carbon fibers	Potassium anode	160	500	1000	31
Yolk−shell FeS2@C	Potassium anode	203	10,000	1500	32
FeS2/C@C	Potassium anode	262	100	100	33
SnS_2/N-rGO	Potassium anode	402	1000	100	34
Sn_3P_4/GO		116	800	60	35
$Co_{0.85}$Se-QDs/C	Potassium anode	402	50	100	36
CQDs/hollow C	Potassium anode	160	1000	800	37
N-rich Cu_2Se/C	Potassium anode	190	100	200	38
rGO@MCSe	Potassium anode	310	5000	120	39
$K_{0.220}$Fe[Fe(CN)$_6$]$_{0.805}$·$4.01H_2O$	Potassium cathode	160	120	120	40
Nitrogen/oxygen co-doped mesoporous carbon octahedrons (MCOs)	Potassium anode	80	2000	3000	41
Nitrogen-doped carbon nanofibers	Potassium anode	146	2000	4000	42
Red P into nitrogen-doped porous hollow carbon nanofibers (red P@N-PHCNFs)	Potassium anode	342	5000	100	43
Sb_2S_3 nanodots/carbon	Potassium anode	312.03	1000	200	44
3D-NS/C nanosheets (3D-NSCNs)	Potassium anode	254.9	100	200	45
Bi-nanorods/NS/Carbon matrix	Potassium anode	289	6000	1000	46
N-PC		135	50	60	47
Octahedral $CoSe_2$/N-CNT	Potassium anode	253	2 A/g	600	48

C, mesoporous Carbon, *CNT*, carbon nanotubes, *GO*, Graphene, *QD*, quantum dots, *rGO*, reduced graphene oxide.

coulombic efficiency up to 95%.[68] Moreover, advancement in the energy density of SIBs firmly counts on the evolution of efficacious anode materials. Therefore, exploration for anode materials capable of operating at low potentials, large storage, and reversible capacity along with morphological sustainability constitute a contemporary challenge for SIBs.[69]

Moreover, the chelation among citric acid and transition metal ions under hydrolysis and co-precipitation methods (Fig. 7.3A). The Series of the metal oxides were

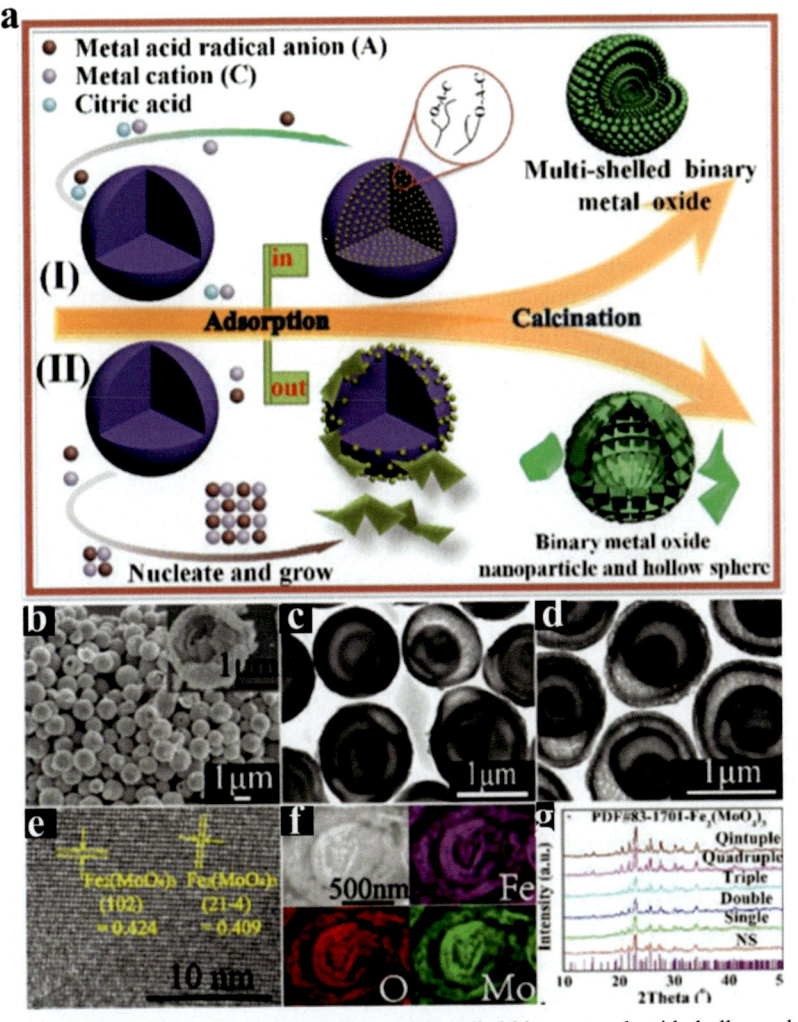

Figure 7.3 (A) The process for synthesizing multi-shelled binary metal oxide hollow sphere. (B) The characterization of multi-shelled $Fe_2(MoO_4)_3$ hollow sphere. (C) SEM and (D) TEM images of quadruple-shelled hollow sphere. (E) TEM, HRTEM and (F) Mapping images. (G) XRD pattern. Copyright 2018, American Chemical Society.[70]

produced and applied for the energy-storage properties. The morphology of the multi-shell $Fe_2(MoO_4)_3$ hollow sphere was observed under SEM and TEM analyses (Fig. 7.3B–E). These shells are consisting of porous nanoparticles, which promote the transport of electrolyte during charge/discharge process in batteries. Furthermore, STEM mapping and XRD pattern confirm the elemental distribution in structure and crystal structure, respectively. The hollow sphere morphology provides a void gap for the ion storage and control the volume expansion during ion storage in battery devices.

One-dimensional Tubular Nanostructure: One-dimensional (1D) structures have found deep interest for sodium-ion storage in their dominant structure, which facilitate the charge transport because of the interconnected networks and control the volume expansion inside the hollow tubules. The number of the tubular morphology such as carbon nanotubes, metal oxides, metal selenides and phosphides could also facilitate the sodium-ion storage. Huang et al., fabricated a product of $TiO_2(B)$ tubular structure presented long cyclic stability and good capacity retentions. The electrical conductivity of these product can also be enhanced through carbon coating such as coating on the MoO_3 nanorods by solvothermal methods.[71,72] However, hollow structure is exhibited decent characteristics for energy-storage devices. The heteroatom (N and S)-doped carbon sphere with special void gap and graphitic layers spacing presented the synergistic effects for obtaining promising charge storage capacity, rate capability, and improved capacity retentions (94% after 2000 cycles). There is another example of the hollow structure of $Na_2Ti_3O_7$ nanosheets coupled in spherical morphology (Fig. 7.4A–B) which exhibited storage capacity of 63 mAhg^{-1} at current density of 8.8 Ag^{-1} for long cycles over the 1000 cycles. The Ti-based anode materials and chalcogenide-based anode materials presented hollow structures and presented high storage capacity. Additionally, composites of hollow structures were applied as active materials to enhance the storage capacity of secondary metal-ion batteries (Fig. 7.4C–D). Similarly, the SnS sheets like morphology were also fabricated on MoO_3 nanorods in the presence of ammonia and proper glucose treatment under solvothermal reaction and annealing processes (Fig. 7.4E–H). The respective products presented the high storage capacity of 465 mAhg^{-1} for 100 cycles (Fig. 7.4I–K). Because of the carbon conductive layers on Sb (Sb@C) presented high electrochemical performance of 310 mAhg^{-1} at 20 Ag^{-1} for long cyclic life of 2000 cycles (Fig. 7.4L).

These materials have improved the reaction kinetics and reduced the volume changes to some extent. The huge volume expansion of electrode material is a major cause for the failure of the alloy-based anode, particularly in KIBs due to large atomic size. Until now, the conductive matrix has shown enough potential to control volume changes during the intercalation and deintercalation of potassium ions as a most promising strategy.[36] Various forms of carbon matrix such as carbon nanotube,[76] doped carbon and graphene[77] and reduced graphene[44] have been explored as an anode material. Of all, reduced garphene oxide is considered as soft carbon matrix, which offered high electrical conductivity and flexible 2D structure. The cycling capability and rate capacity can be enhanced with the heteroatom doping of N,[38,44] P[27] and F[78] which tune the electronic features and provide favourable bonding sites and increase the electrical conductivity of active material. The doping of elements can be done at high temperatures during the carbonization of organic material. N doping in pyridine (pyridinic N PN) causes greater adsorption of

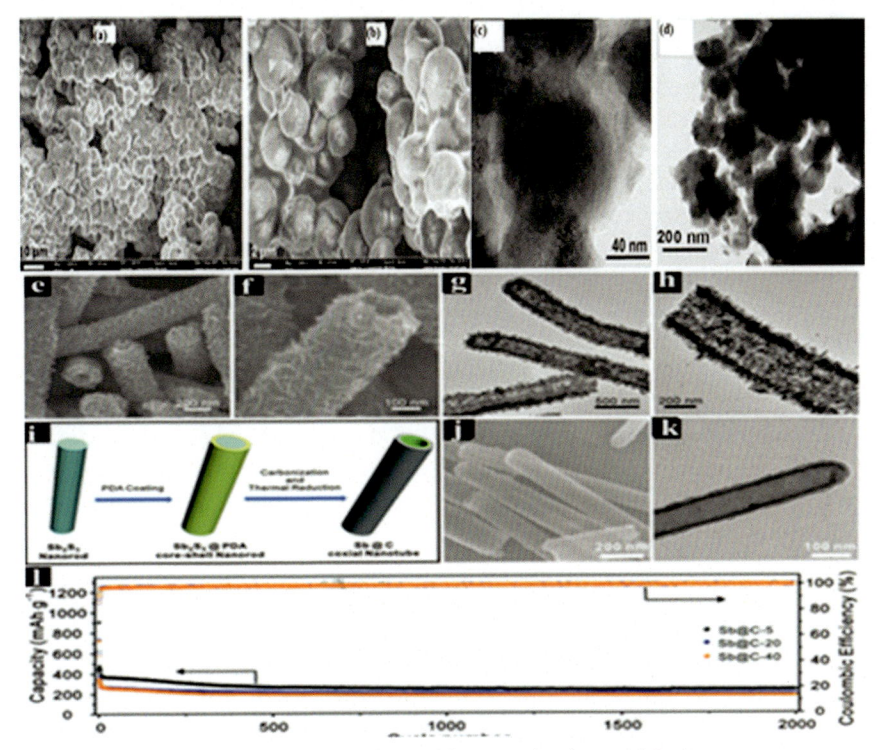

Figure 7.4 Synthesis Strategy, Morphological Characterization, and Performance Investigations of Single-Shelled Hollow Spheres and 1D Tubular Nanostructures (A−B) FESEM image and (C−D) TEM image of N-HSDC/NVP-HMs-2 hollow spheres. Copyright 2020, Elesvier.[73] FESEM images and (G and H) TEM images of SnS@C nanotubes. Copyright 2017, Wiley-VCH.[74] (I) Schematic illustration of the formation of Sb@C coaxial nanotubes. (J) FESEM image and (K) TEM image of Sb@C coaxial nanotubes with an annealing time of 20 min (Sb@C-20). (L) Cycling performance of Sb@C-20 and Sb@C coaxial nanotubes with annealing time of 5 and 40 min (denoted as Sb@C-5 and Sb@C-40, respectively) at a current density of 1.0 A g^{-1}, respectively. Copyright 2016, Royal Society of Chemistry.[75]

potassium ions than graphitic N (GN), while nitrogen doping in graphite enriches active sites with electrons, leading to enhanced electrical conductivity.[17,79] The template-derived nitrogen-doped nonporous carbon has extended surface area and large interspacing due to graphitic carbon provide numerous voids for the insertion/desertion of K$^+$ ions. In another report, the doping of fluorine atom into carbon, alter the hybridization of carbon from sp^2 to sp^3 state, which is responsible for the high electrical conductivity and stability.[80] Bi-nanorods has been taking into account as promising replacement of carbon, onus to its similar size to carbon atom.[46] Efforts have extended to multiple-element doping such as N/S,[39] P/O[30]-doped carbon. It is reported that oxygen doping into the carbon framework give a way to store high energy density and capacity perk up the properties of electrode material(wettability) and enhancing the active surface area.[30]

However, layered oxides materials are exhibited efficacious charge storage capacity with different redox reactions during charge/discharge processes. The transmutation of the complex structure with prominent contraction/expansion and reduced/enlarge inter-layer spacing and structure reduction. Furthermore, the insertion of metal insertion into the structure proceeds with irreversible phase transmutation for the storage of sodium ions. Nevertheless, the storage of layered oxides also promoted electron transport. For example, the $Na_{2/3}Fe_{1/2}Mn_{1/2}O_2$ nanofibers exhibited a stacked structure and were produced through the electrospinning process. The promising electrical conductivity promotes the diffusion paths in $Na_{2/3}Fe_{1/2}Mn_{1/2}O_2$ nanofibers in sodium-ion storage. Similarly, the fabrications of $Na_{0.85}Li_{0.1}Ni_{0.18}Mn_{0.54}Fe_{0.18}O_{1.98}$ with crystalline structure through template method (Fig. 7.5A−C and E−G). The high crystallinity of $Na_{0.85}Li_{0.1}Ni_{0.18}Mn_{0.54}Fe_{0.18}O_{1.98}$ microflowers exhibited a high storage capacity of 55 mAhg^{-1} and cyclic stability (Fig. 7.5D and H). The polyanionic components had a crystalline structure and volume expansion during electrochemical reactions in SIBs. There is another product $NaMN/Mg_3(PO_4)_2$ nanospheres with a diameter of 50 nm. The structural and electrical properties were tuned through RuO_2 layers on

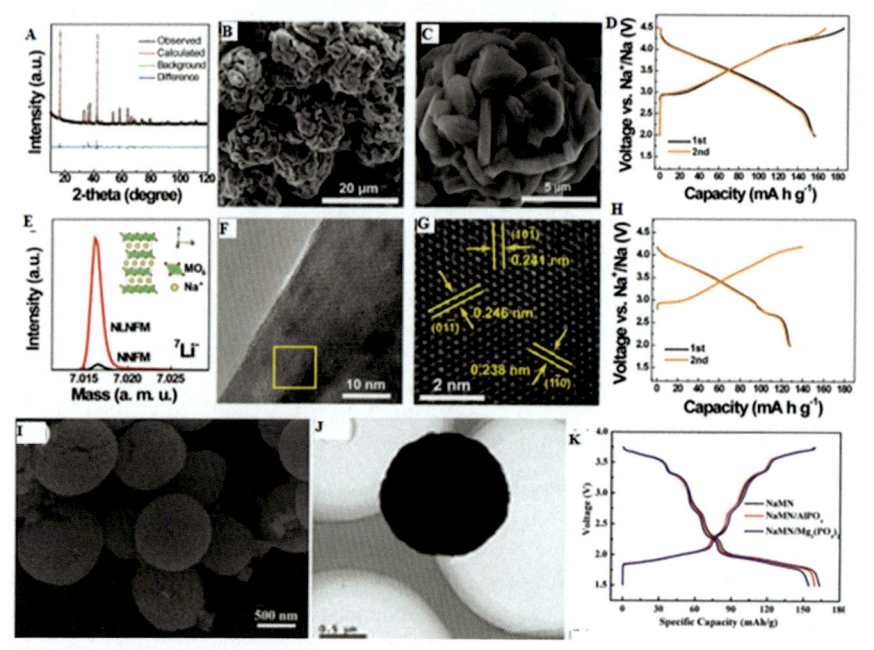

Figure 7.5 Morphologies and Electrochemical charge storage performances (A) XRD pattern (B−C) FESEM image and (D) Galvanostatic charge and discharge curves of the $Na_{0.85}Li_{0.1}$ $Ni_{0.18}Mn_{0.54}Fe_{0.18}O_{1.98}$ electrode at different current densities. (E) TOF-SIMS plots of seven Li signal in NLNMF and the control sample $Na_{0.9}Ni_{0.3}Mn_{0.4}Fe_{0.3}O_2$ (NNMF) integrated over 1000 s of Cs + sputtering (10 s sampling step). The inset shows a structural illustration of O_3-type TM oxide. and (F−G) HRTEM image. (H) Rate capability of the $Na_{0.85}Li_{0.1}Ni_{0.18}$ $Mn_{0.54}Fe_{0.18}O_{1.98}$ nanoparticles (denoted as NNMF) electrodes. Copyright 2018, Elsevier.[81] (I) FESEM images of the $NaMN/Mg_3(PO_4)_2$. (J) HRTEM Image (K) Galvanostatic charge and discharge curves of the $NaMN/Mg_3(PO_4)_2$ electrode between 1.5 and 3.75 V at 0.1C.[82]

NaMN/Mg$_3$(PO$_4$)$_2$ nanospheres. These electrode materials presented the storage capacity of 126 mAhg^{-1} at a rate of 0.1C and a stable capacity of 59 mAg^{-1} at a high rate of 40C for long cycling of 1000 cycles. Cao et al. fabricated products of NaMN/Mg$_3$(PO$_4$)$_2$ nanoparticles coated on nanospheres and coated with carbon and denoted as NaMN/Mg$_3$(PO$_4$)$_2$ (Fig. 7.5I). Nevertheless, the carbon coating presented the regular lattice fringes like graphene structure with an amorphous carbon layer (Fig. 7.5J). The NaMN/Mg$_3$(PO$_4$)$_2$ presented high storage capacity of 115 mAhg^{-1} at a current density of 117 mAg^{-1} and was stable for a long cyclic life of 20,000 cycles (Fig. 7.5K).

Thus far, lots of micro structural designs with new various compositions of atoms and molecules have considerably improves the properties of SIBs. In this regard, carbon-based electrodes have proven efficient electrochemical activity base on the insertion-alloying type mechanism.[103] Although these materials come up with low energy density, but alternate alloying-conversion with other metals and nonmetals, permitting the insertion/desertion storage process.[86] For several years, nanotechnology has revolutionize the material science offering promising candidates with morphologies and synergist effect of different metals have presented remarkable electrode designs beyond insertion-storage mechanism along with enhanced desired properties.[95] A comparison of rate capacity after performing a cycling stability test for various electrodes is presented in Table 7.3.

Alloying of materials with sodium metal demonstrated the much better performance of anodes for SIBs. Indeed, numerous metals have been alloyed with metals and nonmetals such as tin,[104] antimony,[105] phosphorous, nickel,[106] and iron[106] delivering the high specific capacity along with good cyclability and reversible capacity rate attributed to the exchange of multivalent electrons when alloying process occurred. Nevertheless, structural degradation of electrodes has been observed upon continuous cycling of electrode material that can be linked to the fact that when sodium ions intercalate or deintercalate, anodic material undergo huge volume expansion. Most tin-based electrodes offer the problem of agglomeration during electrochemical performances. Encapsulating the carbon matrixes improves the mechanical strength and electrical conductivity of tin, iron, antimony, and phosphorus-based materials.[95] Lan et al. reported the copper-tin phosphide with more phosphorous contents, Cu content stabilizes the electrode material and limits the aggregation of tin, giving the high capacity 811 mA h g^{-1} at a current density of 25 mA g^{-1}.[107] The volume expansion and electrolyte instability are improved by introducing the carbon nanotubes, which improves the cyclability of anode material, delivering a capacity of 512 mA h g^{-1} after the 100th cycle. Tuning of the structure is also important for finding efficacious properties for energy-storage technology. The MOF-derived interconnected beehive-like structure consisted of porous carbon shells of diameter 10 nm with a particular lattice structure for the storage of Na$^+$/K$^+$ ions. Similarly, the heteroatom (S, N)-doped graphitic shells consisted of interlayer spacing of 0.375 nm with a high surface area. These products were exhibited the storage capacity of 448 mAhg^{-1} at a current density of 100 mAg^{-1} for SIBs. However, these products were also presented for potassium ion storage capacity of 320 mAhg^{-1} at a current density of 50 mAg^{-1} in KIBs (Fig. 7.6).[83]

Yong et al. synthesized the nitrogen-doped carbon-coated WS$_2$ nanosheets derived by carbonization of polypyrrole, demonstrating a capacity rate of 236 and 238.1 mAh g^{-1} at 50 and 2000 mA/g, respectively, connected with nitrogen-coated carbon to induce the synergistic effect of WS$_2$ nanosheets[96] (Fig. 7.7). Li et al. reported Sb$_2$O$_3$ nanowires

Table 7.3 Comparison of rate capacity at particular current density after cycling stability test for sodium ion batteries.

Sample	Application	Rate capability	Current density	No. of cycles	References
Sb_2S_3/rGO	Sodium-anode	237	20	30	
Sb_2S_3 NPs	Sodium-anode	605	2400	500	84
Sn-rGO/ $NaLi_{0.2}Ni_{0.25}Mn_{0.75}O_\delta$	Sodium-anode	173			85
Sb_2S_3@C rods	Sodium-anode	699	100	100	86
Copper hexacyanoferrate (CuHCF)	Sodium-anode	51	15C	300	87
NVP@C/rGO-U	Sodium-anode	106			88
$NiSe_2$/C porous nanofiber	Sodium-anode				89
Sb@C NPs	Sodium-anode	553	—	50	90
Sb@(N, S—C)	Sodium-anode	390	1000	1000	86
(MoO_2/C)	Sodium-anode	367	50	100	91
Phosphorus-doped carbon nanosheets (P-CNSs)		149	5 A/g	5000	92
Fe_7S_8/N-doped graphene nanosheets	Sodium-anode	393	400	5000	93
Porous S and N co-doped thin carbon (S/N@C)	Sodium-anode	169	32,000	4500	94
Tin(II) sulfide—carbon (SnS—C)	Sodium-anode	568	20	80	95
Nitrogen-doped conductive carbon/ WS_2 (WS_2-NC)	Sodium-anode	369	100		96
Trogtalite $CoSe_2$ encapsulated into (BCN) nanotubes ($CoSe_2$@BCN-750)	Sodium-anode	580	100	100	97
Ni_2P/ZnP_4 embedded in P-doped/C microspheres	Sodium-anode	175	100	100	98
Solvothermal-derived S-Doped graphene	Sodium-anode	217	2000	1000	99
Hard carbon/Sb composite	Sodium-anode	360		200	100
Sb/graphitic carbon	Sodium-anode	280	0.1	160	101
Carbon coating on the SnO_2 nanoparticles	Sodium-anode	302	1C	200	102

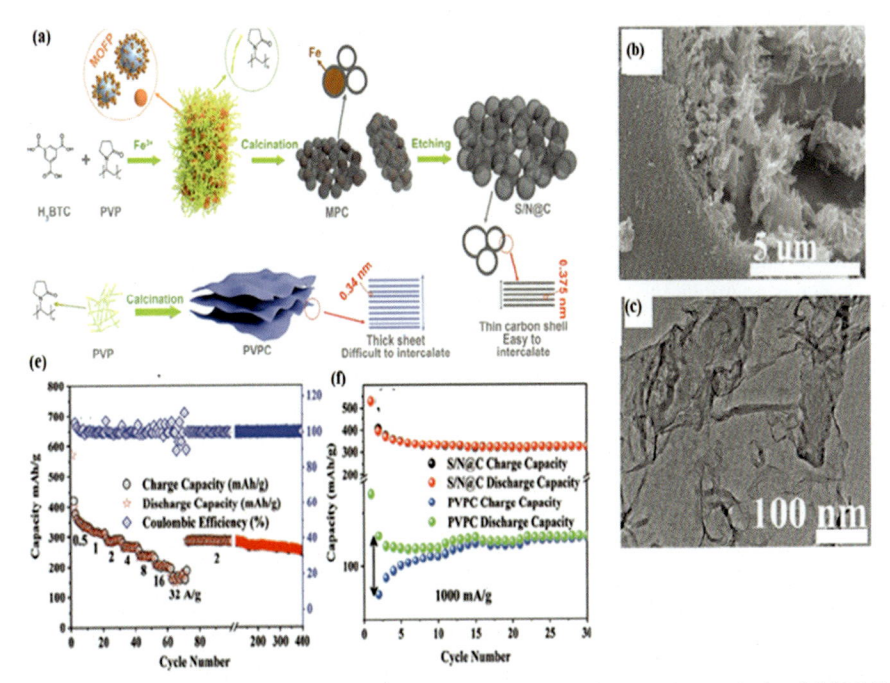

Figure 7.6 Schematic representation of S/N@C synthesis. (B,C) FESEM analysis of S/N@C. Comparative analysis of Na storage capability of different products at the current density of 1000 mA g^{-1}, (D) cyclic life at the current density of 16,000 mA g^{-1}, Copyright 2019, Wiley-VCH.[83]

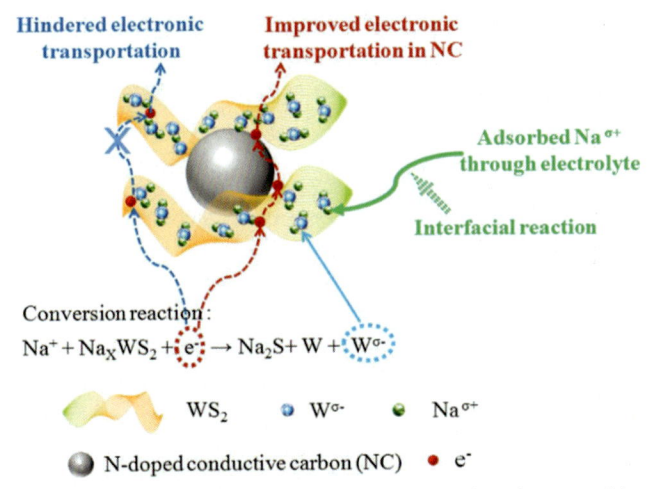

Figure 7.7 Schematic illustration of the mechanism of involving nitrogen with carbon (WS$_2$ nanosheets).[108]

exhibiting a reversible capacity of 273 mAh/g, attributed to alloying with sodium ions.[105] Kostiantyn et al. successfully prepared colloidal Sb_2S_3 nanoparticles (NPs) as Sb_2S_3 producing high specific capacity. Crystalline colloidal particles having the uniform size of crystallites (18−200 nm) offer good cycling stability material.[109] The electrical conductivity of antimony-based materials can also be enhanced by anchoring nanoparticles onto the reduced graphene oxide. The conversion of antimony sulfide nanorods has disclosed a large tolerance for structural modifications during alloying with sodium metal. Antimony sulfide (stibnite) on graphene exhibited excellent electrochemical properties in which graphene act as a template for the nanoparticles with an energy density of 80 Wh/g.[84]

Sodium-sulfur battery offer agglomeration formation of polysulfides which dissolved in the electrolyte and block the active sites. Mostly sodium sulfur electrodes suffer from the shuttling of polysulfides is a major issue in RT-Na-S batteries. The PIN coating of electrode material improves the ionic interfacial properties between sodium anodes. Song et al. reviewed several the NASICON-type electrode for instance, $Na_3Zr_2Si\ 2PO_{12}$ is used as ion separator.[110,111]

Iron sulfides and Molybdenum sulfides has been recognized as an efficient anode although low its low conductivity contributed to its volume expansion following the sodiation/desodiation process hindered its utilization. Anyhow, carbonization and nitrogen doping of Carbon improves the electrochemical properties of Fe_7S_8 and reduces the volume expansion up to 50%.[93] Carbon-based anode materials confronting the numerous issues of facile synthesis and uneven distribution of heteroatom doping faded the actual performance of Molybdenum sulfides layer. In a report, nitrogen-doped amorphous micron-sized carbon ribbons derived from biomass decorated on few layers of MoS2 nanosheets using carboxymethylcellulose (CMC) and polyvinylidene fluoride (PVDF) as binders. The respective anode material have shown excellent initial coulombic efficiency of 75% and high specific capacity of 366 $mAhg^{-1}$ at a current density of 1 Ag^{-1}.[112] In recent report, Tabassum and coworkers analyze the encapsulation of $CoSe_2$ into the boron, nitrogen-doped carbon nanotube (Fig. 7.8). The ultrafast conduction of BCN tubes and opens porous structure of material offer fast sodiation and desodiation process with negligible volume variations even after long term cycling stability test.

7.3 Metal−air batteries

Despite the great success of LIBs, another technology is needed capable of providing high energy and power density to long driving ranges (e.g., 500 km). Unfortunately, LIBs have reached its limit—i.e., LIBs can provide minimally 30% of demand in global energy. Therefore, unable to equip the transportation and promote grid scale stationary energy storage in future. This demand for high energy density and power has increasing exponentially in order to fulfill the global energy demand. To fulfill global energy demand, environmental friendly and low cost new technology and system are indispensable for long range driving of almost 500 Wh/kg.[114] As numerous technologies have been explored, among them metal−air batteries have been most sought after.

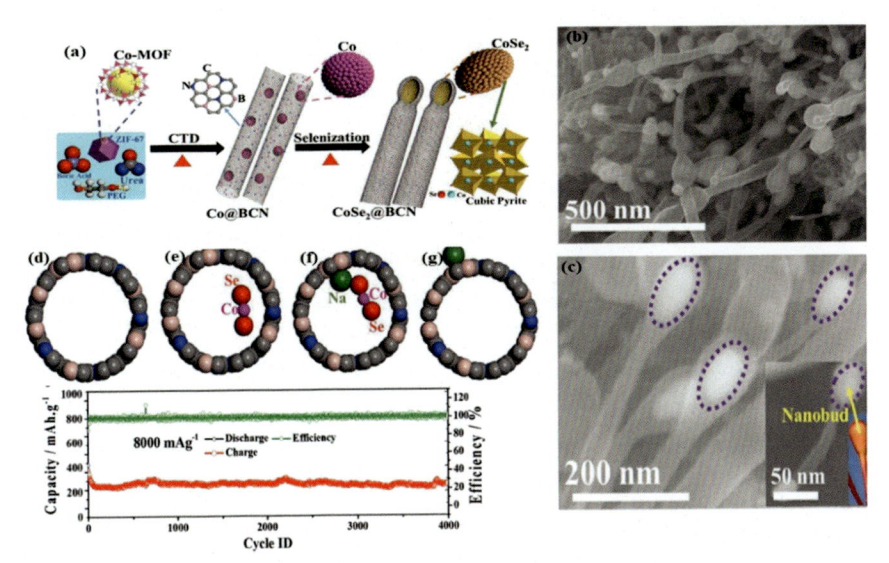

Figure 7.8 Schematic representation for the synthesis of CoSe$_2$@BCN-750 nanotubes. (A,B) FESEM images of CoSe$_2$@BCN-750 nanotubes, CoSe$_2$ NB inside the BCN nanotubes and nano-bud of eucalyptus flower. Schematic structure of (D) BCN, (E) CoSe$_2$@BCN, (F) Na adsorbed CoSe$_2$@BCN, and (G) Na adsorbed BCN nanotubes, (H) cyclic performance at high current density of CoSe$_2$@BCN-750.[97]

Generally, metal−air batteries consist of cathode feed on oxygen air and a metal anode soaked in an appropriate electrolyte. The metal anode is made of any alkali metals such as Li, Na or transition metal (Fe[115] and Zn).[116,117] There are four types of metal−air batteries depending on the type of electrolyte. The electrolyte can be aqueous, nonaqueous or aprotic solvent or protic solvents lies on the nature of anodic material being utilized. The cathode is air breathing made of open porous structure permit the flow of Oxygen from air. They have large theoretical energy density, i.e., 3−30 times more than that of LIBs.[114]

Research on metal−air batteries have begun a few decades ago. First Zn−air batteries were reported by Maiche in 1878 and its commercial products made its way to market in 1932.[118] Despite the early development in metal−air batteries, none of them have viability to replace Li-ion technology. Even though metal−air batteries have been explored for several years, numerous challenges need to consider for large-scale applications. Metal anode during self-charging confronted with various issues such as hydrogen generation, corrosion, dendritic formation, electrode decomposition lead up to poor reversible cycling stability.[119] At the same time metal−air cathode efficiency, get reduced due to lack of bifunctional OER and ORR catalyst. Side reactions generate gas diffusion which is responsible for the blockage of active sites of catalyst.[120] Therefore, in this chapter we provide the overview of metal−air batteries in aqueous or nonaqueous electrolyte and major technical hindrances for large-scale applications and discuss the electrochemistry of metal−air batteries in different electrolytes. There is another example for the electrocatalysts are

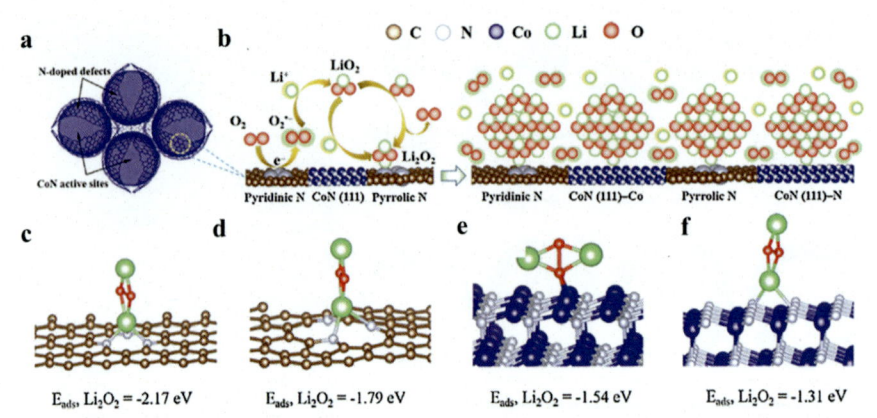

Figure 7.9 Density functional calculations and electrochemical mechanism study of BND-Co@ G-MCH. (A) Schematic of the biphasic N-doped cobalt@graphene multiple-capsule heterostructures. (B) Schematic of the proposed mechanism for reactions occurring during the discharge process. (C–F) Density functional theory (DFT) calculations showing the binding energies between Li_2O_2 monomer and active sites on the support catalyst surfaces. The pyridinic (C) and pyrrolic (D) N-doping sites on graphenes, and the CoN (111) surface with the Co (E) and N (F) terminations were considered for calculations. Copyright 2017, American Chemical Society.[113]

MOF-derived metal nitrides, oxides, and carbides for the excellent catalytic performance for Li–oxygen batteries such as oxygen reduction reaction (ORR). Similarly, the morphology of the nanohybrids of Co@G-MCH was produced in bi-steps under the flow of Argon and then ammonia and nitrogen gases, which helped for the insertion of N in carbon structure. Besides, the homogeneous morphology of the product provides sufficient space for the storage of discharge products. However, the carbon inclusion in product enhances the cyclic stability because of the large electrical conductivity. Furthermore, the Co@G-MCH was loaded on the flexible substrate of Ni foam that exhibited the high storage capacity and stability and confirmed through DFT calculation (Fig. 7.9).[113,122]

7.3.1 Aqueous Zn–air batteries

Transition metals such as Zn and Fe are thermodynamically unstable in aqueous electrolyte but due to passivation of electrode surface with their corresponding oxides and hydroxides, can be used in aqueous electrolyte. During discharge process oxidation of metal occur at the anode and oxygen molecule captured from air is diffused at cathode and get reduced to hydroxides.[123,124]

Anode: M M^{n+} + ne$^-$

Cathode: O_2 + ne$^-$ + H_2O 4OH$^-$

Where M shows the metal and n is the oxidation number of metal ions. The reverse reactions occur during charging process. On recharging, the anodic electrode plated with metal and O_2 release at the cathode. Among the aqueous metal—air batteries, Zn—air so far has been listed as most promising technology. Zn—air has high theoretical energy density (1353 Wh/kg) without including Oxygen mass with a theoretical working voltage of 1.65 V and potentially low fabrication cost.[125] But practically, its approachable energy density is 350 to 500 wh/kg very low than theoretical energy density.[123] Their applied applications are limitized because of inappropriate rate capability and inefficiency of air catalyst. Moreover, during recharging process, Zn—air batteries require large overpotential that reduces the efficiency of air cathode up to 60%. The poor cycle life of rechargeable batteries rest on the unavailability of favourable bifunctional electrocatalyst and anode.[123]

Optimization of air catalyst and cathode morphology plays significant role in accelerating battery efficiency. Air-cathode acting as limiting electrode on which oxygen reduction reaction carried out on the triple phase boundary when solid electrode is interplayed with liquid electrolyte and O_2 gas. As ORR is a very sluggish reaction. Therefore, designing the efficient bifunctional air catalyst along with favourable morphology providing large surface area would greatly enhance the efficiency of Zn—air batteries.[119,126]

The other major issue lies with the rechargeable metal—air batteries is poor cycling stability. Many-reported bifunctional electrocatalyst have low tolerance under reduction-oxidative environment during charge-discharge cycles. The cycling causes serious detrimental effect on the strategy of alloying the Zn metal with Al,[127] Si and Fe[128] or inducing additives such as organic linkers, acids, polymers and silicates.[129] These strategies have reduced the parasitic generation of hydrogen gas at Zn-anode and lessen the solubility of discharge product in the electrolyte and depress the dendritic deposition of zinc during charging ORR active site at the cathode together with electrochemical corrosion of the carbon support.[130,131] Furthermore the anode of Zn—air battery needs particularly engineering to circumvent the electrochemical corrosion and also prevent the deposition and dissolution.[132] To tackle this issue, many researchers have employed.[112]

During early development of Zinc—air batteries, Zinc was used as anodic electrode. Later, researchers have focused on the development of efficient zinc electrode by increasing the surface area of electrodes such as spheres, flakes, and fibers offers large surface area that produce large discharging currents. Owing to large surface and fibrous type zinc electrode enhances the capacity of electrode up to 40% and generate more power during discharging currents. But large surface area enhances the corrosion rate as well as provide surface for the parasitic reactions during charging and discharging reactions.[133]

In rechargeable Zinc—air batteries, main issue is the dissolution and accumulation of zinc during charging process as well as dendritic formation affect the morphology of electrode and overall reduces the life cyclibility of electrode. The above-mentioned issues reduce the contact between anode and electrolyte and lessen the ion conductivity of electrode. To enhance the life cycle of Zinc—air batteries, the formation of Zn—Ni alloy on the surface plays a significant role in reduction dendrites formation.[133,134]

Moreover the addition of additives also improves the life cycles and reduces the corrosion rate of electrodes.[135] Banik and coworkers reported the utilization of polyethyleneimine as an electrolyte additive, blocks the growth of dendrites at tip upon activation process in alkaline electrolyte.[136]

7.3.2 Nonaqueous Li–air batteries

Li air batteries, of all the rechargeable batteries, have attracted the attention of researchers, owing to its high theoretical energy density (approximately 3500 $Whkg^{-1}$) inculcating Li metal and Li_2O_2 mass. Such remarkable energy density can be attributed to "environmental oxygen" as a cathode material ultimately lessens the weight of the overall cell. Moreover, the capability of Li anode to transform high capacity at extremely low potential (-3.04 V vs. SHE), produce certain favourable discharge capacity and high operational voltage, respectively. The anode is made of Li-metal and cathode is a carbon matrix coated with catalyst.[137] The Li–air batteries can be grouped into four types of $Li–O_2$ solid-state batteries based on their electrolytes: aprotic $Li–O_2$, aqueous $Li–O_2$, mixed $Li–O_2$, and all $Li–O_2$. Of these, the aprotic assemblage has gained the most attention because it protects the Li-metal by impeding the redox reaction with electrolyte and anode while on charge and discharge.[138] In principle, the $Li–O_2$ chemistry is based on the following conversion reaction:

$$2Li + O_2 \leftrightarrow Li_2O_2 \; E0 = 2.96V \text{ versus } Li^+/Li$$

$$O_2 + Li^+ + e- \rightarrow LiO_2$$

$$LiO_2^* + Li^+ + e- \rightarrow Li_2O_2$$

$$2LiO_2 \rightarrow Li_2O_2 + O_2$$

During discharging, Li_2O_2 is produced by oxygen reduction reaction with an electromotive force of 2.9 V. Moreover, easiness in synthesizing carbon composite material and economical availability make it more preferable to other LIBs.[116] The open structure of MOF displays remarkable specific capacities for example, HKUST-1 and M-MOF-74 can reach specific capacities of 4000 to 9420 $mAhg^{-1}$.[139] The bimetallic MOF-74, induces into the $Li–O_2$ cathodes and can easily access the catalytic sites by oxides species. The Bimetallic MOF permits the double cyclibility generating 1000 $mAhg^{-1}$ at a current rate of 200 mAg^{-1}.[140] Li–air batteries work under both aqueous and nonaqueous electrolytes. Under aqueous electrolytes, the formation of LiOH hinders the clogging of cathodic electrode during discharging, while decomposition of LiOH require large overpotentials consequently lower the energy density and cycliability of Battery. Based on utilization of aprotic solvent, it must be stable against active redox reactions to produce reduced oxygen species, low volatility and high boiling point as well as high diffusivity of oxygen gas to increase the mass transport to the cathode.[141,142]

7.3.3 Sodium—air batteries

The upsurge in sodium—air batteries is related to the high abundance of sodium metal, comparatively easy extraction and processing, and low cost of sodium salts. Moreover the physiochemical similarities to the electrochemistry of Li batteries give a way toward advance development of sodium—air batteries.[116] The argument for sodium—air batteries is as elementary as that for Li—air cells. Particularly, saleable feasible batteries for transportation and grid storage piercingly stipulate specific energies near to 1700 Whkg^{-1} to be equate with gasoline. Therefore, the development of effective cathode and anode reaching high energy density is a major critical hitch for advance technology.[120]

The efficiency of sodium—air battery is tone down because electrodes cannot store sodium ions beyond their theoretical capacities, which limitize the reversible capacity of hosting and releasing the sodium ions from the anode and cathode. Notably, owing to the lower intercalation potential of sodium ion, the specific energy of SIBs is approximately near to modern state-of-the-art LIBs (ranges between 75 and 200 Whkg^{-1}) as the redox potential of the SIB is higher (-2.71 V vs. standard reduction potential) than that of the LIB (-3.01 V vs. standard reduction potential). In variance with the density of sodium is 1.7 times more than that of Li which suggest storage capacity per unit volume of sodium cells is akin to or higher than that of LIB.[143–145]

The design of Na—air batteries consists of cathode and anode made up of porous carbonaceous materials and pristine sodium metal, respectively. An aluminum foil can be employed as a separator of two electrodes holding an aprotic solvent. The oxygen molecule, while discharging, as a reactant diffuses from air to the porous carbon.[146]

Therefore, possible conversion reactions involve the following electrochemical reaction.

$$O_2 + 2e^- + 2Na^+ \rightarrow Na_2O_2$$

Owing to involvement of electrochemical reaction of Na-metal and O_2 molecule at cathode and anode, formation of superoxide/peroxides of sodium have been reported as a major discharge product. Overpotential is another limiting factor for all types of high energy and power density applied battery. In case of Na—O_2 battery it requires only 200 mV which is three-fourth times lower than the previously reported Li—O_2 cell.[147] It is previously reported that formation of superoxide (NaO_2) as major discharge product, have more storage capacity and high reversible rate requiring low overpotential (<200 mV) even at higher current densities. While the peroxide (Na_2O_2) as a major discharge product, require large overpotential ascribed to its more insulating nature than that of superoxide. Moreover the ab initio studies indicate that decomposition of superoxide and evolution of O_2 molecule during charging requires low overpotential compared with Na_2O_2, Li_2O_2, and Na_2O.[148] This can be attributed to the high electrical conductivity of NaO_2.[149,150]

Moreover, the formation of superoxide is more kinetically favored, and peroxide is thermodynamically favored reaction. Therefore, a room temperature Na—air battery performs best when electrode design favors the formation of superoxide over peroxide.[151] But still several challenges are need to address for state of art Na−air battery even when the discharge product is superoxide. For instance, dissolution of NaO_2 in the electrolyte, conversion of NaO_2 to Na_2O_2 in the presence of O^{2-} radical. Further the parasitic reactions of superoxide with air cathode leads to increase the overpotential of battery considerably and consequently reduces the cycling stability and round-trip efficiency. The choice of electrolyte and solvent greatly influences the discharge product in Na−air batteries.[147]

Hartman observe the formation of superoxide NaO_2 as a paramount emitting product. The superoxide NaO_2 is more stable than peroxide and exhibits lower Gibbs free energy (-229 kJ/mol), which corresponds to a lower current of 2.27 V.[116] Hence another route for the conversion reaction is

$$O_2 + e^- + Na^+ \rightarrow NaO_2.$$

Hartmann observed the oxygen reduction reaction at 2.25 V via cyclic voltametry technique indicating the formation of NaO_2 Even though the Na_2O_2 is thermodynamically favored contrarily the formation of superoxide is more kinetically favored as it involves the formation of one electron per unit.[116]

The formation of air cathode is pivotal in electrochemical reactions. As the discussed above, during discharging the deposition of products Na_2O_2/NaO_2 get entangled in the pores of carbonaceous electrode which in return terminates the feasible discharge reaction sterically.[152] Therefore, an optimized air electrode is needed for the enhanced electrochemical reactions. Inspite of lower specific capacity, abundant and cheap cost sodium resources have pushed scientists to make worthwhile advancements in sodium−air technology. As the matter of the fact, it is only metal−air battery hitherto exhibiting recharge ability. Hence, promotion of efficient Na−O_2 battery is indispensable to make use of its vast potential. This new class of metal−air battery was not demonstrated so far 2010 and its full potential needs to be explored.[144]

7.4 Electrode material design

In order to utilize renewable energy sources, a reliable, stable and long-term devices are prerequisite to store various types of energy in the form of chemical energy. To mitigate the limitations of energy-storage-and-conversion devices, optimization of nanostructures designs is crucial. Typically, an electrode must be able to provide large power densities at low current densities and can deliver high voltage. Therefore an electrode must possess high surface area, optimized pore size and its distribution in nanostructure material which affect the credibility of the nanoporous electrode.[69] Myriads of materials have tried to get to the bottom of the efficient electrode. However, acquiring an efficient electrode is still a challenge. Meta organic frameworks are a porous class of compounds made up of inorganic nodes twinkled in the organic linkers. Commutable nature of MOF permits the immense synthetic adjustments, rendering physical and chemical flexibilities.[153] Many properties of MOF many properties of

MOF can be tuned such as porosity, stability and particle morphology and conductivity for particular application.[154] The facile tenability of MOF also enables their utilization as a precursor or templates in creating the functional materials. The MOF can be tuned to various structures to obtain desirable characteristics with customizing of the intended synthesis process.[155] The efficiency of electrochemical energy-storage devices depends on the configuration of nanostructure including particle size, active surface area, pore size distribution, crystallinity and availability of functional groups for efficient conductivity and observance of variations in MOF structures due to reversible insertion mechanisms of metal-ion batteries(for example; Li ions (Li^+) for LIBs). The coulombic efficiency of electrochemical energy-storage devices mainly depends on the level of porosity in the MOF. As highly porous structure leads to the crossover of the fuel in the cell which ultimately harm the cell components.[156]

Most of the efficiency of electrochemical energy-storage devices is limited due to high internal resistence. MOF can modify the rigid and flexible parts through their electronic structures providing solution to mitigate the internal resistence. One of the simplest strategies to enhance the ionic conductivity of MOF is to incorporate the guest species that can assist the transport. Additionally, integration of counter ions can significantly increase the conductivity.[156] Lots of investigations have been conducted on modifying the abundant carbon nanostructures for storage purposes. Significant work has been done to highlight the crucial factors (pore size and surface area) that require adjusting the material of the desired electrode to increase the storage capacity. Also, the effect of pore size and wall thicknesses on the reversible insertion of Li^+ for high performance LIBs was examined further.[157]

Another major issue is the availability of larger pore apertures which results in generation of interpenetrating networks. The open structure of MOFs provides routs to easy diffusion of ions. But for the enhanced transport of ions, the structure of MOF should be less compact aiding facile transfer of electrons instead of high crystallinity. Additionally, the conductivity can also be augmented by introducing various functional groups creating network of different functionalities.

7.4.1 Template-derived mesoporous carbon

Mesoporous carbon has attracted the attention of researchers due to its abundant availability as well as structural stability. Production of carboneous material not only provides large capacitive storage but also increase the rate capability of electrode material to intercalate/deintercalate Li-ions for longer timer period. It also demonstrates the large surface area of carbon material, which is essential for the active redox reactions for energy-storage devices. Different strategies have employed to produce various types of mesoporous carbon.[52] As MOF consist of organic and inorganic entities, their decompositions at high temperatures produce wide variety of carbon materials, which can directly be utilized as electrode material. There are several ways to either derive mesoporous carbon by carbonization by pretreatment or post treatment. In this case, yang et al. reported formation of ordered mesoporous carbon with polypyrrole as precursor. The ordered mesoporous carbon is derived by pyrolysis of host silica on which control amount of polypyrrole is loaded. The obtained material has large surface area and pore volume.[158] NiMoO has been prepared via pyrolysis of polyvinyl alcohol as template material. The resulting material has many mesopores exhibiting large surface area (Fig. 7.10). By fabricating various structures through several synthesis techniques can also increase the surface area of carbon dramatically. Wilgosz et al. synthesize the mesoporous carbon spheres (high surface area of 666.8 m^2/g) using chemical vapor deposition technique with ethylene as a precursor. The introduction of secondary metal ions into the large cavities of porous material significantly improves the porosity and conductivity of mesoporous carbon materials. In LIBs, the conductivity and porosity can be improved by establishing the relation of material structure, pore size and electrochemical performance. Hu and its coworkers develop mesoporous carbon nanowires by soft-hard dual template approach. The comparison of the formation of various mesopores of carbon nanowires/Fe_2O_3 composites through soft, hard, and dual template method. The resulting synergistic material mesoporous carbon nanowires/Fe_2O_3 composites enhanced the conductivity and reversible Li storage capacity as high as 819 $mAhg^{-1}$ at a current density of 0.5 Ag^{-1}.[159]

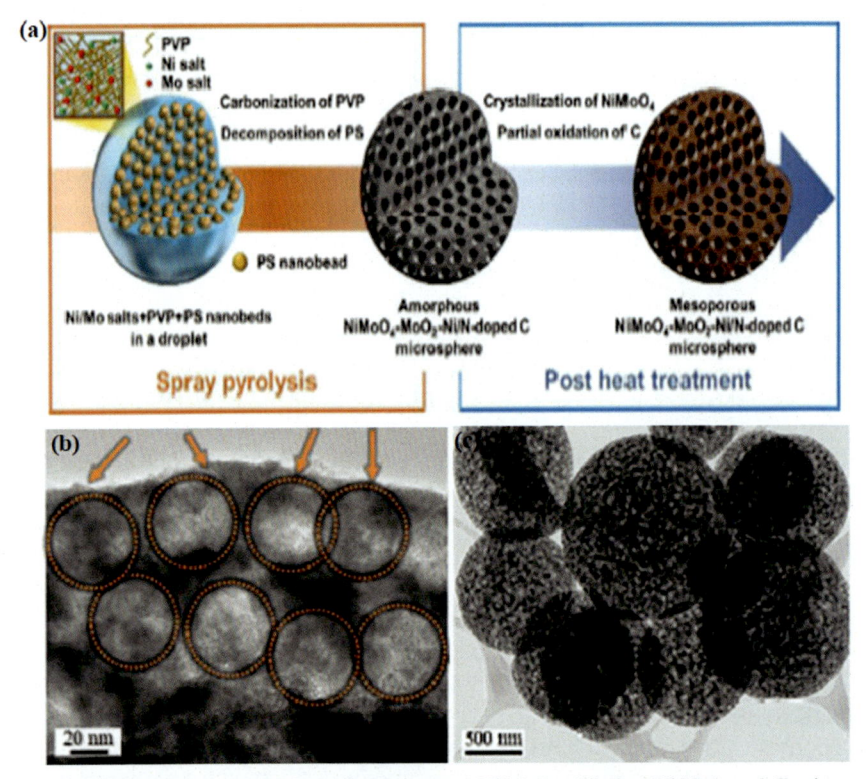

Figure 7.10 Formation mechanism of mesoporous $NiMoO_4$–MoO_2–Ni/N-doped C microsphere. Morphologies, of NiMo/C-300 microspheres prepared by spray pyrolysis and oxidation at 300°C (B) SEM images (C) TEM image showing pores of microspheres.[121]

Zhu et al. reported the scalable synthesis process for the porous carbon electrodes. They synthesized the mesoporous carbon from sustainable biomass cotton cellulose material. A MgO is incorporated to design various macropores and mesopores in carbon material following acid washing produce a mesoporous carbon with specific area of 1260 m^2/g. The cost-effective sample synthesis approach has shown remarkable electrochemical properties as anode delivered a high reversible capacity of 793 mAhg^{-1} at a current density of 0.5 Ag^{-1} after 500 cycles (Table 7.4). The carbon anode also showed a high-rate capability and a capacity of 355 mAhg^{-1} can be obtained at a current density of 4 Ag^{-1}.[179]

Carbon materials synthesized by direct carbonization of precursors or physiochemical vaporization tend to offer low surface area, disordered structures, nonuniform pore size distribution in derived carbon material. MOF-derived carbon material mitigate these issues to some extent owing to presence of high surface area and open framework, MOF-derived carbon materials have shown remarkable surface area and electrochemical properties.[155] The MOF-driven technique comprises single step facile process, and can produce various derived structures by controlling synthesis duration and concentration ratios of precursors at low cost.[178] Conversion of MOF into porous carbon material in inert atmosphere at high temperatures affords myriads of uniform distribution pores in LIBs. MOF-based materials have open up new pathways to produce various porous structures at molecular level, which ease the facile transportation of mass across the electrodes-electrolyte interphase. Liu et al. synthesized the hollow cobalt sulfide nanoparticles embedded in the graphitic nanocages using Co-based zeolitic imidazolate framework (ZIF-67) as the template.[168] Entanglement of sulfur in porous carbon nanoplate increases the surface area as well as cycling ability and delivered a discharge capacity of 730 mAh/g after 50 cycles. This can be ascribed to the conductive carbon open structure and limit the formation of polysulfides and other pores provides the channel for the fast transport of Li$^+$ ion across the separator. One of major issue for the metal-ion batteries (Li/S, Li/Se) is the dissolution of sulfides by forming polysulfide/polyselenides species into the electrolyte. Metal-derived mesoporous carbon can be employed to increase the mass transport and strict the dissolution of polysulfide species.[61] Yan et al. recently explored the sulfur-doped ordered mesoporous bimodal carbon using SiO$_2$ as template as high capacity Li/S cathode material. The material has large surface area following heat treatment and infiltration of sulfur facilitates the restriction of polysulfide formation in the electrolyte. 3D ordered bimodal mesoporous carbon with a high specific surface area of 1368.7 m^2/g. The obtained S/O BMC cathode demonstrate large discharge capacity of 1590 mAh/g and maintains 989 mAh/g after 100 cycles.[167] Similarly, various MOF-derived structures have been reported for the superior performance of the SIB. Wang et al. explored the nitrogen-doped carbon-coated Co$_3$O$_4$ nanoparticles having high capacity for Na-storage ions.[177]

7.4.2 Transition metal-based derived carbon material

Transition metal chalcogenides have attracted the huge attention owing to its feasible, low cost, thermally stable as well as high energy density and power density attributed to the alloy-conversion type chemistry for various metal-ion batteries than that of intercalation process of graphitic/graphene materials.[180,181] Moreover, the metal transition

Table 7.4 Comparison of template- and MOF-derived mesoporous carbon-based electrodes.

Sample	Precursor	Application	Rate capacity (mAh/g)	Cycle. no.	References
Hierarchical mesoporous carbon foam (ECF)	Melamine-formaldehyde foam (MF)	Li/ion- anode	215	300	160
Ordered mesoporous carbon	Polypyrrole	Li/ion- anode			158
Ordered mesoporous carbon Nanowire (OMCNW)/Fe_2O_3	FDU-15, CMK-8 and OMCNW	Li/ion- anode	819	1200	159
NiMo/C-300 composite)	Polystyrene	Li/ion- anode	693	1000	121
Ordered mesoporous $NiCo_2O_4$ microspheres	Mesoporous silica (KIT-6)	Li/ion- anode	430	100	161
Mesoporous hard carbon	Sucrose and nanoscaled calcium carbonate ($CaCO_3$)	Li/ion- anode	633	50	162
Fe_3O_4@CNT	CNT	Li/ion- anode	358	50	163
SnO_2@ordered mesoporous car	Organic tin	Li/ion- anode	787	60	164
CoxP@NC	ZIF-67	Li/ion- anode	928	100	165
Sn@NPC	1,4-BDC	Li/ion- anode	575	500	166
Si@ZIF-67	ZIF-67	Li/ion- anode	1230	500	
Ni/Co_3O_4-doped NiO	{$CoxNi_2$-x(tza)$_2$(Htza) (μ_3-O) (H_2O)]·H_2O}n	Li-anode	479	85	
S/OBMC	SiO_2 nanoarrays	Li/ion- anode	989	100	167
MGN@MC/S	Al-MIL-101-NH2	Li/S-cathode	475	50	32
N-doped carbon honeycomb (NCH nanofilms)	N-rich MOF	Li/ion-anode	609	500	29
Hollow Co_9S_8 NP embedded In GC nanocages (HCSP/GCC)	ZIF-67	Li/ion-anode	365	150	168
Nanoporous CuxS-C/C	MOF-199	Na/ion-cathode	372	110	169
Mg/MgO-embedded mesoporous carbon	Magnesium 1,4-benzenedicarboxylate	Li/S-cathode	302	200	170

Core/shell CoP@ C Polyhedrons anchored on 3D-RGO/NF	ZIF-67	Na/ion-anode	473	100	171
MnCo-MOF-74	MOF-74	Li/ion-cathode	1000	44	172
N,S-co-doped/C	2,5disulfanylterephthalic acid (BDC-(SH)2)	Li/S-cathode		100	61
Co/N-Carbon nanocubes	Co-ZIF-67	Li/ion-anode	688	100	61
Yolk–shell Co$_3$O$_4$/C dodecahedrons	ZIF-67	Li-anode	1100	120	173
CuS@Cu-BTC	Cu-BTC	Li/ion-anode	1609	200	174
C@MnO2@S@PPy	CNT	Li/S-cathode	587	500	32
Nitrogen-doped nanoporous carbons (N–NPCs)	Al-2,5-pyridinedicarboxylic acid (PDC)	Li/ion-anode	207	80	175
Co$_3$O$_4$ mesoporous NW/NF	Co-ZIFs	Li/ion-anode	1609	240	176
Co$_3$O$_4$@NC	ZIF-67	Na/ion-anode	263	1000	177
Co/S/C	ZIF-67	Li/S-anode	780	140	61
Fe$_2$O$_3$ nanotubes@Co$_3$O$_4$ composite	MIL-88B@ZIF-67	Li/ion-anode	791	30	178
N/CoO-NW	ZIF	Li/ion-anode			61

chalcogenides have more capacity to store metal ions; substantial redox activities lead to higher electrochemical properties compared with metal oxides. In order to develop transition metal-based state of art electrode material, researchers need to control the variations in volume expansion of electrode material which lead to high agglomeration of sulfides/selenide/phosphides and low conductivity. To tackle these issues, MOF-derived materials have open up a new pathway to develop highly porous, flexible and enhancing the large surface area of the desired material and reduces the pulverization of sulfides, selenides and phosphides by shortening the ion diffusion pathway.[157] Moreover, designing the various nanostructures in 1D/2D/0D decorated on the MOF-derived carbon materials is promising approach to improve the efficiency of battery. ZIF-67 has been utilized to derive highly porous carbon because it is highly porous MOF with excellent crystallinity. The large 3D structures have numerous interlamellar spacing which facilitate the alloying/extraction chemistry.[182,183] Ge et al. reported the formation of MOF-derived core shells CoPx/C polyhedron on the reduced graphene oxide/NF exhibiting the large surface area and improved electrochemical performances.[171] In another report, Liu et al. demonstrated the synthesis of hybrid tube consist of CoP particles encased in the ZIF-derived N-doped Carbon as Li-ion anode. The CoP/NC has shown remarkable electrochemical activities, capable of giving high cycling stability and rate performances even at higher current densities (1 A/g).[184] Heteroatom doping of MOF-derived material further modify the electronic states of carbon and provide electrons for feasible mass transport therefore increases the storage capacity of LIBs. Han et al. demonstrated the formation of nitrogen-doped honeycomb like structure decorated on 2D nanosheets as Li-ion anode material. The electrode material possesses the high cycling stability and storage capacity.[29] Wang et al. synthesize the Co nanoparticles entangled in the MOF-derived mesopours carbon nanocubes. Increment in capacity during electrochemical performance can be ascribed to the activation and pseudo-capacitance of material.[61] Similarly, Dai et al. reported the Tin nanodots embedded in MOF-derived N-doped Carbon as Li-ion anode. Owing to the large capacity of Co-based material for the Li-storage, as well as earth abundant material have made this element attractive toward energy-storage-and-conversion devices.[166]

Another approach to combat volumetric changes as well as improving the electrochemical properties of Li-ion battery is the development of heteroatom MOF-derived bimetallic metal oxides, offering more porous structures to buffer volume variations. The selection of ligand metal ions influences the many properties of MOF-derived material. Hu and coworkers synthesize the two new isomorphous MOFs via solvothermal approach as Li-ion anode exhibiting the extraordinary cycling stability and rate performances (Table 7.4).[185] Kim et al. synthesize the bimetallic MnCo-MOF-74 as efficient cathode material for Li–air battery facilitating the improved reverse capability and coulombic efficency.[172] Several other carboxylate-based organic ligands have been explored to derive various nanostructures of Carbon. According to reported facts that M-BTC have large capacity for Li storage due to availability of multivacant redox sites which enhance the insertion/extraction capacity of Li ions into and from the benzene ring of carboxylate ions of MOF. Recently Wang et al. prepare CuS/Cu-BTC via direct sulfidation of Cu-BTC. The synergistic effect of CuS and Cu-BTC provide large tolerance of volume variations during insertion/desertion mechanism of Li-ions and also facile transport of Li-ions is promoted owing to BTC ligands.[174]

7.5 Opportunities and challenges

For the production of highly efficient, reliable and undeleted source of energy, metal-ion batteries have provided great opportunity to fulfill the deficiency of cheap power production largely. The Nanotechnology has open up gate for pollution-free and environmentally friendly energy devices. For this purpose, MOF-derived materials have great potential to develop efficient anodic and cathodic material of battery. One of the biggest challenges in metal-ion batteries is to enhance their specific capacitance and rate cyclibility. The performance is also determined by capacity retention after prolong cycling tests. Mostly, the efficiency of LIBs is reduced by the formation of dendrites that produce thermal runaway and cell destruction, and the significant volume expansion in PIBs poses serious threats. MOF-derived anodes can play a significant role in enhancing surface area and porous nature of MOF can boost the insertion and extraction mechanism of batteries without destroying crystal structure of material that ultimately enhances the rate capability and cycling stability of metal-ion batteries. Owing to the flexible structure, redox active entities and excellent insertion/extraction chemistry of MOF are regarded as "next generation materials."

7.6 Summary and conclusions

The finding of high storage capacity and cost-effective electrodes has been an important part of the advanced research in electrochemical energy-storage technologies. Fabrication of electrodes materials for the efficient renewable energy-storage devices has been a deep interest for the materials scientists and chemists. An ideal electrode should be large storage capacity, costless, abundant resources, long cyclic stability, high capacity retention, and high coulombic efficiency. Major type of electrodes consisting of porous carbon, CNTs and metal-based nanohybrid materials. Among the several high storage capacity electrodes, the hierarchical carbon-based metal species nanohybrids have been presented tremendous growth in last decades and exhibited large electrochemical energy-storage properties. All of the above presented electrodes materials have shown suitable properties for the metal-ion storage into interlayer spacing of graphitic materials. The various methods were studied for the large ionic size intercalations into expanded interlayer spacing of 0.37 nm of graphitic-based electrodes.

In this chapter, we have summarized the progress of carbon/metal-based nanohybrids for secondary metal-ion batteries. However, the designing of efficient stable electrodes for the sodium and potassium ion storage is quite challenging and smart and facile synthetic techniques are quite needed. The major challenge with sodium and KIBs' low coulombic efficiency, stability, and storage capacity. In addition, the use of carbon@metal-based nanohybrids electrode materials can be further extended to reach the demands of modern world. These nanohybrids should be highly conductive, porous surface for the transport of electrolytes, structural stability, and active moieties. Moreover, fabrication of the electrode materials for metal−air batteries is difficult and some suitable methods are to be explored. The metal−air batteries are consisting of electrocatalytic reactions during charge/discharge reactions. For the designing of these electrocatalysts, key parameters of high catalytic activity, large mass diffusion, conductive surface, large number of active sites and mass transport should be addressed. Conclusively, the electrodes and electrocatalysts are always needed for the smart secondary metal-ion batteries which play import role for the improving the human lives through smart appliances and vehicles.

References

1. Linda ED, Barden JL, Mellish ML, Murphy BT, Hojjati PGB, Zaretskaya V, Lindstrom P. In: Administration EI, editor. *International energy outlook 2013 with projections to 2040 (IEO 2013)*. Washington, DC: U.S. DoE; 2013.
2. Tabassum H, et al. A universal strategy for hollow metal oxide nanoparticles encapsulated into B/N Co-doped graphitic nanotubes as high-performance lithium-ion battery anodes. *Adv Mater* 2018;**30**(8):1705441.
3. Kong D, et al. Seed-assisted growth of α-Fe2O3 nanorod arrays on reduced graphene oxide: a superior anode for high-performance Li-ion and Na-ion batteries. *J Mater Chem* 2016;**4**(30):11800−11.
4. Li W, et al. Carbon nanofiber-based nanostructures for lithium-ion and sodium-ion batteries. *J Mater Chem* 2017;**5**(27):13882−906.
5. Qu C, et al. MOF-derived α-NiS nanorods on graphene as an electrode for high-energy-density supercapacitors. *J Mater Chem* 2018;**6**(9):4003−12.
6. Park J, et al. Three-dimensional aligned mesoporous carbon nanotubes filled with Co3O4 nanoparticles for Li-ion battery anode applications. *Electrochim Acta* 2013;**105**:110−4.
7. El Kharbachi A, et al. Exploits, advances and challenges benefiting beyond Li-ion battery technologies. *J Alloys Compd* 2020;**817**:153261.
8. Zhang X, et al. Advanced cathodes for potassium-ion battery. *Curr Opin Electrochem* 2019;**18**:24−30.
9. Pramudita JC, et al. An initial review of the status of electrode materials for potassium-ion batteries. *Adv Energy Mater* 2017;**7**(24):1602911.
10. Sha M, et al. Anode materials for potassium-ion batteries: current status and prospects. *Carbon Energy* 2020;**2**(3):350−69.
11. Yabuuchi N, et al. Research development on sodium-ion batteries. *Chem Rev* 2014; **114**(23):11636−82.
12. Zheng X, et al. Sodium metal anodes for room-temperature sodium-ion batteries: applications, challenges and solutions. *Energy Storage Mater* 2019;**16**:6−23.
13. Fan X, et al. Battery technologies for grid-level large-scale electrical energy storage. *Trans Tianjin Univ* 2020;**26**(2):92−103.
14. Jian Z, Luo W, Ji X. Carbon electrodes for K-ion batteries. *J Am Chem Soc* 2015;**137**(36): 11566−9.
15. Skundin A, Kulova T, Andrey Y. Sodium-ion batteries (a review). *Russ J Electrochem* 2018;**54**:113−52.
16. Chen M, et al. Building high power density of sodium-ion batteries: importance of multidimensional diffusion pathways in cathode materials. *Front Chem* 2020;**8**(152).
17. Chen C, et al. Nitrogen-rich hard carbon as a highly durable anode for high-power potassium-ion batteries. *Energy Storage Mater* 2017;**8**:161−8.

18. Luo B, Ye D, Wang L. Recent progress on integrated energy conversion and storage systems. *Adv Sci* 2017;**4**(9):1700104.
19. Tabassum H, et al. Recent advances in confining metal-based nanoparticles into carbon nanotubes for electrochemical energy conversion and storage devices. *Energy Environ Sci* 2019;**12**:2924−56.
20. Tang C, et al. Ternary FexCo1−xP nanowire array as a robust hydrogen evolution reaction electrocatalyst with Pt-like activity: experimental and theoretical insight. *Nano Lett* 2016;**16**(10):6617−21.
21. Liu J, et al. Advanced energy storage devices: basic principles, analytical methods, and rational materials design. *Adv Sci* 2018;**5**(1):1700322.
22. Onar OC, Khaligh A. Chapter 2−energy sources. In: Rashid MH, editor. *Alternative energy in power electronics*. Boston: Butterworth-Heinemann; 2015. p. 81−154.
23. Yao H-R, et al. Rechargeable dual-metal-ion batteries for advanced energy storage. *Phys Chem Chem Phys* 2016;**18**(14):9326−33.
24. Nishi Y. 2-past, present and future of lithium-ion batteries: can new technologies open up new horizons? In: Pistoia G, editor. *Lithium-ion batteries*. Amsterdam: Elsevier; 2014. p. 21−39.
25. Zhang W, Liu Y, Guo Z. Approaching high-performance potassium-ion batteries via advanced design strategies and engineering. *Sci Adv* 2019;**5**(5):eaav7412.
26. Tao X-S, et al. Facile synthesis of hollow carbon nanospheres and their potential as stable Anode materials in potassium-ion batteries. *ACS Appl Mater Interfaces* 2020;**12**(11): 13182−8.
27. Alvin S, Chandra C, Kim J. Extended plateau capacity of phosphorus-doped hard carbon used as an anode in Na- and K-ion batteries. *Chem Eng J* 2020;**391**:123576.
28. Ge J, et al. MoSe2/N-Doped carbon as anodes for potassium-ion batteries. *Adv Energy Mater* 2018;**8**.
29. Han X, et al. MOF-derived honeycomb-like N-doped carbon structures assembled from mesoporous nanosheets with superior performance in lithium-ion batteries. *J Mater Chem* 2018;**6**(39):18891−7.
30. Ma G, et al. Phosphorus and oxygen dual-doped graphene as superior anode material for room-temperature potassium-ion batteries. *J Mater Chem A* 2017;**5**(17):7854−61.
31. Zhang P, et al. Dendritic core-shell nickel-iron-copper metal/metal oxide electrode for efficient electrocatalytic water oxidation. *Nat Commun* 2018;**9**(1):381.
32. Zhang X, et al. Nanoporous sulfur-doped copper oxide (Cu2OxS1−x) for overall water splitting. *ACS Appl Mater Interfaces* 2018;**10**(1):745−52.
33. Li X, et al. S-doped carbon-coated FeS2/C@C nanorods for potassium storage. *Acta Metall Sin. (Engl. Lett.)* 2021;**34**:321−8.
34. Sheng C, et al. SnS2/N-Doped Graphene as a superior stability Anode for potassium-ion Batteries by inhibiting "shuttle effect". *Batter Supercaps* 2020;**3**(1):56−9.
35. Yang W, et al. Facile synthesis of tin phosphide/reduced graphene oxide composites as anode material for potassium-ion batteries. *Ionics* 2019;**25**(10):4795−803.
36. Liu Z, et al. Tuning metallic Co0.85Se quantum dots/carbon hollow polyhedrons with tertiary hierarchical structure for high-performance potassium ion batteries. *Nano-Micro Lett* 2019;**11**(1):96.
37. Hong W, et al. Carbon quantum dot micelles tailored hollow carbon anode for fast potassium and sodium storage. *Nano Energy* 2019;**65**:104038.
38. Zhu X, et al. Self-supporting N-rich Cu2Se/C nanowires for highly reversible, long-life potassium-ion storage. *Sustain Energy Fuels* 2020;**4**(5):2453−61.

39. Sun Z, et al. Construction of bimetallic selenides encapsulated in nitrogen/sulfur Co-doped hollow carbon nanospheres for high-performance sodium/potassium-ion half/full batteries. *Small* 2020;**16**(19):1907670.

40. Zhang C, et al. Potassium prussian blue nanoparticles: a low-cost cathode material for potassium-ion batteries. *Adv Funct Mater* 2017;**27**(4):1604307.

41. Xia G, et al. Nitrogen/oxygen co-doped mesoporous carbon octahedrons for high-performance potassium-ion batteries. *J Mater Chem* 2019;**7**(19):12317−24.

42. Xu Y, et al. Highly nitrogen doped carbon nanofibers with superior rate capability and cyclability for potassium ion batteries. *Nat Commun* 2018;**9**(1). 1720-1720.

43. Yu Y, et al. Boosting potassium-ion battery performance by encapsulating red phosphorus in free-standing nitrogen-doped porous hollow carbon nanofibers. *Nano Lett* 2019;**19**.

44. Lu Y, Chen J. Robust self-supported anode by integrating Sb2S3 nanoparticles with S,N-codoped graphene to enhance K-storage performance. *Sci China Chem* 2017;**60**(12): 1533−9.

45. Ma X-X, et al. Electrochemical performance evaluation of CuO@Cu2O nanowires array on Cu foam as bifunctional electrocatalyst for efficient water splitting. *Chin J Anal Chem* 2020;**48**:e20001−12.

46. Jiao T, et al. Bismuth nanorod networks confined in a robust carbon matrix as long-cycling and high-rate potassium-ion battery anodes. *J Mater Chem* 2020;**8**(17):8440−6.

47. Wang H, et al. Lotus root-like porous carbon for potassium ion battery with high stability and rate performance. *J Power Sources* 2020;**466**:228303.

48. Yu Q, et al. Metallic octahedral CoSe(2) threaded by N-doped carbon nanotubes: a flexible framework for high-performance potassium-ion batteries. *Adv Sci* 2018;**5**(10). 1800782-1800782.

49. Hosaka T, et al. Research development on K-ion batteries. *Chem Rev* 2020;**120**(14): 6358−466.

50. Yang D, et al. Embracing high performance potassium-ion batteries with phosphorus-based electrodes: a review. *Nanoscale* 2019;**11**(33):15402−17.

51. Rajagopalan R, et al. Advancements and challenges in potassium ion batteries: a comprehensive review. *Adv Funct Mater* 2020;**30**(12):1909486.

52. Hou H, et al. Carbon anode materials for advanced sodium-ion batteries. *Adv Energy Mater* 2017:1602898.

53. Do J, et al. Towards stable Na-rich layered transition metal oxides for high energy density sodium-ion batteries. *Energy Storage Mater* 2020;**25**:62−9.

54. Liu Q, et al. Recent progress of layered transition metal oxide cathodes for sodium-ion batteries. *Small* 2019;**15**(32):1805381.

55. Jin T, et al. Polyanion-type cathode materials for sodium-ion batteries. *Chem Soc Rev* 2020;**49**(8):2342−77.

56. Zhao LN, et al. Polyanion-type electrode materials for advanced sodium-ion batteries. *Materials Today Nano* 2020;**10**:100072.

57. Wessells CD, et al. Nickel hexacyanoferrate nanoparticle electrodes for aqueous sodium and potassium ion batteries. *Nano Lett* 2011;**11**(12):5421−5.

58. Wessells CD, Huggins RA, Cui Y. Copper hexacyanoferrate battery electrodes with long cycle life and high power. *Nat Commun* 2011;**2**(1):550.

59. Luo W, et al. Encapsulating segment-like antimony nanorod in hollow carbon tube as long-lifespan, high-rate anodes for rechargeable K-ion batteries. *Nano Res* 2019;**12**(5): 1025−31.

60. Reddy ALM, et al. Synthesis of nitrogen-doped graphene films for lithium battery application. *ACS Nano* 2010;**4**(11):6337−42.

61. Wang L, et al. ZIF-67-Derived N-doped Co/C nanocubes as high-performance anode materials for lithium-ion batteries. *ACS Appl Mater Interfaces* 2019;**11**(18):16619−28.

62. Gabaudan V, et al. Double-walled carbon nanotubes, a performing additive to enhance capacity retention of antimony anode in potassium-ion batteries. *Electrochem Commun* 2019;**105**:106493.

63. Liu Y, et al. Nitrogen-doped bamboo-like carbon nanotubes as anode material for high performance potassium ion batteries. *J Mater Chem* 2018;**6**(31):15162−9.

64. Shen Q, et al. Encapsulation of MoSe2 in carbon fibers as anodes for potassium ion batteries and nonaqueous battery−supercapacitor hybrid devices. *Nanoscale* 2019;**11**(28): 13511−20.

65. Yang S, Hetterscheid DGH. Redefinition of the active species and the mechanism of the oxygen evolution reaction on gold oxide. *ACS Catal* 2020;**10**(21):12582−9.

66. Khan MA, et al. Copper ion beam irradiation-induced effects on structural, morphological and optical properties of tin dioxide nanowires. *Chin Phys Lett* 2016;**33**(7).

67. Deng J, et al. Facile synthesis of Cu2S nanoplates as anode for potassium ion batteries. *Mater Lett* 2020;**262**:127048.

68. Yuan Z, Si L, Zhu X. Three-dimensional hard carbon matrix for sodium-ion battery anode with superior-rate performance and ultralong cycle life. *J Mater Chem A* 2015;**3**(46): 23403−11.

69. Zhu J, et al. High energy batteries based on sulfur cathode. *Green Energy Environ* 2019; **4**(4):345−59.

70. Zhao X, et al. Construction of multishelled binary metal oxides via coabsorption of positive and negative ions as a superior cathode for sodium-ion batteries. *J Am Chem Soc* 2018;**140**(49):17114−9.

71. Wang N, et al. Double-walled Sb@TiO2−x nanotubes as a superior high-rate and ultralong-lifespan anode material for Na-ion and Li-ion batteries. *Adv Mater* 2016;**28**(21): 4126−33.

72. Wang S, et al. Rational design of three-layered TiO2@Carbon@MoS2 hierarchical nanotubes for enhanced lithium storage. *Adv Mater* 2017;**29**(37):1702724.

73. Sun K, et al. N-doped hard/soft double-carbon-coated Na3V2(PO4)3 hybrid-porous microspheres with pseudocapacitive behaviour for ultrahigh power sodium-ion batteries. *Electrochim Acta* 2020;**335**:135680.

74. He P, et al. Hierarchical nanotubes constructed by carbon-coated ultrathin SnS nanosheets for fast capacitive sodium storage. *Angew Chem Int Ed* 2017;**56**(40):12202−5.

75. Liu Z, et al. Sb@C coaxial nanotubes as a superior long-life and high-rate anode for sodium ion batteries. *Energy Environ Sci* 2016;**9**(7):2314−8.

76. Liu X, et al. Graphene/N-doped carbon sandwiched nanosheets with ultrahigh nitrogen doping for boosting lithium-ion batteries. *J Mater Chem* 2016;**4**(4):1423−31.

77. Share K, et al. Role of nitrogen-doped graphene for improved high-capacity potassium ion battery anodes. *ACS Nano* 2016;**10**(10):9738−44.

78. Ju Z, et al. Directly synthesis few-layer F-doped graphene foam and its lithium/potassium storage properties. *ACS Appl Mater Interfaces* 2016;**8**.

79. Komaba S, et al. Potassium intercalation into graphite to realize high-voltage/high-power potassium-ion batteries and potassium-ion capacitors. *Electrochem Commun* 2015;**60**: 172−5.

80. Liu S, et al. Novel sponge-like structure of a high-capacity nitrogen−fluorine codoped reduced graphene (N, F-rGO) film electrode and electrochemical performance in lithium-ion batteries. *J Phys Chem C* 2020;**124**(31):16739−47.

81. You Y, et al. Insights into the improved high-voltage performance of Li-incorporated layered oxide cathodes for sodium-ion batteries. *Inside Chem* 2018;**4**(9):2124–39.
82. Wang Y, et al. Improved cycle and air stability of P3-Na0.65Mn0.75Ni0.25O2 electrode for sodium-ion batteries coated with metal phosphates. *Chem Eng J* 2019;**372**:1066–76.
83. Mahmood A, et al. Ultrafast sodium/potassium-ion intercalation into hierarchically porous thin carbon shells. *Adv Mater* 2019;**31**(2):1805430.
84. Yu DYW, et al. High-capacity antimony sulphide nanoparticle-decorated graphene composite as anode for sodium-ion batteries. *Nat Commun* 2013;**4**(1):2922.
85. Prosini PP, et al. Tin-decorated reduced graphene oxide and NaLi(0.2)Ni(0.25)Mn(0.75)O(δ) as electrode materials for sodium-ion batteries. *Materials* 2019;**12**(7):1074.
86. Cui C, et al. Antimony nanorod encapsulated in cross-linked carbon for high-performance sodium ion battery anodes. *Nano Lett* 2019;**19**(1):538–44.
87. Ma X-H, et al. Synthesis of copper hexacyanoferrate nanoflake as a cathode for sodium-ion batteries. *Ceram Int* 2018;**45**.
88. Park S, et al. Phase-pure Na3V2(PO4)2F3 embedded in carbon matrix through a facile polyol synthesis as a potential cathode for high performance sodium-ion batteries. *Nano Res* 2019;**12**(4):911–7.
89. Cho JS, Lee SY, Kang YC. First introduction of NiSe2 to anode material for sodium-ion batteries: a hybrid of graphene-wrapped NiSe2/C porous nanofiber. *Sci Rep* 2016;**6**(1):23338.
90. Feng J, et al. Enhanced electrochemical stability of carbon-coated antimony nanoparticles with sodium alginate binder for sodium-ion batteries. *Prog Nat Sci Mater Int* 2018;**28**(2):205–11.
91. He H, et al. MoO2 nanosheets embedded in amorphous carbon matrix for sodium-ion batteries. *R Soc Open Sci* 2017;**4**:170892. https://doi.org/10.1098/rsos.170892.
92. Hou H, et al. Large-area carbon nanosheets doped with phosphorus: a high-performance anode material for sodium-ion batteries. *Adv Sci* 2017;**4**(1):1600243.
93. Jin A, et al. Spindle-like Fe7S8/N-doped carbon nanohybrids for high-performance sodium ion battery anodes. *Nano Res* 2019;**12**(3):695–700.
94. Zhang H-w, et al. N, S co-doped porous carbon nanospheres with a high cycling stability for sodium ion batteries. *N Carbon Mater* 2017;**32**:517–26.
95. Wu L, et al. A tin(ii) sulfide–carbon anode material based on combined conversion and alloying reactions for sodium-ion batteries. *J Mater Chem* 2014;**2**(39):16424–8.
96. Liu Y, et al. Nitrogen-doped carbon coated WS2 nanosheets as anode for high-performance sodium-ion batteries. *Front Chem* 2018;**6**(236).
97. Tabassum H, et al. Encapsulating trogtalite CoSe2 nanobuds into BCN nanotubes as high storage capacity sodium ion battery anodes. *Adv Energy Mater* 2019;**9**(39):1901778.
98. Huang L, et al. Bimetallic phosphides embedded in hierarchical P-doped carbon for sodium ion battery and hydrogen evolution reaction applications. *Sci China Mater* 2019;**62**(12):1857–67.
99. Quan B, et al. Solvothermal-derived S-doped graphene as an anode material for sodium-ion batteries. *Adv Sci* 2018;**5**(5):1700880.
100. Liu X, et al. Enhancing the anode performance of antimony through nitrogen-doped carbon and carbon nanotubes. *J Phys Chem C* 2016:120.
101. Zhao X, et al. Antimony/graphitic carbon composite anode for high-performance sodium-ion batteries. *ACS Appl Mater Interfaces* 2016;**8**(22):13871–8.
102. Kalubarme R, Lee J-Y, Park C-J. Carbon encapsulated tin oxide nanocomposites: an efficient anode for high performance sodium-ion batteries. *ACS Appl Mater Interfaces* 2015;**7**.

103. Hou H, et al. One-dimensional rod-like Sb2S3-based anode for high-performance sodium-ion batteries. *ACS Appl Mater Interfaces* 2015;**7**(34):19362−9.

104. Xiong X, et al. Enhancing sodium ion battery performance by strongly binding nano-structured Sb2S3 on sulfur-doped graphene sheets. *ACS Nano* 2016;**10**(12):10953−9.

105. Li K, Liu H, Wang G. Sb2O3 nanowires as anode material for sodium-ion battery. *Arabian J Sci Eng* 2014;**39**(9):6589−93.

106. He Q, et al. Fe(7)S(8) nanoparticles anchored on nitrogen-doped graphene nanosheets as anode materials for high-performance sodium-ion batteries. *ACS Appl Mater Interfaces* 2018;**10**(35):29476−85.

107. Lan D, Wang W, Li Q. Cu4SnP10 as a promising anode material for sodium ion batteries. *Nano Energy* 2017;**39**:506−12.

108. Wang X, et al. Improved Na storage performance with the involvement of nitrogen-doped conductive carbon into WS2 nanosheets. *ACS Appl Mater Interfaces* 2016;**8**(36):23899−908.

109. Kravchyk KV, Kovalenko MV, Bodnarchuk MI. Colloidal antimony sulfide nanoparticles as a high-performance anode material for Li-ion and Na-ion batteries. *Sci Rep* 2020;**10**(1):2554.

110. Song S, et al. A Na(+) superionic conductor for room-temperature sodium batteries. *Sci Rep* 2016;**6**. 32330-32330.

111. Kimpa MI, et al. *Review on material synthesis and characterization of sodium (Na) super-ionic conductor (NASICON).* 2018.

112. Pang Y, et al. Few-layer MoS2 anchored at nitrogen-doped carbon ribbons for sodium-ion battery anodes with high rate performance. *J Mater Chem* 2017;**5**(34):17963−72.

113. Tan G, et al. Toward highly efficient electrocatalyst for Li−O2 batteries using biphasic N-doping Cobalt@Graphene multiple-capsule heterostructures. *Nano Lett* 2017;**17**(5):2959−66.

114. Li Y, Lu J. Metal−air batteries: will they Be the future electrochemical energy storage device of choice? *ACS Energy Lett* 2017;**2**(6):1370−7.

115. McKerracher R, et al. A review of the iron−air secondary battery for energy storage. *ChemPlusChem* 2014;**80**.

116. Hartmann P, et al. A comprehensive study on the cell chemistry of the sodium superoxide (NaO2) battery. *Phys Chem Chem Phys* 2013;**15**(28):11661−72.

117. Gilligan GE, Qu D. Chapter 12−zinc-air and other types of metal-air batteries. In: Menictas C, Skyllas-Kazacos M, Lim TM, editors. *Advances in batteries for medium and large-scale energy storage.* Woodhead Publishing; 2015. p. 441−61.

118. Zhang J, et al. Zinc−air batteries: are they ready for prime time? *Chem Sci* 2019;**10**(39):8924−9.

119. Yang G, et al. Carbon-based alloy-type composite anode materials toward sodium-ion batteries. *Small* 2019;**15**(22):1900628.

120. Wang C, et al. Recent progress of metal−air batteries—a mini review. *Appl Sci* 2019;**9**(14).

121. Oh SH, et al. Three-dimensionally ordered mesoporous multicomponent (Ni, Mo) metal oxide/N-doped carbon composite with superior Li-ion storage performance. *Nanoscale* 2018;**10**(39):18734−41.

122. Dong Y, et al. Metal-organic frameworks and their derivatives for Li−air batteries. *Chin Chem Lett* 2020;**31**(3):635−42.

123. Caramia V, Bozzini B. Materials science aspects of zinc-air batteries: a review. *Mater Renew Sustain Energy* 2014;**3**.

124. Liu J, Liu X-W. Two-dimensional nanoarchitectures for lithium storage. *Adv Mater* 2012; **24**(30):4097−111.
125. Shin J, et al. Aqueous zinc ion batteries: focus on zinc metal anodes. *Chem Sci* 2020;**11**(8): 2028−44.
126. Cai X, et al. Recent advances in air electrodes for Zn−air batteries: electrocatalysis and structural design. *Mater Horiz* 2017;**4**(6):945−76.
127. Durmus YE, et al. Influence of Al alloying on the electrochemical behavior of Zn electrodes for Zn−air batteries with neutral sodium chloride electrolyte. *Front Chem* 2019; **7**(800).
128. Weinrich H, et al. Silicon and iron as resource-efficient anode materials for ambienttemperature metal-air batteries: a review. *Materials* 2019;**12**(13):2134.
129. Mainar AR, et al. Enhancing the cycle life of a zinc−air battery by means of electrolyte additives and zinc surface protection. *Batteries* 2018;**4**(3):46.
130. Yang D, et al. Electrode materials for rechargeable zinc-ion and zinc-air batteries: current status and future perspectives. *Electrochem Energy Rev* 2019;**2**(3):395−427.
131. Pan J, et al. Advanced architectures and relatives of air electrodes in Zn−air batteries. *Adv Sci* 2018;**5**(4):1700691.
132. Tomboc GM, et al. Ideal design of air electrode—a step closer toward robust rechargeable Zn−air battery. *Apl Mater* 2020;**8**(5):050905.
133. Gu P, et al. Rechargeable zinc−air batteries: a promising way to green energy. *J Mater Chem A* 2017;**5**(17):7651−66.
134. Yi J, et al. Challenges, mitigation strategies and perspectives in development of zincelectrode materials and fabrication for rechargeable zinc−air batteries. *Energy Environ Sci* 2018;**11**(11):3075−95.
135. Trudgeon DP, et al. Screening of effective electrolyte additives for zinc-based redox flow battery systems. *J Power Sources* 2019;**412**:44−54.
136. Banik S, Akolkar R. Suppressing dendritic growth during alkaline zinc electrodeposition using polyethylenimine additive. *Electrochim Acta* 2014:179.
137. Imanishi N, Yamamoto O. Perspectives and challenges of rechargeable lithium−air batteries. *Mater Today Adv* 2019;**4**:100031.
138. Balaish M, Kraytsberg A, Ein-Eli Y. A critical review on lithium−air battery electrolytes. *Phys Chem Chem Phys* 2014;**16**(7):2801−22.
139. Wu D, et al. Metal-organic frameworks as cathode materials for Li-O2 batteries. *Adv Mater* 2014;**26**(20):3258−62.
140. Yan W, et al. Downsizing metal−organic frameworks with distinct morphologies as cathode materials for high-capacity Li−O2 batteries. *Mater Chem Front* 2017;**1**(7): 1324−30.
141. Lai J, et al. Electrolytes for rechargeable lithium−air batteries. *Angew Chem Int Ed* 2020; **59**(8):2974−97.
142. Li OL, Ishizaki T. Chapter 4 - development, challenges, and prospects of carbon-based electrode for lithium-air batteries. In: Cheong KY, Impellizzeri G, Fraga MA, editors. *Emerging materials for energy conversion and storage*. Elsevier; 2018. p. 115−52.
143. Das S, Lau S, Archer L. Sodium-oxygen batteries: a new class of metal-air batteries. *J Mater Chem A* 2014;**2**.
144. Yin W-W, Fu Z-W. The potential of Na−air batteries. *ChemCatChem* 2017;**9**(9): 1545−53.
145. Yadegari H, Sun X. Recent advances on sodium−oxygen batteries: a chemical perspective. *Acc Chem Res* 2018;**51**(6):1532−40.

146. Sahgong SH, et al. Rechargeable aqueous Na−air batteries: highly improved voltage efficiency by use of catalysts. *Electrochem Commun* 2015;**61**:53−6.

147. Bi X, et al. A critical review on superoxide-based sodium−oxygen batteries. *Small Methods* 2019;**3**(4):1800247.

148. Schröder D, et al. How to control the discharge product in sodium−oxygen batteries: proposing new pathways for sodium peroxide formation. *Energy Technol* 2017;**5**(8): 1242−9.

149. Yang S, Siegel DJ. Intrinsic conductivity in sodium−air battery discharge phases: sodium superoxide vs sodium peroxide. *Chem Mater* 2015;**27**(11):3852−60.

150. Mekonnen Y, et al. Thermodynamic and kinetic limitations for peroxide and superoxide formation in Na−O2 batteries. *J Phys Chem Lett* 2018;**9**.

151. Sun Q, Yang Y, Fu Z-W. Electrochemical properties of room temperature sodium−air batteries with non-aqueous electrolyte. *Electrochem Commun* 2012;**16**(1):22−5.

152. Liu S, Liu S-S, Luo J-Y. Carbon-based cathodes for sodium-air batteries. *N Carbon Mater* 2016;**31**(3):264−70.

153. Xiao X, et al. Synthesis of micro/nanoscaled metal−organic frameworks and their direct electrochemical applications. *Chem Soc Rev* 2020;**49**(1):301−31.

154. Wu C-W, et al. Importance of cobalt-doping for the preparation of hollow CuBr/Co@CuO nanocorals on copper foils with enhanced electrocatalytic activity and stability for oxygen evolution reaction. *ACS Sustainable Chem Eng* 2020;**8**(26):9794−802.

155. Fominykh K, et al. Iron-doped nickel oxide nanocrystals as highly efficient electrocatalysts for alkaline water splitting. *ACS Nano* 2015;**9**(5):5180−8.

156. Li C, et al. Heteroatomic interface engineering in MOF-derived carbon heterostructures with built-in electric-field effects for high performance Al-ion batteries. *Energy Environ Sci* 2018;**11**(11):3201−11.

157. Zhang Q, et al. All-metal-organic framework-derived battery materials on carbon nanotube fibers for wearable energy-storage device. *Adv Sci* 2018;**5**(12):1801462.

158. Yang C-M, et al. Facile template synthesis of ordered mesoporous carbon with polypyrrole as carbon precursor. *Chem Mater* 2005;**17**(2):355−8.

159. Hu J, et al. Dual-template ordered mesoporous carbon/Fe2O3 nanowires as lithium-ion battery anodes. *Nanoscale* 2016;**8**(26):12958−69.

160. Chen Y, et al. Nitrogen-doped carbon for sodium-ion battery anode by self-etching and graphitization of bimetallic MOF-based composite. *Inside Chem* 2017;**3**(1):152−63.

161. Ren Q, et al. Highly ordered mesoporous NiCo2O4 as a high performance anode material for Li-ion batteries. *Front Chem* 2019;**7**(521).

162. Xu B, et al. Nano-CaCO3 templated mesoporous carbon as anode material for Li-ion batteries. *Electrochim Acta* 2011;**56**:6464−8.

163. Yang L, et al. Novel Fe3O4-CNTs nanocomposite for Li-ion batteries with enhanced electrochemical performance. *Electrochim Acta* 2014;**144**:235−42.

164. Wang X, Zhaoqiang L, Yin L. Nanocomposites of SnO2@ordered mesoporous carbon (OMC) as anode materials for lithium-ion batteries with improved electrochemical performance. *CrystEngComm* 2013;**15**:7589.

165. Ge X, Li Z, Yin L. Metal-organic frameworks derived porous core/shellCoP@C polyhedrons anchored on 3D reduced graphene oxide networks as anode for sodium-ion battery. *Nano Energy* 2017;**32**:117−24.

166. Dai R, Sun W, Wang Y. Ultrasmall tin nanodots embedded in nitrogen-doped mesoporous carbon: metal-organic-framework derivation and electrochemical application as highly stable Anode for lithium ion batteries. *Electrochim Acta* 2016;**217**:123−31.

167. Song Y, et al. High capacity and rate capability of S/3D ordered bimodal mesoporous carbon cathode for lithium/sulfur batteries. *J Mater Res* 2019;**34**(4):600−7.

168. Liu J, et al. MOF-derived hollow Co9S8 nanoparticles embedded in graphitic carbon nanocages with superior Li-ion storage. *Small* 2016;**12**.

169. Kang C, et al. Highly efficient nanocarbon coating layer on the nanostructured copper sulfide-metal organic framework derived carbon for advanced sodium-ion battery anode. *Materials* 2019;**12**(8):1324.

170. Dhawa T, et al. In situ Mg/MgO-embedded mesoporous carbon derived from magnesium 1,4-benzenedicarboxylate metal organic framework as sustainable Li-S battery cathode support. *ACS Omega* 2017;**2**(10):6481−91.

171. Ge X, Zhaoqiang L, Yin L. Metal-organic frameworks derived porous core/ShellCoP@C polyhedrons anchored on 3D reduced graphene oxide networks as anode for sodium-ion battery. *Nano Energy* 2016;**32**.

172. Kim SH, et al. Bimetallic metal−organic frameworks as efficient cathode catalysts for Li−O2 batteries. *ACS Appl Mater Interfaces* 2018;**10**(1):660−7.

173. Wu Y, et al. Interface-modulated fabrication of hierarchical yolk−shell Co3O4/C dodecahedrons as stable anodes for lithium and sodium storage. *Nano Res* 2017;**10**: 2364−76.

174. Wang P, et al. MOF-derived CuS@Cu-BTC composites as high-performance anodes for lithium-ion batteries. *Small* 2019;**15**(47):1903522.

175. Shi X, et al. Multifunctional nitrogen-doped nanoporous carbons derived from metal− organic frameworks for efficient CO2 storage and high-performance lithium-ion batteries. *New J Chem* 2019;**43**(26):10405−12.

176. Tabassum H, et al. Large-scale fabrication of BCN nanotube architecture entangled on a three-dimensional carbon skeleton for energy storage. *J Mater Chem* 2018;**6**(42): 21225−30.

177. Wang Y, et al. Superior sodium-ion storage performance of Co3O4@nitrogen-doped carbon: derived from a metal−organic framework. *J Mater Chem* 2016;**4**(15):5428−35.

178. Zhang SL, et al. Metal−organic framework-assisted synthesis of compact Fe2O3 nanotubes in Co3O4 host with enhanced lithium storage properties. *Nano-Micro Lett* 2018; **10**(3):44.

179. Zhu C, Akiyama T. Cotton derived porous carbon via an MgO template method for high performance lithium ion battery anode. *Green Chem* 2015;**18**.

180. Iqbal S, et al. Recent development of carbon based materials for energy storage devices. *Mater Sci Energy Technol* 2019;**2**(3):417−28.

181. Nishihara H, Kyotani T. Templated nanocarbons for energy storage. *Adv Mater* 2012; **24**(33):4473−98.

182. Zhang P, et al. ZIF-derived porous carbon: a promising supercapacitor electrode material. *J Mater Chem A* 2014;**2**(32):12873−80.

183. Wang F, et al. From ZIF nanoparticles to hierarchically porous carbon: toward very high surface area and high-performance supercapacitor electrode materials. *Inorg Chem Front* 2019;**6**(1):32−9.

184. Liu Y, et al. ZIF-67 derived carbon wrapped discontinuous CoxP nanotube as anode material in high-performance Li-ion battery. *Mater Today Chem* 2020;**17**:100284.

185. Hu B-W, et al. Heterometallic Metal-Organic Frameworks approach to enhancing lithium storage for their derivatives as anodes materials. *Inorg Chim Acta* 2019;**494**:1−7.

Metal oxide–carbon nanofibers based composites for supercapacitors and batteries

Sadia Khalid[1], Ashir Saeed[1,2], Mohammad Azad Malik[3] and Muhammad Saeed Akhtar[4]

[1]Nanosciences & Technology Department, National Centre for Physics, Quaid-i-Azam University Campus, Islamabad, Pakistan; [2]Department of Physics, Khwaja Fareed University of Engineering and Information Technology, Rahim Yar Khan, Punjab, Pakistan; [3]School of Materials, The University of Manchester, Manchester, United Kingdom; [4]Department of Physics, Division of Science and Technology, University of Education Lahore, Lahore, Punjab, Pakistan

Abbreviations

ACNFs	activated carbon nanofibers
BET	Brunauer–Emmett–Teller
CNFs	carbon nanofibers
CNTs	carbon nanotubes
CV	cyclic voltammetry
EIS	electrochemical impedance spectroscopy
ESR	electrochemical series resistance
GCD	galvanostatic charge–discharge test
LABs	lithium-air batteries
LIBs	lithium-ion batteries
LSBs	lithium-sulfur batteries
MO–CNFs	metal oxide-based carbon nanofibers
MOs	metal oxides
NCNFs	nitrogen-enriched carbon nanofibers
PAN	polyacrylonitrile
PCNFs	porous carbon nanofibers
SCB	sugarcane bagasse
SCs	supercapacitors
SIBs	sodium-ion batteries
VGCFs	vapor-grown carbon nanofibers

8.1 Introduction

Metal oxides are attractive, potentially ideal electrode materials owing to their promising electrochemical properties for use in energy storage and conversion devices,

Metal Oxide-Carbon Hybrid Materials. https://doi.org/10.1016/B978-0-12-822694-0.00019-3

including lithium-air batteries (LABs), lithium-ion batteries (LIBs), lithium-sulfur batteries (LSBs), sodium-ion batteries (SIBs), and supercapacitors (SCs). Batteries exhibit well-defined recharging properties, high storage capacity, high power density, cycling stability, and environmentally friendly behavior, whereas SCs exhibit high specific capacitance, cyclic stability, and greater power density. Metal oxides have limited use as battery electrodes because of the variations in surface volume charge and aggregation issues during the charging/discharging cycle. Therefore, it has been a great challenge to fabricate an electrode for energy-storing devices that possesses good performance with stability over many charging/discharging cycles, retains capacity maintenance, and can increase its capability rate.

To date, various materials and approaches have been employed to address the key issues in batteries and SCs. One-dimensional nanostructures, e.g., carbon nanofibers (CNFs), are attractive materials for energy-storage devices and can be produced commercially. The large surface-to-volume ratio, good mechanical stability, and high specific surface area result in good performance over many charging/discharging cycles, thermal stability, and environmentally friendly interaction owing to attractive electrochemical characteristics.[1] Moreover, the efficiency of energy-storage devices can be improved by incorporating metal oxides in CNFs. These metal oxide-based CNFs (MO−CNFs) enhance electrode performance with favorable characteristics that are not feasible with individual materials. These MO−CNFs have excellent performance when used as an electrode in LIBs and SCs. Metal oxide-based CNFs periodically embedded in the matrices of CNFs can produce stable lithium superoxides without transformation between gaseous and solid phases during the charging/discharging cycle.[2]

Diverse metal oxide-based CNF composites have been studied to advance energy-storage devices, including silicon oxide, iron oxide, zinc oxide, manganese oxide, tin oxide, titanium oxide, and nickel oxide. Many techniques have been employed to synthesize the metal oxide-based CNFs, such as photoablation, physical vapor deposition, chemical vapor deposition, electrospinning, the template-assisted method, sol−gel, and the hydrothermal deposition method. Among these techniques, electrospinning exhibits many advantages because it is a simple, cost-effective, and commercial mass production technique. Electrospinning can produce metal oxide-based CNFs with high volume-to-surface ratios. Various materials can be used to synthesize a variety of nanofibers.

The modification in conductivity of metal oxides by reinforcing with CNFs has effectively boosted the performance of these materials. This chapter discusses the recent advancement in metal oxide−CNF-based composites for energy-storage devices such as SCs and batteries. The current issues, challenges, improvements, and future perspectives of metal oxide−CNFs are demonstrated in detail.

8.2 Metal oxides

Metal oxides are candidates for electronics,[3] photoelectrochemical water splitting,[4] photocatalysis,[5]   piezoelectricity,[6]   biological,[7]   energy,   and   environmental applications.[8] Closely packed metal oxides structures can be formed by the coordination of metal ions with oxides. MOs exhibit exceptional physical, structural, chemical, optical, electrical, and electronic properties owing to their electronic configuration. Manganese oxide (MnO_2), titanium oxide (TiO_2), iron oxide (Fe_2O_3, Fe_3O_4), cobalt oxide (Co_3O_4, CoO), nickel oxide (NiO), copper oxide (CuO, Cu_2O), zinc oxide (ZnO), tin oxide (SnO_2), vanadium oxide (VO_x), molybdenum oxide (MoO_x), silicon oxide (SiO_2), tungsten oxide (WO_3), zirconium oxide (ZrO_2), and cerium oxide (CeO_2) exhibit diverse properties depending on their crystal structures and oxidation states.[9]

Besides inherent characteristics, metal oxide nanostructures can be synthesized with unique features unlike those of their bulk counterparts. The high surface-to-volume ratio, quantum confinement effect, and more active sites enhance the reactivity and ultimately the material's performance.

8.2.1 Synthesis of metal oxides

Generally, the synthesis methods of metal oxide nanoparticles can be divided into three groups: (1) physical methods, (2) chemical, and (3) biological methods (Table 8.1). Chemical methods allow rational material design based on phase, morphology, dimensions, and composition according to the area of implication. Coprecipitation, hydrothermal/solvothermal, and sol—gel methods are the most widely used wet chemical methods.[10,11]

The synthesis method defines the quality, reproducibility, and scalability of the material accompanied by environmentally friendly precursors. For example, the synthesis of metal oxides through sol—gel is one of the most economical and resource-saving approaches in time and energy. The primary steps involved in a typical sol—gel method are hydrolysis, condensation, aging, drying, and calcination. The synthesis of metal oxides with various dimensions (0-D, 1-D, 2-D, 3-D) and forms (powder, porous, mesoporous) at low temperatures is a fascinating feature of this procedure. The physiochemical properties can be easily tailored using a variety of molecular precursors, experimental conditions (pH, time, temperature, concentration, etc.), templates, and chelating agents.[12]

Table 8.1 Synthesis methods for metal oxides.

Physical methods	Chemical methods	Biological methods
Ball Milling	Sol—gel	Plants
Sputtering	Polyol	Algae
Electron Beam Deposition	Hydrothermal	Fungi
Laser Ablation	Coprecipitation	Bacteria
Thermal Evaporation	Microemulsion	
Electrospraying	Chemical Vapor Deposition	

8.3   Carbon nanofibers

The carbon family (fullerenes, carbon nanotubes, graphene, nanodiamonds, etc.) displays exceptional physicochemical properties with nano dimensions. One-dimensional fibrous carbon nanostructures such as carbon nanotubes and CNFs exhibit high aspect (length-to-diameter) ratios, more surface area, and exceptional transport properties.[13] CNFs are sp2-based noncontinuous linear filaments (ca. diameter: 100 nm, aspect ratio >100), unlike conventional carbon fibers with far greater diameters (Fig. 8.1).[14]

CNFs are attractive candidates for sustainable energy-storage applications with reference to their facile and economical fabrication methods. On the other side, their high electrical conductivity, chemical, mechanical, and thermal stability make them favorable electrode materials for batteries and SCs. CNF nanomaterials have been applied in various applications such as energy storage/conversion devices, catalysis, gas sensors, microwave/radiation adsorption, water purification, and biomedical engineering.[15-17]

8.3.1   Synthesis of carbon nanofibers

Numerous fabrication methods are being used to produce CNFs, such as electrospinning, catalytic chemical vapor deposition, arc discharge, and laser ablation.[18]

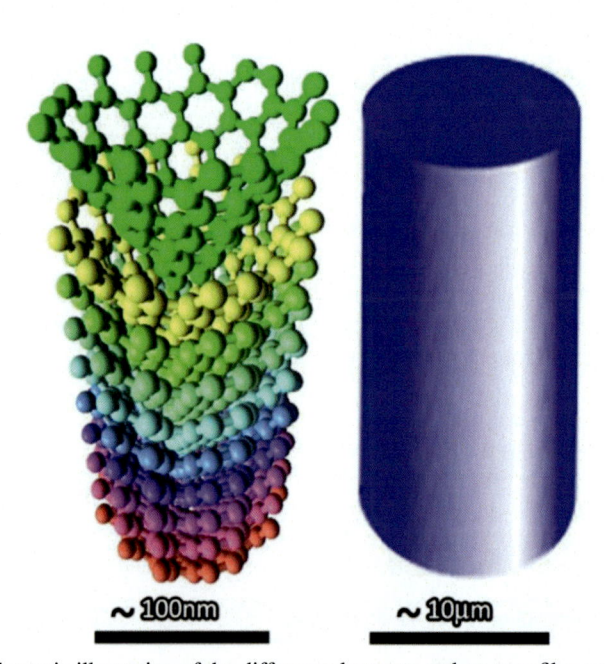

Figure 8.1 Schematic illustration of the difference between carbon nanofiber and conventional carbon fiber.[14]
Reproduced from Ref. 14 Copyright (2014), with permission from MDPI under the Creative Commons Attribution license.

Electrospinning is the most widely used flexible technique for producing nanofibers with a diameter ranging from 10 to 1000 nm. A high voltage is applied between the grounded substrate and syringe to charge the polymer solution in a typical electrospinning approach. The electric force is due to applied voltage causing a jet of the solution. The diameter of the moving jet narrows and is continuously drawn as the electric force and charge repulsion. The jet is converted into a solid phase by evaporation, resulting in the deposition of thin fibers onto the targeted substrate (Fig. 8.2).[19]

These fibers can be deposited as a web network using mesh or metal plate as a target and can also be collected in a well-arranged pattern by applying constant speed. The dimensions of fibers can be controlled by optimizing the distance between the targeted substrate and syringe tip, concentration of materials, and adjusting the voltage. The major issue in the fabrication of CNF by electrospinning is the control over morphology, pore size distribution, and surface area. These issues can be addressed by using optimized experimental conditions. Mass production through inexpensive electrospinning is also a challenging task because of low yield.[20]

Carbon-based polymeric precursors like polyacrylonitrile (PAN), polyvinyl alcohol, polyimide, phenolic resin, propylene glycol monomethyl ether acetate, pitch, and cellulose are used in the electrospinning technique.[21] PAN is the most common precursor because of the product's high carbon yield and superior properties.[22]

For example, Chen et al. recently reported the synthesis of CNFs using sugarcane bagasse (SCB) by electrospinning technique. SCB powder was pretreated with toluene-ethanol for 4 h followed by air-drying in appropriate conditions. Then dispersion of SCB powder in a dual solvent system containing dimethyl sulfoxide/

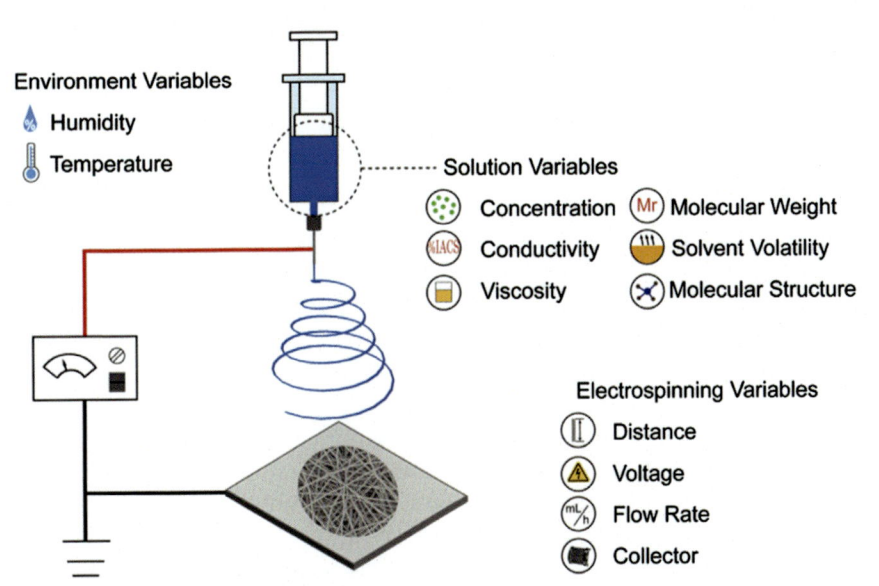

Figure 8.2 Schematic diagram of a conventional electrospinning setup, as well as environment, solution, and electrospinning variables.[19]
Reproduced from Ref. 19 Copyright (2019), with permission from Elsevier.

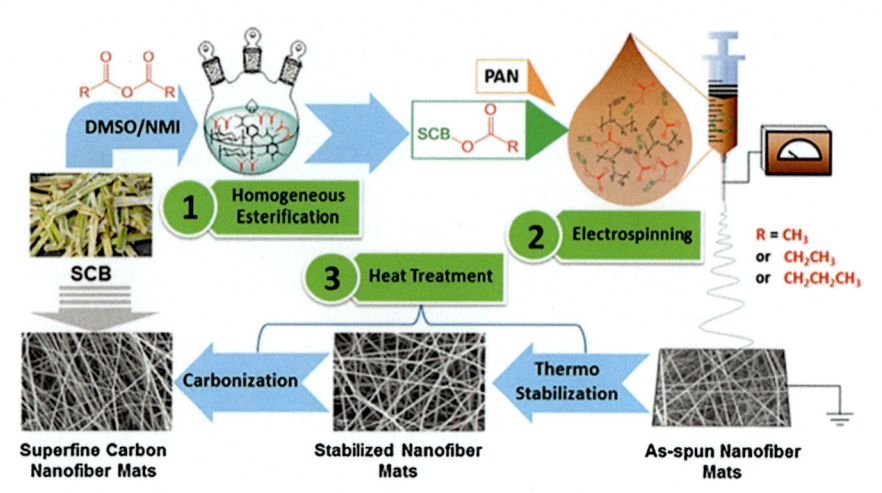

Figure 8.3 Schematic description of the preparation of superfine carbon nanofiber mats derived from sugarcane bagasse.[23]
Reproduced from Ref. 23. Copyright 2019, with permission from MDPI Creative Common CC BY license.

N-methylimidazole was used for esterification. The CNFs were prepared by electrospinning followed by carbonization process at 800°C (Fig. 8.3).[23]

8.4   Metal oxide–carbon nanofiber based composites

Metals and metal oxides have been used extensively as electrode materials for batteries and SCs owing to their promising electrochemical properties. Nevertheless, the limiting performance issues that need to be resolved are low electronic and electrical conductivity, hindered ion permeation, and diffusion. Consequently, poor cyclic performance and low-rate capability lead to short storage device lifetimes.

CNF enhances the reactivity of the electrode materials by providing more active surface area. It also provides fast diffusion of the ions/electrons, hence increasing the composite material's permeability by providing fast transport channels. One of the basic advantages of using CNF is the flexibility of designing practical devices. The interactive forces of CNF couple the composite material to give it strength and structural stability.

CNF-based materials are captivating more attention in the recent research and development (R&D) of energy materials.[2] Nanocomposites combine the features of individual materials with the ability to design a highly efficient product. Metal oxide–CNF-based composites have numerous applications (Fig. 8.4). CNFs provide active support to the metal oxides and offers the flexible modification of basic functional features. MO–CNF composites have been widely applied in SCs,[24] lithium–sulfur batteries (LSBs),[25] SIBs,[26] photocatalysis,[27] fuel cells,[28] sensing,[29] and biomedical applications.[30]

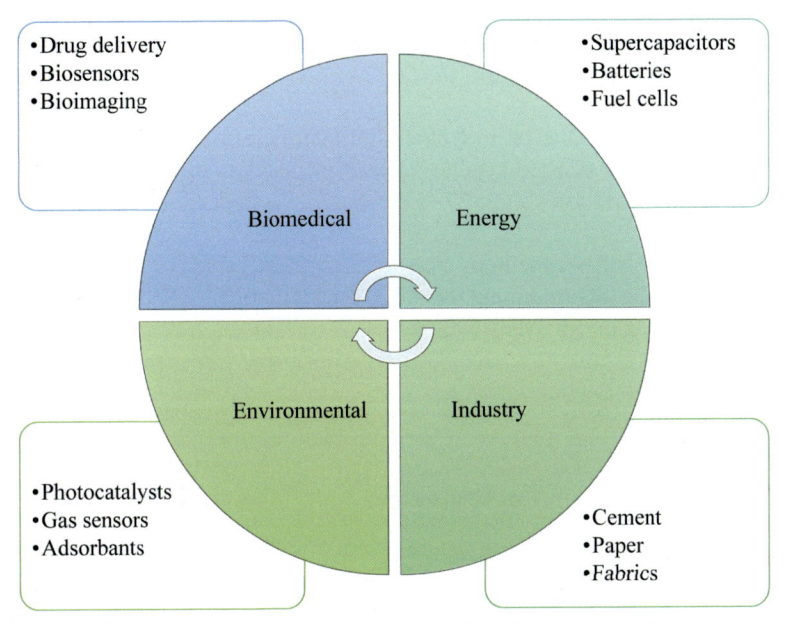

Figure 8.4 Applications of metal oxide—carbon nanofiber-based composites.

Nanocomposites offer a fascinating system to couple metal oxides with CNFs for the better performance of devices.[31] Basic features of CNFs such as porous architectures, larger specific surface areas, more active sites, and structural designability can resolve the major issues.[26] Recently, Hu et al. reported the preparation of hollow CNFs (specific surface area: 2628.10 m^2/g, pore volume: 2.32 cm^3/g) by NaOH activation of the electrospinning silicon oxy carbonitride (SiCNO) fibers, followed by carbonization at 800°C. The porous hollow CNFs exhibited a specific capacity of 259.86 F/g; capacitance retention of 95.3% after 10,000 cycles; and energy density of 12.99 W h/kg at 1.0 A/g of current density.[32]

Metal oxide - CNFs composites

- Inhibit the agglomeration of active materials
- Enhanced electronic conductivity
- Larger specific surface area
- More electrochemical active sites
- Effective electrolyte permeation
- Improved chemical and thermal stability
- Superior mechanical strength
- Structural flexibility

8.5   Synthesis of metal oxide—carbon nanofiber based composites

MO—CNF composites have been synthesized using electrospinning, hydrothermal/solvothermal, template assisted synthesis, electrodeposition, electrospray, etc. Recent R&D in synthesis procedures and MO—CNF-based electrodes for SCs and batteries is discussed next.

Some of these synthesis methods are summarized in Table 8.2.

For example, the incorporation of Co_3O_4 nanoparticles (10 ± 5 nm) in multichannel CNFs (P—Co—MCNFs) was achieved by oxygen plasma treatment at an

Table 8.2 Synthesis methods for metal oxide—carbon nanofiber composites.

Sr. No.	Composite	Synthesis method	Application	Year	Ref.
1	MnO_2 NW/CNF	Electrospinning, carbonization, and electrodeposition	SCs	2017	[33]
2	Fe_3O_4 doped CNF/ MnO_2	Electrospinning, carbonization, and electrospray	SCs	2017	[34]
3	CNF—CNT-Co_3O_4	Preoxidation, carbonization, and CO_2 activation	SCs	2019	[35]
4	Co_3O_4- MCNFs	Oxygen plasma exposure	SCs	2020	[36]
5	CNF—Ni@Fe_2O_3 and CNF—Ni@ MnO_2	Electrospinning, carbonization, and electrodeposition Solution-phase assembly method	SCs	2020	[37]
6	ZnO@CNF	Electrospinning carbonization	SIBs	2020	[38]
7	CoO_x@NCF	Electrospinning followed by heat treatment	LABs	2019	[39]
8	MoO_3	Electrospinning and carbonization solvothermal	LSBs	2020	[40]
9	ZnO—Co_3O_4/PCNF	Electrospinning-catalyzed-graphitization	LIBs	2017	[41]
10	V_2O_3/CNF	Electrospinning and carbonization	LIBs	2017	[42]
11	CNF—SiO_2	Electrospinning	LIBs	2019	[43]
12	p-SiO_2@N—CNF	Sol—gel electrospinning carbonization	LIBs	2019	[44]
13	CNFs@S/MnO_2	Electrospinning-in situ assembly	LSBs	2020	[45]

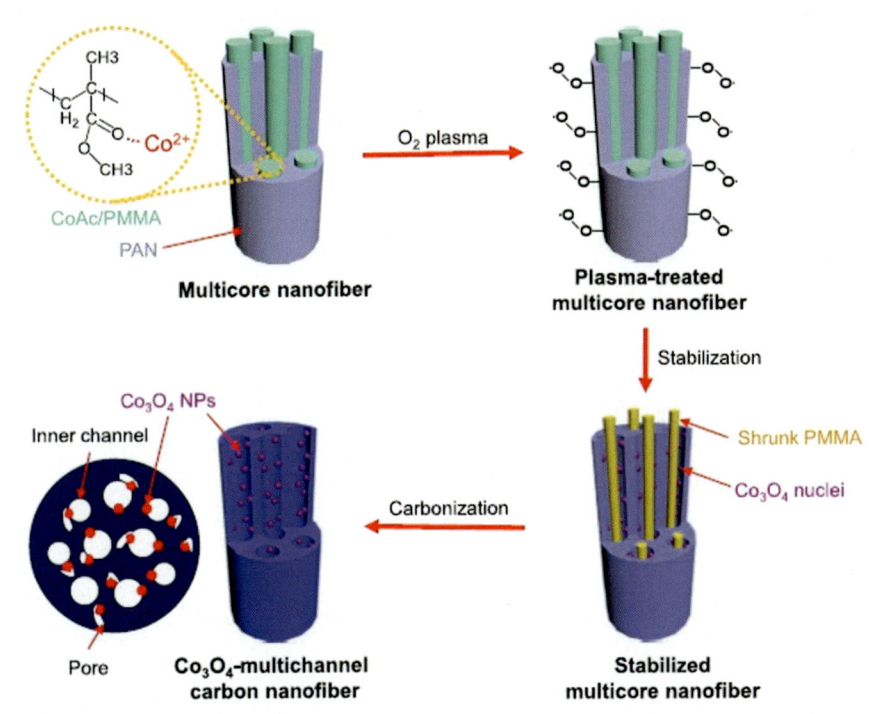

Figure 8.5 Illustration of the sequential synthesis of the cobalt oxide-incorporated multichannel carbon nanofibers.[36]
Reproduced from Ref. 36 Copyright (2020), with permission from American Chemical Society.

exposure of 100–190 W for 5 min (Fig. 8.5). Polyacrylonitrile (PAN), poly(methyl methacrylate), and cobalt (II) acetate were used as starting materials to fabricate multi-core polymer nanofibers by a coelectrospinning method. The precursor fibers were used for oxygen plasma treatment, followed by stabilization and carbonization. The stabilization process was completed at 270°C (air-controlled environment) to convert Co^{2+} into Co_3O_4, while carbonization was performed at 900°C (inert environment) to convert PAN into carbon. The Brunauer–Emmett–Teller (BET) surface area of P–Co–MCNF was calculated as 1187 m^2/g. The P–Co–MCNF-based SC electrode exhibited excellent electrochemical performance that was attributed to the porous structure of CNFs.[36]

8.6 Recent research and development: metal oxide–carbon nanofiber based electrodes

8.6.1 Metal oxide–carbon nanofiber composite based supercapacitor electrodes

MO–CNF composites for SC electrodes were explored based on vapor-grown nano-fibers of carbon and cobalt-manganese oxide (VGCF- $CoMnO_2$). The electrochemical output was tested by cyclic voltammetry (CV) analysis in 1M KOH. They reported the highest 630 F/g specific capacitance of VFCF- $CoMnO_2$ composite at 5 mV/s. The stability test for 10,000 cycles resulted in only 5% degradation of the primary capacitance. The superior performance of the composite was attributed to the incorporation of $CoMnO_2$ and synergetic effect between a double layer of VGCF and pseudo nature of $CoMnO_2$.[46]

Activated CNF—Zn oxide (ACNF—ZnO) nanocomposites were fabricated by an electrospinning technique. A two-electrode system tested the electrochemical efficiency in 6M KOH. They reported that ACNF—ZnO (20) had the highest specific capacitance at 178.2 F/g, energy density at 22.7 Wh/kg, and power density at 400 W/kg among all composites. The ACNF—ZnO composite showed enhanced energy density and specific capacitance values over pure ACNF (6.6 Wh/kg, 125.8 F/g). The stability test resulted in only 25% degradation of the initial capacitance after 1000 cycles. The superior performance of the ACNF—ZnO nanocomposite-based electrodes was attributed to the enhanced surface area, active sites due to ZnO, cooperative effect between the pseudo nature of ZnO, and double-layer behavior of ACNF.[47]

The pyrolysis technique was used to fabricate CNFs and dilute NiO-based composites for SC electrodes. The NiO decorated nanofibers of carbon (CNF—NiO) were derived from Zn and Ni-based fibers of the metal-organic framework. Through BET analysis, they measured a maximum 598 m^2/g surface area of CNF—NiO(0.100) among all other composites. The electrochemical performance was tested by three electrode schemes in 6M KOH. They reported 14,926 F/g specific capacitance and 33.74 Wh/kg energy density of CNF—NiO composite when the 0.43 weight percentage of NiO alone contributed 234 F/g specific capacitance. The stability test resulted in only 10% degradation of the initial capacitance of the CNF—NiO composite after 5000 cycles. The dominant outcome of the composite was attributed to the interactive effect between the pseudo nature of NiO and a double layer of CNF.[48]

Binder-free SC electrodes based on CNFs incorporated with vanadium monoxide and amorphous vanadium (CNF—VO—VO_x) were prepared by electrospinning process. The reported 651 m^2/g surface area of CNF—VO—VOx through BET analysis was highest among other composites of vanadium oxide with CNFs. The electrochemical efficiency of CNF—VO—VO_x-based symmetric electrodes was tested in 6M KOH using two electrode mechanisms. The composite's maximums were observed as 325.7 F/g specific capacitance, 10.0 Wh/kg, and 964.5 W/kg power density at 1 A/g. The EIS test resulted in 0.28 Ω electrochemical series resistance (ESR) of the composite. The stability test resulted in 92% capacitance retention after 5000 cycles at 4 A/g. The eminent output of the symmetric CNF—VO—VO_x-based electrodes was attributed to the conductivity enhancement, pore structure, and faster ion transfer capability due to the addition of quasi-metallic vanadium oxide and redox rich VO—VO_x vanadium couples.[49]

Cai et al. reported efficient symmetric micromesoporous structure electrodes for SCs based on $MnCo_2O_4$-incorporated N-doped nanofibers of carbon (CNF—N— $MnCo_2O_4$). The BET analysis resulted in a maximum 388.7 m^2/g surface area of CNF—N— $MnCo_2O_4$-1 (1 mmol loading of $Mn(CH_3COO)_2$.$4H_2O$) among all the composites.

The electrochemical output was tested in 6M KOH by two symmetric electrode approaches. The composite exhibited the maximum 871.5 F/g specific capacitance and 30.26 Wh/kg energy density at 0.5 A/g. The stability test was carried out for 5000 cycles, resulting in only a 10.3% loss of primary capacitance. However, the electrochemical impedance spectroscopy (EIS) analysis showed that the ESR of CNF–N–MnCo$_2$O$_4$-1 (0.68 Ω) was tolerably larger than other composites. The overall performance of the CNF–N–MnCo$_2$O$_4$-1 composite was attributed to the micromesoporous structure and pseudocapacitive nature of MnCo$_2$O$_4$, which assist redox reactions.[50]

Yu et al. reported iron oxide decorated in lignin-based nanofibers of hollow carbon for SCs (HCNF–Fe oxide) by coaxial electrospinning mechanism. They synthesized the nanocomposites by varying the percentage of iron oxide to observe its effect on the performance of HCNFs. Through BET analysis, they reported the surface area of 401, 301, and 281 m^2/g for HCNF–15, HCNF–20, and HCNF–25, respectively. The electrochemical efficiency was tested in three electrode systems using CV, galvanostatic charge–discharge test (GCD), and EIS analysis. The HCNF–25 composite exhibited 121.5 F/g specific capacitance at 0.5 A/g, the highest among the other composites. Through EIS analysis, they observed that the increasing percentage of iron oxide in HCNF increased ESR. The CV stability test at 100 mV/s resulted in 27% degradation of the output capacitance of the HCNF–25 composite after 1000 cycles. The performance of the HCNF–Fe oxide base composite was superior to that of CNFs. The superb performance of HCNF–Fe oxide composites was attributed to electrolytic ion–iron oxide interactions, surface morphology, meso/microporous structure, and hollow structure of CNFs, which facilitated the rapid transportation of the electrolytic ions.[51]

Recently, Su et al. fabricated a 3-D nickel-metalized CNF network by electrospinning, carbonization followed by an electrodeposition process. CNF–Ni@Fe$_2$O$_3$ anode was prepared by hydrolysis of urea via solution-phase assembly method while CNF–Ni@MnO$_2$ cathode was prepared by electrodeposition method. A pseudocapacitor device was assembled using CNF–Ni@Fe$_2$O$_3$ and CNF–Ni@MnO$_2$ that showed a maximum energy density of 4.32 mW h/cm^3 at the power density of 10.29 mW/cm^3. The nanoarchitecture of metal oxides, surface metallization by nickel metal, and surface area of CNF enhanced the performance of electrodes and increased the device's stability (Fig. 8.6).[37] Fig. 8.7 represents the summary of CNF–MO-based SC electrodes.

8.6.2 Metal oxide–carbon nanofiber composite based electrodes for batteries

Q. Si et al. reported ball-milled SiO–CNF-based composite anode to improve the cyclic performance of Li-ion batteries. The electrochemical test resulted in 700 mAh/g reversible capacity after 200 cycles. The predominant cyclic performance of SiO–CNF-based composite was attributed to the modification in the valence state of Si due to the ball milling mechanism.[52]

Lee et al. reported flexible electrodes based on carbon and tungsten oxide (CNF–WO$_x$) nanofibers by electrospinning for lithium-ion batteries. The resulting composite

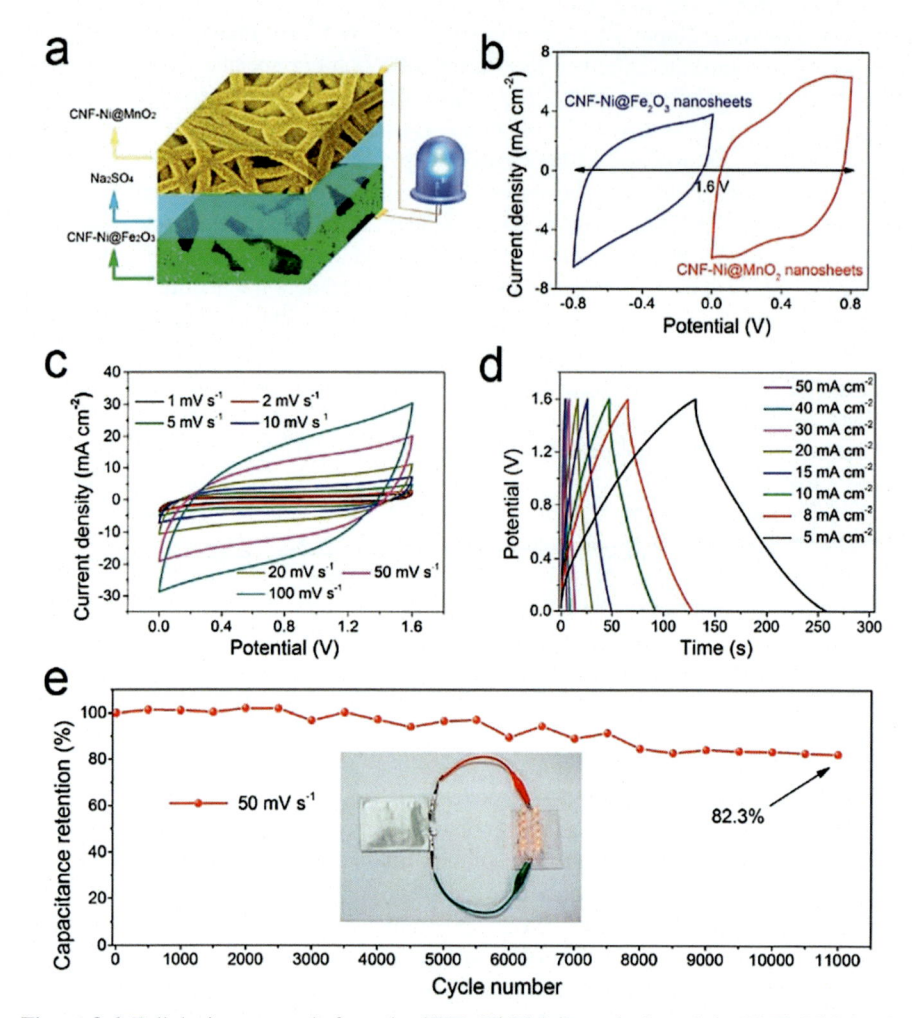

Figure 8.6 Full device test made from the CNF–Ni@MnO$_2$ cathode and the CNF–Ni@Fe$_2$O$_3$ anode. (A) Schematic image illustrating the pseudocapacitor working in a 0.5 M Na$_2$SO$_4$ aqueous electrolyte. (B) CV curves of the cathode and anode within the potential window (0.8–0.8 V). (C) CV curves at different scan rates from 1 to 100 mV/s. (D) GCD curves at different current densities ranging from 5 to 50 mA cm^2. (E) Cycling stability test of the device at 50 mV/s showing 82.3% capacitance retention after 11,000 cycles. The inset panel shows a soft pack pseudocapacitor that can light up 12 LEDs in parallel connection.[37]
Reproduced from Ref. 37 Copyright (2020), with permission from The Royal Society of Chemistry under Creative Commons Attribution-NonCommercial 3.0 Unported License.

consists of a high percentage of tungsten oxide (80% weightage). The CNF–WO$_x$ composite exhibited a maximum of 481 mAh/g reversible capacity and is utilized directly as an anode. The stability test after 85 cycles resulted in 65.6% (321 mAh/g) retention of its maximum reversible capacity. The superior performance was attributed

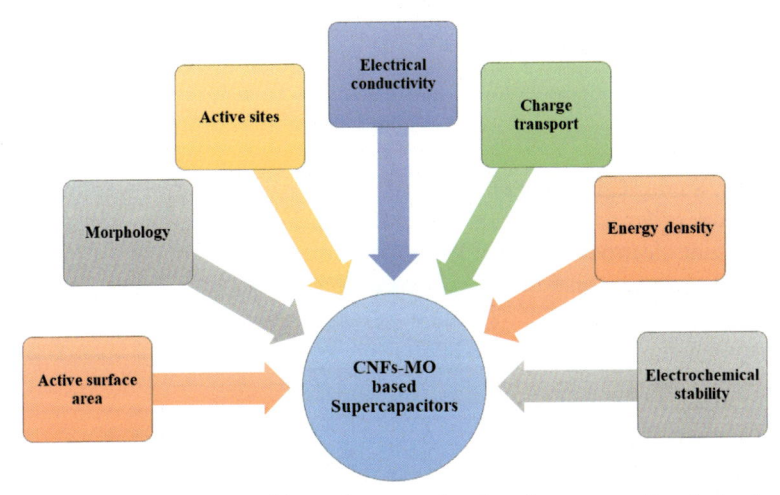

Figure 8.7 Summary of metal oxide—carbon nanofiber-based supercapacitor electrodes.

to the evenly trapping of W-oxide particles inside nanofibers of carbon, which enhances stability and provides better electrical contact.[53]

M J Song et al. reported cathode for LABs based on binder-free CNF—Co3O4 composites by electrospinning mechanism. The CNF—Co3O4 composite resulted in a maximum 159 m^2/g specific surface area 2.15 times that of pristine CNF. The electrochemical performance resulted in a 760 mAh/g discharge capacity of CNF—Co3O4 nanocomposite at 500 mA/g, 10.6 times that of pristine CNF. Moreover, the stability test of CNF—Co3O4 resulted in 55% of capacity retention after 10 charge—discharge cycles, which was again higher than that of pristine CNF (45%). The superior performance of CNF—Co3O4 composite was attributed to the catalytic activity and evenly distribution of Co3O4 in the CNF structure.[54]

Dirican et al. reported high-performance anode material for Na-ion batteries by electrodeposition of Sn oxide on porous CNFs (SnO2—PCNFs). They coated the SnO2—PCNF composite with amorphous carbon to enhance the cycling stability (C—SnO2—PCNF). The electrochemical efficiency of the composite was tested by utilizing coin cells in 1M NaClO4. They reported 374 mAh/g and 188 mAh/g (at 100th cycle) capacity for the C—SnO2—PCNF and SnO2—PCNF composites, respectively. Moreover, the C—SnO2—PCNF exhibited 82.7% capacity retention and 98.8% coulombic efficiency at the 100th cycle. The superior performance of the C—SnO2—PCNF composite was attributed to the incorporation of SnO2 and coating of amorphous carbon, which improved the cyclic stability.[55]

Zhang et al. reported high-performance 3D N-enriched nanofibers of carbon and Co3O4-based composites (NCNF—Co3O4) for Li/Na-batteries. They measured 395.8 m^2/g surface area of the (NCNF—Co3O4) by using BET analysis. The BET adsorption curve of the sample indicated the presence of mesoporous structure with pore size in the range of 1.7—60 nm. The NCNF—Co3O4 composite-based electrode exhibited the highest 1199 mAh/g (at 200 mA/g) and 645 mAh/g (at 100 mA/g) for Li- and Na-based batteries, respectively. The stability test for Li and Na resulted in 721 and 301 mAh/g (at 1000 mA/g) capacity retention after 400 cycles, respectively. The superior electrochemical performance was attributed to the structure of 3-D N-doped CNF, the mesoporous structure of composite, and the incorporation of Co_3O_4.[56]

Recently, Susca et al. modified CNFs with ZnO nanoparticles for Na electroplating. CNFs network offers a high surface area for the mechanically stable electrode. ZnO@ CNF composite electrode delivered 1500 cycles at 1 mA/cm^2 with lower local current density and volume fluctuations. The symmetric battery based on ZnO@CNF composite anodes delivered stable plating/stripping cycles for >1000 h by effectively inhibiting the growth of dendrites and Na agglomerates (Fig. 8.8).[38]

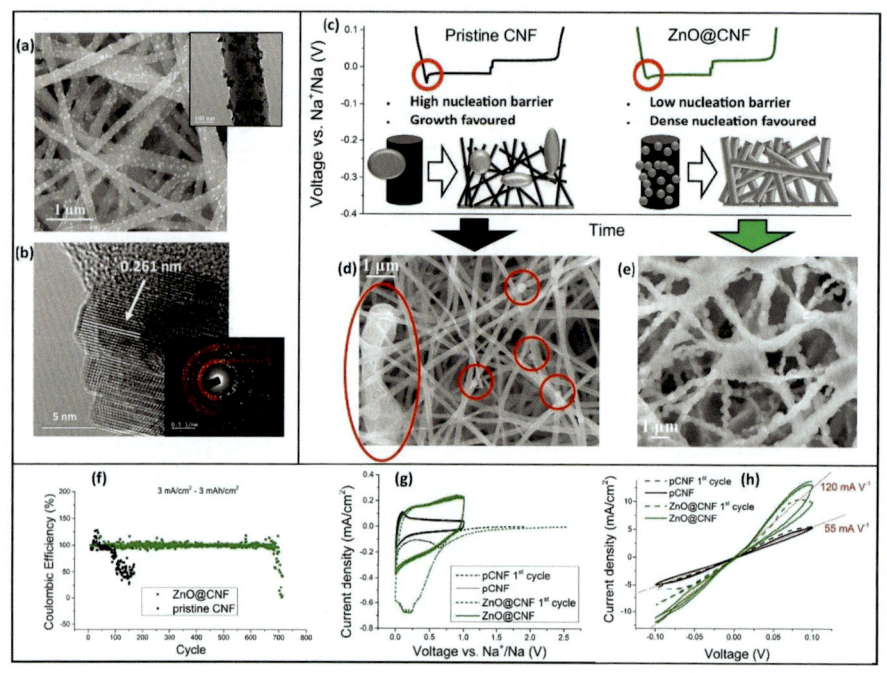

Figure 8.8 (A) SEM and TEM (inset) images of ZnO@CNF; (B) HRTEM image and SAED (inset) of a ZnO nanoparticle embedded in ZnO@CNF. D-spacing and reflection planes are highlighted. (C) Typical voltage profiles of a plating/stripping cycle for the pristine CNF and ZnO@CNF cells, highlighting the difference in overpotential and illustrating its implications; (D and E) SEM images of PCNFs and ZnO@CNFs, respectively, after discharge to 0 V versus metallic Na. (F) Coulombic efficiencies of the PCNF and ZnO@CNF half-cells with a loading of 3 mAh/cm^2, measured at a current density of 3 mA/cm^2; (G) Cyclic voltammetry of half cells made from PCNFs and ZnO@CNFs at a scan rate of 0.1 mV/s, where the first cycle is represented by a dashed line; (H) Cyclic voltammetry of symmetric cells made from PCNFs and ZnO@CNFs at a scan rate of 0.1 mV/s where a dashed line represents the first cycle for both materials, while the second and third cycles by solid lines, and the red lines signify the linear fits and the corresponding slopes.[38]

8.7 Outlook and future perspectives

Although metal oxides are the most used electrode materials because of their promising properties, their low conductivity has proven a great hurdle in exceptional storage energy devices. On the other hand, the fabrication of efficient CNFs is a huge challenge in terms of balance between cost and quality. The incorporation of the metal oxides with CNF can overcome these drawbacks. MO—CNF-based composite nanomaterials combine the properties of metal oxide and CNFs and thus enhance the material's overall properties for particular applications (Figs. 8.9 and 8.10).

MO—CNF-based SC and battery electrodes are very exciting nanomaterials for electrochemical energy-storage applications. Several synthesis methods have been utilized to prepare efficient electrodes based on MO—CNFs, but it is puzzling to identify the most promising synthesis procedure. Rational designing of the MO—CNF nanocomposites has shown encouraging results according to the desired application area. CNF allows fabricating the nanocomposites in numerous forms such as membranes and aerogels for practical applications. MO—CNF-based electrodes for SCs and batteries have elevated the performance of the devices.

Nevertheless, it is difficult to fabricate metal oxide-based CNFs because nanofiber morphology changes during the calcination process and result in the nanocrystals continuously grow in size. It was observed that the morphology of fibers and the spin ability of solutions strongly depend upon three main factors: solution properties, processing condition, and atmosphere condition. Some parameters need to be controlled during the fabrication of metal oxide-based CNFs, such as applied voltage, the distance between the syringe and targeted substrate, concentration, and flow rate during the combination of precursors.

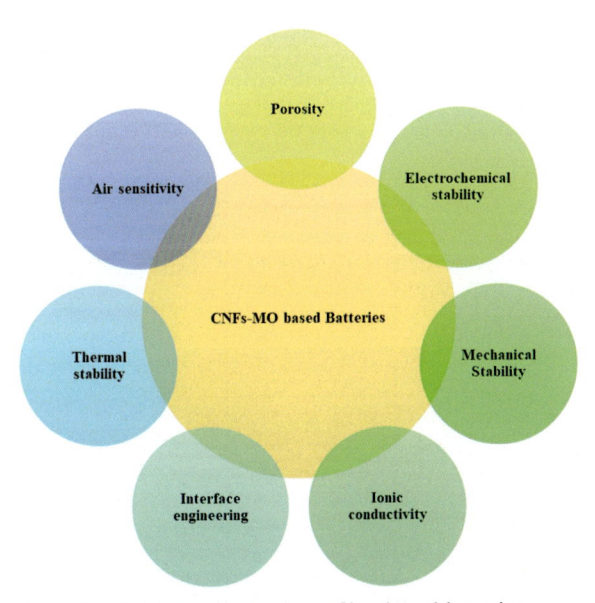

Figure 8.9 Summary of metal oxide—carbon nanofiber-based batteries.

Figure 8.10 Key facts of metal oxide–carbon nanofiber-based electrodes.

Table 8.3 Summary of metal oxide–carbon nanofiber composite-based supercapacitor electrodes.

MO–CNF composites	Synthesis technique	Capacitance (F/g)	Cyclic life	Year	References
VGCF-CoMnO$_2$	Vapor-grown	630 at 5 mV/s	95% (at 200 mV/s) after 10,000 cycles	2009	[46]
ACNF–ZnO	One-step electrospinning	178.2 at 25 mV/s	75% (at 1 mA/cm^2) after 1000 cycles	2014	[47]
CNF–NiO	Pyrolysis	14,926 at 1 mV/s	90% after 5000 cycles	2017	[48]
CNF–VO–VO$_x$	Electrospinning	325.7 at 1 A/g	92% (4 A/g) after 5000 cycles	2016	[49]
CNF–N–MnCo$_2$O$_4$	–	871.5 at 0.5 A/g	89.7% (at 5 A/g) after 5000 cycles	2018	[50]
HCNF–FeO	Coaxial electrospinning	121.5 at 0.5 A/g	73% (at 100 mV/s) after 1000 cycles	2018	[51]

Table 8.4 Summary of metal oxide−carbon nanofiber composite-based electrodes for batteries.

MO−CNF composites	Synthesis technique	Reversible capacity (mAh/g)	Cyclic life	year	References
SiO−CNF	Ball-milling	700	<1% after 100 cycles	2011	[52]
CNF−WO$_x$	Electrospinning	481 at 50 mA/g	65.6% (at 50 mA/g) after 85 cycles	2014	[53]
CNF−Co$_3$O$_4$	Electrospinning	760 at 500 mA/g	55% (at 500 mA/g) after 10 cycles	2015	[54]
C−SnO$_2$−PCNF	Electrodeposition	374 at 50 mA/g	82.7% after (at 50 mA/g) 100 cycles	2015	[55]
3D−NCNF−Co$_3$O$_4$	−	1199 at 200 mA/g	60% (at 1000 mA/g) after 400 cycles	2019	[56]

Further, conductivity issues of the MO−CNF composites can be resolved by ternary composites with conducting polymers. In addition, metal oxide nanoarchitectures (quantum dots, nanoparticles, nanowires, nanowebs, nanosheets, nanoflakes, and nanoribbons) with multidimensional features can boost performance factors. Nanotechnology has enabled us to fabricate multifunctional energy materials, a significant aspect for sustainable energy-storage devices. Tables 8.3 and 8.4 summarize MO−CNF composite-based SC and battery electrodes.

References

1. Soltani S, et al. Recent progress in the design and synthesis of nanofibers with diverse synthetic methodologies: characterization and potential applications. *New J Chem* 2020; **44**(23):9581−606.
2. Zhou X, et al. Carbon nanofiber-based three-dimensional nanomaterials for energy and environmental applications. *Mater Adv* 2020;**1**(7):2163−81.
3. Troughton J, Atkinson D. Amorphous InGaZnO and metal oxide semiconductor devices: an overview and current status. *J Mater Chem C* 2019;**7**(40):12388−414.
4. Ros C, Andreu T, Morante JR. Photoelectrochemical water splitting: a road from stable metal oxides to protected thin film solar cells. *J Mater Chem* 2020;**8**(21):10625−69.
5. Karthikeyan C, et al. Recent advances in semiconductor metal oxides with enhanced methods for solar photocatalytic applications. *J Alloys Compd* 2020;**828**:154281.

6. Apte A, et al. 2D electrets of ultrathin MoO_2 with apparent piezoelectricity. *Adv Mater* 2020;**32**(24):2000006.

7. Ren B, Wang Y, Ou JZ. Engineering two-dimensional metal oxides via surface functionalization for biological applications. *J Mater Chem B* 2020;**8**(6):1108−27.

8. Chen L, et al. Regulation of intrinsic physicochemical properties of metal oxide nanomaterials for energy conversion and environmental detection applications. *J Mater Chem A* 2020;**8**(34):17326−59.

9. Prakash J, et al. Metal oxide-nanoparticles and liquid crystal composites: a review of recent progress. *J Mol Liq* 2020;**297**:112052.

10. Nikam AV, Prasad BLV, Kulkarni AA. Wet chemical synthesis of metal oxide nanoparticles: a review. *CrystEngComm* 2018;**20**(35):5091−107.

11. Yuliarto B, et al. Green synthesis of metal oxide nanostructures using naturally occurring compounds for energy, environmental, and bio-related applications. *New J Chem* 2019; **43**(40):15846−56.

12. Parashar M, Shukla VK, Singh R. Metal oxides nanoparticles via sol−gel method: a review on synthesis, characterization and applications. *J Mater Sci Mater Electron* 2020;**31**(5):3729−49.

13. Kim YA, et al. Carbon nanofibers. In: *Springer handbook of nanomaterials*. Springer; 2013. p. 233−62.

14. Feng L, Xie N, Zhong J. Carbon nanofibers and their composites: a review of synthesizing, properties and applications. *Materials* 2014;**7**(5):3919−45.

15. Gogotsi Y. Not just graphene: the wonderful world of carbon and related nanomaterials. *MRS Bull* 2015;**40**(12):1110−21.

16. Zhang J, et al. Carbon science in 2016: status, challenges and perspectives. *Carbon* 2016; **98**:708−32.

17. Wang Z, et al. Carbon nanofiber-based functional nanomaterials for sensor applications. *Nanomaterials* 2019;**9**(7):1045.

18. Lu W, et al. Progress in catalytic synthesis of advanced carbon nanofibers. *J Mater Chem* 2017;**5**(27):13863−81.

19. Long YZ, Yan X. Chapter 2 - Electrospinning: the setup and procedure. In: Ding B, Wang X, Yu J, editors. *Electrospinning: nanofabrication and applications*. William Andrew Publishing; 2019. p. 21−52.

20. Nie G, et al. Key issues facing electrospun carbon nanofibers in energy applications: ongoing approaches and challenges. *Nanoscale* 2020;**12**(25):13225−48.

21. Lee B-S, Yu W-R. Electrospun carbon nanofibers as a functional composite platform: a review of highly tunable microstructures and morphologies for versatile applications. *Funct Compos Struct* 2020;**2**(1):012001.

22. Zhang L, et al. A review: carbon nanofibers from electrospun polyacrylonitrile and their applications. *J Mater Sci* 2014;**49**(2):463−80.

23. Chen W, et al. A feasible way to produce carbon nanofiber by electrospinning from sugarcane bagasse. *Polymers* 2019;**11**(12):1968.

24. Liang J, et al. Recent advances in electrospun nanofibers for supercapacitors. *J Mater Chem A* 2020;**8**:16747−89.

25. Tong Z, et al. Carbon-containing electrospun nanofibers for lithium−sulfur battery: current status and future directions. *J Energy Chem* 2021;**54**:254−73.

26. Kim J-G, et al. N-doped hierarchical porous hollow carbon nanofibers based on PAN/ PVP@SAN structure for high performance supercapacitor. *Compos B Eng* 2020;**186**: 107825.

27. Dai Z, et al. Synthesis of TiO_2@ lignin based carbon nanofibers composite materials with highly efficient photocatalytic to methylene blue dye. *J Polym Res* 2020;**27**:1−12.

28. Verma S, Sinha-Ray S, Sinha-Ray S. Electrospun CNF supported ceramics as electrochemical catalysts for water splitting and fuel cell: a review. *Polymers* 2020;**12**(1):238.

29. Said RAM, et al. Review—insights into the developments of nanocomposites for its processing and application as sensing materials. *J Electrochem Soc* 2020;**167**(3):037549.

30. Sriram B, et al. A ternary nanocomposite based on nickel(iii) oxide@f-CNF/rGO for efficient electrochemical detection of an antipsychotic drug (Klonopin) in biological samples. *New J Chem* 2020;**44**(25):10250–7.

31. Seok D, et al. Recent progress of electrochemical energy devices: metal oxide–carbon nanocomposites as materials for next-generation chemical storage for renewable energy. *Sustainability* 2019;**11**(13):3694.

32. Liu Y, et al. Advanced supercapacitors based on porous hollow carbon nanofiber electrodes with high specific capacitance and large energy density. *ACS Appl Mater Interfaces* 2020; **12**(4):4777–86.

33. Saito Y, et al. Manganese dioxide nanowires on carbon nanofiber frameworks for efficient electrochemical device electrodes. *RSC Adv* 2017;**7**(20):12351–8.

34. Iqbal N, et al. Flexible Fe_3O_4@Carbon nanofibers hierarchically assembled with MnO_2 particles for high-performance supercapacitor electrodes. *Sci Rep* 2017;**7**(1):15153.

35. Zhang M, et al. Elastic and hierarchical carbon nanofiber aerogels and their hybrids with carbon nanotubes and cobalt oxide nanoparticles for high-performance asymmetric supercapacitors. *Carbon* 2020;**158**:873–84.

36. Kim SG, et al. Facile synthesis of Co_3O_4-incorporated multichannel carbon nanofibers for electrochemical applications. *ACS Appl Mater Interfaces* 2020;**12**(18):20613–22.

37. Su S, et al. Interface metallization enabled an ultra-stable Fe_2O_3 hierarchical anode for pseudocapacitors. *RSC Adv* 2020;**10**(15):8636–44.

38. Susca A, et al. Affinity-engineered carbon nanofibers as a scaffold for Na metal anodes. *J Mater Chem* 2020;**8**(29):14757–68.

39. Zhang X, et al. An efficient, bifunctional catalyst for lithium–oxygen batteries obtained through tuning the exterior Co 2+/Co 3+ ratio of CoO x on N-doped carbon nanofibers. *Catal Sci Technol* 2019;**9**(8):1998–2007.

40. Li H, et al. Self-assembly of MoO_3-decorated carbon nanofiber interlayers for high-performance lithium–sulfur batteries. *Phys Chem Chem Phys* 2020;**22**(4):2157–63.

41. Chen R, et al. Highly mesoporous C nanofibers with graphitized pore walls fabricated via $ZnCo_2O_4$-induced activating-catalyzed-graphitization for long-lifespan lithium-ion batteries. *J Mater Chem* 2017;**5**(41):21679–87.

42. Liu X, et al. Facile preparation of a V_2O_3/carbon fiber composite and its application for long-term performance lithium-ion batteries. *New J Chem* 2017;**41**(13):5380–6.

43. Jayabalan AD, et al. Electrospun 3D CNF–SiO_2 fabricated using non-biodegradable silica gel as prospective anode for lithium–ion batteries. *Ionics* 2019;**25**(11):5305–13.

44. Aboalhassan AA, et al. Self-assembled porous-silica within N-doped carbon nanofibers as ultra-flexible anodes for soft lithium batteries. *iScience* 2019;**16**:122–32.

45. Hu J, et al. In situ assembly of MnO_2 nanosheets on sulfur-embedded multichannel carbon nanofiber composites as cathodes for lithium-sulfur batteries. *Sci China Mater* 2020;**63**(5): 728–38.

46. Kim SH, et al. Cobalt-manganese oxide/carbon-nanofiber composite electrodes for supercapacitors. *Int J Electrochem Sci* 2009;**4**:1489–96.

47. Kim CH, Kim B-H. Zinc oxide/activated carbon nanofiber composites for high-performance supercapacitor electrodes. *J Power Sources* 2015;**274**:512–20.

48. Yang Y, et al. Dilute NiO/carbon nanofiber composites derived from metal organic framework fibers as electrode materials for supercapacitors. *Chem Eng J* 2017;**307**:583–92.

49. Tang K, et al. Self-reduced VO/VOx/carbon nanofiber composite as binder-free electrode for supercapacitors. *Electrochim Acta* 2016;**209**:709−18.

50. Cai N, et al. MnCo$_2$O$_4$@nitrogen-doped carbon nanofiber composites with meso-microporous structure for high-performance symmetric supercapacitors. *J Alloys Compd* 2019;**782**:251−62.

51. Yu B, Gele A, Wang L. Iron oxide/lignin-based hollow carbon nanofibers nanocomposite as an application electrode materials for supercapacitors. *Int J Biol Macromol* 2018;**118**: 478−84.

52. Si Q, et al. Improvement of cyclic behavior of a ball-milled SiO and carbon nanofiber composite anode for lithium-ion batteries. *J Power Sources* 2011;**196**(22):9774−9.

53. Lee J, et al. Simple fabrication of flexible electrodes with high metal-oxide content: electrospun reduced tungsten oxide/carbon nanofibers for lithium ion battery applications. *Nanoscale* 2014;**6**(17):10147−55.

54. Song MJ, et al. Self-standing, binder-free electrospun Co$_3$O$_4$/carbon nanofiber composites for non-aqueous Li-air batteries. *Electrochim Acta* 2015;**182**:289−96.

55. Dirican M, et al. Carbon-confined SnO$_2$-electrodeposited porous carbon nanofiber composite as high-capacity sodium-ion battery anode material. *ACS Appl Mater Interfaces* 2015;**7**(33):18387−96.

56. Zhang K, et al. Universal construction of ultrafine metal oxides coupled in N-enriched 3D carbon nanofibers for high-performance lithium/sodium storage. *Nanomater Energy* 2020; **67**:104222.

Metal oxide–carbon composite electrode materials for rechargeable batteries

Ghulam Ali[1] and Faiza Jan Iftikhar[2]
[1]U.S.-Pakistan Center for Advanced Studies in Energy (USPCAS-E), National University of Sciences and Technology (NUST), Islamabad, Pakistan; [2]NUTECH School of Applied Sciences and Humanities, National University of Technology, Islamabad, Pakistan

9.1 Introduction

Contemporary society is plagued by energy issues concomitant with the environmental pollution that has become rampant in the 21st century. Fossil fuels have been the fuel of choice for hundreds of years despite wreaking havoc on the environment in which we live and creating serious sustainability issues for our planet.[1] Furthermore, there has been an exponential increase in the supply–demand gap—even though this gap is being met in bulk using fossil fuels, its depletion and triggering of environmental hazards has been unprecedented. This critical situation has encouraged the global scientific community to look for alternatives to fossil fuels in the form of renewable energy.[2] Among them, solar and wind energy is generated intermittently without an efficient storage method, thus limiting their efficient use. Thus, electrical energy-storage (EES) devices are being sought as an effective way to harness renewable energy sources more efficiently.[3] Lithium-ion batteries (LIBs), sodium-ion batteries (SIBs), magnesium-ion batteries (MIBs), zinc-ion batteries (ZIBs), metal–sulfur batteries, and redox flow batteries (RFBs) have found a fair share of the market for consumer-based applications. Hence, these specific EES cells, with special reference to metal oxide-based anode and cathodes composited with carbonaceous material, are discussed (vide infra).

9.1.1 Lithium-ion batteries

Lithium-based electrochemical energy-storage devices with a theoretical capacity of ~3860 mAh/g have made tremendous progress, yet they are not suitable for applications that require extended stability and large energy capacity. Graphite has been used for de/intercalation of Li ions as the negative electrode, but their theoretical capacity is quite low, i.e., 372 mAh/g[4] Thus, it is imperative to pursue a search for new materials that can overcome the shortcomings of the graphite electrode as the negative electrode.

Metal oxides have received considerable attention owing to their high storage capacity due to conversion reaction represented by Eq. (9.1); nevertheless, they undergo drastic volume changes during charge/discharge that reduce the electrical contact and

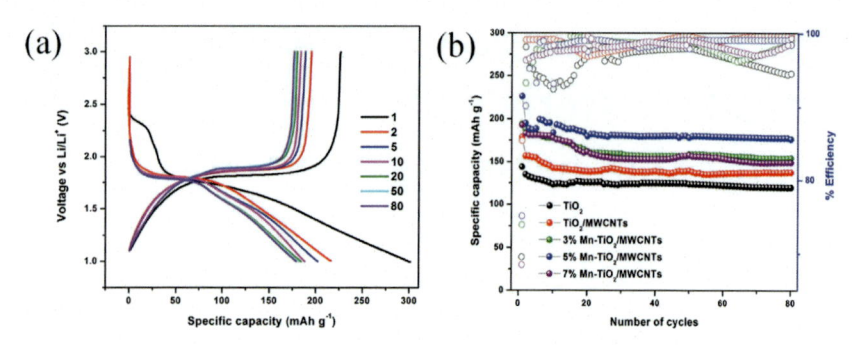

Figure 9.1 (A) Charge–discharge cycles for Mn-doped TiO_2/multiwalled carbon nanotubes and (B) specific capacities and coulombic efficiency measured in a potential window of 1.0–3.0 V versus Li/Li$^+$ at a current density of 0.1 C.
Reproduced by permission of the Royal Society of Chemistry Ref. 6.

ultimately affect the stability of the electrode apart from being resistant to charge transfer; this also results in reduced capacity.[5] If metal oxides are used with carbonaceous material, theoretical capacity is increased (see Fig. 9.1).[6,7] Hence, studies have been conducted to composite these metal oxides with carbonaceous material and develop a high-performance electrode material based on synergistic materials.[8]

Metal oxides undergo conversion reaction during charge/discharge of Li ions, given by

$$MO_x + 2xe^- + 2xLi^+ \leftrightarrow M° + xLi_2O \tag{9.1}$$

CNTs have been used to facilitate electron transport by compositing them with metal oxides, which offers advantages such as compensating for the low electrical conductivity of metal oxides, acting as a scaffold by binding with metal oxides, and increasing their mechanical strength, which is a provision of the electronic highway for effective transport and the formation of a percolation network that facilitates electrolyte diffusion, high energy density, and high specific surface area.[5]

Co_3O_4 nanoparticles (NPs) with a high theoretical capacity of 890 mAh/g decorated onto 3-D mesoporous CNTs to improve conductivity have been reported as anodes for LIBs. The formation of a conductive network of CNTs with finely dispersed Co_3O_4 NPs demonstrated a high reversible capacity of 627 mAh/g after 50 cycles and significantly higher electrical conductivity than pristine Co_3O_4, thus indicating improved electrochemical performance, as shown in Fig. 9.2.[9]

Many studies have attempted to alleviate the stress caused by volume changes resulting from the charge/discharge process by anchoring Co_3O_4 with a high mass loading onto conductive nanotubes, and such CNTs have led to improved electrochemical performance and stability.[10,11] The sidewalls of polyvinyl alcohol-modified CNTs were immobilized with Fe_3O_4 NPs by the coprecipitation method. The highest discharge capacity of 656 mAh/g was reported at the 145th cycle with 67% weight of Fe_3O_4.[12] Similarly, a composite of Mn_3O_4 with CNT has been shown to act as

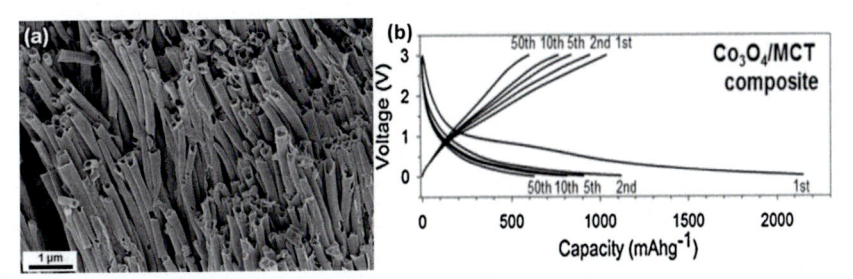

Figure 9.2 (A) SEM image for Co_3O_4/MCT showing a tubular morphology, (B) charge—discharge curves for Co_3O_4/MCT for selected cycles at a current density of 100 mA/g in a potential range of 0.01—3 V (vs. Li+/Li).[9]

an excellent matrix providing an efficient route for conduction of electrons during de/lithiation in LIBs, exhibiting a charge capacity of 342 mAh/g at 10.0 C without the need for any binder.[13] Similarly, Fe_2O_3 NPs composited with single-walled CNTs show an exceptionally high reversible capacity of 1243 mAh/g at a current density of 50 mA/g due to shortening the path length of Li ions and mitigation of strain caused by volume change resulting from prolonged capacity retention at high energy densities.[14]

Graphene with a 2-D honeycomb structure has been used due to its significantly high surface area, electrical conductivity, and tensile strength.[15] However, despite its many advantages, its performance deteriorates during electrochemical reactions due to its restacking. It has been contended that compositing graphene with metal oxide can enhance material stability by connecting the sites and improving its electrochemical performance.

Fe_3O_4—graphene nanocomposite exhibits the reversible capacity of 1048 mAh/g at the 90th cycle retaining 99% of the initial capacity. The enhanced performance results from the graphene sheets, which prevented aggregation of Fe_3O_4 NPs and compensated for volume changes during charge/discharge by providing buffered spaces to accommodate changes (see Fig. 9.3).[16] In another study, graphene enwrapped

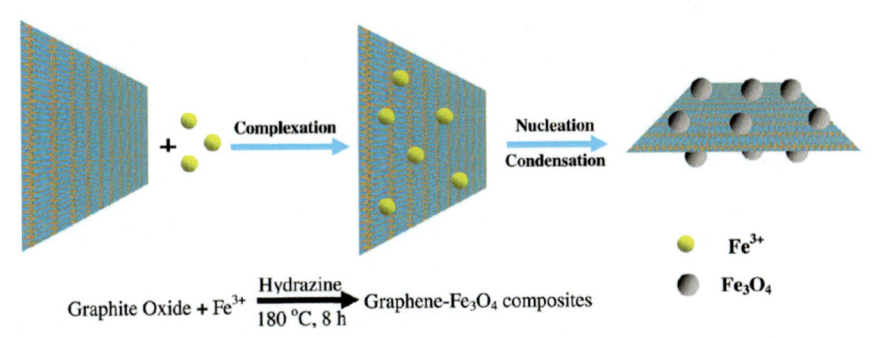

Figure 9.3 Modification of graphene with metal oxide showing active sites to bind metal oxides to graphene.[16]

Fe_3O_4 and improved the cycling stability of the electrode by preventing the agglomeration of NPs while improving the conductive pathway for electrons. The enwrapped composite confers extra structural stability to the electrode.[17] Similarly, CuO NPs of 30 nm dimensions are homogeneously distributed over the graphene sheets providing a conducting platform and buffered spaces to accommodate expansion and contraction of Li ions. It exhibits a reversible capacity of 583.5 mAh/g and retention of 75.5% after the 50th cycle.[18] Fe_2O_3 and CuO embedded onto porous graphene as a composite paper electrode exhibited enhanced specific capacity at high current densities attributed to increased Li-ion storage due to diffusion.[19] Holey graphene increases the ion transport and electrolyte penetration than nonporous graphene and can easily accommodate volume strains due to the charge/discharge process even after 4000 cycles.[20] The porous nature of graphene is responsible for the flexible nature of the electrode confining the metal oxides between the layers and preventing restacking of the layers. The Co_3O_4 hybridized to graphene sheets have been suggested to act as a high-performance anode in LIBs by virtue of their unique architecture and can deliver a reversible capacity of 800 mAh/g in a voltage range of $0.001-3.0$ V at a current density of 200 mA/g.[21] A composite of 1-D nanowalled Co_3O_4 on to 2-D graphene sheets is synthesized sequentially integrating the composite on the graphene acting as the conducting substrate, and hence the need for a separate current collector or binder is eliminated, resulting in excellent stability of the anodic material without losing capacity even after 500 cycles and boosted energy density.[22] In another study, porous nanofibers of Co_3O_4 coating on the reduced graphene oxide have been shown to facilitate ion transport and charge transfer and act as stress buffer to accommodate volume changes. The work presents preliminary results with a reversible capacity of ~ 900 mAh/g at a current density of 1 A/g.[23]

Due to the hierarchy of 3-D structured porous carbon, improved electrochemical performance is observed. The large surface area enables sufficient electrolyte contacts to realize the high energy and power density observed for TiO_2 nanocrystal embedded onto a mesoporous carbon framework.[24] Functionalized MWCNTs (see Fig. 9.4) have shown substantially improved lithium storage and high rate capability of up to 20 C of Ni-doped TiO_2 nanocomposite.[25] The mesopore structure provides a short pathway for ions, while the micropores confine the Li ions and are effective for electrolyte transport. The optimized ratio of SnO_2 embedded onto mesoporous carbon is reported to demonstrate outstanding rate capabilities. Even at very high current densities, the charge$-$discharge capacity is maintained, and hence the material is promising as an anode material for LIBs.[26] Similarly, SnO_2 NPs with a size of $4-5$ nm encapsulated in the ordered tube-like mesoporous carbon walls of 2 nm were reported to show well-maintained morphology after 100 cycles with the possibility of max loading of SnO_2 due to the presence of nanospaces. These nanovoids allow the changes in volume due to lithiation to be compensated, and thus the unique tubular architecture with a high pore volume is suggested as the LIB anode.[27] Monodispersed MnO NPs were inserted into a carbonaceous matrix to form a mesoscale hybrid, resulting in a hierarchical lotus-like structure with excellent stability and rate capability.[28]

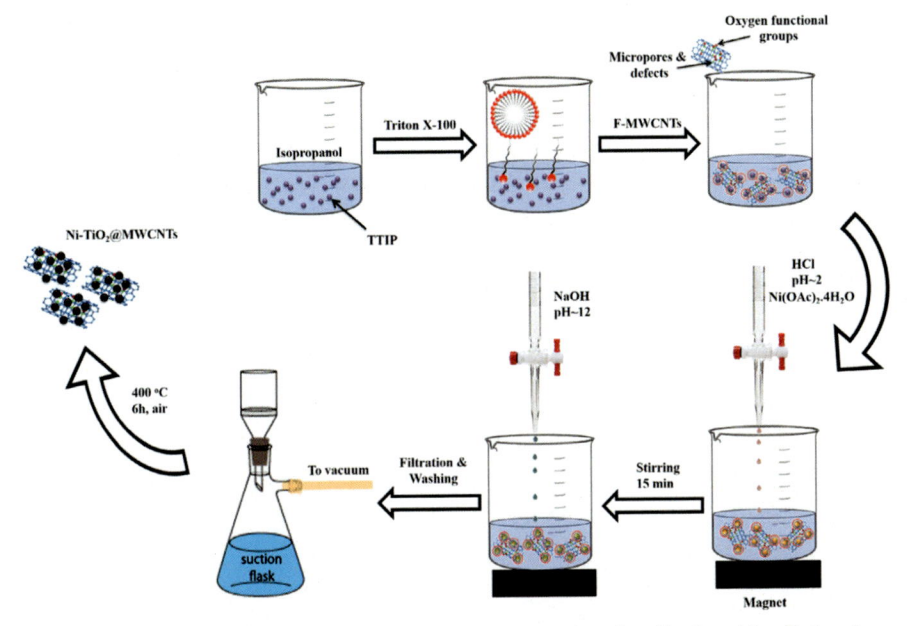

Figure 9.4 Growth of nickel-doped titanium dioxide over functionalized multiwalled carbon nanotubes.[25]

9.1.2 Sodium-ion batteries

Lithium is expensive and must be replaced by more economical sodium; however, sodium replacement does not come without challenges. With a larger size than Li^+, sodium induces greater active material stress during the charge—discharge of Na ions, resulting in reduced energy density.[29]

One-dimensional CNT acts as a pathway for electron transduction when metal oxide NPs are dispersed onto the carbon matrix and cause a synergistic effect by decreasing the hindrance to battery capacity due to volume changes in metal oxides; hence, it shows excellent electrochemical performance. It is reported that SnO_2 modified MWCNT demonstrates high storage capacity for Na ions in the first cycle while coulombic efficiency is boosted with subsequent cycling.[30] In another study, Mn-doped TiO_2 NPs are enwrapped by MWCNT to improve the contact between active materials and improve electrical conductivity, as indicated by high rate performance studies.[6] A nanocomposite consisting of a CNT core to improve electrical conductivity, an increased number of active sites for enhanced storage capacity, and a shell of synergistically coupled TiO_2 and MoO_2 has been reported as a high-performance anode with high capacity and rate capability for SIBs with boosted electrodics.[31]

Fe_2O_3 nanocrystals embedded onto graphene nanosheets as well Fe_2O_3 anchored over rGO has been reported to exhibit excellent rate performance capability and charge transfer reactions as a consequence of synergism offered by graphene and metal oxide where graphene acts as an excellent carbon matrix as well in addition to increasing the conductivity of the active material.[32] Ultrafine SnO_2 dispersed homogenously on GO follows an alloying and dealloying reaction that expands the volume upon sodiation and dealloying when desodiation occurs, while a stable solid electrolyte interface is maintained, resulting in improved battery performance.[33] Dawei et al. have reported a composite of SnO_2/graphene with a structural advantage over SnO_2 and graphene alone, demonstrating a high sodium storage capacity of 700 mAh/g at the second cycle and high reversibility that retained its structure at high current density (see Fig. 9.5).[34]

Anatase TiO_2 offers low electrical conductivity and poor diffusion of ions across the electrode and electrolyte interface that can be overcome by encapsulating it into a 3-D graphene aerogel to provide more active sites to facilitate charge transfer reactions, improving the storage capacity and uptake of Na ions. The TiO_2@3D graphene aerogel delivers a reversible capacity of 164.9 mAh/g at a current density of 5 A/g.[35] In another study, 2.9 wt.% carbon-coated anatase TiO_2 showed high rate capability up to 10 A/g while delivering a capacity of 82 mAh/g, compared with 20 mAh/g for the bare electrode (see Fig. 9.6).[36] SnO_2 nanosheets are reported to grow onto a 3-D carbonized

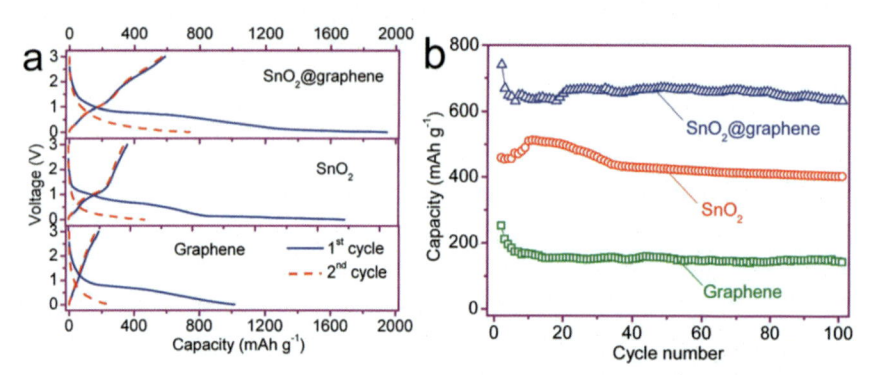

Figure 9.5 (A) Galvanostatic discharge and charge curves for pristine and modified SnO_2@ graphene nanocomposites where its reversible discharge capacity is given as 700 mAh/g at a current density of 20 mA/g (B) Cycling performance of pristine and modified electrodes at 20 mA/g.[34]

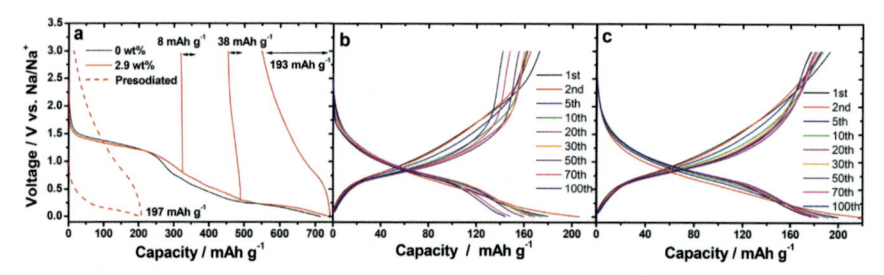

Figure 9.6 (A) Comparison of first charge–discharge curves of 0 wt.% and 2.9 wt.% carbon-coated anatase TiO_2. Long cycling charge–discharge curves of (B) bare and (C) 2.9 wt.% carbon-coated TiO_2.[36]

eggshell membrane with a porous massively interconnected fiber network facilitating electrolyte penetration and improving contact resistance, thus reducing the volume-change stress that is a major problem in SIBs. This active material exhibits a reversible capacity of 656 mAh/g at 0.1 A/g at the fifth cycle and retained at the 200th cycle, and a reversible capacity of 420 mAh/g at 0.2 A/g.[37]

9.1.3 Zinc and magnesium-ion batteries

Recently, much work has focused on monovalent Na^+ and K^+ ions and di/trivalent ions such as $Zn^{2+,}$ Mg^{2+}, and Al^{3+}.[38] Among them, ZIBs are more appealing and have garnered considerable attention for their natural abundance, cost-effectiveness, stability, and high safety compared with LIBs and have found their way into applications such as large-scale EES devices.[38–40] It has a similar structure to LIBs, comprising a cathode to house the Zn ions and an anode for Zn oxidation with the electrolyte required for the discharge/charge process. Much attention has been diverted to Zn anode and cathode material in recent years.[41,42] Superior materials in terms of electrochemical performance, durability, and flexibility have been sought for wearable electronic devices. Thus, solid-state rechargeable ZIBs are promising candidates in terms of safety and mechanical flexibility while sustaining high storage capacities for wearable devices.[40] Solid-state ZIBs have striking advantages over aqueous electrolyte-based ZIBs in circumventing the issue of electrolyte leakage, yet they face challenges, and improvements must be made for practical applications.[40] Considerable research has been dedicated to exploring high-performance cathode ZIBs.

9.1.4 Manganese oxide cathode materials

Mn-based oxides have been very popular due to their excellent electrochemical performance enabling storage and conversion reactions, eco-benevolence, and natural abundance. Based on redox chemistry, MnO_2 displays a high specific capacity supporting one-electron redox transfer. Three different crystallographic polymorphs of MnO_2 built on octahedral units such tunnel forms including $\alpha-$, $\beta-$, $\gamma-MnO_2$, layered forms and spinel form including $\delta-$ and $\lambda-MnO_2$ respectively, and $\delta-MnO_2$ have been reported as an intercalation electrode. MnO_2 cathodes are known to suffer capacity fading due to transformation of structure and disproportionation reaction of Mn^{3+} leading to active material dissolution. Additionally, its storage mechanism is still vague, and further studies are required for its complete mechanistic elucidation. This could be significantly improved by adding $MnSO_4$ in an electrolyte of $ZnSO_4$, which improves electrode integrity forming an amorphous layer at the cathode. However, engineering the structure by conjunction with carbon such as carbon fiber paper under slightly acidic conditions has led to an improvement in electrochemical performance and shown maintenance of high stability due to Mn^{4+}/Mn^{2+}.[43]

MnO_2-based electrodes have low conductivity and lead to the poor performance of the battery.[44] This could be rectified by compositing the MnO_2 electrode with carbon for H^+ and Zn^{2+} intercalation/deintercalation. Thus, graphene has been shown to integrate with MnO_2 nanostructure to demonstrate excellent electrochemical performance

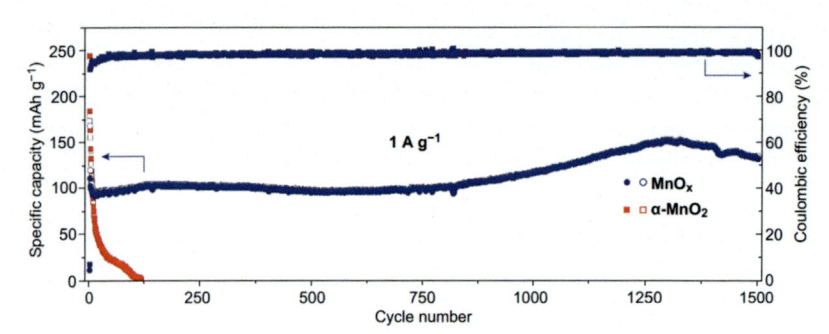

Figure 9.7 Zn/MnOx ZIBs with excellent cycling performance and structural variation during discharge.[46]

by forming a nanoflower-sandwiched structure. The integration of MnO_2 with a carbonaceous matrix increases the surface area substantially to boost charge transport.[45] Active low crystalline of MnO_x NPs coated over a carbon substrate results in high energy density and specific capacity due to facile contact with electrolyte and ease of formation of double redox couple of Mn^{2+} and Mn^{4+} where Mn^{2+} is being supplied from the active material as well as the electrolyte as shown by Fig. 9.7.[46] Similarly, a yolk-shell structure comprising of a carbon shell with MnO_2 NPs for its electrochemical reaction and cycling stability results in high performance for ZIBs.[47]

Nanoflower-layered δ-MnO_2 on graphite has been synthesized as an insertion ion electrode for Zn ions and has displayed high specific capacity compared with pristine MnO_2, as well as extraordinary conductivity and stability.[48] MnO modified with nitrogen-doped carbon has delivered a current density of 500 mA/g, a reversible capacity of 176.3 mAh/g at a graphite current collector. It was demonstrated that diffusion-controlled and surface-controlled process leads to a capacitive behavior for the modified cathode for ZIBs. Phase evolution studies confirmed an insertion/extraction of Zn^{2+} ions accompanied by dissolution/deposition of Mn^{2+} ions at the cathode, contributing to its reversibility and stability.[49] Similar results for MnO_2 as the active material in ZIBs with carbon sphere doped with nitrogen showed improved discharge capacity of 349 mAh/g and capacity retention of 78.7% even after 2000 cycles compared with MnO_2 deposited onto hollow carbon spheres. This choice proves itself as a promising candidate for future economical energy applications.[50] Nanoflakes of α-MnO_2 are deposited over the conductive scaffold of CNT by a simple facile electrodeposition method to assemble a flexible battery for applications in wearable electronics. It is found to deliver a high charge capacity, rate capability, and cycle stability even after 1000 cycles, as shown in Fig. 9.8.[51] Similarly, Khamsanga et al. have reported a versatile method for depositing heterostructured MnO_2 on MWCNT that results in boosting the discharge capacity of Zn^{2+} ions due to diffusion, to 236 mAh/g at a current density of 400 mA/g in ZIBs.[52]

9.1.5 Other cathode materials

Due to structural instability resulting in poor cyclability, a 2-D heterostructure is reported comprising an amorphous V2O5 on graphene with a highly stable structure

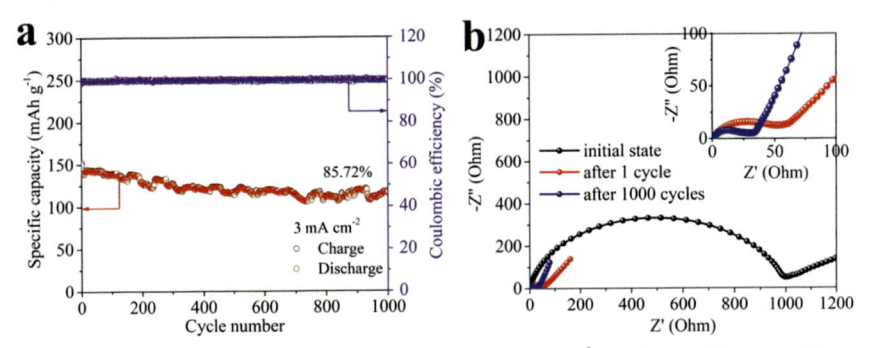

Figure 9.8 Cycling performance of α-MnO$_2$/CNT at 3 mA/cm^2, retaining high reversible capacity of 85.72% after 1000 cycles, (B) Nyquist plots at the initial state, after 1 cycle and 1000 cycles.[51]

as cathode material for aqueous ZIBs compared with a highly stable structure for aqueous pristine V$_2$O$_5$-based ZIBs. The flexible cathode material has shown enhanced specific capacity of 447 mAh/g, resulting in fast electron and ion transfer for ZIBs.[53]

Polyionic materials such as the crystal structure for LiV$_2$(PO$_4$)$_3$ as cathode material exhibit high voltage and a stable structural framework with excellent redox performance due to the robust intercalation of Zn^{2+} ions into available vacancies in the active material during the discharge process. Carbon nanosheets wrapped around Na$_3$V$_2$(PO$_4$)$_3$ have acted as host electrodes for Zn^{2+} ions and have demonstrated a discharge capacity of 97 mAh/g at more than 100 cycles.[54]

MIBs came about after LIBs and have proven to be promising candidates in terms of cost-effectiveness, eco-friendliness, and safety.[55] Nevertheless, the large ionic size of magnesium has restricted the advancement of this technology because of its sluggish diffusion. This has been overcome by using an electrolyte of magnesium organo-haloaluminate in tetrahydrofuran showing extraordinary voltage stability with Mg as the anode and Mg$_x$Mo$_3$S$_4$ as the cathode. V$_2$O$_5$, a positive electrode material with V^{5+}/V^{4+} redox, has been extensively used in LIBs and supports a layered structure connected by weak van der Waals forces. However, the intercalation of Mg^{2+} into the host structure of V$_2$O$_5$ is quite sluggish due to ionic size, and various strategies have been used to achieve fast diffusion of ions into the structure.[55,56] Hydrated V$_2$O$_5$ on graphene sheets has been reported as cathode materials in the organic electrolyte and exhibits high specific capacity, even accommodating Mg^{2+} ion in a temperature range of $-30-55°C$. The high reversible capacity is attributed to fast diffusion of Mg^{2+} ions promoted by H$_2$O crystals present between the bilayer of V$_2$O$_5$ that induce charge shielding effect and an increased interlayer distance.[57] Still, water can react with Mg metal, resulting in poor battery performance in terms of capacity retention.[58] This could be improved by using nano dimensioned cathode material instead of bulk material, resulting in delocalization of the electrons.[59] Thus, porous carbon modified nanocrystals of V$_2$O$_5$ were used as the cathode in MIBs and resulted in high storage capacity, working voltage, and rate capability.[59] Furthermore, the hollow sphere in Li$_3$VO$_4$ is coated with carbon to improve conductivity and ion transport in acetonitrile-based electrolyte with high specific capacity but low retentions[60] attributed to passivation layer on the metal Mg anode, and hence Mg ions become trapped.[61]

9.1.6 Metal–air batteries

Metal–air batteries have been endorsed as the next-generation high-performance EES device, as LIBs are fast approaching their performance limit and are expected to increase it by only 30%,[62] which is insufficient to interface them to grid scale and electric vehicles.[63] Yet that too is laden with challenges that must be overcome concerning the electrolyte used as well as the anode and cathode. Thus, to employ metal–air batteries for large-scale applications, we still have some way to go. Metal–air batteries are an amalgamation of architectural features present in fuel cells and conventional batteries comprising breathable highly porous cathode for the oxygen reduction reaction (ORR), while the anode could be reactive metals such Li, Na, or K in nonaqueous media or Mg, Zn, Al, or Fe in aqueous media.[64−68] Li, Na, and K are too reactive in aqueous electrolytes and must be protected by a conductive film in order to perform, which increases the cost of fabrication and offers a process that is too complex to fabricate.[69] The surfaces of Mg, Zn, Al, and Fe become passivated by the electrolyte and are oxidized at the electrode to the metal cation, while oxygen is reduced at the cathode to OH^- in aqueous media during discharge reactions. For the Zn–air battery, practical energy density is achieved at 350−500 Wh/kg compared with 1353 Wh/kg of theoretical energy density, and this is attributed to the low efficiency of the air catalyst used at the cathode. Hence, research has focused on engineering the cathode materials, as they play a decisive role in improving battery performance.[70] ORR is an uphill reaction and shares design principles with hydrogen fuel cells. Hence, primary attention has been given to designing catalysts to speed up the sluggish ORR reaction and amplify the triple-phase boundary between the solid electrode, O_2 gas, and liquid electrolyte. Precious and nonprecious metals have been employed. Pt acts as the benchmark catalyst, but its activity is suppressed by high coverage of the electrode surface with OH species. Hence, CoO grown on nitrogen-doped CNT has been proposed to outperform the gold standard of Pt/C with the same percentage of loading in the alkaline medium as demonstrated in Fig. 9.9.[71]

While for rechargeable batteries, the catalyst should be dually functional to catalyze both ORR and oxygen evolution reactions such as OER. However, even though huge strides have been taken in the development of air-breathable cathode, the polarization of the electrode during dis/charge is so large that the energy efficiency of rechargeable Zn-air batteries cannot go beyond the 65% required for practical applications. Although such bifunctional catalysts can be made durable, many show a low tolerance for cycling stability because of the varied oxidizing and reducing environments during charge–discharge, which is severely detrimental to active sites at the cathode responsible for ORR, additionally leading to corrosion of the carbon matrix. The inception of such an electrode with high activity and stability is rare. This issue could be circumvented by using a bi-cathode that involves employing two separate electrodes by decoupling the catalyst for ORR and OER during charge–discharge.[72] The same issue has been observed with other aqueous metal–air batteries. Fe air batteries are fairly

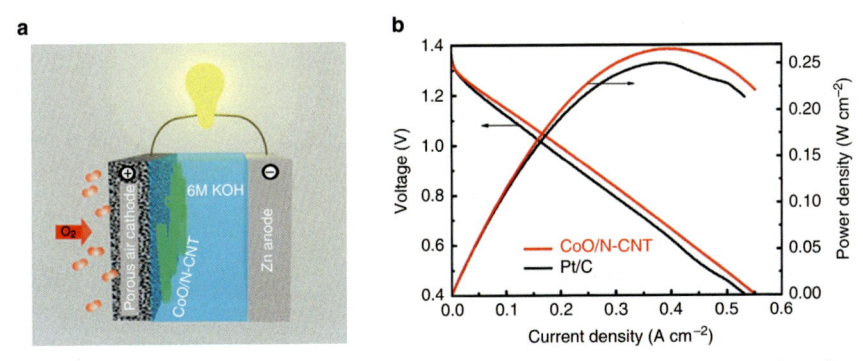

Figure 9.9 (A) schematic showing a primary Zn-air battery with CoO/N-CNT used as electrocatalyst in 6 M KOH (B) shows polarization curve and corresponding power density plot Zn-air battery using CoO/N-CNT compared with the commercial one Pt/C catalyst with peak power density of 265 mW/cm^2.[71]

stable with low energy density, however, and are environmentally friendly and cost-effective, whereas the anodes of Al and Mg air batteries are severely affected by corrosion when in contact with the electrolyte and hence cannot be employed for EVs.[73] Nonaqueous metal—air batteries suffer from serious challenges such as the use of aprotic solvents, which involves a mechanism different from aqueous metal—air batteries and thus involves complex superoxide chemistry that leads to compromised discharge capacity from blockage of the cathode by superoxides.[64,65,74] Nonaqueous metal—air batteries are still in their infancy and have a long way to go to overcome the challenges offered by the electrolytes and cathode used, which will not be discussed further in this chapter.

9.1.7 Metal—sulfur batteries

In rechargeable metal—sulfur batteries (RMSBs), metal anodes and sulfur cathodes exhibit high specific capacity. RMSBs still have challenges to overcome, including dissolution of sulfur compounds in the electrolyte and high reactivity of the anodes under consideration. The cathode is more focused on where the scaffolding and architecture of sulfur cathode plays an important role. Porous Cr_2O_3/C nanocomposite derived from MOF with size dimension <100 nm allows for electrochemical reactions in Li/S battery with Li_2S_8 dissolved in derivatized ether. It shows excellent stability and a discharge capacity of 900—780 mAh/g at different charge rates and 100 charges/discharge cycles versus Li^+/Li couple.[75]

9.1.8 Redox flow batteries

It is a flow battery in the sense that for some configurations, there is no physical transfer of material, while some use electrochemical redox couple for reversible reaction in the solution phase at the anode and cathode for energy storage. The electrodes do not

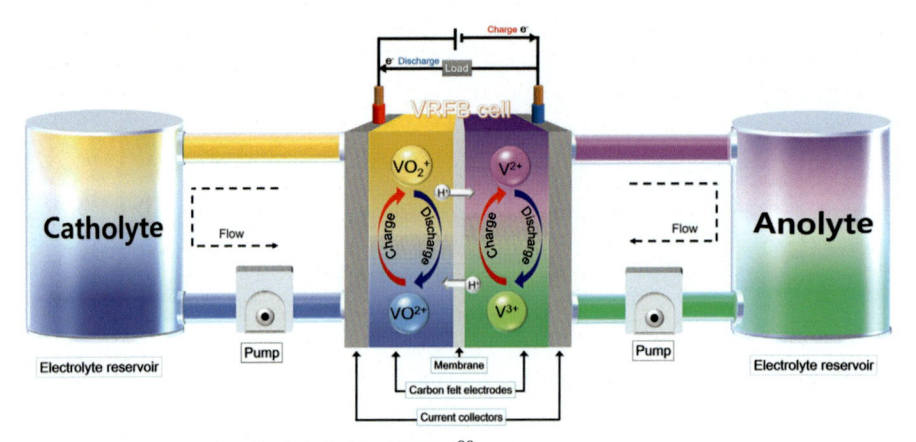

Figure 9.10 The basic principle behind RFBs.[80]

phase change during reaction or undergo intercalation/deintercalation as redox reaction happens in the solution phase where the reactants are dissolved, and hence degradation of the electrode is not a matter of concern.[76] All vanadium system is comprised of vanadium in two oxidations states to compensate for an efficiency loss by preventing crossing over and its irreversible reaction.[77] V^{4+}/V^{5+} acts as the redox couple at the positive electrode, where the active electrolyte is stored in tanks away from the electrodes. Thus, during discharge reactions, the catholyte (positive electrolyte) and anolyte (negative electrolyte) contain the V^{3+} and V^{4+}. During charging, a reduction of catholyte to V^{2+} and oxidation of anolyte to V^{5+} take place.[78] A redox flow battery based on vanadium (VRFB) has shown to be a promising candidate for ESS that employs conversion reactions and finds applications where a few kW to MW are required. However, issues such as cycling stability and charge−discharge cycle rate have hindered its commercialization acceptance. Electrocatalyst can be used to increase the efficiency of the anode and cathode in VRFB. The VRFB have shown an efficiency of more than 80%.[79] A basic schematic for the VRFB is shown in Fig. 9.10.

It has been demonstrated that modifying the graphite felt (GF) with Mn_3O_4,[81] NiO/Ni,[82] WO_3,[83] W-doped Nb_2O_5,[84] and PbO_2[85] results in improved kinetics for reaction to happen at the electrode by increasing the porosity of the carbon felt and likewise engineering microscaled GF with nanosized pores by a simple method with outstanding catalytic activity toward VO^{2+}/VO_2^+ and V^{2+}/V^{3+}.[86] A novel method was devised to modify GF with etched cobalt oxide on CNT, enhancing the wettability of the GF compared with unetched GF.[87] TiO_2 NPs added to negative carbon electrode improves conductivity and wettability of the electrode and thus improves the kinetics. This has been shown to boost the efficiency to store energy in VRFB.[88]

Similarly, GF is modified by zirconium oxide NPs to improve the accessibility of the electrolyte and increase the active sites to give the best electrochemical performance and cycling stability.[89] Likewise, $W_{18}O_{49}$ nanowires were employed to modify graphite felt as an electrocatalyst by a solvothermal process leading to improved energy efficiency providing oxygen vacancies as active sites for redox actions.[90] Thus, in another study, the carbon felt was modified with SnO_2 as an electrocatalyst and showed remarkable retention of capacity at $150\,mA/cm^2$ at 77.3% and improved cycling stability.[80]

Electrospun carbon nanofibers (ECNF) based on polyacrylonitrile decorated with CeO_2 NPs increase the electrochemically active surface area compared with pristine ECNF, which have poor hydrophilicity and hence affect the electrochemical performance. Thus, CeO_2 NPs embedded on the ECNF have been shown to improve the wettability of the negative electrode in VFB compared with the positive electrode.[91]

9.2 Conclusion

The chapter gives an overview of the applications of carbon-based materials composited with metal oxides as the electrode of choice used in different batteries. New materials need to be sought to improve the storage capacity and electrochemical performance of the electrodes in batteries. Hence, various batteries such LIBs, SIBs, Mg and ZIBs, and RFBs have been investigated, and it has been found that integrating carbon with metal oxides improves the drastic volume changes that happen during charge/discharge of the batteries with improvement in conductivity. This leads to improved battery performances. Thus, the synergy of carbon-based materials and metal oxides results in overcoming the challenges presented by either pure carbon-based material or metal oxides. This leads to improved stability, electroconductivity, battery performance, rate capability and endows the composite with modified, enhanced properties. However, while using transition-based metals with carbon material, the issue of hole size needs to be considered for efficient electrolyte permeation in the electrode material. Furthermore, hollowed architecture is more advantageous for electrodes due to their buffering effect of any volume change during the charge/discharge process.

Compositing carbonaceous material with metal oxides has been explored for some time; however, this domain is still rife with challenges. Thus, other than using conductive carbon-based materials, another possibility for developing high-end battery technology to overcome extensive volume changes and conductivity issues during de/lithiation and de/sodiation, specifically for LIBs and SIBs, respectively, self-healing binders tailored to promote electron conductivity. Additionally, electrolyte composition is another field of research that can be explored for achieving the best performance of batteries. More importantly, we need to understand the reaction mechanism in detail, which has not garnered much attention because of its complex reactions. The mechanism with regards to nascent battery technology such as RFB has still to be explored. Similarly, to overcome the shortcoming of aqueous ZIBs in terms of energy density that can be achieved with organic electrolytes, hybrid aqueous ZIBs should be explored as high-performance storage batteries. Still, the integration of materials such as carbon and metals oxides has a great prospect of producing high-performance batteries that are sustainable, greener, and more efficient.

References

1. Owusu PA, Asumadu-Sarkodie S. A review of renewable energy sources, sustainability issues and climate change mitigation. *Cogent Eng* 2016;**3**:1167990.
2. McPherson M, Johnson N, Strubegger M. The role of electricity storage and hydrogen technologies in enabling global low-carbon energy transitions. *Appl Energy* 2018;**216**:649−61.
3. Larcher D, Tarascon J-M. Towards greener and more sustainable batteries for electrical energy storage. *Nat Chem* 2015;**7**:19−29.
4. Ali G, Patil SA, Mehboob S, Ahmad M, Ha HY, Kim H-S, Chung KY. Determination of lithium diffusion coefficient and reaction mechanism into ultra-small nanocrystalline SnO_2 particles. *J Power Sources* 2019;**419**:229−36.
5. Ali G, Mehboob S, Ahmad M, Akbar M, Kim S-O, Ha HY, Chung KY. High-rate lithium storage and kinetic investigations of a cubic Mn_2SnO_4@Carbon nanotube composite anode. *J Alloys Compd* 2020;**823**:153789.
6. Ata Ur R, Ali G, Badshah A, Chung KY, Nam K-W, Jawad M, Arshad M, Abbas SM. Superior shuttling of lithium and sodium ions in manganese-doped titania @ functionalized multiwall carbon nanotube anodes. *Nanoscale* 2017;**9**:9859−71.
7. Jayaprakash N, Jones WD, Moganty SS, Archer LA. Composite lithium battery anodes based on carbon@Co_3O_4 nanostructures: synthesis and characterization. *J Power Sources* 2012;**200**:53−8.
8. Seok D, Jeong Y, Han K, Yoon DY, Sohn H. Recent progress of electrochemical energy devices: metal oxide−carbon nanocomposites as materials for next-generation chemical storage for renewable energy. *Sustainability* 2019;**11**:3694.
9. Park J, Moon WG, Kim G-P, Nam I, Park S, Kim Y, Yi J. Three-dimensional aligned mesoporous carbon nanotubes filled with Co_3O_4 nanoparticles for Li-ion battery anode applications. *Electrochim Acta* 2013;**105**:110−4.
10. Abbas SM, Hussain ST, Ali S, Ahmad N, Ali N, Munawar KS. Synthesis of carbon nanotubes anchored with mesoporous Co_3O_4 nanoparticles as anode material for lithium-ion batteries. *Electrochim Acta* 2013;**105**:481−8.
11. He X, Wu Y, Zhao F, Wang J, Jiang K, Fan S. Enhanced rate capabilities of Co_3O_4/carbon nanotube anodes for lithium ion battery applications. *J Mater Chem* 2013;**1**:11121−5.
12. He Y, Huang L, Cai J-S, Zheng X-M, Sun S-G. Structure and electrochemical performance of nanostructured Fe_3O_4/carbon nanotube composites as anodes for lithium ion batteries. *Electrochim Acta* 2010;**55**:1140−4.
13. Luo S, Wu H, Wu Y, Jiang K, Wang J, Fan S. Mn_3O_4 nanoparticles anchored on continuous carbon nanotube network as superior anodes for lithium ion batteries. *J Power Sources* 2014;**249**:463−9.
14. Li X, Gu H, Liu J, Wei H, Qiu S, Fu Y, Lv H, Lu G, Wang Y, Guo Z. Multi-walled carbon nanotubes composited with nanomagnetite for anodes in lithium ion batteries. *RSC Adv* 2015;**5**:7237−44.
15. Ali G, Lee JH, Chang W, Cho B-W, Jung H-G, Nam K-W, Chung KY. Lithium intercalation mechanism into $FeF_3 \cdot 0.5H_2O$ as a highly stable composite cathode material. *Sci Rep* 2017;**7**:42237.
16. Su J, Cao M, Ren L, Hu C. Fe_3O_4−Graphene nanocomposites with improved lithium storage and magnetism properties. *J Phys Chem C* 2011;**115**:14469−77.
17. Zhou G, Wang D-W, Li F, Zhang L, Li N, Wu Z-S, Wen L, Lu GQ, Cheng H-M. Graphene-wrapped Fe_3O_4 anode material with improved reversible capacity and cyclic stability for lithium ion batteries. *Chem Mater* 2010;**22**:5306−13.

18. Mai YJ, Wang XL, Xiang JY, Qiao YQ, Zhang D, Gu CD, Tu JP. CuO/graphene composite as anode materials for lithium-ion batteries. *Electrochim Acta* 2011;**56**:2306—11.

19. Zhang X, Zhou J, Chen X, Song H. Pliable embedded-type paper electrode of hollow metal oxide@porous graphene with abnormal but superior rate capability for lithium-ion storage. *ACS Appl Energy Mater* 2018;**1**:48—55.

20. Peng L, Fang Z, Zhu Y, Yan C, Yu G. Holey 2D nanomaterials for electrochemical energy storage. *Adv Energy Mater* 2018;**8**:1702179.

21. Kim H, Seo D-H, Kim S-W, Kim J, Kang K. Highly reversible Co_3O_4/graphene hybrid anode for lithium rechargeable batteries. *Carbon* 2011;**49**:326—32.

22. Li L, Zhou G, Shan X-Y, Pei S, Li F, Cheng H-M. Co_3O_4 mesoporous nanostructures@ graphene membrane as an integrated anode for long-life lithium-ion batteries. *J Power Sources* 2014;**255**:52—8.

23. Hu R, Zhang H, Bu Y, Zhang H, Zhao B, Yang C. Porous Co_3O_4 nanofibers surface-modified by reduced graphene oxide as a durable, high-rate anode for lithium ion battery. *Electrochim Acta* 2017;**228**:241—50.

24. Chang P-y, Huang C-h, Doong R-a. Ordered mesoporous carbon—TiO_2 materials for improved electrochemical performance of lithium ion battery. *Carbon* 2012;**50**: 4259—68.

25. Ata ur R, Ali G, Abbas SM, Iftikhar M, Zahid M, Yaseen S, Saleem S, Haider S, Arshad M, Badshah A. Axial expansion of Ni-doped TiO_2 nanorods grown on carbon nanotubes for favourable lithium-ion intercalation. *Chem Eng J* 2019;**375**:122021.

26. Wang X, Li Z, Yin L. Nanocomposites of SnO_2@ordered mesoporous carbon (OMC) as anode materials for lithium-ion batteries with improved electrochemical performance. *CrystEngComm* 2013;**15**:7589—97.

27. Han F, Li W-C, Li M-R, Lu A-H. Fabrication of superior-performance SnO_2@C composites for lithium-ion anodes using tubular mesoporous carbon with thin carbon walls and high pore volume. *J Mater Chem* 2012;**22**:9645—51.

28. Cao Z, Shi M, Ding Y, Zhang J, Wang Z, Dong H, Yin Y, Yang S. Lotus-root-like MnO/C hybrids as anode materials for high-performance lithium-ion batteries. *J Phys Chem C* 2017; **121**:2546—55.

29. Bhange DS, Ali G, Kim J-Y, Chung KY, Nam K-W. Improving the sodium storage capacity of tunnel structured $NaxFexTi_{2-x}O_4$ (x = 1, 0.9 & 0.8) anode materials by tuning sodium deficiency. *J Power Sources* 2017;**366**:115—22.

30. Wang Y, Su D, Wang C, Wang G. SnO_2@MWCNT nanocomposite as a high capacity anode material for sodium-ion batteries. *Electrochem Commun* 2013;**29**:8—11.

31. Ma C, Li X, Deng C, Hu Y-Y, Lee S, Liao X-Z, He Y-S, Ma Z-F, Xiong H. Coaxial carbon nanotube supported TiO_2@MoO_2@carbon core—shell anode for ultrafast and high-capacity sodium ion storage. *ACS Nano* 2019;**13**:671—80.

32. Liu X, Chen T, Chu H, Niu L, Sun Z, Pan L, Sun CQ. Fe_2O_3-reduced graphene oxide composites synthesized via microwave-assisted method for sodium ion batteries. *Electrochim Acta* 2015;**166**:12—6.

33. Zhang Y, Xie J, Zhang S, Zhu P, Cao G, Zhao X. Ultrafine tin oxide on reduced graphene oxide as high-performance anode for sodium-ion batteries. *Electrochim Acta* 2015;**151**: 8—15.

34. Su D, Ahn H-J, Wang G. SnO_2@graphene nanocomposites as anode materials for Na-ion batteries with superior electrochemical performance. *Chem Commun* 2013;**49**:3131—3.

35. Luo R, Ma Y, Qu W, Qian J, Li L, Wu F, Chen R. High pseudocapacitance boosts ultrafast, high-capacity sodium storage of 3D graphene foam-encapsulated TiO_2 architecture. *ACS Appl Mater Interfaces* 2020;**12**:23939—50.

36. Kim K-T, Ali G, Chung KY, Yoon CS, Yashiro H, Sun Y-K, Lu J, Amine K, Myung S-T. Anatase titania nanorods as an intercalation anode material for rechargeable sodium batteries. *Nano Lett* 2014;**14**:416−22.

37. Zhao X, Luo M, Zhao W, Xu R, Liu Y, Shen H. SnO_2 nanosheets anchored on a 3D, bicontinuous electron and ion transport carbon network for high-performance sodium-ion batteries. *ACS Appl Mater Interfaces* 2018;**10**:38006−14.

38. Fang G, Zhou J, Pan A, Liang S. Recent advances in aqueous zinc-ion batteries. *ACS Energy Lett* 2018;**3**:2480−501.

39. Jia X, Liu C, Neale ZG, Yang J, Cao G. Active materials for aqueous zinc ion batteries: synthesis, crystal structure, morphology, and electrochemistry. *Chem Rev* 2020;**120**: 7795−866.

40. Li Y, Fu J, Zhong C, Wu T, Chen Z, Hu W, Amine K, Lu J. Batteries: recent advances in flexible zinc-based rechargeable batteries (adv. energy mater. 1/2019). *Adv Energy Mater* 2019;**9**:1970001.

41. Wu B, Zhang G, Yan M, Xiong T, He P, He L, Xu X, Mai L. Graphene scroll-coated α-MnO_2 nanowires as high-performance cathode materials for aqueous Zn-ion battery. *Small* 2018;**14**:1703850.

42. Yuan C, Zhang Y, Pan Y, Liu X, Wang G, Cao D. Investigation of the intercalation of polyvalent cations (Mg^{2+}, Zn^{2+}) into λ-MnO_2 for rechargeable aqueous battery. *Electrochim Acta* 2014;**116**:404−12.

43. Chao D, Zhou W, Ye C, Zhang Q, Chen Y, Gu L, Davey K, Qiao SZ. An electrolytic Zn−MnO_2 battery for high-voltage and scalable energy storage. *Angew Chem Int Ed* 2019; **58**:7823−8.

44. Toupin M, Brousse T, Bélanger D. Charge storage mechanism of MnO_2 electrode used in aqueous electrochemical capacitor. *Chem Mater* 2004;**16**:3184−90.

45. Liu J, Zhang Y, Li Y, Li J, Chen Z, Feng H, Li J, Jiang J, Qian D. In situ chemical synthesis of sandwich-structured MnO_2/graphene nanoflowers and their supercapacitive behavior. *Electrochim Acta* 2015;**173**:148−55.

46. Huang J, Zeng J, Zhu K, Zhang R, Liu J. High-performance aqueous zinc−manganese battery with reversible Mn 2+/Mn 4+ double redox achieved by carbon coated MnO x nanoparticles. *Nano-Micro Lett* 2020;**12**:1−12.

47. Liu W, Liu P, Hao R, Huang Y, Chen X, Cai R, Yan J, Liu K. One-Dimensional MnO_2 nanowires space-confined in hollow mesoporous carbon nanotubes for enhanced Zn^{2+} storage performance. *ChemElectroChem* 2020;**7**:1166−71.

48. Khamsanga S, Pornprasertsuk R, Yonezawa T, Mohamad AA, Kheawhom S. δ-MnO_2 nanoflower/graphite cathode for rechargeable aqueous zinc ion batteries. *Sci Rep* 2019;**9**: 8441.

49. Tang F, He T, Zhang H, Wu X, Li Y, Long F, Xiang Y, Zhu L, Wu J, Wu X. The MnO@N-doped carbon composite derived from electrospinning as cathode material for aqueous zinc ion battery. *J Electroanal Chem* 2020;**873**:114368.

50. Chen L, Yang Z, Cui F, Meng J, Jiang Y, Long J, Zeng X. Ultrathin MnO2 nanoflakes grown on N-doped hollow carbon spheres for high-performance aqueous zinc ion batteries. *Mater Chem Front* 2020;**4**:213−21.

51. Huang A, Chen J, Zhou W, Wang A, Chen M, Tian Q, Xu J. Electrodeposition of MnO_2 nanoflakes onto carbon nanotube film towards high-performance flexible quasi-solid-state Zn-MnO_2 batteries. *J Electroanal Chem* 2020;**873**:114392.

52. Khamsanga S, Nguyen MT, Yonezawa T, Thamyongkit P, Pornprasertsuk R, Pattananuwat P, Tuantranont A, Siwamogsatham S, Kheawhom S. MnO_2 heterostructure on carbon nanotubes as cathode material for aqueous zinc-ion batteries. *Int J Mol Sci* 2020;**21**: 4689.

53. Wang X, Li Y, Das P, Zheng S, Zhou F, Wu Z-S. Layer-by-layer stacked amorphous V_2O_5/graphene 2D heterostructures with strong-coupling effect for high-capacity aqueous zinc-ion batteries with ultra-long cycle life. *Energy Storage Mater* 2020;**31**:156–63.

54. Li G, Yang Z, Jiang Y, Zhang W, Huang Y. Hybrid aqueous battery based on Na_3V_2 (PO4) 3/C cathode and zinc anode for potential large-scale energy storage. *J Power Sources* 2016;**308**:52–7.

55. Saha P, Datta MK, Velikokhatnyi OI, Manivannan A, Alman D, Kumta PN. Rechargeable magnesium battery: current status and key challenges for the future. *Prog Mater Sci* 2014;**66**:1–86.

56. Muldoon J, Bucur CB, Gregory T. Quest for nonaqueous multivalent secondary batteries: magnesium and beyond. *Chem Rev* 2014;**114**:11683–720.

57. An Q, Li Y, Deog Yoo H, Chen S, Ru Q, Mai L, Yao Y. Graphene decorated vanadium oxide nanowire aerogel for long-cycle-life magnesium battery cathodes. *Nano Energy* 2015;**18**:265–72.

58. Novak P, Desilvestro J. Electrochemical insertion of magnesium in metal oxides and sulfides from aprotic electrolytes. *J Electrochem Soc* 1993;**140**:140.

59. Cheng Y, Shao Y, Raju V, Ji X, Mehdi BL, Han KS, Engelhard MH, Li G, Browning ND, Mueller KT. Molecular storage of Mg ions with vanadium oxide nanoclusters. *Adv Funct Mater* 2016;**26**:3446–53.

60. Yang Y, Li J, He X, Wang J, Sun D, Zhao J. A facile spray drying route for mesoporous Li_3VO_4/C hollow spheres as an anode for long life lithium ion batteries. *J Mater Chem* 2016;**4**:7165–8.

61. Levi E, Levi M, Chasid O, Aurbach D. A review on the problems of the solid state ions diffusion in cathodes for rechargeable Mg batteries. *J Electroceram* 2009;**22**:13–9.

62. Van Noorden R. The rechargeable revolution: a better battery. *Nat News* 2014;**507**:26.

63. Dunn B, Kamath H, Tarascon J-M. Electrical energy storage for the grid: a battery of choices. *Science* 2011;**334**:928–35.

64. Ren X, Wu Y. A low-overpotential potassium–oxygen battery based on potassium superoxide. *J Am Chem Soc* 2013;**135**:2923–6.

65. Hartmann P, Bender CL, Vračar M, Dürr AK, Garsuch A, Janek J, Adelhelm P. A rechargeable room-temperature sodium superoxide (NaO 2) battery. *Nat Mater* 2013;**12**:228–32.

66. Abraham K, Jiang Z. A polymer electrolyte-based rechargeable lithium/oxygen battery. *J Electrochem Soc* 1996;**143**:1.

67. Zaromb S. The use and behavior of aluminum anodes in alkaline primary batteries. *J Electrochem Soc* 1962;**109**:1125.

68. Öjefors L, Carlsson L. An iron—air vehicle battery. *J Power Sources* 1978;**2**:287–96.

69. Grande L, Paillard E, Hassoun J, Park JB, Lee YJ, Sun YK, Passerini S, Scrosati B. The lithium/air battery: still an emerging system or a practical reality? *Adv Mater* 2015;**27**:784–800.

70. Li Y, Lu J. Metal—air batteries: will they be the future electrochemical energy storage device of choice? *ACS Energy Lett* 2017;**2**:1370–7.

71. Li Y, Gong M, Liang Y, Feng J, Kim J-E, Wang H, Hong G, Zhang B, Dai H. Advanced zinc-air batteries based on high-performance hybrid electrocatalysts. *Nat Commun* 2013;**4**:1–7.

72. Lee CW, Sathiyanarayanan K, Eom SW, Yun MS. Novel alloys to improve the electrochemical behavior of zinc anodes for zinc/air battery. *J Power Sources* 2006;**160**:1436–41.

73. Goldstein J, Brown I, Koretz B. New developments in the Electric Fuel Ltd. zinc/air system. *J Power Sources* 1999;**80**:171–9.

74. Johnson L, Li C, Liu Z, Chen Y, Freunberger SA, Ashok PC, Praveen BB, Dholakia K, Tarascon J-M, Bruce PG. The role of LiO 2 solubility in O 2 reduction in aprotic solvents and its consequences for Li−O 2 batteries. *Nat Chem* 2014;**6**:1091.

75. Benítez A, Marangon V, Hernández-Rentero C, Caballero Á, Morales J, Hassoun J. Porous Cr_2O_3@C composite derived from metal organic framework in efficient semi-liquid lithium-sulfur battery. *Mater Chem Phys* 2020;**255**:123484.

76. Weber AZ, Mench MM, Meyers JP, Ross PN, Gostick JT, Liu Q. Redox flow batteries: a review. *J Appl Electrochem* 2011;**41**:1137.

77. Mehboob S, Ali G, Abbas S, Chung KY, Ha HY. Elucidating the performance-limiting electrode for all-vanadium redox flow batteries through in-depth physical and electrochemical analyses. *J Ind Eng Chem* 2019;**80**:450−60.

78. Gundlapalli R, Kumar S, Jayanti S. Stack design considerations for vanadium redox flow battery. *INAE Lett* 2018;**3**:149−57.

79. Soloveichik GL. Flow batteries: current status and trends. *Chem Rev* 2015;**115**:11533−58.

80. Mehboob S, Ali G, Shin H-J, Hwang J, Abbas S, Chung KY, Ha HY. Enhancing the performance of all-vanadium redox flow batteries by decorating carbon felt electrodes with SnO_2 nanoparticles. *Appl Energy* 2018;**229**:910−21.

81. Kim KJ, Park M-S, Kim J-H, Hwang U, Lee NJ, Jeong G, Kim Y-J. Novel catalytic effects of Mn_3O_4 for all vanadium redox flow batteries. *Chem Commun* 2012;**48**:5455−7.

82. Park JJ, Park JH, Park OO, Yang JH. Highly porous graphenated graphite felt electrodes with catalytic defects for high-performance vanadium redox flow batteries produced via NiO/Ni redox reactions. *Carbon* 2016;**110**:17−26.

83. Shen Y, Xu H, Xu P, Wu X, Dong Y, Lu L. Electrochemical catalytic activity of tungsten trioxide- modified graphite felt toward VO^{2+}/VO^{2+} redox reaction. *Electrochim Acta* 2014;**132**:37−41.

84. Li B, Gu M, Nie Z, Wei X, Wang C, Sprenkle V, Wang W. Nanorod niobium oxide as powerful catalysts for an all vanadium redox flow battery. *Nano Lett* 2014;**14**:158−65.

85. Wu X, Xu H, Lu L, Zhao H, Fu J, Shen Y, Xu P, Dong Y. PbO_2-modified graphite felt as the positive electrode for an all-vanadium redox flow battery. *J Power Sources* 2014;**250**:274−8.

86. Liu Y, Shen Y, Yu L, Liu L, Liang F, Qiu X, Xi J. Holey-engineered electrodes for advanced vanadium flow batteries. *Nano Energy* 2018;**43**:55−62.

87. Abbas S, Lee H, Hwang J, Mehmood A, Shin H-J, Mehboob S, Lee J-Y, Ha HY. A novel approach for forming carbon nanorods on the surface of carbon felt electrode by catalytic etching for high-performance vanadium redox flow battery. *Carbon* 2018;**128**:31−7.

88. Tseng T-M, Huang R-H, Huang C-Y, Hsueh K-L, Shieu F-S. Improvement of titanium dioxide addition on carbon black composite for negative electrode in vanadium redox flow battery. *J Electrochem Soc* 2013;**160**:A1269.

89. Zhou H, Shen Y, Xi J, Qiu X, Chen L. ZrO_2-Nanoparticle-Modified graphite felt: bifunctional effects on vanadium flow batteries. *ACS Appl Mater Interfaces* 2016;**8**:15369−78.

90. Bayeh AW, Kabtamu DM, Chang Y-C, Chen G-C, Chen H-Y, Liu T-R, Wondimu TH, Wang K-C, Wang C-H. Hydrogen-treated defect-rich W18O49 nanowire-modified graphite felt as high-performance electrode for vanadium redox flow battery. *ACS Appl Energy Mater* 2019;**2**:2541−51.

91. Jing M, Zhang X, Fan X, Zhao L, Liu J, Yan C. CeO_2 embedded electrospun carbon nanofibers as the advanced electrode with high effective surface area for vanadium flow battery. *Electrochim Acta* 2016;**215**:57−65.

Two-dimensional transition metal carbide (MXene) for enhanced energy storage

Jameela Fatheema[1,2], Deji Akinwande[2] and Syed Rizwan[1]
[1]Physics Characterization and Simulations Lab (PCSL), Department of Physics, School of Natural Sciences (SNS), National University of Sciences and Technology (NUST), Islamabad, Pakistan; [2]Microelectronics Research Center, The University of Texas at Austin, Austin, TX, United States

10.1 Introduction

The rapid development of daily human life has led to the modernization of energy-storage devices. Renewable energy has attracted considerable attention for its potential to meet daily energy requirements while maintaining environmental cleanliness and avoiding pollution. Wind and solar energy are the most accessible sources. It is essential to have highly functional energy-storage devices to exploit all the energy obtained from these sources. For this purpose, many studies are available about different materials for use in batteries and supercapacitors for rechargeable energy storage.

Since the award of the Nobel Prize in 2004 for studies on graphene, two-dimensional (2-D) materials have inspired much research for to their distinguished properties and applications in energy-storage devices. Two-dimensional materials have a structure that extends in two dimensions, while the third dimension is limited. The research since then has mainly focused on exploring further 2-D materials such as germanene, phosphorene, bismuthine, and silicone.[1−4] Two-dimensional materials, because of their large interlayer distances and exposed surface areas, provide many opportunities for manipulating them to the desired requirements. In addition to the above-mentioned 2-D materials based on a single element, some 2-D materials consist of more than one element, including h-BN, transition metal dichalcogenides (TMDs), and MXenes.[5,6] In molecular 2-D compounds, MXenes have shown great potential for many applications due to their unique properties and remarkable structure.

MXenes are early transition metal carbides or nitrides exfoliated from their parent compound MAX, which has a layered hexagonal structure. Here, "M" is an early transition metal like Ti, V, Ta, Cr, "A" is a member of group 13 or 14 like Al or Si, and X is C, N, or both. The chemical formula for MAX is M_nAX_{n+1}, where $n = 1, 2,$ and 3.[7] More than 70 members are reported within the MAX family, since MAX has a variety of phases with different elemental compositions. Thus, a wide variety of MXenes can be obtained by removing the A layer, with each compound having distinct properties and a structure like 2-D graphene. The crystal structures of MAX phases for n = 1, 2,

Metal Oxide-Carbon Hybrid Materials. **https://doi.org/10.1016/B978-0-12-822694-0.00002-8**

and 3, along with corresponding MXene phases after etching out the A layer, are shown in Fig. 10.1. MAX and MXene phases generally show a hexagonal symmetry belonging to a space group—194 P63/mmc.[7] Some examples of MAX are Ti_3SiC_2, Ti_3AlN_2, Nb_2AlC, V_2AlC, Ta_4AlC_3, and Mo_2AlC, which after exfoliation form MXene Ti_3C_2, Nb_2C, V_2C, and Mo_2C, respectively.[8,9]

MXene has been studied extensively both experimentally and theoretically—via density functional theory (DFT) calculations and other means—and shows exceptional electronic, mechanical, magnetic, and optical properties.[10] As transition metals have

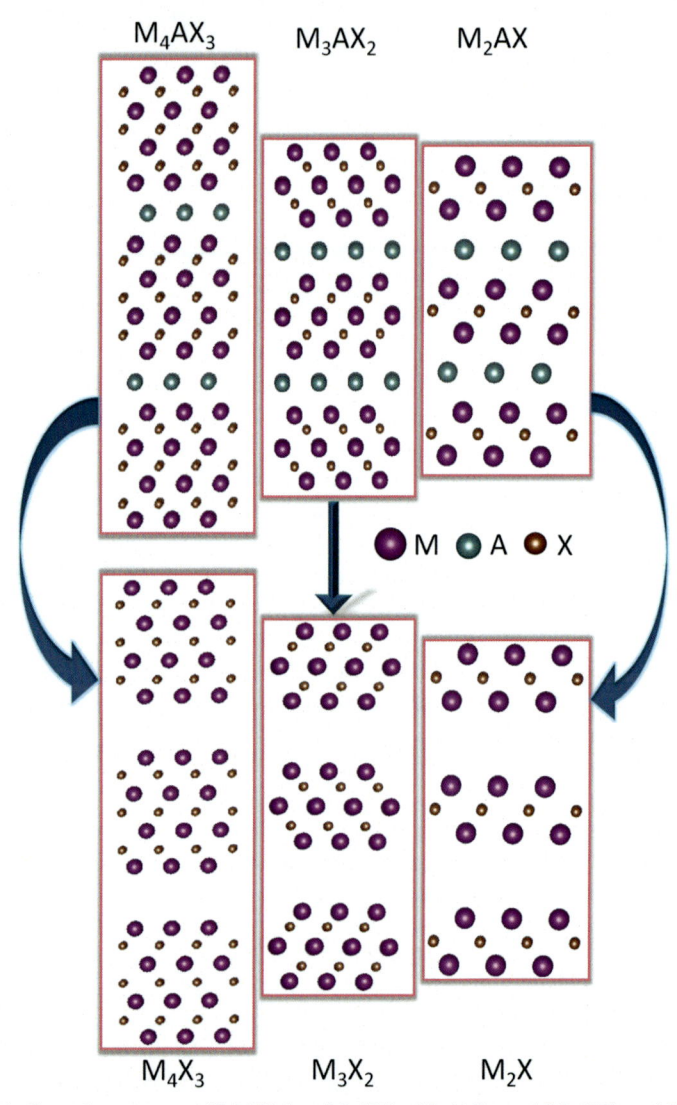

Figure 10.1 Crystal structures of MAX, i.e., M_4AlC_3, M_3AlC_2, and M_2AlC, and MXene, i.e., M_4C_3, M_3C_2, and M_2C, representing the three possible phases for $n = 1, 2$, and 3.

an open d-orbital, many possibilities exist for engineering the structure and properties of MXene to the desired requirements. Since its discovery in 2011, MXenes have shown application in the fields of supercapacitors, topological insulators, batteries, resistive random access memory, photocatalysts, organic dyes, spintronics, etc.[11–20] Their excellent properties in conduction and storage provide a pathway for electrochemical energy storage, and as time has passed, many studies have been conducted. Compared with other 2-D materials like graphene and TMDs, the MXene family presents a range of compounds with enhanced performance in energy-storage applications.[21] Many MXene compounds have demonstrated high storage capacities and endurance that can be altered and modified by defect engineering. This chapter provides a brief review and understanding of recent developments of MXene for batteries, capacitors, and hybrid storage. Before exploring the applications, it is crucial to provide an overview of the synthesis and structure of MXene and its different phases.

10.2   Synthesis and structure

MXene is synthesized from its parent compound MAX by removing the A layer through etching. First, MXene is prepared by a wet etching method using a solution of MAX powder and 50wt% hydrofluoric acid (HF) at room temperature for 2 h,[22] resulting in the exfoliation of Ti_3AlC_2 to $Ti_3C_2T_x$, where T represents functional groups, i.e., $-F$, $-OH$, and $-O$, that become attached on the MXene surface after removing the A layer. A MAX powder and HF solution (10−50 wt%) is prepared to synthesize MXene. The solution is then stirred continuously at a fixed temperature for a time; both parameters depend on the MAX phase and wt% of HF. Afterward, the solution is washed until the pH value is maintained and dried to obtain MXene in powder form. MXene obtained from etching has functional groups at the surface, and it is nearly impossible to attain bare MXene (without functional groups). The sonication of certain MXene solvents like dimethyl sulfoxide is carried out and results in delaminated MXene (few-layered MXene). Delamination is necessary to explore the 2-D material properties of MXene, as the increase in the c-lattice parameter (c-LP) results in increased possibilities for intercalation of several elements and molecules in MXene. Fig. 10.2 shows a schematic for synthesizing etched and delaminated MXene.

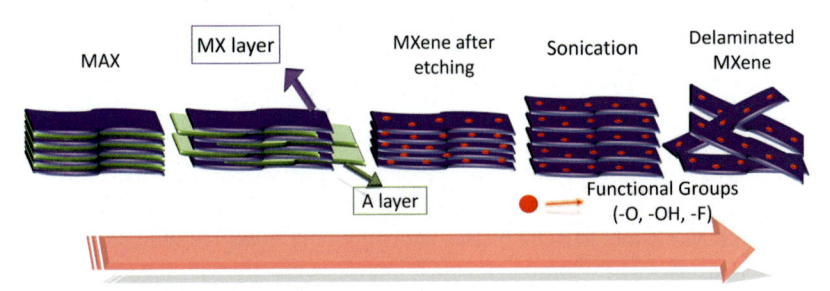

Figure 10.2 Schematic representation for the synthesis of MXene from MAX, showing etched and delaminated MXene.

Many studies have shown the synthesis of MXene through wet chemical etching, mainly using HF. V_2C has been etched from 49 wt% HF and then delaminated using deionized water and tetramethylammonium hydroxide.[23] Shan et al. synthesized V_2C under varying temperatures and times and concluded that 90°C at 5 h for HF reaction are optimum conditions.[24] To avoid health hazards, it is imperative to follow techniques of HF-free MXene synthesis using other etchants like ammonium fluorite salts, lithium fluoride, and HCl.[25,26] Wu et al. reported V_2C MXene synthesis from V_2AlC MAX using different etchants among which the solution of NaF + HCl gave highly pure etched V_2C MXene.[27] Two-dimensional molybdenum carbide (Mo_2C) MXene was prepared from MAX, i.e., Mo_2GaC using LiF and HCl solutions, and then delaminated with tetrabutylammonium hydroxide.[28] For enhanced divacancy properties, molybdenum carbide $Mo_{1.33}C$ was prepared from $(Mo_{2/3}Sc_{1/3})_2AlC$ by etching out Sc and Al using the HF etching technique.[29] Hence, 2-D MXene can be synthesized through various techniques to enhance already present properties. V_4C_3, Nb_4C_3, and Ta_4C_3 were prepared from V_4AlC_3, Nb_4AlC_3, and Ta_4AlC_3 using the HF synthesis technique. While the etchant remains the same, the temperature and time of reaction change with the MAX phase depending on transition metal and bond characteristics.[30–32] Similarly, double-transition-element-based MXenes can also be prepared such that there are two different M elements while the basic structure is maintained.[33] Different etchants have been introduced for synthesizing MXenes, like HF, NaF + HCl, LiF + HCl, and NH_4F + HCl, each having different temperatures and etching times—corresponding to MXenes as well—that result in varying physical properties that can be tuned accordingly.

Another technique for synthesizing of MXene is the bottom-up strategy. Halim et al. reported the formation of MXene thin films from MAX Ti_3AlC_2 deposited on sapphire substrate with varying thickness using Ti Al and C targets. MAX thin film was then etched using HF and NH_4HF separately. The films obtained through HF etching were MXene, while NH_4HF etched MXene was intercalated with NH_4^+ ions.[34] The resistivity and transmittance of MXene film slightly varied with changing etching time, deposition time, and temperature. Furthermore, proceeding with the previously mentioned research, several more studies have focused on the formation of MXene through bottom-up synthesis. Molybdenum carbide has been synthesized using chemical vapor deposition (CVD) while using copper as an etchant.[35,36] The process has been optimized by changing the temperature and time for deposition, resulting in thin films of diverse thickness. In addition, CVD can be used to prepare thin films of MAX phases that can be converted to MXene by selective etching of "A."[37,38] MXene inks can also be prepared using various additives that can be then stamped or printed onto the substrates.[39,40]

The synthesis process is crucial for MXene, as it has a determining role in the 2-D structure of MXene that is of utmost importance—more specifically, the c-LP and the distance between two consecutive layers—in the number of surface terminations. Although both MAX and MXene phases are commonly observed with a crystal structure in space group 194 P63/mmc, the c-LP increases greatly with the exfoliation and delamination process. Additionally, doping and intercalation of other atoms, molecules, and compounds affect the c-LP. Due to the large c-LP, there is an increase in the storage capabilities of other ions and an increase in defect sites. These defect sites can be chemically engineered to enhance the properties like conduction. Defects in 2-D MXene can appear as vacancy sites, additional adsorption sites due to large c-LP and surface terminations, i.e., $-O$, $-H$, $-F$. The number of surface terminations, structure and particle size of MXene, crystallinity, and intercalating material between the MXene layers are the factors affected by the synthesis route taken.[41,42] The change in etchants, dopants, temperature, and time can result in MXene with different storage capacities. For V_2CT_x, the synthesis through $NH_4HF + HCl$ results in higher capacitance with higher interlayer distance and increased number of sites for intercalation. Thus, the synthesis procedure and impurities have an impact on material capacities, indicating that fewer impurities result in higher and more stable capacitance. The advantages and disadvantages of certain etchants or dopants used in the synthesis depend on the composition of MAX and MXene, the nature of the bonding between M and X as well as M and A, and the intercalating agent. In addition, synergistic effects in heterostructures of MXenes with other 2-D materials are another advantageous feature. Overall, the synthesis and structure of MXene can be used to tune its properties by careful planning and execution of experiments.

10.3 Energy storage in MXene

Energy storage in MXene has been vastly studied, both through experimentation and DFT analysis. A review is given of storage in MXene for metal-ion batteries (MIBs), supercapacitors, hybrid/asymmetric devices, and other energy-storage devices.

10.3.1 Metal-ion batteries

Due to environmental changes, the world is tending toward rechargeable batteries, as they can be reused many times, whereas a nonrechargeable battery becomes useless after its first use, emitting harmful gases or materials after disposal. Hence, rechargeable MIBs like lithium-ion batteries (LIBs), sodium-ion batteries (SIBs), and potassium-ion batteries (PIBs) are being studied and are evolving as alternative devices. Their qualities like high energy storage, scalability, cost-effectiveness, and rechargeability have motivated the development of batteries like LIBs, SIBs, and others. Typically, the LIB cathode is made of lithium phosphates or cobaltates, and the anode is usually a 2-D material like graphene or MXene.[43–45] Hence, numerous studies are being conducted on MXene for applications in rechargeable batteries by analyzing, tuning, and modifying their structures through DFT and experiments. The performance of a material is measured in terms of capacitance (mAh/g) and current density (mA/g). DFT analyses are essential to understanding the inner mechanisms in MXene. Several structural compositions of 2-D tantalum carbide MXene studied through DFT have shown that TaC has the higher storage capacity of 556 mAh/g, whereas TaC_2 and Ta_2C have storage capacities of 523 mAh/g and 264 mAh/g, respectively.[46] From the analysis, the TaC structure was metastable, while

TaC_2 and Ta_2C were stable and more likely to be synthesized because of their lower formation energies. As a result, TaC_2 has excellent lithium storage capacity and optimum behavior for synthesis and application as an anode for LIBs.

Hafnium-based MXene was studied with and without the functional groups to analyze the storage capabilities and stability for anode applications in LIBs and SIBs.[47] It was observed that Hf_3C_2 has a higher storage capacity for Li$^+$ ions that decreases when functional groups are attached due to the change in diffusion pathway and charge on the surface. Additionally, Yu et al. showed that several structures of 2-D molybdenum carbide monolayers have a prominent storage capacity of Li and Na ions in the absence of functional groups.[48] Similarly, many theoretical studies have reported promising characteristics of MXene as an anode in LIBs and SIBs.[49−53]

Sun et al. analyzed the lithiation capacities and structural stabilities in oxygen-functionalized 2-D MXenes V_2CO_2, Cr_2CO_2, Sc_2CO_2, Nb_2CO_2, Hf_2CO_2, and Ta_2CO_2, among which V_2CO_2 shows the highest lithiation capacity of 735 mAh/g.[54] Then, the exploration of V_2C—etched from MAX using NaF + HCl—as an anode for LIBs showed a capacity of 250 mAh/g at 50 mA/g current density.[55] Furthermore, Wang et al. prepared V_2CT_x through a detailed optimized synthesis process and analyzed the storage capacity in the sample prepared through four different etchants, i.e., LiF + HCl, NaF + HCl, KF + HCl, and NH_4HF + HCl.[24] The storage capacity was 943 mAh/g at a current density of 50 mA/g for the MXene etched via NH_4HF + HCl.

Beyond the good performance of MXenes as potential electrodes for MIBs, doped structures and heterostructures of MXenes have been presented as enhanced electrodes with greater capacitance, stability, and retention characteristics. A DFT study showed that S-functionalized Ti_3C_2 has better characteristics for Na storage than bare MXene.[50] In addition, the DFT study predicted that S and O-functionalized Ti_3C_2 shows promising behavior for LIBs.[50,56] Kajiyama et al. observed the sodiation and desodiation process in $Ti_3C_2T_x$, which revealed a sodiation capacity of 150 mAh/g at 20 mA/g.[57] Moreover, a detailed study of three samples, i.e., $Ti_3C_2T_x$, S-decorated $Ti_3C_2T_x$, and sodium-intercalated $Ti_3C_2T_x$, showed that S-decorated $Ti_3C_2T_x$ has better performance in SIBs among the three samples, as predicted by theoretical studies.[58]

Nb-based MXenes have also shown decent results for LIBs and SIBs. Rui et al. calculated the capacitance of bare Nb_2CT_x and N-doped Nb_2CT_x for performance as an anode in LIBs. The performance was observed to have increased by 90% through doping N in MXene.[59] Du et al. studied half-cell and full-cell performance of Nb_2C anodes for SIBs.[60] Nb_4C_3 etched from Nb_4AlC_3 in HF for 10 days was investigated by Ping et al. along with Nb_4C_3, $Nb_{3.5}Ta_{0.5}C_3$, and $Nb_{3.9}Ta_{0.1}C_3$ synthesized in HF for 5 days.[33] The sample prepared through a 10-day synthesis process demonstrated higher capacities for LIBs of 140 mAh/g at 100 mA/g. Ti_3CNT_x etched from Ti_3AlCN using HF showed notable performance as an SIB anode, while a DFT study showed its potential in LIBs as well.[61,62] Overall, different approaches to synthesizing the various MXenes produce properties tuned according to the optimized etching process.

MXene modified into three-dimensional structures by doping or treatment with certain solutions showed an increment in capacities. In three-dimensional (3-D) structures, MXene porous foam prepared from treatment with sulfur thiosulfate demonstrated excellent behavior as an anode in LIBs.[63] Another example of a 3-D MXene structure is vanadium dioxide-doped Ti_3C_2, which has a flowerlike 3-D arrangement while MXene sheets are not destroyed.[64] The 3-D flowerlike arrangement of MXene and VO_2 heterostructures revealed a storage capacity of 323 mAh/g at 50 mA/g as an SIB anode. A $Ti_3C_2T_x$–SnS_2/Sn_3S_4 hybrid was prepared through the solvothermal calcination method. A schematic of the layered structure is given in Fig. 10.3A. Cyclic voltammetry (CV) analysis showed 707 mAh/g capacitance at 5000 mA/g for first-charging capacity, which is three times the capacitance observed for the pristine sample of MXene, i.e., $Ti_3C_2T_x$ MXene of 264 mAh/g.[65] At 100 mA/g, 462 mAh/g capacitance was observed after the first 100 cycles, while MXene showed 114 mAh/g. The cyclic stability of the hybrid structure is shown in Fig. 10.3B. The combination of two highly capacitive materials results in the added and improved properties in the Ti_3C_2–SnS_2/Sn_3S_4 hybrid as well as stability for application as an LIB anode. Hence, combining another material with 2-D MXene can increase its potential and stability as an anode.[66]

Another type of heterostructure is the hybrid structure of two different 2-D materials, which is a 2-D material intercalating between the MXene sheets. Chao et al. demonstrated that TMDs and MXene heterostructures are good storage materials for SIB anodes in a theoretical study.[67] Experimentally, meanwhile, $MoS_2/Ti_3C_2T_x$ and $MoSe_2/Ti_3C_2T_x$ have been reported to have higher and more stable capacities than pristine samples, i.e., MoS_2, MoS_2, and $Ti_3C_2T_x$, as anodes in SIBs and LIBs.[68,69] $MoSe_2/Ti_3C_2T_x$, and $MoSe_2/Ti_3C_2T_x$@C hybrid structures were synthesized, and observations showed their possible application in PIBs.[70] $MoSe_2/Ti_3C_2T_x$ and $MoSe_2/Ti_3C_2T_x$@C displayed a storage capacity of ~340 mAh/g and 347 mAh/g, which becomes 102 mAh/g and 355 mAh/g after 100 cycles at 200 mA/g, respectively. The addition of C in $MoSe_2/Ti_3C_2T_x$ improved the cyclic stability of the samples. Black phosphorous and $Ti_3C_2T_x$ composites have shown a capacity of 559 mAh/g at 500 mA/g in SIBs; however, the discharge capacity is reduced very rapidly from 1064 mAh/g to 80 mAh/g after 100 cycles.[71] Study conducted by Zhao et al. shows that addition of polymer poly diallyldimethylammonium chloride (PDDA) to the black phosphorous nanoparticles before synthesizing the heterostructures improves the cyclic stability as well as the storage capacity.[72] Fig. 10.4A and B shows, respectively,

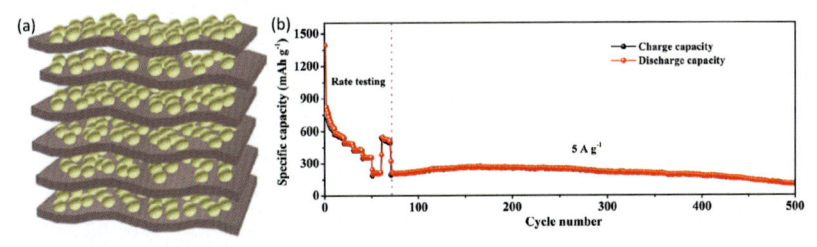

Figure 10.3 (A) Schematic for $Ti_3C_2T_x$-SnS_2/Sn_3S_4 hybrid (B) Charge–discharge capacity of $Ti_3C_2T_x$-SnS_2/Sn_3S_4 hybrid at 5000 mA/g.
Copyrights.[65]

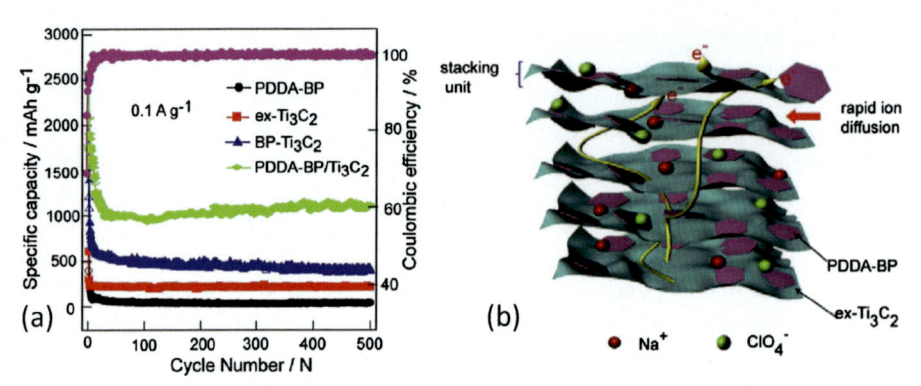

Figure 10.4 (A) Cyclic stability of poly diallyldimethylammonium chloride (PDDA)-black phosphorene-MXene hybrid structure and (B) schematic showing the stacking arrangement of MXene and PDDA-black phosphorene.
Copyrights.[72]

the cyclic performance of PPDA-black phosphorene-MXene heterostructure and a schematic showing the presence of black phosphorene on the surface of MXene. Table 10.1 enlists various MXenes, derivatives of several MXenes, and their applications, etchants and dopants, capacitance, and current density. From these reports, it is clear that the different phases of MXenes for possible surface tuning and available adsorption sites make it an interesting candidate for exploration and application in MIBs.

10.3.2 Supercapacitors

Supercapacitors are rechargeable energy-storage devices that have attracted quite a bit of attention for their high power density, high volumetric capacitance, and long lifetimes.[74] Supercapacitors are electrochemical capacitors that possess a larger storage capacity than a usual capacitor. There are mainly two types, i.e., electric double-layer capacitors (EDLCs) and pseudocapacitors.[21] In EDLCs, the charge is stored in the electrolyte−electrode interface, forming double layers, i.e., the charge of the electrode and opposite ions from the electrolyte forms two layers at the interface. The charge storage in EDLC do not involve faradaic reactions. Whereas for pseudocapacitor, the charge storage occurs due to faradaic redox reactions at the electrode.[75] CV measurements are carried out to assess the capacitance of a material involving the parameters like voltage window and scan rate. The capacitance is observed to vary with scan rate for different materials calculated in terms of F/g, F/cm^2, and F/cm^3 for gravimetric, areal, and volumetric capacitance, respectively.[21] Recently, 2-D materials have shown great potential for supercapacitance applications.[75,76] More specifically, numerous reports have shown different phases of MXene with varying structures and morphology as EDLC and pseudocapacitor.[76,77]

Insights into the mechanism of MXene as a supercapacitor electrode can be gained through DFT studies.[78−80] Ando et al. gave a detailed description of capacitive and

Table 10.1 Summary of applications, etchants and dopants (utilized during synthesis of respective MXenes), current density, and storage capacity of MXene and MXene-derived electrodes.

MXene electrode	Application in MIBs	Etchant and dopants	Current density	Storage capacity	Ref
V_2CT_x	LIBs	NaF + HCl	50 mA/g	250 mAh/g	55
V_2CT_x	LIBs	NH_4F + HCl	50 mA/g	943 mAh/g	24
$Ti_3C_2T_x$	SIBs	HF	20 mA/g	150 mAh/g	57
$S-Ti_3C_2T_x$	SIBs	HF and Na	2 A/g	200 mAh/g	58
Nb_2CT_x	LIBs	HF	120 mA/g	190 mAh/g	59
$N-Nb_2CT_x$	LIBs	HF and urea	120 mA/g	360 mAh/g	59
Few-layered Nb_2CT_x	LIBs	HF	200 mA/g	524 mAh/g	73
Nb_2CT_x	Half-cell SIBs		1 A/g	102 mAh/g after 500 cycles	60
Nb_2CT_x	Full-cell SIBs		100 mA/g	181 mAh/g after 50 cycles	60
$Nb_4C_3T_x$	LIBs	HF	100 mA/g	140 mAh/g	33
Ti_3CNT_x	SIBs	HF	20 mA/g	507 mAh/g	61
Porous $Ti_3C_2T_x$-foam	LIBs	LiF + HCL and sulfur thiosulfate	50 mA/g	455.5 mAh/g	63
$VO_2- Ti_3C_2T_x$	SIBs	HF and vanadyl acetylacetone	50 mA/g	323 mAh/g	64
$Ti_3C_2T_x$-SnS_2/Sn_3S_4 hybrid	LIBs	HF and thioacetamide + K_2SnO_3	5 A/g	707 mAh/g	65
$NH_2-Si/Ti_3C_2T_x$	LIBs	LiF + HCl and NH_2-Si nanoparticle	60 mA/g	1020 mAh/g	66
$MoS_2/Ti_3C_2T_x$	LIBs		1 A/g	451.3 mAh/g	68
$MoSe_2/Ti_3C_2T_x$@C	PIBs	LiF + HCl and $MoSe_2$ + dopamine hydrochloride	200 mA/g	347 mAh/g	70
$MoSe_2/Ti_3C_2T_x$	SIBs	HF	1 A/g	578 mAh/g	69
Black phosphorous/$Ti_3C_2T_x$	SIBs	HF and black phosphorene + N-dimethyl formamide	500 mA/g	559 mAh/g	71
PDDA-black phosphorene/ $Ti_3C_2T_x$	SIBs	LiF + HCl and PDDA + black phosphorene	0.1 A/g	1773 mAh/g	72

pseudocapacitive behavior in MXene. Charge transfer induced by the redistribution of charge—because of orbital couplings between the ions from electrolyte and MXene or surface terminations of MXene—results in the decreases of potential at the electrode—electrolyte interface, which in turn gives rise to the pseudocapacitive behavior.[74] Cheng et al. performed a thorough calculation of M_2X, M_3X_2, and M_4X_3 structures, where M is a transition metal and X is either C or N. They concluded that for a promising MXene pseudocapacitor, the compound should have a large hydrogen adsorption energy and low potential at the point of zero charge.[79] Additionally, zirconium carbide showed the most prominent pseudocapacitive behavior. Intercalation of H^+ ions in $Ti_3C_2T_x$ is considered an important aspect to understand and illustrate the effect of H_2SO_4 as an electrolyte for pseudocapacitive and EDLC storage in MXene.[80,81]

$V_4C_3T_x$ MXene has shown a noticeable behavior as a supercapacitor for EDLC as well as pseudocapacitance. The total specific capacitance, pseudocapacitance, and double-layer capacitance is calculated to be 268.5 F/g, 100.1 F/g, and 168.4 F/g, respectively.[30] Fig. 10.5A—C shows the transmission electron microscopy image of the 2-D structure of $V_4C_3T_x$, CV curves for a 2 mV/s to 100 mV/s scan rate and cyclic capability at a 10 A/g current density showing 97.23% retention of capacitance for 10,000 cycles. In addition, another vanadium-based MXene, V_2CT_x, has shown

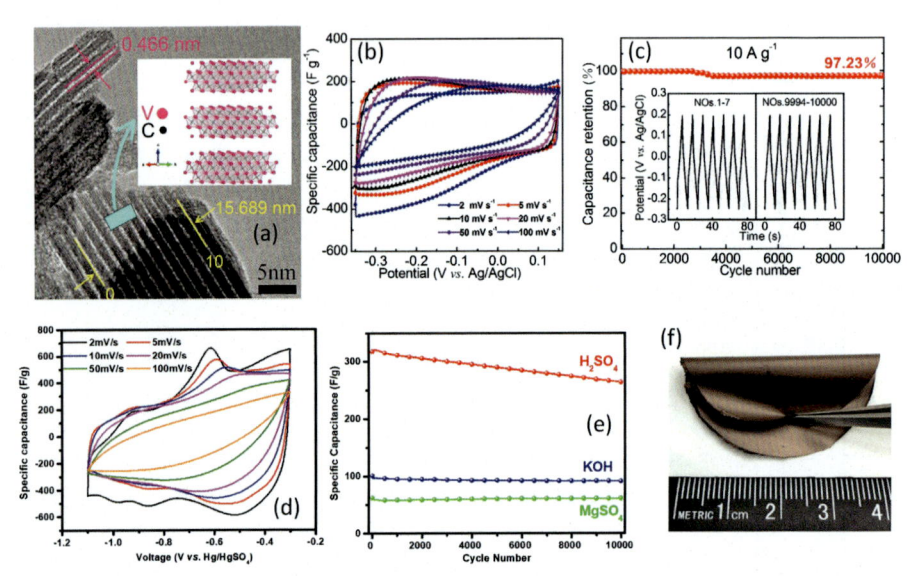

Figure 10.5 (A) Transmission electron microscopy (TEM) image of $V_4C_3T_x$ MXene (B) Specific capacitance of $V_4C_3T_x$ demonstrated through CV curves at different scan rates, i.e., 2 mV/s, 5 mV/s, 10 mV/s, 20 mV/s, 50 mV/s and 100 mV/s (C) 97.23% retention of specific capacitance in $V_4C_3T_x$ electrode for 10,000 cycles at 10 A/g. (D) CV curves for V_2CT_x in 1M H_2SO_4 electrolyte at scan rates 2 mV/s, 5 mV/s, 10 mV/s, 20 mV/s, 50 mV/s and 100 mV/s (E) Cyclic performance of V_2CT_x in 1M H_2SO_4, 1M KOH and 1M $MgSO_4$ electrolyte (F) flexible V_2CT_x film.
(A—C) Copyrights.[30] (D—F) Copyrights.[23]

favorable behavior as a supercapacitor electrode. Shan et al. reported 487 F/g specific capacitance at a scan rate of 2 mV/s in V_2CT_x in 1M H_2SO_4 electrolyte. Pseudocapacitive behavior was observed in three electrolytes, i.e., H_2SO_4, KOH, and $MgSO_4$. For H_2SO_4 as an electrolyte, vanadium carbide showed a higher capacitance; however, $MgSO_4$ showed greater stability by comparison. Fig. 10.5D—F shows the CV curve, cyclic performance, and flexible electrode of vanadium carbide, respectively. The higher capacitance loss in H_2SO_4 and KOH compared with $MgSO_4$ might be because of the irreversible redox reaction, which can lead to degradation of the vanadium carbide electrode.[23] Armin et al. demonstrated that V_2CT_x electrodes lose stability and shape after a month in an open environment. However, when Li-ions are intercalated into MXene, the structure and shape of the film remain the same.[82] The study reports the supercapacitance of Li- and Na-intercalated MXene (Li— V_2CT_x and Na—V_2CT_x) electrodes in different electrolytes. Na— V_2CT_x showed the highest capacitance of 420 F/g at a 5 mV/s scan rate in 3M H_2SO_4. Furthermore, enhanced capacitance was observed in Mo_2CT_x, $Mo_{1.33}CT_x$, and $Ti_3C_2T_x$ for thicknesses of 2, 5, and 3 μm, respectively.[28,29,83] In addition, Ta_4C_3, Ti_2C, Ti_2N, and i_3C_2 have been reported for supercapacitors with different capacitances for various scan rates and electrolytes.[31,84—86] Table 10.2 summarizes MXenes and their derivatives by capacitance, electrolyte, and scan rate.

Numerous materials such as nitrogen, niobium, ammonium citrate, bismuth oxychloride, KOH, and polypyrrole have been hybridized in MXenes, specifically $Ti_3C_2T_x$, to enhance the performance and stability of MXene electrodes[87—92] Additionally, a few studies have performed electrochemical characterization of symmetric supercapacitor (SSC) devices such that both electrodes are the same composite. Xia et al. doped bismuth oxychloride into $Ti_3C_2T_x$, then analyzed the SSC device of BiO-CL-$Ti_3C_2T_x$.[91] A similar approach has been observed in numerous reports.[87,90,92,93] $Ti_3C_2T_x$ MXene modified by $Fe(OH)_3$ exhibits a nanoporous characteristic that enhances flexibility, capacitance, and ion transport in the electrode.[94] Graphene and MXene composites have also been studied in many different arrangements to increase the capacitance, stability, and durability of MXene-based electrodes for supercapacitors and SSC devices.[95—98]

Flexible MXene fibers fabricated by loading $Ti_3C_2T_x$ suspension on 5 cm silver-plated nylon fibers were demonstrated as an all-solid-state supercapacitor. The SSC device retained its capacitance with bending at 30, 45, and 100 degrees as well as twisting and knotting of the fibers.[99] Flexible and stretchable devices are being synthesized and fabricated for potential applications in wearable electronics. In another study, MXene/reduced graphene oxide (rGO) hybrid fibers were observed with bending of 0, 45, 90, and 180 degrees.[100] Recently, MXene/rGO composites have shown excellent capacitance retention of under 300% uniaxial strain and 200% biaxial strain in composite electrodes and SSC devices.[101]

Another important aspect is MXene inks that can be used to fabricate microsupercapacitor (MSC) devices by stamping or printing.[40,102,103] Nicolosi et al. stamped ying-yang (Y— $Ti_3C_2T_x$), spiral, and interdigital (I—$Ti_3C_2T_x$) designs of $Ti_3C_2T_x$ inks on PET, glass, and paper. The interdigital design showed a higher capacitance of 57 mF/cm^2 at a 5 mV/s scan rate.[103] Tian et al. demonstrated a $Ti_3C_2T_x$ MXene

Table 10.2 Summary of performance of electrodes as supercapacitors and symmetric supercapacitor (SSC) devices of MXene and its derivatives.

Electrode/SSC device of MXene or MXene composite	Electrolyte	Scan rate	Capacitance	Ref
$V_4C_3T_x$	1M H_2SO_4	2 mV/s	209 F/g	30
V_2CT_x	1M H_2SO_4	2 mV/s	487 F/g	23
Na$-V_2CT_x$	3M H_2SO_4	5 mV/s	420 F/g	82
5 μm thick $Ti_3C_2T_x$	1M H_2SO_4	2 mV/s	245 F/g	83
2 μm thick Mo_2CT_x	1M H_2SO_4	2 mV/s	196 F/g	28
3 μm thick $Mo_{1.33}CT_x$	1M H_2SO_4	2 mV/s	339 F/g	29
$Ti_3C_2T_x$	6M KOH	1 mV/s	450 F/g	84
Ti_2CT_x	6M KOH	1 mV/s	250 F/g	84
Ti_2NT_x	1M $MgSO_4$	2 mV/s	201 F/g	85
N_2/H_2 annealed Ti_2CT_x	-	5 mV/s	51 F/g	86
$Ta_4C_3T_x$	0.1M H_2SO_4	5 mV/s	481 F/g	31
PPY/$Ti_3C_2T_x$	0.5M H_2SO_4	30 mV/s	406 F/cm^3	87
SSC device		0.3 mA/cm^2	35.6 F/cm^3	
N$-Ti_3C_2T_x$	1M H_2SO_4	1 mV/s	192 F/g	88
Nb$-Ti_3C_2T_x$	6M KOH		442.7 F/g	89
N$-$O$-$ C@$Ti_3C_2T_x$	6M KOH	1 A/g	250.6 F/g	90
SSC-device		1 A/g	53.8 F/g	
BiOCl$- Ti_3C_2T_x$	1M KOH	1 A/g	247 F/g	91
SSC-device		0.5 A/g	64.3 F/g	
Alkalinized $Ti_3C_2T_x$	1M H_2SO_4	2 mV/s	496 F/g	92
SSC-device		1 A/g	294 F/g	
Ti_2CT_x SSC-device	1M KOH	2 mV/s	517 F/cm^3	93
Nanoporous $Ti_3C_2T_x$	3M H_2SO_4	0.5 A/g	351 F/g	94
SCNT/$Ti_3C_2T_x$	1M KOH	2 mV/s	314 F/cm^3	95
$Ti_3C_2T_x$/PAN carbonized	1M H_2SO_4	50 mV/s	205 mF/cm^2	96
PDDA-rGO/$Ti_3C_2T_x$	3M H_2SO_4	2 mV/s	335.4 F/g	97
SSC device		2 V/s	24 F/g	
$Ti_3C_2T_x$/AC$-$SSC device		4000 mV/s	73 F/g	98
$Ti_3C_2T_x$/silver plated nylon fiber SSC devices	PVA H_2SO_4 gel	2 mV/s	328 mF/cm^2	99
SSC device $Ti_3C_2T_x$/rGO fibers	PVA H_2SO_4 gel	10 mV/s	565.4 mF/cm^2	100
	PVA H_3PO_4 gel	10 mV/s	372.2 mF/cm^2	
$Ti_3C_2T_x$/rGO 300% strain	1M H_2SO_4	0.5 A/g	40 mF/cm^2	101
SSC device 300% strain	PVA H_2SO_4 gel	0.1 A/g	18.6 mF/cm^2	
3-D printed $Ti_3C_2T_x$ (MSC)	PVA H_2SO_4 gel	2 mV/s	1035 mF/cm^2	102
Y$- Ti_3C_2T_x$ (MSC)	PVA H_2SO_4 gel	5 mV/s	28.3 mF/cm^2	103
Y$- Ti_3CNT_x$ (MSC)		5 mV/s	33.8 mF/cm^2	
I$- Ti_3C_2T_x$ (MSC)		5 mV/s	57 mF/cm^2	
$Ti_3C_2T_x$/CNF electrode	3M H_2SO_4	5 mV/s	396 F/g	104
$Ti_3C_2T_x$ SSC device		2 mV/s	28.6 mF/cm^2	
$Ti_3C_2T_x$/CNF SSC device		2 mV/s	25.3 mF/cm^2	
$Ti_3C_2T_x$ (MSC) on Si	PVA H_2SO_4 gel	1 μA/cm^2	470 μF/cm^2	105
p- $Ti_3C_2T_x$	PVA H_2SO_4 gel	1 A/g	48.4 mF/cm^2	39
SA- $Ti_3C_2T_x$	PVA H_2SO_4 gel	1 A/g	108.1 mF/cm^2	39

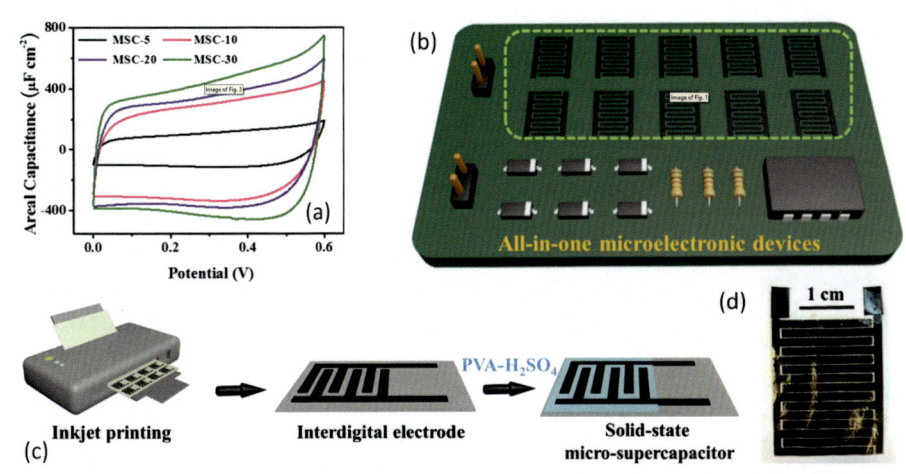

Figure 10.6 (A) Cyclic voltammetry curved for Ti_3C_2 microsupercapacitor (MSC) device on a silicon substrate (B) Schematic for MSC fabricated on-chip for electronic devices (C) Schematic for inkjet printing of MXene Ti_3C_2 MSC on paper and (D) Photograph of MXene MSC printed on paper.
(A and B) Copyright.[105] (C and D) Copyrights.[39]

and carbon nanofibril (CNF) composite as an electrode with 396 F/g capacitance at a 5 mV/s scan rate and then proceeded to fabricate an SSC device of pristine MXene and MXene/CNF composite with a 900 μm width, 1 cm length, and 130 μm gap between electrodes with 1.7 μm thickness. At a scan rate of 2 mV/s, the pristine MXene capacitor and MXene/CNF capacitor showed 28.6 mF/cm^3 and 25.3 mF/cm^3 capacitance, respectively, for application in flexible electronics.[104] With the advancement of technology and portable energy, it is necessary that energy storage of on-chip devices is also increased. Huang et al. studied the CV curves of interdigital MXene inks based on the chip and presented a schematic for an on-chip symmetric MSC of MXene (Fig. 10.6A and B).[105] Moreover, Wu et al. studied the consistency of MXene inks and observed that after 80 days, pristine MXene becomes oxidized and converts to titanium dioxide. Hence, they used several ligands in the suspension and observed that sodium ascorbate (SA)-based MXene ink was oxidation resistive.[39] SA-MXene ink was then used for the inkjet printer system prepared by adding propylene glycol. The interdigital design was printed on paper, then PVA-H$_2$SO$_4$ was pasted as shown in Fig. 10.6C. Electrochemical performance of MSC given in Fig. 10.6D was observed. The results showed 108.1 mF/cm^2 capacitance for SA-MXene ink at 1 A/g. Irrefutably, MXene shows great potential for use in SSC and MSC devices for use as an electrode deposited through spin-coating and ink-printing.

10.3.3 Hybrid and asymmetric energy storage

Symmetric supercapacitors are high-power, low-energy devices, whereas batteries are high-energy, low-power devices. The aim of hybrid energy storage is to combine these two characteristics for a high-power, high-energy device. The mechanism, simply states, is the combination of a capacitor electrode and a battery electrode. The capacitor electrode is considered to have EDLC properties, while the battery electrode is considered to work on the principle of faradaic redox reactions as a pseudocapacitive storage material.[106] Another option is to fabricate an electrode from a composite material containing elements from battery and capacitor electrodes.[107] Such a device for which both electrodes are of different materials with different characteristics is known as an asymmetric supercapacitor (ASC) or hybrid supercapacitor. The initial reports were based on activated carbon as a positive electrode and lithium titanate, manganese oxide, nickel oxide, etc. as a negative electrode.[106,108] Recently, graphene and other 2-D materials also showed potential for hybrid energy storage.[109–112] In an approach toward enhancing the storage capacities, MXenes have also been used to fabricate ASCs in many different assortments. The electrodes range from pristine MXenes to composites and heterostructures used as both positive and negative electrodes in a distinct manner. A brief review is presented in Table 10.3.

Fig. 10.7A provides a schematic illustration of an ASC device with AC and MXene as electrodes. Xie et al. studied an ASC of AC//$Ti_3C_2T_x$ deposited on Si substrate initially and suggested its application in microelectronic chips.[113] Fig. 10.7B shows the CV curves at different scan rates, and Fig. 10.7C shows the detailed procedure of fabricating said ASC and its packaging. Chang et al. fabricated a similar ASC device and showed that it retains its capacitance when stretched by 50% and 100%.[114] Several heterostructures of MXene with other compounds have also been studied for ASC like Nb_2CT_x/CNT, MnO_2/$Ti_3C_2T_x$/Carbon Cloth, NiCo-metal−organic framework-$Ti_3C_2T_x$.[73,115−117] The aim remains to enhance retention characteristics by combining materials that have shown good capacitive capabilities but low structural stability and retention. NiCo-layered double hydroxide (LDH) electrodes show small retention of 17%, as the structure shatters during the charge−discharge process; however, a heterostructure of MXene and Ni_2Co-LDH results in increased capacitance as well as increased retention capabilities.[118] Lu et al. then proceeded to assemble the ASC of graphene//Ni_2Co-LDH- $Ti_3C_2T_x$, which showed a high storage capacity of 140 F/g at 2 A/g with 90% retention.

Beyond AC, graphene is also considered a good electrode for ASC. rGO-coated conducting polymers were used as an anode where the conducting polymers were polyaniline (PANI), poly 3-4 ethylenedioxythiophene, and polypyrrole.[119] The negative electrode was pristine $Ti_3C_2T_x$, and 3M sulfuric acid was used as an electrolyte to assemble ASC. Among the three devices, $Ti_3C_2T_x$//rGO@PANI showed the highest retention and capacitance. Furthermore, PANI has been studied with MXene and graphene with increased capacitance. Fu et al. studied Ti_2CT_x @ polyaniline (PANI) @ graphene heterostructure combined with graphene electrode as an ASC.[120] Additionally, Li et al. synthesized 3-D heterostructures of $Ti_3C_2T_x$ MXene using PANI molecules, as they filled up the interlayer spaces in MXene and assembled ASC using MXene as an anode and an MXene@PANI 3-D structure as a cathode.[121] In another study, wavy $Ti_3C_2T_x$ MXene as a cathode was synthesized using a polystyrene suspension that was removed by annealing at 450°C, and the sheets were then compressed mechanically. For anodes, rGO was modified in PANI suspension, and then CNT was added to form a heterostructure of rGO@CNT@PANI. The assembled all-solid-state ASC and the CV curves of the separate electrode and the ASC are shown

Table 10.3 Brief list of asymmetric supercapacitors (ASCs) based on MXene and MXene derivatives, with capacitance (at scan rate) and energy density (at power density).

Asymmetric capacitor	Electrolyte	Capacitance at scan rate / Energy density at power density	Ref
AC//$Ti_3C_2T_x$	PVA-Na_2SO_4	7.8 mF/cm^2 at 0.2 mA/cm^2 / 3.5 mWh/cm^3 at 100 mW/cm^3	113
AC//$Ti_3C_2T_x$	PVA-Li_2SO_4	46 F/g at 25 mV/s / 5.5 Wh/kg at 0.5 kW/kg	114
AC//NiCoAl-LDH- $Ti_3C_2T_x$	PVA-KOH	128.9 F/g at 0.5 A/g / 45.8 Wh/kg at 346 W/kg	115
AC//NiCo-MOF- $Ti_3C_2T_x$		126.3 F/g at 0.5 A/g / 39.5 Wh/kg at 562.5 W/kg	116
AC//MnO_2−$Ti_3C_2T_x$- CC		94.7 F/g at 1 A/g / 29.8 Wh/kg at 749.2 W/kg	117
AC//Nb_2CT_x/CNT	1M H_2SO_4	462 mF/cm^2 at 2 mV/s / 154.1 µWh/cm^2 at 3750.4 µW/cm^2	73
Graphene//$NiCo_2$-LDH- $Ti_3C_2T_x$		140 F/g at 2 A/g / 68 Wh/kg at 388 W/kg	118
$Ti_3C_2T_x$//rGO@PANI	3M H_2SO_4	57 F/g at 5 mV/s / 17 Wh/kg	119
$Ti_3C_2T_x$//rGO@PPY	3M H_2SO_4	59 F/g at 5 mV/s / 16 Wh/kg	119
$Ti_3C_2T_x$//rGO@PEDOT	3M H_2SO_4	47 F/g at 5 mV/s / 13 Wh/kg	119
Graphene//Ti_2CT_x/PANI/ Graphene	1M H_2SO_4	94.5 F/g at 1 A/g / 42.3 Wh/kg at 950 W/kg	120
$Ti_3C_2T_x$//$Ti_3C_2T_x$ @PANI	3M H_2SO_4	- / 14.8 Wh/kg at 0.49 kW/kg	121
$Ti_3C_2T_x$//rGO@CNT@PANI	PVA H_2SO_4	116.9 F/g at 10 mV/s / 590 Wh/kg at 28.6 W/kg	122
$Ti_3C_2T_x$//RuO_2	PVA H_2SO_4	93 F/g at 50 mV/s / 29 Wh/kg at 3.8 kW/kg	123
$Ti_3C_2T_x$−FeOOH//MnO_2-CC	$LiSO_4$	115 mF/cm^2 at 2 mA/cm^2 / 40 µWh/cm^2 at 1.6 mW/cm^2	124
BMCY//BRCY	PVA H_2SO_4	123 F/g at 2 mA/cm^2 / 168 µWh/cm^2 at 975 µW/cm^2	125
$Ti_3C_2T_x$//VN-PC	6M KOH	105 F/g at 1 A/g / 12.8 Wh/kg at 985.8 kW/kg	126
$Ti_3C_2T_x$//$ZnCo_2O_4$	PVA-KOH	281.25 F/g at 0.5 A/g / 99.948 Wh/kg at 800 W/kg	127
$Ti_3C_2T_x$//Ni−S−$Ti_3C_2T_x$	6M KOH	69.4 C/g at 0.5 A/g / 20 Wh/kg at 0.5 kW/kg	128
$Ti_3C_2T_x$//CuS−$Ti_3C_2T_x$	1M KOH	49.3 F/g at 1 A/g / 15.4 Wh/kg at 750.2 W/kg	129
$Ti_3C_2T_x$//MoO_2−$Ti_3C_2T_x$	PVA-LiCl	19 mF/cm^2 at 2 mV/s / 9.7 mWh/cm^3 at 198 mW/cm^3	130

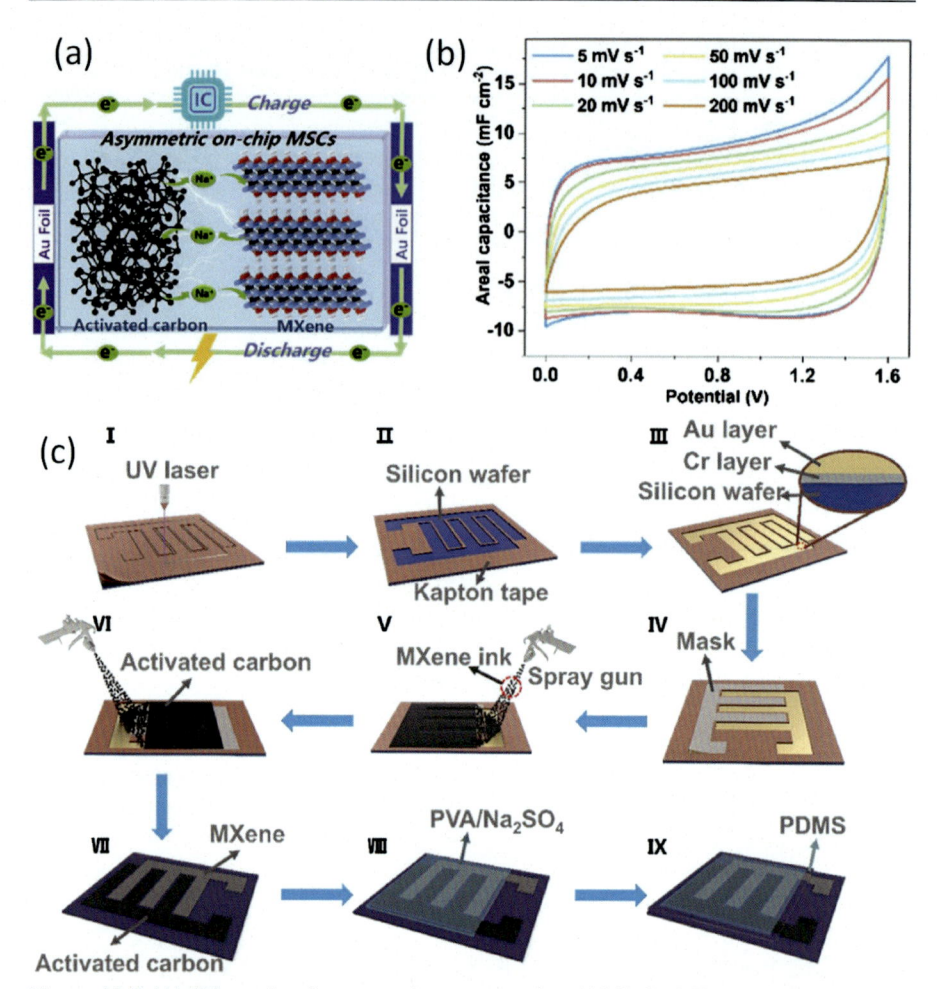

Figure 10.7 (A) Schematic of asymmetric supercapacitor (ASC) for MXene//AC (B) Cyclic voltammetry (CV) curves for ASC of AC//$Ti_3C_2T_x$ (C) fabrication process for ASC of AC// $Ti_3C_2T_x$ using MXene inks on a silicon chip. Copyrights.[113]

in Fig. 10.8.[122] The device showed high volumetric energy density and power density of 70 Wh/L and 111 kW/L with 82.2% retention after 10,000 cycles.

Beyond graphene, other oxide materials are used as counter electrodes, such as ruthenium oxide and manganese oxide. Jiang et al. studied an all-pseudocapacitive ASC with $Ti_3C_2T_x$ as the cathode and RuO_2 as the anode, where the electrodes were fabricated on carbon fabric and sealed in plastic with 1M sulfuric acid as the electrolyte.[123] The device showed retention of 86% for 20,000 cycles at 20 A/g. Lin et al. decorated $Ti_3C_2T_x$ MXene with 1-D quantum dots of FeO−OH, optimized the addition ratio of FeO−OH, and then proceeded to fabricate the ASC device.[124] $Ti_3C_2T_x$−FeO-OH//MnO_2-CC (MnO_2 electrodeposited on CC) maintained its

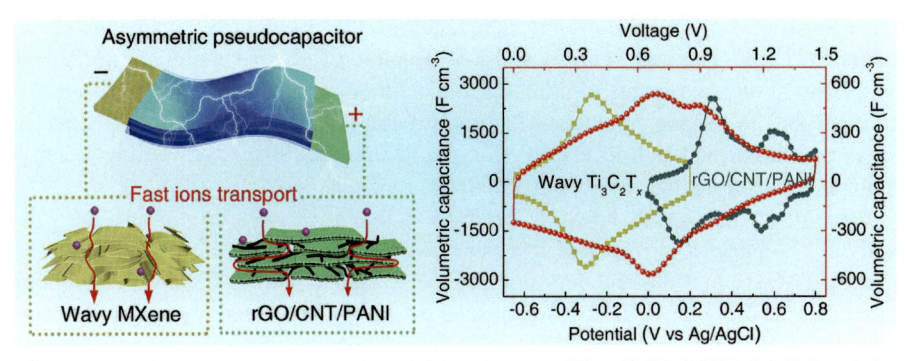

Figure 10.8 Asymmetric supercapacitor (ASC) of wavy MXene//rGO@CNT@PANI and the CV curves for wavy MXene electrode, rGO@CNT@PANI electrode, and ASC MXene// rGO@CNT@PANI.
Copyrights.[122]

capacitance at 60, 120, and 180 degrees bend with a larger voltage window than $Ti_3C_2T_x$//RuO_2. An important study used biscrolled $Ti_3C_2T_x$ MXene CNT on yarn (BMCY) as an electrode paired with a biscrolled RuO_2 CNT on yarn (BRCY) electrode for flexible, bendable, and wearable electronics.[125] Here, BMCYs are anodes, and BRCYs are cathodes. The devices showed 100% retention for ASC at 90 degrees bend for 1000 cycles and retained their capacitive characteristics for 45, 135, and 180 degrees bend. This is a considerable advancement toward stable wearable electronics.

Moreover, other 3-D compounds have been utilized in an ASC study with MXene as the negative electrode, effectively increasing the capacitance compared with the symmetric devices.[126,127] Javed et al. fabricated an ASC of $ZnCo_2O_4$//$Ti_3C_2T_x$ that displayed 281.25 F/g capacitance 0.5 A/g with retention of 87.44% for 500 cycles at 180 degrees bend,[127] as well as the ability to light up 14 LEDs when three ASCs were connected in series. In addition, several studies have demonstrated MXene// modified MXene ASCs as enhanced supercapacitor energy-storage devices.[128−130] Liang et al. decorated used $Ti_3C_2T_x$ MXene with molybdenum oxide MoO_2 electrode for an ASC with an MXene counter electrode.[130] The ASC was assembled on both chip and cloth and showed 88% retention for 10,000 cycles at 0.5 mA/cm². The device showed good integration capability and compatibility for various substrates. Hence, MXene-based asymmetric storage devices open up a new front for stable electronics with good performance under conditions like bending, stretching, and strain, thereby providing an opportunity for fabrication of high energy density, high power density, and high stability materials to accomplish the enhanced requirements of rechargeable energy-storage devices.

10.3.4   Other energy-storage applications

In addition to its use in MIB applications, MXene has shown encouraging behavior as an electrode for other rechargeable batteries like Li−S and Zn−air and exceptional characteristics for applications in other batteries like Li—S, Li—O, and metal—air batteries.[131-135] The high stability and 2-D structure of MXene and good mechanical properties allow the process of lithiation and delithiation or adsorption and desorption of other ions to happen smoothly while the structure is maintained. Beyond batteries, supercapacitors, and hybrid energy storage, MXene has displayed various energy-storage applications like solar cells, catalysis, fuel cells, etc.[136-139]

Recent reports have shown MXene as a potential material for use in solar cells.[138,140] Yang et al. reported the enhanced performance of solar cells after integrating $Ti_3C_2T_x$ in SnO_2 and deposition in the layered structure.[141] Power conversion efficiency of the solar device was enhanced by the high conductivity of 2-D MXene. Chen et al. inserted a pristine MXene ($Ti_3C_2T_x$) layer into their device and reported that MXene helps increase hole extraction and stability, as the functional groups reduce the trap defects from $CsPbBr_3$ perovskite material.[142]

Various 2-D MXenes have reportedly shown promising electrocatalytic behavior like oxygen evolution reaction, hydrogen evolution reaction, and oxygen reduction reaction.[143] Considering the basic advantages of 2-D materials, catalytic activity can be improved greatly by defect engineering.[144,145] Due to defect engineering, the possibility of manipulating materials to meet technology requirements—e.g., thin devices, fast processing, and high energy storage—increases. A hybrid structure of MXene with MoS_2 shows an improved hydrogen evolution reaction having interface contacts of MoS_2 between MXene.[146]

For clean energy, thermoelectricity is an important factor; hence, the search for highly efficient and stable thermoelectric material is of keen interest. MXenes, particularly those that are Mo-based, exhibit prominent characteristics as a thermoelectric material that have been validated through experimental and DFT studies.[147,148] Sarikurt et al. reported that functional groups attached to MXene surfaces further enhance thermoelectric properties.[149] Beyond energy storage, MXene has displayed applications in medicines, cancer therapy, sensors, organic dyes, and data storage.[150-158]

10.4 Conclusion and outlook

In conclusion, 2-D MXenes have noteworthy performance that can be attributed to some of their key characteristics:

1. High surface areas are present and available for adsorption.
2. Edge boundaries present additional sites for adsorbent or dopant.
3. Synergistic composites provide structure with increased opportunities for catalytic activities.
4. Thin structure enables technological miniaturization.
5. Functional groups can play an additional role in enhancing surface adsorption and desorption.

In addition, the structure of MXene can be tuned through numerous processes like doping, hybridizing the structure, and surface engineering. Hence, MXene presents itself as a prominent candidate for MIBs, supercapacitors, and hybrid energy-storage systems. However, certain aspects block the path to commercializing MXene in energy-storage applications. First, MXene synthesis is a key aspect. For stable electro-chemical performance, fluoride-free compounds have shown better performance than MXenes synthesized through HF etching. The increased number of sites available on the surface leads to higher ion diffusion and better storage. Secondly, an important aspect is the maintenance of the 2-D structure and fabrication of ultrathin films of 2-D materials. As the MXene structure deteriorates with time, long-term stability of the ink and electrode is essential for its commercialization in energy-storage devices. Based on its high storage capacities, MXene can be employed in various applications, but it is important to come forward with a methodical process that will help in the real-ization of using high-storage-capacity MXene on a commercial basis. In addition, MXene has efficiently demonstrated its integration into various substrates and chips, the deposition of MXene inks on flexible substrates, and maintenance of capacitance and energy density at different bending angles and stretched states. The various research conducted to date has overcome various limitations, but numerous challenges remain to be investigated and elaborated by further research. The essential factors for energy storage using MXene to meet modern requirements are cost-effectiveness, long lifetimes, retention of high specific capacities, and high energy densities. Furthermore, defining safe and easy synthesis and fabrication paths for a diverse range of MXenes is essential. Decisively, further studies to work out novel synthesis methods and then fabricate thin films through several options, e.g., stamping, inkjet printing, and others, must be optimized. This will present an optimum roadway for introducing MXene to the energy-storage industry. In addition, synthesizing different MXenes, heterostruc-tures of MXenes with other 2-D material, or decorating it with 1-D materials present many approaches to increasing its energy-storage potential and applications.

Abbreviations

2-D	two-dimensional
AC	activated carbon
ASC	asymmetric supercapacitor
CC	carbon cloth
CNFs	carbon nanofibrils
CNTs	carbon nanotubes
CV	cyclic voltammetry
CVD	chemical vapor deposition
DFT	density functional theory
EDLC	electric double-layer capacitor
HF	hydrofluoric acid
LDH	layered double hydroxide
LIBs	lithium-ion batteries
MIBs	metal-ion batteries
MOF	metal–organic framework
MSC	microsupercapacitor
PANI	polyaniline
PC	porous carbon

PDDA	poly diallyldimethylammonium chloride
PEDOT	poly 3-4 ethylenedioxythiophene
PIBs	potassium-ion batteries
PPY	polypyrrole
PVA	polyvinyl alcohol
PVDF	polyvinylidene difluoride
rGO	reduced graphene oxide
SA	sodium ascorbate
SSC	symmetric supercapacitor
SIBs	sodium-ion batteries
TEM	transmission electron microscopy
TMDs	transition metal dichalcogenides

Acknowledgments

The authors would like to acknowledge the Higher Education Commission (HEC) of Pakistan, United States Agency for International Development (USAID) and Department of State, USA for providing research funding under the Project No. HEC/R&D/PAKUS/2017/783.

References

1. Das S, Robinson JA, Dubey M, Terrones H, Terrones M. Beyond graphene: progress in novel two-dimensional materials and van der Waals solids. *Annu Rev Mater Res* 2015;**45**: 1−27.
2. Ke Q, Wang J. Graphene-based materials for supercapacitor electrodes−a review. *J Materiomics* 2016;**2**:37−54.
3. Pumera M, Sofer Z. 2D monoelemental arsenene, antimonene, and bismuthene: beyond black phosphorus. *Adv Mater* 2017;**29**:1605299.
4. Ren X, Lian P, Xie D, Yang Y, Mei Y, Huang X, Wang Z, Yin X. Properties, preparation and application of black phosphorus/phosphorene for energy storage: a review. *J Mater Sci* 2017;**52**:10364−86.
5. Donarelli M, Ottaviano L. 2D materials for gas sensing applications: a review on graphene oxide, MoS_2, WS_2 and phosphorene. *Sensors* 2018;**18**:3638.
6. Khazaei M, Ranjbar A, Arai M, Sasaki T, Yunoki S. Electronic properties and applications of MXenes: a theoretical review. *J Mater Chem C* 2017;**5**:2488−503.
7. Naguib M, Mochalin VN, Barsoum MW, Gogotsi Y. 25th anniversary article: MXenes: a new family of two-dimensional materials. *Adv Mater* 2014;**26**:992−1005.
8. Naguib M, Mashtalir O, Carle J, Presser V, Lu J, Hultman L, Gogotsi Y, Barsoum MW. Two-dimensional transition metal carbides. *ACS Nano* 2012;**6**:1322−31.
9. Naguib M, Gogotsi Y. Synthesis of two-dimensional materials by selective extraction. *Acc Chem Res* 2015;**48**:128−35.
10. Khazaei M, Mishra A, Venkataramanan NS, Singh AK, Yunoki S. Recent advances in MXenes: from fundamentals to applications. *Curr Opin Solid State Mater Sci* 2019;**23**: 164−78.

11. Fatheema J, Fatima M, Monir NB, Khan AS, Rizwan S. A comprehensive computational and experimental analysis of stable ferromagnetism in layered 2D Nb-doped Ti_3C_2 MXene. *Phys E Low-dimens Syst Nanostruct* 2020:114253.

12. Yan X, Wang K, Zhao J, Zhou Z, Wang H, Wang J, Zhang L, Li X, Xiao Z, Zhao Q. A new memristor with 2D $Ti_3C_2T_x$ MXene flakes as an artificial bio-synapse. *Small* 2019;**15**: 1900107.

13. Jiang H, Wang Z, Yang Q, Hanif M, Wang Z, Dong L, Dong M. A novel $MnO_2/Ti_3C_2T_x$ MXene nanocomposite as high performance electrode materials for flexible supercapacitors. *Electrochim Acta* 2018;**290**:695−703.

14. Ke C, Fan D, Chen C, Li X, Jiang M-Y, Hu XJ. Two-dimension tetragonal transition-metal carbides anodes for non-lithium-ion batteries. *Phys Chem Chem Phys* 2020;**22**:21208−21.

15. Iqbal MA, Ali SI, Amin F, Tariq A, Iqbal MZ, Rizwan S. La-and Mn-codoped bismuth ferrite/Ti_3C_2 MXene composites for efficient photocatalytic degradation of Congo red dye. *ACS Omega* 2019;**4**:8661−8.

16. Babar ZUD, Anwar S, Mumtaz M, Iqbal M, Zheng R-K, Akinwande D, Rizwan S. Peculiar magnetic behaviour and Meissner effect in two-dimensional layered Nb_2C MXene. *2D Mater* 2020. https://doi.org/10.1088/2053-1583/ab86d2.

17. Huang Z-Q, Xu M-L, Macam G, Hsu C-H, Chuang F-C. Large-gap topological insulators in functionalized ordered double transition metal carbide MXenes. *Phys Rev B* 2020;**102**: 75306.

18. Rizwan S, Awan SU, Irfan S. Room-temperature ferromagnetism in Gd and Sn co-doped bismuth ferrite nanoparticles and co-doped $BiFeO_3$/MXene (Ti_3C_2) nanohybrids for spintronics applications. *Ceram Int* 2020;**46**(18 A):29011−21.

19. Tariq A, Ali SI, Akinwande D, Rizwan S. Efficient visible-light photocatalysis of 2D-MXene nanohybrids with Gd^{3+} and Sn^{4+} Codoped bismuth ferrite. *ACS Omega* 2018;**3**: 13828−36.

20. Rafiq S, Awan S, Zheng R-K, Wen Z, Rani M, Akinwande D, Rizwan S. Novel room-temperature ferromagnetism in Gd-doped 2-dimensional $Ti_3C_2T_x$ MXene semiconductor for spintronics. *J Magn Magn Mater* 2019:165954.

21. Zhang C, Ma Y, Zhang X, Abdolhosseinzadeh S, Sheng H, Lan W, Pakdel A, Heier J, Nüesch F. Two-dimensional transition metal carbides and nitrides (MXenes): synthesis, properties, and electrochemical energy storage applications. *Energy Environ Mater* 2020; **3**:29−55.

22. Naguib M, Kurtoglu M, Presser V, Lu J, Niu J, Heon M, Hultman L, Gogotsi Y, Barsoum MW. Two-dimensional nanocrystals produced by exfoliation of Ti_3AlC_2. *Adv Mater* 2011;**23**:4248−53.

23. Shan Q, Mu X, Alhabeb M, Shuck CE, Pang Di, Zhao X, Chu X-F, Wei Y, Du F, Chen G, Gogotsi Y, Gao Y, Dall'Agnese Y. Two-dimensional vanadium carbide (V_2C) MXene as electrode for supercapacitors with aqueous electrolytes. *Electrochem Commun* 2018;**96**: 103−7.

24. Wang L, Liu D, Lian W, Hu Q, Liu X, Zhou A. The preparation of V_2CT_x by facile hydrothermal-assisted etching processing and its performance in lithium-ion battery. *J Mater Res Technol* 2020;**9**:984−93.

25. Guan Y, Jiang S, Cong Y, Wang J, Dong Z, Zhang Q, Yuan G, Li Y, Li X. A hydrofluoric acid-free synthesis of 2D vanadium carbide (V 2 C) MXene for supercapacitor electrodes. *2D Mater* 2020;**7**:25010.

26. Alhabeb M, Maleski K, Anasori B, Lelyukh P, Clark L, Sin S, Gogotsi Y. Guidelines for synthesis and processing of two-dimensional titanium carbide (Ti_3C_2T x MXene). *Chem Mater* 2017;**29**:7633−44.

27. Wu M, Wang B, Hu Q, Wang L, Zhou A. The synthesis process and thermal stability of V_2C MXene. *Materials* 2018;**11**. https://doi.org/10.3390/ma11112112.

28. Halim J, Kota S, Lukatskaya MR, Naguib M, Zhao M-Q, Moon EJ, Pitock J, Nanda J, May SJ, Gogotsi Y. Synthesis and characterization of 2D molybdenum carbide (MXene). *Adv Funct Mater* 2016;**26**:3118−27.

29. Tao Q, Dahlqvist M, Lu J, Kota S, Meshkian R, Halim J, Palisaitis J, Hultman L, Barsoum MW, Persson POÅ. Two-dimensional Mo 1.33 C MXene with divacancy ordering prepared from parent 3D laminate with in-plane chemical ordering. *Nat Commun* 2017;**8**:1−7.

30. Wang X, Lin S, Tong H, Huang Y, Tong P, Zhao B, Dai J, Liang C, Wang H, Zhu X, Sun Y, Dou S. Two-dimensional V_4C_3 MXene as high performance electrode materials for supercapacitors. *Electrochim Acta* 2019;**307**:414−21.

31. Syamsai R, Grace AN. Ta_4C_3 MXene as supercapacitor electrodes. *J Alloys Compd* 2019; **792**:1230−8.

32. Ghidiu M, Naguib M, Shi C, Mashtalir O, Pan LM, Zhang B, Yang J, Gogotsi Y, Billinge SJL, Barsoum MW. Synthesis and characterization of two-dimensional Nb 4 C 3 (MXene). *Chem Commun* 2014;**50**:9517−20.

33. Cai P, He Q, Wang L, Liu X, Yin J, Liu Y, Huang Y, Huang Z. Two-dimensional Nb-based $M_4C_3T_x$ MXenes and their sodium storage performances. *Ceram Int* 2019;**45**:5761−7.

34. Halim J, Lukatskaya MR, Cook KM, Lu J, Smith CR, Näslund L-Å, May SJ, Hultman L, Gogotsi Y, Eklund P. Transparent conductive two-dimensional titanium carbide epitaxial thin films. *Chem Mater* 2014;**26**:2374−81.

35. Liu Z, Xu C, Kang N, Wang L, Jiang Y, Du J, Liu Y, Ma X-L, Cheng H-M, Ren W. Unique domain structure of two-dimensional α-Mo_2C superconducting crystals. *Nano Lett* 2016;**16**:4243−50.

36. Xu C, Wang L, Liu Z, Chen L, Guo J, Kang N, Ma X-L, Cheng H-M, Ren W. Large-area high-quality 2D ultrathin Mo 2 C superconducting crystals. *Nat Mater* 2015;**14**:1135−41.

37. Lai C-C, Fashandi H, Lu J, Palisaitis J, Persson POÅ, Hultman L, Eklund P, Rosen J. Phase formation of nanolaminated Mo 2 AuC and Mo 2 (Au 1− x Ga x) 2 C by a substitutional reaction within Au-capped Mo 2 GaC and Mo 2 Ga 2 C thin films. *Nanoscale* 2017;**9**: 17681−7.

38. Gao L, Li C, Huang W, Mei S, Lin H, Ou Q, Zhang Y, Guo J, Zhang F, Xu S. MXene/polymer membranes: synthesis, properties, and emerging applications. *Chem Mater* 2020; **32**:1703−47.

39. Wu C-W, Unnikrishnan B, Chen I-WP, Harroun SG, Chang H-T, Huang C-C. Excellent oxidation resistive MXene aqueous ink for micro-supercapacitor application. *Energy Storage Mater* 2020;**25**:563−71.

40. Zhang CJ, McKeon L, Kremer MP, Park S-H, Ronan O, Seral-Ascaso A, Barwich S, Coileáin CÓ, McEvoy N, Nerl HC. Additive-free MXene inks and direct printing of micro-supercapacitors. *Nat Commun* 2019;**10**:1−9.

41. Ronchi RM, Arantes JT, Santos SF. Synthesis, structure, properties and applications of MXenes: current status and perspectives. *Ceram Int* 2019;**45**:18167−88.

42. Verger L, Xu C, Natu V, Cheng H-M, Ren W, Barsoum MW. Overview of the synthesis of MXenes and other ultrathin 2D transition metal carbides and nitrides. *Curr Opin Solid State Mater Sci* 2019;**23**:149−63.

43. Aierken Y, Sevik C, Gülseren O, Peeters FM, Çakır D. MXenes/graphene heterostructures for Li battery applications: a first principles study. *J Mater Chem A* 2018;**6**:2337−45.

44. Nitta N, Wu F, Lee JT, Yushin G. Li-ion battery materials: present and future. *Mater Today* 2015;**18**:252−64.

45. Tarascon J-M, Armand M. *Materials for sustainable energy: a collection of peer-reviewed research and review articles from nature publishing group*. World Scientific; 2011. p. 171−9.

46. Yu T, Zhang S, Li F, Zhao Z, Liu L, Xu H, Yang G. Stable and metallic two-dimensional TaC 2 as an anode material for lithium-ion battery. *J Mater Chem A* 2017;**5**:18698−706.

47. Yang Z, Zheng Y, Li W, Zhang J. Investigation of two-dimensional hf-based MXenes as the anode materials for li/na-ion batteries: a DFT study. *J Comput Chem* 2019;**40**:1352−9.

48. Yu Y, Guo Z, Peng Q, Zhou J, Sun Z. Novel two-dimensional molybdenum carbides as high capacity anodes for lithium/sodium-ion batteries. *J Mater Chem A* 2019;**7**:12145−53.

49. Meng Q, Ma J, Zhang Y, Li Z, Hu A, Kai J-J, Fan J. Theoretical investigation of zirconium carbide MXenes as prospective high capacity anode materials for Na-ion batteries. *J Mater Chem A* 2018;**6**:13652−60.

50. Meng Q, Ma J, Zhang Y, Li Z, Zhi C, Hu A, Fan J. The S-functionalized Ti3C2 Mxene as a high capacity electrode material for Na-ion batteries: a DFT study. *Nanoscale* 2018;**10**:3385−92.

51. Fan K, Ying Y, Li X, Luo X, Huang H. Theoretical investigation of V 3 C 2 MXene as prospective high-capacity anode material for metal-ion (Li, Na, K, and Ca) batteries. *J Phys Chem C* 2019;**123**:18207−14.

52. Hu J, Xu B, Ouyang C, Yang SA, Yao Y. Investigations on V_2C and V_2CX_2 (X= F, OH) monolayer as a promising anode material for Li ion batteries from first-principles calculations. *J Phys Chem C* 2014;**118**:24274−81.

53. Hu J, Xu B, Ouyang C, Zhang Y, Yang SA. Investigations on Nb 2 C monolayer as promising anode material for Li or non-Li ion batteries from first-principles calculations. *RSC Adv* 2016;**6**:27467−74.

54. Sun D, Hu Q, Chen J, Zhang X, Wang L, Wu Q, Zhou A. Structural transformation of MXene (V_2C, Cr_2C, and Ta_2C) with O groups during lithiation: a first-principles investigation. *ACS Appl Mater Interfaces* 2016;**8**:74−81.

55. Liu F, Zhou J, Wang S, Wang B, Shen C, Wang L, Hu Q, Huang Q, Zhou A. Preparation of high-purity V_2C MXene and electrochemical properties as Li-ion batteries. *J Electrochem Soc* 2017;**164**:A709.

56. Wang X, Cai Y, Wu S, Li B. Sulfur functions as the activity centers for high-capacity lithium ion batteries in S- and O-bifunctionalized MXenes: a density functional theory (DFT) study. *Appl Surf Sci* 2020;**525**:146501.

57. Kajiyama S, Szabova L, Sodeyama K, Iinuma H, Morita R, Gotoh K, Tateyama Y, Okubo M, Yamada A. Sodium-ion intercalation mechanism in MXene nanosheets. *ACS Nano* 2016;**10**:3334−41.

58. Sun S, Xie Z, Yan Y, Wu S. Hybrid energy storage mechanisms for sulfur-decorated Ti_3C_2 MXene anode material for high-rate and long-life sodium-ion batteries. *Chem Eng J* 2019;**366**:460−7.

59. Liu R, Cao W, Han D, Mo Y, Zeng H, Yang H, Li W. Nitrogen-doped Nb_2CT_x MXene as anode materials for lithium ion batteries. *J Alloys Compd* 2019;**793**:505−11.

60. Du L, Duan H, Xia Q, Jiang C, Yan Y, Wu S. Hybrid charge-storage route to Nb 2 CT x MXene as anode for sodium-ion batteries. *Chemistry* 2020;**5**:1186−92.

61. Zhu J, Wang M, Lyu M, Jiao Y, Du A, Luo B, Gentle I, Wang L. Two-dimensional titanium carbonitride mxene for high-performance sodium ion batteries. *ACS Appl Nano Mater* 2018;**1**:6854−63.

62. Chen X, Kong Z, Li N, Zhao X, Sun C. Proposing the prospects of Ti_3CN transition metal carbides (MXenes) as anodes of Li-ion batteries: a DFT study. *Phys Chem Chem Phys* 2016;**18**:32937−43.

63. Zhao Q, Zhu Q, Miao J, Zhang P, Wan P, He L, Xu B. Flexible 3D porous MXene foam for high-performance lithium-ion batteries. *Small* 2019;**15**:e1904293.

64. Wu F, Jiang Y, Ye Z, Huang Y, Wang Z, Li S, Mei Y, Xie M, Li L, Chen R. A 3D flower-like VO 2/MXene hybrid architecture with superior anode performance for sodium ion batteries. *J Mater Chem A* 2019;**7**:1315−22.

65. Li J, Han L, Li Y, Li J, Zhu G, Zhang X, Lu T, Pan L. MXene-decorated SnS_2/Sn_3S_4 hybrid as anode material for high-rate lithium-ion batteries. *Chem Eng J* 2020;**380**:122590.

66. Cui Y, Wang J, Wang X, Qin J, Cao M. A hybrid assembly of MXene with NH_2- Si nanoparticles boosting lithium storage performance. *Chem - Asian J* 2020;**15**:1376−83.

67. Tang C, Min Y, Chen C, Xu W, Xu L. Potential applications of heterostructures of TMDs with MXenes in sodium-ion and $Na-O_2$ batteries. *Nano Lett* 2019;**19**:5577−86.

68. Luan S, Han M, Xi Y, Wei K, Wang Y, Zhou J, Hou L, Gao F. MoS_2-decorated 2D Ti_3C_2 (MXene): a high-performance anode material for lithium-ion batteries. *Ionics* 2020;**26**: 51−9.

69. Xu E, Zhang Y, Wang H, Zhu Z, Quan J, Chang Y, Li P, Yu D, Jiang Y. Ultrafast kinetics net electrode assembled via $MoSe_2$/MXene heterojunction for high-performance sodium-ion batteries. *Chem Eng J* 2020;**385**:123839.

70. Huang H, Cui J, Liu G, Bi R, Zhang L. Carbon-coated $MoSe_2$/MXene hybrid nanosheets for superior potassium storage. *ACS Nano* 2019;**13**:3448−56.

71. Li H, Liu A, Ren X, Yang Y, Gao L, Fan M, Ma T. A black phosphorus/Ti 3 C 2 MXene nanocomposite for sodium-ion batteries: a combined experimental and theoretical study. *Nanoscale* 2019;**11**:19862−9.

72. Zhao R, Qian Z, Liu Z, Zhao D, Hui X, Jiang G, Wang C, Yin L. Molecular-level heterostructures assembled from layered black phosphorene and Ti_3C_2 MXene as superior anodes for high-performance sodium ion batteries. *Nano Energy* 2019;**65**:104037.

73. Xiao J, Wen J, Zhao J, Ma X, Gao H, Zhang X. A safe etching route to synthesize highly crystalline Nb_2CT_x MXene for high performance asymmetric supercapacitor applications. *Electrochim Acta* 2020;**337**:135803.

74. Ando Y, Okubo M, Yamada A, Otani M. Capacitive versus pseudocapacitive storage in MXene. *Adv Funct Mater* 2020:2000820.

75. Kumar KS, Choudhary N, Jung Y, Thomas J. Recent advances in two-dimensional nanomaterials for supercapacitor electrode applications. *ACS Energy Lett* 2018;**3**:482−95.

76. Das P, Wu Z-S. MXene for energy storage: present status and future perspectives. *J Phys Energy* 2020;**2**:03204.

77. Zhang CJ, Nicolosi V. Graphene and MXene-based transparent conductive electrodes and supercapacitors. *Energy Storage Mater* 2019;**16**:102−25.

78. Osti NC, Naguib M, Ostadhossein A, Xie Y, Kent PRC, Dyatkin B, Rother G, Heller WT, van Duin ACT, Gogotsi Y. Effect of metal ion intercalation on the structure of MXene and water dynamics on its internal surfaces. *ACS Appl Mater Interfaces* 2016;**8**:8859−63.

79. Zhan C, Sun W, Kent PRC, Naguib M, Gogotsi Y, Jiang D. Computational screening of MXene electrodes for pseudocapacitive energy storage. *J Phys Chem C* 2018;**123**:315−21.

80. Mu X, Wang D, Du F, Chen G, Wang C, Wei Y, Gogotsi Y, Gao Y, Dall'Agnese Y. Revealing the pseudo-intercalation charge storage mechanism of MXenes in acidic electrolyte. *Adv Funct Mater* 2019;**29**:1902953.

81. Zhan C, Naguib M, Lukatskaya M, Kent PRC, Gogotsi Y, Jiang D. Understanding the MXene pseudocapacitance. *J Phys Chem Lett* 2018;**9**:1223−8.

82. VahidMohammadi A, Mojtabavi M, Caffrey NM, Wanunu M, Beidaghi M. Assembling 2D MXenes into highly stable pseudocapacitive electrodes with high power and energy densities. *Adv Mater* 2019;**31**:1806931.

83. Ghidiu M, Lukatskaya MR, Zhao M-Q, Gogotsi Y, Barsoum MW. Conductive two-dimensional titanium carbide 'clay'with high volumetric capacitance. *Nature* 2014;**516**: 78−81.
84. Syamsai R, Kollu P, Jeong SK, Grace AN. Synthesis and properties of 2D-titanium carbide MXene sheets towards electrochemical energy storage applications. *Ceram Int* 2017;**43**: 13119−26.
85. Djire A, Bos A, Liu J, Zhang H, Miller EM, Neale NR. Pseudocapacitive storage in nanolayered Ti_2NT x MXene using Mg-ion electrolyte. *ACS Appl Nano Mater* 2019;**2**: 2785−95.
86. Rakhi RB, Ahmed B, Hedhili MN, Anjum DH, Alshareef HN. Effect of postetch annealing gas composition on the structural and electrochemical properties of Ti_2CT x MXene electrodes for supercapacitor applications. *Chem Mater* 2015;**27**:5314−23.
87. Zhu M, Huang Y, Deng Q, Zhou J, Pei Z, Xue Q, Huang Y, Wang Z, Li H, Huang Q. Highly flexible, freestanding supercapacitor electrode with enhanced performance obtained by hybridizing polypyrrole chains with MXene. *Adv Energy Mater* 2016;**6**: 1600969.
88. Wen Y, Rufford TE, Chen X, Li N, Lyu M, Dai L, Wang L. Nitrogen-doped $Ti_3C_2T_x$ MXene electrodes for high-performance supercapacitors. *Nano Energy* 2017;**38**:368−76.
89. Fatima M, Fatheema J, Monir NB, Siddique AH, Khan B, Islam A, Akinwande D, Rizwan S. Nb-doped MXene with enhanced energy storage capacity and stability. *Front Chem* 2020;**8**:168.
90. Pan Z, Ji X. Facile synthesis of nitrogen and oxygen co-doped C@ Ti_3C_2 MXene for high performance symmetric supercapacitors. *J Power Sources* 2019;**439**:227068.
91. Xia QX, Shinde NM, Yun JM, Zhang T, Mane RS, Mathur S, Kim KH. Bismuth oxychloride/MXene symmetric supercapacitor with high volumetric energy density. *Electrochim Acta* 2018;**271**:351−60.
92. Zhang X, Liu Y, Dong S, Yang J, Liu X. Surface modified MXene film as flexible electrode with ultrahigh volumetric capacitance. *Electrochim Acta* 2019;**294**:233−9.
93. Zhu K, Jin Y, Du F, Gao S, Gao Z, Meng X, Chen G, Wei Y, Gao Y. Synthesis of Ti_2CT_x MXene as electrode materials for symmetric supercapacitor with capable volumetric capacitance. *J Energy Chem* 2019;**31**:11−8.
94. Fan Z, Wang Y, Xie Z, Xu X, Yuan Y, Cheng Z, Liu Y. A nanoporous MXene film enables flexible supercapacitors with high energy storage. *Nanoscale* 2018;**10**:9642−52.
95. Fu Q, Wang X, Zhang N, Wen J, Li L, Gao H, Zhang X. Self-assembled $Ti_3C_2T_x$/SCNT composite electrode with improved electrochemical performance for supercapacitor. *J Colloid Interface Sci* 2018;**511**:128−34.
96. Levitt AS, Alhabeb M, Hatter CB, Sarycheva A, Dion G, Gogotsi Y. Electrospun MXene/carbon nanofibers as supercapacitor electrodes. *J Mater Chem A* 2019;**7**:269−77.
97. Yan J, Ren CE, Maleski K, Hatter CB, Anasori B, Urbankowski P, Sarycheva A, Gogotsi Y. Flexible MXene/graphene films for ultrafast supercapacitors with outstanding volumetric capacitance. *Adv Funct Mater* 2017;**27**:1701264.
98. Yu L, Hu L, Anasori B, Liu Y-T, Zhu Q, Zhang P, Gogotsi Y, Xu B. MXene-bonded activated carbon as a flexible electrode for high-performance supercapacitors. *ACS Energy Lett* 2018;**3**:1597−603.
99. Hu M, Li Z, Li G, Hu T, Zhang C, Wang X. All-solid-state flexible fiber-based MXene supercapacitors. *Adv Mater Technol* 2017;**2**:1700143.
100. Yang Q, Xu Z, Fang B, Huang T, Cai S, Chen H, Liu Y, Gopalsamy K, Gao W, Gao C. MXene/graphene hybrid fibers for high performance flexible supercapacitors. *J Mater Chem A* 2017;**5**:22113−9.

101. Zhou Y, Maleski K, Anasori B, Thostenson JO, Pang Y, Feng Y, Zeng K, Parker CB, Zauscher S, Gogotsi Y. Ti$_3$C$_2$T x MXene-reduced graphene oxide composite electrodes for stretchable supercapacitors. *ACS Nano* 2020;**14**:3576−86.

102. Orangi J, Hamade F, Davis VA, Beidaghi M. 3D printing of additive-free 2D Ti$_3$C$_2$T x (MXene) ink for fabrication of micro-supercapacitors with ultra-high energy densities. *ACS Nano* 2019;**14**:640−50.

103. Zhang C, Kremer MP, Seral-Ascaso A, Park S-H, McEvoy N, Anasori B, Gogotsi Y, Nicolosi V. Stamping of flexible, coplanar micro-supercapacitors using MXene inks. *Adv Funct Mater* 2018;**28**:1705506.

104. Tian W, VahidMohammadi A, Reid MS, Wang Z, Ouyang L, Erlandsson J, Pettersson T, Wågberg L, Beidaghi M, Hamedi MM. Multifunctional nanocomposites with high strength and capacitance using 2d mxene and 1d nanocellulose. *Adv Mater* 2019;**31**: 1902977.

105. Huang H, He J, Wang Z, Zhang H, Jin L, Chen N, Xie Y, Chu X, Gu B, Deng W. Scalable, and low-cost treating-cutting-coating manufacture platform for MXene-based on-chip micro-supercapacitors. *Nano Energy* 2020;**69**:104431.

106. Cericola D, Kötz R. Hybridization of rechargeable batteries and electrochemical capacitors: principles and limits. *Electrochim Acta* 2012;**72**:1−17.

107. Dubal DP, Ayyad O, Ruiz V, Gomez-Romero P. Hybrid energy storage: the merging of battery and supercapacitor chemistries. *Chem Soc Rev* 2015;**44**:1777−90.

108. Amatucci GG, Badway F, Du Pasquier A, Zheng T. An asymmetric hybrid nonaqueous energy storage cell. *J Electrochem Soc* 2001;**148**:A930.

109. Liu W, Zhang M, Li M, Li B, Zhang W, Li G, Xiao M, Zhu J, Yu A, Chen Z. Advanced electrode materials comprising of structure-engineered quantum dots for high-performance asymmetric micro-supercapacitors. *Adv Energy Mater* 2020;**10**:1903724.

110. Shi S, Xu C, Yang C, Chen Y, Liu J, Kang F. Flexible asymmetric supercapacitors based on ultrathin two-dimensional nanosheets with outstanding electrochemical performance and aesthetic property. *Sci Rep* 2013;**3**:2598.

111. Zheng S, Wu Z-S, Wang S, Xiao H, Zhou F, Sun C, Bao X, Cheng H-M. Graphene-based materials for high-voltage and high-energy asymmetric supercapacitors. *Energy Storage Mater* 2017;**6**:70−97.

112. Sahoo R, Pal A, Pal T. 2D materials for renewable energy storage devices: outlook and challenges. *Chem Commun* 2016;**52**:13528−42.

113. Xie Y, Zhang H, Huang H, Wang Z, Xu Z, Zhao H, Wang Y, Chen N, Yang W. High-voltage asymmetric MXene-based on-chip micro-supercapacitors. *Nano Energy* 2020: 104928.

114. Chang T-H, Zhang T, Yang H, Li K, Tian Y, Lee JY, Chen P-Y. Controlled crumpling of two-dimensional titanium carbide (MXene) for highly stretchable, bendable, efficient supercapacitors. *ACS Nano* 2018;**12**:8048−59.

115. Zhao R, Wang M, Zhao D, Li H, Wang C, Yin L. Molecular-level heterostructures assembled from titanium carbide MXene and Ni−Co−Al layered double-hydroxide nanosheets for all-solid-state flexible asymmetric high-energy supercapacitors. *ACS Energy Lett* 2017;**3**:132−40.

116. Wang Y, Liu Y, Wang C, Liu H, Zhang J, Lin J, Fan J, Ding T, Ryu JE, Guo Z. Significantly enhanced ultrathin NiCo-based MOF nanosheet electrodes hybrided with Ti$_3$C$_2$T$_x$ MXene for high performance asymmetric supercapacitor. *Eng Sci* 2020;**9**:50−9.

117. Zhou H, Lu Y, Wu F, Fang L, Luo H, Zhang Y, Zhou M. MnO2 nanorods/MXene/CC composite electrode for flexible supercapacitors with enhanced electrochemical performance. *J Alloys Compd* 2019;**802**:259−68.

118. Lu C, Li A, Zhai T, Niu C, Duan H, Guo L, Zhou W. Interface design based on Ti_3C_2 MXene atomic layers of advanced battery-type material for supercapacitors. *Energy Storage Mater* 2020;**26**:472−82.

119. Boota M, Gogotsi Y. MXene—conducting polymer asymmetric pseudocapacitors. *Adv Energy Mater* 2019;**9**:1802917.

120. Fu J, Yun J, Wu S, Li L, Yu L, Kim KH. Architecturally robust graphene-encapsulated MXene Ti_2CT x@ polyaniline composite for high-performance pouch-type asymmetric supercapacitor. *ACS Appl Mater Interfaces* 2018;**10**:34212−21.

121. Li K, Wang X, Li S, Urbankowski P, Li J, Xu Y, Gogotsi Y. An ultrafast conducting polymer@ MXene positive electrode with high volumetric capacitance for advanced asymmetric supercapacitors. *Small* 2020;**16**:1906851.

122. Li K, Wang X, Wang X, Liang M, Nicolosi V, Xu Y, Gogotsi Y. All-pseudocapacitive asymmetric MXene-carbon-conducting polymer supercapacitors. *Nano Energy* 2020: 104971.

123. Jiang Q, Kurra N, Alhabeb M, Gogotsi Y, Alshareef HN. All pseudocapacitive MXene-RuO_2 asymmetric supercapacitors. *Adv Energy Mater* 2018;**8**:1703043.

124. Zhao K, Wang H, Zhu C, Lin S, Xu Z, Zhang X. Free-standing MXene film modified by amorphous FeOOH quantum dots for high-performance asymmetric supercapacitor. *Electrochim Acta* 2019;**308**:1−8.

125. Wang Z, Qin S, Seyedin S, Zhang J, Wang J, Levitt A, Li N, Haines C, Ovalle-Robles R, Lei W. High-performance biscrolled MXene/carbon nanotube yarn supercapacitors. *Small* 2018;**14**:1802225.

126. Venkateshalu S, Grace AN. $Ti_3C_2T_x$ MXene and Vanadium nitride/porous carbon as electrodes for asymmetric supercapacitors. *Electrochim Acta* 2020:136035.

127. Javed MS, Lei H, Shah HU, Asim S, Raza R, Mai W. Achieving high rate and high energy density in an all-solid-state flexible asymmetric pseudocapacitor through the synergistic design of binder-free 3D ZnCo 2 O 4 nano polyhedra and 2D layered Ti 3 C 2 T x-MXenes. *J Mater Chem* 2019;**7**:24543−56.

128. Luo Y, Yang C, Tian Y, Tang Y, Yin X, Que W. A long cycle life asymmetric super-capacitor based on advanced nickel-sulfide/titanium carbide (MXene) nanohybrid and MXene electrodes. *J Power Sources* 2020;**450**:227694.

129. Pan Z, Cao F, Hu X, Ji X. A facile method for synthesizing CuS decorated Ti_3C_2 MXene with enhanced performance for asymmetric supercapacitors. *J Mater Chem* 2019;**7**: 8984−92.

130. Zhang L, Yang G, Chen Z, Liu D, Wang J, Qian Y, Chen C, Liu Y, Wang L, Razal J. MXene coupled with molybdenum dioxide nanoparticles as 2D-0D pseudocapacitive electrode for high performance flexible asymmetric micro-supercapacitors. *J Materiomics* 2020;**6**:138−44.

131. Pang J, Mendes RG, Bachmatiuk A, Zhao L, Ta HQ, Gemming T, Liu H, Liu Z, Rummeli MH. Applications of 2D MXenes in energy conversion and storage systems. *Chem Soc Rev* 2019;**48**:72−133.

132. Yang J, Bao W, Jaumaux P, Zhang S, Wang C, Wang G. MXene-based composites: synthesis and applications in rechargeable batteries and supercapacitors. *Adv Mater Interfaces* 2019;**6**:1802004.

133. Li N, Meng Q, Zhu X, Li Z, Ma J, Huang C, Song J, Fan J. Lattice constant-dependent anchoring effect of MXenes for lithium-sulfur (Li-S) batteries: a DFT study. *Nanoscale* 2019;**11**:8485−93.

134. Li N, Chen X, Ong W-J, MacFarlane DR, Zhao X, Cheetham AK, Sun C. Understanding of electrochemical mechanisms for CO_2 capture and conversion into hydrocarbon fuels in transition-metal carbides (MXenes). *ACS Nano* 2017;**11**:10825−33.

135. Liu F, Liu Y, Zhao X, Liu K, Yin H, Fan L-Z. Prelithiated V2 C MXene: a high-performance electrode for hybrid magnesium/lithium-ion batteries by ion cointercalation. *Small* 2020;**16**:e1906076.

136. Yadav A, Dashora A, Patel N, Miotello A, Press M, Kothari DC. Study of 2D MXene Cr_2C material for hydrogen storage using density functional theory. *Appl Surf Sci* 2016; **389**:88−95.

137. Li Z, Wu Y. 2D early transition metal carbides (MXenes) for catalysis. *Small* 2019;**15**: 1804736.

138. Agresti A, Pazniak A, Pescetelli S, Di Vito A, Rossi D, Pecchia A, auf der Maur M, Liedl A, Larciprete R, Kuznetsov DV. Titanium-carbide MXenes for work function and interface engineering in perovskite solar cells. *Nat Mater* 2019;**18**:1228−34.

139. Liu D, Wang R, Chang W, Zhang L, Peng B, Li H, Liu S, Yan M, Guo C. Ti 3 C 2 MXene as an excellent anode material for high-performance microbial fuel cells. *J Mater Chem* 2018;**6**:20887−95.

140. Fu H-C, Ramalingam V, Kim H, Lin C-H, Fang X, Alshareef HN, He J-H. MXene-contacted silicon solar cells with 11.5% efficiency. *Adv Energy Mater* 2019;**9**:1900180.

141. Yang L, Dall'Agnese Y, Hantanasirisakul K, Shuck CE, Maleski K, Alhabeb M, Chen G, Gao Y, Sanehira Y, Jena AK. SnO 2−Ti 3 C 2 MXene electron transport layers for perovskite solar cells. *J Mater Chem* 2019;**7**:5635−42.

142. Chen T, Tong G, Xu E, Li H, Li P, Zhu Z, Tang J, Qi Y, Jiang Y. Accelerating hole extraction by inserting 2D Ti 3 C 2-MXene interlayer to all inorganic perovskite solar cells with long-term stability. *J Mater Chem* 2019;**7**:20597−603.

143. Liu J, Peng W, Li Y, Zhang F, Fan X. *2D MXene-based materials for electrocatalysis.* Transactions of Tianjin University; 2020. p. 1−23.

144. Sun T, Zhang G, Xu D, Lian X, Li H, Chen W, Su C. Defect chemistry in 2D materials for electrocatalysis. *Mater Today Energy* 2019;**12**:215−38.

145. Kuznetsov DA, Chen Z, Kumar PV, Tsoukalou A, Kierzkowska A, Abdala PM, Safonova OV, Fedorov A, Müller CR. Single site cobalt substitution in 2D molybdenum carbide (MXene) enhances catalytic activity in the hydrogen evolution reaction. *J Am Chem Soc* 2019;**141**:17809−16.

146. You J, Si C, Zhou J, Sun Z. Contacting MoS_2 to MXene: vanishing p-type Schottky barrier and enhanced hydrogen evolution catalysis. *J Phys Chem C* 2019;**123**:3719−26.

147. Khazaei M, Arai M, Sasaki T, Estili M, Sakka Y. Two-dimensional molybdenum carbides: potential thermoelectric materials of the MXene family. *Phys Chem Chem Phys* 2014;**16**: 7841−9.

148. Kim H, Anasori B, Gogotsi Y, Alshareef HN. Thermoelectric properties of two-dimensional molybdenum-based MXenes. *Chem Mater* 2017;**29**:6472−9.

149. Sarikurt S, Çakır D, Keçeli M, Sevik C. The influence of surface functionalization on thermal transport and thermoelectric properties of MXene monolayers. *Nanoscale* 2018; **10**:8859−68.

150. Zhu J, Ha E, Zhao G, Zhou Y, Huang D, Yue G, Hu L, Sun N, Wang Y, Lee LYS. Recent advance in MXenes: a promising 2D material for catalysis, sensor and chemical adsorption. *Coord Chem Rev* 2017;**352**:306−27.

151. Tariq A, Iqbal MA, Ali SI, Iqbal MZ, Akinwande D, Rizwan S. Ti_3C_2-MXene/Bismuth ferrite nanohybrids for efficient degradation of organic dye and colorless pollutant. *ChemRxiv* 2019:1−33.

152. Sundaram A, Ponraj JS, Wang C, Peng WK, Manavalan RK, Dhanabalan SC, et al. Engineering of 2D transition metal carbides and nitrides MXenes for cancer therapeutics and diagnostics. *J Mater Chem B* 2020;**8**:4990−5013.
153. Sinha A, Zhao H, Huang Y, Lu X, Chen J, Jain R. MXene: an emerging material for sensing and biosensing. *Trends Anal Chem* 2018;**105**:424−35.
154. Rafieerad A, Yan W, Sequiera GL, Sareen N, Abu-El-Rub E, Moudgil M, Dhingra S. Application of Ti_3C_2 MXene quantum dots for immunomodulation and regenerative medicine. *Adv Healthc Mater* 2019;**8**:1900569.
155. Noor Q, Zahra SA, Serna MI, Abuoudah CK, Iqbal MZ, Akinwande D, et al. Silicon carbide-assisted co-existence of magnetic phases in well-optimized Ti_3SiC_2-etched MXene. *Ceram Int* 2020;**46**(17):27419−25.
156. Iqbal M, Fatheema J, Noor Q, Rani M, Mumtaz M, Zheng R-K, Khan SA, Rizwan S. Co-existence of magnetic phases in two-dimensional MXene. *Mater Toady Chem* 2020;**16**: 100271.
157. George SM, Kandasubramanian B. Advancements in MXene-Polymer composites for various biomedical applications. *Ceram Int* 2020;**46**:8522−35.
158. Ding G, Zeng K, Zhou K, Li Z, Zhou Y, Zhai Y, Zhou L, Chen X, Han S-T. Configurable multi-state non-volatile memory behaviors in Ti 3 C 2 nanosheets. *Nanoscale* 2019;**11**: 7102−10.

Vanadium oxide–carbon composites and their energy storage applications

Zeeshan Tariq[1,2,3], Sajid Ur Rehman[1,2], Xiaoming Zhang[1,2] and Chuanbo Li[1,2]
[1]School of Science, Minzu University of China, Beijing, PR China; [2]Optoelectronics Research Center, Minzu University of China, Beijing, PR China; [3]State Key Laboratory on Integrated Optoelectronics, Institute of Semiconductors, Chinese Academy of Sciences, Beijing, PR China

11.1 Introduction

Technological advancements and the invention of new portable electronic devices (laptops, camcorders, and cell phones) are fueling energy demand.[1] In addition, electric and hybrid-electric vehicles use energy storage and conversion devices, thus further boosting energy requirements.[2] One of the most important challenges in recent decades has been to overcome mounting energy demand. Mostly, fossil fuels (oil, coal, natural gas) are used to generate energy to meet the needs of humankind.[3] These fossil fuels have limited reserves, and with continued use, they will one day become depleted. In addition, these fossil fuel resources cause substantial environmental damage by releasing toxic gases into the atmosphere.[4–6] Alternative renewable energy resources that are environmentally friendly have received considerable attention in recent decades.[2] In this scenario, batteries,[7–9] supercapacitors,[6,10] and fuel cells[11,12] serve as efficient and capable renewable energy storage systems.[13] There is an urgent need for material that can be used in renewable energy systems for better, more efficient energy conversion and storage.

In recent decades, vanadium oxides (VOs) have received considerable attention as energy storage and conversion material. VOs exhibit high energy density, high capacity, and low cost.[14–16] V_2O_5 is the most stable oxide among VOs, such as V_2O_3, V_3O_5, V_2O_5, VO_2, and VO. In addition, it owns a high state of oxidation for use as an oxidizing agent and as amphoteric oxide.[17,18] The low electric conductivity of VOs usually hinders their applications as electrode material. These VOs are combined with carbonaceous materials like carbon fiber, graphene, and carbon nanotubes because of their better processing ability, low electrical resistivity, high surface area, and low cyclability.[19–22]

11.2 Vanadium oxide–carbon composite applications

VO/carbon composites are used as electrode material in metal-ion batteries and[23] supercapacitors[19] as well as in electrocatalysis.[24] Fig. 11.1 is a pictorial representation of the potential applications of vanadium oxide/carbon composites.

Metal Oxide-Carbon Hybrid Materials. https://doi.org/10.1016/B978-0-12-822694-0.00014-4

11.2.1 Application in metal-ion batteries

Nowadays, rechargeable batteries are receiving considerable attention because of their safety, low cost, and high capacity. Rechargeable batteries are vital for electric/hybrid vehicles and mobile computing devices that consume high amounts of electric energy. Using metals oxidized to multivalent cations as cathode materials is an auspicious strategy.[23]

Ihsan et al.[25] synthesized V_2O_5/mesoporous carbon (mc) composites by sol method at room temperature and tested them for electrochemical performance. A TEM image of V_2O_5/mc is displayed in Fig. 11.2A. Fig. 11.2B and C depicts scanning transmission electron microscope images for V_2O_5/mc without bright field (BF) sintering and annular dark field. The pore size distribution is shown in Fig. 11.2D. The pore size of mc is around 3 nm and decreases to 2—3 nm when mixed with V_2O_5. Rate capabilities, cyclic performance, and charge/discharge profile are displayed in Fig. 11.3A—C, respectively. Rate capabilities and cyclic performance of V_2O_5/mc are measured against V_2O_5 nanoparticle (np) and V_2O_5 composite with mc by simple mixing (sm). Cyclic performance of V_2O_5/mc composite is improved over V_2O_5 np and V_2O_5/mc sm at a 500 mA g^{-1} current density. V_2O_5/mc displays a reversible capacity of 198 mAh g^{-1} after 100 cycles. Nyquist plots are displayed in Fig. 11.3D, and the

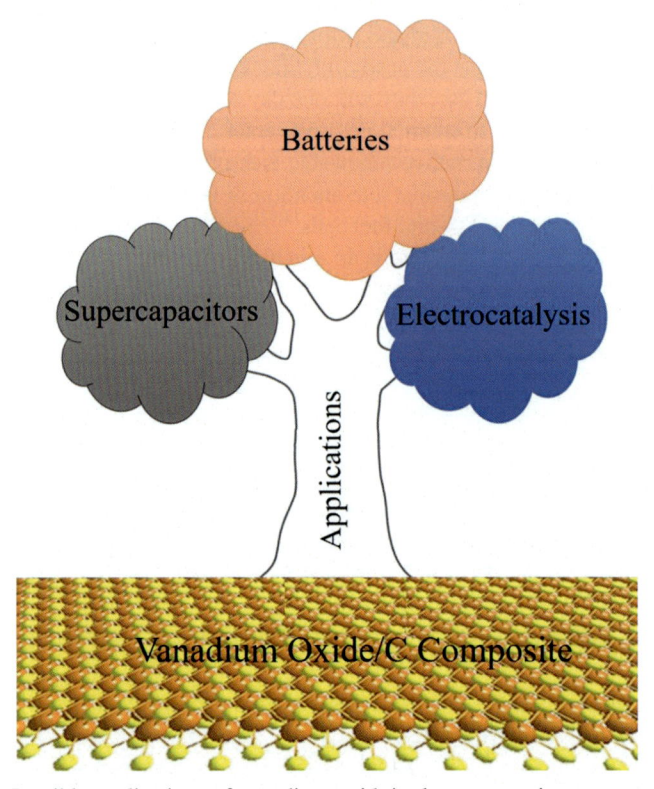

Figure 11.1 Possible applications of vanadium oxide/carbon composites.

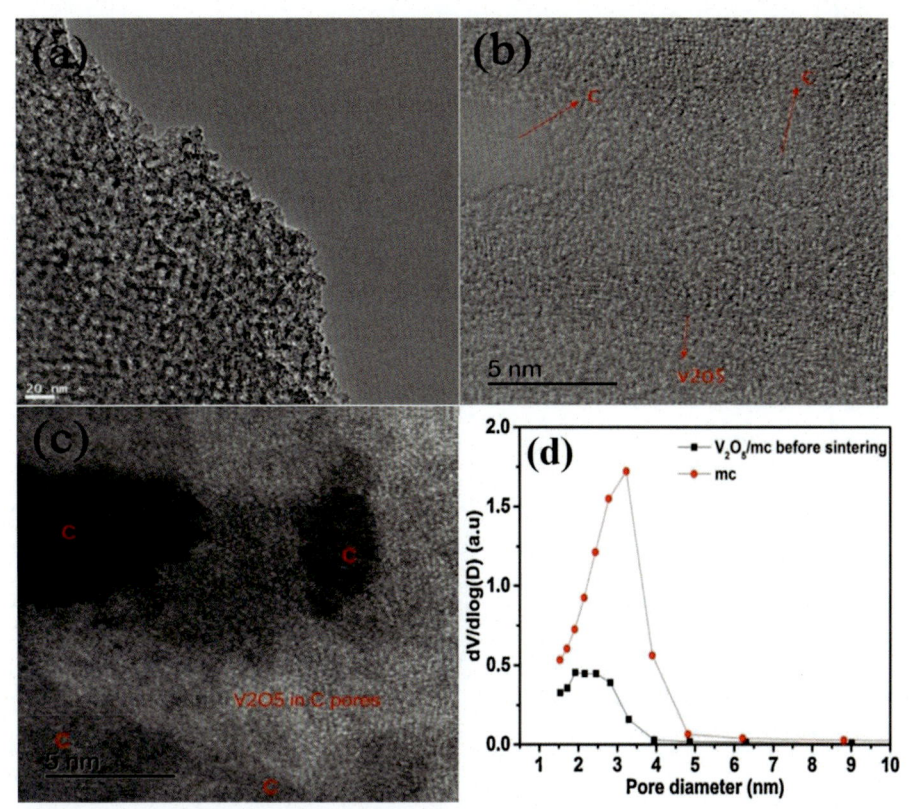

Figure 11.2 Depicting (A) TEM image of mc (B) STEM bright-field mapping image of V_2O_5/mc (C) STEM annular dark-field mapping image of V_2O_5/mc and (D) distribution of pore size for mesoporous carbon (mc) and before sintering V_2O_5/mc.
Reprinted with permission from Ref. 25. Copyright (2015), Elsevier.

charge transfer resistance (R_{ct}) is better for V_2O_5/mc (397 Ω cm^{-2}) than for V_2O_5-np (604 Ω cm^{-2}), which might be due to the mesoporous structure of carbon, which enables fast transport channels for Li$^+$.

De Juan-Corpuz et al.[26] grew V_2O_5 on carbon cloth (CC) using a single-step hydrothermal method and achieved nanofiber morphology. The electrochemical performance of the synthesized V_2O_5/CC composite was tested for a binder-free aqueous zinc battery. Scanning electron microscopy (SEM), magnified SEM, and transmission electron microscopy (TEM) are displayed in Fig. 11.4A—C, respectively.

Galvanometric charge-discharge performance of synthesized nanofibers of V_2O_5 on carbon cloth is tested in the presence of two electrolytes, i.e., 3 M zinc sulfate (ZnSO$_4$) and 3 M zinc trifluoromethanesulfonate Zn (OTf)$_2$ at 100 mA g^{-1} current density. It is observed that the battery's performance is not good with the ZnSO$_4$ electrolyte, which might be associated with the dissolution of the V_2O_5/CC composite in ZnSO4. Capacity increased rapidly (211 mAh g^{-1}) to the 20th cycle, capacity then dropped sharply, and the battery failed. However, the battery with Zn (OTf)$_2$ electrolyte displayed good

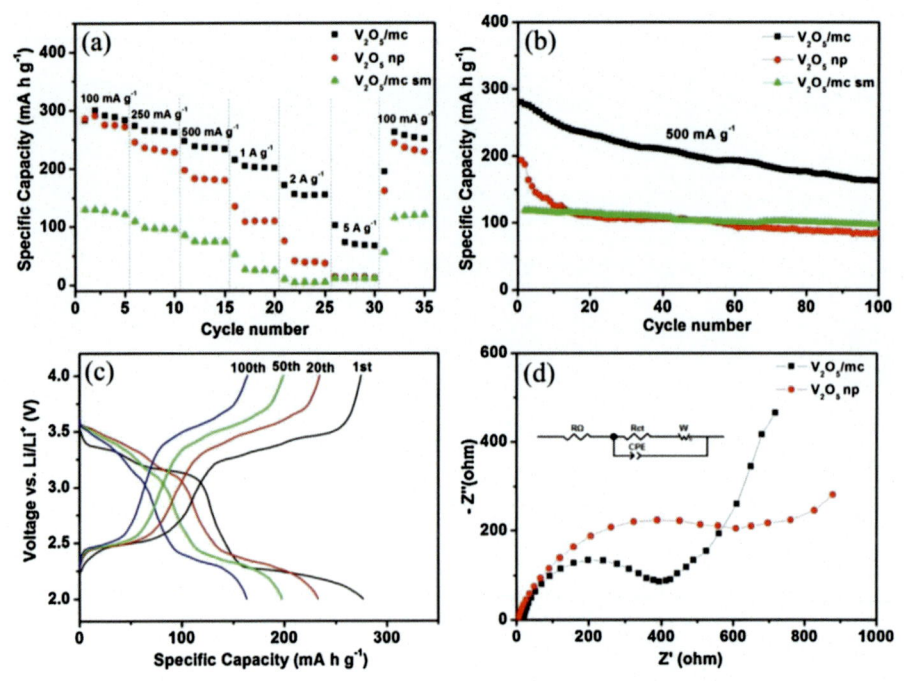

Figure 11.3 Displaying (A) rate capabilities at different current densities for V_2O_5/mc, V_2O_5-np and V_2O_5/mc sm (B) cyclic performance at 500 mA g^{-1} of V_2O_5/mc, V_2O_5-np and V_2O_5/mc sm (C) charge/discharge profile at 100 mA g^{-1} for V_2O_5/mc and (D) Nyquist plots and equivalent circuit (inset) for V_2O_5/mc and V_2O_5-np.
Reprinted with permission from Ref. 25. Copyright (2015), Elsevier.

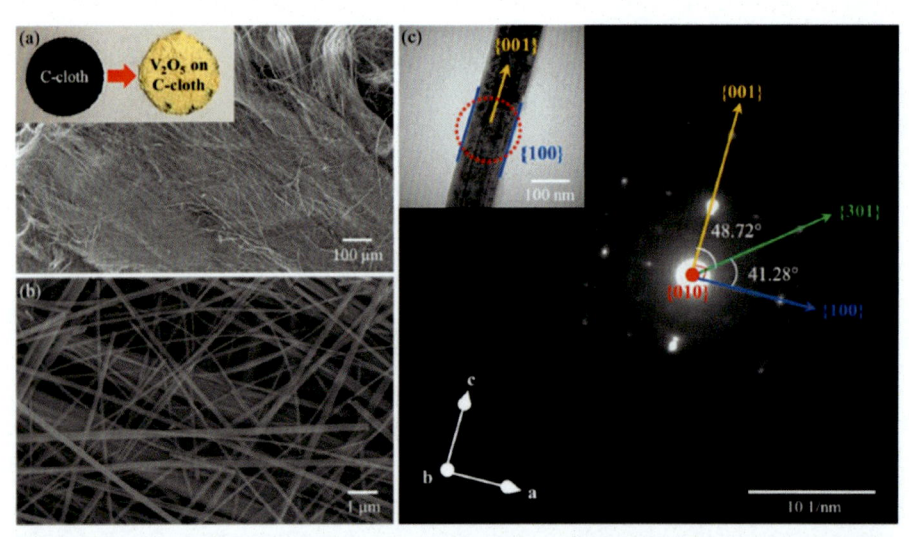

Figure 11.4 Displaying (A) SEM image of synthesized material (inset displaying bare carbon cloth and carbon cloth coated with V_2O_5 nanofibers (B) magnified SEM of image (A) and (C) SAED pattern with TEM image (inset) of selected part for SAED.
Reprinted with permission of Ref. 26. Copyright (2019), MDPI.

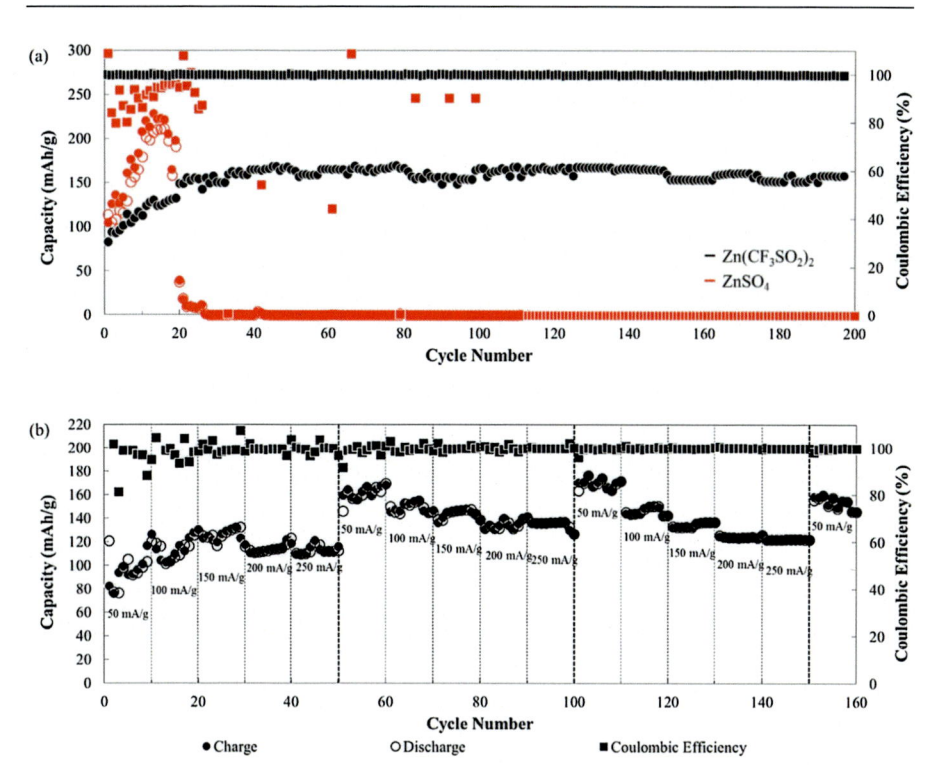

Figure 11.5 (A) Cyclic performance of V_2O_5/CC composite in 3 M $ZnSO_4$ and 3 M Zn $(OTf)_2$ electrolyte (B) Rate capabilities at various current densities with three repetitions. Reprinted with permission of Ref. 26. Copyright (2019), MDPI.

results, and capacity showed an increasing trend to the 30th cycle and then stabilized at around 155 mAh g^{-1} (Fig. 11.5A).

Rate capability is depicted in Fig. 11.5B at different current densities ranging from 50 mA g^{-1} to 250 mA g^{-1} with three repetitions in 3 M Zn $(OTf)_2$ electrolytes. In the first set of rate capability, capacity continues to increase regardless of increasing current densities; this test looked erroneous. The second and third sets of rate capability tests looked more stable and thus used for further analysis. The capacity retention in these two tests is more than 90%, which indicates stable material for negative electrodes even at different applied current densities.

Chiku et al.[23] synthesized amorphous V_2O_5/C composite by solution method in an ice bath using Ketjenblack as the carbon source. SEM and TEM carried morphological characterization. Electrochemical performance of as-synthesized VO composite as the positive electrode was investigated for an aluminum ion battery. Mo is used as a current collector, and Al is used as a counter electrode. Fig. 11.6A and B display KB SEM images and pristine V_2O_5/C, respectively. KB has a porous structure from accumulating spherical particles with diameters near 50 nm (Fig. 11.6A), and Fig. 11.6B displays KB particles uniformly covered by V_2O_5. TEM images are displayed in Fig. 11.6C and D, and lattice fringes of KB can be seen in (d). V_2O_5 lattice fringes are not clear in TEM, implying that V_2O_5 particles are amorphous, not crystalline.

Figure 11.6 (A) SEM image of KB (B) SEM image of pristine V_2O_5/C (B, D) TEM images of V_2O_5/C.
Reprinted with permission of Ref. 23. Copyright (2015), American Chemical Society.

Charge/discharge curves, capacities, cyclic performance, and charge/discharge curves of KB electrode are displayed in Fig. 11.7A—D, respectively. The initial discharge capacity is 150 mAh g^{-1}, but after the 15th cycle, the capacity falls below 100 mAh g^{-1} for the C/20 charge/discharge rate. Charge and discharge capacities are almost the same, suggesting that Al coin cells have higher coulombic efficiency. Cyclic performance at various current densities is displayed in Fig. 11.7C. When discharge current is decreased to C/40, the capacity for initial discharge is 200 mAh $g-1$ and when it is C/10, the capacity is below 100 mAh g^{-1}. Amorphous V_2O_5/C is a good material for the anode in aluminum ion batteries.

Wang et al.[27] synthesized VO/carbon (VOx@C) hierarchical porous composites by refluxing and annealing at different temperatures. There are two phases of VOs present in the synthesized sample, i.e., V_2O_3 and VO_2. Morphological characterization was carried out using SEM, TEM, and HRTEM EDX mapping to examine the distribution of different elements. The synthesized carbon composite material's electrochemical performance was tested for high-performance lithium-ion batteries. The SEM image (Fig. 11.8A) of the sample VO_x-500 comprises hierarchical microspheres formed by quadrangular nanoprisms. Fig. 11.8B is a magnified SEM image revealing that each nanoprism comprises nanoparticles interconnected into a porous structure. Seemingly, it smoothens electrolyte infusion and wetting on the surface of the electrode, Li^+ circulation with shortened paths, and enhances Li^+ storage capacity.

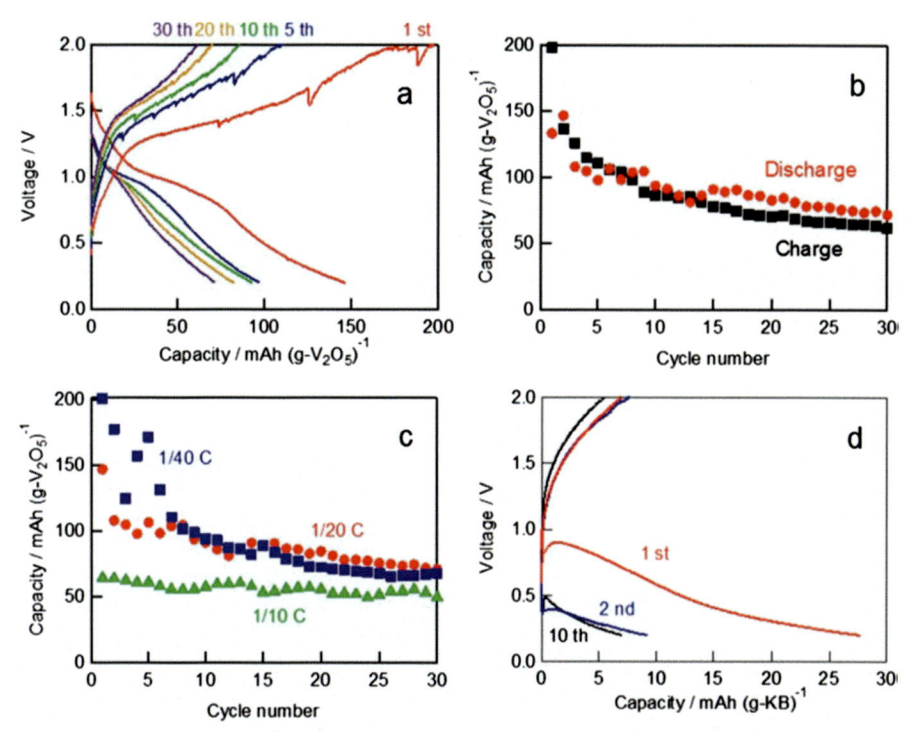

Figure 11.7 (A) Charge/discharge curves for selected cycles (B) Charge/discharge capacities at C/20 (C) Cyclic performance at various current densities like C/10, C/20, C/40 (D) charge/discharge curves electrode fabricated with KB.
Reprinted with permission of Ref. 23. Copyright (2015), American Chemical Society.

EDX mapping of C, V, and O is illustrated in Fig. 11.8D—F, revealing that all elements are uniformly distributed throughout the sample.

TEM and HRTEM images of the VO_x-500 sample are depicted in Fig. 11.9. Micrographs of TEM (Fig. 11.9A) confirms the spherical nature of the particles of sample VO_x-500. Some nanoprism hollow structures are observed in Fig. 11.9B and C. HRTEM results are shown in Fig. 11.9D, which reveals the interconnected nanoparticles of porous nanoprisms. Lattice fringes with distances of 0.37 and 0.32 nm correspond to (012) and (110) planes of V_2O_3 and VO_2, respectively.

The CV profile, charge/discharge curves, rate capability, and cyclic performance of a VO_x-500 sample are depicted in Fig. 11.10A—D, respectively. Various pairs of redox peaks with multistep redox reactions for Li^+ insertion/extraction can be seen in the CV profile of the electrode containing V_2O_3 and VO_2. Galvanostatic charge/discharge curves at different current rates are displayed in Fig. 11.10B. Higher reversible capacities at 100 mA g^{-1}, 200 mA g^{-1}, and 500 mA g^{-1} are 503 mAh g^{-1}, 487 mAh g^{-1}, and 453 mAh g^{-1}, respectively, and displayed in Fig. 11.10C. The cyclic performance of electrodes containing VO_x-400, VO_x-500, VO_x-600, and VO_x-700 samples at a

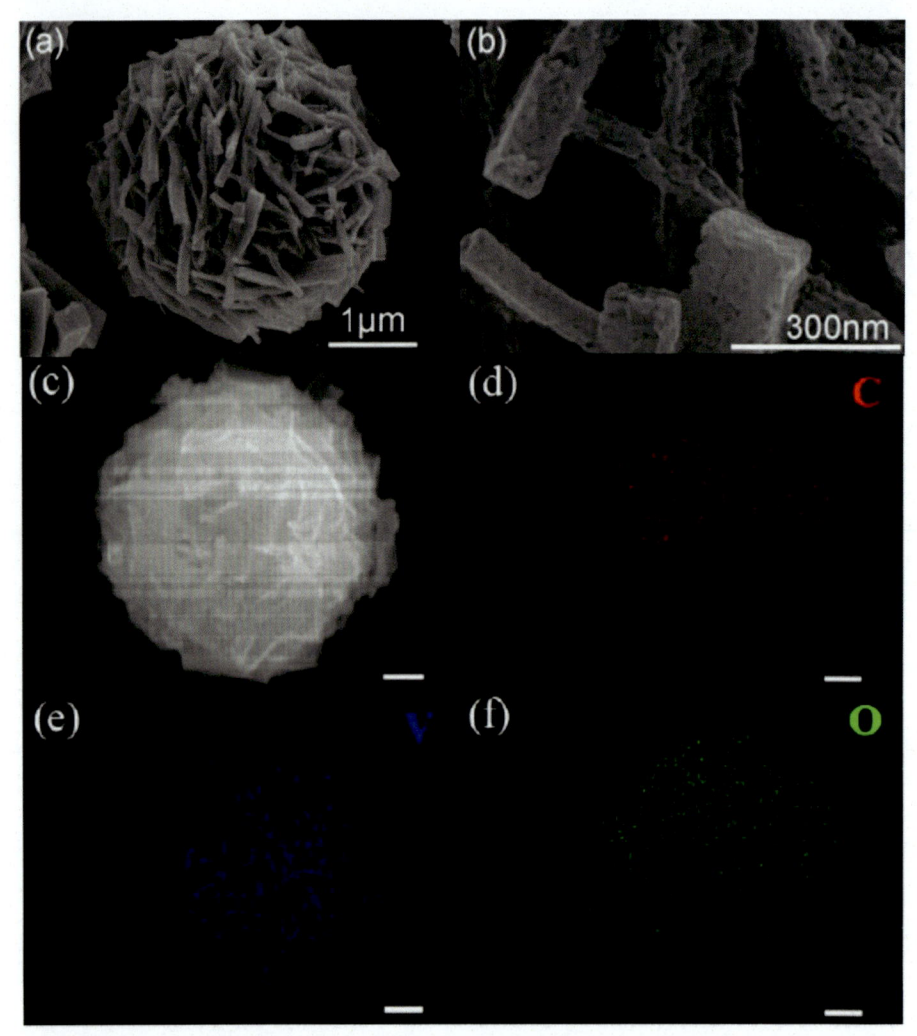

Figure 11.8 (A and C) SEM image (B) Magnified SEM image (D–F) EDX mapping of C, V, and O in VO_x-500 sample. All scales for (C–F) are 0.5 μm.
Reprinted with permission from Ref. 27. Copyright (2017), American Chemical Society.

current density of 100 mA g^{-1} is presented in Fig. 11.10D. The VO_x-500 sample exhibits a maximum capacity of 569 mAh g^{-1}, and its capacity performance is superior to the other samples. The VO_x-400, VO_x-600, and VO_x-700 samples display deprived capacity of Li^+ storage.

Among different morphologies, hierarchical microspheres of VO/carbon composites exhibit good cyclic performance for lithium-ion batteries.

11.2.2 Application in supercapacitors

Supercapacitors are also ascribed as electrochemical capacitors and are used as devices to store electrochemical energy. Supercapacitors exhibit high power density and long

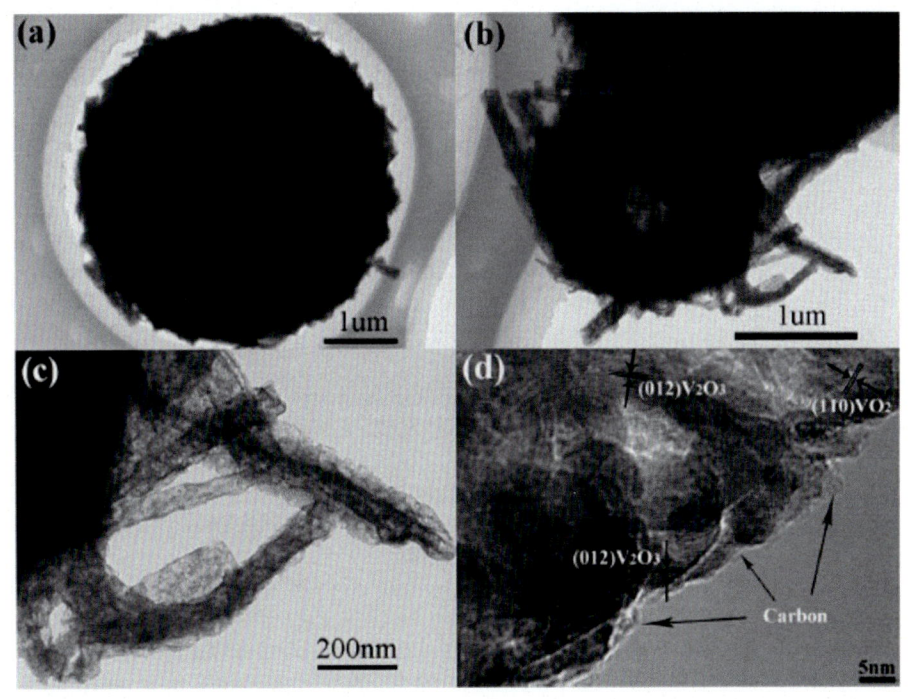

Figure 11.9 (A−C) TEM images (D) HRTEM micrograph for VO_x-500 sample. Reprinted with permission from Ref. 27. Copyright (2017), American Chemical Society.

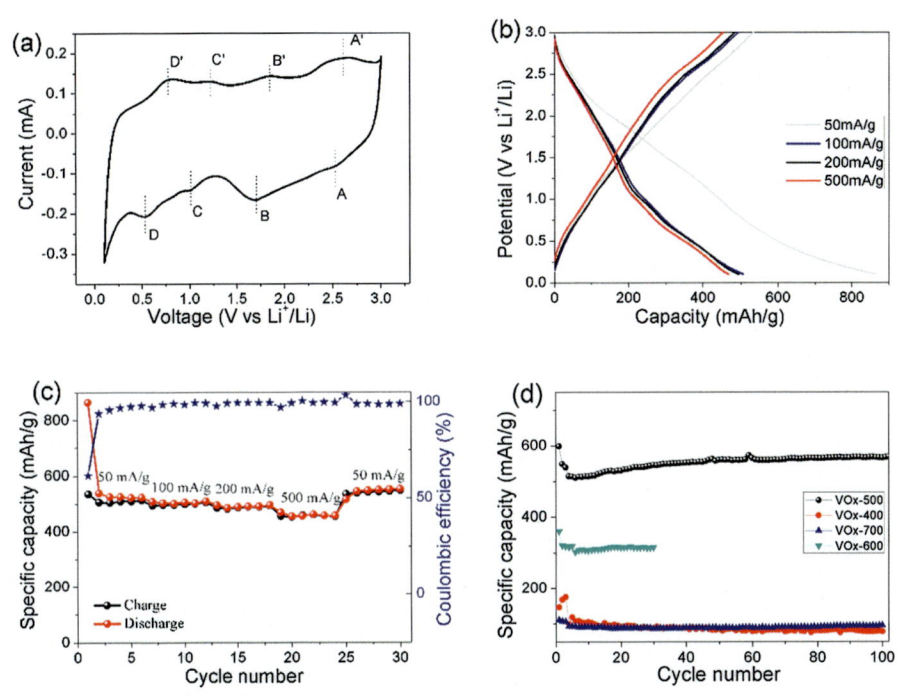

Figure 11.10 (A) CV profile of VO_x-500 (B) Charge/discharge curves of VO_x-500 (C) Rate capabilities at different current densities for VO_x-500 (D) Cyclic performance of VO_x-400, VO_x-500, VO_x-600, and VO_x-700 samples at a current density of 100 mA g^{-1}. Reprinted with permission from Ref. 27. Copyright (2017), American Chemical Society.

cyclic life compared with batteries.[28] Transition metal oxides (TMOs) revealed electrochemically reversible and fast faradaic redox reactions for charge storage in supercapacitors, which results in high capacitance. Among different TMOs, VOs have received considerable attention because they exhibit a wide range of oxidation states that result in a wide range of redox reactions for supercapacitor operations. In addition, VOs are cheap and abundant.[29−31] Supercapacitors have broad applications ranging from hybrid vehicles, electric vehicles, and defense and aerospace.[31]

Chen et al.[32] hydrothermally synthesized V_2O_5/CNT composites by mixing aqueous VO with hydrophilic CNTs. They tested the synthesized composites as anode material for asymmetric supercapacitors. SEM, TEM, and HRTEM images are displayed in Fig. 11.11. SEM images show the fibrous morphology of the synthesized V_2O_5/CNT composites.

Anode capacitance versus maximum energy density at different cathode capacitance (100, 120, 150 Fg^{-1}) of V_2O_5/CNT and cyclic voltammograms at a scan rate of 2 mV s^{-1} for (a) V_2O_5 nanowires (b) V_2O_5/CNT composites, and (c) CNTs are displayed in Fig. 11.12A and B, respectively.

A supercapacitor with anode material of V_2O_5/CNT composite at a power density of 210 W kg^{-1} exhibits an energy density of 40 Wh Kg^{-1} and a maximum power density of 20 kW kg^{-1}. The overall performance of asymmetric supercapacitors is better than electric double-layer capacitors. Fig. 11.12B displays cyclic voltammograms comparison of V_2O_5 nanowires, V_2O_5/CNT composite, and CNTs. Electrode composed of nanowires depicts two pairs of symmetric, broad, and well-separated redox peaks, revealing slow insertion/disinsertion kinetics of Li^+.[33] Nanocomposite displayed well-defined two pairs of redox peaks. This improved performance of electrode consists of nanocomposite might be associated with the fibrous structure.

Zhou et al.[34] fabricated nanocomposites of V_2O_5/Polyindole (V_2O_5/PIn) decorated on activated carbon cloth (ACC). Performance of synthesized V_2O_5/PIn@ACC

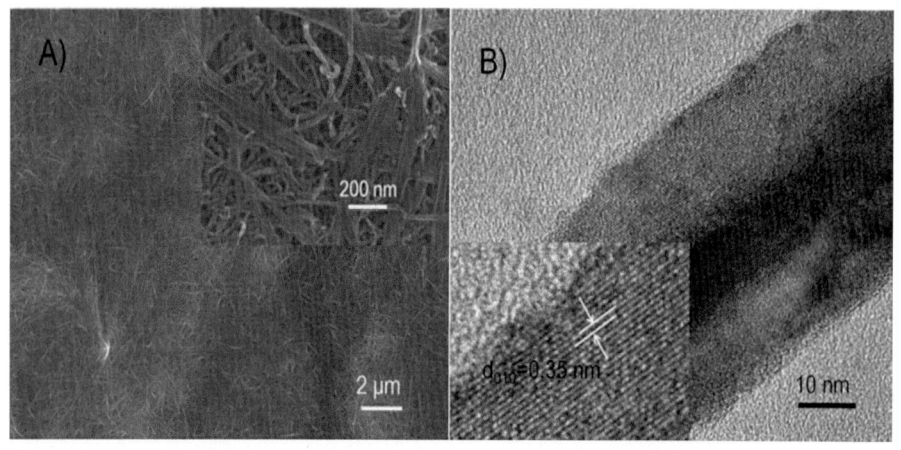

Figure 11.11 (A) SEM and magnified SEM (inset) images of V_2O_5/CNT (B) TEM and HRTEM images of V_2O_5/CNT composite.
Reprinted with permission from Ref. 32. Copyright (2011), John Wiley & Sons, Inc.

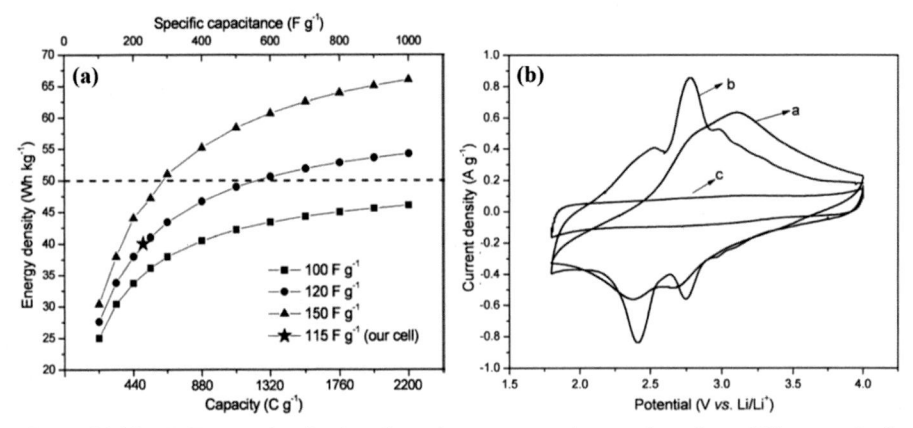

Figure 11.12 (A) Energy density (maximum) versus capacitance of anode at different cathode capacitance (B) Cyclic voltammograms at 2 mV s^{-1} scan rate for (A) V_2O_5 nanowires (B) V_2O_5/CNT composite and (C) CNTs.
Reprinted with permission from Ref. 32. Copyright (2011), John Wiley & Sons, Inc.

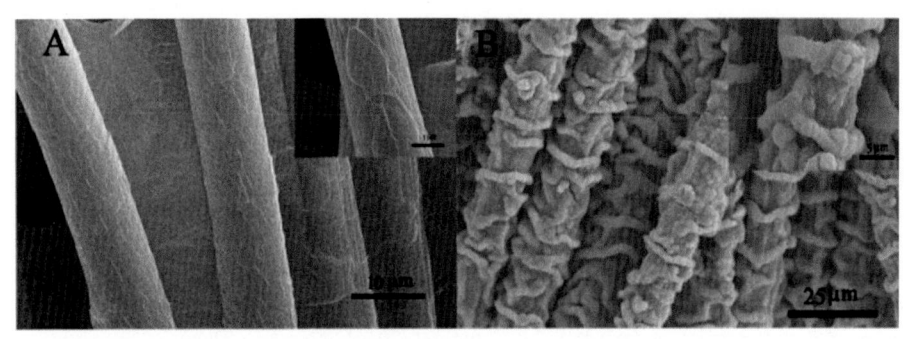

Figure 11.13 Depicting SEM images (A) rGO@ACC (b) V_2O_5/PIn@ACC. Insets in (A) and (B) are magnified images of rGO@ACC and V_2O_5/PIn@ACC, respectively.
Reprinted with permission from Ref. 34. Copyright (2016), American Chemical Society.

composites is tested for asymmetric supercapacitor. Morphology of fabricated carbon composites is displayed in Fig. 11.13. SEM images show that V_2O_5/PIn decorated on ACC displays a bamboo-like morphology, whereas reduced graphene oxide (rGO) on ACC (rGO@ACC) covers the carbon fibers with thin layers. V_2O_5/PIn@ACC is used as an anode, and rGO@ACC is used as a cathode in the supercapacitor.

The CV curve (Fig. 11.14A) shows that the rGO@ACC electrode has a quasi-rectangular shape lacking noticeable redox peaks at scan rates ranging from 5 mV/s to 50 mV/s, indicating admirable double-layer charge storage performance. V_2O_5/PIn@ACC exhibits distorted rectangular shapes at scan rates with a range of 5 to 50 mV/s because of the pseudocapacitive performance of PIn and V_2O_5 (Fig. 11.14B). At 1 A/g, V_2O_5/PIn@ACC exhibit a specific capacitance of 535.3 F/g, much higher than V_2O_5@ACC (304.7 F/g) at alike measuring conditions

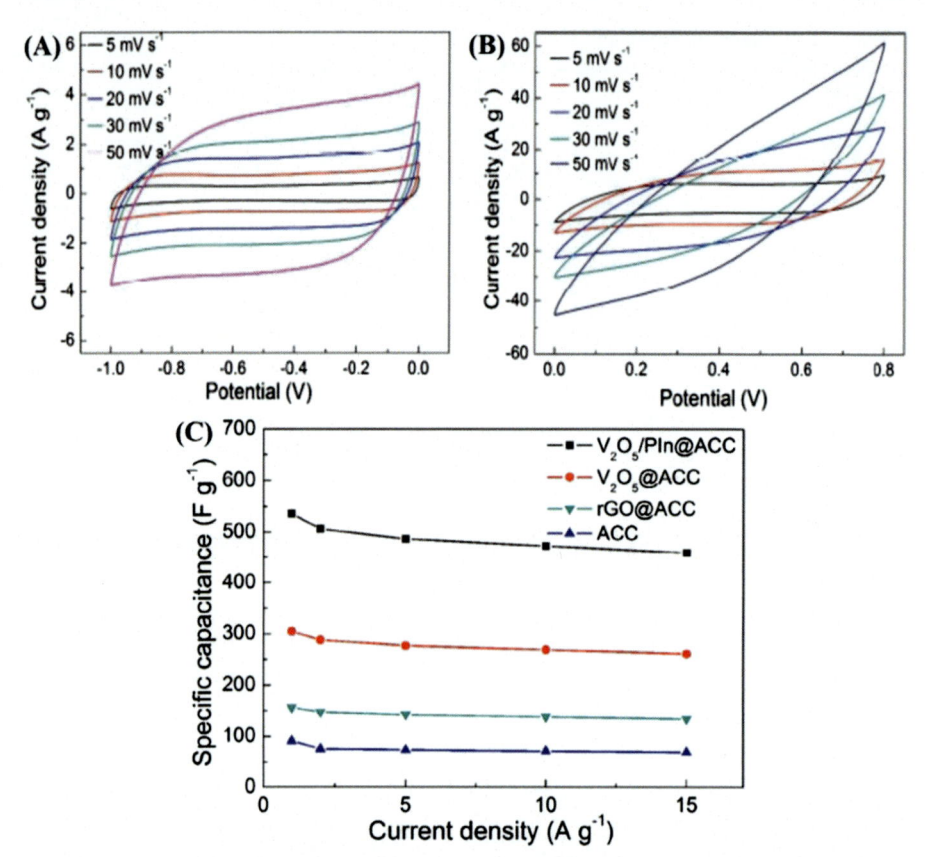

Figure 11.14 CV curves for (A) rGO@ACC (B) V_2O_5/PIn@ACC (C) Specific capacitance of V_2O_5/PIn@ACC, V_2O_5@ACC, rGO@ACC, and ACC, respectively.
Reprinted with permission from Ref. 34. Copyright (2016), American Chemical Society.

(Fig. 11.14C). This is probably due to improved electronic conductivity because of PIn and the amalgamation with V_2O_5 for faster transport of ions. These results are better than reported in the literature, such as rGO/V_2O_5 xerogels (195.4 F/g, 1 A/g)[35] and V_2O_5 nanosheets (253 F/g, 1 A/g).[36]

Perera et al.[19] fabricated V_2O_5 nanowires (VNWs) and Li^+ doped VNWs ((Li) VNWs) by hydrothermal method. Carbon composite of VNWs and (Li)VNWs were prepared by mixing carbon nanotube (CNT) and with vanadium nanowire, and then bath sonicated was carried out for 1 h and probe sonicated for 2 h. The as-synthesized samples are named as CNT-VNWs and CNT-(Li)VNWs. The CNT-VNWs composites are fabricated in different v/v ratios such as 10:2 mL, 10:4 mL, 6:6 mL, 4:10 mL, and 2:10 mL. Composites with a 1:1 ratio display better performance than other composites. These composites are used to make binder-free flexible paper electrodes and then studied their performance in supercapacitors. TEM and SEM (Fig. 11.15) are used to explore the morphology of the synthesized samples. The diameter of the synthesized VNW is ~24 nm (Fig. 11.15C).

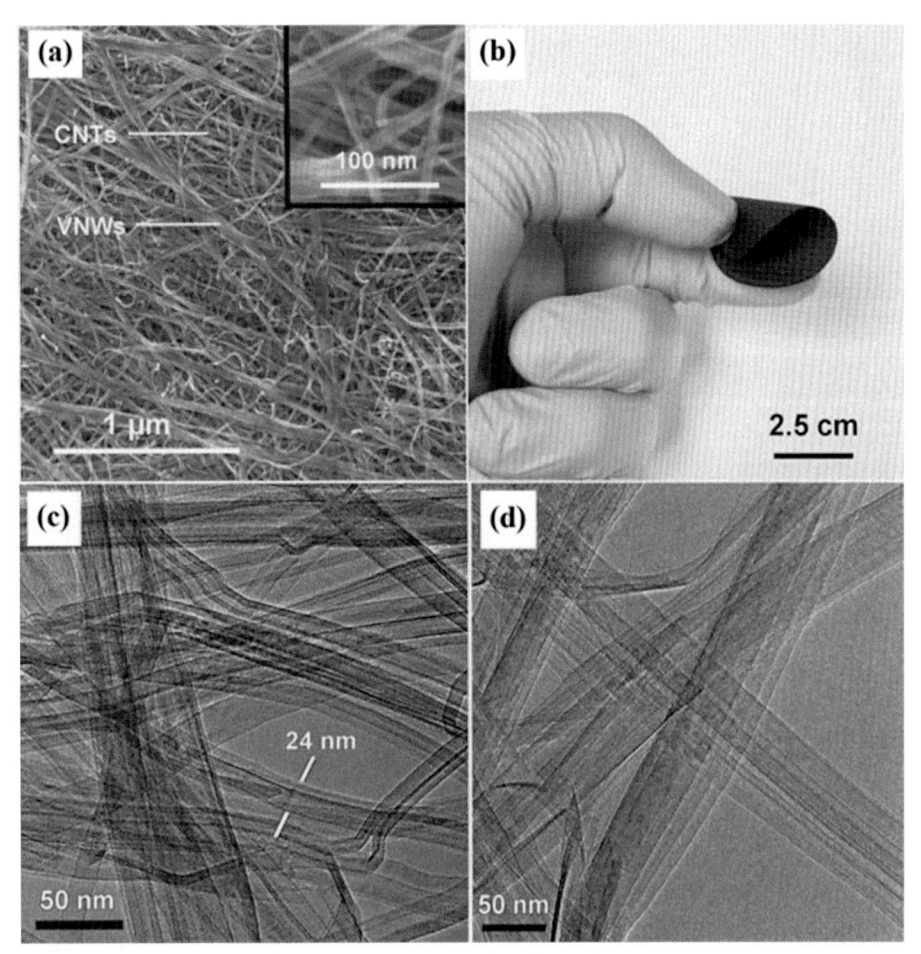

Figure 11.15 Displaying (A) SEM image of a flexible paper electrode of CNT-VNWs (B) Digital image of a freestanding electrode of CNT-VNWs (C—D) TEM images of CNT-VNWs displaying the diameter ~ 24 nm.
Reprinted with permission from Ref. 19. Copyright (2011), John Wiley & Sons, Inc.

The electrochemical performance was investigated in voltages ranging from -0.5 to 2.5 V, and electrolyte LiTFSI in acetonitrile was used. Fig. 11.16A illustrating quasi rectangular shape for cyclic voltammograms of CNT-VNW$_{1:1}$ demonstrating ideal capacitive behavior. It can be seen that the electrode with the least amount of CNTs exhibits the least current output (Fig. 11.16B). The capacitive property of the electrode composed of CNT-VNW$_{1:1}$ is displayed in Fig. 11.16C at various sweep rates—for instance, 10 mVs^{-1}, 25 mVs^{-1}, 50 mVs^{-1}, and 75 mVs^{-1}. A comparison of cyclic voltammogram between CNT-VNW$_{1:1}$ and CNT-(Li)VNW$_{1:1}$ is given in Fig. 11.16D. Eq. (11.1) [19] is employed to evaluate specific capacitance (C_{sp}) and found to be 48.5 Fg^{-1} for CNT-VNW$_{1:1}$ and 57.3 Fg^{-1} for CNT-(Li)VNW$_{1:1}$. This high

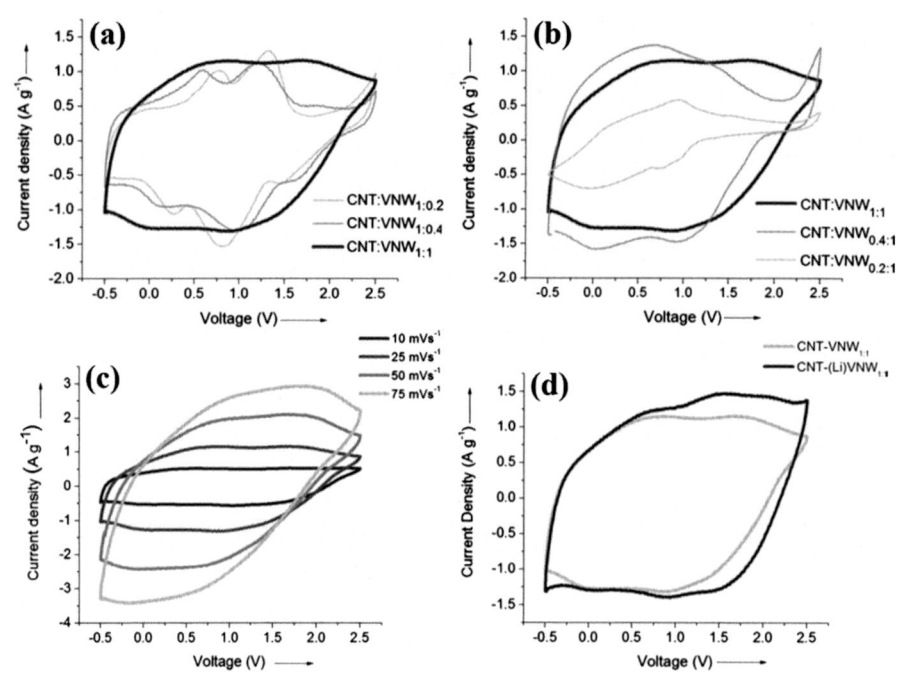

Figure 11.16 (A–B) CV curves for electrodes with different compositions (C) Cyclic voltammogram for CNT-VNW$_{1:1}$ at different scan rates (D) Comparison between CNT-VNW$_{1:1}$ CNT-(Li)VNW$_{1:1}$ electrode in a voltage range of −0.5 to 2.5 V. Reprinted with permission from Ref. 19. Copyright (2011), John Wiley & Sons, Inc.

capacitance might be because of the higher faradaic activity of Li^+ in nanowires. Power density and Energy of CNT-(Li)VNW$_{1:1}$ is also high:

$$C_{sp} = (I \times \Delta t)/(m \times \Delta V) \tag{11.1}$$

where C_{sp} is specific capacitance, I represents constant current discharge, Δt is discharge time, m is the electrode mass, and ΔV is the voltage drop during discharge.

Wang et al.[37] synthesized 3-D graphene/VO$_2$ nanobelts hydrogel by a facile hydrothermal method using graphene oxide and commercial V$_2$O$_5$ as precursors. Two-dimensional flexible graphene sheets and 1-D VO$_2$ nanobelts self-assembled during architecture formation of graphene/VO$_2$. This interconnected porous structure is formed through hydrogen bonding which also assists in transporting ions and charges in the electrode. Graphene hydrogel was also synthesized without V$_2$O$_5$ for comparison. The formation process for 3-D graphene/VO$_2$ nanobelts hydrogel is depicted in Fig. 11.17. Wang et al. tested the electrochemical performance of as-synthesized composite for supercapacitor in K$_2$SO$_4$ electrolyte.

The surface morphologies of the fabricated hydrogel of 3-D graphene/VO$_2$ are explored by field emission scanning electron microscopy (FESEM). High-magnification FESEM images reveal that the hydrogel of graphene/VO$_2$ has a beltlike

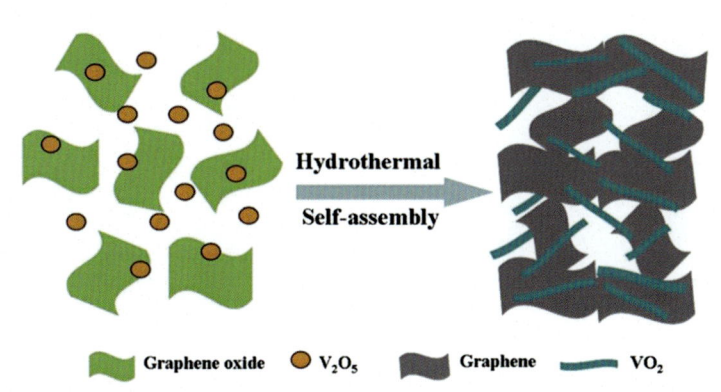

Figure 11.17 Formation process for a hydrogel of 3-D graphene/VO$_2$ nanobelts. Reprinted with permission from Ref. 37. Copyright (2014), Royal Society of Chemistry.

structure forming hierarchical architecture composed of intertwined belts like VO$_2$ and graphene sheets (Fig. 11.18A—B). Irreversible agglomeration was lessened by the interaction between graphene and VO$_2$, thus improving the electrical conductivity of the composite. Fig. 11.18C—D displays TEM images of composites, revealing that VO$_2$ belts are grown directly on highly conductive graphene sheets. The TEM and SAED images are shown in Fig. 11.18E and F, respectively, display the lattice fringes with d-spacing 0.58 nm, which agree with the (200) VO2(B) plane and reveal the single crystalline structure of VO$_2$ nanobelts.

CV curves for a hydrogel of graphene, VO$_2$ nanobelt, and graphene/VO$_2$ composites are shown in Fig. 11.19A at a sweep rate of 5 mVs^{-1} for a −0.6 to 0.6 V potential range. The CV curve for VO$_2$ nanobelts and graphene/VO^2 hydrogels displays two pairs of redox peaks ascribed to electrochemical insertion of the K$^+$ process as follows:

$$VO_2 + xK^+ + xe^- \leftrightarrow K_xVO_2$$

where x represents the inserted K$^+$ mole fraction. The CV curves of graphene/VO$_2$ at various sweep rates, such as at 5 mVs^{-1}, 10 mVs^{-1}, 20 mVs^{-1}, and 50 mVs^{-1}, are shown in Fig. 11.19B. Comparison study of specific capacitance and energy density for a hydrogel of graphene, nanobelts of VO$_2$, and composites of graphene/VO$_2$ are shown in Fig. 11.19C and D, respectively. Specific capacitance is calculated by using Eq. (11.1).

Calculated specific capacitance in two-electrode arrangements for a hydrogel of graphene/VO$_2$ composite is 426 Fg^{-1} at 1 A g^{-1} in potential range −0.6−0.6 V, which is way better than the 191 Fg^{-1} and 243 Fg^{-1} for nanobelts of VO$_2$ and hydrogel of graphene respectively. The findings of Wang et al.[37] show the great potential and importance of graphene composite hydrogels in fabricating devices for energy storage with high energy densities and high power.

The bamboo-like morphology of VO/carbon composite displays very good capacitance compared with the other morphologies discussed in this section. According to Zhou et al.[34] this is probably due to the improved electronic conductivity of PIn and the amalgamation with V$_2$O$_5$ for faster transport of ions.

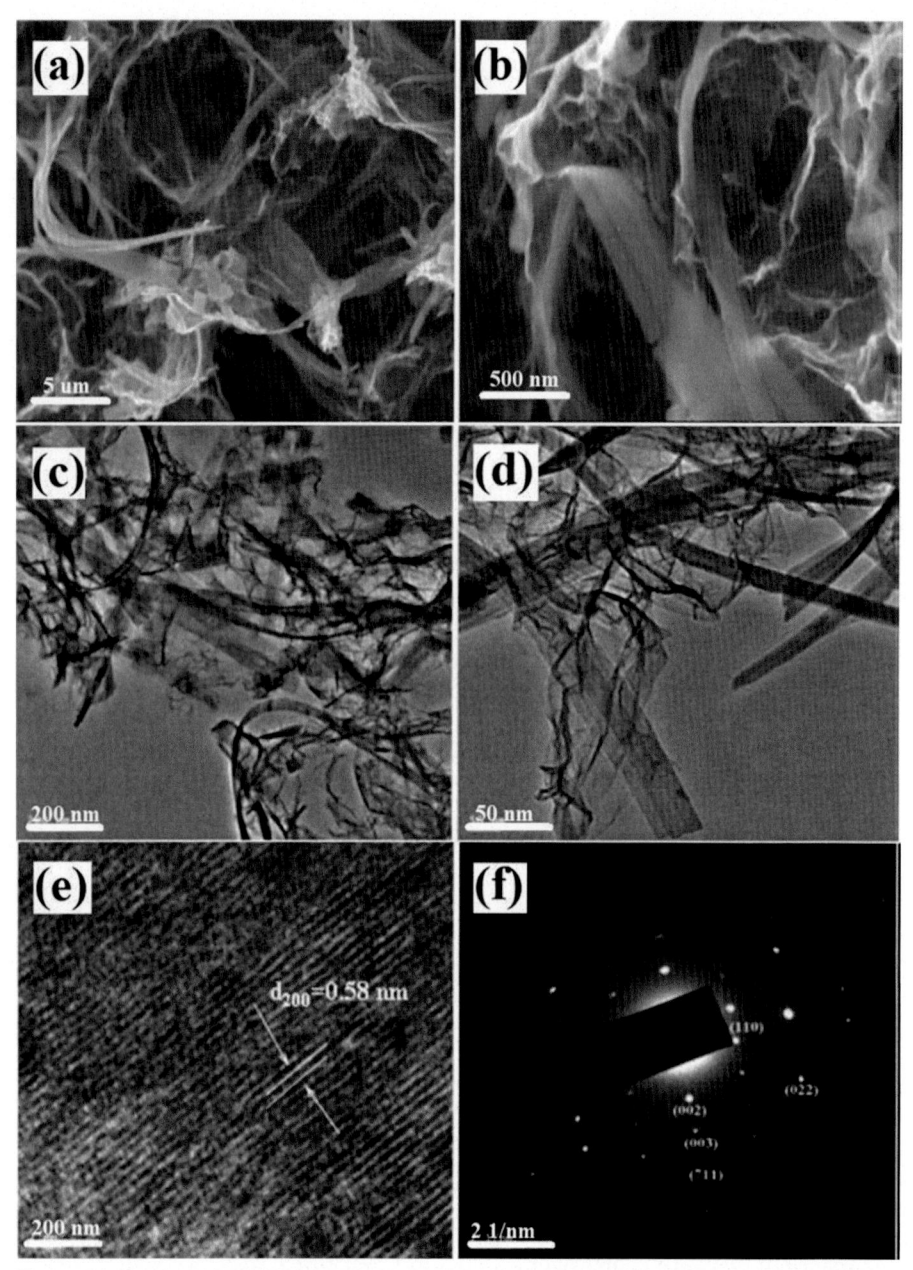

Figure 11.18 (A−B) FESEM images (C−E) TEM images (F) SAED pattern of a hydrogel of 3-D graphene/VO$_2$ nanobelt composites.
Reprinted with permission from Ref. 37. Copyright (2014), Royal Society of Chemistry.

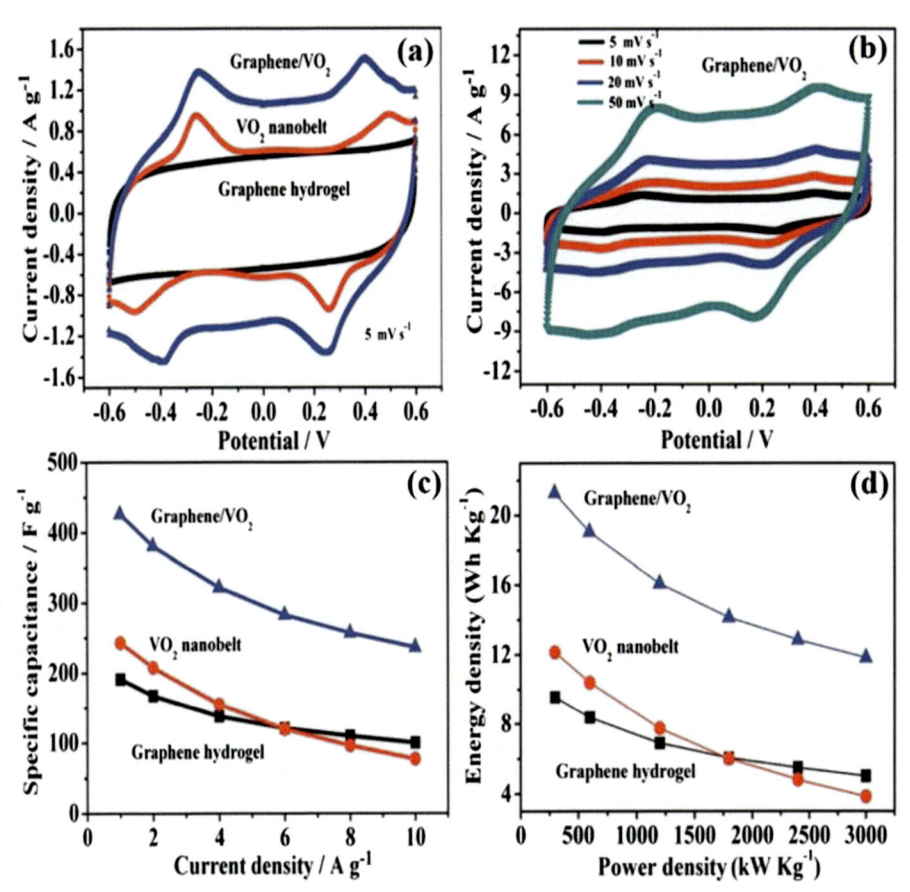

Figure 11.19 (A) CV curves for a hydrogel of graphene, nanobelts of VO$_2$, and composites of graphene/VO$_2$ at a scan rate 5 mVs^{-1} (B) CV curve of graphene/VO$_2$ composite at various scan rates (C) Specific capacitance and (D) Ragone plots of supercapacitors composed of the hydrogel of graphene, nanobelts of VO$_2$ and composites of graphene/VO$_2$.
Reprinted with permission from Ref. 37. Copyright (2014), Royal Society of Chemistry.

11.2.3 Application in electrocatalysis

The search for renewable and clean energy sources is a tough task for the sustainable development of society when concerns are increasing because of pollution, decreasing availability of fossil fuels, and climate change.[38,39] Need for advanced energy systems like fuel cells and water electrolysis is vital for this development. A deep understanding of electrocatalysis's important concepts and principles is required for understanding these next-generation energy devices.[40–42] VOs are a potential candidate to be used in electrocatalysis because they are cheap, abundant, and less toxic, and vanadium provides a wide range of oxidation states which gives a wide range of redox reactions.[29,30]

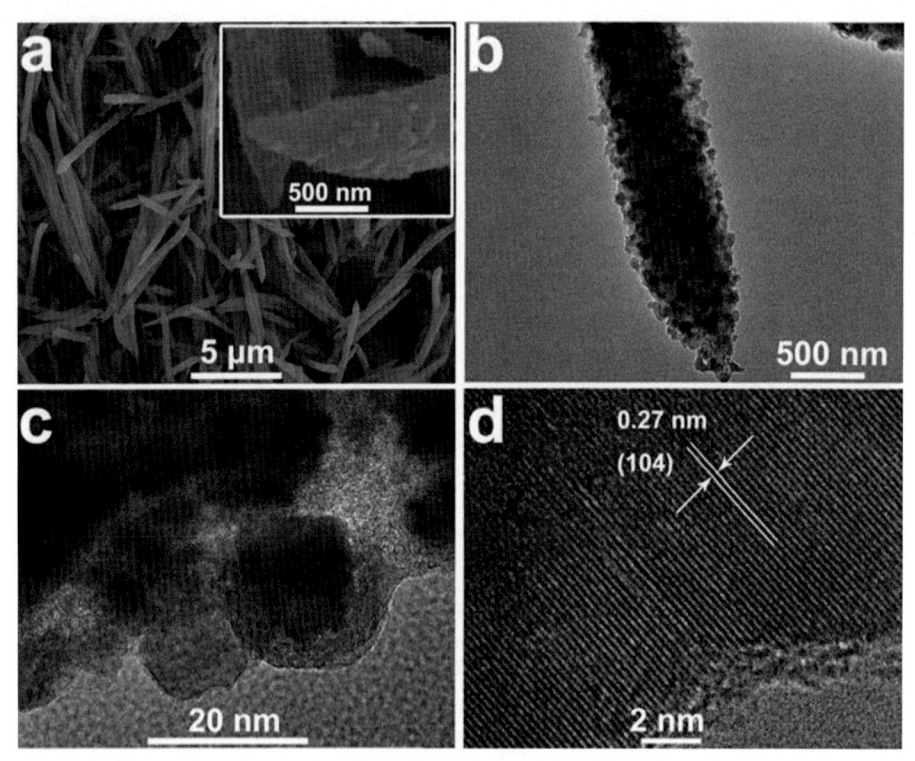

Figure 11.20 (A) SEM image, inset is high-magnification SEM (B) Low magnification (C) High-magnification TEM image (D) HRTEM image of V_2O_3/C composite.
Reprinted with permission from Ref. 24. Copyright (2019), Royal Society of Chemistry.

Zhang et al.[24] prepared V_2O_3/C by solvothermal and Ar annealing. First, they prepared a metallic oxide framework (MOF) of vanadium by the solvothermal method, and by annealing it in the presence of Ar gas, they synthesized V_2O_3/C composite. Morphology of the synthesized composite material is studied by using SEM, TEM, and HRTEM. V_2O_3/C composite SEM image displays shuttle-like morphology with a rough surface, as shown in Fig. 11.20A. Low- and high-magnification TEM images are given in Fig. 11.20B–C, respectively, further endorsing the SEM results. HRTEM image shows the d-spacing of lattice fringes at d = 0.27 nm, which matches the (104) plane of V_2O_3 (Fig. 11.20D).

Electrochemical measurements of as-synthesized V_2O_3/C composites are carried out for N_2 reduction reaction (NRR) in a solution of 0.1 M Na_2SO_4 for an electrochemical cell with two compartments. V_2O_3/C is pasted on carbon paper (CP) and employed as a working electrode (V2O3/C-CP). In the cathodic compartment, N_2 is supplied in the feed gas stream where transportation of protons takes place through the electrolyte and reacts with N_2 to give NH_3 the following mechanism: $N_2 + 6H^+ + 6e^- \rightarrow 2NH_3$. In N_2-saturated electrolyte, NRR electrocatalytic activity of V_2O_3/C-CP is measured for 2 h.

Chronoamperometric curves are shown in Fig. 11.21A. When the potential changes from −0.5 V to −1.0 V, current density (total) rises from 0.1 mA cm^{-2} to more than

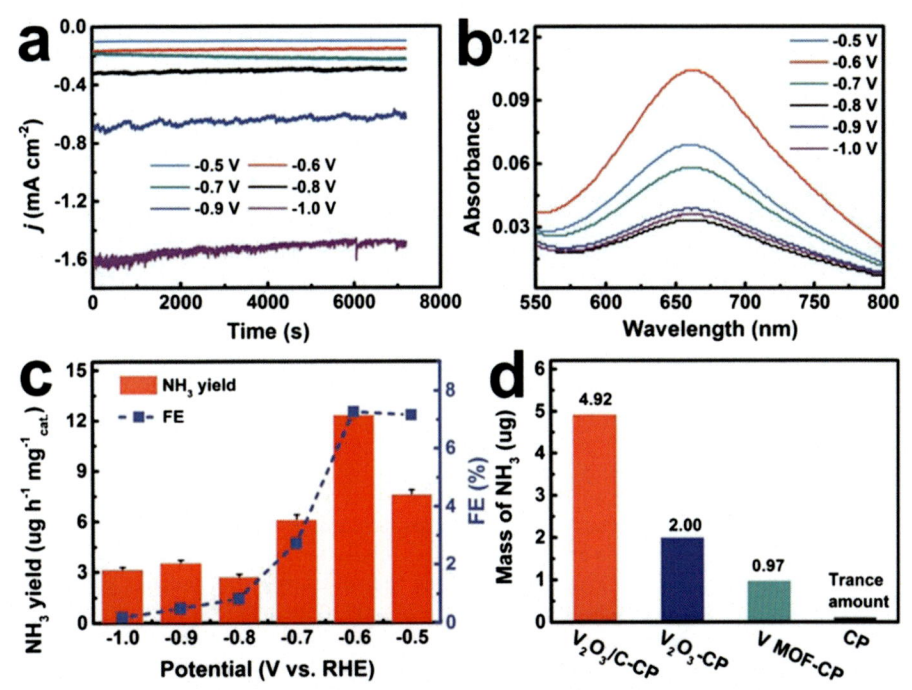

Figure 11.21 (A) Chronoamperometric curve of V_2O_3/C-CP at different potentials for 2 h for NRR (B) UV-Vis spectra for absorption at different potentials for indophenol stained electrolytes after electrolysis of 2 h (C) Yield of NH3 and Fes for V_2O_3/C-CP at different potentials (D) Comparison of different electrodes after 2 h electrolysis for the amount of NH3 generated at -0.6 V.
Reprinted with permission from Ref. 24. Copyright (2019), Royal Society of Chemistry.

1.5 mA cm^{-2} is observed. Electrolytes UV-Vis absorption spectra illustrate characteristic absorption at wavelength 660 nm ascribed to indophenol, representing the occurrence of the electrocatalytic process for NRR in the potential range of -0.5 to -1.1 V (Fig. 11.21B). The yield of NH_3 and Fes at different potentials are given in Fig. 11.21C. At -0.6 V, the highest yield of NH_3 (12.3 $\mu g\ h^{-1}\ mg^{-1}$ cat.) and highest FE (7.28%) was attained. Comparative study of different electrodes (V_2O_3/C-CP, V_2O_3-CP, V MOF-CP) was also carried out, as displayed in Fig. 11.21D. The best catalytic performance of V_2O_3/C-CP is attributed to the high active surface area. Thus, for N_2 reduction to NH_3, V_2O_3/C experimentally proved as a high-performance electrocatalyst at ambient conditions. In addition, it displays high structural and electrochemical stability.

11.3 Conclusions

This chapter has presented an overview of various syntheses for fabricating various VO/carbon composites and their potential applications in innovative energy storage and conversion systems, such as metal-ion batteries, supercapacitors, and electrocatalysis. VO/carbon composites are promising materials because of their high energy density, high capacity, and low cost. The toxicity of VOs varies significantly due to the nature of the compound and oxidation states of V, and pentavalent vanadium is the most toxic and stable form. SEM and TEM results of VO/C composites are also discussed along with the electrochemical performance, which shows that different carbon-based morphologies help in the better transportation of ions in energy storage and conversion applications which richly improve the performance of VOs. In a nutshell, this inclusive review will trigger the interest in VO/C composites having high power and energy density for the next-generation energy storage and conversion device, which ultimately leads to the solution for energy shortage.

References

1. Gao M-R, et al. Nanostructured metal chalcogenides: synthesis, modification, and applications in energy conversion and storage devices. *Chem Soc Rev* 2013;**42**(7):2986–3017.
2. Wang Y-J, Wilkinson DP, Zhang J. Noncarbon support materials for polymer electrolyte membrane fuel cell electrocatalysts. *Chem Rev* 2011;**111**(12):7625–51.
3. Chen G, et al. Nanochemistry and nanomaterials for photovoltaics. *Chem Soc Rev* 2013; **42**(21):8304–38.
4. Kalyanasundaram K, Grätzel M. Themed issue: nanomaterials for energy conversion and storage. *J Mater Chem* 2012;**22**(46):24190–4.
5. Nozik AJ, Miller J. *Introduction to solar photon conversion*. ACS Publications; 2010.
6. Mao J, et al. Graphene aerogels for efficient energy storage and conversion. *Energy Environ Sci* 2018;**11**(4):772–99.
7. Kang W, et al. A review of recent developments in rechargeable lithium–sulfur batteries. *Nanoscale* 2016;**8**(37):16541–88.
8. Yang Q, et al. Ionic liquids and derived materials for lithium and sodium batteries. *Chem Soc Rev* 2018;**47**(6):2020–64.
9. Fang R, et al. More reliable lithium-sulfur batteries: status, solutions and prospects. *Adv Mater* 2017;**29**(48):1606823.
10. Chen X, et al. Smart, stretchable supercapacitors. *Adv Mater* 2014;**26**(26):4444–9.
11. Jiang C, et al. Challenges in developing direct carbon fuel cells. *Chem Soc Rev* 2017;**46**(10): 2889–912.
12. Cao T, et al. Recent advances in high-temperature carbon–air fuel cells. *Energy Environ Sci* 2017;**10**(2):460–90.
13. Kurinec SK. *Emerging photovoltaic materials: silicon and beyond*. John Wiley & Sons; 2018.
14. Aliahmad N, et al. V2O5/graphene hybrid supported on paper current collectors for flexible ultrahigh-capacity electrodes for lithium-ion batteries. *ACS Appl Mater Interfaces* 2018; **10**(19):16490–9.
15. Hu P, et al. Porous V_2O_5 microspheres: a high-capacity cathode material for aqueous zinc–ion batteries. *Chem Commun* 2019;**55**(58):8486–9.
16. Liu P, et al. Ultrathin nanoribbons of in situ carbon-coated $V_3O_7 \cdot H_2O$ for high-energy and long-life Li-ion batteries: synthesis, electrochemical performance, and charge–discharge behavior. *ACS Appl Mater Interfaces* 2017;**9**(20):17002–12.

17. Wang Y, Cao G. Synthesis and enhanced intercalation properties of nanostructured vanadium oxides. *Chem Mater* 2006;**18**(12):2787—804.

18. Pol VG, et al. Core— Shell vanadium oxide— carbon nanoparticles: synthesis, characterization, and luminescence properties. *J Phys Chem C* 2009;**113**(24):10500—4.

19. Perera SD, et al. Vanadium oxide nanowire—carbon nanotube binder-free flexible electrodes for supercapacitors. *Adv Ener Mater* 2011;**1**(5):936—45.

20. Zhao X, et al. The role of nanomaterials in redox-based supercapacitors for next generation energy storage devices. *Nanoscale* 2011;**3**(3):839—55.

21. Balducci A, et al. High temperature carbon—carbon supercapacitor using ionic liquid as electrolyte. *J Power Sources* 2007;**165**(2):922—7.

22. Raut AS, Parker CB, Glass JT. A method to obtain a Ragone plot for evaluation of carbon nanotube supercapacitor electrodes. *J Mater Res* 2010;**25**(8):1500.

23. Chiku M, et al. Amorphous vanadium oxide/carbon composite positive electrode for rechargeable aluminum battery. *ACS Appl Mater Interfaces* 2015;**7**(44):24385—9.

24. Zhang R, et al. Metal—organic framework-derived shuttle-like V_2O_3/C for electrocatalytic N_2 reduction under ambient conditions. *Inorg Chem Front* 2019;**6**(2):391—5.

25. Ihsan M, et al. V_2O_5/mesoporous carbon composite as a cathode material for lithium-ion batteries. *Electrochim Acta* 2015;**173**:172—7.

26. De Juan-Corpuz LM, et al. Binder-free centimeter-long V_2O_5 nanofibers on carbon cloth as cathode material for zinc-ion batteries. *Energies* 2019;**13**(1):1—13.

27. Wang H-E, et al. Superior pseudocapacitive lithium-ion storage in porous vanadium oxides@ C heterostructure composite. *ACS Appl Mater Interfaces* 2017;**9**(50):43665—73.

28. Simon P,Y. Gogotsi-materials for electrochemical capacitors. *Nat Mater* 2008;**7**:845.

29. Kim I-H, et al. Synthesis and electrochemical characterization of vanadium oxide on carbon nanotube film substrate for pseudocapacitor applications. *J Electrochem Soc* 2006;**153**(6): A989.

30. Sun D, et al. The relationship between nanoscale structure and electrochemical properties of vanadium oxide nanorolls. *Adv Funct Mater* 2004;**14**(12):1197—204.

31. Boukhalfa S, Evanoff K, Yushin G. Atomic layer deposition of vanadium oxide on carbon nanotubes for high-power supercapacitor electrodes. *Energy Environ Sci* 2012;**5**(5): 6872—9.

32. Chen Z, et al. High-performance supercapacitors based on intertwined CNT/V_2O_5 nanowire nanocomposites. *Adv Mater* 2011;**23**(6):791—5.

33. Dong W, Mansour A, Dunn B. Structural and electrochemical properties of amorphous and crystalline molybdenum oxide aerogels. *Solid State Ionics* 2001;**144**(1—2):31—40.

34. Zhou X, et al. Bamboo-like composites of V_2O_5/polyindole and activated carbon cloth as electrodes for all-solid-state flexible asymmetric supercapacitors. *ACS Appl Mater Interfaces* 2016;**8**(6):3776—83.

35. Xu J, et al. Synthesis and electrochemical properties of graphene/V_2O_5 xerogels nanocomposites as supercapacitor electrodes. *Solid State Ionics* 2014;**262**:234—7.

36. Nagaraju DH, et al. Two-dimensional heterostructures of V_2O_5 and reduced graphene oxide as electrodes for high energy density asymmetric supercapacitors. *J Mater Chem* 2014; **2**(40):17146—52.

37. Wang H, et al. One-step strategy to three-dimensional graphene/VO_2 nanobelt composite hydrogels for high performance supercapacitors. *J Mater Chem* 2014;**2**(4):1165—73.

38. Seh ZW, et al. Combining theory and experiment in electrocatalysis: insights into materials design. *Science* 2017;**355**(6321).

39. Chu S, Majumdar A. Opportunities and challenges for a sustainable energy future. *Nature* 2012;**488**(7411):294—303.

40. Stamenkovic VR, et al. Energy and fuels from electrochemical interfaces. *Nat Mater* 2017; **16**(1):57−69.
41. Sun Y, et al. Atomically-thin two-dimensional sheets for understanding active sites in catalysis. *Chem Soc Rev* 2015;**44**(3):623−36.
42. Dai L, et al. Metal-free catalysts for oxygen reduction reaction. *Chem Rev* 2015;**115**(11): 4823−92.

Metal oxide-carbon composites in biomedical, catalytic, and other applications

Metal oxide–carbon composites and their applications in optoelectronics and electrochemical energy devices

12

Asadullah Dawood [1,3], Junaid Ahmad [2], Saif Ali [2], Sami Ullah [2], Zeeshan Asghar [2] and Matiullah Shah [1]
[1]Department of Physics, University of Wah, Wah Cantt, Punjab, Pakistan; [2]Department of Physics, Division of Science and Technology, University of Education Lahore, Lahore, Punjab, Pakistan; [3]Department of Physics, School of Sciences and Humanities, Nazarbayev University, Nur Sultan, Kazakhstan

12.1 Introduction

Owing to the global economy's exponential growth, unconventional energy resources (e.g., fossil fuels) have been employed on a massive scale, with significant implications for our environment. This has prompted an urgent call for renewable and reliable green energy sources (e.g., solar and wind) and low-cost, highly efficient, and environmentally friendly energy conversion and storage systems. Electrochemical energy-storage systems that can store electricity as chemical energy have many benefits, such as high power and energy densities, low costs, long lives, and environmental friendliness, and are seen as promising energy-storage systems. In recent years, many metal oxide–carbon (MO–C) composites have been considered promising candidates for optoelectronic and electrochemical energy-storage systems such as solar panels, photovoltaic solar cells, Li-ion batteries (LIBs), photodiodes, fuel cells, and supercapacitors (SCs). Metal oxides (MOs) such as SnO_2,[1] $COMn_2O_4$[2] Fe_2O_3,[3] NiO,[4] NiC_2O_4,[5] $ZnMn_2O_4$,[6] and MnO have been used as ideal electrodes for LIBs and electrochemical capacitors (ECs). Usually, during electrochemical reactions, these MOs can use multiple-electron transfer per formula unit but have a potential basic capacity/capacitance many times greater than that of traditional carbon-based materials. Unfortunately, there are many problems facing the practical usage of these MO-based electrodes, like rapid capacity/capacitance fading upon cycling and unsatisfactory high-rate efficiency. These disadvantages of MOs may be due to the destruction of structures during repetitive processes of charge/discharge, especially for materials of battery types that experience tremendous volume expansion upon the injection of lithium, and slow electronic/ionic transport. Therefore, the development of successful methods to improve the electrochemistry of MO-based electrodes and realize their future applications in energy-storage devices is of great importance. Nanostructure

Metal Oxide-Carbon Hybrid Materials. https://doi.org/10.1016/B978-0-12-822694-0.00007-7

engineering has successfully improved reversible capacity/capacitance, cycling stability, and rate efficiency compared with their bulk equivalents, among numerous techniques to achieve performance improvement for MO-based electrodes. To date, with some progress in electrochemical efficiency for LIBs and ECs, a wide range of MO-based nanostructures has been documented.[7] One-dimensional (1-D) nanowires/tubes,[8] two-dimensional (2-D) nanosheets,[9] and three-dimensional (3-D) hollow spheres/cubes,[10] are representative of these complex nanostructures.[11,12] On the other hand, the ability to modify carbon nanotube (CNT) composites by diameter control offers unique opportunities to customize optical and optoelectronic properties.[13] Direct bandgap materials developed into a number of optoelectronic instruments, such as light detectors, light emitters, and transparent conductors, are CNTs.[14,15] It has been previously reported that fluorescence quantum efficiency (QF) of scattered CNTs was shown in early studies to be in the range of 10^{-3} with an efficient radiative lifetime of $1-10$ ns at room temperature. Limited QFs in CNTs may be due to many nonradiographic mechanisms, namely exciton—exciton annihilation, the existence of low-energy dark excitons that cannot relax radiatively to the field, and effective photometric nonradiative decline.[16] It is important to remind ourselves that carbon composites in optoelectronics are a favorite choice.[17,18] Optoelectronic devices that have been developed based on carbon range from passive devices to modulators, detectors, light amplifiers, and sources. In this contribution, we give an overview of MO—C composites and their applications in optoelectronics and electrochemical energy devices.

12.2   Types of carbon composites

The property of an element to exist in more than one form while retaining the same physical state is known as allotropy. The arrangement of atoms and the nature of chemical bonds define various of these materials' physical and chemical properties. Carbon is one of the most fascinating elements in this regard, with the potential to shape a broad variety of structures with profoundly distinct properties. It is the basic building unit of about 20 million identified molecules, and the whole of organic chemistry is based on it. "Soft" graphite and "hard" diamond are classical allotropic carbon forms used in diverse scientific and technological applications and a wide variety of products, including consumer goods in daily human living.[19]

The novel allotropic forms of carbon have been introduced over the last few decades—i.e., fullerenes (1985),[20] CNTs (1991),[21] and graphene (2004).[22] These innovative materials have attracted considerable research interest for many applications because of their extraordinary and novel properties. Therefore, carbon-based materials are frequently considered "wonder materials" among scientists seeking to know increasingly more from observations and experimentation with carbon forms.[23,24]

Curiosity is a crucial component that enables groundbreaking discoveries and innovations that make life on the planet sustainable. So scientists from all over the world are greatly committed to attaining the ultimate potential of different technologies. In this perspective, they introduced various techniques to optimize the performance of materials. The preparation of nanocomposites is among one of these techniques.

Composites are materials comprise of two or more distinct materials that mutually contribute to the entire structure to combine the leading properties of each component however simultaneously preserve their own identities. The fabrication of composite material has dynamic technological significance because it combines the distinctive properties of its components and also induces certain novel extraordinary characteristics due to their interaction.

MOs are a fascinating class of semiconducting materials that are intensively studied these days in electrochemistry and optoelectronics due to their distinctive electronic and optical properties. But there are still certain limitations that constrain the optimal performance of MO, such as poor stability, low electrical conductivity,[25] and porosity tailoring problems.[26] From this perspective, extensive efforts have been made to boost MO efficiency in optoelectronics and electrochemical devices and to resolve low-capacity problems of carbon nanomaterial.[27] So the formation of MO−C composites produce the synergistic effects that increase the electrical conductivity of MO and solve the low capacity problem of carbon-based materials.[27]

On the basis of carbon dimensionality, MO−C composites can be classified into the following types:

1. MO−0-D carbon composites
2. MO−1-D carbon composites
3. MO−2-D carbon composites
4. MO−3-D carbon composites

12.2.1 Metal oxide−zero-dimensional carbon composites

Significant development has been made in the area of 0-D nanostructured materials over the last few years. Especially, 0-D carbon nanomaterials, including carbon quantum dots (QDs), graphene quantum dots (GQDs), and carbon nanoparticles, have been applied in SCs in addition to their extensive use in light-emitting diodes (LEDs), solar cells, lasers, and single-electron transistors. GQDs are small particles with single or several layers of graphene exhibiting the special properties of both QDs and graphene.[28,29] GQDs have emerged as attractive materials for energy-storage systems and fuel cells because of their elevated specific surface area, good mobility,[30] and extraordinary electrical conductivity.[31]

GQDs can be simply synthesized by chemical exfoliation of graphite into GO, followed by GO reduction in regulated environments. By applying a one-step solvothermal process, Liu et al.[32] synthesized GQDs from GO powder and identified exceptional power density, the longer life cycle for SCs, and high rate capability, but their weak energy density and poor specific capacitance remained the key constraints. Consequently, Zhu et al.[33] reported the synthesis of a hybrid composite of thermally reduced carbon QDs and RuO_2 (RCQD/RuO_2) via the sol−gel technique combined with the impregnation method. RCQD/RuO_2 composites presented a specific capacitance of 594 F/g at 1 A/g and a better capacity retention rate (77.4%) compared with RuO_2 (58.2%). Furthermore, the prepared electrode exhibited outstanding cycling stability of 96.9% to 5000 cycles, which was higher than RuO_2 (73.9%). The exceptional electrochemical performance of the prepared electrode was primarily due to the development of the hybrid network based on RCQD, which facilitates fast electron/ion transfer during the charging/discharging and further enhances the use of RuO_2, contributing to rapid redox reactions. These promising findings gave a fresh insight into the synthesis and design of high-performance SC hybrid electrode materials based on cost-effective QD because the toxic nature and the high price of RuO_2 toxic nature hampers its practical use in SCs. In another report, the same group presented the fabrication of $CQDs/NiCo_2O_4$ composites by a simple one-step reflux route and subsequent calcination treatment.[34] The composites obtained show outstanding electrochemical properties, especially cycle stability (98.75% up to 10,000 cycles), capability rate (60.8% up to 100 A/g), and high specific capacitance (856 F/g at 1 A/g). High mesoporosity, elevated surface areas, and higher electronic conductivity of as-fabricated composites are factors responsible for superior electrochemical efficiency. Yuan et al.[35] prepared carbon nanoparticles/MnO_2 nanorods hybrid structure on carbon fabric by a simple procedure of flame synthesis accompanied by the process of electrochemical deposition and acquired a maximum capacitance of 109 mF/cm^2. In fact, the prepared hybrid material not only enables cation diffusion between the electrode and the electrolyte but also helps to beat the weak electrical conductivity of MnO_2.[36] In addition, Fan et al.[37] documented the synthesis of carbon nanoparticles encapsulated in hollow NiO nanostructures by calcination from carbon-coated $Ni(OH)_2$ precursor, which exhibited a better specific capacitance (988.7 F/g at 0.5 A/g) than that of bare NiO hollow nanospheres (594.4 F/g). Various potential applications of GQDs are shown in Fig. 12.1.

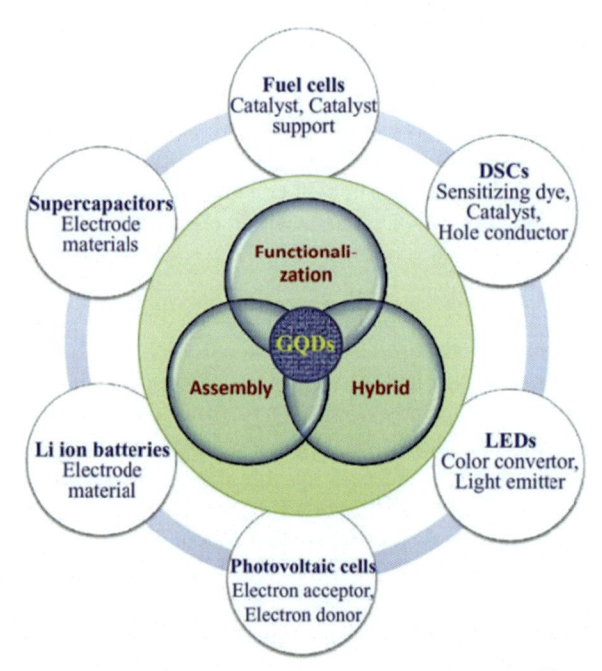

Figure 12.1 Various promising applications of graphene quantum dots.[28]
Retrieved with permission from Ref. 28.

12.2.2 Metal oxide—one-dimensional carbon composites

1-D carbon/metal oxide composites have been researched extensively due to their special structure. One-dimensional nanomaterials of carbon nanofibers or CNTs commonly form a composite structure by a homogeneous decoration of MOs around them. As-formed composites show better features that come from their distinctive structure. Especially, CNTs with a high aspect ratio in a continuous conductive network may promote charge transport due to the decreased contact resistance with neighboring nanoparticles higher than that of a pure MO nanomaterial. Therefore, we can say that CNTs may act as electron highways, allowing the transport of charges along the longitudinal direction.[38,39]

MO—1-D carbon nanocomposites provide many advantageous features, such as elevated specific surface area, improved mechanical toughness, and high electrical conductivity compared with the pristine MOs highlighted in Fig. 12.2. So to increase electrochemical efficiency, various MOs, such as NiO,[40] MnO_2, TiO_2, Co_3O_4,[41] and SnO_2, have been coupled with CNTs. In a report, Zhao et al.[42] compared the electrical conductivity of CO_3O_4 with $Co@Co_3O_4/CNTs$ at 18 MPa to analyze the impact of CNT compositing on the material properties. The nanocomposite $Co@Co_3O_4/CNT$ was fabricated by arc discharge and low-temperature oxidation. Due to the significant role of CNTs as an electronic conductive highway, the electrical conductivity of the fabricated nanocomposite was increased from 10^4 to 7.6 S/m. It is notable that the fabricated nanocomposite experiences nearly 10,000 times greater electrical conductivity than that of pristine.CO_3O_4[43] This increased electrical conductivity of the MO/CNT composite facilitates the transport of charges, contributing to improved electrochemical efficiency of composite electrode-based energy-storage systems.[44]

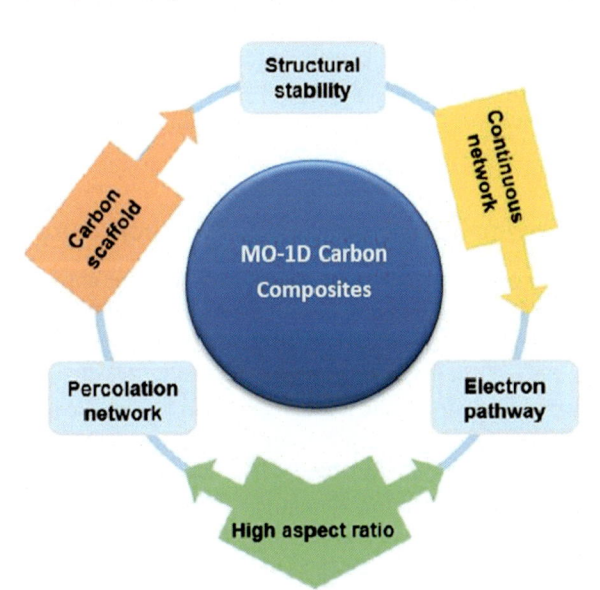

Figure 12.2 Some promising properties of metal oxide—one-dimensional carbon composites. Reproduced with permission from Ref. 46.

A new approach to fabricate high-efficient stretchable CNT/MnO_2 nanosheets via a simple biscrolling and wrapping process has been recently developed by Xianhong Zheng et al.[45] The fabricated nanosheets exhibited excellent stretchability along with extraordinary bending stability, elevated specific capacitance of 658 mF/cm^2, and 15.2 $\mu Wh/cm^2$ of energy density. Their research work will open the way for assembling high-performance stretchable energy-storage systems.

12.2.3 Metal oxide—two-dimensional carbon composites

Graphene, an atomically of carbon thin two-dimensional allotropic form, has captured considerable interest in the scientific community.[47] Many promising features make graphene desirable for optoelectronics and electrochemical applications.[48] First, graphene provides numerous active sites due to its elevated surface area along with very high surface-to-volume ratio; second, it fastens the electrochemical kinetics because of its charge carrier mobility and extraordinary electrical conductivity; third, it possesses high tensile strength (1.0 TPa) that improves cycle and device stability. As an electrode material, even though graphene shows several features, but it does not attain excellent performance because of restacking during electrochemical reactions.

A number of research publications highlighted that the coupling of MO with 2-D graphene has increased overall electrochemical efficiency. For instance, Wang and co-workers[49] reported the synthesis of $Fe_3O_4/2$-D rGO composites via hydrothermal treatment. The fabricated nanocomposite displays elevated specific capacitance (220.1 F/g). The decreased agglomeration between nanoparticles of Fe_3O_4 in presence of rGO and chemical interaction between them are responsible for improved electrochemical performance. Zhang et al.[50] documented the fabrication of $NiCo_2O_4$@rGO hybrid nanomaterials through deposition of GO on the $NiCo_2O_4$ nanosheets by nickel foam with any polymer binder. The porous and interconnected structure and formation of the Schottky electric field in the fabricated hybrid nanostructure are responsible for exceptional cycling stability, boosted specific capacitance, and high-rate capability, as portrayed in Fig. 12.3.

12.2.4 Metal oxide—three-dimensional carbon composites

Although improving cyclic stability and the capacitance of composites electrode materials by the incorporation of 2-D graphene is a gifted output for researchers but there are some gray areas described in the literature that require more advancement. For instance, restacking and irreversible graphene accumulation disrupts the porosity of the carbon network that obstructs the ion transport and electrolyte access in it. In addition, chemically inert binders and additives are usually required to fabricate graphene-based electrodes, and the 2-D structure hinders the decoration of other pseudochemical materials on graphene. These are few limitations that need to be addressed. MO—3-D carbon composites play a crucial role to overcome the aforementioned limitations because of their captivating features, such as highly porous structure, improved surface area, wall thickness, and tunable pore size. Fig. 12.4 summarizes the promising features of MO—3-D carbon composites. In a report, Dong and co-workers[52] documented

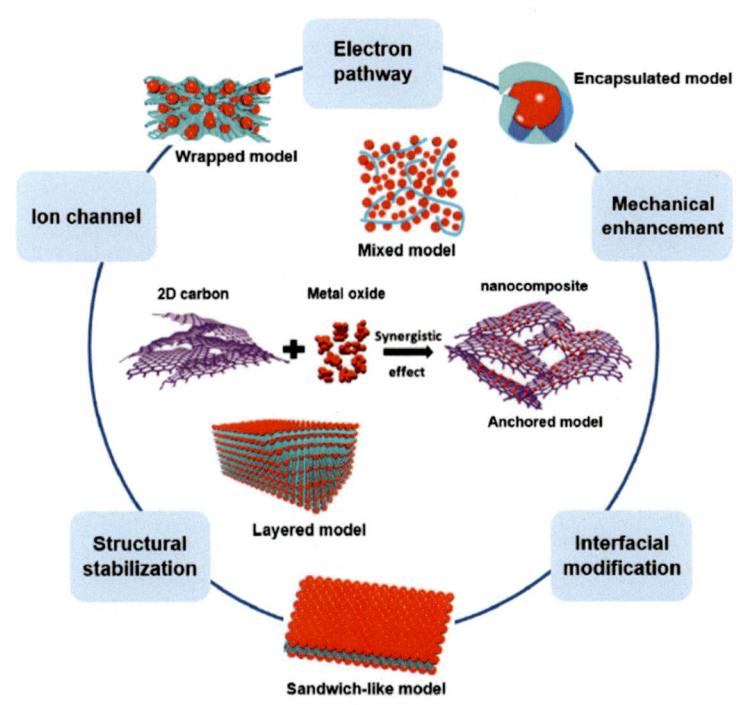

Figure 12.3 Some fascinating structural features of metal oxide—two-dimensional carbon composites.[51]
Reprinted with permission from Ref. 51.

the fabrication of 3-D graphene/Co3O4 nanocomposites with the help of a simple hydrothermal method. The fabricated composite material was used as an electrode for the SC and exhibited high cyclic stability along with good specific capacitance of about 1100 F/g @ 10 A/g. Recently, Zhou et al.[53] reported the fabrication of 3-D PC/Co3O4 hierarchical nanocomposites by pyrolysis technique. The fabricated nanocomposite showed enhanced electrochemical efficiency because of its better ability to accelerate ion diffusion, transfer electrons, and increase the number of active sites. Hence, we can conclude that MO—3-D carbon composite materials can be successfully applied in many potential applications owing to their extraordinary properties.

12.3 Why metal oxide—carbon composites?

The various composite conversions from MO reduction add carbon to the electrocatalytic process that enhances the electrical conductivity of the oxide catalyst, thus improving the bifunctional air electrode quality by the catalytic activity of various transition MOs.[54] The electrochemical properties of the material of MO were improved by the carbon material. MOs have been vigorously examined, and their carbon nanocomposites repeatedly display better performance in galvanostatic.[55]

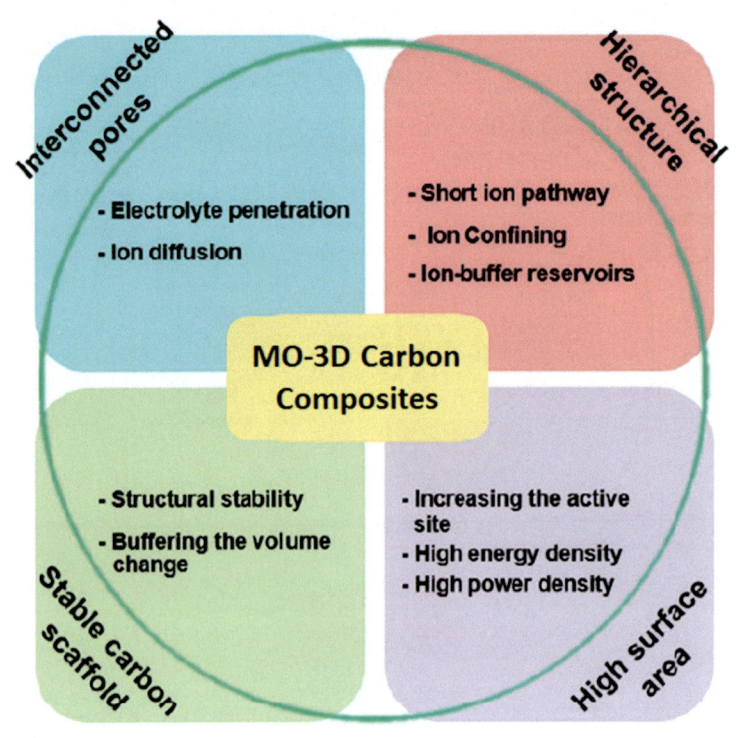

Figure 12.4 Some fascinating characteristics of metal oxide—three-dimensional carbon composites.
Adapted with permission from Ref. 46.

MOs play a significant role in upcoming energy invention and storage technologies. Improved cycling stability and improved high rate capacity for carbon composites containing transition MOs have also been reported.[56] The identification of the correct electrode materials to achieve improved electrochemical performance in terms of cycle life and delivery capacity has played an important role. MO—C composites are sodium-ion battery conversion anodes.[57] In maintaining the structure of iron oxide, the carbon layer plays a significant role and also enhances electronic conductivity. During lithium storage, this carbon layer can store the embedded iron oxide nanoparticles and provide conductivity between graphene sheets.[58] Redox-active composites of MOs/carbons play an important part in future technologies with energy transformation and storage, such as batteries and electrolysis by water. It is reasonable to follow novel electrode materials demonstrating their application in electrochemical devices with enhanced quality. Among the many materials for the capacitive candidate, metal due to their high electron storage space and emissivity, oxides have received significant attention through a chemical redox reaction in an energy-storage device. It is sensible to explore novel electrode materials demonstrating their application in electrochemical devices with improved performance. Among the many materials for the capacitive candidate, metal due to their high electron storage capacity and emissivity, oxides have received significant attention through a chemical redox reaction in an energy-storage device.[46] SCs, also related to it as ultracapacitors, are increasing applications for automobiles, large industrial appliances, hold-up memory, and renewable energy devices power plants high power density is given by SCs compared with rapid charge-discharge.[38]

MO increases the high power and long life cycle of SCs. The SC has high energy through MO and more strength of carbon material than the polymer material.[59] MOs have been widely used in biomedical including applications are magnetic resonance, ablation therapy, and drug delivery. There is a wide range of applications, such as gas sensors, catalysis, sunglasses, cosmetics, paints, and UV detectors.[60] MO has high sensitivity, high dielectric properties, and greater rates of chemical and mechanical stability. The magnetic properties of MOs cover a wide range of applications—magnetic recording media, magnetic resonance imaging, data storage, ferrofluids, and cancer treatment. They are also quite important for the environment and as well as food analysis in organic, bioscience, and industrial water treatment.[61] Therefore, MO carbon composite is so much important in the daily routine of our lives.

12.4 Synthesis techniques of metal oxide–carbon composites

Numerous MO–C composites are promising candidates for optoelectronic and electrochemical energy-storage devices, such as solar cells, photovoltaic solar cells, LIBs, photodiodes, fuel cells, and SCs. MOs, such as SnO_2, $CoMn_2O_4$, Fe_2O_3, NiO, $NiCo_2O_4$, $ZnMn_2O_4$, and MnO, have been utilized as capable electrodes suitable for LIBs, and electrochemical SCs. During electrochemical reactions, MOs utilize multiple-electron transfer per unit formula and achieve much greater theoretical capacitance than conventional carbon-based materials. But MO-based electrodes are practically limited because of several challenges—unsatisfactory high performance and fast capacitance fading upon cycling. Therefore, scientists are trying to overcome these drawbacks by making composites of MOs with carbon-based materials.[7,55,62]

Several synthesis methods have been adopted by the researchers to synthesize the desired MO–C composite having enhanced properties for optoelectronic and electrochemical energy-storage devices. The synthesis process is quite a challenging task to synthesize the MO–C composite in a minimum time and at the least cost. Several synthesis techniques including hydrothermal, sol–gel, solvothermal, Microwave-assisted method, thermal decomposition, and direct mixing are employed among other synthesis techniques. The abovementioned techniques are reported by most of the researchers to synthesize MO–C composite in a large quantity and they are nontoxic to the environment. Herein this section different synthesis techniques that are recently reported by the researchers for MO–C are in the following comparative table and most adopted techniques are discussed briefly.[63]

12.4.1 Microwave-assisted method

The microwave-assisted method has been commonly used in the preparation of nano-materials and chemical reactions. In comparison with other conventional heat conduction methods, this method is energy effective heating method. Heat is transferred by the convection process in the conventional method when the vessel is heated. On other hand, microwave heating is considerably more rapid and more suitable in terms of used energy. Moreover, it produced higher temperature homogeneity.[59]

Rather than heat transfer, Microwave heating is an energy conversion phenomenon, as in this method electromagnetic rays are converted to thermal energy. In this method, heat is produced by the electric component of the electromagnetic field. There are mainly two mechanisms that cause heat one is the dipolar mechanism and the other is the conduction. In the polarization process, heat is generated by a substance when microwaves are irradiated and as the result, it possesses a dipole moment. The dipole will try to bring into the line itself in the direction of the field by rotating itself. Molecules in the substance try to track the field vainly and they strike with one another that causes heat. High- and low-frequency radiation does not increase the heating process. In a high-frequency case, molecules do not respond because the field oscillates too quickly and in a low-frequency case, there is no random motion as molecules obey the field. On the other hand, the conduction is very stronger interaction concerning the generation heat capacity. In that case, under the influence of the electric field of microwave mobile charge carriers move comparatively easily through substance. Due to electric resistance, heat is produced by the induced currents in the sample. The differences between microwave and conventional methods illustrated via plot demonstration by Sarah et al. are shown in Fig. 12.5 and Table 12.1.[59,62]

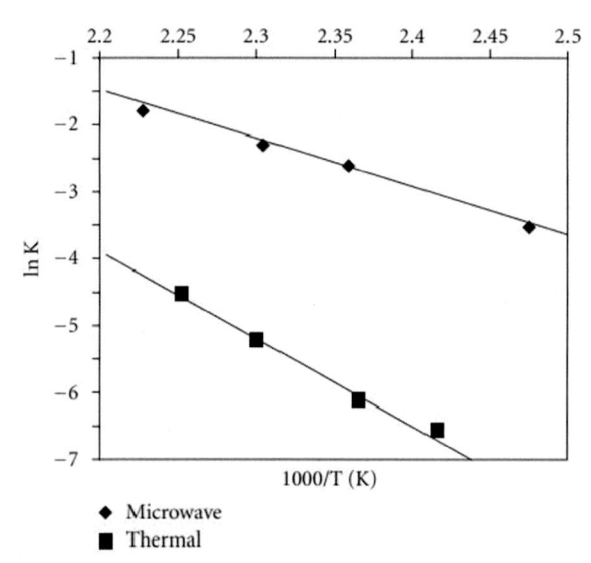

Figure 12.5 The different kinetic plots (between the first order) for thermal activation and microwave for the imidization reaction.
Reused with permission from Ref. 62.

Table 12.1 Results from Fig. 12.1.

Heating mode	ΔH (kJ/mol)	log A
MW	57 ± 5	13 ± 1
Δ	105 ± 14	24 ± 4

Reused with the permission of Ref. 62.

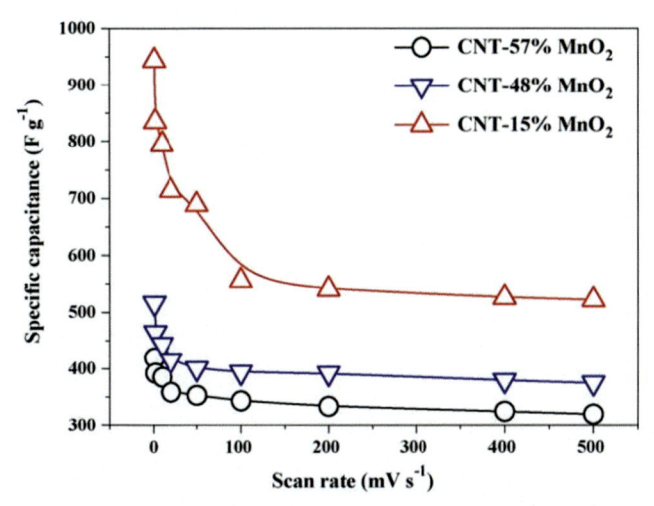

Figure 12.6 Carbon nanotube/manganese dioxide (MnO_2) composite specific capacitances. Reused with the permission of Ref. 65.

W. Zheng et al. developed a Scalable and fast synthesis method to synthesize 3-D MXene derived MO$-$C nanotubes via microwave irradiation. The reported that TEM and SEM images showed the vertically grown CNTs on the MXene nanosheets and obtained hierarchical hybrid structures with both nanotubes and nanosheets that are decorated by the TiO_2/Fe_2O_3 nanoparticles. The obtained hybrid structures showed good electrochemical performance when these structures were tested in LIBs.[64] Jun Yan et al. prepared CNT/MnO_2 (CNT/MnO_2) nanocomposites by microwave heating process and examine the prepared sample by SEM, cyclic voltammetry, TEM, and XPS. They used the galvanostatic charge/discharge to study electrochemical properties. They reported 944 specific capacitance that is 85% of theoretical capacitance when they made 15% MnO_2 composite. They obtained maximum specific capacitance and energy density at 57% content of MnO_2,[65] as shown in Fig. 12.6.

12.4.2 Sol$-$gel synthesis method

The sol$-$gel is the preparation method for ceramics and inorganic polymers from a solution that undergoes a conversion from liquid precursors to sol and then transformed to another form known as a gel. The creation of sol arises through the condensation and hydrolysis process. More generally, the sol can be well-defined and colloidal suspension. In that method, there are several forms of gel. Generally, the gel is a non-fluid 3-D network. Flory in 1947 grouped the gel into four types that are covalent polymer networks, lamellar gels, networks of physically aggregated polymers, ordered, and disordered particulate gels. A more useful classification was made by Kakihana in 1996 for gels that are given in Table 12.2.[66]

The sol−gel technique can be described briefly as follows:

- Partial condensation and hydrolysis of alkoxides are used to form sol−gel preparation by using a polycondensation process to achieve a metal−oxo−metal bond.

Table 12.2 Types of gels.

Type of gel	Bonding	Source	Gel schematic
Colloidal[7]	Particles connected by Van der Waals or hydrogen bonding	Metal oxide or hydroxide sols	
Metal−oxane polymer[4]	Inorganic polymers interconnected via covalent or intermolecular bonding	Hydrolysis and condensation of metal alkoxides e.g., SiO_2 from tetramethyl orthosilicate	
Metal complex[8]	Weakly interconnected metal complexes	Concentrated metal complex solution e.g., aqueous metal−citrate or ethanolic metal urea often form resins or glassy solids rather than gels	
Polymer complex I in situ polymerizable complex ("Pechini" method)[9,10]	Organic polymers interconnected by covalent and coordinate bonding	Polyesterification between polyhydroxy alcohol (e.g., ethylene glycol) and carboxylic acid with metal complex (e.g., metal−citrate)	
Polymer complex II coordinating and crosslinking polymers[11]	Organic polymers interconnected by coordinate and intermolecular bonding	Coordinating polymer (e.g., alginate) and metal salt solution (typically aqueous)	

- For the continuous condensation within the gel network, aging or syneresis is done by shrinking to get results in the expulsion of solvent.
- The gel is dried using the collapse of the porous network (aerogel), i.e., through supercritical drying to prepare dense/thicker xerogel.
- A calcination process is used to remove the M-OH group surface at high temperatures up to $800°C$.[66]

Magda et al. prepared MO (V, W, Zn)−C composites as Al-ion electrodes for electrochemical cells that showed very excellent results because of higher theoretical capacity than LIBs and much lower toxicity at low cost. They studied physicochemical properties of MO−C composites using thermogravimetry, Differential thermal analysis, XRD, SEM.[67] Mingsong Liu et al. used the sol−gel method to prepare TiO_2-Carbon composite via carbonization of the gel. They investigated the gas sensing properties of the composite and found the enhanced adsorption of the gas molecule can modify sensor response effectively. They also reported the enhanced photocatalytic activity of the as-prepared composites.[68]

E. Thauer et al. synthesized Li_3VO_4/C composites via the sol−gel technique and then post-annealing at 650°C in a flow of N_2 for 1 h using either malic acid, tartaric acid, or glucose. They used the prepared material as LIB anodes. They reported that the electrochemical properties were influenced by the texture, morphology, and carbon content. They obtained 400 mAh/g capacity initially. The prepared composite exhibits 299 mAh/g discharge capacity.[69]

12.4.3 Hydrothermal/solvothermal process

Hydrothermal is a suitable and mostly employed method by the researchers to synthesize a single crystal material that depends on the solubility of minerals in hot water under maintaining the condition of high pressure. The best advantage of this method over other methods is to create or synthesize crystalline phases that show unstable behavior at the melting point. This method is also suitable to create good quality crystals and larger crystals of the nanocomposite. It refers to reactions that are heterogeneous. So that preparation of inorganic materials is possible in an aqueous medium above the ambient pressure and temperature.[70,71]

Marina et al. worked on the TiO_2/carbon composite and synthesized this composite by hydrothermal method. They used glucose and isopropoxide solution as a precursor to getting molar ratios of Ti/C from 0.05 to 0.30 in the composite. They reported that the composite prepared from the highest concentrated solution of glucose and represented the enhanced, valuable photocatalytic activity toward both selected pharmaceuticals and methylene blue under UV light. This composite also showed a higher than 81% degradation ratio of methylene blue with high recycling ability,[72] as illustrated in Fig. 12.7.

Yifu et al. performed the hydrothermal synthesis of vanadium dioxide/carbon composite. They measured the electrochemical properties of the surface and used the composite as SC electrodes. Prepared nanocomposite showed specific capacitance (406 F/g) and 0.2 A/g current density. They concluded that V_2O_5 nanoparticles were ideal materials for application in SCs,[73] as summarized in Table 12.3.

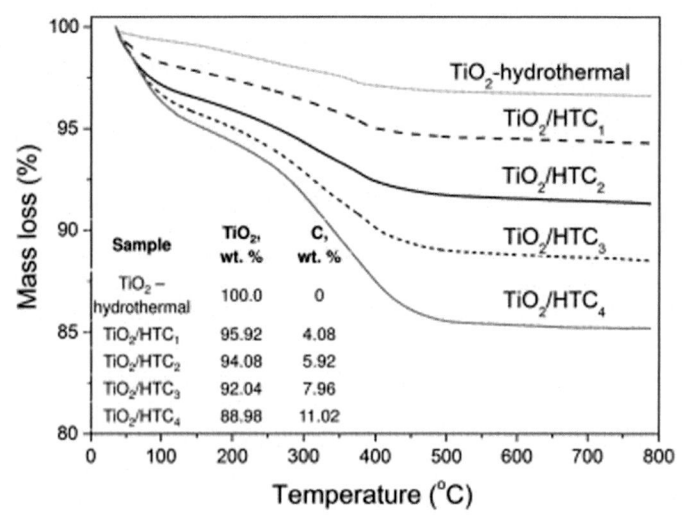

Figure 12.7 Thermogravimetric analysis of composites.
Reused with permission from Ref. 72.

Table 12.3 Comparison of synthesis methods for preparing metal oxide−carbon composites for optoelectronic and electrochemical energy-storage devices.

No	Composites	Method	Application	References
1	Vanadium dioxide/carbon	Facile one-step hydrothermal	Electrode for supercapacitor	[73]
2	Fe_3O_4/carbon	Hydrothermal	Good material for removing toxic chemicals	[74]
3	TiO_2/carbon	Direct mixing	Anode material for LIBs	[75]
4	MnO_2/carbon nanotubes	Coprecipitation	Electrochemical electrodes	[76]
5	Graphene-TiO_2	Solution mixing	Photocatalyst, optoelectronic devices	[77]
6	Graphene/TiO_2	Sol−gel	Photocatalyst, Hydrogen evolution, water splitting	[77]
7	Graphene-MnO_2	Microwave irradiation	Supercapacitor electrode	[78]
8	Fe_3O_4-carbon	Emulsion polymerization	Anode for LIBs	[55]
9	NiO−MCMB (mesocarbon microbeads)	High-energy ball milling	Anode for LIBs	[79]
10	CuO−MCMB	High-energy ball milling	Electrochemical devices, LIBs	[80]

12.5 Applications of metal oxide–carbon composites in optoelectronic devices

Optoelectronics is the communication between optics and electronics that requires the research, design, and manufacture of a hardware system that, by semiconductors, transforms electrical energy into light and light into energy. This system is made of solid crystalline materials that are thinner and heavier than insulators than metals. Optoelectronics products are essentially light-containing electronic devices.[14] In several optoelectronics uses, such as armed forces, telecommunications, automated access control systems, and medical devices, this product can be used. A broad range of instruments, including LEDs and modules, image collection devices, information displays, optical communication systems, optical storage, and remote sensing systems, are covered by this academic field. Telecommunication lasers, blue lasers, optical fibers, LED traffic lights, photodiodes, and solar cells are examples of optoelectronic products. LEDs, laser diodes, photodiodes, and solar cells are the bulk of optoelectronic systems.

Graphene is a monolayer two-dimensional hexagonal honeycomb lattice structure and is the building block of graphite. In 2004, A. Geim and K. Novoselov separated graphene flakes from graphite samples using the mechanical exfoliation method. The resultant graphene structure showed astounding electrical, optical, and thermal characteristics.[1]

Graphene monolayers stack upon each other to form a crystalline lattice. Graphene has three σ-bonds and one π-bond. One π-bond in its structure offers an important role in a half-filled band that makes electrons move freely.[1]

12.5.1 Application of graphene in lasers

Q-switching and mode-locking are two primary ways of understanding that laser ultrashort pulse. With Q-switching, output pulses could be produced having a width in the ms and ns regimes with strong pulse energy characteristics. The ps pulse, fs pulse, and higher peak power can be produced by the mode-locking technology. Both active and passive alternatives are employed to achieve Q-switching and mode-locking. Relatively, passive mode-locking and passive Q-switching techniques do not require modulation of an electric field or light field; they only require the introduction into the laser cavity of nonlinear optical components known as saturable absorbers, so that they are more convenient and effective and can be easily recognized.

Saturable absorber efficiency can be calculated by the following factors: the range of working wavelengths, the depth of modulation, the loss of unsaturated light, the strength of saturated light, and the threshold value of thermal damage. Since graphene has the characteristics of a superbroad working range spectrum, an ultrafast electronic relaxation time measured in femtoseconds (about 150 fs), and a high thermal damage threshold, as a new saturable absorber material, it has garnered great interest. This phenomenon happens when graphene is illuminated, electrons are excited to the conduction band from the valence band of graphene. These hot electrons are cooled in a very short period (hundreds of fs) to form the Fermi-Dirac distribution within the band with the temperature of electrons T_e. Some interband optical transitions that were originally possible could be blocked within the range of $K_B T_e$ from the Fermi energy E_F by the newly formed electron-hole pairs. As a consequence, photon absorption is reduced by approximately $K_B T_e$.[81] In the next 1 ps, the emitted hot electron would continue to be cooled with the dispersive effect of the photon in the band and would dominate the electron-hole compound process. This process describes linear optical transition at a lower intensity of incident light. However, the number of photoexcited electrons and holes significantly increases with the increase in incident light intensity, and its concentration at room temperature is higher than the intrinsic graphene concentration. These excited electrons and holes would fill the energy state between the graphene conduction band and the valence band, the energy level will not be able to accommodate more electrons or holes as per the Pauli Exclusion Principle, thereby blocking the further absorption of light in graphene, expressing the characteristics of saturated absorption.

12.5.1.1 Application of graphene in the mode-locked fiber laser

There are various benefits of fiber lasers, such as good beam efficiency, miniaturization and device structure intensification, ease of heating, and ease of incorporation into the optical fiber communication system. Therefore, it has become the center of light for scientific study and development. Many methods are possible to insert a graphene saturable absorber into the laser cavity; the sandwich configuration is the most widely used. This technique involves transporting the graphene to the fiber end face of FC/PC and then attaching it through the connector to another FC/PC and integrating it into the laser cavity. This technique is simple and reduces the cavity length induced by mode-locking devices being inserted. This helps to create a mode-locked pulse with a high degree of repetition. The preparation method of graphene mode-locking devices with a sandwich configuration is seen in Fig. 12.8. Even then, in this method, the graphene saturable absorber is located vertically in the light path; this requires a sample with a threshold of high thermal impact. On the other hand, too much power can quickly melt the graphene down when faced with a high-power pulse.

Different research groups have suggested various ways to pair graphene into the fiber laser cavity to solve these problems, including the use of D-shaped fiber or tapered fiber coupling side instantaneous field, hollow graphene-filled fiber absorber, and

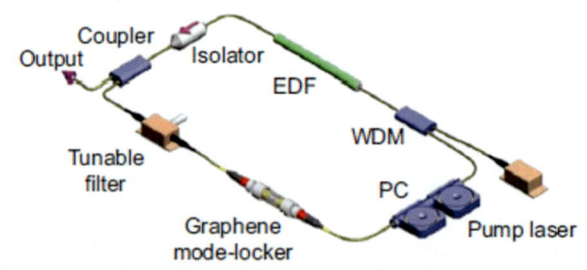

Figure 12.8 Circular laser cavity based on graphene mode-locked device. Reused with permission from Ref. 82.

graphene nanoparticle embedding in the photonic crystal fiber.[83,84] It is reported that negative dispersion erbium-doped fiber laser cavity usually produces just a few tens of mill watts of pulse and pulse power is too low in terms of output power. Currently, the maximum energy of a single pulse is actually 7.3 nJ, and the effect is 3.5 W of high pump capacity. As a consequence, a ytterbium-doped fiber with higher efficiency of light conversion is used.[85] The positive dispersion cavity, in which the Sulzberger-Landau double gold mechanism replaces the conventional nonlinear Schrodinger mechanism, allows for higher chirp pulse power and substantially improved mode-locked pulse output power. The mode-locked pulse energy obtained by this approach has currently achieved 10.2 nJ at best. Another important way to produce high-energy output pulses is to amplify the chirp pulse to the mode-locked pulse with a peripheral circuit.[84,85] To achieve chirp pulse amplification of the graphene mode-locked femtosecond laser output, Cunning et al.[86] used an alternative light route without affecting the beam efficiency and they obtained 1 W average output power and 20 nJ single pulse energy. To give maximum exposure to the graphene ultrafast electronic relaxation and obtain ultrashort mode-locked pulses, Cunning et al.[86] developed a low loss graphene saturable absorption mirror by depositing pure graphene polymer on the mirror plated with a 250 nm gold film and using the linear optical fiber cavity architecture, which shortened the width of the graphene mode-locked laser to below 200 fs. The harmonic mode-locking can be tens of times to increase the output pulse repetition rate. Currently, harmonic mode-locked graphene fibers of 1 and 1.5 μm band have been successfully applied and the elevated repetition rate has reached 2.22 GHz.

12.5.1.2 Application of graphene in the Q-switched fiber laser

The optical pulse length can reach the level of ns by Q-switching technology and the peak power can reach the level of MW. Using the fiber laser, Q-switching technology can produce more fast bursts of energy than with the optical fiber, mode-locking technology, and graphene has significant depth of modulation as well as wideband working characteristics that completely satisfy the criteria for producing Q-switching devices. The sandwich configuration is the main way to incorporate graphene into the Q-switching fiber laser cavity since it is identical to a fiber mode-locked laser. Graphene is usually placed between two FC/PC optical fiber interfaces, to achieve high-power pulse output using the tapered fiber coupling graphene. At present, graphene Q-switching fiber lasers of 1, 1.5, 2, and even 2.78 μm have been documented although wavelength-tunable graphene Q-switching fiber lasers were successfully constructed at the same time, with the highest tuning range up to 50 nm.[87] The efficiency of the graphene Q-switching fiber laser has gradually improved with in-depth testing. For example, Liu et al.[87] deposited oxidized graphene on a tapered optical fiber, then fixed it in a u-shaped base, positioning it as a high-energy saturable absorber in a linear laser cavity, with double-cladding thulium-doped fiber as the medium of gain, to attain a high-power pulse output. It is also obvious that in the optical Q-switching fiber laser, Graphene demonstrated superior performance, which completely demonstrates its advantages in the preparation of a Q-switching laser and provides a new way of creating a practical high-energy laser application.[87]

12.5.1.3   Application of graphene in the solid-state laser

In 2010 Tan et al.[88] performed research on graphene mode-locked solid-state laser by depositing the graphene on the quartz substrate and as a saturable absorber into stable Nd: YAG lasers, and they obtained a mode-locked pulse of 4 ps width at 1064 nm. Around the same time, a study has made progress on the graphene Q-switching solid laser. Yu et al.[89] achieved Q-switching, using graphene generated with the SiC epitaxial method in the Nd: YAG laser, obtaining an output pulse of 159.2 nJ and a pulse width of 161 ns. Xu et al.[90] used the chemical vapor deposition process to prepare the modulation depth of graphene near 100% and stated that the average output power is up to 1.6 W. In the ultrafast optical sector, the research findings undeniably indicate a great potential for graphene. Graphene has currently been subjected to solid-state lasers such as Ti: Sapphire, Nd:KLu $(WO_4)_2$,[90] Nd:YAG, and Cr:ZnS, and wavelengths of 800 nm, 1 μm, 1.25 μm, 1.4 μm, 1.5 μm, and 2 μm are covered by the output pulse. Therefore, the overall output power approached the watt level, and the pulse width of the output was less than 100 fs. The massive performance of the application of graphene in solid-state laser fields has been shown by these parameters.

Both optical fiber lasers and solid lasers have their own benefits and drawbacks. The characteristics of fiber lasers are small volume and good beam quality, and it is simple to incorporate the fiber design into the device. Solid-state lasers, on the other hand, have higher peak power and output energy and smaller pulse widths. Optical fiber laser uses fiber doped with rare earth elements as the medium of gain, while solid-state laser typically uses doping ions or other clear crystal active substances as the medium of gain. The quantum fluorescence efficiency of this type of gain medium is higher and is more likely to achieve a laser output of high intensity and improved beam quality. Thus, solid-state lasers, which are stable and require better beam efficiency, are more suited for the environment.

12.5.2   Application of graphene in photodetectors

1. In photodetectors, photodiodes are very critical components that are used. They generate current through the sensing of light signals. Photons transfer their energy to electrons in photodiodes, which in turn allows them to jump from the band of valance to the band of conduction. Graphene tends to interact with the electromagnetic range that covers most of the region that is visible. With a theoretical limit of 500 GHz, the frequency response of graphene can reach more than 40 GHz. The application of graphene at a wavelength of 1.5 μm on fast communication links generates a strong photonic response.[91]
2. The relation of graphene to the semiconductor is another means of producing current from the photodetector. Current leakage, one of the disadvantages of the graphene photodetector, can be minimized by reducing the bandgap or by coating a dielectric film on the graphene surface.[91] The allocation of metallic nanoparticles on the graphene film is another process. This generates a junction, so the small photodetectors and sensitivity enhancers behave like metallic nanoparticles.[91]

12.5.3   Application of graphene in optical modulators

In current optical systems, optical modulators (OMs) are a key device. A vital aspect of optical circuits is integrated OMs with high bandwidth, limited size, and broadband optical spectrum. Therefore, in the last few decades, semiconductor OMs have been studied extensively although it has faced several inherent obstacles. In electronics and photonics, Graphene has recently been used in the manufacture of OMs because of its unusual optical properties, which offer good performance, such as broadband response, high modulation speed, and high depth of modulation. Here, we present some recent advances in OMs based on graphene. It has been well known that characteristic parameters, such as strength, amplitude, wavelength, phase, and polarization, occur in optical signals. The so-called optical modulation is to change one or more of these parameters and simultaneously encrypt them with different external sources of energy. Optical modulation, accomplished by electro-optics, thermo-optics, and acousto-optics, is a way of altering the effective refractive index of a substance that accounts for the light propagation behavior. In addition, electro-optic modulation, through which an electric field can be applied across graphene to change the effective linear susceptibility and triggering changes in both the actual and imaginary refractive indices. Electro-refraction concerns the difference in the actual portion of the refractive index, whereas electroabsorption concerns the change in the theoretical equivalent. The modulators may be categorized as either electro refractive or electro-absorptive, based on which effect is manifested.

12.5.3.1 Application of graphene in waveguide optical modulators

Since its discovery in 2004,[5–10] Graphene has drawn rising interest due to its outstanding optical and electrical properties. In addition, large optical absorption can be controlled by electrical gating, and optical transitions can be controlled by changing the electronic Fermi level. Due to its ultrahigh carrier mobility,[14,15] Graphene has a wide potential to be used for OMs, and compatibility with CMOS techniques gives it a promising future for OMs. The authors manufactured another system based on double-layer graphene isolated by a thin layer of Al_2O_3[19] on the basis of previous fabrications. To form a modulating electric field, the voltage can be applied to the graphene such that the degree of Fermi can be modified. Optical absorption occurs only when the degree of Fermi falls between the $\pm hv_0/2$ levels; otherwise, the graphene is translucent. This form of design eliminates drawbacks inherited from silicon photonics, such as high loss of insertion and reduced stability of the carrier. Meanwhile, due to greater interaction between the light field and double-layer graphene, the extinction ratio increases dramatically. In Fig. 12.9, the electroabsorption modulator structure is schematically illustrated.

12.5.4 Graphene-based polarizers

Two types of graphene-based polarizers can be used. One is an in-line fiber polarizer, and the other is graphene nanoribbon arrays. The in-line fiber polarizer is based on the polarization-selective coupling between the evanescent field and graphene while in graphene nanoribbon arrays the light can couple to the localized plasmons of the ribbon only when polarized perpendicular to the ribbon axis.[85]

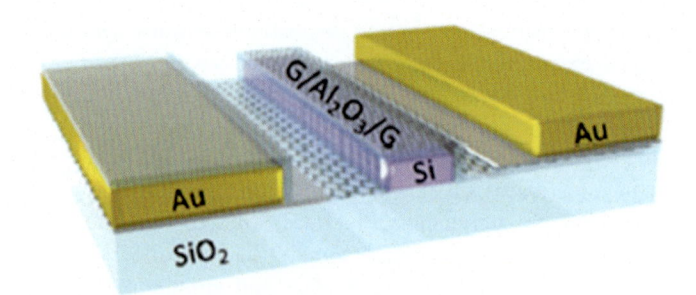

Figure 12.9 Three-dimensional schematic of the device.
Reused with permission from Ref. 92

12.6   Applications of metal oxide—carbon composites in electrochemical energy devices

12.6.1   Batteries

Flexible batteries contribute a vital role in electrochemical energy-storage systems. Generally, we have three kinds of batteries: (1) secondary batteries, (2) ultrabatteries, and (3) flow batteries. Secondary batteries are also known as rechargeable batteries because these batteries can be charged and discharged many times. Energy stored by these batteries is due to well-defined electrochemical reactions. These batteries have gained much attention in extensive applications owing to their unique features, such as long lifetimes and environmentally friendly behavior. Large size, heavy weight, and high cost are a few barriers to large-scale applications. The ultrabattery is another magnificent alternative for the storage and conversion of energy. It is the mixture of lead-acid battery and ultracapacitor in a single cell and an electrolyte. Flow batteries work by the motion of liquid across the membrane. The voltage of these batteries can be determined with the help of the Nernst equation. These batteries have some advantages, such as quick response time and flexible configuration, but the energy densities of such batteries are less than as compared with the secondary batteries.[93] Many researchers designed carbon-based MO composites to improve the efficiency of batteries.

Zhou et al.[94] successfully fabricated the graphene nanosheets (GNSs) and Fe_3O_4 nanocomposite by using in situ reduction technique. They used this composite as an anode for LIBs. This composite revealed the reversible capacity of about 1026 mAh/g after 39 cycles of charging/discharging at a rate of 35 mA/g and 580 mAh/g after 1000 cycles of charging/discharging at the rate of 700 mA/g, as shown in Fig. 12.10A and B. Furthermore, it showed improved cyclic stability and remarkably improved rate ability. The improved results were due to features such as (1) GNSs playing a vital role as a highway for the transportation of electrons, (2) the porous structure of the composite, which facilitates the transport of ions, and (3) the presence of GNSs, which prevents the agglomeration of Fe_3O_4 nanoparticles.[94]

Zhang's group[95] also fabricated carbon-coated Fe_3O_4 nanospindles and characterizations using SEM, TEM, XRD, and electrochemical techniques. This material was

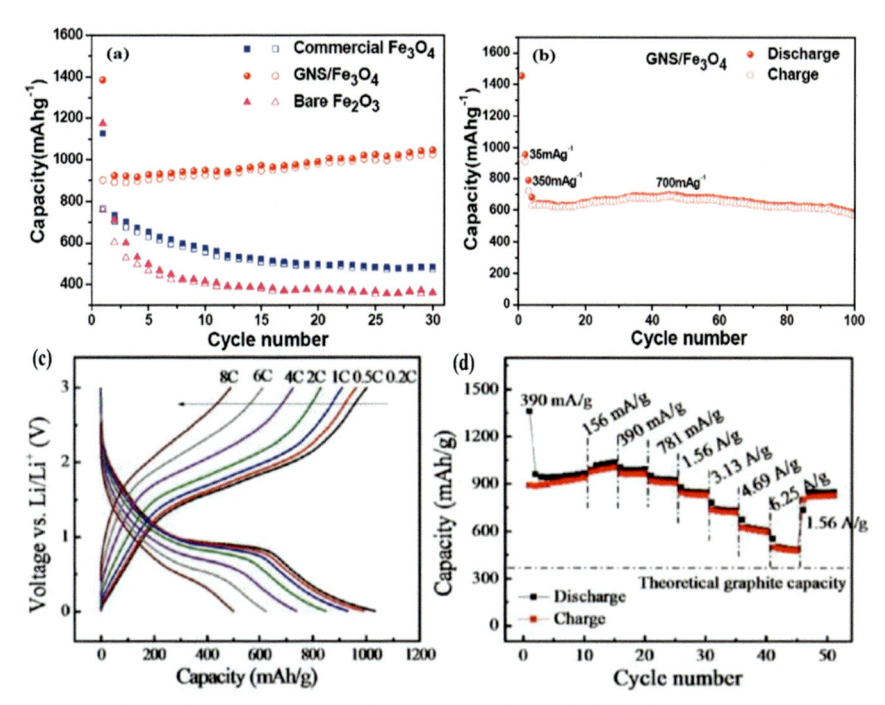

Figure 12.10 (A) Cycling activity of Fe_3O_4 nanoparticles, GNS/Fe_3O_4 nanocomposite and pristine Fe_2O_3 nanoparticles (solid symbols = discharge, hollow symbols = charge), (B) Cycling activity of GNS/Fe_3O_4 nanocomposite at 700 mA/g for 100 cycles, (C) Charging/discharging curves and time (0.2−8 C, 1C = 780 mA/g) and (D) Rate ability of electrode for various current densities (0.156−6.25 A/g).
(A, B) Reprinted with permission from Ref. 94. Copyright 2010, American Chemical Society; (C, D) Reprinted with permission from Ref. 96. Copyright 2012, American Chemical Society.

used as an anode for LIBs and exhibited excellent reversible specific capacities of 745 mAh/g and 600 mAh/g at C/5 and C/2, respectively. Furthermore, high coulombic efficiency, improved cycling ability, and better rate capability were observed over the pure spindles of hematite and particles of magnetite. The results were improved due to the coating of carbon on Fe_3O_4 nanospindles.[95] Jia et al.[96] synthesized the nanocomposites of Fe_3O_4 spheres and CNTs using the aerosol spray and vacuum filtration technique. The synthesized composite showed many characteristics that are essential for the excellent performance of anodes, such as fast transport of ions, the durability of the structure, and excellent conductivity. The electrode having a thickness of ~35 μm can show a capacity of 994 mAh/g and excellent recharging rates as shown in Fig. 12.10C and D.[96] Yang et al.[55] successfully fabricated the nanocomposites of Fe_3O_4 and carbon by using the in situ technique. The synthesized material consists of same-sized nanoparticles of Fe_3O_4 that were embedded in the matrix of carbon and the size is compatible with the average structures of crystalline. They used this material as an anode for LIBs and observed that the composite exhibited improved cycling activity at both (high and low) values of current densities.[55]

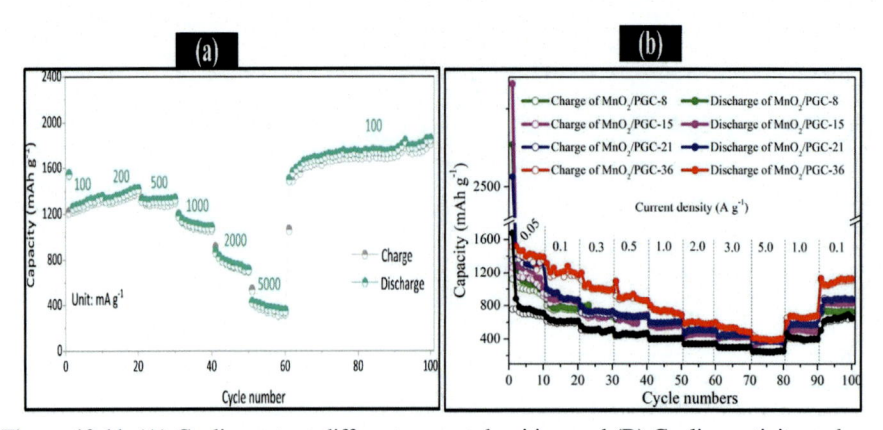

Figure 12.11 (A) Cycling rate at different current densities, and (B) Cycling activity and coulombic ability for 400 cycles.
(A) Reprinted with permission from Ref. 97. Copyright 2020, Elsevier; (B) Reprinted with permission from Ref. 98. Copyright 2020, American Chemical Society.

Ding et al.[97] fabricated the hollow spheres of nitrogen-doped multi-shelled Co_3O_4 and carbon. The morphology of the Co- Metal-organic framework was spherical and uniform having a diameter of 1 μm. It can be seen through SEM images of MS–Co_3O_4@PC that the morphology remains the same, but the diameter of microspheres reduced to 500 nm. They used it as an anode for LIBs and reported that the material exhibited excellent electrochemical activity. The material showed the reversible capacity of 1701 mAh/g and retained it after 50 cycles at the current density of 100 mA/g. A reversible capacity of 427 mAh/g was retained at the large current density of 5000 mA/g as depicted in Fig. 12.11A.[97] Zeng et al.[98] successfully fabricated the composite of MnO_2 and porous graphitic carbon (PGC) with the help of in situ precipitation technique. The synthesized composites maintained a well-proportioned attached porous geometry and appropriate specific surface areas of 190–229 m^2/g. The synthesized nanocomposite revealed enhanced reversible range, cycling stability, and rate performance as an electrode for LIBs due to the collaborative effect of MnO_2 and PGC. Particularly, the MnO_2/PGC-36 composite exhibited the reversible capacity of 1516 mAh/g at 0.05 A/g. Furthermore, it showed prolonged cycling stability with 90% retention after 400 cycles as depicted in Fig. 12.11B.[98]

12.6.2 Supercapacitors

In recent years, renewable energy-storage devices have attracted many researchers around the globe. Particularly, LIBs were successfully utilized for energy-storage systems.[99] But, the use of LIBs was limited owing to its high cost and shorter life cycle. A long life cycle, large power density, and economical and safe energy-storage devices will be required for the forthcoming generation of hybrid vehicles and high-power electronic instruments. SCs are considered potential alternatives to LIBs owing to their

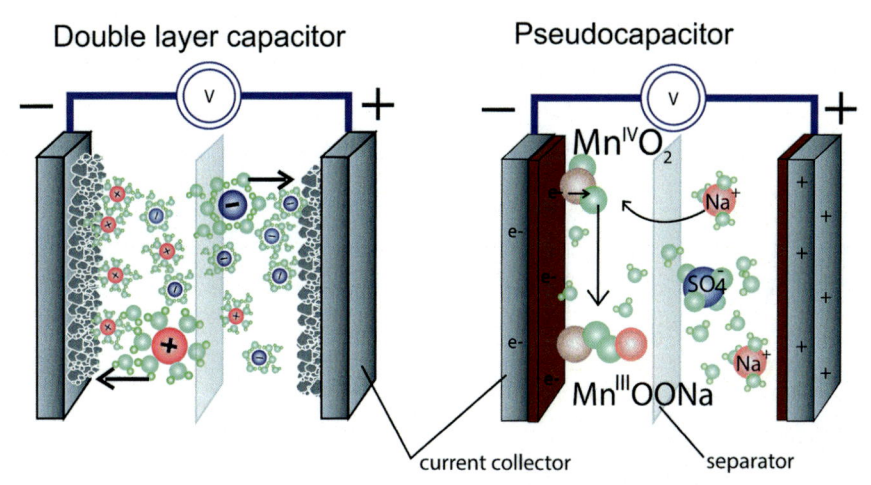

Figure 12.12 Comparison of double-layered capacitor and pseudocapacitor. Reprinted with permission from Ref. 100. Copyright 2013, Royal Society of Chemistry.

unique features, such as high power density, a broad temperature range for performance, and a quick charging/discharging mechanism.[93] The differentiation between pseudocapacitor and double-layer capacitor is illustrated in Fig. 12.12.

Nanostructured materials of carbon owing to their excellent electrochemical performance are normally used as electric double-layer capacitor electrodes. Therefore, many attempts were made to revamp the efficiency of SCs by fabricating the composites of transition MOs and carbon materials. It is expected that the synthesized nanocomposites can improve the efficiency of SCs.[101]

Kumar et al.[102] successfully fabricated the 3-D hierarchical morphology of 0-D Co_3O_4 nanobeads, 1-D CNTs, and 2-D graphene by utilizing the microwave irradiation technique. They used this material as an SC electrode. The distinctive structure of the synthesized nanocomposite exhibited excellent electrochemical activity. This material showed a high specific capacitance of 600 F/g at 0.7 A/g (30 wt.% KOH) as shown in Fig. 12.13A. Moreover, the material exhibited 94.5% retention after 5000 cycles of charging/discharging at 10 A/g that proves its good cycling ability as shown in Fig. 12.13B.

Yu et al.[103] synthesized the nanostructured sponges of hierarchical graphene and MnO_2 by utilizing the facile "dip and dry" technique as shown in Fig. 12.13C. They used commercial sponges as a framework during the synthesis process. They reported that the fabricated nanocomposite can be driven at a maximum scan scale of 200 V/s and also exhibited exceptional cycling activity with 90% retention after 10,0000 cycles at 10 A/g as shown in Fig. 12.13D. The developed SC could be used at a large scale owing to the unique features of this material, such as economic, easy preparation, excellent cycling activity, and high energy and power density.

Zhang et al.[100] synthesized the MnO_2/CF fibers by the electrodeposition of ultrathin nanosheets of MnO_2 on yarns of carbon fiber. The synthesized material as an electrode

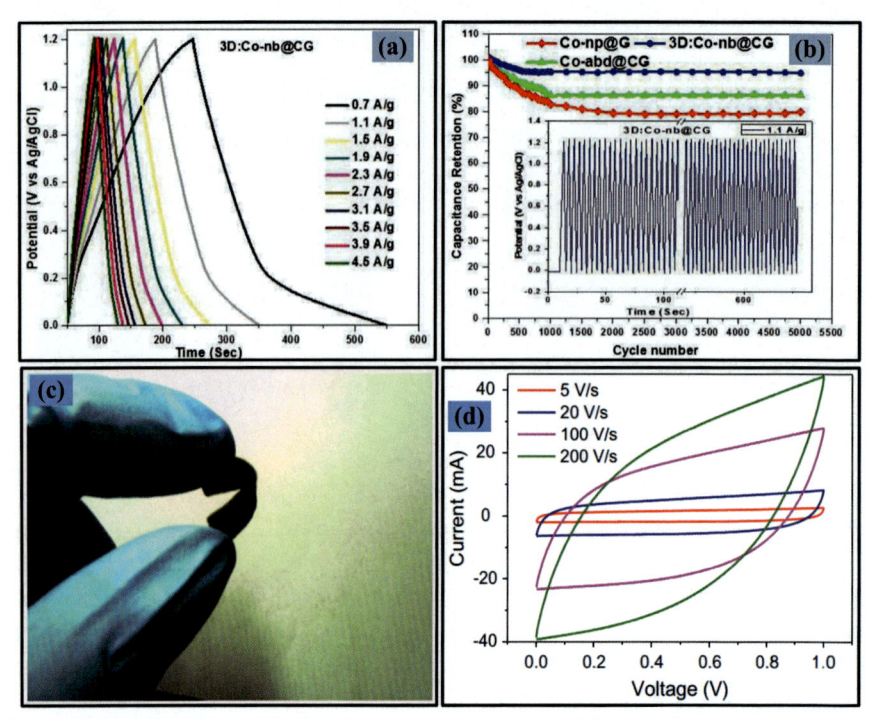

Figure 12.13 (A) Charging/discharging curves of 3-D:Co-nb/CG at various current densities, (B) Capacitance retention of various nanocomposites at 1.1 A/g; (inset) charging/discharging graph of 3-D:Co-nb/CG based capacitor electrode at 1.1 A/g, (C) Image of synthesized RGO/ MnO_2 sponges and (D) Cyclic voltammetry graphs of sponge@RGO based device at various voltage scan scales (5−200 V/s).
(A, B) Reprinted with permission from Ref. 102. Copyright 2015, American Chemical Society, and (C, D) Reprinted with permission from Ref. 103. Copyright 2013, Elsevier.

exhibited a wide volumetric capacitance of 58.7 F/cm^3 owing to the collaborative effect among the MnO_2 and CFs. They developed an SC by utilizing the MnO_2/CFs fiber as an anode and cathode and reported excellent volumetric energy density of 3.8 mWh/cm^3 at a power density of 89 mW/cm^3, enhanced flexibility and extraordinary cycling activity with capacitance fade of 14.2% after 10,000 cycles. Different samples of MnO_2/CFs showed triangular graphs at 0.1 A/g (Fig. 12.14A), which revealed that it has better electrochemical capacitive features. Hu et al.[104] directly synthesized the nanocone forest of hierarchical mesoporous $NiFe_2O_4$ (NFO) on carbon textile (Fig. 12.14B and C) using the hydrothermal technique. The dense bunch of fibers exhibited an average diameter of 5 μm and a specific surface area of 593.60 m^2/g. As an electrode, the as-synthesized material presented a specific capacitance of 697 and 303 F/g at 5 and 75 mV/s. They developed an SC device by utilizing the as-synthesized material and reported a high value of capacitance (584 F/g at 5 mV/s) and improved cycle performance with capacitance detention (93.57% after 10,000 charging/discharging cycles) as depicted in Fig. 12.14D. Furthermore, this material

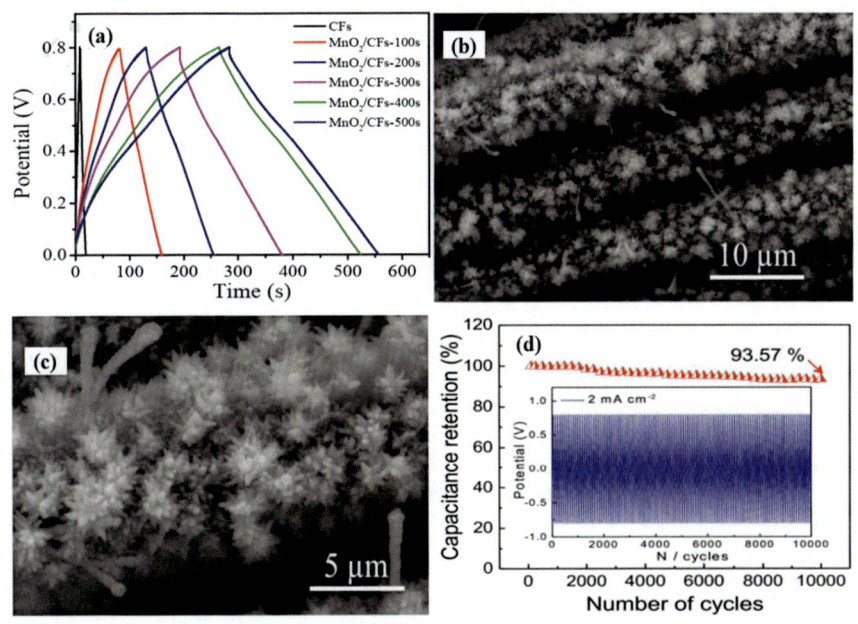

Figure 12.14 (A) Charging/discharging curves of various samples of MnO$_2$/CFs at 0.1 A/g, (B) Growth of NFO nanoforests on carbon textiles (CTs), (C) Growth of NFO nanoforests on the solo fiber of CT, and (D) Graph between capacitance retention and cycle numbers NFO nanoforest-based SC, (Inset: 10,000 cycles of charging/discharging at a uniform current density of 1 mA/cm^2.
(A) Reprinted with permission from Ref. 100. Copyright 2016, Elsevier, and (B—D) Reprinted with permission from Ref. 104. Copyright 2016, Royal Society of Chemistry.

can attain an energy density of 25.5 Wh/kg at a power density of 1372 W/kg that is significantly more than other metal oxide-based electrodes.

12.7 Conclusion

MO—C-based nanocomposites have emerged as a new family of nanomaterials due to their impressive skeletal and controllable physical characteristics that are higher as compared with pristine carbon and MOs. Researchers carried out a great deal of research on MO—C-based nanocomposites to develop advanced materials, methods, formulations, and systems that are flourishing in attempts to resolve the problems in synthesis, characterization, etc. MO—C-based nanocomposites have been developed for electrochemical energy storage and conversion (such as batteries and SCs) and optoelectronic devices (such as lasers). Despite the fact that MO—C-based nanocomposites were utilized efficiently for electrochemical energy storage and optoelectronic applications. It is expected that MO—C-based nanocomposites will provide a huge impact in electrochemical energy storage and optoelectronics fields at a large scale due to their controllable properties with considerable challenges and outlook ahead.

References

1. Chen JS, Lou XW. SnO$_2$-based nanomaterials: synthesis and application in lithium-ion batteries. *Small* 2013;**9**:1877−93.
2. Yunyun F, Xu L, Wankun Z, Yuxuan Z, Yunhan Y, Honglin Q, Xuetang X, Fan W. Spinel CoMn$_2$O$_4$ nanosheet arrays grown on nickel foam for high-performance supercapacitor electrode. *Appl Surf Sci* 2015;**357**:2013−21.
3. Reddy M, Yu T, Sow C-H, Shen ZX, Lim CT, Subba Rao G, Chowdari B. α-Fe$_2$O$_3$ nanoflakes as an anode material for Li-ion batteries. *Adv Funct Mater* 2007;**17**: 2792−9.
4. Wang X, Li X, Sun X, Li F, Liu Q, Wang Q, He D. Nanostructured NiO electrode for high rate Li-ion batteries. *J Mater Chem* 2011;**21**:3571−3.
5. Yang YJ, Li W. Hierarchical nanoflake-assembled flower-like NiCo double hydroxide@ NiC$_2$O$_4$ microspheres for high-performance supercapacitor. *Mater Technol* 2019;**34**: 571−80.
6. Senthilkumar N, Venkatachalam V, Kandiban M, Vigneshwaran P, Jayavel R, Potheher IV. Studies on electrochemical properties of hetarolite (ZnMn$_2$O$_4$) nanostructure for supercapacitor application. *Physic E Low Dimens Syst Nanostruct* 2019;**106**:121−6.
7. Wu HB, Zhang G, Yu L, Lou XWD. One-dimensional metal oxide−carbon hybrid nanostructures for electrochemical energy storage. *Nanoscale Horiz* 2016;**1**:27−40.
8. Li Y, Zhou Z, Shen P, Zhang S, Chen Z. Computational studies on hydrogen storage in aluminum nitride nanowires/tubes. *Nanotechnology* 2009;**20**:215701.
9. Yao Y, Lin Z, Li Z, Song X, Moon K-S, Wong C-p. Large-scale production of two-dimensional nanosheets. *J Mater Chem* 2012;**22**:13494−9.
10. Ansari SA, Parveen N, Kotb HM, Alshoaibi A. Hydrothermally derived three-dimensional porous hollow double-walled Mn$_2$O$_3$ nanocubes as superior electrode materials for supercapacitor applications. *Electrochim Acta* 2020;**355**:136783.
11. Mingabudinova LR, Zalogina AS, Krasilin AA, Petrova MI, Trofimov P, Mezenov YA, Ubyivovk EV, Lönnecke P, Nominé A, Ghanbaja J. Laser printing of optically resonant hollow crystalline carbon nanostructures from 1D and 2D metal−organic frameworks. *Nanoscale* 2019;**11**:10155−9.
12. Liu F, Xue D. Chemical design of complex nanostructured metal oxides in solution. *Int J Nanosci* 2009;**8**:571−88.
13. Esawi A, Morsi K, Sayed A, Taher M, Lanka S. The influence of carbon nanotube (CNT) morphology and diameter on the processing and properties of CNT-reinforced aluminium composites. *Compos A Appl Sci Manuf* 2011;**42**:234−43.
14. Yu X, Marks TJ, Facchetti A. Metal oxides for optoelectronic applications. *Nat Mater* 2016;**15**:383−96.
15. Avouris P, Freitag M. Graphene photonics, plasmonics, and optoelectronics. *IEEE J Sel Top Quant Electron* 2013;**20**:72−83.
16. Smalley RE. *Carbon nanotubes: synthesis, structure, properties, and applications.* 2003.
17. Avouris P, Chen Z, Perebeinos V. Carbon-based electronics. *Nanosci Technol* 2010: 174−84.
18. Avouris P, Freitag M, Perebeinos V. Carbon-nanotube photonics and optoelectronics. *Nat Photonics* 2008;**2**:341−50.
19. Roston E. *The carbon age: how life's core element has become civilization's greatest threat.* USA: Bloomsbury Publishing; 2010.
20. Kroto HW, Heath JR, O'Brien SC, Curl RF, Smalley RE. C60: buckminsterfullerene. *Nature* 1985;**318**:162−3.

21. Iijima S. Helical microtubules of graphitic carbon. *Nature* 1991;**354**:56—8.
22. Novoselov KS, Geim AK, Morozov SV, Jiang D, Zhang Y, Dubonos SV, Grigorieva IV, Firsov AA. Electric field effect in atomically thin carbon films. *Science* 2004;**306**:666—9.
23. Kah M, Hofmann T. The challenge: carbon nanomaterials in the environment: new threats or wonder materials? *Environ Toxicol Chem* 2015;**34**:954.
24. Peng Q, Dearden AK, Crean J, Han L, Liu S, Wen X, De S. New materials graphyne, graphdiyne, graphone, and graphane: review of properties, synthesis, and application in nanotechnology. *Nanotechnol Sci Appl* 2014;**7**:1.
25. Conder KJPSI. *Electronic and ionic conductivity in metal oxides.* 2012. p. 1—44. Switzerland.
26. Barton TJ, Bull LM, Klemperer WG, Loy DA, McEnaney B, Misono M, Monson PA, Pez G, Scherer GW, Vartuli JC. Tailored porous materials. *Chem Mater* 1999;**11**: 2633—56.
27. Guo J, Xu Y, Wang C. Sulfur-impregnated disordered carbon nanotubes cathode for lithium—sulfur batteries. *Nano Lett* 2011;**11**:4288—94.
28. Zhang Z, Zhang J, Chen N, Qu L. Graphene quantum dots: an emerging material for energy-related applications and beyond. *Energy Environ Sci* 2012;**5**:8869—90.
29. Ricciardulli AG, Blom PW. Solution-processable 2D materials applied in light-emitting diodes and solar cells. *Adv Mater Technol* 2020;**5**:1900972.
30. Surya C, Hsiang T. Surface mobility fluctuations in metal-oxide-semiconductor field-effect transistors. *Phys Rev B* 1987;**35**:6343.
31. Mehandru R, Luo B, Kim J, Ren F, Gila B, Onstine A, Abernathy C, Pearton S, Gotthold D, Birkhahn R. AlGaN/GaN metal—oxide—semiconductor high electron mobility transistors using Sc_2O_3 as the gate oxide and surface passivation. *Appl Phys Lett* 2003;**82**:2530—2.
32. Liu WW, Feng YQ, Yan XB, Chen JT, Xue QJ. Supercapacitors: superior micro-supercapacitors based on graphene quantum dots (Adv. Funct. Mater. 33/2013) *Adv Funct Mater* 2013;**23**:4164.
33. Zhu Y, Ji X, Pan C, Sun Q, Song W, Fang L, Chen Q, Banks CE. A carbon quantum dot decorated RuO 2 network: outstanding supercapacitances under ultrafast charge and discharge. *Energy Environ Sci* 2013;**6**:3665—75.
34. Zhu Y, Wu Z, Jing M, Hou H, Yang Y, Zhang Y, Yang X, Song W, Jia X, Ji X. Porous $NiCo_2O_4$ spheres tuned through carbon quantum dots utilised as advanced materials for an asymmetric supercapacitor. *J Mater Chem* 2015;**3**:866—77.
35. Yuan L, Lu X-H, Xiao X, Zhai T, Dai J, Zhang F, Hu B, Wang X, Gong L, Chen J. Flexible solid-state supercapacitors based on carbon nanoparticles/MnO2 nanorods hybrid structure. *ACS Nano* 2012;**6**:656—61.
36. Lan B, Zheng X, Cheng G, Han J, Li W, Sun M, Yu L. The art of balance: engineering of structure defects and electrical conductivity of α-MnO_2 for oxygen reduction reaction. *Electrochim Acta* 2018;**283**:459—66.
37. Fan L, Tang L, Gong H, Yao Z, Guo R. Carbon-nanoparticles encapsulated in hollow nickel oxides for supercapacitor application. *J Mater Chem* 2012;**22**:16376—81.
38. Zhi M, Xiang C, Li J, Li M, Wu N. Nanostructured carbon—metal oxide composite electrodes for supercapacitors: a review. *Nanoscale* 2013;**5**:72—88.
39. Zhang W-D, Xu B, Jiang L-C. Functional hybrid materials based on carbon nanotubes and metal oxides. *J Mater Chem* 2010;**20**:6383—91.
40. Gangwar J, Dey KK, Tripathi SK, Wan M, Yadav RR, Singh RK, Srivastava AK. NiO-based nanostructures with efficient optical and electrochemical properties for high-performance nanofluids. *Nanotechnology* 2013;**24**:415705.

41. Mujtaba J, Sun H, Zhao Y, Xiang G, Xu S, Zhu J. High-performance lithium storage based on the synergy of atomic-thickness nanosheets of TiO_2 (B) and ultrafine Co_3O_4 nanoparticles. *J Power Sources* 2017;**363**:110−6.

42. Zhao Y, Dong W, Riaz MS, Ge H, Wang X, Liu Z, Huang F. "Electron-sharing" mechanism promotes Co@ Co3O4/CNTs composite as the high-capacity anode material of lithium-ion battery. *ACS Appl Mater Interfaces* 2018;**10**:43641−9.

43. Xu L, Jiang Q, Xiao Z, Li X, Huo J, Wang S, Dai L. Plasma-engraved Co_3O_4 nanosheets with oxygen vacancies and high surface area for the oxygen evolution reaction. *Angew Chem* 2016;**128**:5363−7.

44. Santos C, Lado JJ, García-Quismondo E, Rodríguez IV, Palma J, Anderson MA, Vilatela JJ. Interconnected metal oxide CNT fibre hybrid networks for current collector-free asymmetric capacitive deionization. *J Mater Chem* 2018;**6**:10898−908.

45. Zheng X, Zhou X, Xu J, Zou L, Nie W, Hu X, Dai S, Qiu Y, Yuan N. Highly stretchable CNT/MnO_2 nanosheets fiber supercapacitors with high energy density. *J Mater Sci* 2020; **55**:8251−63.

46. Seok D, Jeong Y, Han K, Yoon DY, Sohn H. Recent progress of electrochemical energy devices: metal oxide−carbon nanocomposites as materials for next-generation chemical storage for renewable energy. *Sustainability* 2019;**11**:3694.

47. Mannix AJ, Zhou X-F, Kiraly B, Wood JD, Alducin D, Myers BD, Liu X, Fisher BL, Santiago U, Guest JR. Synthesis of borophenes: anisotropic, two-dimensional boron polymorphs. *Science* 2015;**350**:1513−6.

48. Nawz T, Safdar A, Hussain M, Sung Lee D, Siyar M. Graphene to advanced MoS_2: a review of structure, synthesis, and optoelectronic device application. *Crystals* 2020;**10**: 902.

49. Wang Q, Jiao L, Du H, Wang Y, Yuan H. Fe_3O_4 nanoparticles grown on graphene as advanced electrode materials for supercapacitors. *J Power Sources* 2014;**245**:101−6.

50. Zhang C, Geng X, Tang S, Deng M, Du Y. $NiCo_2O_4$@ rGO hybrid nanostructures on Ni foam as high-performance supercapacitor electrodes. *J Mater Chem* 2017;**5**:5912−9.

51. Wu Z-S, Zhou G, Yin L-C, Ren W, Li F, Cheng H-M. Graphene/metal oxide composite electrode materials for energy storage. *Nanomater Energy* 2012;**1**:107−31.

52. Dong X-C, Xu H, Wang X-W, Huang Y-X, Chan-Park MB, Zhang H, Wang L-H, Huang W, Chen P. 3D graphene−cobalt oxide electrode for high-performance supercapacitor and enzymeless glucose detection. *ACS Nano* 2012;**6**:3206−13.

53. Li S, Yang K, Ye P, Ma K, Zhang Z, Huang Q. Three-dimensional porous carbon/Co_3O_4 composites derived from graphene/Co-MOF for high performance supercapacitor electrodes. *Appl Surf Sci* 2020;**503**:144090.

54. Malkhandi S, Trinh P, Manohar AK, Jayachandrababu K, Kindler A, Prakash GS, Narayanan S. Electrocatalytic activity of transition metal oxide-carbon composites for oxygen reduction in alkaline batteries and fuel cells. *J Electrochem Soc* 2013;**160**: F943.

55. Yang Z, Shen J, Archer LAJJOMC. An in situ method of creating metal oxide−carbon composites and their application as anode materials for lithium-ion batteries. *J Mater Chem* 2011;**21**:11092−7.

56. Chou S-L, Lu L, Wang J-Z, Rahman MM, Zhong C, Liu H-K. The compatibility of transition metal oxide/carbon composite anode and ionic liquid electrolyte for the lithium-ion battery. *J Appl Electrochem* 2011;**41**:1261.

57. Hasa I, Verrelli R, Hassoun J. Transition metal oxide-carbon composites as conversion anodes for sodium-ion battery. *Electrochim Acta* 2015;**173**:613−8.

58. Cai Y. *Carbon/metal oxide composites and their application in lithium-ion batteries*. 2013.

59. Faraji S, Ani FN. Microwave-assisted synthesis of metal oxide/hydroxide composite electrodes for high power supercapacitors—a review. *J Power Sources* 2014;**263**:338—60.

60. Sengul AB, Asmatulu EJECL. *Toxicity of metal and metal oxide nanoparticles: a review.* 2020. p. 1—25.

61. Prakash J, Khan S, Chauhan S, Biradar A. Metal oxide-nanoparticles and liquid crystal composites: a review of recent progress. *J Mol Liq* 2020;**297**:112052.

62. Motshekga SC, Pillai SK, Sinha Ray S, Jalama K, Krause R. Recent trends in the microwave-assisted synthesis of metal oxide nanoparticles supported on carbon nanotubes and their applications. *J Nanomater* 2012;**2012**.

63. Das D, Plazas-Tuttle J, Sabaraya IV, Jain SS, Sabo-Attwood T, Saleh NB. An elegant method for large scale synthesis of metal oxide—carbon nanotube nanohybrids for nano-environmental application and implication studies. *Environ Sci Nano* 2017;**4**:60—8.

64. Zheng W, Zhang P, Chen J, Tian W, Zhang Y, Sun Z. Microwave-assisted synthesis of three-dimensional MXene derived metal oxide/carbon nanotube/iron hybrids for enhanced lithium-ions storage. *J Electroanal Chem* 2019;**835**:205—11.

65. Yan J, Fan Z, Wei T, Cheng J, Shao B, Wang K, Song L, Zhang M. Carbon nanotube/MnO_2 composites synthesized by microwave-assisted method for supercapacitors with high power and energy densities. *J Power Sources* 2009;**194**:1202—7.

66. Danks AE, Hall SR, Schnepp Z. The evolution of 'sol—gel'chemistry as a technique for materials synthesis. *Mater Horiz* 2016;**3**:91—112.

67. Mączka M, Pasierb P. Sol-gel synthesis of metal (V, W, Zn) oxide-carbon nanocomposites as cathode materials for Al-ion batteries. *Ceram Int* 2019;**45**:11041—9.

68. Wang M, Jin C, Luo Q, Kim EJ. Sol-gel derived TiO_2—carbon composites with adsorption-enhanced photocatalytic activity and gas sensing performance. *Ceram Int* 2020;**46**:18608—13.

69. Thauer E, Zakharova G, Wegener S, Zhu Q, Klingeler R. Sol-gel synthesis of Li_3VO_4/C composites as anode materials for lithium-ion batteries. *J Alloys Compd* 2021;**853**:157364.

70. Huang G, Lu C-H, Yang H-H. *Magnetic nanomaterials for magnetic bioanalysis.* Elsevier; 2019.

71. Senapati S, Maiti P. *Emerging bio-applications of two-dimensional nanoheterostructure materials.* Elsevier; 2020.

72. Maletić M, Vukčević M, Kalijadis A, Janković-Častvan I, Dapčević A, Laušević Z, Laušević M. Hydrothermal synthesis of TiO_2/carbon composites and their application for removal of organic pollutants. *Arab J Chem* 2019;**12**:4388—97.

73. Zhang Y, Zheng J, Wang Q, Hu T, Meng C. Hydrothermal synthesis of vanadium di-oxides/carbon composites and their transformation to surface-uneven V_2O_5 nanoparticles with high electrochemical properties. *RSC Adv* 2016;**6**:93741—52.

74. Chen M, Shao L-L, Li J-J, Pei W-J, Chen M-K, Xie X-H. One-step hydrothermal synthesis of hydrophilic Fe_3O_4/carbon composites and their application in removing toxic chemicals. *RSC Adv* 2016;**6**:35228—38.

75. Lee J, Jung YS, Warren SC, Kamperman M, Oh SM, DiSalvo FJ, Wiesner U. Direct access to mesoporous crystalline TiO_2/carbon composites with large and uniform pores for use as anode materials in lithium ion batteries. *Macromol Chem Phys* 2011;**212**:383—90.

76. Hou Y, Cheng Y, Hobson T, Liu J. Design and synthesis of hierarchical MnO_2 nano-spheres/carbon nanotubes/conducting polymer ternary composite for high performance electrochemical electrodes. *Nano Lett* 2010;**10**:2727—33.

77. Williams G, Seger B, Kamat PV. TiO_2-graphene nanocomposites. UV-assisted photo-catalytic reduction of graphene oxide. *ACS Nano* 2008;**2**:1487—91.

78. Yan J, Fan Z, Wei T, Qian W, Zhang M, Wei F. Fast and reversible surface redox reaction of graphene$-MnO_2$ composites as supercapacitor electrodes. *Carbon* 2010;**48**:3825−33.

79. Verrelli R, Hassoun J. High-capacity NiO−(mesocarbon microbeads) conversion anode for lithium-ion battery. *ChemElectroChem* 2015;**2**:988−94.

80. Verrelli R, Hassoun J, Farkas A, Jacob T, Scrosati B. A new, high performance CuO/$LiNi_{0.5}Mn_{1.5}O_4$ lithium-ion battery. *J Mater Chem A* 2013;**1**:15329−33.

81. Bao Q, Zhang H, Wang Y, Ni Z, Yan Y, Shen ZX, Loh KP, Tang DY. Atomic-layer graphene as a saturable absorber for ultrafast pulsed lasers. *Adv Funct Mater* 2009;**19**: 3077−83.

82. Lin Y-H, Yang C-Y, Liou J-H, Yu C-P, Lin G-R. Using graphene nano-particle embedded in photonic crystal fiber for evanescent wave mode-locking of fiber laser. *Opt Express* 2013;**21**:16763−76.

83. Song Y-W, Jang S-Y, Han W-S, Bae M-K. Graphene mode-lockers for fiber lasers functioned with evanescent field interaction. *Appl Phys Lett* 2010;**96**:051122.

84. Choi SY, Cho DK, Song Y-W, Oh K, Kim K, Rotermund F, Yeom D-I. Graphene-filled hollow optical fiber saturable absorber for efficient soliton fiber laser mode-locking. *Opt Express* 2012;**20**:5652−7.

85. Bao Q, Hoh H, Zhang Y. *Graphene photonics, optoelectronics, and plasmonics*. CRC Press; 2017.

86. Cunning B, Brown C, Kielpinski D. Low-loss flake-graphene saturable absorber mirror for laser mode-locking at sub-200-fs pulse duration. *Appl Phys Lett* 2011;**99**:261109.

87. Liu C, Ye C, Luo Z, Cheng H, Wu D, Zheng Y, Liu Z, Qu B. High-energy passively Q-switched 2 μm Tm 3+-doped double-clad fiber laser using graphene-oxide-deposited fiber taper. *Opt Express* 2013;**21**:204−9.

88. Tan W, Su C, Knize R, Xie G, Li L-J, Tang D. Mode locking of ceramic Nd: yttrium aluminum garnet with graphene as a saturable absorber. *Appl Phys Lett* 2010;**96**:031106.

89. Yu WJ, Liu Y, Zhou H, Yin A, Li Z, Huang Y, Duan X. Highly efficient gate-tunable photocurrent generation in vertical heterostructures of layered materials. *Nat Nanotechnol* 2013;**8**:952−8.

90. Xu J-L, Li X-L, Wu Y-Z, Hao X-P, He J-L, Yang K-J. Graphene saturable absorber mirror for ultra-fast-pulse solid-state laser. *Opt Lett* 2011;**36**:1948−50.

91. Kusmartsev F, Wu W, Pierpoint M, Yung K. *Application of graphene within optoelectronic devices and transistors*. Springer; 2015.

92. Liu M, Yin X, Zhang X. Double-layer graphene optical modulator. *Nano Letters* 2012;**12**: 1482−5.

93. Butt FK, Ullah S, Ahmad J, Rehman SU, Tariq Z. *Graphitic carbon nitride/metal oxides nanocomposites and their applications in engineering*. Springer; 2020.

94. Zhou G, Wang D-W, Li F, Zhang L, Li N, Wu Z-S, Wen L, Lu GQ, Cheng H-M. Graphene-wrapped Fe_3O_4 anode material with improved reversible capacity and cyclic stability for lithium ion batteries. *Chem Mater* 2010;**22**:5306−13.

95. Zhang WM, Wu XL, Hu JS, Guo YG, Wan LJ. Carbon coated Fe_3O_4 nanospindles as a superior anode material for lithium-ion batteries. *Adv Funct Mater* 2008;**18**:3941−6.

96. Jia X, Chen Z, Cui X, Peng Y, Wang X, Wang G, Wei F, Lu Y. Building robust architectures of carbon and metal oxide nanocrystals toward high-performance anodes for lithium-ion batteries. *ACS Nano* 2012;**6**:9911−9.

97. Ding Y, Hu L, He D, Peng Y, Niu Y, Li Z, Zhang X, Chen S. Design of multishell microsphere of transition metal oxides/carbon composites for lithium ion battery. *Chem Eng J* 2020;**380**:122489.

98. Zeng H, Xing B, Zhang C, Chen L, Zhao H, Han X, Yi G, Huang G, Zhang C, Cao Y. In situ synthesis of MnO2/porous graphitic carbon composites as high-capacity anode materials for lithium-ion batteries. *Energy Fuel* 2020;**34**:2480−91.

99. Lee S, Hong J, Kang K. Redox-active organic compounds for future sustainable energy storage system. *Adv Energy Mater* 2020;**10**:2001445.

100. Li K, Xie X, Zhang WD. Porous graphitic carbon nitride derived from melamine−ammonium oxalate stacking sheets with excellent photocatalytic hydrogen evolution activity. *ChemCatChem* 2016;**8**:2128−35.

101. Zheng M, Xiao X, Li L, Gu P, Dai X, Tang H, Hu Q, Xue H, Pang H. Hierarchically nanostructured transition metal oxides for supercapacitors. *Sci China Mat* 2018;**61**: 185−209.

102. Kumar R, Singh RK, Dubey PK, Singh DP, Yadav RM. Self-assembled hierarchical formation of conjugated 3D cobalt oxide nanobead−CNT−graphene nanostructure using microwaves for high-performance supercapacitor electrode. *ACS Appl Mater Interfaces* 2015;**7**:15042−51.

103. Ge J, Yao H-B, Hu W, Yu X-F, Yan Y-X, Mao L-B, Li H-H, Li S-S, Yu S-H. Facile dip coating processed graphene/MnO2 nanostructured sponges as high performance supercapacitor electrodes. *Nanomater Energy* 2013;**2**:505−13.

104. Javed MS, Zhang C, Chen L, Xi Y, Hu C. Hierarchical mesoporous NiFe$_2$O$_4$ nanocone forest directly growing on carbon textile for high performance flexible supercapacitors. *J Mater Chem* 2016;**4**:8851−9.

Graphene oxide–metal oxide composites, syntheses, and applications in water purification

Kiran Aftab[1], Jianhua Hou[2] and Zia Ur Rehman[2,3]
[1]Department of Chemistry, Government College University Faisalabad, Faisalabad, Punjab, Pakistan; [2]Jiangsu Key Laboratory of Environmental Material and Engineering, School of Environmental Science and Engineering, Yangzhou University, Yangzhou, Jiangsu, PR China; [3]School of Physics, College of Physical Science and Technology and School of Environmental Science and Engineering, Yangzhou University, Yangzhou, Jiangsu, PR China

13.1 Overview of graphene oxides and metal oxides

13.1.1 History

A global concern in the present era is water scarceness. Growth in population and industrialization are the main causes for the rising scarceness of water. Among the many causes for shortages in potable water, the foremost is increasing industrial development leading to toxic wastes in water bodies. Industrial runoff includes coloring agents, toxic metals, high biological/chemical oxygen demand, turbidity values, and salinity, which greatly reduce the natural balance of the aquatic ecosystem. Water purification technology has advanced worldwide to attempt to address this present and future dilemma. The chemistry approach is used to fabricate and design new functional compounds/materials with efficient adsorption, high redox, and photocatalytic properties to treat surface water, groundwater, and industrial wastewater reservoirs.

Composites have been substantiated as favorable materials in this regard. Composites are made by amalgamating two or more constituents with different properties into a single material with an integrated set of properties. Up to now, the research on synthesis and fabrication with high porosity, smaller pore sizes, and density has been of simple conventional materials (AC, clay, zeolite, MIP, etc.) with movement toward reusable advanced composites.

13.1.2 Introduction

In earlier times, materials made up of carbon, such as activated carbon with a high surface area, were used widely to purify water. But nowadays, carbon-based composites are preferable for water treatment. Although extensive research has been conducted on inorganic materials, carbon-based materials are becoming more popular among researchers and scientists because of their novel size-dependent properties. Carbon-based compounds and their derivatives are simple, less hazardous, and possess mass-scale availability. The mind-blowing properties of sp^2 hybridized carbon derivatives such as C_{60}, i.e., buckminsterfullerene (zero-dimensional, or 0-D), carbon nanotubes (one-dimensional, or 1-D), graphene sheets (two-dimensional, or 2-D), and graphite (three-dimensional, or 3-D) have ensured new directions in research. Of all these compounds, graphene is the most popular because of its novel characteristics.

Graphene comprises 2-D tunable nanomaterials consisting of a single layer of sp^2 hybridized carbon atoms with carbon hexagons appearing like a honeycomb. It can be used as a better alternative to activated carbon because of its double surface area. Graphene is a "miracle material" discovered in 2004 by Andre Geim and Kostya Novoselov at the University of Manchester using a mechanical exfoliation method. Graphene shows novel and intriguing properties such as high-temperature charge-carrying mobility ($\sim 100,000$ cm^2 V^{-1} s^{-1}), great surface area (~ 2630 m^2 g^{-1}), high mechanical strength ($2.4 + 0.4$ TPa), high thermal conductivity ($\sim 2000-5000$ W m K^{-1}), large electric current density (10^8 A cm^{-1}), and high optical transparency.[1]

Though graphene is a useful material, it is expensive and relatively hard to produce. Efforts are ongoing to access effective and economical ways to synthesize and use graphene derivatives, including graphene oxide (GO), an oxidized form of graphene. GO consists of many oxygen-containing functional groups such as hydroxyl, epoxide, and carboxyl groups and is a monolayered material that is both economical and abundant. GO is mainly produced by three methods, namely, Brodie, Staudenmaier, and Hummers. These methods involve graphite oxidation in the presence of either strong acids (e.g., nitric acid or its mixture of sulfuric acid) or oxidizing agents (e.g., $KMnO_4$, $NaNO_3$, $KClO_3$).[2]

Historically, metal oxides (MOs) have been extensively explored because of their novel photocatalytic characteristics, and the process began after the initial study of electrochemical photolysis of water by Fujishima and Honda. Besides water photolysis, MOs have been used as photocatalysts for numerous water pollutants with effective outcomes. Though MOs have novel features, devices using them have failed to compete with methods like reverse osmosis and filtration because of some deficiencies of MOs. The major reasons for MO failure are that its activity is limited to the ultraviolet region solely because of its wider bandgap, and the chance of recombination of charge carriers reduces the lifetime of active MO components, thus rendering it incapable of continuing to degrade pollutants. Many potential solutions have been explored to overcome this deficiency, such as doping, surface modification, and amalgam formation. Of these methods, the best known for enhancing the photocatalytic ability of MO is to combine it with electron-scavenging agents. Photocatalytic capability is boosted by lessening the impact of the wider bandgap and electron-hole recombination on the process of photocatalysis. Of the methods mentioned, the best one used is a combination of metal oxide with electron acceptor carbonaceous species, specifically GO.

As discussed, GO's novel properties have proven it an efficient material for various purposes. Recently, novel graphene oxide—metal oxide nanocomposites have been fabricated for water remediation because of their degradation properties, ability to eliminate lethal organic contaminants and inorganic toxic metal ions, and disinfectant properties. Compared with GO alone, GO—MO nanocomposites are considered best for water treatment because of their specific structural morphology, robustness, long-lasting and environmentally friendly nature, easy recycling, and photochemical properties. According to recent studies, GO—MO composites can play various but not limited roles as photocatalysts, adsorbents, and antimicrobial agents (disinfectants), making them an effective material against major water pollutants, i.e., organic and inorganic pollutants, and waterborne pathogens.

According to the current state of the art, various GO—MO composites are utilized for different purposes. For example, the composite of zero-valent iron with GO is highly efficient in removing tiny concentrations of chromium ions (Cr^{+3}, Cr^{+6}) from polluted water. TiO_2 exhibits disinfectant properties when used in the form of nanocomposites with GO. Zirconium dioxide (ZrO_2) is less expensive and insoluble in water than titanium dioxide (TiO_2). The role of GO—MO composites as adsorbents, photocatalyst, and disinfectant is given below.

Adsorption is one of the best-known methods that are easy to use and cost-effective. It can remove almost all types of pollutants. GOs exhibit a larger surface area than conventional adsorbents such as activated carbons. The adsorption capacity of the GO can be greatly increased by making their nanocomposites with nanomaterial because the nanomaterial possesses a larger surface area to volume ratio. So, as a result of the nanocomposite, the surface area potentially increases and thereby increases the adsorption capacity of the composite material. The GO sheet increases the mechanical robustness and surface area of the adsorbent. GO inhibits the leaching of metal oxide during water treatment and improves the adsorption of organic molecules (especially the aromatic structures) by $\pi - \pi$ interaction.[3] Composite of TiO_2 with GO was firstly used for water purification. Other MOs such as SiO_2, ZnO, $CoFe_2O_4$, MnO_2, Fe_2O_3, and ZrO_2 have also been hybridized separately with GO for water purification as nanoadsorbents.

Various MOs are exploited for the photolysis of water, and when these catalysts are exploited as nanocomposites with GO, they act as photocatalysts. These nanocomposites possess extraordinary properties such as higher conductivity, adsorption, tunable optical character, stability, and long-lasting sustainability. In GO—MO composites used as photocatalysts, GO acts as an electron acceptor and prevents the recombination of electrons and holes while enhancing the absorption range of metal oxide from UV to the visible region of electromagnetic radiation. GO also prevents the cluster formation of metal oxide nanomaterials by dispersing them evenly. The graphene in GO helps the metal oxide interact with organic pollutants by $\pi - \pi$ stacking.[3] Different nanocomposites are used as photocatalysts, i.e., composites of GO with different MOs and complex oxides such as TiO_2, ZnO, $CuFe_2O_4$, $ZnFe_2O_4$, Cu_2O, Bi_2WO_6, SnO_2, $ZnWO_4$, $BiMoO_6$, and BiO_4 are used as remarkable photocatalysts for the degradation of synthetic dyes.

GO nanocomposites with MOs such as Fe_3O_4 and TiO_2 are used as disinfectant agents for water purification. Composites with nanotubes are more efficient than those with nanoparticles. Recently, Gollavelli and coworkers synthesized a nanocomposite of Fe_3O_4 with GO. This nanocomposite possesses outstanding disinfectant properties. The major function of GO in GO—MO composites is to damage the microbe's cell membrane with its sharp edges and impair the metabolic pathway of the microbe cell.

13.1.3 Advances

Although GO–MO composites are quite useful for water purification, further efforts are underway to make these composites even more beneficial. In working to make more advanced composites, materials are added, or their mechanical structures are changed. For example, as we already know that the composite of TiO_2 and graphene is the best-known photocatalyst. The nonmetal doping of this composite makes it more efficient. The sulfur and nitrogen (S–N) doped TiO_2/GO exhibit enhanced photocatalytic degradation rates on various dyes, especially methyl orange. These doped GO/TiO_2 composites are prepared using thiourea ($CS(NH_2)_2$) where thiourea and GO work together to boost the photocatalytic activity or photosensitivity of TiO_2.[3] Another example of the nanocomposite of GO–MO, which is becoming more advanced over time, is the Fe_3O_4/GO composite modified by polyaniline (PANI). This hierarchical Fe_3O_4/GO/PANI composite shows greater adsorption capacity that is superior to other adsorbents.[4] Complete understating of GO–MO composite usage in the water pollution problem is still a challenge. Therefore, exhaustive studies are encountered presently considering their structure, formation, size, viability, reproducibility, preparation methods, and properties.

13.2 General routes of graphene oxide–metal oxide composites for wastewater treatment

Synthesis approaches used to design graphene oxide–metal oxide composites ultimately define the resultant composite's physical, mechanical, and structural properties.[5] Let us assume that the graphene composite's mechanical properties rely on the organization of sheets, the loading content of graphene materials, and the specific surface area. The affinity of components, interfacial strength, and dispersion control the final strength, toughness elongation of composites, and stiffness. Therefore, the synthesis route is selected by surface-functionalized integrated chemical species. An important task in the preparation of composites is to achieve a uniform dispersion of MOs in the polymer matrix of GO.[6]

In general, MOs are bonded/loaded on graphene nanocomposites either by postimmobilization (ex situ hybridization) or in situ binding (in situ crystallization),[7] as illustrated in Fig. 13.1.

13.2.1 Ex situ crystallization

13.2.1.1 Solution mixing process

Solution mixing processing is an ex situ hybridization (postimmobilization) technique implicates by mixing GO solution into presynthesized MOs (Fe_2O_3, Al_2O_3, SnO_2, NiO, MnO_2, Cu_2O, and Co_3O_4) solution. The resulting suspension can then be precipitated, precipitated composite can then be extracted, dried, and further processed for testing and application.[7] In the synthesis of GO–MO composites, the solution mixing

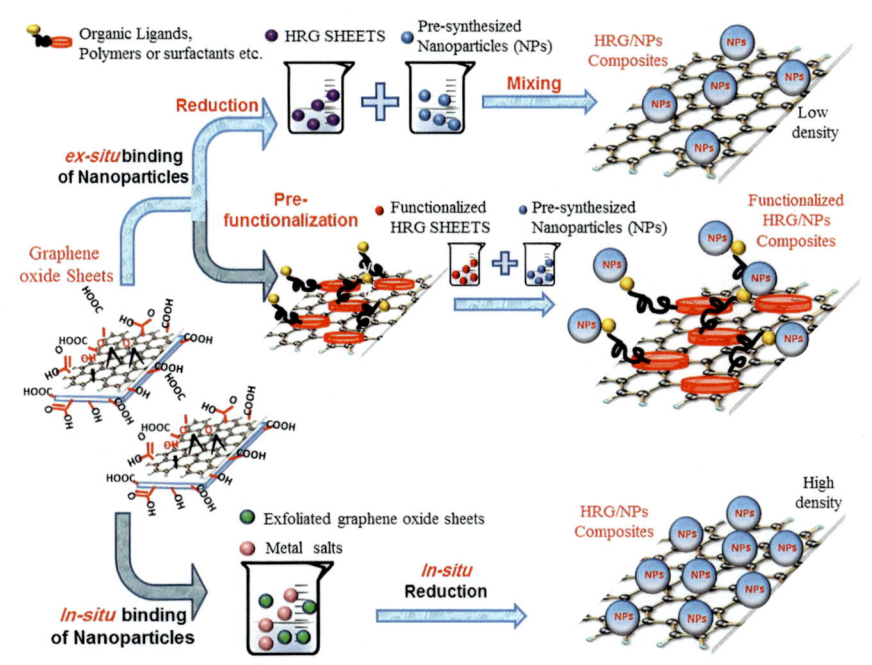

Figure 13.1 Schematic representation of different methods for synthesis of graphene oxide−metal oxide composites.
Reprinted with permission.[7]

approach disperses the GO platelets either by phase transfer, lyophilization methods, and surfactants techniques.[8] Conjugated GO is functionalized by noncovalent p-p stacking or covalent C−C coupling reactions. MOs can be prepared with controlled size in excellent yield from metal-organic precursors or to start with metallic powders. Such as, very fine zinc oxide nanoparticles can be synthesized from metallic zinc powder (source precursor) in an alkaline medium (KOH) at room temperature. Metal oxide deposited on the graphene sheets immersed in the solution to form the graphene oxide−metal oxide composite. Choosing the oxides and optimizing the physicochemical conditions are significant in synthesizing oxides and the resulting composites.

Solution mixing has been widely reported in the literature, as the method consists of thorough mixing of separately prepared NPs or salt precursors ($HAuCl_4$, K_2PtCl_4, K_2PdCl_4, and $AgNO_3$) with GO suspension followed by hydrazine monohydrate/sodium borohydride reduction.[8] Bell et al.[9] integrated reduced graphene oxide (rGO) into TiO_2 film by solution-phase photocatalytic reduction method that bared 10-fold increase in photogenerated current than simply TiO_2. Akhavan et al. used the same practice for GO−TiO_2 thin film formation and studied its antibacterial activity against *Escherichia coli*.[10] Peak et al. used ethylene glycol to disperse chemically reduced graphene nanocomposite and later reassembled it in the presence of SnO_2 nanoparticles to synthesize a graphene-SnO_2 nanocomposite with superior cyclic performance.[11]

Composite production with a uniform dispersion of reinforced material at low temperature is the main advantage of the solution mixing method; however, choice of common solvents, difficulties in solvent removal, aggregation issues, and thin-film limitations are key challenges.[7]

13.2.1.2 Self-assembly method

Another efficient, economical method for GO–MO preparation is self-assembly, which offers switching of size and structure through component design and is applicable in many fields. The sensitivity to environmental dynamics, low mechanical strength, and need for surfactants are major disadvantages. Surfactants not only improve the nanoparticle loading but also facilitate the dispersion of GO sheets. Self-assembly based on the electrostatic attraction among GO sheets (negatively charged) and nanoparticles (positively charged) has been used for the simple and cost-effective synthesis of graphene-based semiconductors. Some other strategies in the self-assembly method that can produce graphene-based nanocomposites are two-phase interface and template-free self-assembly.

13.2.1.3 Electrospinning process

Electrospinning is an efficient and reasonably priced popular ex situ method for graphene-based photocatalysts synthesis. In a typical electrospinning process, graphene-containing polymeric solution and photocatalyst precursor is electrospun into a nanofiber mat. The nanofiber is then sintered at high temperatures, yielding the photocatalyst in its final form. Compared with other synthesis methods, the need for postprocessing presents a major disadvantage of the electrospinning method of synthesis.

13.2.2 In situ crystallization

13.2.2.1 Hydrothermal/solvothermal method

Hydrothermal and solvothermal syntheses produce GO–MO composites, usually at high temperatures/pressures in an autoclave.[12] In the hydrothermal method, an aqueous solution is used at a temperature higher than $100°C$ and 1 bar of pressure, while in the solvothermal method nonaqueous solution is used. Experimental method of hydrothermal/solvothermal synthesis after choosing suitable reagents, define the mole ratio of reagents to mix/add the reagents. A mixture of reagents is sealed in an autoclave under a suitable reaction temperature, reaction time, and state (dynamic or static crystallization). After the reaction is completed, the products are washed and dried for further characterization.

The solvothermal method was used to prepare TiO_2-GO nanocomposites, and the microanalytical structure demonstrates that in the prepared nanocomposites on the GO sheets, TiO2 NPs are uniformly dispersed. These synthesized nanocomposites exhibit more catalytic activity than that of TiO2 nanoparticles alone in pollutant photodegradation.[13] GO/MFe$_2$O$_4$[14] nanocomposites were synthesized by the solvothermal process when depositing the dispersed MFe2O4 microspheres on the GO nanosheets and showed a high removal efficiency of Remazol Black B from the wastewater. ZnO[15] and $Ag_2CO_3-TiO_2$[16] nanoparticles were decorated over the sheets of GO by a hydrothermal process and exhibited excellent adsorption behavior toward heavy metals and organic dyes (rhodamine B, methyl orange, and methylene blue) as well as exhibiting high catalytic degradation under irradiation of visible light. GO and Ni—Al layered double hydroxide (GO@LDH) nanocomposites were prepared by the hydrothermal process and proved to be efficient adsorbents for the removal of U(VI) from wastewater.[17]

13.2.2.2 Sol—gel method

The sol—gel method is a controllable multistep process for GO—MO composite synthesis first used in the late 1980s. Sequential hydrolysis of metal precursor produces a solution of metal hydroxide that changes into a 3-D gel by immediate condensation reaction in the sol—gel method. The gel is dried either by evaporation or supercritical drying to obtain xerogel or aerogel, respectively, and then the desired product.[18] Aqueous/nonaqueous solvents allow GO/rGO to react covalently with other compounds and other functionalities with high dispersive ability in the sol—gel method.

Zhang et al.[19] synthesized graphene-TiO_2 nanocomposites using tetrabutyl titanate and GO by the sol—gel method and discovered that the graphene content and calcination atmosphere affect composite photocatalytic activity. Li et al.[20] grew ultradispersed TiO_2 nanoparticles on GO by the sol—gel method on graphene showed the specific capacity twice of mechanically mixed composites. Giampiccolo et al.[21] synthesized graphene-TiO_2 nanocomposite with a 50 ppb detection limit for sensing the NO_2 via the sol—gel route.

13.2.3 In situ reduction

In situ method is preferably used to synthesize nanocrystals of graphene oxide—noble metal (Ag, Pt, Au) graphene oxide—metal oxide ($SnO2$, Mn_3O_4) composites.[22] The negatively charged functional units (hydroxyl, carbonyl, and carboxyl) present on the surface of GO electrostatic interact with metal ions. Afterward, metal ions are reduced by reducing agents, and nanoparticles grow on the GO surface. CuO/GO nanocomposites were prepared under different temperatures (120 C, 150 C, and 180 C) using GO starting materials and cupric acetate.[23]

Chen et al.[24] changed the existent method by in situ growth of nanocrystals on the surface of crumpling GO sheets and solvent evaporation process. During the process, dissolved precursor ions (e.g., MnO, Ag_2O) are decorated on CG using an ultrasonic nebulizer to form ball-based hybrids and serve as a binder-free electrode in energy storage and conversion devices.[24] Zhang et al. prepared mesoporous TiO_2-GO hybrid by one-step hydrothermal method using glucose as crosslinkers. The resultant hybrids hindered the aggregation of nanocrystals because of exposure of many glucose owing hydroxyl groups adsorb on the GO.[25]

Xianjun et al. and Ting et al. used microwave irradiation to fuse FeCl3 and zinc oxide precipitates in GO suspension.[26] Shortly, highly photocatalytic active GO-based composite preparation is possible by simple/cheap solution processes with efficient degradation of organic dyes and pesticides under ultraviolet light irradiation, e.g., graphene hybrid with multi-branched Au—ZnO hybrid nanocrystals.[27]

13.3 Synthesis and specific properties of graphene oxide—metal oxide composites for wastewater treatment

GO is coupled with a wide variety of transition and noble metal oxide to enhance the photocatalytic activity of GO-based composites. The coupling of GO with metal oxide enhances electron transfer, the interfacial contact area, and ultimate photocatalytic activity. The dimensionality of material plays a starring role in defining the graphene oxide—metal oxide properties. Based on the morphology of MOs, nanoparticles/quantum dots (0-D), nanorods/nanotubes/nanofibers (1-D), nanosheets/nanoplates (2-D), and nanoflower/nanospheres (3-D), graphene oxide—metal oxide composites are categorized as GO/0-D MO, GO/1-D MO, GO/2-D MO, and GO/3-D MO composites.

13.3.1 Graphene oxide/zero-dimensional metal oxide composites

GO—0-D metal oxide nanocomposites are developed by different techniques—solvothermal, hydrothermal, flame spray pyrolysis, etc. The SnCl4, HCl, H2O, and GO mixture was heated at $120°C$ and $180°C$ for 12 h to synthesize the SnO_2/rGO nanocomposite. Au is added in the form of $HAuCl_4$ salt and ultrasonicated to obtain uniform dispersion. Then the sample was dropped on the platinum electrode to reduce GO and get SnO_2/rGO hybrid. Pt-activated SnO_2/rGO composites with a high concentration of oxygen vacancy and NiOs/rGO nanohybrid were fabricated by two-step hydrothermal treatment. NiO nanoparticles powder was prepared using $NiCl_4.6H_2O$ as the source of Ni and then calcined at $400°C$ and mixed with rGO solution with various ratios for NiO/rGO of 2:1, 4:1, and 8:1. Undoped and Ni-doped SnO_2 nanoparticles and graphene were developed by flame spray pyrolysis and graphite by the electrolytic exfoliation technique, respectively. Finally, the spin coating method was used to deposit a film to use the composite for a gas-sensing application.[28]

ZnO/rGO composite was prepared through the solvothermal method using Zn $(NO_3)_2$ and NaOH in ethanol at $120°C$ for low-temperature acetylene sensing. Ag-loaded ZnO/GO hybrid was synthesized by adding $AgNO_3$ with a 2:1 ratio by chemical route. Hydrazine hydrate is then added to the mixer to reduce GO at $110°C$ for 8 h.

13.3.2 Graphene oxide/one-dimensional metal oxide composites

One-dimensional nanostructures (nanotubes, nanorods, nanofibers, nanowires) due to large surface-to-volume ratio and open porosity are considered the most capable of detecting analytes in gaseous phases. Graphene sheets were synthesized using the chemical vapor deposition method and transferred to the top of the ZnO nanorods using a polymethyl methacrylate treatment. Dense nanorods of ZnO were grown by hydrothermal method and double side coverage of reduced graphene sheets to form heterostructures of ZnO/graphene/ZnO for efficient ethanol detection.

Nanorods with single-crystal WO_3 on graphene surface were synthesized through a one-step hydrothermal method[7] having 25 times improved NO_2 sensitivity. $GO-TiO_2$/ZnO nanotubes composite were synthesized by electrodeposition of rGO on TiO_2/ZnO nanotubes.[29] The rGO/SnO_2 quantum wire was also synthesized for H_2S sensing at room temperature. Single crystal SnO_2 nanowire for efficient NO_2 sensing was also directly grown on the platinum electrode by thermal evaporation.

13.3.3 Graphene oxide/two-dimensional metal oxide composite

The metal oxide nanosheets are two plane nanostructured materials studied because of their extensive presentations in water treatment. The production of 2-D nanomaterials is mostly done by mechanical exfoliation, self-assembly, bottom-up wet chemistry method, and ultrafacile technique. The bottom-up wet chemistry method was used by Xie and coworkers[30] to prepare several ultrathin metal oxide nanosheets. But the bottom-up method required severe synthetic environments of pressure, temperature, and pH values. For large-scale preparation of metal oxide nanosheets (such as Fe_2O_3, WO_3, TiO_2, Co_3O_4, and ZnO), a simplistic production method is an ultrafacile technique using metal salts, glucose, and urea as the raw constituents.

The thickness and thinness of the small sheets differ with their chemical structure that is about 50 nm for TiO_2, about 25 nm for Fe_2O_3, 30 nm for WO_3, and 10 nm for Co_3O_4.[31] This oxide TiO_2 had diverse stages of rutile and anatase, but all the remaining samples can be allotted with only one phase. The Brunauer—Emmett—Teller method was used to identify the pores of particles with diameters around 5 nm, and the observation also described that nanosheets contained small particles with a usual size of 9.6 nm. The enlarged vision of the folded nanosheet edge gave the 15 nm thickness of the nanosheet. TEM was used to describe the edge parts of the ZnO. Properties such as low cost and great degradation efficiency of aluminum oxide nanosheets make it quite attractive as a metal adsorbent for heavy metals removal in water treatment.

13.3.4 Graphene oxide/three-dimensional metal oxide composite

Three-dimensional graphene-based hybrid materials show significant improvement in both adsorptive and catalytic activity. Methods such as template-based synthesis, 3-D printing, self-assembly, and freeze-drying were used to produce 3-D graphene-based hybrid materials. The formation of smart composite materials with numerous metal arrangements is more favored from the environmental point of view because no additional greater temperature reactions and lethal solvents were accepted in this technique. Moreover, the formation of 3-D flexible and conductive graphene complexes by chemical vapor deposition can be changed for other small particles to made 3-D graphene-based hybrid materials more effective. Cao et al.[36] described a chemical vapor deposition technique to formulate MoS_2 covered 3-D graphene (MoS_2/3-DGN) syntheses by retaining 3-D graphene as a template for the direct growth of MoS_2. The graphene foams constructed on metal and polymer foams are used to make 3-D graphene-based hybrid materials by mixing several small arrangements. Recently, the self-assembly technique was also used to formulate 3-D networks in which the accumulation of other smaller building blocks can facilitate the preparation of 3-D graphene-based composite materials by involving metal ion induction and chemical reduction method through electrostatic interactions, $\pi-\pi$ stacking, and hydrogen bonding.

The small metal oxides in the composition of 3-D GO—MO composites retain numerous unique properties, such as the production of iron oxide-based small particles goethite (α-FeOOH) and magnetite (Fe_3O_4), that have received attention for eliminating toxic metals from wastewater resources. The small-scale α-FeOOH had greater elimination capability toward uranium than a non-nanoscale α-FeOOH. Maghemite-based GO composite easily disconnected from aqueous solution during the water treatment by applying a magnetic field.[29] Various 0-D, 1-D, 2-D, and 3-D composites are described in Table 13.1.

13.4   Water purification methods using graphene oxide—metal oxide composites

GOs are remarkably positioned in carbon-based adsorptive materials because of the large basal plane with vivid loading positions for metal oxide nanoparticles to produce composites. Adsorption properties of graphene-based composites depend upon pore size, structural area ($2600\ m^2/g$), and individual adsorption site contribution.[32] For adsorption processes, there are two major contributors:

- negatively charged surfaces of GOs
- positively charged heavy metals

Electrostatic forces concerning GO (negatively charged) and metal ions (positively charged) result in an ion exchange process. Hence, the metal ion adsorption capacity on the surface of GO/rGO is subject to their standard reduction potential, stability constant, electronegativity, and the relative positions of metals in electrochemical series but not on the ionic radius of metals. On the other hand, depending on the process, exothermicity or endothermicity adsorption capacity decreases or increases with temperature,[33] as illustrated in Fig. 13.2.

Table 13.1 Comparison of different zero-, one-, two-, and three-dimensional composites.

Zero-dimensional (0 − D)	One-dimensional (1 − D)	Two-dimensional (2 − D)	Three-dimensional (3 − D)
Clusters, quantum dots, atomic aggregates, metal nanoparticles, graphene quantum dots, fullerenes Dimensions in xyz are <100 nm.	Nanobars, nanowires, carbon nanotubes, nanoribbons Dimensions in xy are <100 nm.	Nanofilms, nanolayers, graphene, graphene oxide two layered graphene Dimensions in one direction, is <100 nm.	Graphite, polycrystals, diamond, graphite oxide, MOF, pillared graphene, aerogels Dimensions in xyz are >100 nm.

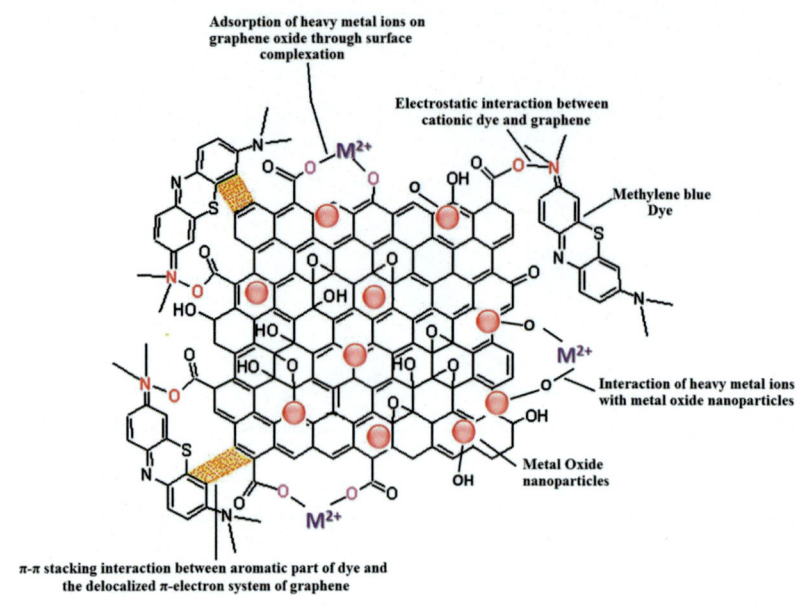

Figure 13.2 Various interactions involved in pollutants and graphene oxide—metal oxide composite copied from.[2]
Reused with permission.[2]

Along with adsorption, the involvement of the unique feature "capacitive deionization" allows GO—MO composites to ascend over other water desalination techniques. Capacitive deionization adsorbs ions onto the electrode surface based on electric potential differece. Ions present in the water form a double layer with opposite charges onto the surface of the electrode. On applying the external field, the adsorbed ionic layer desorbs easily from the electrode without generating secondary waste, making it an eco-friendly material.

13.4.1 Graphene oxide—metal oxide for adsorption method

13.4.1.1 Adsorption of inorganic contaminants

A variety of MOs, such as Fe_3O_4, TiO_2, ZnO, Cu_2O, and $ZnFe_2O_4$, in combination with GO, have been found to be adsorbents for removing toxic inorganic pollutants from water. Among GO—MO composites, Fe_3O_4 is a frequently used metal oxide because of its magnetic properties, which simplifies the posttreatment collection of adsorbents in a batch system.[33] Moreover, the formation of GO—Fe_3O_4 composites enhances their stability and lifetime, which were the major hurdles for its application, even in continuous flow systems. The most common anionic contaminants present in wastewater are chromium, arsenic, and fluoride. Inorganic arsenic varieties— arsenite [As(III)] and arsenate [As(V)]—are more contaminated than the organic pattern. As(V) is dominant in surface water, while As(III) is prevailing in groundwater systems under most conditions of pH, As(III) is present in anionic form ($H_2AsO_3^-$). As(V) in solution exist mainly as H_3AsO_4. Among the two oxidation states of chromium, Cr (III) is more toxic than Cr (VI). At low pH, the prevailing Cr (VI) species is $HCrO_4^-$, as increases in pH transformed it into $Cr_2O_7^-/Cr_2O_4^-$. Chandra et al.[34] used a positively charged magnetite reduced GO composite to remove arsenic/arsenous acid (H_3AsO_4) by electrostatic attraction. Magnetite-reduced GO (M-rGO) charged positively below the point of zero charge (pH pzc) led to accelerated adsorption of As(V) anions. Beyond the point of zero charge as the solution pH increases, adsorption of As(III) on M-rGO by surface complexation also increases, unlike the electrostatic adsorption interactions of As(V).

Outer-sphere complexation capacity decreases with reduced ionic strength, but toward Co(II) ion adsorption, a slight change in composite sorption ability was observed. That attributes the involvement of inner-sphere surface complexation of Co(II) ion on M-rGO rather than outer-sphere complexation or ion exchange.[34] Apart from Fe_3O_4, MOs such as SiO_2, ZnO, MnO, and ZrO_2 have been fabricated with graphene to design new adsorbents with higher selectivity toward targeted adsorbate for large scale application of composites as adsorbents. The SiO_2/graphene composites prepared by Hao et al. verified high selectivity for Pb(II) out of several available divalent ions of nickel, cobalt and cadmium, chromium, and copper.[35] Zong and coworkers obtained ZrO_2 functionalized graphite oxide by a postgrafting method and employed it as a sorbent material to remove phosphate ions.[36]

GO—TiO_2 hybrids synthesized by the self-assembly of TiO_2 on exfoliated GO were used for removing Cd^{2+}, Zn^{2+}, and Pb^{2+} from contaminated water. The oxygen moiety on exfoliated GO with TiO_2 NPs was significantly responsible for efficiently removing cationic pollutants such as Zn^{+2}, Pb^{+2}, and Cd^{+2}. It has the highest absorption capacity for the ions i-e Zn^{+2} (92.2 mg/g), Pb^{+2} (74.4 mg/g), and Cd^{+2} (68.3 mg/g).[20] Similarly, for the removal of Cr^{5+}, the hybrid material used is GO—TiO_2, rGO—TiO_2 by the process of adsorption or photocatalytic reduction.[37]

Cr^{6+} was removed by rGO—TiO_2 hybrids with a removal rate of 86.5%. Light adsorption strength is enhanced by adding rGO, and it minimizes the electron-hole pair recombination in TiO_2. Uranium (U^{6+}) can be removed from water by using hierarchical 3-D rGO/LDH (layered double hydroxide) composites as the adsorbents. The adsorption capacity of 3-D rGO/LDH obtained experimentally was 277.80 mg/g, this revealed excellent adsorption for U^{6+} removal from contaminated water.[38] The mercury ion Hg^{2+} is removed from contaminated water using a 3-D porous graphene composite constituted of graphene nanosheets assembled with α-FeOOH and silica microspheres and functionalized by thiol. A maximum adsorption capacity was more than 800 mg/g at 400 mg/L of conc. of mercury ion,[39] as illustrated in Fig. 13.3.

13.4.1.2 Adsorption of organic contaminants

Applications of GO—MO composite in various forms have been applied for the sorption of organic pollutants. GO adsorption of polycyclic aromatic hydrocarbons, pesticides, drugs, phenolic compounds, and humic acids shows excellent results. Although

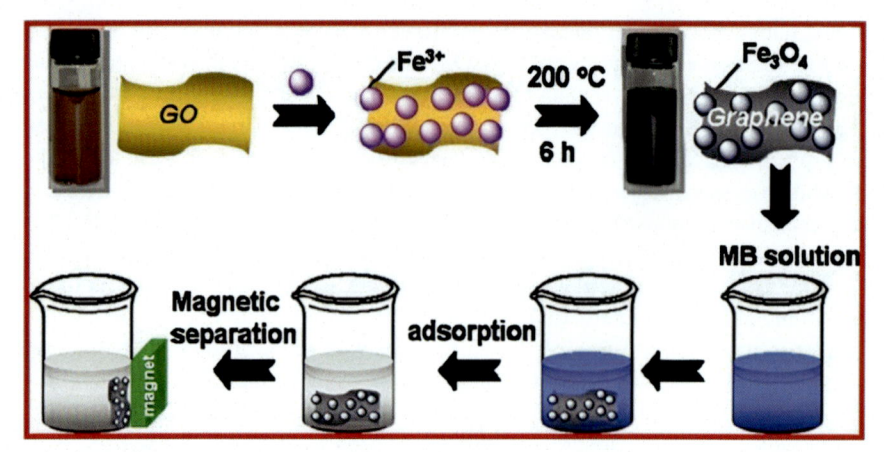

Figure 13.3 Schematic of the preparation of GNS/Fe_3O_4 composites and their methylene blue adsorption.[42]
Reused with permission. Copyright 2011, Elsevier.

several techniques are present to remove organic pollutants from water, adsorption has been adopted as the handiest. In carbon materials, π-π bonds, hydrophobic effect, electrostatic bond hydrogen, and the covalent bond can be identified. All these factors play a vital role in the adsorption of organic compounds of very small-sized carbon particles. It was noticed that the GO has the highest adsorption capability (190 mg g^{-1}) compared with graphite for humic acid. Organic contaminants like benzene, toluene, and dimethylbenzene from aqueous solutions can be removed using rGO–Ni nanocomposite material. Graphene nanosheets and GO showed high efficiency to remove 1-naphthylamine than 2-naphthol,[40] and composites prepared like GO/camphor sulfonic acid (CSA) hybrids have greater adsorption capacity than that of pure CSA and pure GO because of the presence of carboxylic groups (COOH) in GO/CSA composites. Recently it has been published that the adsorptive removal of anti-inflammatory drugs such as ketoprofen, naproxen, and AIDS from contaminated water using (GO/MOF) composites.[41]

Another example is the adsorption of norfloxacin (NOR) and ciprofloxacin (CIP) by reduced GO/magnetite (rGO–M) composites. With instruments such as the transmission electron microscope (TEM), scanning electron microscope (SEM), and XRD, adsorbents have been characterized. The adsorption of CIP and NOR depends upon the pH, π–π interactions, and electrostatic repulsions. The adsorption data were fitted, and results were investigated using adsorption isotherms and many others. The highest single layer uptake of CIP and NOR onto the graphene adsorbent showed 18.22–22.20 mg g^{-1}, respectively.[42] CoFe2O4-GNs removed methyl orange with an adsorption capacity of 71.5 mg/g.[43]

Dye adsorption is also being investigated for different graphene composites. For carcinogenic dyes, porous GO foams are applied. For practical application, 3-D GO is prepared by facile and other techniques. The foam produced by these techniques can be used directly without any pretreatment like ultrasonication. At the same time, these foams also have antibacterial activities against bacteria (*E. coli*) in nutrient and aqueous media, which again shows its prospects for water treatment[43].[43–45] Among various graphene composites, Fe_3O_4-GO composite has been widely used for adsorption of methylene blue (167.2 mg g^{-1}),[3] Congo red (33.7 mg g^{-1}), malachite green (22.2 mg g^{-1}),[31] and Remazol Blue (133.6 mg g^{-1})[44] dyes. MnFe2O4—GNs adsorption capacity toward methylene blue was 25 mg g^{-1}.[45]

13.4.2 Photocatalysis method

13.4.2.1 Graphene oxide—metal oxide composites for catalytic oxidative degradation of organic contaminants

In GO, the ratio between carbon and oxygen varies from 2.1 to 2.9. This material has a brownish-gray color and comprises lightly bound layers. Every layer has a 2-D arrangement of C atoms, and its thickness is almost 1.1 nm.[46] This layered material is attached with O containing several functional groups. This type of surface structure permits GO to swell freely and dissolve effortlessly in water. The attractive structure of GO makes it a faultless and suitable candidate for catalysis.[47] Because of these interesting properties of GO, much research work has been conducted and continues for catalytic processes on GOs and their composites with other materials, especially metallic oxides such as titanium oxide (TiO_2), iron oxide (Fe_3O_4), cobalt oxide (Co_3O_4), zinc oxide (ZnO), manganese oxide (MnO_2), and $Fe(OH)_3$. These composites with GO show good catalytic performance. Because of these amazing properties of GO—MO composites, they have become suitable candidates for oxidative degradation of organic contaminants.[48]

Composite GO—Co_3O_4 catalytic activity varies with different loading of Co_3O_4. A 50% Co_3O_4/GO composite showed maximum oxidative degradation of Orange II compared with the Co_3O_4 nanoparticles or similar Co_2^+ particles alone.[49]

In industries that release wastewater, many toxic materials and compounds of nitroarene are found and form a major category of pollutants. Many of them are not easy to remove from water in the existence of a catalyst. The toxic pollutant 4-nitrophenol is dangerous for both humans and animals, can exist for a long time in the environment, and is not easily biodegradable. So finding an easy and effective method for removing 4-nitrophenol and other toxic pollutants is difficult. Chemical reduction is an economical method for removing these toxic materials from water, but this process is very slow. Among noble metals (NPs), Pd has great potential and has already been used to remove these pollutants from wastewater. To improve catalytic activity, GO/Fe3O4/Pd was developed to degrade 4-nitrophenol, Congo red, methyl blue, and methyl orange. This nanocomposite is used first time for the degradation of organic containments from the wastewater as recycled catalyst.[50]

Azo dyes are also a big source of water pollution. Various methods have been used to remove azo dyes from wastewater and protect the environment, but typically used techniques have some problems that decrease material efficiency. For this purpose, several materials are used as catalysts, such as Ti, Zn, and Mn, and MnO_2 is a suitable candidate because of its economic cost, good stability, excellent compatibility, and easy approach. There is a separation issue while using MnO_2 in aquatic solution. GO is a promising material for degradation ability to reduce this problem. So GO—MnO_2 composite has good application for catalytic ability due to good electrical conductivity and electron transformation.[51] In this research work, GO—MnO_2 composite is used as a catalyst in RB5 to degrade organic pollutants in water. The GO—MnO_2 composite was discussed for RB5 with an increment of H_2O_2 for degradation. The results showed that a pulp density of 60 mg L^{-1} of GO—MnO_2 provides 70% efficiency, but when 0.15 M of H_2O_2 is added, it provides about 99% efficiency.[52]

13.4.2.2 Graphene oxide—metal oxide composites for photocatalytic degradation of organic contaminants

The photocatalytic deprivation of organic pollutants from water has collected great interest, as it utilizes renewable solar light and yields nontoxic materials in the reaction. Light irradiation on metal oxide catalysts is a key advantage (solar light, visible light, UV light sources, etc.) to run the catalysis reaction. The technique's long history began in 1960 and has developed into an attractive field over the last few decades while researchers have achieved many valuable results in this area of research. In 1972, Fujishima—Honda discovered how sunlight acts with semiconductors for water oxidation. Although much work has been done on photocatalytic materials, their reaction rate, stability, and selectivity must be improved. In this opinion, composites of 2-D GO with metal oxides (ZnO, SnO_2, WO_3, a-Fe_2O_3, TiO_2, etc.) provide a good platform for electron transport because of their small junctions and grain boundaries among different nanomaterials.

MOs are irradiated by light, generate charge carriers' (e−) and holes (h+) to drive the photocatalytic reaction. The holes generated by MOs act as oxidizing agents and oxidize the organic species. Graphene here acts as a very efficient sink for the transportation of electrons. Moreover, the Fermi level of graphene, 0 V versus nanomembrane hybrid electronics, is less negative than most semiconductor constituents, resulting in the promising bandgap placement of MOs and graphene. It is known that electrons transfer takes place at down potential while holes transfer lead to the up potential; therefore, in GO—MO composites, the electron transfer occurs from the conduction band of metal oxides to graphene, and active species radicals are formed, which results in the deprivation of organic pollutants.[53]

Large numbers of MOs (ZnO, WO_3, CuO, Cu_2O, Mn_3O_4, Mn_2O_3, and SnO_2) and complex oxides (Bi_2WO_6, $ZnWO_4$, Bi_2MoO_6, $BiVO_4$, $BaCrO_4$, and $CoFeO_4$) are combined with GO as a photocatalyst [84] with graphene content and particle size diversity.

Many research articles have been published on graphene and TiO_2 composite for photocatalysis. TiO_2 NPs for photocatalysts have been used to degrade organic pollutants because of attractive properties of TiO_2 such as economical price, good stability, nontoxic material, and high efficiency. The synthesized methods are complex because of the use of high temperatures and templates to assemble TiO_2. There is a need for material assistance to increase composite efficiency and be synthesized at low temperatures for a new TiO_2 photocatalyst. For this, graphene has an attractive material for its unique properties. The results showed that composite gives 95% efficiency in the first 9 min, but neat TiO_2 showed only 40% efficiency of this dye after 18 min. Supplementary addition of Au, Ag, Pt, Nd, Fe onto TiO_2 graphene composite surface also increases composite photocatalytic activity by extending the lifetime of charge carriers.[54] Dharwadkar et al.[55] studied the photodegradation of sulfolane, phenol, and methyl orange using powdered TiO_2 and titanium oxide—graphene composite formed by hydrothermal method. Sulfolane was irradiated with 365 nm, and that is the maximum absorption wavelength of TiO_2. TiO_2 and titanium/GO were tested at 460 nm for phenol, methyl orange, and sulfolane. Sulfolane showed no degradation under 460 nm, but it showed 89% degradation at 365 nm with TiO_2 powder while 34% degradation with graphene composite. Phenol and methyl orange were better degraded with graphene/TiO_2 composite at 460 nm. This study suggests that graphene/titanium oxide is best for pollutants with high absorption in visible light and more interactions with graphene.

ZnO—GO and ZnO—GO—Pd composites showed a maximum degradation rate for methylene blue at $\lambda = 365$ and a minimum degradation rate at $\lambda = 254$ nm.[50] Luo et al.[56] reported a 67% increase in ZnO—rGO photodegradation efficiency compared with pristine ZnO by suppressing the photocorrosion of hollow spheres and upgrading the stability and endurance photocatalyst.

The combination of the GO with magnetic Fe_3O_4 has become a promising new composite because of its additional benefit of recycling catalyst, which decreases the cost of separation. Problems while using GO—Fe_3O_4 composites are strong solid binding between GO and magnetic NPs that reduce adsorption ability. A unique and novel composite of GO and Fe_3O_4 has been fabricated to overcome these problems by a facile and economical method, Fe_3O_4@GO nanoclusters. This material has a core-shell structure with many advantages over the previously mentioned nanocomposite. This Fe_3O_4@GO composite was investigated to remove methylene blue (99.5%) and rhodamine B (88%) dyes in just 5 minutes from an aqueous solution. The results showed an increase in pulp density enhanced the efficiency of Fe_3O_4@GO due to the availability of a large surface area of composite.[57]

Mengting et al.[58] investigated the use of $BaTiO_3$/GO for photocatalytic degradation of methylene blue with 95% efficiency by UV/Visible light irradiation for 3 h; however, the byproducts formed by degradation were still required for degradation. To enhance the performance of composite BaTiO3/GO, the composition varied by changing the weight of $BaTiO_3$ and characterized by XRD, BET, FTIR, SEM, and TEM, as illustrated in Fig. 13.4.

13.4.3 Electrochemical purification method

In the past 2 decades, electro-Fenton, anodic oxidation, solar photoelectro-Fenton have been used as technically advanced techniques for wastewater treatment. In anodic oxidation, pollutants are destructed at the anode using Fenton's reagent. In electro-Fenton, iron catalyst H2O2 is added at the cathode with O2/air feeding, while in ECAOPs, the Fenton reaction between Fe^{2+} and electrogenerated H_2O_2 produces

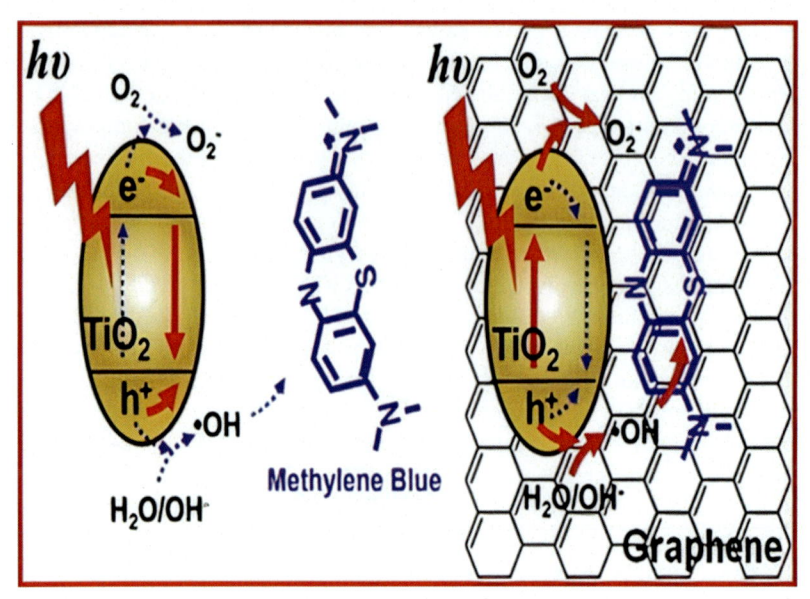

Figure 13.4 Schematic sketch of the methylene blue photodegradation process with pristine TiO_2 nanorods and TiO_2 nanorod-decorated graphene sheets.[79]
Reprinted with permission. Copyright 2012, Elsevier.

hydroxyl radicals ($^{\bullet}OH$) as a powerful oxidizing agent for removal of contaminants.[59] Efficiency of electrochemical water treatment is affected greatly by electrode material. Carbonaceous/graphene-based materials are considered good in electrochemical electrode materials because of their high conductivity and economical commercial availability.

13.4.3.1 Graphene oxide–metal oxide composites for electrocatalytic degradation of organic contaminants

In electrocatalytic activity, electrodes such as Pt, SnO_2, Fe_2O_3, MnO_2, CoO, RuO_2, TiO_2, and PbO_2, and C-based materials like graphene and active carbon, have much importance in conditions f perform reaction and degradation performance. Among these, PbO_2 has good potential for electrocatalytic activity material, but electrode fouling is a problem while using PbO_2 to remove organic containments from the wastewater. The nanocomposite PbO_2-rGO is considered good electrode material for degrading organic containments from wastewater because of its combined MO and GO properties. Govindaraj et al.[60] equipped a graphene/Fe_3O_4 electrode with quinone for constant H_2O_2 and $^{\bullet}OH$ electrogeneration. Complete removal of bisphenol A was detected with less than 1% iron leaching using new electrode material.

Ti/Sb–SnO_2, Ti/Sb–SnO_2-GNS, and Ti/Sb–SnO_2-NGNS are used as electrode materials for electrocatalytic activity in methylene blue and orange II. The decolorization performance of Ti/Sb–SnO_2–NGNS is better over time compared with the other electrode materials used in this research. The efficiency of Ti/Sb—SnO$_2$-NGNS is 97.7% after 100 min for methylene blue decolorization, and the efficiencies of Ti/Sb—SnO$_2$ and Ti/Sb—SnO$_2$-GNS are 43.8% and 93.6%, respectively.[61]

13.4.3.2 Graphene oxide—metal oxide composites for the photoelectrocatalytic degradation of organic containments

Photoelectrocatalysis is the combination of two processes, the photocatalytic method and the electrochemical oxidation method. There is a need for photoactive electrode material with electrical current and light of appropriate wavelength to remove organic pollutants in this process. In this method, the photocatalytic method's recombination of charge carriers (electrons, holes) is slow with the given potential. Due to this, the life of charge carriers increases, and as a result, degradation ability increases in the presence of light. In this process, when light with sufficient energy falls on the semiconductor material (S), the electron-hole pair generates. In these charge carriers, the holes interact with the organic pollutants, and the molecules of water adsorbed on the catalyst material, and degradation of pollutants occurs.[62]

In photoelectrocatalytic activity through photocatalytic and electrolytic methods, oxidant species can be produced. The recombination process of charge carriers is reduced by applied potential, and the electrons are guided away by the opposite electrode. During the reaction, the transfer of electrons from the containment to the +ve electrode is also possible. So during the different routes of oxidation, more organic pollutants are degraded in photoelectrocatalytic activity. The photoelectrocatalytic process is illustrated in Fig. 13.5.

Many materials, such as titanium oxide (TiO$_2$), zinc oxide (ZnO), WO$_3$, bismuth vanadium oxide (BiVO$_4$), and their composites, have shown good performance for photoelectrocatalysis to degrade organic pollutants. But MOs such as TiO$_2$ face some problems: the combination rate of electron-hole pairs is high, and electron transport ability is not good for photoelectrocatalytic efficiency. Some strategies have been adopted to improve TiO$_2$ performance and reduce these problems.[63] One of these is to combine TiO$_2$ with carbon-based materials. In this approach, graphene, or rGO, has more importance due to its unique and interesting properties, such as high SA, good electrical conductivity, and excellent interaction with nanoparticles. Briefly, a conclusion can be made that the combination of MO with GO is a promising combination for reducing the organic containments from the wastewater. Nanocomposite rGO/TiO$_2$ film degradation of organic contaminants showed good efficiency in photoelectrocatalytic activity. In nanocomposite rGO/TiO2, different concentrations of rGO for the photoelectrocatalytic reaction in RhB showed that with the addition of rGO, its performance increased and gained its highest efficiency at a 1.0% concentration. But when additional amounts were added, further performance improvements were not seen because of the decrease in photocatalytic activity and high recombination rate.[64,65]

Some articles have proposed synthesizing Ag/ZnO/rGO for sensors and photocatalysts to remove containments.[66,67] From this, the Ag—ZnO—rGO composite has become a promising material as an electrode for the degradation of organic

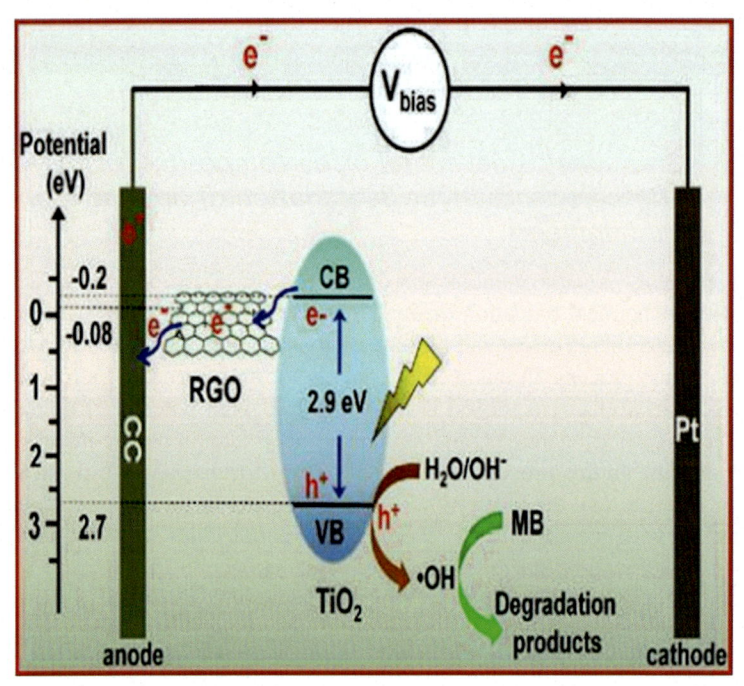

Figure 13.5 Schematic diagram of the reduced graphene oxide/TiO_2/Carbon coated electrode system and their photoelectrocatalytic degradation process of methylene blue in the presence of visible light.
Reprinted with permission.[81] Copyright 2013, Elsevier.

containments from wastewater. In this study, using this composite for degradation of orange II showed very good degradation performance results. The photoelectrocatalytic performance of the different materials rGO, ZnO−rGO, and Ag−ZnO−rGO was evident as electrode material for removing orange II dye. Other results for the composite Ag−ZnO−rGO, studied for degradation under different methods such as EC and PC and photoelectrochemical, showed maximum results for degradation with photoelectrocatalytic activity.[68] So Ag−ZnO−rGO is a promising material for degrading organic containments from wastewater.

13.4.4 Graphene oxide−metal oxide composites for disinfection

GO−MO nanomaterials also emerged as antimicrobial because of their high surface-to-volume ratio, crystallographic structure, and adaptability to various substances. Liu et al. reported that GO shows more bactericidal activity than graphite (Gt), graphite oxide (GtO), and rGO. Moreover, GO coupling and compositing with other materials such as MOs results increase in its antimicrobial activity. These composites may increase antimicrobial activity by producing reactive oxidative species,[69] which damage the cell membrane, DNA, and proteins and ultimately lead to cell destruction. They may also increase the adherence of cells toward graphene by enhancing the graphene surface area. The aggregation of the GO sheets (GRS) exhibits antimicrobial activity, too, but dispersed GO sheets show better activity because of their higher surface area. Some MOs also show antimicrobial activity, so their composites with graphene/GO increase their antimicrobial activity due to synergistic effects,[2] as illustrated in Fig. 13.6.

Graphene oxide—titanium oxide composite thin films can be used for disinfection purposes. Akhavan et al.[70] prepared and examined the antibacterial activity of GO—TiO2 by using an antibacterial drop test. In his experiment, he increased the antimicrobial activity of TiO2 by coupling it with GO film. The annealed GO—TiO_2 film shows 25% more efficiency against *E. coli* compared with bare TiO2. And after the photocatalytic reduction of this film, this photocatalytic activity enhanced to about 60%. Further, by increasing the photocatalytic reduction time of this thin film to about 4 h, the antibacterial activity becomes even better. As during the photocatalytic reduction process, the exposure of TiO2 to UV light resulted in the production of photoinduced electron-hole pairs, and ethanol acts to scavenge these electro hole pairs to produce ethoxy radicles as given by the following equation:

$$TiO_2(h- + e-) + C_2H_5OH \rightarrow TiO_2(e-) + \cdot C_2H_4OH + H+$$

So these photoexcited electrons get accumulated on the surface of thin TiO_2. These accumulated electrons reduce the functional groups present on the GO plates, thereby decreasing the time required to reduce functional groups to 100 ns from milliseconds

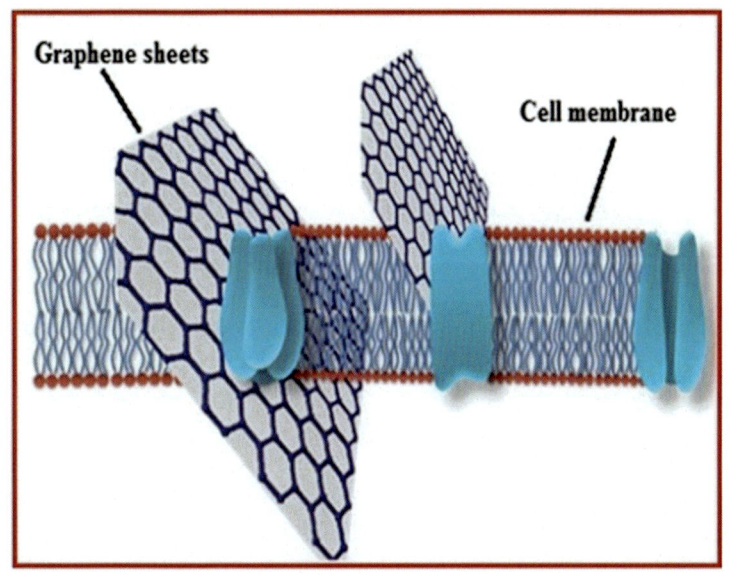

Figure 13.6 Schematic illustration of cell membrane damage by graphene sheets.[2] Reused with permission. Copyright 2014, Royal Society of Chemistry.

(by oxidant reduction), which results in increased photocatalytic of TiO2 and leads to increased antibacterial activity. The general mechanism may be that[71] a very slight increase in the C−OH group occurs after the process, showing that these thin films are chemically stable and thus can be conveniently applied for this process. Liu et al. also synthesized and examined the bactericidal effect of graphene oxide−titanium oxide nanorod composite (GO−TiO$_2$ NRC) by exposing the common water pathogen *E. coli* to it under solar irradiation. Similarly, it showed significantly greater antimicrobial activity than TiO2 and killed about 90% of the *E. coli* bacterium within 27 min. It was further confirmed that GO−TiO2 NRC has more and better activity than GO−TiO2 nanoparticle composites, which might be because of more facets on rods than nanoparticles. (7).

GO enwrapped Ag$_3$PO$_4$ composite can be used for the complete removal of *E. coli* bacterium as reported by Lei Liu.[72] Primarily when this composite is dissolved in the water, some of the Ag+ ion gets dissolved in water because Ag$_3$PO$_4$ is slightly soluble in water and destroys some of the membrane proteins. Secondly, GO gathers the microorganisms in combination with Ag3PO4 due to redox reactions, thus increasing the interaction between bacterial cells and catalysts. When visible light is irradiated on the composite during photocatalytic activity, the photogenerated electrons from Ag$_3$PO$_4$ move toward GO. The electron pairs on GO absorb surface O$_2$ and form various reactive oxidative species that destroy the microbes. This charge transfer between GO and Ag$_3$PO$_4$ also makes the complex more stable. Despite the stability of the composite, it is destroyed during the process by bacterial secretions—chloride ion, thiol ion, etc., thus limiting its reusability. In addition, Ag is a costly material.

Ham Raj Pant et al.[73] have suggested the utilization of silver-zinc oxide/reduced GO (Ag−ZnO/rGO) as he prepared and examined its antimicrobial activity. The photoexcited electrons are similarly trapped by the O2 molecules to form reactive oxidative species that destruct the microbes. Furthermore, due to the larger surface area of ZnO/GO, more Ag particles are present on the surface, which will destroy the bacteria even in the absence of UV light. It has been reported that Ag+ stops DNA replication and hinders ribosomal protein expression. Moreover, this material is economical, environmentally friendly, and can be recovered easily thus can be reused, as summarized in Table 13.2.

13.5   Challenges and future perspective for graphene oxide−metal oxide composites

Despite many promising features, GO−MO composites still cannot compete with traditional processes such as filtration and reverse osmosis. This failure can be attributed to shortcomings such as the wide bandgap of MOs, which limits the composite activity to only the UV region. This increases the charge carrier recombination possibility that shortens active site lifetimes and is the major reason for degrading the pollutants from the water. Several strategies have been practiced to solve these issues (surface modification, doping, amalgamation of MOs, and electron-scavenging

Table 13.2 Comparison of the activities of different graphene oxide—metal oxide composites for water decontamination.

Composite	Pollutant	Irradiation	Efficiency	Reference
TiO$_2$/Gr	Sulfolane	LED	34%	[55]
	Phenol		48%	
	Methyl orange		25%	
TiO$_2$/GrO	Dichloroacetic acid	UV/VIS	87%	[75]
TiO$_2$/GrO	Phenol	UV	100%	[76]
Ba—TiO$_3$-Go	Methylene blue	UV	95%	[58]
N—TiO$_2$-G	Pesticides	VIS	80%	[77]
TiO$_2$-rGO—CoO	2-Chlorophenol	VIS	98%	[78]
Mn-doped ZnO/GrO	Methylene blue	VIS	99%	[79]
G/ZnO	Methylene blue	UV	23%	[80]
	Methyl orange			
Mn-doped ZnO/GrO	2,4-Dichlorophenox yacetic acid	LED	66%	[81]
G-ZnO NCs	Methylene blue	VIS	89%	[82]
	Methyl orange			

Reused with permission.[74]

agents). However, graphene itself can promote GO—MO composite visible light response and tune semiconductor optical characteristics by the collaborative effect of doping. Then again, the major issues are the defective and imperfect graphene structure. Adverse effects are exhibited on composite photocatalytic response due to the structural defects in graphene.

Graphene with minimum structural defects rendered a greater electronic mean free path for the electrons and allowed them to flow farther from that of metal oxide—graphene interface that makes the recombination difficult. The properties and structure of the graphene are dictated by the pathway of reaction adopted for the preparation of the metal oxide—graphene composites. That is why selecting the appropriate route/ way and reaction conditions is highly recommended with the desired properties to obtain graphene.

The defect in the metal oxide can trap the excited electrons and can also cause the reduction in electron recombination, so for improving metal oxide—graphene composite photocatalytic activity, defect engineering could be the important strategy.

A second crucial factor is the graphene loading amount in the composite, which can significantly affect photocatalytic activity and must be adjusted accordingly. Photocatalytic activity can be increase by enhancing the content of graphene in the composite, but apart from this, graphene can hinder photocatalytic activity by preventing adequate synergy between the metal oxide and light.

Other than in photocatalysis, GO—MO composites have become more attractive than traditional adsorbents (i.e., mesoporous and activated carbons) because of their much higher graphene surface area (2600 m^{2}g^{-1}). Adsorption capacity can also be reduced when there is an increase in the amount of metal oxide above a critical limit. MOs are comparatively heavy and own a lower surface area/weight ratio than graphene, so excessive increases in the quantity of metal oxide in composites result in reduced adsorption capacity.

The adsorption capacity of these absorptive materials depends greatly on the pH of some organic molecules, e.g., dyes, heavy metal ions, and many other pollutants; hence, it is crucial to maintain a favorable adsorption pH. Hydroxyl group concentration is enhanced on the surface with increases in the value of pH, which hinders metal ion adsorption; variously, H+ ion concentration increases in the solution when the pH is highly acidic, which for the sorption site competes with the metal ions available at the adsorbent. So it is highly desirable to maintain the pH at which the performance hindering side reactions could be prevented. Some of the toxic metallic ions, i.e., chromium and arsenic, form metastable species at different pH values, and only a few of these species (metastable) may be adsorbed on the adsorbent.

Composite durability is another issue for practical application. Metal oxide leaching in the treated water could be due to labile binding and weak/loose physical adhesion that could be the reason for water cross-contamination as well as reduced adsorbent lifetime. Thus, strength is a matter of great concern between the bonding of the metal oxide and graphene, and also the reactants nature, conditions of reaction, and route of synthesis must be chosen that facilitates the firm bonding between the graphene and metal oxide. Metal oxide and graphene tight bonding can be attained by modifying the surface of metal oxide and graphene with the functional groups.

The metal and MO composites of hybrids, i.e., Fe_3O_4, Fe, TiO_2, and Ag, have been explored for disinfecting and sanitizing water as antimicrobial agents. Some perspectives on the challenges are discussed below.

Graphene sheet size is the main factor on which the antimicrobial activity of GO and pristine graphene depends directly. Larger graphene sheets are recommended, as the cell can be enveloped by it, and also, the larger graphene sheets show greater antimicrobial activity than the smaller graphene sheets. Thus, the fabrication method could be chosen as is; for the antibacterial applications, the method that can produce larger graphene sheets, i.e., burn–quench method, thermocatalytic decomposition, chemical vapor deposition, etc.

Graphene sheets also can form agglomerates and reduce their size, enhancing their efficiency. Hence, the crucial factor is the graphene surface modification with functional groups and molecules that can effectively prevent the folding of sheets and help maintain the stabilize dispersion. In addition, OH and −COOH presence in certain active functional groups can increase surface roughness. That is why increasing the antimicrobial activity that, through rubbing, can damage the cell's outer membranes. It could be an effective way to extemporize the antimicrobial activity of the graphene composites by increasing the active group numbers on the graphene surface. Sharp extremities of the graphene sheets also contribute to its antimicrobial activity, wherein the atomically sharp edges can slice through the cell membrane. Therefore, particular morphologies and structures that provide a very high degree of edge-planes are significant.

A frequently tried approach is also the microbial cell inactivation by the mediation of reactive oxygen species. Finally, use of lower defect graphene, use of graphene in composite with optimized content, interfacial contact between graphene and metal oxide, on the graphene sheets uniform distribution of the nanoparticles of metal oxide would be hot future working strategies for efficient and durable graphene oxide—metal oxide composites designing to solve water pollution problems.

References

1. Kroto HW, et al. C60: buckminsterfullerene. *Nature* 1985;**318**(6042):162—3.
2. Upadhyay RK, Soin N, Roy SS. Role of graphene/metal oxide composites as photocatalysts, adsorbents and disinfectants in water treatment: a review. *RSC Adv* 2014;**4**(8):3823—51.
3. Shang X, et al. Sulphur, nitrogen-doped TiO2/graphene oxide composites as a high performance photocatalyst. *J Exp Nanosci* 2014;**9**(7):749—61.
4. Li J, et al. Hierarchical GOs/Fe3O4/PANI magnetic composites as adsorbent for ionic dye pollution treatment. *RSC Adv* 2014;**4**(72):38192—8.
5. Singh V, et al. Graphene based materials: past, present and future. *Prog Mater Sci* 2011; **56**(8):1178—271.
6. Guo F, et al. A review of the synthesis and applications of polymer—nanoclay composites. *Appl Sci* 2018;**8**(9):1696.
7. Khan M, et al. Graphene based metal and metal oxide nanocomposites: synthesis, properties and their applications. *J Mater Chem* 2015;**3**(37):18753—808.
8. Paek S-M, Yoo E, Honma I. Enhanced cyclic performance and lithium storage capacity of SnO2/graphene nanoporous electrodes with three-dimensionally delaminated flexible structure. *Nano Letters* 2009;**9**(1):72—5.
9. Shin HJ, et al. Efficient reduction of graphite oxide by sodium borohydride and its effect on electrical conductance. *Adv Funct Mater* 2009;**19**(12):1987—92.
10. Bell NJ, et al. Understanding the enhancement in photoelectrochemical properties of photocatalytically prepared TiO$_2$-reduced graphene oxide composite. *J Phys Chem C* 2011; **115**(13):6004—9.
11. Uddin AI, Phan D-T, Chung G-S. Low temperature acetylene gas sensor based on Ag nanoparticles-loaded ZnO-reduced graphene oxide hybrid. *Sensor Actuator B Chem* 2015; **207**:362—9.
12. Feng S-H, Li G-H. Hydrothermal and solvothermal syntheses. In: *Modern inorganic synthetic chemistry*. Elsevier; 2017. p. 73—104.
13. Yadav HM, Kim J-S. Solvothermal synthesis of anatase TiO$_2$-graphene oxide nanocomposites and their photocatalytic performance. *J Alloys Compd* 2016;**688**:123—9.
14. Sheshmani S, Falahat B, Nikmaram FR. Preparation of magnetic graphene oxide-ferrite nanocomposites for oxidative decomposition of Remazol Black B. *Int J Biol Macromol* 2017;**97**:671—8.
15. Ranjith KS, et al. Multifunctional ZnO nanorod-reduced graphene oxide hybrids nanocomposites for effective water remediation: effective sunlight driven degradation of organic dyes and rapid heavy metal adsorption. *Chem Eng J* 2017;**325**:588—600.
16. Pant B, Park M, Park S-J. Hydrothermal synthesis of Ag$_2$CO$_3$-TiO$_2$ loaded reduced graphene oxide nanocomposites with highly efficient photocatalytic activity. *Chem Eng Commun* 2020;**207**(5):688—95.

17. Yu S, et al. One-pot synthesis of graphene oxide and Ni-Al layered double hydroxides nanocomposites for the efficient removal of U(VI) from wastewater. *Sci China Chem* 2017; **60**(3):415−22.

18. Yaqoob AA, Serrà A, Ibrahim MNM. Advances and challenges in developing efficient graphene oxide-based ZnO photocatalysts for dye photo-oxidation. *Nanomaterials* 2020; **10**(5):932.

19. Zhang X, et al. Show compounds show chemical terms show biomedical terms graphene/TiO_2 nanocomposites: synthesis, characterization and application in hydrogen evolution from water photocatalytic splitting. *J Mater Chem* 2010;**20**:2801−6.

20. Zhao Y, et al. Enhanced photo-reduction and removal of Cr (VI) on reduced graphene oxide decorated with TiO_2 nanoparticles. *J Colloid Interface Sci* 2013;**405**:211−7.

21. Giampiccolo A, et al. Sol gel graphene/TiO_2 nanoparticles for the photocatalytic-assisted sensing and abatement of NO_2. *Appl Catal B Environ* 2019;**243**:183−94.

22. Khalil I, et al. Graphene−gold nanoparticles hybrid−synthesis, functionalization, and application in a electrochemical and surface-enhanced Raman scattering biosensor. *Materials* 2016;**9**(6):406.

23. Song J, et al. Synthesis of graphene oxide based CuO nanoparticles composite electrode for highly enhanced nonenzymatic glucose detection. *ACS Appl Mater Interfaces* 2013;**5**(24): 12928−34.

24. Mao S, et al. A general approach to one-pot fabrication of crumpled graphene-based nanohybrids for energy applications. *ACS Nano* 2012;**6**(8):7505−13.

25. Qiu B, Xing M, Zhang J. Mesoporous TiO_2 nanocrystals grown in situ on graphene aerogels for high photocatalysis and lithium-ion batteries. *J Am Chem Soc* 2014;**136**(16):5852−5.

26. Ren Z, et al. Mechanical properties of nickel-graphene composites synthesized by electrochemical deposition. *Nanotechnology* 2015;**26**(6):065706.

27. Zeng D, et al. Preparation of multi-branched Au−ZnO hybrid nanocrystals on graphene for enhanced photocatalytic performance. *Mater Lett* 2015;**161**:379−83.

28. Hazra A, Samane N, Basu S. A review on metal oxide-graphene derivative nano-composite thin film gas sensors. In: *Multilayer thin films-versatile applications for materials engineering*. IntechOpen; 2020.

29. Hu C, et al. A brief review of graphene−metal oxide composites synthesis and applications in photocatalysis. *J Chin Adv Mat Soc* 2013;**1**(1):21−39.

30. Lee J-W, et al. Fouling-tolerant nanofibrous polymer membranes for water treatment. *ACS Appl Mater Interfaces* 2014;**6**(16):14600−7.

31. Xi Y, Mallavarapu M, Naidu R. Preparation, characterization of surfactants modified clay minerals and nitrate adsorption. *Appl Clay Sci* 2010;**48**(1−2):92−6.

32. Abu-Nada A, McKay G, Abdala A. Recent advances in applications of hybrid graphene materials for metals removal from wastewater. *Nanomaterials* 2020;**10**(3):595.

33. Lü W, et al. Facile preparation of graphene−Fe_3O_4 nanocomposites for extraction of dye from aqueous solution. *CrystEngComm* 2014;**16**(4):609−15.

34. Chandra V, et al. Water-dispersible magnetite-reduced graphene oxide composites for arsenic removal. *ACS Nano* 2010;**4**:3979−86.

35. Hao L, et al. SiO_2/graphene composite for highly selective adsorption of Pb (II) ion. *J Colloid Interface Sci* 2012;**369**(1):381−7.

36. Zong E, et al. Adsorptive removal of phosphate ions from aqueous solution using zirconia-functionalized graphite oxide. *Chem Eng J* 2013;**221**:193−203.

37. Lee Y-C, Yang J-W. Self-assembled flower-like TiO_2 on exfoliated graphite oxide for heavy metal removal. *J Ind Eng Chem* 2012;**18**(3):1178−85.

38. Tan L, et al. Enhanced adsorption of uranium (VI) using a three-dimensional layered double hydroxide/graphene hybrid material. *Chem Eng J* 2015;**259**:752–60.
39. Kabiri S, et al. Functionalized three-dimensional (3D) graphene composite for high efficiency removal of mercury. *Environ Sci* 2016;**2**(2):390–402.
40. Ji L, et al. Graphene nanosheets and graphite oxide as promising adsorbents for removal of organic contaminants from aqueous solution. *J Environ Qual* 2013;**42**(1):191–8.
41. Wang Y, et al. Synthesis of three-dimensional graphene-based hybrid materials for water purification: a review. *Nanomaterials* 2019;**9**(8):1123.
42. Carmalin Sophia A, et al. Application of graphene based materials for adsorption of pharmaceutical traces from water and wastewater-a review. *Desal Water Treat* 2016;**57**(57):27573–86.
43. Nupearachchi C, Mahatantila K, Vithanage M. Application of graphene for decontamination of water; implications for sorptive removal. *Groundwat Sustain Dev* 2017;**5**:206–15.
44. Qi P, Pichler T. Competitive adsorption of as (III), as (V), Sb (III) and Sb (V) onto ferrihydrite in multi-component systems: implications for mobility and distribution. *J Hazard Mater* 2017;**330**:142–8.
45. Bai S, et al. One-pot solvothermal preparation of magnetic reduced graphene oxide-ferrite hybrids for organic dye removal. *Carbon* 2012;**50**(6):2337–46.
46. Zhang K, et al. Graphene oxide/ferric hydroxide composites for efficient arsenate removal from drinking water. *J Hazard Mater* 2010;**182**(1–3):162–8.
47. Xu X, et al. Interfacial engineering in graphene bandgap. *Chem Soc Rev* 2018;**47**(9):3059–99.
48. Min Y, et al. Enhanced chemical interaction between TiO_2 and graphene oxide for photocatalytic decolorization of methylene blue. *Chem Eng J* 2012;**193**:203–10.
49. Shi P, et al. Co_3O_4 nanocrystals on graphene oxide as a synergistic catalyst for degradation of Orange II in water by advanced oxidation technology based on sulfate radicals. *Appl Catal B Environ* 2012;**123**:265–72.
50. Yang X, et al. Superparamagnetic graphene oxide–Fe_3O_4 nanoparticles hybrid for controlled targeted drug carriers. *J Mater Chem* 2009;**19**(18):2710–4.
51. Hu M, Hui K, Hui K. Role of graphene in MnO_2/graphene composite for catalytic ozonation of gaseous toluene. *Chem Eng J* 2014;**254**:237–44.
52. Wolfenden BS, Willson RL. Radical-cations as reference chromogens in kinetic studies of ono-electron transfer reactions: pulse radiolysis studies of 2, 2′-azinobis-(3-ethylbenzthiazoline-6-sulphonate). *J Chem Soc Perkin Transact* 1982;**2**(7):805–12.
53. Kumar S, et al. Photocatalytic degradation of organic pollutants in water using graphene oxide composite. In: *A new generation material graphene: applications in water technology*. Springer; 2019. p. 413–38.
54. Geim AK, Novoselov KS. The rise of graphene. In: *Nanoscience and technology: a collection of reviews from nature journals*. World Scientific; 2010. p. 11–9.
55. Dharwadkar S, Yu L, Achari G. *Graphene-Titanium dioxide composite: a novel photocatalyst for the degradation of organic contaminants*. 2019.
56. Kumar SG, Rao KK. Zinc oxide based photocatalysis: tailoring surface-bulk structure and related interfacial charge carrier dynamics for better environmental applications. *RSC Adv* 2015;**5**(5):3306–51.
57. Luo Q-P, et al. Reduced graphene oxide-hierarchical ZnO hollow sphere composites with enhanced photocurrent and photocatalytic activity. *J Phys Chem C* 2012;**116**(14):8111–7.
58. Mengting Z, et al. Applicability of $BaTiO_3$/graphene oxide (GO) composite for enhanced photodegradation of methylene blue (MB) in synthetic wastewater under UV–vis irradiation. *Environ Pollut* 2019;**255**:113182.

59. Brillas E, Sirés I, Oturan MA. Electro-Fenton process and related electrochemical technologies based on Fenton's reaction chemistry. *Chem Rev* 2009;**109**(12):6570−631.
60. Divyapriya G, Nambi IM, Senthilnathan J. An innate quinone functionalized electrochemically exfoliated graphene/Fe$_3$O$_4$ composite electrode for the continuous generation of reactive oxygen species. *Chem Eng J* 2017;**316**:964−77.
61. Ng YH, et al. To what extent do graphene scaffolds improve the photovoltaic and photocatalytic response of TiO$_2$ nanostructured films? *J Phys Chem Lett* 2010;**1**(15):2222−7.
62. Hoffmann MR, et al. Environmental applications of semiconductor photocatalysis. *Chem Rev* 1995;**95**(1):69−96.
63. Gan WY, Zhao H, Amal R. Photoelectrocatalytic activity of mesoporous TiO$_2$ thin film electrodes. *Appl Catal Gen* 2009;**354**(1−2):8−16.
64. Zhu G, et al. Graphene-incorporated nanocrystalline TiO$_2$ films for CdS quantum dot-sensitized solar cells. *J Electroanal Chem* 2011;**650**(2):248−51.
65. Vinodgopal K, Hotchandani S, Kamat PV. Electrochemically assisted photocatalysis: titania particulate film electrodes for photocatalytic degradation of 4-chlorophenol. *J Phys Chem* 1993;**97**(35):9040−4.
66. Meng A, et al. Rapid synthesis of a flower-like ZnO/rGO/Ag micro/nano-composite with enhanced photocatalytic performance by a one-step microwave method. *RSC Adv* 2014;**4**(104):60300−5.
67. Anandan S, Ikuma Y, Niwa K. An overview of semi-conductor photocatalysis: modification of TiO2 nanomaterials. In: *Solid state phenomena*. Trans Tech Publ; 2010.
68. Ntsendwana B, et al. Photoelectrochemical oxidation of p-nitrophenol on an expanded graphite−TiO$_2$ electrode. *Photochem Photobiol Sci* 2013;**12**(6):1091−102.
69. Zeng X, et al. Silver/reduced graphene oxide hydrogel as novel bactericidal filter for point-of-use water disinfection. *Adv Funct Mater* 2015;**25**(27):4344−51.
70. Akhavan O, Ghaderi E. Photocatalytic reduction of graphene oxide nanosheets on TiO$_2$ thin film for photoinactivation of bacteria in solar light irradiation. *J Phys Chem C* 2009;**113**(47):20214−20.
71. Tijani JO, et al. A review of combined advanced oxidation technologies for the removal of organic pollutants from water. *Water Air Soil Pollut* 2014;**225**(9):2102.
72. Liu L, Liu J, Sun DD. Graphene oxide enwrapped Ag$_3$ PO$_4$ composite: towards a highly efficient and stable visible-light-induced photocatalyst for water purification. *Catal Sci Technol* 2012;**2**(12):2525−32.
73. Pant HR, et al. A green and facile one-pot synthesis of Ag−ZnO/RGO nanocomposite with effective photocatalytic activity for removal of organic pollutants. *Ceram Int* 2013;**39**(5):5083−91.
74. García-Betancourt M, et al. *Low dimensional nanostructures: measurement and remediation technologies applied to trace heavy metals in water.* 2020.
75. Ribao P, Rivero MJ, Ortiz I. TiO$_2$ structures doped with noble metals and/or graphene oxide to improve the photocatalytic degradation of dichloroacetic acid. *Environ Sci Pollut Control Ser* 2017;**24**(14):12628−37.
76. Shahbazi R, Payan A, Fattahi M. Preparation, evaluations and operating conditions optimization of nano TiO$_2$ over graphene based materials as the photocatalyst for degradation of phenol. *J Photochem Photobiol Chem* 2018;**364**:564−76.
77. Ayoubi-Feiz B, Mashhadizadeh MH, Sheydaei M. Preparation of reusable nano N-TiO2/ graphene/titanium grid sheet for electrosorption-assisted visible light photoelectrocatalytic degradation of a pesticide: effect of parameters and neural network modeling. *J Electroanal Chem* 2018;**823**:713−22.

78. Sharma A, Lee B-K. Rapid photo-degradation of 2-chlorophenol under visible light irradiation using cobalt oxide-loaded TiO2/reduced graphene oxide nanocomposite from aqueous media. *J Environ Manag* 2016;**165**:1—10.

79. Rouzafzay F, et al. Graphene@ ZnO nanocompound for short-time water treatment under sun-simulated irradiation: effect of shear exfoliation of graphene using kitchen blender. *J Alloys Compd* 2020:154614.

80. Phophayu S, et al. Modified graphene quantum dots-zinc oxide nanocomposites for photocatalytic degradation of organic dyes and commercial herbicide. *J Reinforc Plast Compos* 2020;**39**(3—4):81—94.

81. Ebrahimi R, et al. Photocatalytic degradation of 2, 4-dichlorophenoxyacetic acid in aqueous solution using Mn-doped ZnO/graphene nanocomposite under LED radiation. *J Inorg Organomet Polym Mater* 2020;**30**(3):923—34.

82. Maruthupandy M, et al. Graphene-zinc oxide nanocomposites (G-ZnO NCs): synthesis, characterization and their photocatalytic degradation of dye molecules. *Mater Sci Eng B* 2020;**254**:114516.

Biomedical applications of metal oxide−carbon composites

14

Ammar Z. Alshemary[1,3], Ali Motameni[4] and Zafer Evis[2]
[1]Department of Biomedical Engineering, Karabük University, Karabük, Turkey; [2]Department of Engineering Sciences, Middle East Technical University, Ankara, Turkey; [3]Biomedical Engineering Department, Al-Mustaqbal University College, Hillah, Babil, Iraq; [4]Department of Metallurgical and Materials Engineering, Middle East Technical University, Ankara, Turkey

14.1 Introduction

Metal oxide (MO) nanomaterials have received extensive research attention owing to their remarkable properties and applications in various sectors (Fig. 14.1). However, these materials cannot fulfill all the possible requirements of researchers. This deficiency can be ascribed to the natural brittleness and low fracture toughness of MOs. Both issues continue to drive researchers by encouraging them to explore novel techniques to synthesize metal oxides and upgrade their properties to meet application requirements by incorporating other metallic or nonmetallic elements.[1−3] Similarly,

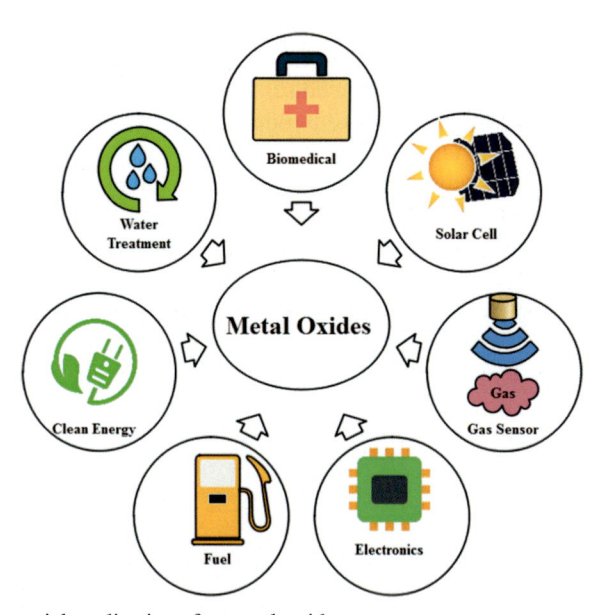

Figure 14.1 Potential applications for metal oxides.

Metal Oxide-Carbon Hybrid Materials. https://doi.org/10.1016/B978-0-12-822694-0.00004-1

their properties can be upgraded, in contrast to individual component upgrading, by combining two or more materials (composite materials).[4]

For the last 20 years, carbon-based nanofillers like graphene oxide (GO), nanodiamonds (NDs), and carbon nanotubes (CNTs) have been the subjects of vast research activity because of the superior and remarkable characteristics they possess over other materials.[4] Although these modern materials have an enormous number of applications in various fields, such as sensors, electronics, and biofuels, here we focus particularly on their application in the biomedical field in dealing with emerging clinical disorders. In recent years, the synthesis of these MO-based composites has been widely considered. Incorporating a carbon-based ceramic matrix may change physicochemical properties while also providing multidisciplinary utilizations in photocatalytic and medical care areas. Several research papers, reviews, and book chapters have already been published regarding biomedical applications. This chapter provides an extensive and updated vision of MO−carbon composites to deal with various clinical diseases.

14.2 Metal oxide nanoparticles

The importance of MO nanomaterials and their wonderful applications in biomedicine has been increasingly recognized over the last 2 decades. The integration of MO-based nanomaterials into the clinical area has opened new horizons for young researchers. This has led to a better understanding of molecular biology, and as a result, new opportunities to develop novel techniques to synthesize these nanomaterials to treat various diseases. Previously, it was difficult to deal with different clinical diseases because of the large sizes of these particles. The synthesis and characterization of diversified nanoparticles are very important in dealing with the medical issues that arise from day to day in current biomedical applications. In this context, various metal-based nanoparticles, such as silver, copper (Cu), and gold, and MOs like titanium oxide (TiO_2), iron oxide (Fe_2O_3), and zinc oxide (ZnO) have grown in importance because of their superior and exceptional physical and chemical properties.[5−10] These MOs are quite attractive in biomedical fields for anticancer, antifungal, and antibacterial applications because of their nontoxic characters in biological fields.[11−13] A vast collection of strategies are used to synthesize MOs and MO−carbon composites. These strategies include hydrothermal (batch and continuous), sol−gel, microwave (batch and continuous), precipitation, microwave-assisted sol−gel, and sonochemical synthesis methods.[14−21] Well-known nanoparticles that have been investigated at a large scale include Fe_2O_3, TiO_2, ZnO, manganese oxide (MnO_2), CeO_2, and zirconia (ZrO_2).

14.2.1 Iron oxides

Iron oxide nanoparticles are the most important materials utilized in various biomedical applications. Their popularity and importance arise from their particular magnetic, chemical, and biological properties. They are chemically highly stable, biocompatible, and nontoxic, and they possess high magnetic susceptibility.[22−28] These particles are available in different oxidation states and include iron(II) oxide (FeO), iron(III) oxide (Fe_2O_3), and iron(II,III) oxide (Fe_3O_4). Although they have many applications, hematite (Fe_2O_3) and magnetite (Fe_3O_4) are most commonly employed in biomedical applications because of their highly biocompatible characters.[29] Fe_3O_4 can oxidize easily, and therefore, its coatings with different biocompatible materials further enhance their scope to overcome various clinical diseases. These materials include some metals, ceramic materials, and a few polymeric materials.[30−36] Such combinations of iron oxide nanomaterials and their coatings help them further overcome agglomeration and facilitate functionalization and conjugation to antibodies and anticancer drugs. They can be used to treat tissue repair, hypothermia, in targeted drug delivery, and as contrast agents in magnetic resonance imaging (MRI).[30,32,36−44] Although hypothermia can speed up radiotherapy and chemotherapy treatments, healthy tissue malignancy, precise energy delivery, and control to the target have remained as big snags in cancer treatment.[45] Currently, the best way to overcome this difficulty medically is to introduce magnetic hypothermia. Magnetic Fe_2O_3 nanoparticles provide precise dose control and truly focused thermal therapy.[45] However, as new challenges occur, the magnetic nanoparticles can be used in injectable form, carried to the tumor site, and then activated using an alternating current (AC) magnetic field to deposit localized heating in the treatment target (42−43°C) (Fig. 14.2).[45,46] Gilchrist et al. first introduced this concept to vaccinate magnetic Fe_2O_3 nanoparticles (20−100 nm) into lymphatic tissues to treat residual cancer cells under an alternating magnetic field.[47] Wust et al. demonstrated that injecting a high volume of magnetic nanoparticles

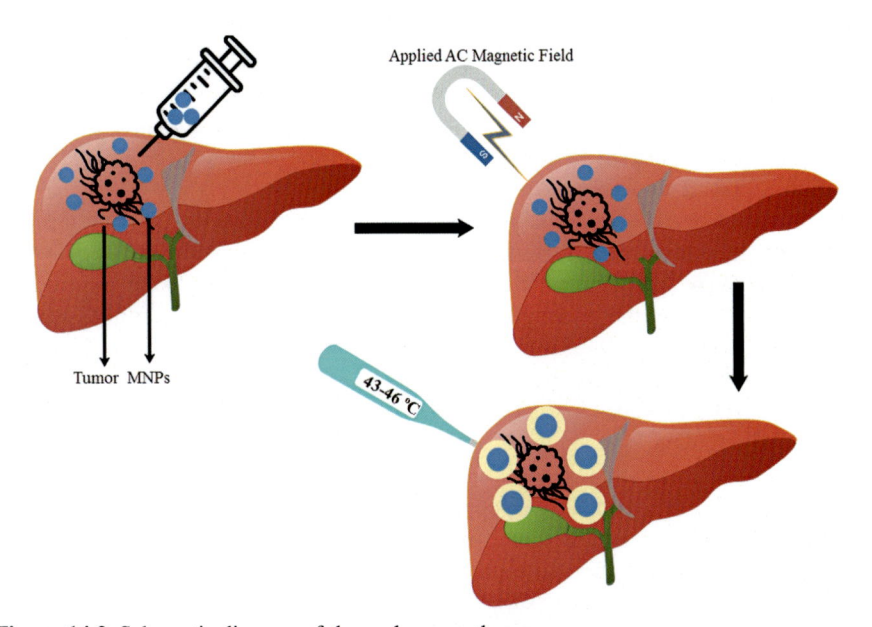

Figure 14.2 Schematic diagram of thermal cancer therapy.

(Fe_2O_3, ~ 10 kg/m^3) precisely into the cancerous tissue and then exposing them to an AC magnetic field boosted the temperature of the tumor tissue to about 43°C.[46,48]

Nanoparticle properties can be upgraded using various elements to make them more versatile for various biomedical applications. For example, $MnFe_2O_4$ is more biocompatible and is regularly employed as a contrast agent in MRI and hypothermia.[49-54] Cu-doped Fe_2O_3 showed diversity in longitudinal relaxivity values, and later, these particles were used in targeted molecular imaging and angiography in vivo.[55] However, cobalt (Co)-doped ferrite particles exhibited better antimicrobial activity.[56] Further studies were also conducted to determine the affinity constants, and similarly, cytotoxicity studies were conducted using these particles. These nanomaterials have different phases, morphologies, and particle sizes, and their different morphologies, sizes, coatings, and stabilities have various uses in the medical field. For instance, 10–20 nm iron oxide nanoparticles have better utilization in varied clinical usage, specifically because supermagnetic properties become more operative for many applications within this range. Growing evidence has accumulated from work performed to date, highlighting their use for biomedical applications, and work continues to further explore their properties to make them increasingly beneficial for the medical field. Major concerns related to toxicology, immunology, and biocompatibility are the current challenges yet to be addressed in detail to improve their medical scope.

14.2.2 Titanium dioxide

Doxorubicin has been known for some time as an effective drug with anticancer potential. It has been found very effective against human cancers, such as ovarian carcinoma, breast carcinoma, acute lymphoblastic leukemia, and hepatocellular carcinoma.[57] Nevertheless, doctors do not recommend long-term medical use because of certain acute side effects like cardiotoxicity, which may contribute to cardiomyopathy and congestive heart failure.[58] Owing to this tangible threat, there has been an earnest need to introduce new delivery methods that may improve its therapeutic efficacy while overcoming its side effects. In this instance, a nanocarrier drug delivery system has played a positive role in cancer chemotherapy.

To further deal with cancer therapy, a method was introduced to entrap drugs with complexing transition metal ions, and it was further demonstrated that the pH of the medium and the nature of the material are important in the delivery of the encapsulated drug.[59] This work was then extended to other inorganic materials. Regarding this view, TiO_2 nanoparticles have been found beneficial for various medical therapies because they are nontoxic, possess chemical stability, and are environmentally friendly.[60,61]

Among MOs, TiO_2 nanoparticles are becoming more prominent because of their possible participation in novel medical therapies (Fig. 14.3).

TiO_2 is a modern inorganic material that has been well known in the material science field because of its remarkable photoactivity. Its most important trait is its potential to produce an array of reactive oxygen species (ROS) after illumination with ultraviolet (UV) light in aqueous media.[62] This ability of TiO_2 makes it an attractive choice for medical use. As a result, these nanoparticles may induce cell death and further utilize photodynamic therapy to treat an extensive range of maladies and

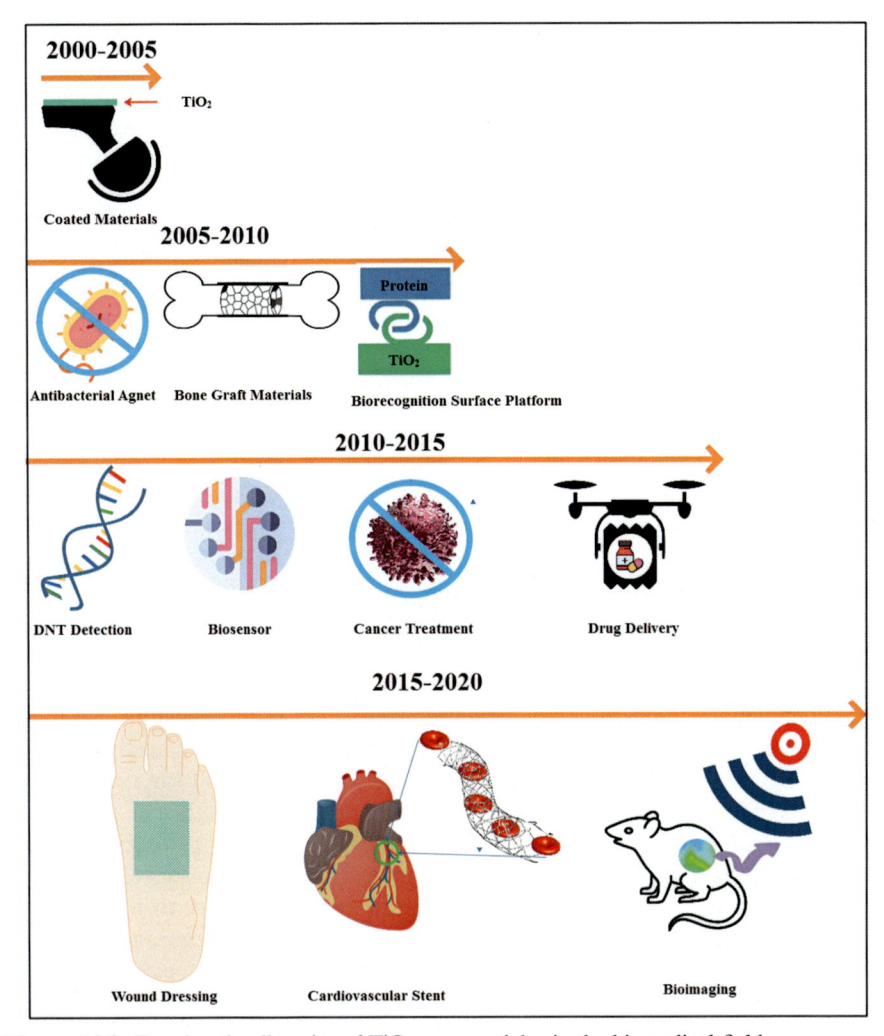

Figure 14.3 Showing the diversity of TiO$_2$ nanoparticles in the biomedical field.

cancer. They have been studied and used as photosensitizing agents to cure photodynamic inactivation of antibiotic-resistant bacteria and malignant tumors.

There are different types of TiO$_2$ nanoparticles according to phase, morphology, and particle nature: anatase, rutile, and brookite. Similarly, they may be porous, mesoporous, highly crystalline, amorphous, or semicrystalline. Morphologically, they may be spherical, flowerlike, rodlike, or platelike. These different physical aspects of TiO$_2$ particles make them an interesting choice for researchers not only for synthesizing TiO$_2$ using different techniques but also for studying their beneficial characteristics for various applications, including medical, photocatalytic, cosmetics, and biomaterials fields.

The surface area of TiO_2 nanoparticles is another important aspect that always remains under consideration in research, as it has a vital role when selected for diverse applications or fields. TiO_2 nanostructures in phase-pure form, composite composition, or combination with biomolecules can be applied in various applications. They can successfully be used as photosensitizers in photodynamic therapy, and similarly, because of their enhanced light absorption capacity, they can be suggested for targeted therapy in the clinical area. Wu et al. synthesized mesoporous nanoparticles via a controlled hydrolysis method and used them for drug delivery uses.[63] The cytotoxicity of these particles was revealed through human breast cancer cells, and results exhibited good compatibility. Also apparent was that mesoporous TiO_2 particles have a high level of attraction for phosphate moieties like DNA and, therefore, can be functionalized with phosphate-containing fluorescent species. Further, they can be utilized for intracellular bioimaging in drug delivery to female breast cancer.[63]

14.2.3 Zinc oxide

Nowadays, nanomaterials receive prominent exposure because of the special biological features that make them more attractive for researchers in biomedical applications, thus allowing them to be prescribed for antibacterial or anticancer infections. ZnO is a versatile nanomaterial gaining importance in various clinical applications. It can be used in antiinflammatory, anticancer, antibacterial, and antioxidant activities. Consequently, it can also be used as a bioimaging agent.[64−66] In the literature, ZnO with a particle size less than 100 nm has significant biocompatibility and bioactivity. Additionally, ZnO particles have long-term utilization as cosmetic and sunscreen agents because of their superior UV properties.[67] In other research, the influence of agglomerated ZnO nanoparticles was studied on the human skin.[68] The particles were suspended in a commercial sunscreen base on the skin of five volunteers (ages 20−30 years) for 6 h daily for 5 days. The results confirmed the presence of nanoparticles in the skin furrows and within the superficial layers of the stratum corneum. This study was conducted with the help of multiphoton tomography with fluorescence lifetime imaging microscopy. Further studies showed that ZnO particles do not penetrate the viable epidermis, and similarly, no toxic cellular effects of the ZnO nanoparticles were observed even after many daily studies in vitro.

Apart from the biomedical field, ZnO nanoparticles have widespread applications as attractive photocatalytic agents and in electrotechnology industries.[69,70] Moreover, Zn is an essential element present in various tissues such as skin, brain, muscles, and bone. Likewise, Zn is the main constituent of various enzyme systems and an active member of different metabolic pathways of the living system and particularly shows a vital role in hematopoiesis, neurogenesis, and protein and nucleic acid syntheses.[69,71−73] Nanosized ZnO can deliver Zn to the body, as it absorbs readily—it is thus reliably used as a food additive. Owing to this, ZnO nanoparticles have been suggested as the safest of materials by the US Food and Drug Administration.[74] This recognition has made ZnO particles more attractive and prominent in biomedical applications. Particularly, in contrast to many other MOs, ZnO nanomaterials are comparatively inexpensive and show excellent healthy results after their use as antiinflammatory, anticancer, antibacterial, antimicrobial, and bioimaging agents.[75–79]

14.2.4 Zirconium dioxide

Teeth play a crucial role in chewing food, and as a result, they ensure good health and quality of life. However, most of the population faces oral diseases, many of which are due to imperfections of the teeth.[80,81] Therefore, there is an intense need to introduce novel materials and techniques to restructure heterogeneous dental anatomical structures.[82–84] In this connection, ceramic crowns may play a productive role in restoring dental prostheses and implants.[85–87] Titanium implants have already been used in ZrO_2 implants have become increasingly popular because of their superior osseointegration character compared with titanium implants.[88–90]

ZrO_2 nanoparticles have become increasingly popular because of their hardness, high fracture toughness, and strength. ZrO_2 shows three different phases, including monoclinic, tetragonal, and cubic, depending on composition and sintering temperature. Monoclinic ZrO_2 is most stable at room temperature; however, it transforms into the tetragonal form above 1170°C and finally to the cubic state above 2370°C.

ZrO_2 is a substantial substance with natural white color, excellent strength, good corrosion resistance, and sound chemical properties. Because of its high biocompatibility has been used as the implanting material to deal with teeth related to surgical problems.[79,91–94] Therefore, small blocks of ZrO_2 like micropowders or nanopowders are used to resolve clinical problems. Even though ZrO_2 has numerous applications, it has issues regarding its fabrication and high cost; despite these issues, nanosized ZrO_2 is impressive in applications as a filling powder, bionics promoter, and enhancer of the mechanical properties of dental ceramics and tissue engineering scaffolds.[95–98] The latest studies have confirmed that ZrO_2 can significantly improve fracture toughness and the shear bond and flexural strength of employed material.[99,100] In another study, nanosized ZrO_2 prepared through a modification method was reported to give nanostructured ZrO_2 with better biocompatibility, and similarly, nanoporous ZrO_2 implants may enhance the osseointegration process.[101]

14.2.5 Manganese dioxide

Nanotechnology developments have opened new pathways to explore potential solutions to medical disorders and effectively influence the medical area.[102–104] Regarding various biomedical materials available for different clinical applications, transition MOs, particularly MnO_2, have gained recognition for their structure-property association in several areas: catalysis, optoelectronics, and energy-based applications.[105–108] Manganese (Mn) is the 12th best-known element on the planet and the third most abundant element after titanium and iron.[109] This abundant element is also an important part of various enzymes and functions as a cofactor. Furthermore, MnO_2 may have various morphological shapes such as nanobelts, nanosheets, nanowires, and nanofibers. Owing to the variable oxidation states of Mn, it may form various oxides such as MnO, MnO_2, Mn_2O_3, Mn_3O_4, and Mn_5O_8.[110–112] Mn has been utilized in biomedical applications in different oxides and shapes. MnO_2 nanosheets with a high surface area (thickness in the range of nanometer to micrometer) were declared good for controlled drug delivery. According to the literature, MnO_2 nanoparticles, particularly MnO_2 nanosheets, are attractive for biological sensing, drug delivery, chemodynamic therapy, and molecular imaging.[113,114]

14.2.6 Cerium oxide

Cerium is the most abundant element from the lanthanide series, and unlike other rare earth metals, it is present in two oxidation states, i.e., $+3$ and $+4$.[115] Cerium is abundant in the oxide forms CeO_2 and Ce_2O_3. These oxides are good UV absorbers and have many applications as chemical mechanical polishing agents, catalysts, sensors, and corrosion-protection agents.[116–120] Their UV-absorbing capabilities have a crucial role in cosmetic products. They have also exhibited a protective role against radiation, cellular damage mediated by toxic agents, and neurological disorders.[121] Pure nano-CeO_2 has poor water solubility and shows difficulties in biological functions. However, numerous studies have revealed that the polymer coating of nanomaterials improves its stability, biocompatibility, and water solubility. For example, nanoceria layered with dextran displays a good antioxidant property.[122] Owing to its many applications, nanoceria is actively released into the environment and affects human health. Humans inhale it because of its presence within the environment, and this is its major concern. However, there is controversy regarding the role of nanoceria. According to some researchers, its toxicity level is low, and it has been found not to mediate cytotoxicity or inflammation.[123–125] At the same time, according to some researchers, nanosized CeO_2 triggers death in cells. This cell death has been ascribed to ROS from the prooxidative effect.[126] Despite this controversy, nanosized CeO_2 has achieved growing importance, and at a large scale, it has been used to treat various biomedical complications; CeO_2 has been found to manifest good activity against both gram-positive and gram-negative bacteria. However, a maximum antimicrobial role was detected against gram-negative strains. Different studies confirmed that chitosan−CeO_2 composites have good antibacterial actions against test pathogens, as evident from the qualitative analysis.[127] Increasing the concentration of chitosan−CeO_2 hybrid increases the inhibition zone against *Bacillus subtilis and Escherichia coli,* as shown in Fig. 14.4. The mentioned study confirmed that hybrid chitosan−CeO_2 nanoparticles with an even 50 μL concentration had good antibacterial activity.[127] This may be due to the impact of direct contact of the synthesized particles on the membrane of *E. coli* (initially, disruption of the cell membrane, and finally, cell death).[127–129]

Thus far, several in vitro and in vivo research studies have been conducted to explore the influence of MOs and mixed MOs in the biomedical field. These studies have confirmed that these nanoparticles have comprehensive utilization in many areas but specifically in the clinical field. An overview of these MOs and their applications is given in Table 14.1.

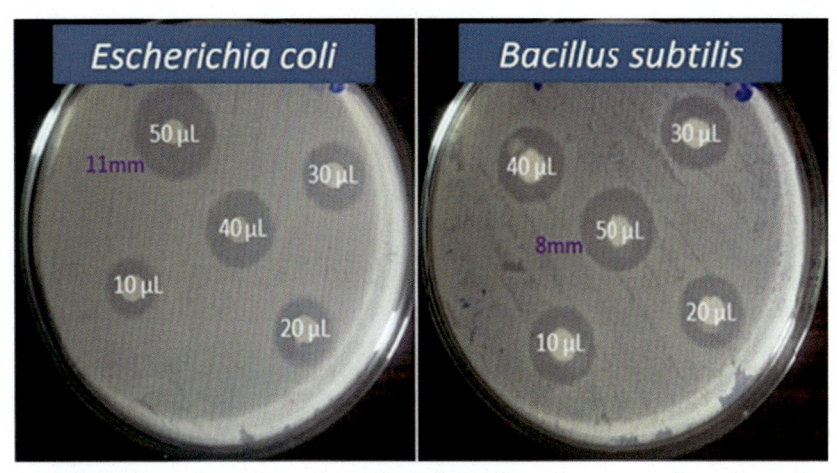

Figure 14.4 Antibacterial performance of hybrid chitosan–CeO_2 nanoparticles against *E. coli* and *B. subtilis*.[127]
This figure has been reproduced with permission from Elsevier (License Number 4984960999264).

14.3 Carbon-based materials

Biomaterials have been constantly employed in biomedical applications. This induction in the medical field is due to their certain specific properties such as high specific surface area, diversification in the morphological images, and transference from micro size to nanoscale level compared with their bulk counterparts and similarly augmentation in their chemical reactivity and modification in their physicochemical properties.[181] For example, nanosized and modified MnO_2, ZrO_2, and CeO_2 are more beneficial for the biomedical field. So far, numerous studies have been performed on a vast number of nanoparticles, including pure metal-based nanoparticles, MO-based nanoparticles, mixed MO-based nanoparticles, substituted materials, and natural resource-derived nanomaterials. However, in present circumstances, researchers are diverting their interest in carbon-based nanomaterials as they have exhibited interesting properties such as good mechanical strength, chemical resistance, and low density. Nevertheless, CNTs, GO, NDs, and carbon-related materials (like few-layer graphene and reduced graphene) received more attention.[182] Over the last 30 years, carbon-based nanofillers have included GO, NDs, and fullerenes, with researchers especially attracted to the superior characteristics of CNTs. Carbon-based materials (CBMs) have been studied considerably for biomedical applications because of their exceptional physicochemical properties. These include electrical, mechanical, optical, thermal, and structural multiplicity. Based on these properties, CBMs comprise CNT, GO, graphene quantum dots, or NDs. All these carbon-based nanomaterials have numerous utilizations in cancer therapy, drug delivery, biosensing, and against various antimicrobial agents.[183–188]

Table 14.1 Biomedical applications of metal oxides.

Metal oxide	Biomedical applications	References
Calcium oxide	Antibacterial agent, drug delivery system	130,131
Strontium oxide	Antibacterial agent, antiresorptive agent, bone graft materials	132–134
Barium oxide	Antigastroduodenal ulcer agent	135
Niobium pentoxide	Enhanced bone healing and used as a coated material on Ti6Al4V (medical implant)	136,137
Bismuth(III) oxide	Radiopaque agent	138
Lanthanum(III) oxide	Enhanced bone regeneration, showed hemocompatibility properties, and was used as an antibacterial agent	139–141
Samarium(III) oxide	Nuclear nanomedicine, antioxidant agent	142,143
Europium(III) oxide	Bioimaging	144,145
Gadolinium(III) oxide	MRI contrast agent, anticancer drug delivery system, image-guided therapy, an antibacterial agent	146–149
Terbium(III, IV) oxide	Antibacterial agent, anticancer agent	150,151
Holmium(III) oxide	MRI	152
Erbium(III) oxide	Antibacterial agent	149
Yttrium(III) oxide	Bioimaging probes, radiopaque agent	153,154
Tellurium dioxide	Antibacterial agent	155
Vanadium(V) oxide	Antibacterial agent, antifungal agent, anticancer agent, enhanced osteogenic effects	156–159
Chromium(III) oxide	Antibacterial agent	160–162
Cobalt(II) oxide	Drug delivery system, MRI contrast agents, anticancer agent, an antibacterial agent	163–166
Copper(II) oxide	Hydrophilic drug delivery model, anticancer agent, an antibacterial agent	167–169
Nickel(II) oxide	Antibacterial agent, drug delivery system	170,171
Gallium(III) oxide	Antibacterial agent, drug delivery system, bioimaging agent	172,173
Bismuth(III) oxide	Antibacterial agent, drug delivery system, biolabel radiopacifier agent	174–177
Silver oxide	Antibacterial, antioxidant agent	178,179
Aluminum oxide	Nanostructured substrates for cell-interface studies	180

Since their introduction, CBMs have gained importance as nanomaterials for diversified applications in various fields. As we know, carbons exist in three allotropic forms, including amorphous carbon, graphite, and diamond.[183,189–192] This class includes newly introduced CNTs, GO, graphene quantum dots, and carbon fullerenes. Each carbon family member reveals different characteristics and shows numerous biological applications such as drug delivery, imaging, diagnosis, cancer therapy, and tissue engineering.[193–195] Consequently, these carbon-based nanomaterials, particularly

GO, CNTs, and NDs, have attained special attention because of their exceptional electrical and structural features, which give them more strength, tractability, and electrical conductivity toward different biological species, and this is beneficial in medical analysis, sensing, and the treatment of several diseases.[196–198] Even though carbon-substituted MOs have many applications in different fields, we shall focus on CNT, GO, and NDs, which are currently considered appealing materials because of their extraordinary intrinsic properties.

14.3.1 Carbon nanotubes

CNTs have exceptional mechanical strength and special electrical and optical properties, and further, they can be loaded with various promising materials and drugs for biomedical use.[199] Important factors like morphology, dispersion, interaction, size, and alignment make CNT attractive for modern researchers working on fabricating them for various applications such as drug delivery, biosensors, cancer therapy, imaging, and tissue engineering.

In 1991, CNTs were identified as the allotropic form of carbon and since then have been widely studied for a wide range of fields like biosensors, drug carriers in medicines, and as materials reinforcement and electrode materials. There are various types of CNTs: single-walled carbon nanotubes (SWCNTs), double-walled carbon nanotubes (DWCNTs), and multiwalled carbon nanotubes (MWCNTs). SWCNTs have a small diameter in the range of 1–2 nm, whereas MWCNTs have a diameter of up to 100 nm. Similarly, DWCNTs are the interfaces between the SWCNTs and MWCNTs and share many features with SWCNTs, such as narrow diameters and significant mechanical properties. Regarding CNT nanoparticles, the study of nanoparticles' chemical structure and intrinsic composition is quite important, as both affect surface properties like catalytic activity, hydrophobicity, hydrophilicity, and possible dissolution rate. In addition, the decoration of surface these nanoparticles can modify surface properties. This may lead to altered biological properties and thus leave a marked influence on their biodistribution.[200,201] Besides this, more requirements are needed for biomedical applications of CNTs and must be considered positively. In this connection, the first one is their high priority which increases their safety level and restricts the release of toxic ions during different biological processes. The synthesis of pure CNTs is a bit difficult task. The second big task is the formulation of CNTs which is also very important for the biomedical field. This is important, as it helps the significant dispersion of CNTs in solvents, especially water. However, the hydrophobic nature of CNTs makes it difficult to disperse them in highly polar solvents. However, this challenge can be resolved by the functionalization of CNTs. For example, two different mediums (covalent and noncovalent) can be dissolved using a surfactant or templating agent. According to the current trend, CNT applications can flourish by supplementing various functional groups in CNTs. In this scenario, two functionalization approaches are very popular to functionalize CNTs, as explained in Fig. 14.5. According to the first one, CNT dispersibility can be increased in aqueous media through the oxidation method using a strong acid like H_2SO_4.[202] In the second approach, based on CNT noncovalent functionalization, various functional groups like proteins, nucleic acids, and other therapeutic agents can be added to the CNTs.[203,204]

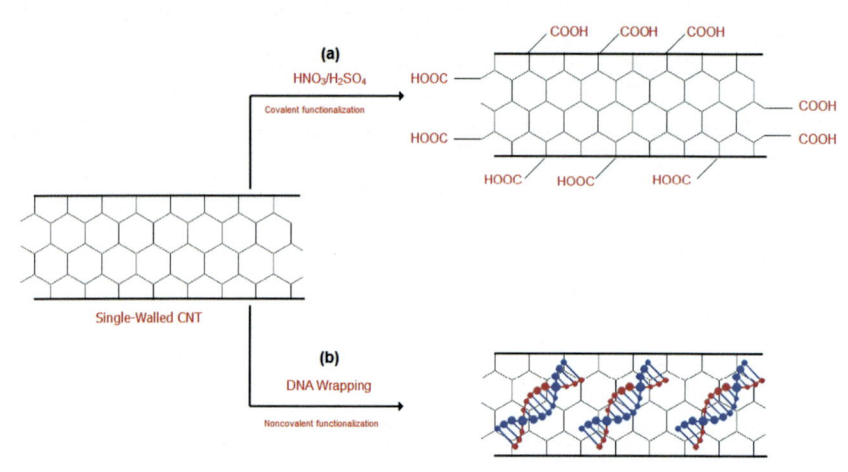

Figure 14.5 Functionalization of carbon nanotubes (A) using oxidation reaction by a strong acid, and (B) using surfactants such as DNA wrapping.

A third big challenge to using CNTs in the biomedical field is their dispersion rate when starting from dry powder. But this can be controlled by avoiding the drying steps during CNT processing. A similar issue is the increased viscosity of CNTs during their dispersion in fluid even at low concentrations, which makes it difficult to prepare well-dispersed CNT nanocomposites.

Despite these challenges, CNTs have shown good biocompatibility and have large surface areas, and such properties make them a good selection for a variety of biomedical fields. A summary of these biomedical applications is provided below.[205]

14.3.1.1 Carbon nanotubes for tissue engineering

The treatment of bone defects, such as eliminating tumors, trauma, and irregular bone growth in humans, faces important restrictions. Present rehabilitation such as metal prostheses, allographs, and autographs do not normally encourage bone-regeneration processes. As an alternative, they substitute the missed bone with artificial bioactive materials. However, using modern tissue engineering, an attempt can be made to create new tissues by tissue replacement. This can be done by culturing bone cells on synthetic three-dimensional scaffolds or live prostheses. In this connection, a good scaffold for bone tissue regeneration should have exceptional mechanical properties. It should have a large degree of porosity, high pore interconnectivity for bone tissue ingrowth, good biocompatibility with surrounding tissue, and large pore size. Similarly, the synthetic scaffold material should either be biodegradable or nonbiodegradable.

All these demands can be met using CNTs. The tensile strength of SWCNTs is about 100 times better than that of stainless steel. Similarly, the specific gravity of SWCNTs is about one-sixth that of the usual materials.[206–208] This means CNTs can find powerful applications in hard tissue engineering. Additionally, MWCNTs have shown high bone tissue compatibility and similarly have exhibited accelerated bone formation.[209] All this discussion demonstrates that CNTs can be the better candidates to deal with hard tissue engineering issues and their solution using CNTs.

The influence of CNTs on living tissues has attracted much interest, as today researchers consider them the basic materials for biomedical applications. This field, including regenerative medicine and tissue engineering, was developed recently using CNTs.[210] Furthermore, CNTs have attracted interest because of their specific characteristics such as a high surface-area-to-volume ratio and good mechanical, thermal, and electrical characteristics. Even though CNTs have remarkable properties and applications in the medical field, the safety of CNTs remains debatable. According to some researchers, CNTs have good surfaces for cellular growth and further show significant effects on neural signal transmission.[211] In another investigation, it was showed that CNTs may directly stimulate brain circuit activity.[212] Similarly, another research confirmed that CNTs have favorable surface texture and better adsorption of culture medium proteins, thus permitting mature patterning and growth of neuron networks in contrast to SiO_2 surfaces.[213] Owing to the remarkable properties of CNTs, particularly as nanocarriers have grown in importance because of their exceptional cell-transfection capabilities. Certain drugs like 10-hydroxycamptothecin can be covalently bonded to CNTs with increased efficiency.[214] Similarly, hydroxypropyl-β-cyclodextrin-modified CNTs display comparatively good delivery and release of anticancer drug formononetin, which might be useful to treat cancer and further have the potential to attack the cancer cells.[215–217]

In short, in this development field, new tissues are encapsulated in suitable biological material.[218–220] In this regard, CNTs play a crucial role. CNTs allow improvement in the mechanoelectrical aspects of scaffolds and further permit the occurrence of chemical reactions inside the cells. Furthermore, CNTs can be successfully employed in cardiac tissue and neural tissue engineering.[220]

14.3.1.2 Carbon nanotubes for drug delivery

Drug delivery represents an effective and attractive alternative to the oral or injectable delivery of drugs. Lately, CNTs have been explored as potential materials because of their nanosized and absorption capacity in drug delivery systems. A research group synthesized electrosensitive CNTs to control drug release using polyethylene oxide and pentaerythritol triacrylate polymers.[221] Kang et al. synthesized self-assembly of highly thermal conductive CNTs hybridized with chitosan.[222] After characterization, they confirmed that this self-assembly has effective drug-loading and releasing properties. Correspondingly, some drug-loaded work has been accomplished based on various CNTs to introduce and develop a novel transdermal drug delivery system.[223] In this connection, to enhance the drug delivery ability of CNTs, a new skin patch technique is also in operation to extend this study further.[224] CNTs have axial symmetry and are tubular; these unique characters help in enhancing drug delivery and diagnosis ability of CNTs to the cancer cells. Additionally, the extensive use of the CNTs in the drug delivery system minimizes the drawbacks of limited solubility and nonselective biodistribution of the other materials. Moreover, drug delivery ability through CNTs has been recognized much better than the existing approaches.[225] There is another prevalent technique helping in delivering selective drug delivery to the cancer cells that are cutting-edge functionalization of CNTs.[226]

14.3.1.3 Carbon nanotubes for bioimaging

Owing to their excellent properties, CNTs can improve the functions of the biomedical imaging field.[227] Regarding the biomedical imaging field, photoacoustic and MRI are important fields. CNTs have been revealed as major materials for these application areas, and their activity can be enhanced using an external magnetic field. Similarly, it was also reported that imaging could be further developed using quantum dots and gold nanoparticles. The carboxyl functionalized SWCNTs adjusted with the various polymeric substances and folic acid conjugate can be useful as a bioimaging probe to target the folate receptor overexpressing tumor cells.[228] In the same way, SWCNTs can be better materials in fluorescent bioimaging probes.[229] Although much detail about CNT applications could be added here, this brief detail is more than enough to highlight that CNTs have opened a new era for practical applications and offer incredible fields, including electronics, chemical applications, therapeutics and biomedicine, and biomedical areas. From the abovementioned details, it can be seen that CNTs play an essential role in every discovery because of CNTs' unique mechanical, electrical, structural, and biomedical features. For example, cancer therapy was a big challenge in the past, but using CNT-targeted drug delivery today is helping in cancer-related treatment therapies. In current circumstances, researchers are diverting their focus to functionalize CNTs to enhance their features and applications. All these efforts should prove fruitful, but CNTs have safety concerns because of their toxicity.

14.3.2 Graphene oxide

Since the advent of carbon materials, CBMs are getting remarkable attention because of their unique properties like excellent flexibility, high transparency, superior charge carrier mobility, superior thermal quality, high strength mechanical characters, and extraordinary electronic qualities.[230–235] These wonderful characteristics include their more environmentally and biologically friendly nature compared with inorganic materials. Carbon is one of the most popular units of our ecosystem and therefore has a widespread affiliation with biological systems, particularly because of its less reported toxic effects. Graphite is a well-known, naturally occurring CBM with several applications. Graphene is a single layer of graphite in a two-dimensional lattice, and therefore it is expected to have many applications in the biological system. It consists of a densely packed network of carbon atoms assembled in a honeycomb manner and is reported to show a high theoretical surface area.[236] Additionally, graphene is highly transparent and shows good absorption of visible light, and similarly has almost negligible reflectance.[230,237,238] Owing to these features, graphene particles hold great potential in many areas like nanoelectronics, sensors, hydrogen storage cells, and supercapacitors.[239–241] Besides these areas, graphene is well known in numerous biomedical applications such as promising biosensing materials and drug delivery agents.[242–244] Furthermore, the proficiency of graphene materials has been recognized to be much greater than that of CNTs.[245]

Likewise, graphene, GO-based nanomaterials have been attracting attention because of their better surface functionalize ability, amphiphilicity, aqueous processability, and fluorescence quenching ability, making them suitable for various biomedically and clinically based applications. GO has been derived from oxidized graphite and has been considered an excellent promising material for various biological applications. These remarkable characteristics of GO are because GO has a unique structure, and this uniqueness is due to the special chemical structure of GO, which is composed of SP^2 hybridized carbon domains further encircled by SP^3 hybridized carbon atoms and at the same time with hydrophilic functional groups embedded with oxygen atoms.[246] In summary, nanofunctionalized biocompatible GO with various morphological features can be used as anticancer drugs, in cellular imaging, as drug load and drug delivery agents, and for several medically oriented disorders.[247–249]

Research conducted by Fallatah et al. exhibited that GO can reduce biofilm viability and detached biofilm cells accompanying membrane death after 48 h treatment, whereas 24 and 72 h treatment did not show appreciable results.[250] Another study, performed by Di Giulio et al., revealed that GO particles may significantly reduce the effect of *Pseudomonas aeruginosa*, *Staphylococcus aureus*, and *Candida albicans* after different treatment intervals.[251]

Regarding antibacterial and antimicrobial activities with graphene and GO, there is ongoing debate among different groups, but a sole approach is accepted that somehow both have the capability of infection caused by these varieties of bacteria. However, it is a commonly established phenomenon that antimicrobial activity of GO nanoparticles could be the result of physicochemical factors.[252,253] How these physical and chemical mechanisms act to finally die the bacterial cells is shown in Fig. 14.6. Following the physical method, it has been revealed that the wrapping pathway of larger GO sheets supports further thwarting of bacterium proliferation. But in contrast to the physical route, the chemical method results in ROS overproduction. Later, these ROS not only oxidize the fatty acids but also facilitate the mechanism of lipid peroxides. These lipid peroxides then stimulate a chain reaction that ultimately disintegrates the cell membrane, and finally, the disease-causing cells die.[254] Even here, there are two opinions about the mode of action of graphene. One group favors that the graphene surface is primarily responsible for antimicrobial activity, whereas according to other groups, surface edges are responsible for this activity. Yet according to another current report, bare GO eradicates the bacterial species, while the screening mechanism of the basal plane of GO makes them sluggish.[255]

14.3.3 Nanodiamonds

NDs are CBMs with large surface areas.[256] NDs can be further functionalized using various ligands that help to conjugate them with drugs.[257] The basic crystal structure

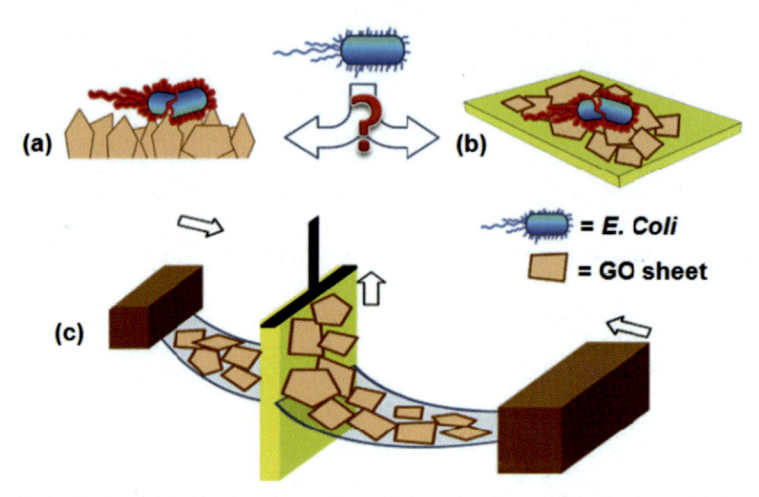

Figure 14.6 (A) A model showing the antibacterial mechanism of GO (B) A model showing the antibacterial activity of GO- Langmuir—Blodgett films and (C) A set up indicating how Langmuir—Blodgett model work out.[253]

This figure has been reproduced with permission from the Royal Society of Chemistry (License Number: ID1089274-1).

of ND is nanocrystalline and has tetrahedrally bonded carbon atoms in the form of a three-dimensional cubic lattice. It imparts the properties of diamond in it and has coatings of the onion-shaped carbon shell functional groups on the surface.[258,259] Nevertheless, NDs are frequently labeled as a crystalline diamond core fenced by an onionlike amorphous graphite shell.[260] The SP^2/SP^3 bonds in NDs are fairly flexible and have two geometrical forms (stretched-faced diamond and puckered-faced diamond). The stretched face of the diamond can act like a graphene plane. Similarly, puckered-faced graphene may become a diamond surface.

ND particles have good hardness, good wear resistance, bioactivity, and significant biocompatibility with living tissues, and because of these properties, NDs have a huge number of applications in the biomedical area.[261] It can be employed to rectify orthopedic surgical disorders, cardiovascular problems, and dental-oriented complaints. There is a different type of NDs particles and have different medical care utilizations.

The most stirring results of NDs have been shown in drug delivery, imaging, and diagnostics.[262−264] Although these two areas are growing quickly, ND particles are gaining popularity in biodegradable bone surgery devices, killing drug-resistant microorganisms and tissue-engineering scaffolds because of enhanced cellular adhesion.[261] Additionally, ND particles facilitate powerful resistance toward viruses and deliver genetic materials into the cell nucleus.[265] All these present prospects stress the need for a greater understanding of NDs to further expand their applications for various biomedical fields.

14.4 Metal oxide—carbon composites: synthesis and biomedical applications

Substantial awareness has been ascended in the research of MOs and mixed MOs nanoparticles for many years, specifically for biomedical uses. The integration of nanotechnology science into the field of the clinical area has opened new potentials for the future. A coherent working in this area allowed better appreciative knowhow about molecular biology. Consequently, modern science has been providing novel approaches to treat various illnesses which were formerly problematic to mark because of size limitations. From a medical point of view, the preparation of biofunctional nanoparticles is essential. In this connection, the synthesis of mixed MOs substituted with carbon atoms has recently attracted the attention of research groups, and this area continues to flourish. In the last part of this chapter, we shall highlight those techniques and carbon-substituted MO nanoparticles with novel properties that today are facilitating physicians to deal with current diseases and their treatment using these modern materials. Besides medical area applications, CBMs have several utilizations for electrical devices like electrodes and supercapacitors because of their excellent electrical conductivity and high surface area.[266] Although MO—carbon composites have many applications in the energy sector, the biomedical care applications area has been rising as a modern field.

Various techniques are constantly being used to fabricate these modern materials for use in the clinical area. In this regard, two popular approaches are in situ and ex situ techniques to synthesize MO-doped carbon composites.[267,268] Metals such as Fe, Cu, Ti, Ce, and Zn have been incorporated into GO, CNTs, and NDs to synthesize metal—carbon composites for various applications like biomedical imaging, biomedical sensing, cancer treatment, targeted drug delivery, and surgical disorders.

Copper is a well-known transition metal because of its antibacterial character and has been used to synthesize phase-pure MOs, mixed MOs, Cu-incorporated hydroxyapatite nanoparticles. Owing to their well-established antimicrobial characteristics, many research efforts have been executed to prepare carbon-incorporated MO nanomaterials to deal with different biological and biomedical applications. Novel Cu-substituted carbon dots have been fabricated using a simple and facile single-step hydrothermal method.[269] These nanodots were found to have outstanding fluorescent properties. Transmission electron microscopy and X-ray diffraction characterizations of Cu carbon dots confirmed that their size is 3.76 nm, and they are spherical and monodispersed in shape. These particles were further used to identify iron in human blood serum and other clinical applications. In another investigation, Cu carbon dots were prepared using a Cu complex of polyacrylic acid coordinating carboxylic group and Cu ions.[270] These nanodots, later on, were inducted for the fluorescence imaging of human cervical cancer and in human neuroblastoma cells because of having bright fluorescence and low toxicity. This study further revealed that Cu carbon dots might be fruitful therapeutic agents for various biological applications.

In 2007, a group of researchers reported the synthesis of carbon substituted TiO_2 micro/nanospheres and nanotubes through a single-source chemical vapor deposition technique.[271] In this work, titanium butoxide was used as a carbon, oxygen, and titanium source. The as-prepared CNT-TiO_2 was well instructed with a diameter around 100 nm, and thickness was reported around 15 nm. Results confirmed that these particles have the best photocatalytic activity. Another research report described the synthesis of flexible, lightweight, and disposable MWCNT−ZnO nanofibers.[272] The nanofibers were prepared using the electrospinning technique through the calcination process. Results confirmed that these biosensors are effective against malaria biomarkers and could be effective for other biomarker detection systems. According to current research studies, one-dimensional semiconducting MO nanomaterials have earned recognition among researchers because of their thermal characteristics, chemical stability, and high-care diagnostic ability.[273,274] Besides these properties, one-dimensional nanomaterial provides rapid electron transport compared with their bulk counterparts.[275] Additionally, in a study ZnO-doped CNTs and MWCNT−ZnO nanoparticles were observed to show high sensing ability in contrast to ZnO or MWCNTs individually.[276,277] Because of these versatilities of the MO-doped CBMs in another project, mesoporous TiO_2@ZnO-GO hybrid nanostructures were synthesized using a facile sonochemical method.[278] These nanocarriers were further inducted for colon-specific drug delivery study.

Finally, these studies confirmed that MOs, and mixed MOs composites have a wide scope, and a great deal of work is required to explore new horizons in different fields regarding the fascinating characteristics of these modern materials. To sum up this discussion, applications of different MOs-Carbon composites are given in Table 14.2.

Table 14.2 Biomedical applications of metal oxide−carbon composites.

Metal oxide−carbon composites			
MO	**Carbon**	**Biomedical application**	**References**
γ-Fe_2O_3	MWCNTs	Nanomedicine system	279
MgO	CNTs	Impedimetric genosensor	280
TiO_2	CNTs	Implant coating materials	281
MgO	MWCNTs	Antibacterial agent	282
ZnO	GO	Drug delivery systems	283
TiO_2+ZnO	GO	Antibacterial agent	284
Fe_3O_4	NDs	Contrast imaging and targeted delivery system	285
TiO_2	NDs	Sonodynamic therapy for cancer treatment	286
Ag/ZnO	rGO	Antibacterial agent	287
ZnO	MWCNTs	Malaria biomarker detection	272
Pd−ZnO	rGO	Antimicrobial and antioxidant agent	288
ZnO	rGO	Antimicrobial agent	289
ZnO	GO	Antimicrobial agent	290

CNT, carbon nanotube; *GO*, graphene oxide; *MO*, metal oxide; *MWCNT*, multiwalled CNT; *NDs*, nanodiamonds; *rGO*, reduced graphene oxide.

14.5 Conclusions

This study demonstrated the importance of MOs, CNTs, GO, NDs, and MO-doped carbon composites. Metallic and MOs nanomaterials including TiO_2, ZnO, ZrO_2, CeO_2, GO, CNTs, and NDs have garnered potential involvement from various fields, particularly modern medical therapies. These modern materials currently have a high rise in the scientific fields because of their interesting and attractive characteristics such as mechanical, electrical, optical, electrochemical, high chemical stability, high thermal stability, and high surface to volume ratio. The aforementioned information confirms that using these nanostructures individually or in composite forms enhances their properties and makes them multidimensional and thus vast their scope for various fields, particularly biomedical areas. Besides, their versatile surface-active properties, morphological diversity make them imminent materials for many medical-oriented applications. Particularly, various materials' ability to produce ROS further induces cell death and has shown utilization in the photodynamic therapies for the medication of a wide range of cancer. Various studies have confirmed that these MOs and carbon composites (hybrid materials) can be studied as photosensitizing agents to treat malignant tumors and inactivate different bacterial species. These materials further revealed their usefulness in targeted medicinal therapies because they can accelerate light absorption, which is helpful in targeted therapy. These studies have confirmed that these modern carbon-based composites could successfully facilitate humanity's efforts to overcome numerous medical-based issues; however, this work cannot be found here. Still, there is a vacuum to work further and investigate these materials' employment in other fields. Many studies indicate that the increased therapeutic efficiencies of phase-pure MOs (TiO_2, ZnO, ZrO_2, CeO_2) and carbon-based composites can be further augmented using targeted drug delivery. Moreover, it has been observed that functionalization with various organic functional groups enhances the scope of these materials. Fortunately, many techniques and functionalized materials have presently been synthesized to deal with emerging day-to-day medical queries.

Although much work has been done and is in operation, we suggest that the fresh researcher focuses on controlling production costs, toxic effects, and the attractiveness of composites for bioimaging in clinical analysis, targeted drug delivery, and photothermal cancer treatment. Additionally, the fabrication of these materials and their characterization are big challenges for implementing them. Modern challenges include a limited lifetime, operational stability, solubility in aqueous media, and reproducibility. In the future, researchers must target these areas to further enhance the use of modern hybrid composite materials for more fascinating applications.

References

1. Soldano C. Hybrid metal-based carbon nanotubes: novel platform for multifunctional applications. *Prog Mater Sci* 2015;**69**:183−212.
2. Liu Y, Chae HG, Choi YH, Kumar S. Effect of carbon nanotubes on sintering behavior of alumina prepared by sol−gel method. *Ceram Int* 2014;**40**:6579−87.

3. Zapata-Solvas E, Gómez-García D, Domínguez-Rodríguez A. Towards physical properties tailoring of carbon nanotubes-reinforced ceramic matrix composites. *J Eur Ceram Soc* 2012;**32**:3001−20.

4. Mallakpour S, Khadem E. Carbon nanotube−metal oxide nanocomposites: fabrication, properties and applications. *Chem Eng J* 2016;**302**:344−67.

5. Robinson I, Ung D, Tan B, Long J, Cooper AI, Fernig DG, Thanh NTK. Size and shape control for water-soluble magnetic cobalt nanoparticles using polymer ligands. *J Mater Chem* 2008;**18**:2453−8.

6. Roca A, Costo R, Rebolledo A, Veintemillas-Verdaguer S, Tartaj P, Gonzalez-Carreno T, Morales M, Serna C. Progress in the preparation of magnetic nanoparticles for applications in biomedicine. *J Phys Appl Phys* 2009;**42**:224002.

7. Gupta AK, Naregalkar RR, Vaidya VD, Gupta M. Recent advances on surface engineering of magnetic iron oxide nanoparticles and their biomedical applications. *Nanomedicine* 2007;**2**:23−39.

8. Dwivedi S, Wahab R, Khan F, Mishra YK, Musarrat J, Al-Khedhairy AA. Reactive oxygen species mediated bacterial biofilm inhibition via zinc oxide nanoparticles and their statistical determination. *PLoS One* 2014;**9**:e111289.

9. Mirzaei H, Darroudi M. Zinc oxide nanoparticles: biological synthesis and biomedical applications. *Ceram Int* 2017;**43**:907−14.

10. Daoud WA, Xin JH, Zhang Y-H. Surface functionalization of cellulose fibers with titanium dioxide nanoparticles and their combined bactericidal activities. *Surf Sci* 2005;**599**: 69−75.

11. Nadeem M, Tungmunnithum D, Hano C, Abbasi BH, Hashmi SS, Ahmad W, Zahir A. The current trends in the green syntheses of titanium oxide nanoparticles and their applications. *Green Chem Lett Rev* 2018;**11**:492−502.

12. Ealias AM, Saravanakumar M. A review on the classification, characterisation, synthesis of nanoparticles and their application. *IOP Conf Ser Mater Sci Eng* 2017:032019.

13. Sangaiya P, Jayaprakash R. A review on iron oxide nanoparticles and their biomedical applications. *J Supercond Nov Magnetism* 2018;**31**:3397−413.

14. Xu C. *Continuous and batch hydrothermal synthesis of metal oxide nanoparticles and metal oxide-activated carbon nanocomposites*. Doctoral dissertation. Georgia Institute of Technology; 2006.

15. Hayashi H, Hakuta Y. Hydrothermal synthesis of metal oxide nanoparticles in supercritical water. *Materials* 2010;**3**:3794−817.

16. Ding Y, Xu L, Chen C, Shen X, Suib SL. Syntheses of nanostructures of cobalt hydrotalcite like compounds and Co_3O_4 via a microwave-assisted reflux method. *J Phys Chem C* 2008;**112**:8177−83.

17. Akram M, Butt FK, Alshemary AZ, Goh Y-F, Ibrahim WAW, Hussain R. Continuous microwave flow synthesis (CMFS) of nanosized titania: structural, optical and photocatalytic properties. *Mater Lett* 2015;**158**:95−8.

18. Akram M, Taha A, Butt FK, Awan AS, Hussain R. Effect of reactant concentration on the physicochemical properties of nanosized titania synthesized by microwave-assisted continuous flow method. *J Mater Sci Mater Electron* 2017;**28**:10449−56.

19. Niederberger M. Nonaqueous sol−gel routes to metal oxide nanoparticles. *Acc Chem Res* 2007;**40**:793−800.

20. Dwivedi R, Maurya A, Verma A, Prasad R, Bartwal K. Microwave assisted sol−gel synthesis of tetragonal zirconia nanoparticles. *J Alloys Compd* 2011;**509**:6848−51.

21. Wang N, Zhu L, Wang M, Wang D, Tang H. Sono-enhanced degradation of dye pollutants with the use of H_2O_2 activated by Fe_3O_4 magnetic nanoparticles as peroxidase mimetic. *Ultrason Sonochem* 2010;**17**:78−83.

22. Senpan A, Caruthers SD, Rhee I, Mauro NA, Pan D, Hu G, Scott MJ, Fuhrhop RW, Gaffney PJ, Wickline SA. Conquering the dark side: colloidal iron oxide nanoparticles. *ACS Nano* 2009;**3**:3917−26.

23. Kim BH, Lee N, Kim H, An K, Park YI, Choi Y, Shin K, Lee Y, Kwon SG, Na HB. Large-scale synthesis of uniform and extremely small-sized iron oxide nanoparticles for high-resolution T 1 magnetic resonance imaging contrast agents. *J Am Chem Soc* 2011;**133**: 12624−31.

24. Bardajee GR, Hooshyar Z. One-pot synthesis of biocompatible superparamagnetic iron oxide nanoparticles/hydrogel based on salep: characterization and drug delivery. *Carbohydr Polym* 2014;**101**:741−51.

25. Ling D, Hyeon T. Chemical design of biocompatible iron oxide nanoparticles for medical applications. *Small* 2013;**9**:1450−66.

26. Kong SD, Lee J, Ramachandran S, Eliceiri BP, Shubayev VI, Lal R, Jin S. Magnetic targeting of nanoparticles across the intact blood−brain barrier. *J Contr Release* 2012;**164**: 49−57.

27. Sun C, Lee JS, Zhang M. Magnetic nanoparticles in MR imaging and drug delivery. *Adv Drug Deliv Rev* 2008;**60**:1252−65.

28. McNamara K, Tofail SA. Nanosystems: the use of nanoalloys, metallic, bimetallic, and magnetic nanoparticles in biomedical applications. *Phys Chem Chem Phys* 2015;**17**: 27981−95.

29. McNamara K, Tofail SA. *10 biomedical applications of nanoalloys, nanoalloys: from fundamentals to emergent applications*. 2013. p. 345.

30. Hong R, Feng B, Chen L, Liu G, Li H, Zheng Y, Wei D. Synthesis, characterization and MRI application of dextran-coated Fe_3O_4 magnetic nanoparticles. *Biochem Eng J* 2008;**42**: 290−300.

31. Ahmad T, Bae H, Rhee I, Chang Y, Jin S-U, Hong S. Gold-coated iron oxide nanoparticles as a T 2 contrast agent in magnetic resonance imaging. *J Nanosci Nanotechnol* 2012;**12**: 5132−7.

32. McNamara K, Tofail SA. Nanoparticles in biomedical applications. *Adv Phys X* 2017;**2**: 54−88.

33. Iqbal MZ, Ma X, Chen T, Ren W, Xiang L, Wu A. Silica-coated super-paramagnetic iron oxide nanoparticles (SPIONPs): a new type contrast agent of T 1 magnetic resonance imaging (MRI). *J Mater Chem B* 2015;**3**:5172−81.

34. Qu J, Liu G, Wang Y, Hong R. Preparation of Fe_3O_4-chitosan nanoparticles used for hyperthermia. *Adv Powder Technol* 2010;**21**:461−7.

35. Mohammadi-Samani S, Miri R, Salmanpour M, Khalighian N, Sotoudeh S, Erfani N. Preparation and assessment of chitosan-coated superparamagnetic Fe_3O_4 nanoparticles for controlled delivery of methotrexate. *Res Pharm Sci* 2013;**8**:25.

36. Ding Y, Shen SZ, Sun H, Sun K, Liu F, Qi Y, Yan J. Design and construction of polymerized-chitosan coated Fe_3O_4 magnetic nanoparticles and its application for hydrophobic drug delivery. *Mater Sci Eng C* 2015;**48**:487−98.

37. Malekzadeh AM, Ramazani A, Rezaei SJT, Niknejad H. Design and construction of multifunctional hyperbranched polymers coated magnetite nanoparticles for both targeting magnetic resonance imaging and cancer therapy. *J Colloid Interface Sci* 2017; **490**:64−73.

38. Barick K, Singh S, Bahadur D, Lawande MA, Patkar DP, Hassan P. Carboxyl decorated Fe_3O_4 nanoparticles for MRI diagnosis and localized hyperthermia. *J Colloid Interface Sci* 2014;**418**:120−5.

39. Zhao D-L, Zeng X-W, Xia Q-S, Tang J-T. Preparation and coercivity and saturation magnetization dependence of inductive heating property of Fe_3O_4 nanoparticles in an alternating current magnetic field for localized hyperthermia. *J Alloys Compd* 2009;**469**: 215—8.

40. Wang Y, Cao X, Liu G, Hong R, Chen Y, Chen X, Li H, Xu B, Wei D. Synthesis of Fe_3O_4 magnetic fluid used for magnetic resonance imaging and hyperthermia. *J Magn Magn Mater* 2011;**323**:2953—9.

41. Shah RR, Davis TP, Glover AL, Nikles DE, Brazel CS. Impact of magnetic field parameters and iron oxide nanoparticle properties on heat generation for use in magnetic hyperthermia. *J Magn Magn Mater* 2015;**387**:96—106.

42. Kut C, Zhang Y, Hedayati M, Zhou H, Cornejo C, Bordelon D, Mihalic J, Wabler M, Burghardt E, Gruettner C. Preliminary study of injury from heating systemically delivered, nontargeted dextran—superparamagnetic iron oxide nanoparticles in mice. *Nanomedicine* 2012;**7**:1697—711.

43. Akbarzadeh A, Samiei M, Joo SW, Anzaby M, Hanifehpour Y, Nasrabadi HT, Davaran S. Retracted article: synthesis, characterization and in vitro studies of doxorubicin-loaded magnetic nanoparticles grafted to smart copolymers on A549 lung cancer cell line. *J Nanobiotechnol* 2012;**10**:1—13.

44. Ntoutoume GMN, Granet R, Mbakidi JP, Brégier F, Léger DY, Fidanzi-Dugas C, Lequart V, Joly N, Liagre B, Chaleix V. Development of curcumin—cyclodextrin/cellulose nanocrystals complexes: new anticancer drug delivery systems. *Bioorg Med Chem Lett* 2016;**26**:941—5.

45. Soetaert F, Korangath P, Serantes D, Fiering S, Ivkov R. Cancer therapy with iron oxide nanoparticles: agents of thermal and immune therapies. *Adv Drug Deliv Rev* 2020; **163—164**:65—83.

46. Mamiya H. Recent advances in understanding magnetic nanoparticles in AC magnetic fields and optimal design for targeted hyperthermia. *J Nanomater* 2013;**2013**:752973.

47. Gilchrist R, Medal R, Shorey WD, Hanselman RC, Parrott JC, Taylor CB. Selective inductive heating of lymph nodes. *Ann Surg* 1957;**146**:596.

48. Wust P, Gneveckow U, Wust P, Gneveckow U, Johannsen M, Böhmer D, Henkel T, Kahmann F, Sehouli J, Felix R. Magnetic nanoparticles for interstitial thermotherapy—feasibility, tolerance and achieved temperatures. *Int J Hyperther* 2006;**22**:673—85.

49. Doaga A, Cojocariu A, Amin W, Heib F, Bender P, Hempelmann R, Caltun O. Synthesis and characterizations of manganese ferrites for hyperthermia applications. *Mater Chem Phys* 2013;**143**:305—10.

50. Shah SA, Asdi M, Hashmi M, Umar M, Awan S-U. Thermo-responsive copolymer coated $MnFe_2O_4$ magnetic nanoparticles for hyperthermia therapy and controlled drug delivery. *Mater Chem Phys* 2012;**137**:365—71.

51. Lu J, Ma S, Sun J, Xia C, Liu C, Wang Z, Zhao X, Gao F, Gong Q, Song B. Manganese ferrite nanoparticle micellar nanocomposites as MRI contrast agent for liver imaging. *Biomaterials* 2009;**30**:2919—28.

52. Yang H, Zhuang Y, Hu H, Du X, Zhang C, Shi X, Wu H, Yang S. Silica-coated manganese oxide nanoparticles as a platform for targeted magnetic resonance and fluorescence imaging of cancer cells. *Adv Funct Mater* 2010;**20**:1733—41.

53. Yang H, Zhang C, Shi X, Hu H, Du X, Fang Y, Ma Y, Wu H, Yang S. Water-soluble superparamagnetic manganese ferrite nanoparticles for magnetic resonance imaging. *Biomaterials* 2010;**31**:3667—73.

54. Sahoo B, Devi KSP, Dutta S, Maiti TK, Pramanik P, Dhara D. Biocompatible mesoporous silica-coated superparamagnetic manganese ferrite nanoparticles for targeted drug delivery and MR imaging applications. *J Colloid Interface Sci* 2014;**431**:31—41.

55. Fernández-Barahona I, Gutiérrez L, Veintemillas-Verdaguer S, Pellico J, Morales MdP, Catala M, del Pozo MA, Ruiz-Cabello Js, Herranz F. Cu-doped extremely small iron oxide nanoparticles with large longitudinal relaxivity: one-pot synthesis and in vivo targeted molecular imaging. *ACS Omega* 2019;**4**:2719—27.

56. Venkatesan K, Babu DR, Bai MPK, Supriya R, Vidya R, Madeswaran S, Anandan P, Arivanandhan M, Hayakawa Y. Structural and magnetic properties of cobalt-doped iron oxide nanoparticles prepared by solution combustion method for biomedical applications. *Int J Nanomed* 2015;**10**:189.

57. Zheng J, Lee HCM, bin Sattar MM, Huang Y, Bian J-S. Cardioprotective effects of epigallocatechin-3-gallate against doxorubicin-induced cardiomyocyte injury. *Eur J Pharmacol* 2011;**652**:82—8.

58. Ibsen S, Zahavy E, Wrasdilo W, Berns M, Chan M, Esener S. A novel doxorubicin prodrug with controllable photolysis activation for cancer chemotherapy. *Pharm Res* 2010; **27**:1848—60.

59. Barick K, Nigam S, Bahadur D. Nanoscale assembly of mesoporous ZnO: a potential drug carrier. *J Mater Chem* 2010;**20**:6446—52.

60. Song M, Zhang R, Dai Y, Gao F, Chi H, Lv G, Chen B, Wang X. The in vitro inhibition of multidrug resistance by combined nanoparticulate titanium dioxide and UV irradition. *Biomaterials* 2006;**27**:4230—8.

61. Li Q, Wang X, Lu X, Tian H, Jiang H, Lv G, Guo D, Wu C, Chen B. The incorporation of daunorubicin in cancer cells through the use of titanium dioxide whiskers. *Biomaterials* 2009;**30**:4708—15.

62. Ziental D, Czarczynska-Goslinska B, Mlynarczyk DT, Glowacka-Sobotta A, Stanisz B, Goslinski T, Sobotta L. Titanium dioxide nanoparticles: prospects and applications in medicine. *Nanomaterials* 2020;**10**:387.

63. Wu KC-W, Yamauchi Y, Hong C-Y, Yang Y-H, Liang Y-H, Funatsu T, Tsunoda M. Biocompatible, surface functionalized mesoporous titania nanoparticles for intracellular imaging and anticancer drug delivery. *Chem Commun* 2011;**47**:5232—4.

64. Barui AK, Kotcherlakota R, Patra CR. *Biomedical applications of zinc oxide nanoparticles, Inorganic frameworks as smart nanomedicines.* Elsevier; 2018. p. 239—78.

65. Mishra PK, Mishra H, Ekielski A, Talegaonkar S, Vaidya B. Zinc oxide nanoparticles: a promising nanomaterial for biomedical applications. *Drug Discov Today* 2017;**22**: 1825—34.

66. Li S, Mou Q, Leung PH. Synthesis of photoluminescent ZnO quantum dots and its application in bioimaging. *Nanosci Nanotechnol Lett* 2017;**9**:1514—9.

67. Newman MD, Stotland M, Ellis JI. The safety of nanosized particles in titanium dioxide—and zinc oxide—based sunscreens. *J Am Acad Dermatol* 2009;**61**:685—92.

68. Mohammed YH, Holmes A, Haridass IN, Sanchez WY, Studier H, Grice JE, Benson HA, Roberts MS. Support for the safe use of zinc oxide nanoparticle sunscreens: lack of skin penetration or cellular toxicity after repeated application in volunteers. *J Invest Dermatol* 2019;**139**:308—15.

69. Kołodziejczak-Radzimska A, Jesionowski T. Zinc oxide—from synthesis to application: a review. *Materials* 2014;**7**:2833—81.

70. Xiao F-X, Hung S-F, Tao HB, Miao J, Yang HB, Liu B. Spatially branched hierarchical ZnO nanorod-TiO_2 nanotube array heterostructures for versatile photocatalytic and photoelectrocatalytic applications: towards intimate integration of 1D—1D hybrid nanostructures. *Nanoscale* 2014;**6**:14950—61.

71. Smijs TG, Pavel S. Titanium dioxide and zinc oxide nanoparticles in sunscreens: focus on their safety and effectiveness. *Nanotechnol Sci Appl* 2011;**4**:95.

72. Ruszkiewicz JA, Pinkas A, Ferrer B, Peres TV, Tsatsakis A, Aschner M. Neurotoxic effect of active ingredients in sunscreen products, a contemporary review. *Toxicol Rep* 2017;**4**: 245−59.

73. Sahoo S, Maiti M, Ganguly A, Jacob George J, Bhowmick AK. Effect of zinc oxide nanoparticles as cure activator on the properties of natural rubber and nitrile rubber. *J Appl Polym Sci* 2007;**105**:2407−15.

74. Rasmussen JW, Martinez E, Louka P, Wingett DG. Zinc oxide nanoparticles for selective destruction of tumor cells and potential for drug delivery applications. *Expet Opin Drug Deliv* 2010;**7**:1063−77.

75. Senthilkumar K, Senthilkumar O, Yamauchi K, Sato M, Morito S, Ohba T, Nakamura M, Fujita Y. Preparation of ZnO nanoparticles for bio-imaging applications. *Physica Status Solidi (B)* 2009;**246**:885−8.

76. Zhang Z-Y, Xiong H-M. Photoluminescent ZnO nanoparticles and their biological applications. *Materials* 2015;**8**:3101−27.

77. Kim S, Lee SY, Cho H-J. Doxorubicin-wrapped zinc oxide nanoclusters for the therapy of colorectal adenocarcinoma. *Nanomaterials* 2017;**7**:354.

78. Nagajyothi P, Cha SJ, Yang IJ, Sreekanth T, Kim KJ, Shin HM. Antioxidant and anti-inflammatory activities of zinc oxide nanoparticles synthesized using Polygala tenuifolia root extract. *J Photochem Photobiol B Biol* 2015;**146**:10−7.

79. Jin S-E, Jin H-E. Synthesis, characterization, and three-dimensional structure generation of zinc oxide-based nanomedicine for biomedical applications. *Pharmaceutics* 2019;**11**:575.

80. Cochran DL. Inflammation and bone loss in periodontal disease. *J Periodontol* 2008;**79**: 1569−76.

81. Simón-Soro A, Mira A. Solving the etiology of dental caries. *Trends Microbiol* 2015;**23**: 76−82.

82. Sloan A, Smith A. Stem cells and the dental pulp: potential roles in dentine regeneration and repair. *Oral Diseases* 2007;**13**:151−7.

83. Lin Y, Zheng L, Fan L, Kuang W, Guo R, Lin J, Wu J, Tan J. The epigenetic regulation in tooth development and regeneration. *Curr Stem Cell Res Ther* 2018;**13**:4−15.

84. Dong Z, Yang Q, Mei M, Liu L, Sun J, Zhao L, Zhou C. Preparation and characterization of fluoride calcium silicate composites with multi-biofunction for clinical application in dentistry. *Compos B Eng* 2018;**143**:243−9.

85. Sailer I, Makarov NA, Thoma DS, Zwahlen M, Pjetursson BE. All-ceramic or metal-ceramic tooth-supported fixed dental prostheses (FDPs)? A systematic review of the survival and complication rates. Part I: single crowns (SCs). *Dent Mater* 2015;**31**:603−23.

86. Azer SS, Ayash GM, Johnston WM, Khalil MF, Rosenstiel SF. Effect of esthetic core shades on the final color of IPS Empress all-ceramic crowns. *J Prosthet Dent* 2006;**96**: 397−401.

87. Morton D, Chen ST, Martin WC, Levine RA, Buser D. Consensus statements and recommended clinical procedures regarding optimizing esthetic outcomes in implant dentistry. *Int J Oral Maxillofac Implants* 2014;**29**:216−20.

88. Sailer I, Zembic A, Jung RE, Siegenthaler D, Holderegger C, Hämmerle CHF. Randomized controlled clinical trial of customized zirconia and titanium implant abutments for canine and posterior single-tooth implant reconstructions: preliminary results at 1 year of function. *Clin Oral Implants Res* 2009;**20**:219−25.

89. Zembic A, Sailer I, Jung RE, Hämmerle CHF. Randomized-controlled clinical trial of customized zirconia and titanium implant abutments for single-tooth implants in canine and posterior regions: 3-year results. *Clin Oral Implants Res* 2009;**20**:802−8.

90. van Brakel R, Noordmans HJ, Frenken J, de Roode R, de Wit GC, Cune MS. The effect of zirconia and titanium implant abutments on light reflection of the supporting soft tissues. *Clin Oral Implants Res* 2011;**22**:1172—8.

91. Sato H, Yamada K, Pezzotti G, Nawa M, Ban S. Mechanical properties of dental zirconia ceramics changed with sandblasting and heat treatment. *Dent Mater J* 2008;**27**:408—14.

92. Theunissen G, Bouma J, Winnubst A, Burggraaf A. Mechanical properties of ultra-fine grained zirconia ceramics. *J Mater Sci* 1992;**27**:4429—38.

93. Gautam C, Joyner J, Gautam A, Rao J, Vajtai R. Zirconia based dental ceramics: structure, mechanical properties, biocompatibility and applications. *Dalton Trans* 2016;**45**: 19194—215.

94. Yang Y, Qin Z, Zeng W, Yang T, Cao Y, Mei C, Kuang Y. Toxicity assessment of nanoparticles in various systems and organs. *Nanotechnol Rev* 2017;**6**:279—89.

95. Balagangadharan K, Chandran SV, Arumugam B, Saravanan S, Venkatasubbu GD, Selvamurugan N. Chitosan/nano-hydroxyapatite/nano-zirconium dioxide scaffolds with miR-590-5p for bone regeneration. *Int J Biol Macromol* 2018;**111**:953—8.

96. Wang J, Huang C, Wan Q, Chen Y, Chao Y. Characterization of fluoridated hydroxyapatite/zirconia nano-composite coating deposited by a modified electrocodeposition technique. *Surf Coating Technol* 2010;**204**:2576—82.

97. Hakim LF, George SM, Weimer AW. Conformal nanocoating of zirconia nanoparticles by atomic layer deposition in a fluidized bed reactor. *Nanotechnology* 2005;**16**:S375.

98. Vasylkiv O, Sakka Y. Synthesis and colloidal processing of zirconia nanopowder. *J Am Ceram Soc* 2001;**84**:2489—94.

99. Gad MM, Rahoma A, Al-Thobity AM, ArRejaie AS. Influence of incorporation of ZrO2 nanoparticles on the repair strength of polymethyl methacrylate denture bases. *Int J Nanomed* 2016;**11**:5633.

100. Lu X, Xia Y, Liu M, Qian Y, Zhou X, Gu N, Zhang F. Improved performance of diatomite-based dental nanocomposite ceramics using layer-by-layer assembly. *Int J Nanomed* 2012;**7**:2153.

101. Aboushelib MN, Salem NA, Taleb ALA, El Moniem NMA. Influence of surface nano-roughness on osseointegration of zirconia implants in rabbit femur heads using selective infiltration etching technique. *J Oral Implantol* 2013;**39**:583—90.

102. Mansouri A, Gattolliat C-H, Asselah T. Mitochondrial dysfunction and signaling in chronic liver diseases. *Gastroenterology* 2018;**155**:629—47.

103. Fruman DA, Chiu H, Hopkins BD, Bagrodia S, Cantley LC, Abraham RT. The PI3K pathway in human disease. *Cell* 2017;**170**:605—35.

104. Nusse R, Clevers H. Wnt/β-catenin signaling, disease, and emerging therapeutic modalities. *Cell* 2017;**169**:985—99.

105. Meyer J, Hamwi S, Kröger M, Kowalsky W, Riedl T, Kahn A. Transition metal oxides for organic electronics: energetics, device physics and applications. *Adv Mater* 2012;**24**: 5408—27.

106. Loh KP, Ho D, Chiu GNC, Leong DT, Pastorin G, Chow EKH. Clinical applications of carbon nanomaterials in diagnostics and therapy. *Adv Mater* 2018;**30**:1802368.

107. Wu W, Qiu G, Wang Y, Wang R, Ye P. Tellurene: its physical properties, scalable nanomanufacturing, and device applications. *Chem Soc Rev* 2018;**47**:7203—12.

108. Chimene D, Alge DL, Gaharwar AK. Two-dimensional nanomaterials for biomedical applications: emerging trends and future prospects. *Adv Mater* 2015;**27**:7261—84.

109. Veeramani H, Aruguete D, Monsegue N, Murayama M, Dippon U, Kappler A, Hochella MF. Low-temperature green synthesis of multivalent manganese oxide nanowires. *ACS Sustainable Chem Eng* 2013;**1**:1070—4.

110. Layfield RA. Manganese (II): the black sheep of the organometallic family. *Chem Soc Rev* 2008;**37**:1098−107.
111. Prasad AS. Green synthesis of nanocrystalline manganese (II, III) oxide. *Mater Sci Semicond Process* 2017;**71**:342−7.
112. Fei J, Cui Y, Yan X, Qi W, Yang Y, Wang K, He Q, Li J. Controlled preparation of MnO_2 hierarchical hollow nanostructures and their application in water treatment. *Adv Mater* 2008;**20**:452−6.
113. Lin LS, Song J, Song L, Ke K, Liu Y, Zhou Z, Shen Z, Li J, Yang Z, Tang W. Simultaneous Fenton-like ion delivery and glutathione depletion by MnO_2-based nanoagent to enhance chemodynamic therapy. *Angew Chem* 2018;**130**:4996−5000.
114. Xu M, Liang T, Shi M, Chen H. Graphene-like two-dimensional materials. *Chem Rev* 2013;**113**:3766−98.
115. Korsvik C, Patil S, Seal S, Self WT. Superoxide dismutase mimetic properties exhibited by vacancy engineered ceria nanoparticles. *Chem Commun* 2007:1056−8.
116. Dao NN, Dai Luu M, Nguyen QK, Kim BS. UV absorption by cerium oxide nanoparticles/epoxy composite thin films. *Adv Nat Sci Nanosci Nanotechnol* 2011;**2**:045013.
117. Zholobak N, Ivanov V, Shcherbakov A, Shaporev A, Polezhaeva O, Baranchikov AY, Spivak NY, Tretyakov YD. UV-shielding property, photocatalytic activity and photo-cytotoxicity of ceria colloid solutions. *J Photochem Photobiol B Biol* 2011;**102**:32−8.
118. Trovarelli A. Catalytic properties of ceria and CeO_2-containing materials. *Catal Rev* 1996;**38**:439−520.
119. Courbiere B, Auffan M, Rollais R, Tassistro V, Bonnefoy A, Botta A, Rose J, Orsière T, Perrin J. Ultrastructural interactions and genotoxicity assay of cerium dioxide nanoparticles on mouse oocytes. *Int J Mol Sci* 2013;**14**:21613−28.
120. Ivanov VK, Shcherbakov AB, Usatenko A. Structure-sensitive properties and biomedical applications of nanodispersed cerium dioxide. *Russian Chem Rev* 2009;**78**:855.
121. Culcasi M, Benameur L, Mercier A, Lucchesi C, Rahmouni H, Asteian A, Casano G, Botta A, Kovacic H, Pietri S. EPR spin trapping evaluation of ROS production in human fibroblasts exposed to cerium oxide nanoparticles: evidence for NADPH oxidase and mitochondrial stimulation. *Chem Biol Interact* 2012;**199**:161−76.
122. Perez JM, Asati A, Nath S, Kaittanis C. Synthesis of biocompatible dextran-coated nanoceria with pH-dependent antioxidant properties. *Small* 2008;**4**:552−6.
123. Urner M, Schlicker A, Z'graggen BR, Stepuk A, Booy C, Buehler KP, Limbach L, Chmiel C, Stark WJ, Beck-Schimmer B. Inflammatory response of lung macrophages and epithelial cells after exposure to redox active nanoparticles: effect of solubility and anti-oxidant treatment. *Environ Sci Technol* 2014;**48**:13960−8.
124. Fisichella M, Berenguer F, Steinmetz G, Auffan M, Rose J, Prat O. Toxicity evaluation of manufactured CeO_2 nanoparticles before and after alteration: combined physicochemical and whole-genome expression analysis in Caco-2 cells. *BMC Genom* 2014;**15**:700.
125. Franchi LP, Manshian BB, de Souza TA, Soenen SJ, Matsubara EY, Rosolen JM, Takahashi CS. Cyto-and genotoxic effects of metallic nanoparticles in untransformed human fibroblast. *Toxicol Vitro* 2015;**29**:1319−31.
126. Pešić M, Podolski-Renić A, Stojković S, Matović B, Zmejkoski D, Kojić V, Bogdanović G, Pavićević A, Mojović M, Savić A. Anti-cancer effects of cerium oxide nanoparticles and its intracellular redox activity. *Chem Biol Interact* 2015;**232**:85−93.
127. Senthilkumar RP, Bhuvaneshwari V, Ranjithkumar R, Sathiyavimal S, Malayaman V, Chandarshekar B. Synthesis, characterization and antibacterial activity of hybrid chitosan-cerium oxide nanoparticles: as a bionanomaterials. *Int J Biol Macromol* 2017;**104**:1746−52.

128. Hameed ASH, Karthikeyan C, Ahamed AP, Thajuddin N, Alharbi NS, Alharbi SA, Ravi G. In vitro antibacterial activity of ZnO and Nd doped ZnO nanoparticles against ESBL producing *Escherichia coli* and *Klebsiella pneumoniae*. *Sci Rep* 2016;**6**:1—11.

129. Magdalane CM, Kaviyarasu K, Vijaya JJ, Siddhardha B, Jeyaraj B. Photocatalytic activity of binary metal oxide nanocomposites of CeO_2/CdO nanospheres: investigation of optical and antimicrobial activity. *J Photochem Photobiol B Biol* 2016;**163**:77—86.

130. Butt A, Ejaz S, Baron J, Ikram M, Ali S. CaO nanoparticles as a potential drug delivery agent for biomedical applications. *Digest J Nanomater Biostruct* 2015;**10**.

131. Gedda G, Pandey S, Lin Y-C, Wu H-F. Antibacterial effect of calcium oxide nano-plates fabricated from shrimp shells. *Green Chemistry* 2015;**17**:3276—80.

132. Apsana G, Devanna N, Yuvasravana R. Biomimetic synthesis and antibacterial properties of strontium oxide nanoparticles using Ocimum sanctum leaf extract. *Asian J Pharmaceut Clin Res* 2018;**11**:384—9.

133. Patel U, Moss RM, Hossain KMZ, Kennedy AR, Barney ER, Ahmed I, Hannon AC. Structural and physico-chemical analysis of calcium/strontium substituted, near-invert phosphate based glasses for biomedical applications. *Acta Biomater* 2017;**60**:109—27.

134. Bodhak S, Bose S, Bandyopadhyay A. Influence of MgO, SrO, and ZnO dopants on electro-thermal polarization behavior and in Vitro biological properties of hydroxyapatite ceramics. *J Am Ceram Soc* 2011;**94**:1281—8.

135. Paliwal P, Kumar AS, Tripathi H, Singh S, Patne SC, Krishnamurthy S. Pharmacological application of barium containing bioactive glass in gastro-duodenal ulcers. *Mater Sci Eng C* 2018;**92**:424—34.

136. Souza L, Lopes JH, Encarnação D, Mazali IO, Martin RA, Camilli JA, Bertran CA. Comprehensive in vitro and in vivo studies of novel melt-derived Nb-substituted 45S5 bioglass reveal its enhanced bioactive properties for bone healing. *Sci Rep* 2018;**8**:1—15.

137. Dinu M, Braic L, Padmanabhan SC, Morris MA, Titorencu I, Pruna V, Parau A, Romanchikova N, Petrik LF, Vladescu A. Characterization of electron beam deposited Nb_2O_5 coatings for biomedical applications. *J Mech Behav Biomed Mater* 2020;**103**:103582.

138. Mohn D, Zehnder M, Imfeld T, Stark WJ. Radio-opaque nanosized bioactive glass for potential root canal application: evaluation of radiopacity, bioactivity and alkaline capacity. *Int Endod J* 2010;**43**:210—7.

139. Khoshsima S, Alshemary AZ, Tezcaner A, Surdem S, Evis Z. Impact of B_2O_3 and La_2O_3 addition on structural, mechanical and biological properties of hydroxyapatite. *Process Appl Ceram* 2018;**12**:143—52.

140. Balusamy B, Kandhasamy YG, Senthamizhan A, Chandrasekaran G, Subramanian MS, Kumaravel TS. Characterization and bacterial toxicity of lanthanum oxide bulk and nanoparticles. *J Rare Earths* 2012;**30**:1298—302.

141. Jing FJ, Huang N, Liu Y, Zhang W, Zhao X, Fu R, Wang J, Shao Z, Chen J, Leng Y. Hemocompatibility and antibacterial properties of lanthanum oxide films synthesized by dual plasma deposition. *J Biomed Mater Res A* 2008;**87**:1027—33.

142. Popova-Kuznetsova E, Tikhonowski G, Popov AA, Duflot V, Deyev S, Klimentov S, Zavestovskaya I, Prasad PN, Kabashin AV. Laser-ablative synthesis of isotope-enriched samarium oxide nanoparticles for nuclear nanomedicine. *Nanomaterials* 2020;**10**:69.

143. Muthulakshmi V, Balaji M, Sundrarajan M. Biomedical applications of ionic liquid mediated samarium oxide nanoparticles by Andrographis paniculata leaves extract. *Mater Chem Phys* 2020;**242**:122483.

144. Kemal E, Peters R, Bourke S, Fairclough S, Bergstrom-Mann P, Owen D, Sandiford L, Dailey L, Green M. Magnetic conjugated polymer nanoparticles doped with a europium complex for biomedical imaging. *Photochem Photobiol Sci* 2018;**17**:718—21.

145. Syamchand S, Sony G. Europium enabled luminescent nanoparticles for biomedical applications. *J Lumin* 2015;**165**:190−215.
146. Das GK, Heng BC, Ng S-C, White T, Loo JSC, D'Silva L, Padmanabhan P, Bhakoo KK, Selvan ST, Tan TTY. Gadolinium oxide ultranarrow nanorods as multimodal contrast agents for optical and magnetic resonance imaging. *Langmuir* 2010;**26**:8959−65.
147. Liu Z, Liu X, Yuan Q, Dong K, Jiang L, Li Z, Ren J, Qu X. Hybrid mesoporous gadolinium oxide nanorods: a platform for multimodal imaging and enhanced insoluble anticancer drug delivery with low systemic toxicity. *J Mater Chem* 2012;**22**:14982−90.
148. Roux S, Faure AC, Celine M, Riviere C, Bridot J-L, Mutelet B, Marquette CA, Josser V, Le Duc G, Pape L. Multifunctional gadolinium oxide nanoparticles: towards image-guided therapy. *Imag Med* 2010;**2**:211−23.
149. Dĕdková K, Kuzníková Ľ, Pavelek L, Matĕjová K, Kupková J, Barabaszová KČ, Váňa R, Burda J, Vlček J, Cvejn D. Daylight induced antibacterial activity of gadolinium oxide, samarium oxide and erbium oxide nanoparticles and their aquatic toxicity. *Mater Chem Phys* 2017;**197**:226−35.
150. Li C, Sun Y, Li X, Fan S, Liu Y, Jiang X, Boudreau MD, Pan Y, Tian X, Yin J-J. Bactericidal effects and accelerated wound healing using Tb_4O_7 nanoparticles with intrinsic oxidase-like activity. *J Nanobiotechnol* 2019;**17**:1−10.
151. Iram S, Khan S, Ansary AA, Arshad M, Siddiqui S, Ahmad E, Khan RH, Khan MS. Biogenic terbium oxide nanoparticles as the vanguard against osteosarcoma. *Spectrochim Acta Mol Biomol Spectrosc* 2016;**168**:123−31.
152. Atabaev TS, Shin YC, Song S-J, Han D-W, Hong NH. Toxicity and T2-weighted magnetic resonance imaging potentials of holmium oxide nanoparticles. *Nanomaterials* 2017;**7**:216.
153. Das GK, Tan TTY. Rare-earth-doped and codoped Y_2O_3 nanomaterials as potential bioimaging probes. *J Phys Chem C* 2008;**112**:11211−7.
154. Costa BC, Guerreiro-Tanomaru JM, Bosso-Martelo R, Rodrigues EM, Bonetti-Filho I, Tanomaru-Filho M. Ytterbium oxide as radiopacifier of calcium silicate-based cements. Physicochemical and biological properties. *Braz Dent J* 2018;**29**:452−8.
155. Gupta PK, Sharma PP, Sharma A, Khan ZH, Solanki PR. Electrochemical and antimicrobial activity of tellurium oxide nanoparticles. *Mater Sci Eng, B* 2016;**211**:166−72.
156. Raj S, Kumar S, Chatterjee K. Facile synthesis of vanadia nanoparticles and assessment of antibacterial activity and cytotoxicity. *Mater Technol* 2016;**31**:562−73.
157. Sridhar C, Gunvanthrao Yernale N, Prasad M. Synthesis, spectral characterization, and antibacterial and antifungal studies of $PANI/V_2O_5$ nanocomposites. *Int J Chem Eng* 2016; **2016**.
158. Li J, Jiang M, Zhou H, Jin P, Cheung KM, Chu PK, Yeung KW. Vanadium dioxide nanocoating induces tumor cell death through mitochondrial electron transport chain interruption. *Global Chall* 2019;**3**:1800058.
159. Guo J, Zhou H, Wang J, Liu W, Cheng M, Peng X, Qin H, Wei J, Jin P, Li J. Nano vanadium dioxide films deposited on biomedical titanium: a novel approach for simultaneously enhanced osteogenic and antibacterial effects. *Artif Cells, Nanomed Biotechnol* 2018;**46**:58−74.
160. Sarin N, Singh KJ, Kaur H, Kaur R, Singh J. Preliminary studies of the effect of doping of chromium oxide in SiO_2-CaO-P_2O_5 bioceramics for bone regeneration applications. *Spectrochim Acta Mol Biomol Spectrosc* 2020;**229**:118000.
161. Rakesh S, Ananda S, Gowda N. Synthesis of chromium (III) oxide nanoparticles by electrochemical method and *Mukia maderaspatana* plant extract, characterization, $KMnO_4$ decomposition and antibacterial study. *Mod Res Catal* 2013;**2**:127−35.

162. Ramesh C, Mohan Kumar K, Latha N, Ragunathan V. Green synthesis of Cr_2O_3 nanoparticles using Tridax procumbens leaf extract and its antibacterial activity on *Escherichia coli*. *Curr Nanosci* 2012;**8**:603—7.

163. Chattopadhyay S, Dash SK, Ghosh T, Das D, Pramanik P, Roy S. Surface modification of cobalt oxide nanoparticles using phosphonomethyl iminodiacetic acid followed by folic acid: a biocompatible vehicle for targeted anticancer drug delivery. *Cancer Nanotechnol* 2013;**4**:103—16.

164. Chattopadhyay S, Chakraborty S, Laha D, Baral R, Pramanik P, Roy S. Surface-modified cobalt oxide nanoparticles: new opportunities for anti-cancer drug development. *Cancer Nanotechnol* 2012;**3**:13—23.

165. Ren Q, Yang K, Zou R, Wan Z, Shen Z, Wu G, Zhou Z, Ni Q, Fan W, Hu J. Biodegradable hollow manganese/cobalt oxide nanoparticles for tumor theranostics. *Nanoscale* 2019;**11**: 23021—6.

166. Khan S, Ansari AA, Khan AA, Ahmad R, Al-Obaid O, Al-Kattan W. In vitro evaluation of anticancer and antibacterial activities of cobalt oxide nanoparticles. *JBIC J Biol Inorgan Chem* 2015;**20**:1319—26.

167. Assadi Z, Emtiazi G, Zarrabi A. Hyperbranched polyglycerol coated on copper oxide nanoparticles as a novel core-shell nano-carrier hydrophilic drug delivery model. *J Mol Liq* 2018;**250**:375—80.

168. Gnanavel V, Palanichamy V, Roopan SM. Biosynthesis and characterization of copper oxide nanoparticles and its anticancer activity on human colon cancer cell lines (HCT-116). *J Photochem Photobiol B Biol* 2017;**171**:133—8.

169. Jadhav S, Gaikwad S, Nimse M, Rajbhoj A. Copper oxide nanoparticles: synthesis, characterization and their antibacterial activity. *J Cluster Sci* 2011;**22**:121—9.

170. Pang H, Lu Q, Li Y, Gao F. Facile synthesis of nickel oxide nanotubes and their antibacterial, electrochemical and magnetic properties. *Chem Commun* 2009:7542—4.

171. Adhikary J, Chakraborty P, Das B, Datta A, Dash SK, Roy S, Chen J-W, Chattopadhyay T. Preparation and characterization of ferromagnetic nickel oxide nanoparticles from three different precursors: application in drug delivery. *RSC Adv* 2015;**5**:35917—28.

172. Murthy PS, Venugopalan V, Sahoo P, Dhara S, Das A, Tyagi A, Saini G. Gallium oxide nanoparticle induced inhibition of bacterial adhesion and biofilm formation. In: *International conference on nanoscience, engineering and technology (ICONSET 2011)*. IEEE; 2011. p. 490—3.

173. Wang X-S, Situ J-Q, Ying X-Y, Chen H, Pan H-f, Jin Y, Du Y-Z. β-Ga_2O_3:Cr^{3+} nanoparticle: a new platform with near infrared photoluminescence for drug targeting delivery and bio-imaging simultaneously. *Acta Biomater* 2015;**22**:164—72.

174. Sakthi PS, Ratha I, Adarsh T, Anand A, Sinha PK, Diwan P, Annapurna K, Biswas K. In vitro bioactivity and antibacterial properties of bismuth oxide modified bioactive glasses. *J Mater Res* 2018;**33**:178—90.

175. Szostak K, Ostaszewski P, Pulit-Prociak J, Banach M. Bismuth oxide nanoparticles in drug delivery systems. *Pharmaceut Chem J* 2019;**53**:48—51.

176. Oviedo MJ, Contreras O, Rosenstein Y, Vazquez-Duhalt R, Macedo Z, Carbajal-Arizaga G, Hirata GA. New bismuth germanate oxide nanoparticle material for biolabel applications in medicine. *J Nanomater* 2016;**2016**:9782625.

177. Chen C, Hsieh S-C, Teng N-C, Kao C-K, Lee S-Y, Lin C-K, Yang J-C. Radiopacity and cytotoxicity of Portland cement containing zirconia doped bismuth oxide radiopacifiers. *J Endod* 2014;**40**:251—4.

178. Gao A, Hang R, Huang X, Zhao L, Zhang X, Wang L, Tang B, Ma S, Chu PK. The effects of titania nanotubes with embedded silver oxide nanoparticles on bacteria and osteoblasts. *Biomaterials* 2014;**35**:4223−35.
179. Karunakaran G, Jagathambal M, Gusev A, Van Minh N, Kolesnikov E, Mandal AR, Kuznetsov D. *Nitrobacter* sp. extract mediated biosynthesis of Ag_2O NPs with excellent antioxidant and antibacterial potential for biomedical application. *IET Nanobiotechnol* 2016;**10**:425−30.
180. Brüggemann D. Nanoporous aluminium oxide membranes as cell interfaces. *J Nanomater* 2013;**2013**:460870.
181. Auffan M, Rose J, Bottero J-Y, Lowry GV, Jolivet J-P, Wiesner MR. Towards a definition of inorganic nanoparticles from an environmental, health and safety perspective. *Nat Nanotechnol* 2009;**4**:634−41.
182. Wick P, Louw-Gaume AE, Kucki M, Krug HF, Kostarelos K, Fadeel B, Dawson KA, Salvati A, Vázquez E, Ballerini L. Classification framework for graphene-based materials. *Angew Chem Int Ed* 2014;**53**:7714−8.
183. Cha C, Shin SR, Annabi N, Dokmeci MR, Khademhosseini A. Carbon-based nanomaterials: multifunctional materials for biomedical engineering. *ACS Nano* 2013;**7**:2891−7.
184. Vardharajula S, Ali SZ, Tiwari PM, Eroğlu E, Vig K, Dennis VA, Singh SR. Functionalized carbon nanotubes: biomedical applications. *Int J Nanomed* 2012;**7**:5361.
185. Molaei MJ. Carbon quantum dots and their biomedical and therapeutic applications: a review. *RSC Adv* 2019;**9**:6460−81.
186. Bianco A, Kostarelos K, Prato M. Opportunities and challenges of carbon-based nanomaterials for cancer therapy. *Expet Opin Drug Deliv* 2008;**5**:331−42.
187. Muhulet A, Miculescu F, Voicu SI, Schütt F, Thakur VK, Mishra YK. Fundamentals and scopes of doped carbon nanotubes towards energy and biosensing applications. *Mater Today Energy* 2018;**9**:154−86.
188. Li J, Cai C, Li J, Li J, Li J, Sun T, Wang L, Wu H, Yu G. Chitosan-based nanomaterials for drug delivery. *Molecules* 2018;**23**:2661.
189. Mukhopadhyay S, Maiti D, Saha A, Devi PS. Shape transition of TiO_2 nanocube to nanospindle embedded on reduced graphene oxide with enhanced photocatalytic activity. *Cryst Growth Des* 2016;**16**:6922−32.
190. Lin G, Mi P, Chu C, Zhang J, Liu G. Inorganic nanocarriers overcoming multidrug resistance for cancer theranostics. *Adv Sci* 2016;**3**:1600134.
191. Tiwari JN, Vij V, Kemp KC, Kim KS. Engineered carbon-nanomaterial-based electrochemical sensors for biomolecules. *ACS Nano* 2016;**10**:46−80.
192. Zhang D-Y, Zheng Y, Tan C-P, Sun J-H, Zhang W, Ji L-N, Mao Z-W. Graphene oxide decorated with Ru (II)−polyethylene glycol complex for lysosome-targeted imaging and photodynamic/photothermal therapy. *ACS Appl Mater Interf* 2017;**9**:6761−71.
193. Mostofizadeh A, Li Y, Song B, Huang Y. Synthesis, properties, and applications of low-dimensional carbon-related nanomaterials. *J Nanomater* 2011;**2011**:685081.
194. Hong G, Diao S, Antaris AL, Dai H. Carbon nanomaterials for biological imaging and nanomedicinal therapy. *Chem Rev* 2015;**115**:10816−906.
195. Vickers NJ. Animal communication: when i'm calling you, will you answer too? *Curr Biol* 2017;**27**:R713−5.
196. Kumawat MK, Thakur M, Gurung RB, Srivastava R. Graphene quantum dots for cell proliferation, nucleus imaging, and photoluminescent sensing applications. *Sci Rep* 2017;**7**:1−16.

197. Roldo M, Fatouros DG. Biomedical applications of carbon nanotubes. *Annu Rep Sec "C"(Phys Chem)* 2013;**109**:10—35.
198. Liu Z, Yang K, Lee S-T. Single-walled carbon nanotubes in biomedical imaging. *J Mater Chem* 2011;**21**:586—98.
199. Neves V, Heister E, Costa S, Tîlmaciu C, Flahaut E, Soula B, Coley HM, Mcfadden J, Silva SRP. Design of double-walled carbon nanotubes for biomedical applications. *Nanotechnology* 2012;**23**:365102.
200. Nel AE, Mädler L, Velegol D, Xia T, Hoek EM, Somasundaran P, Klaessig F, Castranova V, Thompson M. Understanding biophysicochemical interactions at the nano—bio interface. *Nat Mater* 2009;**8**:543—57.
201. Aggarwal P, Hall JB, McLeland CB, Dobrovolskaia MA, McNeil SE. Nanoparticle interaction with plasma proteins as it relates to particle biodistribution, biocompatibility and therapeutic efficacy. *Adv Drug Deliv Rev* 2009;**61**:428—37.
202. Liu J, Rinzler AG, Dai H, Hafner JH, Bradley RK, Boul PJ, Lu A, Iverson T, Shelimov K, Huffman CB. Fullerene pipes. *Science* 1998;**280**:1253—6.
203. Bianco A, Kostarelos K, Prato M. Applications of carbon nanotubes in drug delivery. *Curr Opin Chem Biol* 2005;**9**:674—9.
204. He H, Pham-Huy LA, Dramou P, Xiao D, Zuo P, Pham-Huy C. Carbon nanotubes: applications in pharmacy and medicine. *BioMed Res Int* 2013;**2013**:578290.
205. Mohajeri M, Behnam B, Sahebkar A. Biomedical applications of carbon nanomaterials: drug and gene delivery potentials. *J Cell Physiol* 2019;**234**:298—319.
206. Iijima S, Ichihashi T. Single-shell carbon nanotubes of 1-nm diameter. *Nature* 1993;**363**:603—5.
207. Iijima S. Carbon nanotubes: past, present, and future. *Phys B Condens Matter* 2002;**323**:1—5.
208. Zanello LP, Zhao B, Hu H, Haddon RC. Bone cell proliferation on carbon nanotubes. *Nano Lett* 2006;**6**:562—7.
209. Usui Y, Aoki K, Narita N, Murakami N, Nakamura I, Nakamura K, Ishigaki N, Yamazaki H, Horiuchi H, Kato H. Carbon nanotubes with high bone-tissue compatibility and bone-formation acceleration effects. *Small* 2008;**4**:240—6.
210. Bethune D, Kiang CH, De Vries M, Gorman G, Savoy R, Vazquez J, Beyers R. Cobalt-catalysed growth of carbon nanotubes with single-atomic-layer walls. *Nature* 1993;**363**:605—7.
211. Lovat V, Pantarotto D, Lagostena L, Cacciari B, Grandolfo M, Righi M, Spalluto G, Prato M, Ballerini L. Carbon nanotube substrates boost neuronal electrical signaling. *Nano Lett* 2005;**5**:1107—10.
212. Mazzatenta A, Giugliano M, Campidelli S, Gambazzi L, Businaro L, Markram H, Prato M, Ballerini L. Interfacing neurons with carbon nanotubes: electrical signal transfer and synaptic stimulation in cultured brain circuits. *J Neurosci* 2007;**27**:6931—6.
213. Béduer Al, Seichepine F, Flahaut E, Loubinoux I, Vaysse L, Vieu C. Elucidation of the role of carbon nanotube patterns on the development of cultured neuronal cells. *Langmuir* 2012;**28**:17363—71.
214. Wu W, Li R, Bian X, Zhu Z, Ding D, Li X, Jia Z, Jiang X, Hu Y. Covalently combining carbon nanotubes with anticancer agent: preparation and antitumor activity. *ACS Nano* 2009;**3**:2740—50.
215. Liu X, Xu D, Liao C, Fang Y, Guo B. Development of a promising drug delivery for formononetin: cyclodextrin-modified single-walled carbon nanotubes. *J Drug Deliv Sci Technol* 2018;**43**:461—8.

216. Shi Kam NW, Jessop TC, Wender PA, Dai H. Nanotube molecular transporters: internalization of carbon nanotube— protein conjugates into mammalian cells. *J Am Chem Soc* 2004;**126**:6850—1.

217. Kostarelos K, Lacerda L, Pastorin G, Wu W, Wieckowski S, Luangsivilay J, Godefroy S, Pantarotto D, Briand J-P, Muller S. Cellular uptake of functionalized carbon nanotubes is independent of functional group and cell type. *Nat Nanotechnol* 2007;**2**:108—13.

218. Paul A, Hasan A, Kindi HA, Gaharwar AK, Rao VT, Nikkhah M, Shin SR, Krafft D, Dokmeci MR, Shum-Tim D. Injectable graphene oxide/hydrogel-based angiogenic gene delivery system for vasculogenesis and cardiac repair. *ACS Nano* 2014;**8**:8050—62.

219. Hasan A, Ragaert K, Swieszkowski W, Selimović Š, Paul A, Camci-Unal G, Mofrad MR, Khademhosseini A. Biomechanical properties of native and tissue engineered heart valve constructs. *J Biomech* 2014;**47**:1949—63.

220. Bosi S, Ballerini L, Prato M. *Carbon nanotubes in tissue engineering, Making and exploiting fullerenes, graphene, and carbon nanotubes.* Springer; 2013. p. 181—204.

221. Im JS, Bai BC, Lee Y-S. The effect of carbon nanotubes on drug delivery in an electrosensitive transdermal drug delivery system. *Biomaterials* 2010;**31**:1414—9.

222. Kang J-H, Kim H-S, Shin US. Thermo conductive carbon nanotube-framed membranes for skin heat signal-responsive transdermal drug delivery. *Polym Chem* 2017;**8**:3154—63.

223. Schwengber A, Prado HJ, Zilli DA, Bonelli PR, Cukierman AL. Carbon nanotubes buckypapers for potential transdermal drug delivery. *Mater Sci Eng C* 2015;**57**:7—13.

224. Kaur J, Gill GS, Jeet K. *Applications of carbon nanotubes in drug delivery: a comprehensive review, Characterization and biology of nanomaterials for drug delivery.* Elsevier; 2019. p. 113—35.

225. Madani SY, Naderi N, Dissanayake O, Tan A, Seifalian AM. A new era of cancer treatment: carbon nanotubes as drug delivery tools. *Int J Nanomed* 2011;**6**:2963.

226. Lim E-K, Kim T, Paik S, Haam S, Huh Y-M, Lee K. Nanomaterials for theranostics: recent advances and future challenges. *Chem Rev* 2015;**115**:327—94.

227. Yang W, Thordarson P, Gooding JJ, Ringer SP, Braet F. Carbon nanotubes for biological and biomedical applications. *Nanotechnology* 2007;**18**:412001.

228. Ag D, Seleci M, Bongartz R, Can M, Yurteri S, Cianga I, Stahl F, Timur S, Scheper T, Yagci Y. From invisible structures of SWCNTs toward fluorescent and targeting architectures for cell imaging. *Biomacromolecules* 2013;**14**:3532—41.

229. Gong H, Peng R, Liu Z. Carbon nanotubes for biomedical imaging: the recent advances. *Adv Drug Deliv Rev* 2013;**65**:1951—63.

230. Nair RR, Blake P, Grigorenko AN, Novoselov KS, Booth TJ, Stauber T, Peres NM, Geim AK. Fine structure constant defines visual transparency of graphene. *Science* 2008; **320**. 1308-1308.

231. Geim AK, Novoselov KS. *The rise of graphene, Nanoscience and technology: a collection of reviews from nature journals.* World Scientific; 2010. p. 11—9.

232. Geim AK. Graphene: status and prospects. *Science* 2009;**324**:1530—4.

233. Novoselov KS, Geim AK, Morozov SV, Jiang D, Zhang Y, Dubonos SV, Grigorieva IV, Firsov AA. Electric field effect in atomically thin carbon films. *Science* 2004;**306**:666—9.

234. Balandin AA, Ghosh S, Bao W, Calizo I, Teweldebrhan D, Miao F, Lau CN. Superior thermal conductivity of single-layer graphene. *Nano Lett* 2008;**8**:902—7.

235. Xu Du IS, Barker A, Andrei EY. Approaching ballistic transport in suspended graphene. *Nat Nanotechnol* 2008;**3**:491—5.

236. Zhu Y, Murali S, Cai W, Li X, Suk JW, Potts JR, Ruoff RS. Graphene and graphene oxide: synthesis, properties, and applications. *Adv Mater* 2010;**22**:3906—24.

237. Bruna M, Borini S. Optical constants of graphene layers in the visible range. *Appl Phys Lett* 2009;**94**:031901.
238. Li W, Chen B, Meng C, Fang W, Xiao Y, Li X, Hu Z, Xu Y, Tong L, Wang H. Ultrafast all-optical graphene modulator. *Nano Lett* 2014;**14**:955—9.
239. He Q, Wu S, Yin Z, Zhang H. Graphene-based electronic sensors. *Chem Sci* 2012;**3**:1764—72.
240. Liu C, Yu Z, Neff D, Zhamu A, Jang BZ. Graphene-based supercapacitor with an ultrahigh energy density. *Nano Lett* 2010;**10**:4863—8.
241. Pham CV, Krueger M, Eck M, Weber S, Erdem E. Comparative electron paramagnetic resonance investigation of reduced graphene oxide and carbon nanotubes with different chemical functionalities for quantum dot attachment. *Appl Phys Lett* 2014;**104**:132102.
242. Pumera M. Graphene in biosensing. *Mater Today* 2011;**14**:308—15.
243. Wisitsoraat A, Mensing JP, Karuwan C, Sriprachuabwong C, Jaruwongrungsee K, Phokharatkul D, Daniels T, Liewhiran C, Tuantranont A. Printed organo-functionalized graphene for biosensing applications. *Biosens Bioelectron* 2017;**87**:7—17.
244. Goenka S, Sant V, Sant S. Graphene-based nanomaterials for drug delivery and tissue engineering. *J Contr Release* 2014;**173**:75—88.
245. Machado BF, Serp P. Graphene-based materials for catalysis. *Catal Sci Technol* 2012;**2**:54—75.
246. Loh KP, Bao Q, Eda G, Chhowalla M. Graphene oxide as a chemically tunable platform for optical applications. *Nat Chem* 2010;**2**:1015.
247. Sun X, Liu Z, Welsher K, Robinson JT, Goodwin A, Zaric S, Dai H. Nano-graphene oxide for cellular imaging and drug delivery. *Nano Res* 2008;**1**:203—12.
248. Liu J, Cui L, Losic D. Graphene and graphene oxide as new nanocarriers for drug delivery applications. *Acta Biomater* 2013;**9**:9243—57.
249. Pan Y, Sahoo NG, Li L. The application of graphene oxide in drug delivery. *Expet Opin Drug Deliv* 2012;**9**:1365—76.
250. Fallatah H, Elhaneid M, Ali-Boucetta H, Overton TW, El Kadri H, Gkatzionis K. Antibacterial effect of graphene oxide (GO) nano-particles against Pseudomonas putida biofilm of variable age. *Environ Sci Pollut Control Ser* 2019;**26**:25057—70.
251. Di Giulio M, Zappacosta R, Di Lodovico S, Di Campli E, Siani G, Fontana A, Cellini L. Antimicrobial and antibiofilm efficacy of graphene oxide against chronic wound microorganisms. *Antimicrob Agents Chemother* 2018;**62**. e00547-00518.
252. Hu W, Peng C, Luo W, Lv M, Li X, Li D, Huang Q, Fan C. Graphene-based antibacterial paper. *ACS Nano* 2010;**4**:4317—23.
253. áde Leon A. On the antibacterial mechanism of graphene oxide (GO) Langmuir—Blodgett films. *Chem Commun* 2015;**51**:2886—9.
254. Li J, Wang G, Zhu H, Zhang M, Zheng X, Di Z, Liu X, Wang X. Antibacterial activity of large-area monolayer graphene film manipulated by charge transfer. *Sci Rep* 2014;**4**:1—8.
255. Hui L, Piao J-G, Auletta J, Hu K, Zhu Y, Meyer T, Liu H, Yang L. Availability of the basal planes of graphene oxide determines whether it is antibacterial. *ACS Appl Mater Interf* 2014;**6**:13183—90.
256. Ansari SA, Satar R, Jafri MA, Rasool M, Ahmad W, Zaidi SK. Role of nanodiamonds in drug delivery and stem cell therapy. *Iran J Biotechnol* 2016;**14**:130.
257. Ho D. *Nanodiamonds: applications in biology and nanoscale medicine.* Springer Science & Business Media; 2009.
258. Iakoubovskii K, Baidakova M, Wouters B, Stesmans A, Adriaenssens G, Vul AY, Grobet PJ. Structure and defects of detonation synthesis nanodiamond. *Diamond Relat Mater* 2000;**9**:861—5.

259. Iakoubovskii K, Mitsuishi K, Furuya K. High-resolution electron microscopy of detonation nanodiamond. *Nanotechnology* 2008;**19**:155705.

260. Shengfu J, Tianlai J, Kang X, Shuben L. Optical properties of nanodiamond layers. *Appl Surf Sci* 1998;**133**:231−346.

261. Turcheniuk K, Mochalin VN. Biomedical applications of nanodiamond. *Nanotechnology* 2017;**28**:252001.

262. Perevedentseva E, Lin Y-C, Jani M, Cheng C-L. Biomedical applications of nanodiamonds in imaging and therapy. *Nanomedicine* 2013;**8**:2041−60.

263. Rifai A, Pirogova E, Fox K. Diamond, carbon nanotubes and graphene for biomedical applications. In: Narayan R, editor. *Encyclopedia of biomedical engineering.* Oxford: Elsevier; 2019. p. 97−107.

264. Kazi S. A review article on nanodiamonds discussing their properties and applications. *Int J Pharm Sci Inven* 2014;**3**:40−5.

265. Kaur R, Badea I. Nanodiamonds as novel nanomaterials for biomedical applications: drug delivery and imaging systems. *Int J Nanomed* 2013;**8**:203−20.

266. De B, Banerjee S, Verma KD, Pal T, Manna P, Kar KK. *Transition metal oxide/carbon nanofiber composites as electrode materials for supercapacitors, handbook of nanocomposite supercapacitor materials II.* Springer; 2020. p. 201−27.

267. Tang J, Fan G, Li Z, Li X, Xu R, Li Y, Zhang D, Moon W-J, Kaloshkin SD, Churyukanova M. Synthesis of carbon nanotube/aluminium composite powders by polymer pyrolysis chemical vapor deposition. *Carbon* 2013;**55**:202−8.

268. Lu K. Freeze cast carbon nanotube-alumina nanoparticle green composites. *J Mater Sci* 2008;**43**:652−9.

269. Xu Q, Wei J, Wang J, Liu Y, Li N, Chen Y, Gao C, Zhang W, Sreeprased TS. Facile synthesis of copper doped carbon dots and their application as a "turn-off" fluorescent probe in the detection of Fe^{3+} ions. *RSC Adv* 2016;**6**:28745−50.

270. Wang J, Xu M, Wang D, Li Z, Primo FL, Tedesco AC, Bi H. Copper-doped carbon dots for optical bioimaging and photodynamic therapy. *Inorg Chem* 2019;**58**:13394−402.

271. Wu G, Nishikawa T, Ohtani B, Chen A. Synthesis and characterization of carbon-doped TiO_2 nanostructures with enhanced visible light response. *Chem Mater* 2007;**19**:4530−7.

272. Panigrahi AK, Singh V, Singh SG. A multi-walled carbon nanotube−zinc oxide nanofiber based flexible chemiresistive biosensor for malaria biomarker detection. *Analyst* 2017;**142**: 2128−35.

273. Wang X, Yu J, Sun G, Ding B. Electrospun nanofibrous materials: a versatile medium for effective oil/water separation. *Mater Today* 2016;**19**:403−14.

274. Hsu P-C, Wu H, Carney TJ, McDowell MT, Yang Y, Garnett EC, Li M, Hu L, Cui Y. Passivation coating on electrospun copper nanofibers for stable transparent electrodes. *ACS Nano* 2012;**6**:5150−6.

275. Lee BH, Song MY, Jang S-Y, Jo SM, Kwak S-Y, Kim DY. Charge transport characteristics of high efficiency dye-sensitized solar cells based on electrospun TiO_2 nanorod photoelectrodes. *J Phys Chem C* 2009;**113**:21453−7.

276. Khanderi J, Hoffmann RC, Gurlo A, Schneider JJ. Synthesis and sensoric response of ZnO decorated carbon nanotubes. *J Mater Chem* 2009;**19**:5039−46.

277. Samadi M, Shivaee HA, Zanetti M, Pourjavadi A, Moshfegh A. Visible light photocatalytic activity of novel MWCNT-doped ZnO electrospun nanofibers. *J Mol Catal Chem* 2012;**359**:42−8.

278. Zamani M, Rostami M, Aghajanzadeh M, Manjili HK, Rostamizadeh K, Danafar H. Mesoporous titanium dioxide@ zinc oxide−graphene oxide nanocarriers for colon-specific drug delivery. *J Mater Sci* 2018;**53**:1634−45.

279. Kılınç E. γ-Fe$_2$O$_3$ magnetic nanoparticle functionalized with carboxylated multi walled carbon nanotube: synthesis, characterization, analytical and biomedical application. *J Magn Magn Mater* 2016;**401**:949—55.

280. Patel MK, Ali MA, Srivastava S, Agrawal VV, Ansari S, Malhotra BD. Magnesium oxide grafted carbon nanotubes based impedimetric genosensor for biomedical application. *Biosens Bioelectron* 2013;**50**:406—13.

281. Prodana M, Duta M, Ionita D, Bojin D, Stan MS, Dinischiotu A, Demetrescu I. A new complex ceramic coating with carbon nanotubes, hydroxyapatite and TiO$_2$ nanotubes on Ti surface for biomedical applications. *Ceram Int* 2015;**41**:6318—25.

282. Prabhu Y, Rao KV, Kumari BS, Pavani T. Decoration of magnesium oxide nanoparticles on O-MWCNTs and its antibacterial studies. *Rendiconti Lincei* 2015;**26**:263—70.

283. Afzal H, Ikram M, Ali S, Shahzadi A, Aqeel M, Haider A, Imran M. Enhanced drug efficiency of doped ZnO—GO (graphene oxide) nanocomposites, a new gateway in drug delivery systems (DDSs). *Mater Res Express* 2020;**7**:015405.

284. El-Shafai N, El-Khouly ME, El-Kemary M, Ramadan M, Eldesoukey I, Masoud M. Graphene oxide decorated with zinc oxide nanoflower, silver and titanium dioxide nanoparticles: fabrication, characterization, DNA interaction, and antibacterial activity. *RSC Adv* 2019;**9**:3704—14.

285. Khasraghi SS, Shojaei A, Sundararaj U. *Highly biocompatible multifunctional hybrid nanoparticles based on Fe3O4 decorated nanodiamond with superior superparamagnetic behaviors and photoluminescent properties.* Materials Science and Engineering: C; 2020. p. 110993.

286. Nazarkovsky M, de Mello HL, Bisaggio RC, Alves LA, Zaitsev V. Hybrid suspension of nanodiamonds-nanosilica/titania in cytotoxicity tests on cancer cell lines. *Inorg Chem Commun* 2020;**111**:107673.

287. Hsueh Y-H, Hsieh C-T, Chiu S-T, Tsai P-H, Liu C-Y, Ke W-J. Antibacterial property of composites of reduced graphene oxide with nano-silver and zinc oxide nanoparticles synthesized using a microwave-assisted approach. *Int J Mol Sci* 2019;**20**:5394.

288. Rajeswari R, Gurumallesh PH. Palladium—Decorated reduced graphene oxide/zinc oxide nanocomposite for enhanced antimicrobial, antioxidant and cytotoxicity activities. *Process Biochem* 2020;**93**:36—47.

289. Sandhya P, Jose J, Sreekala M, Padmanabhan M, Kalarikkal N, Thomas S. Reduced graphene oxide and ZnO decorated graphene for biomedical applications. *Ceram Int* 2018;**44**:15092—8.

290. Archana S, Kumar KY, Jayanna B, Olivera S, Anand A, Prashanth M, Muralidhara H. Versatile graphene oxide decorated by star shaped zinc oxide nanocomposites with superior adsorption capacity and antimicrobial activity. *J Sci Adv Mater Dev* 2018;**3**:167—74.

Antimicrobial studies of metal oxide nanomaterials

Fakhra Liaqat[1,2], Mahammed Ilyas Khazi[3], Ahmad Sher Awan[4], Rengin Eltem[5] and Jian Li[2,3]
[1]Department of Biotechnology, Virtual University of Pakistan, Lahore, Punjab, Pakistan; [2]School of Biological and Chemical Engineering, Panzhihua University, Panzhihua, Sichuan, PR China; [3]Panzhihua Gesala Biotechnology Inc., Panzhihua, Sichuan, PR China; [4]Institute of Education and Research, University of the Punjab Qaid-I-Azam Campus, Lahore, Punjab, Pakistan; [5]Faculty of Engineering, Department of Bioengineering, Ege University, Izmir, Turkey

15.1 Introduction

Microorganisms exist in various environments like water, soil, and air.[1] Microbial infections, even today, are among the main causes of global morbidity and mortality. Reports indicate that nearly half the population within the developing world are infected with microbes, and more than three million people die yearly from these pathogens.[2] Antimicrobial agents inhibit the growth and reduce the harmful effects of microorganisms and are critically important in several industries, including water disinfection, textiles, packaging, construction, medicine, and food.[3] The increasing resistance of microbes to available antimicrobial drugs has led to severe health issues in recent times. Most pathogenic bacteria are resistant to at least one of the antibiotics typically used to treat associated infections.[4] Resistant bacteria are mostly present in hospitals, where genetic mutations in bacteria occur in response to uncontrolled administration of antibiotics. Consequently, the bacteria develop various defensive mechanisms to incapacitate applied chemical substances.[5] Efforts have been made to manage this issue by either discovering new antimicrobial drugs or chemically altering existing ones. Unfortunately, there is no guarantee that the new drugs can cope appropriately with the rapid development of microbial resistance.[6] Keeping this severe problem in mind, discovering alternative antimicrobials that can effectively inhibit microbial growth has become imperative.

With the development of nanotechnology, several important industries are expected to be transformed. Pharmaceutical nanotechnology, with several advantages, has been an area of increasing interest for many research groups. The use of nanomaterials (NMs) as drug carriers has been explored for more than 2 decades and has brought about novel dosage forms with enhanced therapeutic potential and physicochemical properties.[7] Nanotechnology has brought nanosized organic and inorganic particles to the fore, with sizes up to 100 nm in one or more dimensions and properties unique from their bulk equivalent.[6] Metal oxide nanoparticles (MO-NPs) are examples of these NMs, with sizes ranging from 1 to 100 nm. MO-NPs possess distinctive physicochemical characteristics due to their small sizes and are available in a variety of shapes. MO-NPs are the main ingredients in catalysis, disease diagnosis, drug delivery systems, semiconductors, and fuel cells.[8] MO-NPs are a unique class of NMs in terms of their electronic, physicochemical, and electromagnetic properties. Nanosized metal oxide materials have large surface areas, great capacity, fast kinetics, and specific affinities for many contaminants.[9] NMs are most commonly applied as antimicrobial agents or biocides in various industries. Inorganic NMs present stronger antimicrobial effects at lower concentrations, attributed to their higher surface-area-to-volume ratio and distinctive physicochemical properties. In addition, they have greater stability under harsh conditions such as high pressure and temperature. Moreover, some MO-NPs are deliberated as nontoxic and even comprise minerals important to the human body.[10]

Many MO-NPs and noble metals have long been suggested as effective antimicrobials. Among metal oxides, silver oxide (Ag_2O), copper oxide (CuO), titanium dioxide (TiO_2), magnesium oxide (MgO), zinc oxide (ZnO), calcium oxide (CaO), silicon dioxide (SiO_2), aluminum oxide (Al_2O_3), iron oxide (Fe_2O_3), yttrium oxide (Y_2O_3), and cerium dioxide (CeO_2) are known to exhibit antimicrobial activities and are frequently used as antibacterial agents. Several reports have examined the bactericidal efficiency of NPs against different strains of gram-negative and gram-positive bacteria.[1,7,11] Similarly, a few studies have reported the antifungal and antiviral potential of MO-NPs.[12–16]

Similarly, fullerenes, carbon nanotubes (CNTs), single-walled carbon nanotubes, and graphene oxide (GO) nanoparticles (NPs) are new allotropic types of carbon with antimicrobial potential discovered during the last 2 decades and have had substantial applications in various fields of science.[4] The availability of many different NMs with diverse physicochemical and functional properties makes them good antimicrobial agents as a substitute to conventional antimicrobial drugs. Disruption of the cell membrane, generation of reactive oxygen species (ROS), and inactivation of enzymes and nucleic acid are the key antimicrobial mechanisms associated with MO-NPs. These modes of antimicrobial action are entirely different from those manifested by antibiotics. Antibiotics generally target the synthesis of the bacterial cell wall, important bacterial enzymes, DNA replication, protein synthesis, and folic acid metabolism. Bacteria can develop resistance against these mechanisms by one or more means, but the mechanism exhibited by MO-NPs is unique compared with antibiotics. MO-NPs induce cellular damage in microbial cells by changing cellular processes at both molecular and biochemical levels.[8] Hence, MO-NPs offer promise as antimicrobial agents. This chapter summarizes the antimicrobial activities of various MO-NPs along with the mechanisms involved in their antimicrobial activity.

15.2 Synthesis of metal oxide nanoparticles

Various approaches, including chemical, physical, and biological methods of MO-NP synthesis, have been used. The synthesis method defines the physicochemical characteristics, toxicities, bioactivities, and use of MO-NPs. Biological or green synthesis of

Table 15.1 Methods used for the synthesis of MO-NPs.

Method	Brief description	Reference
Chemical synthesis	Use of chemical solvents and high temperature	8,19—21
Wet chemical synthesis	Use of chemicals and hydrothermal technique	22,23
Biosynthesis/green synthesis	Use of plant extracts, protein cages, or microorganisms with precursor compounds	24—28
Sol—gel synthesis	Use of gelation, condensation, and hydroxylation	29,30
Electrochemical synthesis	Use of electrolytes, electrode, and anode	31—34
Pyrolytic synthesis	Use of soluble polymers	35—37
Microwave heating	Use of microwave irradiation as an energy source	38—40
Sonochemical synthesis	Use of high-intensity ultrasound sonication of metal salt	41,42

novel antimicrobial NPs is gaining importance nowadays. Some studies have reported the use of plant extracts and microorganisms to produce NPs. As they are cost-effective and environmentally friendly, these methods can be employed in food packaging, pharmaceuticals, and medical applications.[17] Moreover, green synthesis renders NMs more biocompatible than chemically synthesized materials. For example, green synthesis of MO-NPs with naturally occurring carbohydrate polymers results in polymer-coated NPs that are less toxic and more potent in antimicrobial activities.[18] However, traditional chemical methods are still used extensively to synthesize MO-NPs. Various methods reported in the literature for the synthesis of MO-NPs are summarized in Table 15.1. Synthesis of MO-NPs at a large scale in the economic setup is still a distant dream. Therefore, future work should be directed toward designing novel, suitable, and economical methodologies for scaled-up manufacturing of MO-NPs to meet their growing needs.

15.3 Antimicrobial activity of metal oxide nanoparticles

The number of antimicrobial studies of MO-NPs has expanded exponentially during the last decade because the development of new antimicrobial agents is in dire need of effective control of harmful microbes. In addition to their diverse physical, chemical, electrical, and magnetic properties, the significant antimicrobial potential of MO-NPs against a wide range of microorganisms has made them candidates to become next-generation antibiotics for multi-drug-resistant pathogens. Attributed to their remarkable antimicrobial potential, MO-NPs can be used in biomedicine, food, pharmaceuticals, cosmetics, textiles, water treatment, and agriculture.

Target microbes vary greatly in their genetics, structure, and metabolism. In addition, the physiological state of microbes and growth rate can significantly contribute to their sensitivity to antimicrobial agents. Some environmental factors, including aeration, pH, and temperature, can also play a role. The size, structure, composition, and surface charge of NPs are directly linked to their antimicrobial effects. For example, NPs less than 20 nm in size can effortlessly enter into microbial cells and generate toxic metal ions after dissolution. Thus, smaller particles are usually the most efficient antibacterial agents.[21,43] Furthermore, the shape, chemical modifications, surface coating, mixing, and doping with other NPs and solvents greatly affect antimicrobial activity.[44] Consequently, with this complexity, the mechanisms of the antimicrobial mode of action and toxicity associated with them are still obscure, and one may find the reported literature incomplete and contradictory.

Among the antimicrobial activities of MO-NPs, antibacterial activities are reported more extensively than antifungal and antiviral activities. Similarly, the mechanisms of antibacterial action are frequently explored over antifungal and antiviral mechanisms. The most frequently reported antimicrobial MO-NPs include Ag_2O, ZnO, CuO, Fe_2O_3, Ce_2O_3, and TiO_2. Ag_2O NPs have a much wider range of applications thanks to their biocompatibilities and greater potential for functionalization.[2] CuO, ZnO, TiO_2, Y_2O_3, Ce_2O_3, and Fe_2O_3 have also shown outstanding antimicrobial activities for a broad range of applications. The antimicrobial potential of a few important MO-NPs is briefly presented in this section. Various other MO-NPs and their antimicrobial activities, possible mechanisms of antimicrobial action, and applications are detailed in Table 15.2.

15.3.1 Silver oxide

Silver has various forms like Ag_2O, AgO, Ag_3O_4, and Ag_2O_3 due to interactions with oxygen. Accordingly, silver oxide NPs have many interesting applications resulting from this multivalency. The antibacterial activity of Ag has been recognized since antiquity, and Ag has been used as an antibacterial agent in dentistry and wound healing. Recently, silver oxide NPs have been tested against pathogenic bacteria, and the results indicated that silver oxide NPs are as effective as silver NPs against bacteria.[45] In a recent finding, Ag_2O NPs were biologically synthesized via *Lippia citriodora* plant extract, and their applications for wound healing were investigated. The study also proved the antibacterial and antifungal efficiency of Ag_2O NPs against bacterial and fungal pathogens.[46] Another recent study investigated the antibacterial potential of Ag_2O against biofilm-forming bacterial strains including *Salmonella* sp., *Enterobacter* sp., *Vibrio parahaemolyticus*, and *Micrococcus* sp. and reported that Ag_2O NPs play a role in releasing intracellular constituents together with proteins via membrane destruction, in turn causing bacterial cell death.[47]

15.3.2 Cerium oxide

Cerium oxide (CeO_2) NPs are known to exhibit antibacterial, antifungal, and antiviral activities. Recently, CeO_2 NPs have gained great attention as antimicrobial agents owing to their pretty lower toxic effects on mammalian cells. Bellio, Luzi[48] have investigated the effect of CeO_2 NPs on bacterial membrane permeability and their use in increasing the antibacterial effects of antibiotics against multi-drug-resistant

Table 15.2 Antimicrobial MO-NPs and their applications.

Type of MO-NPs	Antimicrobial activity			Proposed mechanism of antimicrobial action	Applications	References
	Antibacterial	Antifungal	Antiviral			
Ag₂O	*Escherichia coli, Streptococcus mutans, Staphylococcus aureus, Pseudomonas aeruginosa, Vibrio parahaemolyticus, Salmonella* sp., *Enterobacter* sp., *Micrococcus* sp.	*Phytophthora capsici, Colletotrichum acutatum, Cladosporium fulvum, Aspergillus aureus*	–	Disruption of cell membrane and reactive oxygen species (ROS) formation	Control of infectious diseases, wound healing, orthopedic implants, dental composites, medicine	44−47,73−76
CeO₂	*E. coli, S. aureus, P. aeruginosa, B. subtilis, Proteus* spp., *S. pneumoniae, E. faecalis, Salmonella* sp., *Klebsiella* sp., *S. mutans*	*Mucor* spp., *A. flavus, A. niger, Fusarium solani*	*Polio virus*	Membrane disruption, generation of ROS, alteration of electron flow and respiration of bacteria and interfering with the nutrient transport functions. Fungal membrane disruption by electromagnetic interaction and ROS generation	Biomedical, dentistry, wound healing, optical and sensor technologies	48−51,77−79

Continued

Table 15.2 Antimicrobial MO-NPs and their applications.—cont'd

Type of MO-NPs	Antimicrobial activity			Proposed mechanism of antimicrobial action	Applications	References
	Antibacterial	Antifungal	Antiviral			
Fe₂O₃	*S. aureus, E. coli, K. pneumoniae, B. subtilis, Streptococcus salivarius, P. mirabilis, Listeria monocytogenes*	*Candida* spp. *A. niger, A. flavus, Trichothecium roseum, Cladosporium herbarum, P. chrysogenum, Alternaria alternata, Mucor piriformis, P. spinulosum*	H1N1 influenza A virus	Membrane disruption, generation of ROS, damage of bacterial DNA, production of metal ions that disturb the cellular homeostasis and protein function For viruses, virion and composites interaction and reaction of iron oxide with SH groups of proteins in the cell and inactivation of proteins	Biomedical, therapeutic drug carrier	52—58
Al₂O₃	*S. aureus, E. coli, P. aeruginosa, K. pneumonia, Serratia marcescens, B. subtilis, S. epidermidis, Micrococcus luteus, L. monocytogenes, Salmonella typhi, Chromobacterium violaceum*	*F. oxysporum, A. flavus, Candida* spp.	—	Membrane disruption, generation of ROS in bacteria. Attachment and penetration into fungal cells	Biomedical, water treatment, drug carrier, medical implants, therapeutic coatings on implantable devices, orthopedics, dentistry, food	80—86

ZnO	E. coli, S. aureus, MRSA, P. aeruginosa, B. subtilis, K. pneumoniae, S. typhi, E. faecalis	C. albicans, Botrytis cinerea, Penicillium expansum	H1N1 influenza virus, herpes simplex virus (HSV-1)	Membrane disruption, generation of ROS, the release of metal ions, NPs internalization in bacteria Inhibition of cell function and increased nucleic acids derived from the hyphae in fungi	Food, pharmaceutical, biomedical, cosmetics, plant disease management	87—95
MgO	E. coli, S. aureus, MRSA, S. epidermidis, P. mirabilis, S. marcescens, S. typhimurium, P. aeruginosa, E. faecalis, K. pneumoniae, Ralstonia solanacearum, Campylobacter jejuni, S. Enteritidis, S. mutans, B. subtilis, M. luteus	C. albicans, C. glabrata, C. glabrata ER, A. niger, Chaetomium spp., Nigrospora oryzae, F. solani	Foot and mouth disease virus	Membrane disruption, generation of ROS	Biomedical, agriculture, food, and dentistry	96—103

Continued

Table 15.2 Antimicrobial MO-NPs and their applications.—cont'd

Type of MO-NPs	Antimicrobial activity			Proposed mechanism of antimicrobial action	Applications	References
	Antibacterial	Antifungal	Antiviral			
CuO	*S. aureus, E. coli, Enterobacter aerogenes, P. aeruginosa, K. pneumonia, E. faecalis, Shigella flexneri, S. typhimurium, Proteus vulgaris, Pseudomonas fluorescens, Ralstonia solanacearum, Erwinia amylovora, Vibrio eltor, B. subtilis*	*Candida tropicalis, C. albicans, F. oxysporum, A. alternate, Pythium ultimum, A. solani, A. niger, Saccharomyces cerevisiae*	HSV-1, hepatitis C virus	Membrane disruption, generation of ROS	Biomedical, agriculture, food, and dentistry	24,59—64,104
CaO	*E. coli, S. mutans, P. vulgaris, E. faecalis*	—	—	Generation of ROS	Biomedical, dentistry, environmental remediation, water treatment	105—109
SiO$_2$	*B. subtilis, E. coli, S. aureus, P. aeruginosa*	*C. albicans*	HSV type 1 and type 2, HIV, and respiratory syncytial virus (RSV)	Membrane disruption, photodynamic inactivation by the singlet oxygen in bacteria. Binding to viruses through hydrophobic interactions and lower their invasion abilities	Biomedical.	110—113

TiO$_2$	S. aureus, MRSA, E. coli, P. aeruginosa, L. monocytogenes, B. cereus Enterococcus hirae, Bacteroides fragilis	C. albicans, Ustilago tritici	NDV	Prevent bacterial adhesion, generation of ROS	Food, pharmaceutical, biomedical, textile, diseases diagnostics, manufacturing of surgical tools, tissue engineering, Water treatment, environmental remediation	87,114—122
NiO	S. aureus, E. coli, B. subtilis, P. aeruginosa, Lactobacillus casei, K. pneumoniae	A. niger, C. albicans, Mucor racemosus, F. solani, A. flavus	—	Generation of ROS, membrane damage, interaction with intracellular proteins of bacteria and fungi	Biomedical, environmental remediation	123—128
Y$_2$O$_3$	S. aureus, E. coli, Shigella, B. licheniformis. B. subtilis, P. aeruginosa	Saccharomyces cerevisiae	—	Generation of ROS, inactivation of enzymes, oxidative stress in yeast	Biomedical	129—132
Fullerenes	E. coli, Salmonella, S. epidermidis Streptococcus spp., S. aureus, Propionibacterium acnes	Malassezia furfur, C. albicans.	HIV and influenza A virus (subtypes H1N1 and H3N2)	Generation of ROS, blockage of respiratory chain, inhibition of cell metabolism	Biomedical, water treatment	4,69,70,133

Continued

Table 15.2 Antimicrobial MO-NPs and their applications.—cont'd

Type of MO-NPs	Antimicrobial activity			Proposed mechanism of antimicrobial action	Applications	References
	Antibacterial	Antifungal	Antiviral			
CNTs	S. aureus, Salmonella enteric, E. coli, E. faecium, L. acidophilus, B. adolescentis, P. aeruginosa, K. pneumoniae	C. albicans, F. graminearum	HIV, retrovirus, betano-davirus	Generation of ROS, disruption of cell membrane	Biomedical, agriculture	4,71,72,134—139
GO	S. aureus, P, aeruginosa, P. syringae, E. coli, M. smegmatis, X. campestris	C. albicans, Fusarium graminearum, A. niger, A. oryzae F. oxysporum.	Pseudorabies virus, HSV-1, RSV, feline corona-virus, porcine epidemic diarrhea virus	Generation of ROS, membrane disruption	Biomedical, agriculture, drug delivery	65—68,140—145

pathogens. CeO_2 NPs increased the efficiency of antibiotics with synergistic effects via the interaction of NPs with the bacterial membrane. In a previous study, nano CeO_2 was an excellent antibacterial agent against the dental bacteria *Streptococcus mutans*.[49] CeO_2 NPs have also shown antibacterial effectiveness against *Escherichia coli, Staphylococcus aureus, Klebsiella pneumoniae*, and *Pseudomonas aeruginosa* along with antifungal effects against *Mucor* sp., *Aspergillus niger, Aspergillus flavus*, and *Fusarium solani*.[50] Although the antiviral potential of CeO_2 has barely been explored thus far, recent antiviral activity of CeO_2 NPs has been reported against poliovirus (type 1), which strongly encourages widespread investigations into their antiviral potential.[51]

15.3.3 Iron oxide

It is well-known that among various types of NPs, magnetic NPs, owing to their chemical stability, magnetic behavior, and biocompatibility, are used extensively within the biomedical sector. Several transition metals, including iron, nickel, cobalt, and their derivatives, belong to magnetic NPs. Hematite (α -Fe_2O_3), maghemite (γ -Fe_2O_3), magnetite (Fe_3O_4) and goethite (FeO(OH)) are different forms of iron oxides that come under magnetic NPs.[52] These magnetic NPs of iron oxides are chemically and physically stable, environmentally safe and biocompatible, therefore, offering exceptional properties for clinical applications.[53] Antibacterial efficacy of Iron oxide NPs has been screened against *E. coli, Proteus mirabilis,* and *Bacillus subtilis.* The results showed noteworthy inhibition of *P. mirabilis* and *E. coli,* paving the way to use magnetic NPs to cure bacterial diseases.[54] A recent study has reported the effectiveness of α-Fe_2O_3 against the bacteria *Listeria monocytogenes* and the fungi *A. flavus* and *Penicillium spinulosum*.[55]

Iron oxide NPs in intact form lose their magnetism and monodispersibility upon contact with the air. To cope with such drawbacks, various researchers have used biological and chemical agents to transform the surfaces. For instance, coating with chitosan caused a substantial increase in the antimicrobial susceptibility of iron oxide NPs when tested against *B. subtilis* and *E. coli*.[52] Antifungal effects of iron oxide NPs have also been reported against various fungal pathogens (*A. niger, Alternaria alternata, Trichothecium roseum, Penicillium chrysogenum, Cladosporium herbarum,* and *Mucor piriformis*).[56,57] However, the antiviral activity of iron oxide NPs has yet to be explored. A recent study opens a new avenue for the use of iron oxide NPs against viral infections by reporting their antiviral activity against H1N1 influenza A virus.[58] Much work is still required before donating these NPs as antiviral agents and establishing their use against viral pathogens.

15.3.4 Copper oxide

Among various metal oxide NPs, CuO is vitally important and has attracted specific attention because of its simple form, widespread availability, low cost, high conductivity, and unique chemical, thermal, and optical features.[59] CuO NPs have displayed promising antimicrobial potential against several bacteria, including *E. coli, S. aureus, P. aeruginosa, Enterococcus faecalis, Shigella flexneri, K. pneumonia, Salmonella typhimurium,* and *Proteus vulgaris* with maximum effectiveness against *E. coli* and least effectiveness against *E. faecalis*.[60] According to a recent finding, CuO NPs showed antimicrobial susceptibilities against both prokaryotic and eukaryotic microbes, including Gram-positive bacteria, negative bacteria, and unicellular and multicellular fungi.[61] The antibacterial and antifungal potential of CuO NPs has also been evaluated against *Pseudomonas fluorescens* and *Candida albicans*.[62]

Little is known about the antiviral mechanisms of CuO NPs; however, their antiviral effectiveness has been reported by a few studies. For example, a finding demonstrated significant antiviral activity of CuO NPs against herpes simplex virus type 1 and proposed its use in topical formulations for the cure of genital herpetic lesions.[63] Another research group evaluated the effects of CuO NPs against Hepatitis C Virus and demonstrated that CuO NPs targeted the binding and entry of infectious virus particles to the hepatic cells in vitro[64].

15.3.5   Carbon-based materials

Carbon-based NMs are of particular interest owing to their high mechanical strength and noticeable physicochemical characteristics.[65] These NMs are fullerenes, CNTs, graphene and its derivatives, carbon-based quantum dots and nanodiamonds. Among carbon-based NMs, graphene and its derivatives have attracted broad attention recently because of their higher surface area, excellent conductivity, and ease of mass production. These unique properties were exploited in the areas of electronics, energy storage, optics, bioscience and biotechnology.[66] Previous findings have suggested that GO, reduced GO, fullerenes, and CNTs exhibited strong antimicrobial activities[5,67−72] and GO NMs are most extensively investigated in recent years for their antibacterial, antifungal and antiviral properties. The seemingly low toxicity qualified GO NMs to be promising candidates as next generation antimicrobial agents. Antimicrobial studies of carbon-based NMs including fullerenes, CNTs and GO along with their applications are incorporated in Table 15.2.

15.4   Proposed mechanisms of antimicrobial activity of metal oxide nanoparticles

15.4.1   Antibacterial mechanisms

The mode of action of MO-NPs is comparable to antibiotics, since both can discriminate prokaryotic microbial cells from eukaryotic mammalian cells via the metal transport system and metalloproteins of bacteria. On the other hand, the antibacterial effects of MO-NMs are caused by multiple mechanisms, a feature that distinguishes them from antibiotics. Therefore, unlike the case with antibiotics, to provoke any form of resistance against MO-NMs, the bacterial cell must complete multiple gene mutations.[146] MO-NPs can demonstrate bacteriostatic or bactericidal effects. As bacteriostatic agents, they can halt the reproduction of bacteria; nevertheless, the cell does

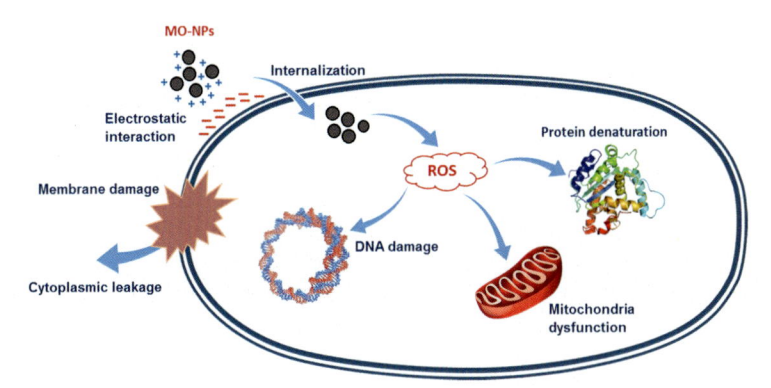

Figure 15.1 Possible mechanisms of MO-NP-mediated antibacterial activity.

not die. If treated bacteria are removed from the solution containing NPs and plated onto fresh NP-free medium, they will begin growing again. As a result of bactericidal action, bacterial cell death takes place, and replating results in no bacterial colonies. Depending on testing conditions, the NPs concentration, and the type of bacteria, certain types of MO-NPs may have bacteriostatic or bactericidal activity through ZnO and TiO$_2$.[43]

Data on the molecular mechanisms of action related to antimicrobial effects of MO-NPs is still in its infancy and explanation of the molecular mechanisms is the subject of demanding research. Many efforts have been made to explore the mechanisms of antibacterial action of MO-NP, however, the existing literature is inadequate. Even though the mechanisms are not comprehensive, literature has revealed two mains extensively reported possible mechanisms responsible for in bactericidal efficiency of MO-NPs (Fig. 15.1).

15.4.1.1 Disruption of cell membrane

NMs act as nanocatalysts, in general, cause membrane damage by interacting electrostatically with the bacterial membrane leads to membrane depolarization, modification of membrane potential and loss of membrane integrity which, in turn, create an imbalance of transport, impaired respiration, disturbance in energy metabolism, cell lysis and ultimately cell death.[44] At biological pH the bacteria have negative charge on their membranes leads to the accumulation of MO-NPs on the cell surface permitting their entry into the cell. Gram-negative bacteria have more negative charge than Gram-positive bacteria, therefore, the electrostatic interaction will be strong in Gram-negative bacteria. Electrostatic attraction depends on the surface area which is greater in case of MO-NPs compared with their native larger sized materials, hence imparting toxicity to the cells. Smaller size particles offer larger surface area and greater antimicrobial efficacy against the growth of harmful bacteria.

The zeta potential shown by MO-NPs promotes the interaction of NPs with cell membranes causing membrane damage and leakage of cytosol contents.[8] Because of disturbance of membrane, ample amount of water is released from the cytosol. Cells try to compensate for their loss by proton efflux pumps and electron transport systems of bacteria. Though, higher demand of these ions leads to serious destruction to these transmembrane systems. This mechanism has been demonstrated by various NPs including ZnO, MgO, and TiO$_2$ NPs. Ag$_2$O NPs explicitly interact with sulfur containing constituents of the cell membrane and the ions produced block the synthesis of cell wall.[146] The interaction of NPs with sulfur groups present in cell wall proteins leads to permanent changes in the cell wall structure which in turns, disrupts the integrity of lipid bilayer and increases the membrane permeability.[2] Occasionally, the formation of pits is seen in the cell walls of bacteria after interacting with a certain type of NPs. Along with cell membrane binding, MO-NPs bind with mesosomes and thus impair cellular respiration, DNA replication and cell division.[8]

Interestingly, metal particles of negative zeta potential can also disrupt the cell membranes as van der Waals and hydrophobic interactions can also occur between the NPs and microbial cells. MO-NPs can specifically bind to phosphate, amine and carboxyl groups in lipids and proteins of membrane barrier surface and subsequently induce cell death. It is important to mention here that MO-NPs remain strongly attached to the surface of damaged or dead cells which may alter their effective concentration in the given solution over time. Since MO-NPs with dissimilar physical and chemical properties show different bactericidal mechanisms and activities, NPs can be developed by mixing two or more metals with greater efficiency for effective removal of different bacterial strains including those highly resistant to conventional drugs.[43]

15.4.1.2 Generation of reactive oxygen species

Second mechanism is the production of ROS also called free radicals, such as hydrogen peroxide or superoxide anions which can also cause severe oxidative stress leading to the formation of pits in the cell membrane, initiating secondary membrane disruption and cell lysis. ROS can be generated indirectly by alterations in the respiratory chain of bacteria or directly by the NPs. Excessive production of ROS resulting into severe oxidative stress, damage of macromolecules, lipid peroxidation, proteolysis, enzyme inactivation and severe damage to nucleic acids.[2,44] After adsorption onto the bacterial cell, CuO NPs were found to produce superoxide anions to induce antibacterial effect in both *S. aureus* and *E. coli*.[147] Some polycationic NMs have exceptional characteristics since they are capable of inducing signal secretion that seems to promote programmed cell death.[44]

15.4.1.3 Other mechanisms

Protein denaturation is another mechanism of antibacterial activity demonstrated by MO-NPs. For instance, Ag may act as weak acid and have ability to interact with sulfur and phosphorus, which are the key constituents of proteins and DNA, respectively. Metals ions can react with these soft bases and damage DNA, hence causing the cell death.[148] Some metal oxides can display genotoxicity and block bacterial cell division by disturbing the replication of chromosomal and plasmid DNA.[52] Furthermore, in some cases photokilling is also observed when MO-NPs interact with bacterial cells in the presence of light. This happens only in case of transition MOs which can be photosensitized. The presence of light can cause photochemical changes of the bacterial membrane, Ca^{2+} permeability, decrease in the activity of superoxide dismutase, inactivation of proteins, DNA damage, and irregular cell division.[8] In a study, Fe_3O_4 NPs were found to inhibit the growth of nosocomial bacteria by exhibiting photochemical reactions.[149]

15.4.2 Antifungal mechanism

In terms of antifungal and antiviral activities iron, zinc and copper oxide NPs are most frequently studied metal oxides. The mechanisms of antifungal effects were demonstrated as membrane disruption and apoptosis.[150−152] Damage of biological macromolecules by generation of ROS was also reported as antifungal mechanism of MO-NPs.[153] The antimicrobial effects of MO-NPs are caused by their greater surface area-to-volume ratio and small size which successfully covers the microbial cells and decrease the supply of oxygen for respiration. Some MO-NPs are capable of releasing metal ions that can be absorbed into the cell followed by interactions with proteins and DNA, which in turn inhibit activity and cell damage.[56] NPs may exhibit different antifungal activities against various types of fungus. NPs can suppress the conidial growth, distort the conidiophores, and deform the surface of fungal hyphae by excessive accumulation of nucleic acid and carbohydrates due to stress response as a self-protecting mechanism of fungal hyphae against the NPs.[91]

15.4.3 Antiviral mechanism

Viruses are intracellular infectious entities that replicate only inside the living cells and can infect all life forms from human, animals, and plants to microorganisms. Viruses contain either DNA or RNA and a protein coat called capsid. The viral infection depends on the entry and attachment of virus onto the host cell by binding of virus ligands to the cell membrane surface proteins. Therefore, the best strategy employed in the development of antiviral agents is the interference of ligand-membrane interactions which leads to blocking of viral attachment and entry into the host cell.[154] Other potential mechanisms of antiviral agents includes anatomization of cell surface receptors, blockage of uncoating, inhibition of transcription of viral nucleic acid and inactivation of enzymes required for virus replication. Very limited literature is present regarding the antiviral mechanism of MO-NPs attributed to vast variety and complex structure of viruses. However, inhibition of ligand-membrane interaction is the commonly reported mechanism of antiviral activity of MO-NP.[155−157] In addition, few mechanisms including damage of viral proteins coat by NP,[158] reduction of DNA formation,[12] and damaging the viral particles[159] were also reported.

15.5 Safety issues

Biggest issue related to the safe use of metal oxides in biological systems is their toxicity, but, like antimicrobial effects, the cytotoxicity of MO-NPs is also dependent on their physical and chemical nature. The toxicity mechanism of MO-NPs in humans remains unclear; however, some features that may be potential risks in using MO-NPs as antimicrobial drugs in humans and other organisms include their size, structure, solubility, system aggregation, oxidant generation, nucleus percolation, and exposure time and dose. Furthermore, the toxicity of some MO-NPs differs with cell and tissue type.[8] MO-NPs with similar sizes but different compositions show varying cytotoxicities. In a study, toxicities of various MO-NPs were compared using A549 cells. The results revealed that CuO induced the highest percentage of cell death and DNA damage, while TiO2 showed the least toxicity. In addition, Fe_3O_4 and Fe_2O_3 were nearly nontoxic compared with the mixed $CuZnFe_2O_4$ NPs, which exhibited severe DNA damage.[160] A study reported that ZnO NPs were the most, TiO_2 NPs the second most, and MgO NPs the least toxic to human cells.[161] A study examined the toxicity of 19 different MO-NPs and proposed that the key factor defining their cytotoxicity is the intrinsic toxicity of the metal ions released.[162] The toxicity of MO-NPs can be modified by using surface coatings, however, the initial surface characteristics of the coating materials should also be taken into consideration.[43] Toxicity can also be controlled using specially designed biological growth media based on the reactivity of metal oxides. Such as, the toxicity of ZnO can be decreased using a medium containing phosphate ions which can help inactivating toxic Zn^{2+} ions by forming Zn-phosphate.[163] Diethylene triamine pentaacetic acid can be used as a metal-ion chelator to decrease the toxicity of MO-NPs.[43] Another method for reducing the toxicity of MO-NPs is doping with other metal ions. For example, ZnO NPs doped with Ta^{5+} ions are less toxic and more influential in antibacterial activity than pristine ZnO NPs.[164]

The use of metal oxides as antimicrobial agents in the food sector needs a serious regulatory framework. Recently, researchers have been particularly concerned about the risk of NMs migrating from packaging materials into food and whether this migration could negatively affect the safety or quality of the packaged product. The US Food and Drug Administration (FDA) developed a Food Contact Notification program within the Center for Food Safety and Applied Nutrition (CFSAN) Office of Food Additive Safety. CFSAN has a research program to discover a framework for safety testing of NMs in foods and cosmetics. It is based on investigating the chance of NPs leaching from food packaging materials, determining whether there is a safety concern, and examining methodologies to study the potential toxicity of NPs. The FDA's National Center for Toxicological Research is establishing analytical tools and procedures to quantify NMs in complex matrices and is conducting toxicity studies. Furthermore, The US National Toxicology Program, through its Nanotechnology Safety Initiative, is doing research on numerous classes of NMs, including MO, fluorescent crystalline semiconductors, fullerenes, CNTs, nanoscale Ag, and nanoscale Au. The European Food Safety Authority, in its scientific belief in the possible hazards arising from nanoscience and nanotechnologies on food and feed safety, affirmed that the risk assessment standard is applicable for engineered NMs.[165] In addition to cytotoxicity, the disruption of beneficial microflora by the antimicrobial effects of MO-NPs is an important concern, and effective distribution must be discovered.[8]

Taking together, the application of MO-NPs as antimicrobial agents depends on their toxicity; therefore, cytotoxicity studies should be conducted with well-characterized NPs using standardized method of toxicity testing. Complete understanding of toxicity mechanisms is equally important to progress in the applications of MO-NPs as antimicrobial agents.

15.6 Stabilization and biocompatibility of metal oxide nanoparticles

MO-NPs dissolve partially in water solutions; the propensity toward dissolution depends on NP composition, structure, and size. In contrast to larger sizes, higher dissolution rates were observed with smaller particles, which formed surface hydroxide and led to partial dissolution and surface faceting.[43] Most MO-NPs have been shown to aggregate once they are hydrated. The high surface energies of NPs are attributed to their large surface-to-volume ratio; however, NPs tend to aggregate to minimize total surface energy.[166] This common aggregation phenomenon of NPs in aqueous solutions is affected by the ionic strength and pH of the solution. NPs can also interact with natural organic matter in various solutions. These interactions disturb the balance between attractive and repulsive forces, which eventually control the aggregation of NPs in solution as well as their attachment to environmental surfaces. Nevertheless, this is the collective effect of pH, ionic strength, ionic composition, nature of organic matter, and other properties of the aqueous solution that lead to either aggregation or stabilization. This also influences NP bioavailability and biocompatibility. Hence, NP size along with aggregation play a vital role in determining their biocompatibility and cytotoxicity.[167]

Control of NP size and dispersion in desired solutions is technically challenging because of complications in manufacturing and biomedical applications, such as accretion, consistency, hydrophilicity, and biocompatibility.[166] The two approaches generally used for NP dissolution in water are electrostatic stabilization and steric repulsion. In the electrostatic stabilization strategy, particles do not aggregate because of their equal charges or electrostatic repulsion. This approach is simple to understand yet demands well-defined pH and ionic strength of the solvent and control of reactive species that can alter the surface charges of MOs. In the case of steric repulsion, stabilizing molecules, such as hydrocarbon polymer or biomacromolecules, modify the NP surface. These polymers and biomacromolecules can be adsorbed or grafted onto the NP surface to avoid direct contact between them, and thus their aggregation can be evaded. Therefore, the NPs remain dispersed in water solution even after a change in pH or salt concentration.[43]

The stabilizing molecules can be coating agents, including protein, ethylene glycol, dextran, and silica.[166] These molecules are known not only for increasing the stability of NPs but also for enhancing their performance. Prevention of NP aggregation and attainment of a stable distribution in an aqueous solution may also be achieved by adding a mild detergent, such as Tween 20 or P-20.[43] In addition, proper functionalization of the NP surface and solvent selection are critical to attaining adequate repelling interactions between NPs to inhibit aggregation and increase the thermodynamic stability of the colloidal solution. For example, the surface of Fe_3O_4 dispersed in aqueous media via citric acid adsorption can be functionalized by applying the coordination of one or two carboxylate functionalities of the citric acid depending on the steric necessity and curvature of the surface.[166] The encapsulation of the MO-NPs by natural polymer chitosan makes them resistant to the unfavorable factors usually present in physiological conditions such as pH and high ionic force. Chitosan can also promote the stimulating effect of MO-NPs on cells by increasing their accumulation.[168] Finally, NPs synthesized by biological methods or green synthesis are more biocompatible than chemically synthesized NPs because natural and nontoxic agents are used.[120]

15.7 Limitations

The complete mechanism behind the antimicrobial effects of MO-NPs is not known. As discussed previously (section number), various mechanisms are being recognized and suggested. Still, this area should be considered more extensively in future studies. Although in vivo experiments are expensive, slow, and ethically questionable, in vitro models cannot fully mimic in vivo conditions because some microorganisms have complex membrane structures. Future work should establish more accurate in vitro assays with increased efficiency and reliable risk assessment and comparability to in vivo conditions. The lack of unified standards to compare antimicrobial activities of metal oxides is another limitation of current studies. Reports are available on assessing MO-NP toxicity both in vitro and in vivo; however, rigorous studies are indispensable. Moreover, applying functionalized carbon NMs as carriers for ordinary antibiotics may help lower associated resistance, improve bioavailability, and provide for targeted delivery. It is important to mention here that no antimicrobial drugs or preparations of metal oxides are clinically available for treating infectious diseases. Some clinical trials are currently being conducted on NP-based antimicrobials; however, none of these trials have progressed to the approval stage. Hence, much remains to consider and investigate before asserting that MO-NPs are novel alternative antimicrobial agents applicable in various industries.

15.8 Conclusion

Because of long-term and inappropriate use of antibiotics, microorganisms are rapidly developing resistance to commonly used antimicrobial drugs. As a result, the discovery of novel and effective antimicrobials is an urgent need. Nanostructured materials can be considered next-generation antibiotics because they possess significant potential to combat the multidrug resistance of pathogens. Recently, efforts have been made to prepare MO NMs with antimicrobial activities against pathogenic microbes. Although no overall conclusion has been reached about the antimicrobial mechanism of MO-NP interactions with mammalian cells and microorganisms, the frequently suggested action mechanisms include ROS formation, cell membrane disruption, NP internalization, and protein or DNA inactivation. MO-based antimicrobials may be applicable in biomedicine, water treatment, agriculture, food, cosmetics, pharmaceuticals, medical devices, dentistry, textiles, and veterinary medicine. The primary impediment to using these antimicrobials, however, is their toxic nature. Methods are now being proposed to minimize their cytotoxicity. With further exploration into their detailed mechanisms of action, safe use strategies, and an economical synthesis methodology, the application of MO-NPs as antimicrobial agents is not far off and offers promise in the near future.

Acknowledgment

This work was supported by Sichuan Province Science and Technology Sastaupport Program in Sichuan, China, with grant number 2019YFN0121 Joint Science and Technology Support Program of Sichuan University and Panzhihua City, with grant number 2019CDPZH-19, Panzhihua Municipal Science and Technology Support Program in Panzhihua, China, with grant number 2018CY-N-5.

References

1. El-Shafai N, El-Khouly ME, El-Kemary M, Ramadan M, Eldesoukey I, Masoud M. Graphene oxide decorated with zinc oxide nanoflower, silver and titanium dioxide nanoparticles: fabrication, characterization, DNA interaction, and antibacterial activity. *RSC Adv* 2019;**9**(7):3704−14.
2. Baranwal A, Srivastava A, Kumar P, Bajpai VK, Maurya PK, Chandra P. Prospects of nanostructure materials and their composites as antimicrobial agents. *Front Microbiol* 2018;**9**:422.
3. Gordon T, Perlstein B, Houbara O, Felner I, Banin E, Margel S. Synthesis and characterization of zinc/iron oxide composite nanoparticles and their antibacterial properties. *Colloid Surface Physicochem Eng Aspect* 2011;**374**(1−3):1−8.
4. Dizaj SM, Mennati A, Jafari S, Khezri K, Adibkia K. Antimicrobial activity of carbon-based nanoparticles. *Adv Pharmaceut Bull* 2015;**5**(1):19.
5. Richtera L, Chudobova D, Cihalova K, Kremplova M, Milosavljevic V, Kopel P, et al. The composites of graphene oxide with metal or semimetal nanoparticles and their effect on pathogenic microorganisms. *Materials* 2015;**8**(6):2994−3011.
6. Khashan KS, Sulaiman GM, Mahdi R. Preparation of iron oxide nanoparticles-decorated carbon nanotube using laser ablation in liquid and their antimicrobial activity. *Artif Cells Nanomed Biotechnol* 2017;**45**(8):1699−709.
7. Dizaj SM, Lotfipour F, Barzegar-Jalali M, Zarrintan MH, Adibkia K. Antimicrobial activity of the metals and metal oxide nanoparticles. *Mater Sci Eng C* 2014;**44**:278−84.
8. Raghunath A, Perumal E. Metal oxide nanoparticles as antimicrobial agents: a promise for the future. *Int J Antimicrob Agents* 2017;**49**(2):137−52.
9. Dontsova TA, Nahirniak SV, Astrelin IM. Metal oxide nanomaterials and nanocomposites of ecological purpose. *J Nanomater* 2019;**2019**.

10. Espitia PJP, Soares NFF, dos Reis Coimbra JS, de Andrade NJ, Cruz RS, Medeiros EAA. Zinc oxide nanoparticles: synthesis, antimicrobial activity and food packaging applications. *Food Bioprocess Technol* 2012;**5**(5):1447−64.

11. Duffy LL, Osmond-McLeod MJ, Judy J, King T. Investigation into the antibacterial activity of silver, zinc oxide and copper oxide nanoparticles against poultry-relevant isolates of Salmonella and Campylobacter. *Food Control* 2018;**92**:293−300.

12. Lu L, Sun R, Chen R, Hui C-K, Ho C-M, Luk JM, et al. Silver nanoparticles inhibit hepatitis B virus replication. *Antivir Ther* 2008;**13**(2):253.

13. El-Sheekh MM, Shabaan MT, Hassan L, Morsi HH. Antiviral activity of algae biosynthesized silver and gold nanoparticles against Herps Simplex (HSV-1) virus in vitro using cell-line culture technique. *Int J Environ Health Res* 2020:1−12.

14. Dung TTN, Nam VN, Nhan TT, Ngoc TTB, Minh LQ, Nga BTT, et al. Silver nanoparticles as potential antiviral agents against African swine fever virus. *Mater Res Express* 2020;**6**(12):1250g9.

15. Broglie JJ, Alston B, Yang C, Ma L, Adcock AF, Chen W, et al. Antiviral activity of gold/copper sulfide core/shell nanoparticles against human norovirus virus-like particles. *PLoS One* 2015;**10**(10):e0141050.

16. Folorunso A, Akintelu S, Oyebamiji AK, Ajayi S, Abiola B, Abdusalam I, et al. Biosynthesis, characterization and antimicrobial activity of gold nanoparticles from leaf extracts of Annona muricata. *J Nanostruct Chem* 2019;**9**(2):111−7.

17. Hoseinnejad M, Jafari SM, Katouzian I. Inorganic and metal nanoparticles and their antimicrobial activity in food packaging applications. *Crit Rev Microbiol* 2018;**44**(2): 161−81.

18. Makvandi P, Wang C, Zare EN, Borzacchiello A, Niu L, Tay FR. Metal-based nanomaterials in biomedical applications: antimicrobial activity and cytotoxicity aspects. *Adv Funct Mater* 2020:1910021.

19. F Hasany S, H Abdurahman N, R Sunarti A, Jose R. Magnetic iron oxide nanoparticles: chemical synthesis and applications review. *Curr Nanosci* 2013;**9**(5):561−75.

20. Kwon SG, Piao Y, Park J, Angappane S, Jo Y, Hwang N-M, et al. Kinetics of monodisperse iron oxide nanocrystal formation by "heating-up" process. *J Am Chem Soc* 2007; **129**(41):12571−84.

21. Yu J, Ju Y, Chen F, Che S, Zhao L, Sheng F, et al. Chemical synthesis and biomedical applications of iron oxide nanoparticles. In: Wang X, Ramalingam M, Kong X, Zhao L, editors. *Nanobiomaterials*; 2017. https://doi.org/10.1002/9783527698646.ch14.

22. Aslani A, Oroojpour V. CO gas sensing of CuO nanostructures, synthesized by an assisted solvothermal wet chemical route. *Phys B Condens Matter* 2011;**406**(2):144−9.

23. Krishnakumar T, Jayaprakash R, Pinna N, Donato N, Bonavita A, Micali G, et al. CO gas sensing of ZnO nanostructures synthesized by an assisted microwave wet chemical route. *Sensor Actuator B Chem* 2009;**143**(1):198−204.

24. Abboud Y, Saffaj T, Chagraoui A, El Bouari A, Brouzi K, Tanane O, et al. Biosynthesis, characterization and antimicrobial activity of copper oxide nanoparticles (CONPs) produced using brown alga extract (Bifurcaria bifurcata). *Appl Nanosci* 2014;**4**(5):571−6.

25. Kulkarni N, Muddapur U. Biosynthesis of metal nanoparticles: a review. *J Nanotechnol* 2014;**2014**.

26. Jeevanandam J, Chan YS, Danquah MK. Biosynthesis of metal and metal oxide nanoparticles. *ChemBioEng Rev* 2016;**3**(2):55−67.

27. Jha AK, Prasad K. Biosynthesis of metal and oxide nanoparticles using Lactobacilli from yoghurt and probiotic spore tablets. *Biotechnol J* 2010;**5**(3):285−91.

28. Ramesh P, Rajendran A, Meenakshisundaram M. Green synthesis of zinc oxide nanoparticles using flower extract cassia auriculata. *J Nanosci Nanotechnol* 2014;**2**(1):41—5.
29. Niederberger M. Nonaqueous sol—gel routes to metal oxide nanoparticles. *Acc Chem Res* 2007;**40**(9):793—800.
30. Bayal N, Jeevanandam P. Synthesis of TiO$_2$— MgO mixed metal oxide nanoparticles via a sol— gel method and studies on their optical properties. *Ceram Int* 2014;**40**(10):15463—77.
31. Pandey P, Merwyn S, Agarwal G, Tripathi B, Pant S. Electrochemical synthesis of multi-armed CuO nanoparticles and their remarkable bactericidal potential against waterborne bacteria. *J Nanoparticle Res* 2012;**14**(1):709.
32. Katwal R, Kaur H, Sharma G, Naushad M, Pathania D. Electrochemical synthesized copper oxide nanoparticles for enhanced photocatalytic and antimicrobial activity. *J Ind Eng Chem* 2015;**31**:173—84.
33. Fajaroh F, Setyawan H, Widiyastuti W, Winardi S. Synthesis of magnetite nanoparticles by surfactant-free electrochemical method in an aqueous system. *Adv Powder Technol* 2012;**23**(3):328—33.
34. Ma H, Yin B, Wang S, Jiao Y, Pan W, Huang S, et al. Synthesis of silver and gold nanoparticles by a novel electrochemical method. *ChemPhysChem* 2004;**5**(1):68—75.
35. Xu X, Guo J, Wang Y. A novel technique by the citrate pyrolysis for preparation of iron oxide nanoparticles. *Mater Sci Eng, B* 2000;**77**(2):207—9.
36. Kim S, Liu B, Zachariah M. Synthesis of nanoporous metal oxide particles by a new inorganic matrix spray pyrolysis method. *Chem Mater* 2002;**14**(7):2889—99.
37. Ghosh T, Dash SK, Chakraborty P, Guha A, Kawaguchi K, Roy S, et al. Preparation of antiferromagnetic Co 3 O 4 nanoparticles from two different precursors by pyrolytic method: in vitro antimicrobial activity. *RSC Adv* 2014;**4**(29):15022—9.
38. Roy A, Bhattacharya J. Microwave-assisted synthesis and characterization of CaO nanoparticles. *Int J Nanosci* 2011;**10**(03):413—8.
39. Lin Y, Baggett DW, Kim J-W, Siochi EJ, Connell JW. Instantaneous formation of metal and metal oxide nanoparticles on carbon nanotubes and graphene via solvent-free microwave heating. *ACS Appl Mater Interfaces* 2011;**3**(5):1652—64.
40. Wang S, Jiang SP, Wang X. Microwave-assisted one-pot synthesis of metal/metal oxide nanoparticles on graphene and their electrochemical applications. *Electrochim Acta* 2011;**56**(9):3338—44.
41. Hassanjani-Roshan A, Vaezi MR, Shokuhfar A, Rajabali Z. Synthesis of iron oxide nanoparticles via sonochemical method and their characterization. *Particuology* 2011;**9**(1):95—9.
42. Xiong HM, Shchukin DG, Möhwald H, Xu Y, Xia YY. Sonochemical synthesis of highly luminescent zinc oxide nanoparticles doped with magnesium (II). *Angew Chem* 2009;**121**(15):2765—9.
43. Stankic S, Suman S, Haque F, Vidic J. Pure and multi metal oxide nanoparticles: synthesis, antibacterial and cytotoxic properties. *J Nanobiotechnol* 2016;**14**(1):1—20.
44. Beyth N, Houri-Haddad Y, Domb A, Khan W, Hazan R. Alternative antimicrobial approach: nano-antimicrobial materials. *Evid Based Complement Alternat Med* 2015;**2015**.
45. Vithiya K, Kumar R, Sen S. Antimicrobial activity of biosynthesized silver oxide nanoparticles. *J Pure Appl Microbiol* 2014;**8**(4):3263—8.
46. Li R, Chen Z, Ren N, Wang Y, Wang Y, Yu F. Biosynthesis of silver oxide nanoparticles and their photocatalytic and antimicrobial activity evaluation for wound healing applications in nursing care. *J Photochem Photobiol B Biol* 2019;**199**:111593.

47. Dharmaraj D, Krishnamoorthy M, Rajendran K, Karuppiah K, Annamalai J, Durairaj KR, et al. Antibacterial and cytotoxicity activities of biosynthesized silver oxide (Ag_2O) nanoparticles using Bacillus paramycoides. *J Drug Deliv Sci Technol* 2020:102111.

48. Bellio P, Luzi C, Mancini A, Cracchiolo S, Passacantando M, Di Pietro L, et al. Cerium oxide nanoparticles as potential antibiotic adjuvant. Effects of CeO_2 nanoparticles on bacterial outer membrane permeability. *Biochim Biophys Acta Biomembr* 2018;**1860**(11): 2428−35.

49. dos Santos CCL, Farias IAP, dos Reis Albuquerque AJ, e Silva PMF, da Costa One GM, Sampaio FC. Antimicrobial activity of nano cerium oxide (IV)(CeO_2) against *Strepto-coccus mutans*. *BMC Proc* 2014;**8**(P48). https://doi.org/10.1186/1753-6561-8-S4-P48.

50. Maqbool Q, Nazar M, Naz S, Hussain T, Jabeen N, Kausar R, et al. Antimicrobial potential of green synthesized CeO_2 nanoparticles from Olea europaea leaf extract. *Int J Nanomed* 2016;**11**:5015.

51. Mohamed HEA, Afridi S, Khalil AT, Ali M, Zohra T, Akhtar R, et al. Promising antiviral, antimicrobial and therapeutic properties of green nanoceria. *Nanomedicine* 2020;**15**(05): 467−88.

52. Arakha M, Pal S, Samantarrai D, Panigrahi TK, Mallick BC, Pramanik K, et al. Antimicrobial activity of iron oxide nanoparticle upon modulation of nanoparticle-bacteria interface. *Sci Rep* 2015;**5**:14813.

53. Arias LS, Pessan JP, Vieira APM, Lima TMT, Delbem ACB, Monteiro DR. Iron oxide nanoparticles for biomedical applications: a perspective on synthesis, drugs, antimicrobial activity, and toxicity. *Antibiotics* 2018;**7**(2):46.

54. Arokiyaraj S, Saravanan M, Prakash NU, Arasu MV, Vijayakumar B, Vincent S. Enhanced antibacterial activity of iron oxide magnetic nanoparticles treated with *Arge-mone mexicana* L. leaf extract: an in vitro study. *Mater Res Bull* 2013;**48**(9):3323−7.

55. Jamzad M, Bidkorpeh MK. Green synthesis of iron oxide nanoparticles by the aqueous extract of *Laurus nobilis* L. leaves and evaluation of the antimicrobial activity. *J Nanostruct Chem* 2020:1−9.

56. Parveen S, Wani AH, Shah MA, Devi HS, Bhat MY, Koka JA. Preparation, characterization and antifungal activity of iron oxide nanoparticles. *Microb Pathog* 2018;**115**: 287−92.

57. Devi HS, Boda MA, Shah MA, Parveen S, Wani AH. Green synthesis of iron oxide nanoparticles using *Platanus orientalis* leaf extract for antifungal activity. *Green Process Synth* 2019;**8**(1):38−45.

58. Kumar R, Nayak M, Sahoo GC, Pandey K, Sarkar MC, Ansari Y, et al. Iron oxide nanoparticles based antiviral activity of H1N1 influenza A virus. *J Infect Chemother* 2019; **25**(5):325−9.

59. El-Batal AI, El-Sayyad GS, Mosallam FM, Fathy RM. Penicillium chrysogenum-mediated mycogenic synthesis of copper oxide nanoparticles using gamma rays for in vitro antimicrobial activity against some plant pathogens. *J Cluster Sci* 2020;**31**(1):79−90.

60. Ahamed M, Alhadlaq HA, Khan M, Karuppiah P, Al-Dhabi NA. Synthesis, characterization, and antimicrobial activity of copper oxide nanoparticles. *J Nanomater* 2014;**2014**.

61. Hassan SE-D, Fouda A, Radwan AA, Salem SS, Barghoth MG, Awad MA, et al. Endophytic actinomycetes Streptomyces spp mediated biosynthesis of copper oxide nanoparticles as a promising tool for biotechnological applications. *J Biol Inorg Chem* 2019; **24**(3):377−93.

62. Sivaraj R, Rahman PK, Rajiv P, Narendhran S, Venckatesh R. Biosynthesis and characterization of Acalypha indica mediated copper oxide nanoparticles and evaluation of its

antimicrobial and anticancer activity. *Spectrochim Acta A Mol Biomol Spectrosc* 2014;**129**: 255−8.

63. Tavakoli A, Hashemzadeh MS. Inhibition of herpes simplex virus type 1 by copper oxide nanoparticles. *J Virol Methods* 2020;**275**:113688.

64. Hang X, Peng H, Song H, Qi Z, Miao X, Xu W. Antiviral activity of cuprous oxide nanoparticles against hepatitis C virus in vitro. *J Virol Methods* 2015;**222**:150−7.

65. Chen J, Sun L, Cheng Y, Lu Z, Shao K, Li T, et al. Graphene oxide-silver nanocomposite: novel agricultural antifungal agent against Fusarium graminearum for crop disease prevention. *ACS Appl Mater Interfaces* 2016;**8**(36):24057−70.

66. Sametband M, Kalt I, Gedanken A, Sarid R. Herpes simplex virus type-1 attachment inhibition by functionalized graphene oxide. *ACS Appl Mater Interfaces* 2014;**6**(2): 1228−35.

67. Zou F, Zhou H, Jeong DY, Kwon J, Eom SU, Park TJ, et al. Wrinkled surface-mediated antibacterial activity of graphene oxide nanosheets. *ACS Appl Mater Interfaces* 2017;**9**(2): 1343−51.

68. Sawangphruk M, Srimuk P, Chiochan P, Sangsri T, Siwayaprahm P. Synthesis and antifungal activity of reduced graphene oxide nanosheets. *Carbon* 2012;**50**(14):5156−61.

69. Brunet L, Lyon DY, Hotze EM, Alvarez PJ, Wiesner MR. Comparative photoactivity and antibacterial properties of C60 fullerenes and titanium dioxide nanoparticles. *Environ Sci Technol* 2009;**43**(12):4355−60.

70. Aoshima H, Kokubo K, Shirakawa S, Ito M, Yamana S, Oshima T. Antimicrobial activity of fullerenes and their hydroxylated derivatives. *Biocontrol Sci* 2009;**14**(2):69−72.

71. Yang C, Mamouni J, Tang Y, Yang L. Antimicrobial activity of single-walled carbon nanotubes: length effect. *Langmuir* 2010;**26**(20):16013−9.

72. Saleemi MA, Fouladi MH, Yong PVC, Wong EH. Elucidation of antimicrobial activity of non-covalently dispersed carbon nanotubes. *Materials* 2020;**13**(7):1676.

73. Maheshwaran G, Bharathi AN, Selvi MM, Kumar MK, Kumar RM, Sudhahar S. Green synthesis of Silver oxide nanoparticles using *Zephyranthes Rosea* flower extract and evaluation of biological activities. *J Environ Chem Eng* 2020:104137.

74. D'Lima L, Phadke M, Ashok VD. Biogenic silver and silver oxide hybrid nanoparticles: a potential antimicrobial against multi drug-resistant *Pseudomonas aeruginosa*. *New J Chem* 2020;**44**(12):4935−41.

75. Boopathi S, Gopinath S, Boopathi T, Balamurugan V, Rajeshkumar R, Sundararaman M. Characterization and antimicrobial properties of silver and silver oxide nanoparticles synthesized by cell-free extract of a mangrove-associated *Pseudomonas aeruginosa* M6 using two different thermal treatments. *Ind Eng Chem Res* 2012;**51**(17):5976−85.

76. Manikandan V, Yi P-I, Velmurugan P, Jayanthi P, Hong S-C, Jang S-H, et al. Production, optimization and characterization of silver oxide nanoparticles using Artocarpus heterophyllus rind extract and their antifungal activity. *Afr J Biotechnol* 2017;**16**(36):1819−25.

77. Zhang M, Zhang C, Zhai X, Luo F, Du Y, Yan C. Antibacterial mechanism and activity of cerium oxide nanoparticles. *Sci China Mater* 2019:1−13.

78. Farias IAP, Santos CCL, Sampaio FC. Antimicrobial activity of cerium oxide nanoparticles on opportunistic microorganisms: a systematic review. *Biomed Res Int* 2018; **2018**.

79. Qi M, Li W, Zheng X, Li X, Sun Y, Wang Y, et al. Cerium and its oxidant-based nanomaterials for antibacterial applications: a state-of-the-art. *Front Mater* 2020. https://doi.org/10.3389/fmats.2020.00213.

80. Manikandan V, Jayanthi P, Priyadharsan A, Vijayaprathap E, Anbarasan P, Velmurugan P. Green synthesis of pH-responsive Al_2O_3 nanoparticles: application to rapid removal of

nitrate ions with enhanced antibacterial activity. *J Photochem Photobiol Chem* 2019;**371**: 205−15.

81. Jalal M, Ansari MA, Shukla AK, Ali SG, Khan HM, Pal R, et al. Green synthesis and antifungal activity of Al 2 O 3 NPs against fluconazole-resistant Candida spp isolated from a tertiary care hospital. *RSC Adv* 2016;**6**(109):107577−90.

82. Sikora P, Augustyniak A, Cendrowski K, Nawrotek P, Mijowska E. Antimicrobial activity of Al_2O_3, CuO, Fe_3O_4, and ZnO nanoparticles in scope of their further application in cement-based building materials. *Nanomaterials* 2018;**8**(4):212.

83. Moshafi MH, Ranjbar M, Ilbeigi G. Eco-friendly and systematic study for synthesis of La3+/α-Al_2O_3 nanoparticles: antibacterial activity against pathogenic microbial strains. *Int J Nanomed* 2019;**14**:10137.

84. Suryavanshi P, Pandit R, Gade A, Derita M, Zachino S, Rai M. Colletotrichum sp.-mediated synthesis of sulphur and aluminium oxide nanoparticles and its in vitro activity against selected food-borne pathogens. *LWT-Food Sci Technol* 2017;**81**:188−94.

85. Shenashen M, Derbalah A, Hamza A, Mohamed A, El Safty S. Antifungal activity of fabricated mesoporous alumina nanoparticles against root rot disease of tomato caused by Fusarium oxysporium. *Pest Manag Sci* 2017;**73**(6):1121−6.

86. Sadiq IM, Chowdhury B, Chandrasekaran N, Mukherjee A. Antimicrobial sensitivity of *Escherichia coli* to alumina nanoparticles. *Nanomed Nanotechnol Biol Med* 2009;**5**(3): 282−6.

87. Jesline A, John NP, Narayanan P, Vani C, Murugan S. Antimicrobial activity of zinc and titanium dioxide nanoparticles against biofilm-producing methicillin-resistant *Staphylococcus aureus*. *Appl Nanosci* 2015;**5**(2):157−62.

88. Venkataraju JL, Sharath R, Chandraprabha M, Neelufar E, Hazra A, Patra M. Synthesis, characterization and evaluation of antimicrobial activity of zinc oxide nanoparticles. *J Biochem Technol* 2014;**3**(5):151−4.

89. Narayanan P, Wilson WS, Abraham AT, Sevanan M. Synthesis, characterization, and antimicrobial activity of zinc oxide nanoparticles against human pathogens. *BioNanoScience* 2012;**2**(4):329−35.

90. da Silva BL, Abuçafy MP, Manaia EB, Junior JAO, Chiari-Andréo BG, Pietro RCR, et al. Relationship between structure and antimicrobial activity of zinc oxide nanoparticles: an Overview. *Int J Nanomed* 2019;**14**:9395.

91. He L, Liu Y, Mustapha A, Lin M. Antifungal activity of zinc oxide nanoparticles against *Botrytis cinerea* and Penicillium expansum. *Microbiol Res* 2011;**166**(3):207−15.

92. Lipovsky A, Nitzan Y, Gedanken A, Lubart R. Antifungal activity of ZnO nanoparticles—the role of ROS mediated cell injury. *Nanotechnology* 2011;**22**(10): 105101.

93. Ghaffari H, Tavakoli A, Moradi A, Tabarraei A, Bokharaei-Salim F, Zahmatkeshan M, et al. Inhibition of H1N1 influenza virus infection by zinc oxide nanoparticles: another emerging application of nanomedicine. *J Biomed Sci* 2019;**26**(1):1−10.

94. Tavakoli A, Ataei-Pirkooh A, Mm Sadeghi G, Bokharaei-Salim F, Sahrapour P, Kiani SJ, et al. Polyethylene glycol-coated zinc oxide nanoparticle: an efficient nanoweapon to fight against herpes simplex virus type 1. *Nanomedicine* 2018;**13**(21):2675−90.

95. Abdelkhalek A, Al-Askar AA. Green synthesized ZnO nanoparticles mediated by *Mentha spicata* extract induce plant systemic resistance against tobacco mosaic virus. *Appl Sci* 2020;**10**(15):5054.

96. El-Sayyad GS, Mosallam FM, El-Batal AI. One-pot green synthesis of magnesium oxide nanoparticles using Penicillium chrysogenum melanin pigment and gamma rays with

antimicrobial activity against multidrug-resistant microbes. *Adv Powder Technol* 2018; **29**(11):2616−25.

97. Pugazhendhi A, Prabhu R, Muruganantham K, Shanmuganathan R, Natarajan S. Anticancer, antimicrobial and photocatalytic activities of green synthesized magnesium oxide nanoparticles (MgONPs) using aqueous extract of Sargassum wightii. *J Photochem Photobiol B Biol* 2019;**190**:86−97.

98. Cai L, Chen J, Liu Z, Wang H, Yang H, Ding W. Magnesium oxide nanoparticles: effective agricultural antibacterial agent against *Ralstonia solanacearum*. *Front Microbiol* 2018;**9**:790.

99. Nguyen N-YT, Grelling N, Wetteland CL, Rosario R, Liu H. Antimicrobial activities and mechanisms of magnesium oxide nanoparticles (nMgO) against pathogenic bacteria, yeasts, and biofilms. *Sci Rep* 2018;**8**(1):1−23.

100. He Y, Ingudam S, Reed S, Gehring A, Strobaugh TP, Irwin P. Study on the mechanism of antibacterial action of magnesium oxide nanoparticles against foodborne pathogens. *J Nanobiotechnol* 2016;**14**(1):54.

101. Naguib GH, Hosny KM, Hassan AH, Al Hazmi F, Al Dharrab A, Alkhalidi HM, et al. Zein based magnesium oxide nanoparticles: assessment of antimicrobial activity for dental implications. *Pak J Pharm Sci* 2018;**31**.

102. Joghee S, Ganeshan P, Vincent A, Hong SI. Ecofriendly biosynthesis of zinc oxide and magnesium oxide particles from medicinal plant *Pisonia grandis* R. Br. leaf extract and their antimicrobial activity. *BioNanoScience* 2019;**9**(1):141−54.

103. Rafiei S, Rezatofighi SE, Ardakani MR, Madadgar O. In vitro anti-foot-and-mouth disease virus activity of magnesium oxide nanoparticles. *IET Nanobiotechnol* 2015;**9**(5):247−51.

104. Antibacterial activity of copper oxide nanoparticles prepared by mechanical milling. In: Amal M, Wibowo J, Nuraini L, Senopati G, Hasbi M, Priyotomo G, editors. *IOP conference series: materials science and engineering*. IOP Publishing; 2019.

105. Ramola B, Joshi NC, Ramola M, Chhabra J, Singh A. Green synthesis, characterisations and antimicrobial activities of CaO nanoparticles. *Orient J Chem* 2019;**35**(3):1154−7.

106. Aguiar AS, Guerreiro-Tanomaru J, Faria G, Leonardo R, Tanomaru-Filho M. Antimicrobial activity and ph of calcium hydroxide and zinc oxide nanoparticles intracanal medication and association with chlorhexidine. *J Contemp Dent Pract* 2015;**16**(8):624−9.

107. Louwakul P, Saelo A, Khemaleelakul S. Efficacy of calcium oxide and calcium hydroxide nanoparticles on the elimination of *Enterococcus faecalis* in human root dentin. *Clin Oral Invest* 2017;**21**(3):865−71.

108. Ijaz U, Bhatti IA, Mirza S, Ashar A. Characterization and evaluation of antibacterial activity of plant mediated calcium oxide (CaO) nanoparticles by employing *Mentha pipertia* extract. *Mater Res Express* 2017;**4**(10):105402.

109. Sato Y, Ishihara M, Nakamura S, Fukuda K, Takayama T, Hiruma S, et al. Preparation and application of bioshell calcium oxide (BiSCaO) nanoparticle-dispersions with bactericidal activity. *Molecules* 2019;**24**(18):3415.

110. Karimiyan A, Najafzadeh H, Ghorbanpour M, Hekmati-Moghaddam SH. Antifungal effect of magnesium oxide, zinc oxide, silicon oxide and copper oxide nanoparticles against *Candida albicans*. *Zahedan J Res Med Sci* 2015;**17**(10):19−23.

111. Maniprasad P, Santra S. Novel copper (Cu) loaded core−shell silica nanoparticles with improved Cu bioavailability: synthesis, characterization and study of antibacterial properties. *J Biomed Nanotechnol* 2012;**8**(4):558−66.

112. Smirnov N, Kudryashov S, Nastulyavichus A, Rudenko A, Saraeva I, Tolordava E, et al. Antibacterial properties of silicon nanoparticles. *Laser Phys Lett* 2018;**15**(10):105602.

113. Chen L, Liang J. An overview of functional nanoparticles as novel emerging antiviral therapeutic agents. *Mater Sci Eng C* 2020:110924.

114. Azizi-Lalabadi M, Ehsani A, Divband B, Alizadeh-Sani M. Antimicrobial activity of Titanium dioxide and Zinc oxide nanoparticles supported in 4A zeolite and evaluation the morphological characteristic. *Sci Rep* 2019;9(1):1−10.

115. El-Naggar ME, Shaheen TI, Zaghloul S, El-Rafie MH, Hebeish A. Antibacterial activities and UV protection of the in situ synthesized titanium oxide nanoparticles on cotton fabrics. *Ind Eng Chem Res* 2016;55(10):2661−8.

116. Nadeem M, Tungmunnithum D, Hano C, Abbasi BH, Hashmi SS, Ahmad W, et al. The current trends in the green syntheses of titanium oxide nanoparticles and their applications. *Green Chem Lett Rev* 2018;11(4):492−502.

117. Rao TN, Babji P, Ahmad N, Khan RA, Hassan I, Shahzad SA, et al. Green synthesis and structural classification of Acacia nilotica mediated-silver doped titanium oxide (Ag/TiO$_2$) spherical nanoparticles: assessment of its antimicrobial and anticancer activity. *Saudi J Biol Sci* 2019;26(7):1385−91.

118. Pavlova EL, Toshkovska RD, Doncheva TE, Ivanova IA. Prooxidant and antimicrobic effects of iron and titanium oxide nanoparticles and thalicarpine. *Arch Microbiol* 2020: 1−8.

119. Akhtar S, Shahzad K, Mushtaq S, Ali I, Rafe MH, Fazal-ul-Karim SM. Antibacterial and antiviral potential of colloidal Titanium dioxide (TiO$_2$) nanoparticles suitable for biological applications. *Mater Res Express* 2019;6(10):105409.

120. Irshad MA, Nawaz R, ur Rehman MZ, Imran M, Ahmad MJ, Ahmad S, et al. Synthesis and characterization of titanium dioxide nanoparticles by chemical and green methods and their antifungal activities against wheat rust. *Chemosphere* 2020:127352.

121. Zhu X, Pathakoti K, Hwang H-M. Green synthesis of titanium dioxide and zinc oxide nanoparticles and their usage for antimicrobial applications and environmental remediation. *Green Synth, Charact Appl Nanopart* 2019:223−63.

122. Daoud WA, Xin JH, Zhang Y-H. Surface functionalization of cellulose fibers with titanium dioxide nanoparticles and their combined bactericidal activities. *Surf Sci* 2005; 599(1−3):69−75.

123. Zinedine A, Mañes J. Occurrence and legislation of mycotoxins in food and feed from Morocco. *Food Control* 2009;20(4):334−44.

124. Din MI, Nabi AG, Rani A, Aihetasham A, Mukhtar M. Single step green synthesis of stable nickel and nickel oxide nanoparticles from *Calotropis gigantea*: catalytic and antimicrobial potentials. *Environ Nanotechnol Monit Manag* 2018;9:29−36.

125. Vaseghi Z, Tavakoli O, Nematollahzadeh A. Rapid biosynthesis of novel Cu/Cr/Ni trimetallic oxide nanoparticles with antimicrobial activity. *J Environ Chem Eng* 2018;6(2): 1898−911.

126. Rakshit S, Ghosh S, Chall S, Mati SS, Moulik S, Bhattacharya SC. Controlled synthesis of spin glass nickel oxide nanoparticles and evaluation of their potential antimicrobial activity: a cost effective and eco friendly approach. *RSC Adv* 2013;3(42):19348−56.

127. Suresh S, Karthikeyan S, Saravanan P, Jayamoorthy K. Comparison of antibacterial and antifungal activities of 5-amino-2-mercaptobenzimidazole and functionalized NiO nanoparticles. *Karbala Int J Mod Sci* 2016;2(3):188−95.

128. Iqbal J, Abbasi BA, Mahmood T, Hameed S, Munir A, Kanwal S. Green synthesis and characterizations of Nickel oxide nanoparticles using leaf extract of *Rhamnus virgata* and their potential biological applications. *Appl Organomet Chem* 2019;33(8):e4950.

129. Vennila raj, Kamaraj P, Sridharan M, Arockiaselvi J. Green synthesis, characterization of yttrium oxide, stannous oxide, yttrium doped tin oxide and tin doped yttrium oxide

nanoparticles and their biological activities. *Mater Today Proc* 2021;**36**(Part 4):920−2. https://doi.org/10.1016/j.matpr.2020.07.032. ISSN 2214-7853.

130. Magdalane CM, Kaviyarasu K, Vijaya JJ, Siddhardha B, Jeyaraj B. Facile synthesis of heterostructured cerium oxide/yttrium oxide nanocomposite in UV light induced photocatalytic degradation and catalytic reduction: synergistic effect of antimicrobial studies. *J Photochem Photobiol B Biol* 2017;**173**:23−34.

131. Mariano-Torres JA, López-Marure A, García-Hernández M, Basurto-Islas G, Domínguez-Sánchez MÁ. Synthesis and characterization of glycerol citrate polymer and yttrium oxide nanoparticles as a potential antibacterial material. *Mater Trans* 2018:M2018248.

132. Moriyama A, Takahashi U, Mizuno Y, Takahashi J, Horie M, Iwahashi H. The truth of toxicity caused by yttrium oxide nanoparticles to yeast cells. *J Nanosci Nanotechnol* 2019;**19**(9):5418−25.

133. Kornev AB, Peregudov AS, Martynenko VM, Balzarini J, Hoorelbeke B, Troshin PA. Synthesis and antiviral activity of highly water-soluble polycarboxylic derivatives of [70] fullerene. *Chem Commun* 2011;**47**(29):8298−300.

134. Mohammed MK, Ahmed DS, Mohammad MR. Studying antimicrobial activity of carbon nanotubes decorated with metal-doped ZnO hybrid materials. *Mater Res Express* 2019;**6**(5):055404.

135. Chen H, Wang B, Gao D, Guan M, Zheng L, Ouyang H, et al. Broad-spectrum antibacterial activity of carbon nanotubes to human gut bacteria. *Small* 2013;**9**(16):2735−46.

136. Iannazzo D, Pistone A, Galvagno S, Ferro S, De Luca L, Monforte AM, et al. Synthesis and anti-HIV activity of carboxylated and drug-conjugated multi-walled carbon nanotubes. *Carbon* 2015;**82**:548−61.

137. Navanietha Krishnaraj R, Chandran S, Pal P, Berchmans S. Investigations on the antiretroviral activity of carbon nanotubes using computational molecular approach. *Comb Chem High Throughput Screen* 2014;**17**(6):531−5.

138. Zhu S, Li J, Huang A-G, Huang J-Q, Huang Y-Q, Wang G-X. Anti-betanodavirus activity of isoprinosine and improved efficacy using carbon nanotubes based drug delivery system. *Aquaculture* 2019;**512**:734377.

139. Wang X, Zhou Z, Chen F. Surface modification of carbon nanotubes with an enhanced antifungal activity for the control of plant fungal pathogen. *Materials* 2017;**10**(12):1375.

140. Li C, Wang X, Chen F, Zhang C, Zhi X, Wang K, et al. The antifungal activity of graphene oxide−silver nanocomposites. *Biomaterials* 2013;**34**(15):3882−90.

141. Chen J, Peng H, Wang X, Shao F, Yuan Z, Han H. Graphene oxide exhibits broad-spectrum antimicrobial activity against bacterial phytopathogens and fungal conidia by intertwining and membrane perturbation. *Nanoscale* 2014;**6**(3):1879−89.

142. Gurunathan S, Han JW, Dayem AA, Eppakayala V, Kim J-H. Oxidative stress-mediated antibacterial activity of graphene oxide and reduced graphene oxide in *Pseudomonas aeruginosa*. *Int J Nanomed* 2012;**7**:5901.

143. Yang XX, Li CM, Li YF, Wang J, Huang CZ. Synergistic antiviral effect of curcumin functionalized graphene oxide against respiratory syncytial virus infection. *Nanoscale* 2017;**9**(41):16086−92.

144. Chen Y-N, Hsueh Y-H, Hsieh C-T, Tzou D-Y, Chang P-L. Antiviral activity of graphene−silver nanocomposites against non-enveloped and enveloped viruses. *Int J Environ Res Publ Health* 2016;**13**(4):430.

145. Ye S, Shao K, Li Z, Guo N, Zuo Y, Li Q, et al. Antiviral activity of graphene oxide: how sharp edged structure and charge matter. *ACS Appl Mater Interfaces* 2015;**7**(38):21571−9.

146. Gold K, Slay B, Knackstedt M, Gaharwar AK. Antimicrobial activity of metal and metal-oxide based nanoparticles. *Adv Therapeut* 2018;**1**(3):1700033.

147. Applerot G, Lellouche J, Lipovsky A, Nitzan Y, Lubart R, Gedanken A, et al. Understanding the antibacterial mechanism of CuO nanoparticles: revealing the route of induced oxidative stress. *Small* 2012;**8**(21):3326—37.

148. Hatchett DW, White HS. Electrochemistry of sulfur adlayers on the low-index faces of silver. *J Phys Chem* 1996;**100**(23):9854—9.

149. Chen W-J, Chen Y-C. Fe_3O_4/TiO_2 core/shell magnetic nanoparticle-based photokilling of pathogenic bacteria. *Nanomedicine* 2010;**5**(10):1585—93.

150. Grumezescu AM. *Antimicrobial nanoarchitectonics: from synthesis to applications.* William Andrew; 2017.

151. Hwang I, Lee J, Hwang JH, Kim KJ, Lee DG. Silver nanoparticles induce apoptotic cell death in *Candida albicans* through the increase of hydroxyl radicals. *FEBS J* 2012;**279**(7): 1327—38.

152. Jo Y-K, Kim BH, Jung G. Antifungal activity of silver ions and nanoparticles on phytopathogenic fungi. *Plant Dis* 2009;**93**(10):1037—43.

153. Nehra P, Chauhan R, Garg N, Verma K. Antibacterial and antifungal activity of chitosan coated iron oxide nanoparticles. *Br J Biomed Sci* 2018;**75**(1):13—8.

154. Salleh A, Naomi R, Utami ND, Mohammad AW, Mahmoudi E, Mustafa N, et al. The potential of silver nanoparticles for antiviral and antibacterial applications: a mechanism of action. *Nanomaterials* 2020;**10**(8):1566.

155. Gaikwad S, Ingle A, Gade A, Rai M, Falanga A, Incoronato N, et al. Antiviral activity of mycosynthesized silver nanoparticles against herpes simplex virus and human parainfluenza virus type 3. *Int J Nanomed* 2013;**8**:4303.

156. Huy TQ, Thanh NTH, Thuy NT, Van Chung P, Hung PN, Le A-T, et al. Cytotoxicity and antiviral activity of electrochemical—synthesized silver nanoparticles against poliovirus. *J Virol Methods* 2017;**241**:52—7.

157. Lara HH, Ayala-Nuñez NV, Ixtepan-Turrent L, Rodriguez-Padilla C. Mode of antiviral action of silver nanoparticles against HIV-1. *J Nanobiotechnol* 2010;**8**(1):1—10.

158. Park S, Park HH, Kim SY, Kim SJ, Woo K, Ko G. Antiviral properties of silver nanoparticles on a magnetic hybrid colloid. *Appl Environ Microbiol* 2014;**80**(8):2343—50.

159. Chen N, Zheng Y, Yin J, Li X, Zheng C. Inhibitory effects of silver nanoparticles against adenovirus type 3 in vitro. *J Virol Methods* 2013;**193**(2):470—7.

160. Karlsson HL, Cronholm P, Gustafsson J, Moller L. Copper oxide nanoparticles are highly toxic: a comparison between metal oxide nanoparticles and carbon nanotubes. *Chem Res Toxicol* 2008;**21**(9):1726—32.

161. Lai JC, Lai MB, Jandhyam S, Dukhande VV, Bhushan A, Daniels CK, et al. Exposure to titanium dioxide and other metallic oxide nanoparticles induces cytotoxicity on human neural cells and fibroblasts. *Int J Nanomed* 2008;**3**(4):533.

162. Horie M, Fujita K, Kato H, Endoh S, Nishio K, Komaba LK, et al. Association of the physical and chemical properties and the cytotoxicity of metal oxide nanoparticles: metal ion release, adsorption ability and specific surface area. *Metall* 2012;**4**(4):350—60.

163. Ng AMC, Chan CMN, Guo MY, Leung YH, Djurišić AB, Hu X, et al. Antibacterial and photocatalytic activity of TiO 2 and ZnO nanomaterials in phosphate buffer and saline solution. *Appl Microbiol Biotechnol* 2013;**97**(12):5565—73.

164. Guo B-L, Han P, Guo L-C, Cao Y-Q, Li A-D, Kong J-Z, et al. The antibacterial activity of Ta-doped ZnO nanoparticles. *Nanoscale Res Lett* 2015;**10**(1):336.

165. Yemmireddy VK, Hung YC. Using photocatalyst metal oxides as antimicrobial surface coatings to ensure food safety—opportunities and challenges. *Compr Rev Food Sci Food Saf* 2017;**16**(4):617—31.

166. Dheyab MA, Aziz AA, Jameel MS, Noqta OA, Khaniabadi PM, Mehrdel B. Simple rapid stabilization method through citric acid modification for magnetite nanoparticles. *Sci Rep* 2020;**10**(1):1−8.

167. Keller AA, Wang H, Zhou D, Lenihan HS, Cherr G, Cardinale BJ, et al. Stability and aggregation of metal oxide nanoparticles in natural aqueous matrices. *Environ Sci Technol* 2010;**44**(6):1962−7.

168. Kurlyandskaya GV, Litvinova LS, Safronov AP, Schupletsova VV, Tyukova IS, Khaziakhmatova OG, et al. Water-Based suspensions of iron oxide nanoparticles with electrostatic or steric stabilization by chitosan: fabrication, characterization and biocompatibility. *Sensors* 2017;**17**(11):2605.

Metal oxide–carbon nanotube composites for photodegradation

Nazia Nasr[1,2,3,4,9], Saadia Mushtaq[5], Hassina Tabassum[6,7], Ishaq Ahmad[1,4], Salem Abdulkarim[8], Tingkai Zhao[1,3] and Muhammad Hassan Sayyad[9]
[1]NPU-NCP Joint International Research Center on Advanced Nanomaterials and Defects Engineering, Northwestern Polytechnical University, Xi'an, Shaanxi, China; [2]Department of Physics, Division of Science and Technology, University of Education Lahore, Lahore, Punjab, Pakistan; [3]School of Materials Science & Engineering, Northwestern Polytechnical University, Xi'an, Shaanxi, China; [4]National Centre for Physics (NCP), Islamabad, Punjab, Pakistan; [5]Department of Physics, Hazara University, Mansehra, KPK, Pakistan; [6]Department of Materials Science and Engineering, College of Engineering, Peking University, Beijing, Haidian District, China; [7]Department of Chemical and Biological Engineering, University at Buffalo, The State University of New York, Buffalo, NY, United States; [8]Department of Physics, Faculty of Science, University of Benghazi, Al Marj City, Libya; [9]Faculty of Engineering Sciences, GIK Institute, Topi, Khyber Pakhtunkhwa, Pakistan

16.1 Introduction

Environmental pollution is a serious global issue that affects millions of lives every day. Industries that use hydrocarbon compounds, heavy metals, and dyes are primary contributors to air and water contamination by discharging these pollutants into untreated water bodies. Over 0.8 million tons of synthetic dyes are produced globally, and most are toxic and carcinogenic. A major fraction of these synthetic dyes flow into the air, rivers, and seas without proper treatment, not only affecting human and aqueous lives but also bioaccumulating in crops and ultimately reducing their photosynthetic activity.

Various methods have been developed to tackle environmental pollution, including ozonation, electrochemical oxidation, bacterial and fungal degradation, coagulation, adsorption and biosorption, chemical precipitation, and ion exchange. But these methods are not completely effective at removing biological and air pollutants, such as microbial growth or volatile organic compounds, that contain benzene, a substance that is carcinogenic to living species. Many of these methods are expensive and destructive; they also create nondegradable by-products that cause further ecosystem agitation and annihilation.[1–4]

Eco-friendly techniques, such as using solar energy to photodegrade pollutants, are alternative methods for removing microbials and toxic organic compounds. Further, they specifically allow complete mineralization of dyes from water.[5]

Advanced oxidation processes (AOPs) that use sunlight and photoactive semiconducting materials are known as proficient photodegradation processes. Metallic nanoparticles are commonly consumed as photocatalysts because of their large surface areas, availability, effectiveness, chemical reactivity, and eco-friendliness.[6] In this chapter, we discuss these photodegradation methods in detail and discuss the roles and performance of metal oxides and metal oxide—carbon nanotube (CNT) hybrid materials in pollutant photodegradation.

16.2   Photodegradation

Photodegradation is a breakdown or dissociation of molecules in the presence of solar radiation, mainly UV-visible (UV-Vis) radiation. Organic compounds comprising hydrocarbons or aromatic groups are mainly photodegradable.[7] This includes polymers, acids, dyes, pharmaceuticals, pesticides, and other pollutants in the air and water. We can degrade any photodegradable material using one process or a combination of two or more during photodegradation. This is possible because the photolysis process can only initiate the materials degradation process while the chromophoric group is being photoactivated, which in turn excites particles within the system to initiate photo-oxidation or AOPs for additional materials degradation.[8]

Photo-oxidation, photo-Fenton, photo-ozonation, and photocatalysis all fall under the umbrella of AOPs that use solar radiation for photodegradation, for which metal oxide—CNT hybrid materials are extensively employed. These are promising methods widely used for the complete remediation and degradation of heavy metals and toxic pollutants, contagious organic compounds, and microbes.[9,10] These processes can effectively destroy and remove specific pollutants, even those present in low concentrations, while leaving no chemical or biological sludge, and thus complete mineralization can be achieved, which is nearly impossible to abolish with other processes. Fig. 16.1 lists categories of AOPs.

16.3   Photocatalytic ozonation

Photocatalytic ozonation is an AOP constructed through the amalgamation of ozonation and photocatalysis. It is a versatile technique for exploiting both photonic processes.

Ozonation: Ozone reacts with various constituents in aqueous solutions in two ways: direct oxidation with molecular ozone or indirect ozonation through the decay of O_3 by OH· radical production. The former involves selective reactions with slow reaction rates, while the latter is nonselective and highly responsive. The highly reactive OH⁻ radicals, when they react with ozone, generate strong chemical oxidants. These chemical oxidants are used as a disinfectant and for the oxidation of contaminants.[11-13]

The overall reaction is given as

$$3O_3 + OH^- + H^+ \rightarrow 2OH^· + 4O_2 \qquad (16.1)$$

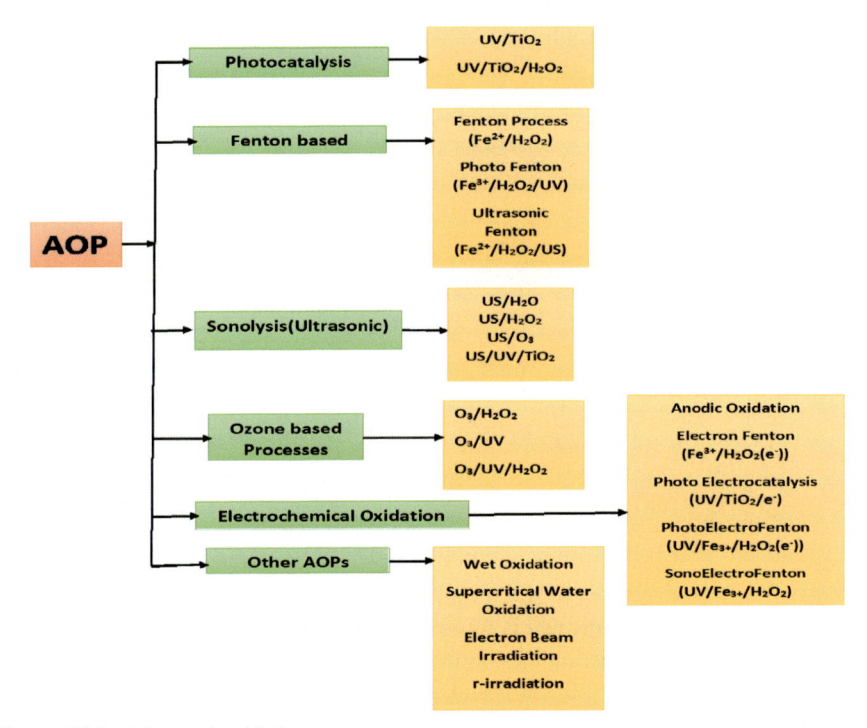

Figure 16.1 Advanced oxidation processes.

Coupling photocatalysis and ozonation: Among AOPs, photocatalytic ozonation—which, as indicated earlier, is a merger of two processes—has received extensive attention. Although the degradation process by ozone oxidation is quite fast and efficient, complete degradation and mineralization through this process is difficult because of the creation of stable and inert by-products, such as carboxylic acids, the important mineralization of the pollutant seldom occurs.[13,14] These carboxylic acids are extremely resilient to becoming oxidized because of the formation of the aromatic ring, and hence complete degradation and mineralization becomes impossible. In contrast, complete mineralization is achievable through photocatalysis. Nevertheless, because of low oxidation rates, a long degradation period is required.[15]

Recently, heterogeneous photocatalysis has been combined with other processes to enhance the photodegradation rate and complete mineralization of pollutants.[16,17] For example, to improve the photocatalytic performance of the TiO_2 oxidant, species such as ozone are added to the photocatalytic process.[18,19] The coupling of photocatalysis with ozonation affects photocatalytic mechanisms by reducing reaction time while increasing efficiency.[16]

16.4 Mechanism of photocatalytic ozonation

The photocatalytic ozonation (PO) mechanism is based on photocatalysis and ozonation mechanisms (shown in Fig. 16.2). When light photons are irradiated on a

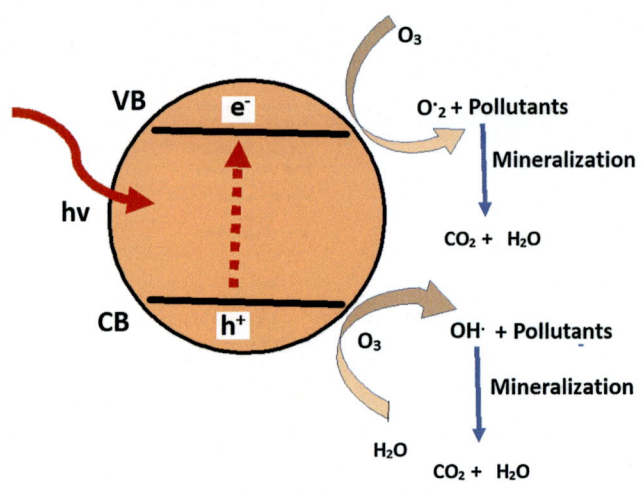

Figure 16.2 Schematic of the mechanism of photocatalytic ozonation.

photocatalyst, an electron-hole pair is formed. These carriers are then scavenged by the ozone, resulting in the formation of OH· and OH⁻ radicals on the surface of a photo-catalyst, which then react with the pollutants, and degradation then occurs. The PO mechanism can be explained by the following reactions:[20]

$$TiO_2 + hv \rightarrow e^- + h^+ \tag{16.2}$$

$$e^- + O_3 \rightarrow O_3^- \tag{16.3}$$

$$O_3^- + H_2O \rightarrow OH^- + OH^\bullet \tag{16.4}$$

$$h^+ OH^- \rightarrow OH^\bullet \tag{16.5}$$

OH· radicals profoundly advance the oxidation rate of PO because of the more elec-trophilic nature of O_3 (e⁻ loving) than O_2 toward photogenerated electrons. This ten-dency causes greater OH⁻ radical production, and hence, an improved mineralization process becomes possible.

<u>Uses:</u> Many pollutants present in water bodies are not biodegradable and are stub-born and contagious to conventional degradation treatments; therefore, advanced remediation and degradation techniques are presently needed to achieve the solo mineralization of pollutants. AOPs, especially ozonation and photocatalysis for removing toxic and refractory organic pollutants in water and wastewater systems, have been extensively reported and well documented in the literature.[21–23]

Photocatalysis is slow because of charge recombination, while ozone-resistant pol-lutants restrict the productivity of the ozonation process. However, the coupling of photocatalysis and the ozonation process through PO can eliminate these shortcomings.[24,25] To prevent the spread of secondary pollutants within the ecosystem, the nanomaterial PO effectively eliminates artificial alkanes. Organic pollutant degradation, wastewater disinfection, and toxicity evaluation of treated wastewater can be achieved through PO. PO is further classified as homogeneous or heterogeneous ozonation. In homogeneous ozonation, some metals in solution form, such as Fe^{2+}, Fe^{3+}, Mn^{2+}, Ni^{2+}, or Co^{2+}, can improve contaminant elimination compared with ozonation alone.[26–28] To generate free radicals O_3 is decomposed by these metallic ions. Moreover, complexes between the catalyst formed by the organic molecule are then oxidized. In heterogeneous ozonation, heterogeneous photocatalysts are employed to produce free radicals during the ozonation process. These heterogeneous photocatalysts include metals and metal oxides—Cu, Pt, Mn—TiO_2, zinc oxide (ZnO), and MgO—on the surfaces of SiO_2, Al_2O_3, CeO_2—CNT, etc.[29–31]

16.5 Metal oxide—carbon nanotubes for photo-ozonation

Metal oxides are used in photo-ozonation, and CNTs and MO—CNT composites have also been observed to play a role. For high efficiency, the unstable oxygen species such as $\cdot OH$ and $\cdot O_2^-$ resulting from the disintegration of O_3 following its reaction with metal oxide are accounted for. TiO_2 metal oxide nanoparticles and thin films are extensively used in this process. Aniline ($C_6H_5NH_2$) and its derivatives carrying a nondegradable aromatic ring are widely used in the chemical industry to manufacture rubber additives, lifesaving drugs, paints, dyes, polymers, chemicals, and pesticide sprays that contribute to environmental toxicity.[32] TiO_2 and CNTs, along with their hybrid forms, have proven themselves as efficient photocatalytic materials for degrading aniline in aqueous solution under solar irradiation when accompanied by O_3.[33–35]

16.6 Fenton and photo-Fenton processes

The Fenton reaction is an environmentally friendly process and does not require any destructive reagents. OH radicals are produced during the activation of hydrogen peroxide (H_2O_2) by the Fe^{2+} ion and form the Fe^{3+} ion. The oxidized Fe^{3+} ion further reacts with H_2O_2 to form the Fe^{2+} ion). The following reaction represents the generation of OH radicals and conversion of Fe^{2+} to Fe^{3+} ions:

$$Fe^{2+} + H_2O_2 \rightarrow Fe^{3+} + OH^- + OH^{\bullet} \tag{16.6}$$

Due to its clarity, the Fenton reaction is commonly used to detach the recalcitrant compounds. A primary disadvantage of the inhomogeneous Fenton process is that mass transfer constraints are disregarded, as the reactants and catalysts exist in the same phase, resulting in the production of iron mud, which requires continuous H_2O_2 supplementing and thus requires extra resources. In addition, dissolved residue Fe catalyst greatly hinders its consumption of Fenton purified water. This drawback gives rise to the evolution of the photo-Fenton process. The photo-Fenton reaction is initiated by the Fenton reagent and UV irradiation of frequency less than 600 nm. Using UV-Vis radiance in the Fenton process can reduce Fe(III) to Fe(II), and hence, extra Fe slurry could be removed this way.

A workable photo-Fenton process for water purification for fertilizing purposes was built at the Solar Platform at Almería, and the technology has been determined to be inexpensive and useable. Similarly, researchers have reported the degradation of reactive dyes (MPG, RR45, etc.), TNT, and their derivatives with more than 60% mineralization through this process.[36]

16.6.1   Heterogeneous photo-Fenton

Although the homogeneous photo-Fenton process is quite useful for the mineralization of many pollutants, some Fe residues limit the use of homogeneous photo-Fenton technology to only some areas. The shortcomings of this technology are removed with an integrated approach of the Fenton process and photocatalysis, making a heterogeneous photo-Fenton process. This allows the use of a semiconductor with a magnetic nanocomposite.

16.7   Metal oxide and carbon-supported nanocatalysts

A variety of iron oxides and hydroxide, including Fe_3O_4, Fe_2O_3, and $FeOOH^-$, have been employed as catalysts in heterogeneous Fenton processes. These metal oxides work within a wide pH range and can be reused.

Fe_2O_3 can act as an N-type semiconductor and has been used as a catalyst in heterogeneous photocatalytic and photo-Fenton processes. Fe_2O_3, which carries magnetic properties, helps separate the reaction medium by applying a magnetic field.[36-38]

Fe-based photocatalysts enable the reaping of photons while preventing unnecessary dispersion of Fe. The high recombination rate of carriers due to the low diffusion length of Fe_2O_3 may be reduced by employing mixed oxides with good photocatalytic properties.

CNT/graphene (G)/graphene oxide-supported catalysts can withstand many reaction cycles than macroscale activated carbon-supported catalysts.[39-41] CNTs are receiving a lot of interest as supporting material. Due to the mesoporous structure and high electrochemical accessibility of CNTs, the mass transfer constraints of CNTs are considerably less than their counterparts. They are stable at any pH level and proficient in inhibiting the agglomeration of catalysts, making them a more proficient choice to be employed. Also, the introduction of functional groups on the surface of CNTs creates covalent bonding between the metal oxide and CNT, which reduces leaching. The use of CNT@Fe catalysts has been reported in the hetero-photo-Fenton for degrading pharmaceutical pollutants and removing organic contaminants such as antibacterial drugs, methylene blue, methyl orange, and azo dye, as well as for the degradation of BPA (polycarbonate).[42,43]

16.8 Photocatalytic degradation

Photocatalytic degradation involves material degradation through photocatalysis. It is a process that involves photons, which produce reactive species by the exposure of a photosensitive catalyst with solar radiation. The photocatalytic degradation process is not possible without a photocatalyst. Usually, a catalyst is used in any reaction to speed up the reaction. In photocatalysis, the catalyst used is also sensitive to light. Usually, a semiconducting material whose energy gap is close to the photons' energy and is easily excited by it is used as a photocatalyst in photocatalytic degradation. Excited by light, semiconducting materials or metal oxides like TiO_2, ZnO nanoparticles, and CNT complexes combine with oxygen and water molecules to produce reactive radicals.[44–46] Those reactive radicals subsequently enter the microorganism's cells, organic dyes, polymers, etc., carrying hydrocarbon compounds that are photodegradable by such photocatalysis.

16.9 Mechanism of photocatalytic oxidation reactions

The photocatalysis was first noticed and explained by Honda-Fujishima, created on the photoelectrochemical water splitting experiment. When light irradiates the semiconductor surface, the photons with energy equal to or greater than the bandgap of the irradiated semiconductor become excited from the valence band (VB) to the conduction band (CB), leaving the holes behind in the VB. Within their lifetime of 10–100 ns, excited electrons recombine, and as a result, heat energy is released as represented below:

$$TiO_2 + hv \rightarrow e^- + h^+ \rightarrow heat \tag{16.7}$$

This takes place in the bulk of the catalyst. If the excited electrons are provided sufficient time to reach the catalyst surface, a subsequent charge transfer initiates redox reactions with the pollutants adsorbed on its surface, as shown in Fig. 16.3.

The holes left behind in the VB also play a crucial role in the photocatalytic degradation of the pollutant. These holes are either trapped in the surface defects making an OH· radical or participates in the oxidation process directly. The mechanism of photocatalytic degradation is given by the following reactions:

$$h^+ + H_2O \rightarrow OH^\bullet + h^+ \tag{16.8}$$

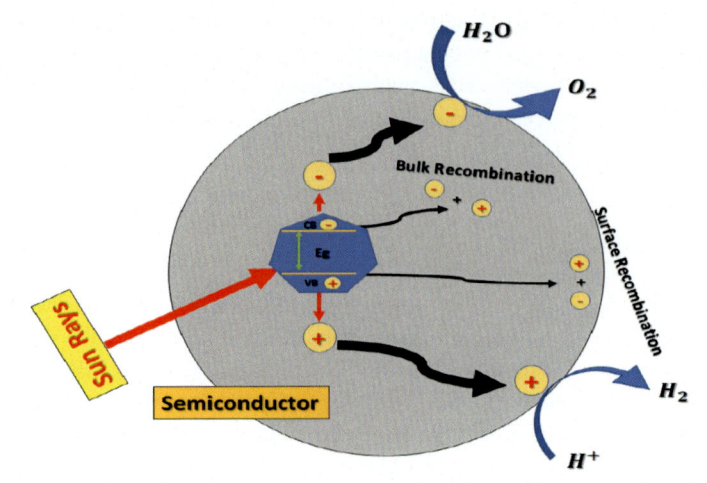

Figure 16.3 Schematic of the mechanism of the semiconductor photocatalytic activity.

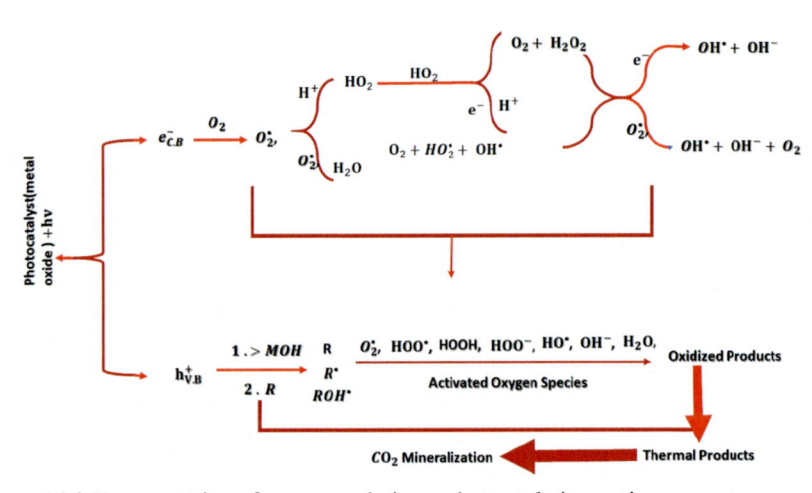

Figure 16.4 Representation of processes during a photocatalytic reaction.

$$h^+ + OH^- \rightarrow OH^{\bullet} \tag{16.9}$$

$$h^+ + M^{n^+} \rightarrow M^{(n+1)^+} \tag{16.10}$$

For an efficient photocatalytic activity, the excited photoelectrons must not accumulate on the surface of the catalyst. For this purpose, oxygen is usually used as an electron scavenger in most cases. Fig. 16.4 shows the chain reaction and formation of oxides because of photocatalysis reaction.

16.10 Measurement of photocatalytic activity

The Beer–Lambert relation is used to measure the photocatalytic performance of any photocatalyst. The degradation percentage is calculated with Eq. (16.1) and plotted against photocatalytic activity:

$$\text{Degradation } (\%) = \frac{C_i - C}{C} \times 100 = \frac{A_i - A}{A} \times 100 \qquad (16.11)$$

where C_i is the original pollutant concentration, A_i is the absorbance of pollutants before irradiation, and C and A are the concentration and absorbance of pollutants at a given time, respectively.

16.11 Features of a photocatalysts

Ideal photocatalysts must be inexpensive, readily available, and eco-friendly, and they should exhibit biological and chemical inertness. In addition, they must be highly photoactive and capable of degrading materials over long periods; they should also be reconsumable. Furthermore, it must be sensitive to the broad UV-Vis solar spectrum, photostable, and nonselective. The most popular photocatalyst is titanium dioxide, as it encompasses nearly all the characteristics just listed.

16.12 Degradation parameters

This section reviews and explains some basic parameters that affect the pollutant photodegradation rate of any pollutants while measuring any photocatalyst's photodegradation activity and efficiency.

This includes

a. concentration of photocatalyst and contaminants
b. photocatalyst surface area and morphology
c. temperature effect
d. pH of the solution effect

16.12.1 Concentrations of photocatalysts and contaminants

The degradation rate of any pollutant is highly dependent and directly related to photocatalyst and contaminant concentrations.

Increases in catalyst concentration lead to increased catalyst loading to an optimal value. A high amount of catalyst loading provides more active sites to generate a large amount of $OH^{\cdot}$ species. Once the saturation level is achieved, the rate of degradation no longer increases with photocatalyst loading, and thus, accumulation occurs and stops or/and scatters solar radiation. This reduces photoexcited electrons and holes that hinder all photocatalytic activity; therefore, the pollution degradation rate begins to decline.[47,48]

Organic compounds due to extensive conjugation are mostly colored. In such types of pollutants, the degradation rate of contaminants rises with the rise in their concentration. In other words, the relation between the concentration of the contaminants and their degradation is directly proportional until the saturation point is achieved. After the saturation level, the degradation rate starts to decline with the increase in the volume of pollutants or contaminants to be degraded. However, this saturation level can be improved by increasing the concentration of the photocatalyst. In this way, the organic compound is substantially adsorbed on the catalyst surface and is effectively decomposed.

Langmuir– Hinshel relation is used to represent the pollutants for the degradation, which follows pseudo-first-order kinetics:

$$\ln\left[\frac{C_o}{C}\right] = k_r K t = kt$$

where "t" is the time for the original concentration of contaminants C_o to decrease to a value C, K is the stability constant for adsorption of the contaminants on photocatalysts, and K_r is the limiting rate of the reaction. For dye degradation, mostly a pollutant concentration of 10–200 mg/L is tested, analogous to the pollutant concentration present under actual wastewater conditions.

16.12.2   Temperature effect

A key factor that influences the degradation rate of a photocatalyst is the surrounding temperature. Oxygen molecules play a crucial role in photoactivity, as it helps in OH radical formation. At elevated temperatures, desorption or disassociation of organic compounds occurs from the photocatalyst surface before reacting with the photoexcited species, which results in poor photocatalytic activity and a low photodegradation rate.

The physical and chemical properties of photocatalysts depend on the calcination temperature as well. With a small rise in the calcination temperature, some lightly bounded impurities and access water molecules are removed, thus creating extra surface sites. Some reports also suggest that calcination temperature negatively affects the photocatalyst surface area, as it adversely affects photocatalyst pore size.

16.12.3   The pH of the solution effects

While studying AOPs, it should be noted that all these reactions are pH vulnerable. The pH level of any solution is associated with the adsorption rate. It regulates the adsorption of the contaminants onto the catalyst surface. The photocatalyst surface attracts positive species with increases in the acidity level of the solution, whereas under alkaline conditions, the photocatalyst attracts negative ions.

The pollutant type and degradation rate greatly influence a solution's pH level.

16.12.4 Surface area and morphology effects

Surface area, crystal size, orientation and structure, atomic distribution, energy gap, and porosity influence a material's chemical and electrical properties. The rate of photodegradation improves with increases in the photocatalyst surface area and pore size. Lin et al. showed this effect by demonstrating a shift in the photocatalytic activity level from high to inert for brookite TiO_2 nanoflowers and nanorods to brookite TiO_2 nanosheets, respectively.

16.13 Metal oxides and other nanocomposites as potential photocatalysts

Semiconductors are mostly employed as photocatalysts. Titanium dioxide is tremendously useful as a photocatalyst and in many fields owing to its nontoxicity, availability, and photosensitivity. Anatase TiO_2 is highly photosensitive because of its elevated Fermi level, better oxygen adsorption ability, and a greater degree of surface hydroxylation. However, the high recombination rate of the e^- hole pair reduces the quantum efficiency of single semiconductors as photocatalysts. This results in slow reaction kinetics. Furthermore, the wide bandgap of semiconductors requires more UV light to produce OH^- radicals during the photocatalytic process. This constitutes a critical energy consumption problem. Many strategies are being adopted to compensate for this shortfall. Recently, numerous metal oxides, metal oxide nanocomposites, complex oxides, and heterojunction-type semiconductors as visible-light-driven photocatalysts have been employed. This includes WO_3, Fe_2O_3, CdSe, WO_3/TiO_2, Nd_2O_3, $Nd_2Sn_2O_7$, $Nd_2Sn_2O_7-SnO_2$, $Nd_2Zr_2O_7$, $Ag@CeO_2$, $Ho_2O_3-SiO_2$, Ho_2O_3, $BiVO_4$, and $CoO_x/BiVO_4$.[49–53]

Graphene—metal oxide composites own high surface area, fast charge transfer rate, and better adsorption capabilities. These features make them desirable for supporting metal oxides to remove metal ions and complex hydrocarbon compound contamination from an aqueous medium. Studies on photoelectrochemical properties and possible applications on graphene with TiO_2 nanocrystal hybrid and composites, graphene—SnO_2, graphene—ZnO, graphene—metal oxides such as RGO—WO_3, $gC_3N_4/TiO_2/CNT$, and RGO—$BiVO_4$ composites have been carried out.[53–58]

16.14   Metal oxide—carbon nanotube nanocomposites

CNT, a carbonaceous material with sp2 hybridization with a length of a few microns and a diameter range of 0.4 to 3 nm, has unique and diverse properties depending on morphology, size, and diameter.

The discovery of CNTs has significantly advanced the science and engineering of carbon materials. CNTs, because of their small dimensions, closed topology, and lattice helicity, possess exceptional electronic properties, strong mechanical strength, chemical, and thermal stability. The high activated surface area and aspect ratio make their catalytic and adsorption properties strong. Fig. 16.5 represents the metal oxide—CNT composite-based photocatalyst. Metal oxide—CNTs are used for photocatalytic activities. Carbonaceous materials like CNTs and graphene provide an electron bridge mediator to a semiconducting photocatalyst and improve the photoactivity of the catalyst. CNTs are used to photodegrade microbes, insecticides, gasoline, dyes, and environmental pollutants containing toxic metallic ions and hydrocarbon compounds, 4-nitrophenol, and many others. CNTs are easily dissolved in organic pollutants accredited by their synergic effect and photoactivity that activates ROS that help in the supplementary degradation of a degradable substance. Carbonaceous materials can act as adsorbents and band tighteners in semiconductor doping, which improves the photocatalyst's absorbance capacity and narrows the semiconductor's energy gap to absorb more light rays within the visible range.

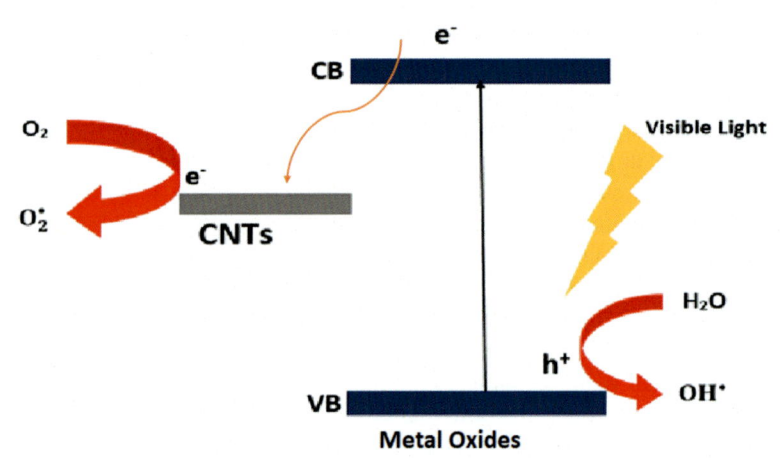

figure 16.5 Representation of photocatalysis of metal oxide—carbon nanotube composite-based catalysts.

16.14.1   Titanium dioxide/carbon nanotube composites

The enhanced photocatalytic properties of CNT—TiO$_2$ composites are explained by various mechanisms each time, as CNT— TiO$_2$ acts in many ways. In one mechanism, TiO$_2$ acts as a sensitizer. Photoexcited electrons are formed in the space charge region of TiO_2 and then shifted to the CNTs, leaving holes behind in TiO_2 to take part in redox reactions.[21,44]

In the second mechanism, CNTs act as sensitizers and transport photoexcited charges to the CB of TiO_2, allowing the formation of superoxide radicals by adsorbed molecular oxygen. After this, the positively charged CNTs take an electron from the VB of the TiO_2, leaving a hole behind. The TiO_2 then reacts with adsorbed water to form OH⁻ radicals. In the case of CNT− TiO_2 composites, CNTs can behave as an impurity additive through Ti−O−C bonds, improving photocatalytic activity by providing more active sites.

16.14.2 Zinc oxide/carbon nanotube composites

ZnO is another metal oxide whose photocatalytic behavior is studied extensively and considered a beneficial agent for degrading many organic pollutants. Like TiO_2, ZnO is also eco-friendly, inert, and stable. It has high electron mobility and thermal conductivity due to its direct wide bandgap energy and considerable free-exciton binding energy. Research has reported photocatalytic activity of ZnO alone and in many other forms and composites, especially with CNTs, which include ZnO/CNT composites, ZnO reduced graphene oxide−CNT composites, Cd-doped ZnO/CNT, Ag−ZnO/CNT nanophotocatalysts, and RGO/ZnO/MoS_2 ternary nanocomposites. ZnO/CNT nanocomposites reportedly have significant electronic and conductive properties, high chemical stability, and unique mesoporous surface that pollutants may be physically adsorbed on their surface.[59−61] The substantial electron-storage capacity of CNTs is reported to improve at the interface of ZnO/CNT nanocomposites.

Zhou et al. for hierarchical ZnO/CNT microsphere composites reported 45% improved degradation for MB under solar irradiation compared with pure ZnO microspheres. Similarly, Cd-doped ZnO/CNT nanocomposites were used for MO dye decomposition with a degradation rate of 93%, attributed to the porous nature, large specific surface area.

TiO_2@ZnO/CNT heteronanostructures applied to photodegradation of organic dye exhibit higher efficiency and better enhancement in UV light decomposition. Czecha et al. explained its improvement and stated that negative charge formation due to excitation migration to the nanotubes improves generation of hydroxyl radicals and free active sites. The e⁻/h⁺ recombination is reduced in the ZnO−TiO_2 structure by the CNTs. Table 16.1 shows some of the MO/CNT composites photocatalysts and their photodegradation efficiencies.

16.15 Conclusion

The role and significance of metal oxide—CNT composites in photodegradation to remove contagious environmental pollutants have been highlighted. Photodegradation is an environmentally friendly technique that uses solar energy and photoactive semiconducting materials to degrade pollutants, microbials, and toxic organic compounds

Table 16.1 Metal oxide—carbon nanotube composite photocatalysts and their photodegradation efficiencies.

Photocatalyst	Synthesized method	Type of pollutants	Pollutant concentration	Degradation efficiency	Ref.
g-C_3N_4/TiO_2/CNT	Facile hydrothermal method	MB6	6 mg/L	Exhibits 5x greater photocatalytic activity related to g-C_3N_4	[62]
Ni, C, N, S multidoped ZrO_2/MWCNT nanocomposites	Ultrasound-assisted hydrothermal method	Indigo carmine anionic dye	3 mg/L	96% compared with Ni, O, C, N, S (48%), and bare ZrO_2 (19%)	[63]
T-ZnO/CNOs hybrid composite	Facile one-step process	2,4-Dinitrophenol DNP	100 mg/50 mL	92%.	[64]
(F-MWCNT)/Co—Ti oxide nanocomposites	Chemical reduction method	Rhodamine B in aqueous medium		93.35% dye while the Co—TiO NPs degraded to about 91.76% dye under the same conditions	[65]

Abbreviations: *F-MWCNT*, functionalized multiwalled carbon nanotube; *MWCNT*, multiwalled carbon nanotube.

from water and air using the various AOPs discussed earlier. The working mechanisms of photo-oxidation, photo-Fenton, photo-ozonation, and photocatalysis have been briefly discussed. An overview of the various synthesis and implementation techniques of MO–CNT hybrid materials for photodegradation are provided. The factors and morphological effects of MO–CNTs on photodegradation have been given in detail. Many MO–CNTs have been shared over time to photodegrade various pollutants.

The significance of MO–CNT composites in PO is evident in the prevention of the spread of secondary pollutants in the ecosystem and makes the nanomaterial PO is an effective technology for the abolition of synthetic alkanes. The effect of concentration of photocatalyst and contaminants, surface area and morphology of a photocatalyst, temperature, and pH of the solution effect on photocatalytic degradation has been discussed.

Reports on the photocatalytic activity of ZnO, TiO_2, and other semiconducting materials and their CNT composites, including ZnO/CNT composites, TiO_2/CNT composites, ZnO reduced graphene oxide–CNT composites, Ni, C, N, and S multidoped ZrO_2/multiwalled CNT nanocomposites, Cd-doped ZnO/CNT, g-C_3N_4/TiO_2/CNT Ag–ZnO/CNT nanophotocatalysts, functionalized, multiwalled CNT/Co–Ti oxide nanocomposites, and RGO/ZnO/MoS_2 ternary nanocomposites, have proven the performance of these MO–CNT composites for the complete degradation of dyes like methyl blue, methyl orange, microbes, and other pollutants. Carbonaceous materials can act as adsorbents and band tighteners upon doping to semiconductors, thus improving photocatalyst absorbance capacity and narrowing the energy gap of the semiconductor to absorb more light rays within the visible range, which is essential for the efficient photodegradation of pollutants.

References

1. Gupta V, Khamparia S, Tyagi I, Jaspal D, Malviya A. Decolorization of mixture of dyes: a critical review. *Global J Environ Sci Manage* 2015;**1**.
2. Pimentel M, Oturan N, Dezotti M, Oturan MA. Phenol degradation by advanced electrochemical oxidation process electro-Fenton using a carbon felt cathode. *Appl Catal B Environ* 2008;**83**:140–9.
3. Putra WP, Kamari A, Yusoff SNM, Ishak CF, Mohamed A, Hashim N, et al. Biosorption of Cu (II), Pb (II) and Zn (II) ions from aqueous solutions using selected waste materials: adsorption and characterisation studies. *J Encapsulation Adsorpt Sci* 2014;**2014**.
4. Gopal M, Pakshirajan K, Swaminathan T. Heavy metal removal by biosorption using Phanerochaete chrysosporium. *Appl Biochem Biotechnol* 2002;**102**:227–37.
5. Halmann MM. *Photodegradation of water pollutants*. CRC Press; 1995.
6. Bora T, Dutta J. Applications of nanotechnology in wastewater treatment—a review. *J Nanosci Nanotechnol* 2014;**14**:613–26.
7. Gontarz JA, Nelson CH. Google Patents. In: *Photodegradable polyolefins containing aryl-substituted 1, 3-diones*; 1976.
8. Yousif E, Haddad R. *Photodegradation and photostabilization of polymers, especially polystyrene*, vol. 2. SpringerPlus; 2013. p. 398.

9. Andreozzi R, Caprio V, Insola A, Marotta R. Advanced oxidation processes (AOP) for water purification and recovery. *Catal Today* 1999;**53**:51−9.

10. Ruppert G, Bauer R, Heisler G. UV-O3, UV-H_2O_2, UV-TiO_2 and the photo-Fenton reaction-comparison of advanced oxidation processes for wastewater treatment. *Chemosphere* 1994;**28**:1447−54.

11. Rosenfeldt EJ, Linden KG, Canonica S, Von Gunten U. Comparison of the efficiency of OH radical formation during ozonation and the advanced oxidation processes O_3/H_2O_2 and UV/H_2O_2. *Water Res* 2006;**40**:3695−704.

12. Elovitz MS, von Gunten U. *Hydroxyl radical/ozone ratios during ozonation processes. I. The Rct concept.* 1999.

13. Qu J, Li H, Liu H, He H. Ozonation of alachlor catalyzed by Cu/Al_2O_3 in water. *Catal Today* 2004;**90**:291−6.

14. Xiao J, Xie Y, Cao H, Wang Y, Zhao Z. g-C_3N_4-triggered super synergy between photocatalysis and ozonation attributed to promoted OH generation. *Catal Commun* 2015;**66**:10−4.

15. Sakthivel S, Neppolian B, Shankar M, Arabindoo B, Palanichamy M, Murugesan V. Solar photocatalytic degradation of azo dye: comparison of photocatalytic efficiency of ZnO and TiO_2. *Sol Energy Mater Sol Cell* 2003;**77**:65−82.

16. Augugliaro V, Litter M, Palmisano L, Soria J. The combination of heterogeneous photocatalysis with chemical and physical operations: a tool for improving the photoprocess performance. *J Photochem Photobiol C Photochem Rev* 2006;**7**:127−44.

17. Gaya UI, Abdullah AH. Heterogeneous photocatalytic degradation of organic contaminants over titanium dioxide: a review of fundamentals, progress and problems. *J Photochem Photobiol C Photochem Rev* 2008;**9**:1−12.

18. Huang X, Shi W, Yuan J, Shi J, Jiang Z, Shangguan W. Synergetic catalytic performance of TiO_2/MCM-41 for ozone-assisted photocatalytic degradation of gaseous acetaldehyde. *Environ Technol* 2011;**32**:307−16.

19. Rajeswari R, Kanmani S. A study on synergistic effect of photocatalytic ozonation for carbaryl degradation. *Desalination* 2009;**242**:277−85.

20. Tanaka K, Abe K, Hisanaga T. Photocatalytic water treatment on immobilized TiO_2 combined with ozonation. *J Photochem Photobiol Chem* 1996;**101**:85−7.

21. Valério A, Wang J, Tong S, de Souza AAU, Hotza D, González SYG. Synergetic effect of photocatalysis and ozonation for enhanced tetracycline degradation using highly macroporous photocatalytic supports. *Chem Eng Process Process Intensif* 2020;**149**:107838.

22. Xiao J, Xie Y, Nawaz F, Jin S, Duan F, Li M, et al. Super synergy between photocatalysis and ozonation using bulk g-C_3N_4 as catalyst: a potential sunlight/O_3/g-C_3N_4 method for efficient water decontamination. *Appl Catal B Environ* 2016;**181**:420−8.

23. Farré MJ, Franch MI, Malato S, Ayllón JA, Peral J, Doménech X. Degradation of some biorecalcitrant pesticides by homogeneous and heterogeneous photocatalytic ozonation. *Chemosphere* 2005;**58**:1127−33.

24. Qian R, Zong H, Schneider J, Zhou G, Zhao T, Li Y, et al. Charge carrier trapping, recombination and transfer during TiO_2 photocatalysis: an overview. *Catal Today* 2019;**335**:78−90.

25. Ikhlaq A, Brown DR, Kasprzyk-Hordern B. Catalytic ozonation for the removal of organic contaminants in water on alumina. *Appl Catal B Environ* 2015;**165**:408−18.

26. Chen M, Zhu L, Liu S, Li R, Wang N, Tang H. Efficient degradation of organic pollutants by low-level Co^{2+} catalyzed homogeneous activation of peroxymonosulfate. *J Hazard Mater* 2019;**371**:456−62.

27. Graedel T, Mandich M, Weschler C. Kinetic model studies of atmospheric droplet chemistry: 2. Homogeneous transition metal chemistry in raindrops. *J Geophys Res: Atmosphere* 1986;**91**:5205—21.

28. Rivas J, Rodríguez E, Beltrán FJ, García-Araya JF, Alvarez P. Homogeneous catalyzed ozonation of simazine. Effect of Mn (II) and Fe (II). *J Environ Sci Health B* 2001;**36**: 317—30.

29. Lang X, Chen X, Zhao J. Heterogeneous visible light photocatalysis for selective organic transformations. *Chem Soc Rev* 2014;**43**:473—86.

30. Védrine JC. Heterogeneous catalysis on metal oxides. *Catalysts* 2017;**7**:341.

31. Védrine JC. Metal oxides in heterogeneous oxidation catalysis: state of the art and challenges for a more sustainable world. *ChemSusChem* 2019;**12**:577—88.

32. Faria P, Monteiro D, Órfão J, Pereira M. Cerium, manganese and cobalt oxides as catalysts for the ozonation of selected organic compounds. *Chemosphere* 2009;**74**:818—24.

33. Liang Y, Wang H, Diao P, Chang W, Hong G, Li Y, et al. Oxygen reduction electrocatalyst based on strongly coupled cobalt oxide nanocrystals and carbon nanotubes. *J Am Chem Soc* 2012;**134**:15849—57.

34. You F-T, Yu G-W, Wang Y, Xing Z-J, Liu X-J, Li J. Study of nitric oxide catalytic oxidation on manganese oxides-loaded activated carbon at low temperature. *Appl Surf Sci* 2017;**413**:387—97.

35. Evans J. The vibrational assignments and configuration of aniline, aniline-NHD and aniline-ND2. *Spectrochim Acta* 1960;**16**:428—42.

36. Sires I, Garrido JA, Rodriguez RM, Brillas E, Oturan N, Oturan MA. Catalytic behavior of the Fe^{3+}/Fe^{2+} system in the electro-Fenton degradation of the antimicrobial chlorophene. *Appl Catal B Environ* 2007;**72**:382—94.

37. Han T, Qu L, Luo Z, Wu X, Zhang D. Enhancement of hydroxyl radical generation of a solid state photo-Fenton reagent based on magnetite/carboxylate-rich carbon composites by embedding carbon nanotubes as electron transfer channels. *New J Chem* 2014;**38**:942—8.

38. Wu K, Xie Y, Zhao J, Hidaka H. Photo-Fenton degradation of a dye under visible light irradiation. *J Mol Catal Chem* 1999;**144**:77—84.

39. Liu Y, Fan Q, Wang J. Zn-Fe-CNTs catalytic in situ generation of H_2O_2 for Fenton-like degradation of sulfamethoxazole. *J Hazard Mater* 2018;**342**:166—76.

40. Almkhelfe H, Li X, Thapa P, Hohn KL, Amama PB. Carbon nanotube-supported catalysts prepared by a modified photo-Fenton process for Fischer—Tropsch synthesis. *J Catal* 2018; **361**:278—89.

41. Kurapati R, Backes C, Ménard-Moyon C, Coleman JN, Bianco A. White graphene undergoes peroxidase degradation. *Angew Chem* 2016;**128**:5596—601.

42. García JC, Pedroza AM, Daza CE. Magnetic Fenton and photo-Fenton-like catalysts supported on carbon nanotubes for wastewater treatment. *Water, Air, Soil Pollut* 2017;**228**:246.

43. Liu X, Zhou Y, Zhang J, Luo L, Yang Y, Huang H, et al. Insight into electro-Fenton and photo-Fenton for the degradation of antibiotics: mechanism study and research gaps. *Chem Eng J* 2018;**347**:379—97.

44. Woan K, Pyrgiotakis G, Sigmund W. Photocatalytic carbon-nanotube—TiO_2 composites. *Adv Mater* 2009;**21**:2233—9.

45. Ahmad M, Ahmed E, Hong Z, Ahmed W, Elhissi A, Khalid N. Photocatalytic, sonocatalytic and sonophotocatalytic degradation of Rhodamine B using ZnO/CNTs composites photocatalysts. *Ultrason Sonochem* 2014;**21**:761—73.

46. Zhang W, Li G, Liu H, Chen J, Ma S, An T. Micro/nano-bubble assisted synthesis of Au/TiO_2@ CNTs composite photocatalyst for photocatalytic degradation of gaseous styrene and its enhanced catalytic mechanism. *Environ Sci* 2019;**6**:948—58.

47. Daneshvar N, Salari D, Khataee A. Photocatalytic degradation of azo dye acid red 14 in water: investigation of the effect of operational parameters. *J Photochem Photobiol Chem* 2003;**157**:111−6.

48. Mozia S, Tomaszewska M, Morawski AW. Photocatalytic degradation of azo-dye acid red 18. *Desalination* 2005;**185**:449−56.

49. Papp J, Soled S, Dwight K, Wold A. Surface acidity and photocatalytic activity of TiO_2, WO_3/TiO_2, and MoO_3/TiO_2 photocatalysts. *Chem Mater* 1994;**6**:496−500.

50. Zinatloo-Ajabshir S, Morassaei MS, Salavati-Niasari M. Facile synthesis of $Nd_2Sn_2O_7$-SnO_2 nanostructures by novel and environment-friendly approach for the photodegradation and removal of organic pollutants in water. *J Environ Manag* 2019;**233**:107−19.

51. Petala A, Noe A, Frontistis Z, Drivas C, Kennou S, Mantzavinos D, et al. Synthesis and characterization of $CoO_x/BiVO_4$ photocatalysts for the degradation of propyl paraben. *J Hazard Mater* 2019;**372**:52−60.

52. Liu M, Suzuki Y. $BiVO_4$ hollow nanoplates with improved photocatalytic water oxidation efficiency. *Curr Nanosci* 2015;**11**:499−503.

53. Khan MM, Ansari SA, Lee J-H, Ansari MO, Lee J, Cho MH. Electrochemically active biofilm assisted synthesis of Ag@ CeO_2 nanocomposites for antimicrobial activity, photocatalysis and photoelectrodes. *J Colloid Interface Sci* 2014;**431**:255−63.

54. Wang T, Li C, Ji J, Wei Y, Zhang P, Wang S, et al. Reduced graphene oxide (rGO)/$BiVO_4$ composites with maximized interfacial coupling for visible lght photocatalysis. *ACS Sustainable Chem Eng* 2014;**2**:2253−8.

55. Zhu W, Sun F, Goei R, Zhou Y. Facile fabrication of RGO-WO_3 composites for effective visible light photocatalytic degradation of sulfamethoxazole. *Appl Catal B Environ* 2017;**207**:93−102.

56. Seema H, Kemp KC, Chandra V, Kim KS. Graphene−SnO_2 composites for highly efficient photocatalytic degradation of methylene blue under sunlight. *Nanotechnology* 2012;**23**:355705.

57. Luo Q-P, Yu X-Y, Lei B-X, Chen H-Y, Kuang D-B, Su C-Y. Reduced graphene oxide-hierarchical ZnO hollow sphere composites with enhanced photocurrent and photocatalytic activity. *J Phys Chem C* 2012;**116**:8111−7.

58. Yu H, Xiao P, Tian J, Wang F, Yu J. Phenylamine-functionalized rGO/TiO_2 photocatalysts: spatially separated adsorption sites and tunable photocatalytic selectivity. *ACS Appl Mater Interfaces* 2016;**8**:29470−7.

59. Moradi M, Haghighi M, Allahyari S. Precipitation dispersion of Ag−ZnO nanocatalyst over functionalized multiwall carbon nanotube used in degradation of acid orange from wastewater. *Process Saf Environ Protect* 2017;**107**:414−27.

60. Ghasemipour P, Fattahi M, Rasekh B, Yazdian F. Developing the ternary ZnO doped MoS 2 nanostructures grafted on CNT and reduced graphene oxide (RGO) for photocatalytic degradation of aniline. *Sci Rep* 2020;**10**:1−16.

61. Azqhandi MHA, Rajabi F, Keramati M. Synthesis of Cd doped ZnO/CNT nanocomposite by using microwave method: photocatalytic behavior, adsorption and kinetic study. *Result Phys* 2017;**7**:1106−14.

62. Chaudhary D, V. D. V, Neeraj K. Noble metal-free g-C_3N_4/TiO_2/CNT ternary nanocomposite with enhanced photocatalytic performance under visible-light irradiation via multi-step charge transfer process. *Solar Energy* 2017;**158**:132−9.

63. Shinde SG, Maheshkumar PP, Kim G-D. Ni, C, N, S multi-doped ZrO_2 decorated on multi-walled carbon nanotubes for effective solar-induced degradation of anionic dye. *J Environ Chemi Eng* 2020;**8**(3). https://doi.org/10.1016/j.jece.2020.103769.

64. Park SJ, Das GS, Schütt F, Adelung R, Mishra YK, Tripathi KM, et al. Visible-light photocatalysis by carbon-nano-onion-functionalized ZnO tetrapods: degradation of 2, 4-dinitrophenol and a plant-model-based ecological assessment. *NPG Asia Mater* 2019; **11**(1):1—13.
65. Zada N, Saeed K, Khan I. Decolorization of Rhodamine B dye by using multiwalled carbon nanotubes/Co—Ti oxides nanocomposite and Co—Ti oxides as photocatalysts. *Appl Water Sci* 2020;**10**(1):1—10.

Metal oxide-carbon—based sensors

Potential carbon nanotube–metal oxide hybrid nanostructures for gas-sensing applications

17

Nuzhat Jamil[1], Farwah Jameel[1], Sadia Zafar Bajwa[1], Asma Rehman[1], Rao F. Hussain Khan[2], Arshad Mahmood[3] and Waheed S. Khan[1]

[1]Nanobiotechnology Group, Industrial Biotechnology Division, National Institute for Biotechnology and Genetic Engineering (NIBGE), Faisalabad, Punjab, Pakistan; [2]Department of Radiation Oncology, Washington University School of Medicine, Missouri, WA, United States; [3]National Institute of Lasers and Optronics, Nilore, Islamabad, Pakistan

17.1   Introduction

Nanoscale systems based on carbon nanotubes (CNTs) have been explored extensively for their biological and medical applications for the past few decades. High sensitivity and selectivity in nanoscaled biosensors have been achieved by applying advanced nanofabrication protocols such as surface-controlled assembly.[1] In recent times, CNTs have been a major subject of interest for gas sensor manufacturers using nanotechnology. The increased surface area of such nanostructures is the main characteristic responsible for large-scale interaction with gas molecules. Various research articles demonstrate that exposing such CNT-based sensors to toxic gases like NH_3, NO_2, CO, ethanol, and benzene vapors leads to changes in electrical conductance of semiconducting CNTs of such instruments, hence resulting in reduced power consumption by such nanosensors.[2]

However, forming a uniformly dispersed smooth semiconducting layer may accompany the problem of the clustering up of CNTs. In recent studies, this problem is perfectly compensated by applying direct current plasma for functionalization of CNTs that helps avoid agglomeration of CNTs. Such plasma-based functionalization not only enhances the surface reactivity of CNTs but also increases their gas detection capability. Therefore, gas sensors fabricated with functionalized multiwalled CNTs (MWCNTs) have exhibited efficient results at ambient temperatures—for instance, depicting efficient NO_2 response even at low concentrations. However, in the case of toxic gases, the detection efficiency of functionalized or untreated MWCNT-based sensors cannot be improved even by applying high temperature values. This limitation is controlled by doping CNTs with nanoclusters of heavy metals, as one research report demonstrated the improved sensitivity of Au- and Ag-doped MWCNTs over untreated MWCNTs in detecting NO_2. An oxygen-based plasma functionalization technique presented a more efficient method

Metal Oxide-Carbon Hybrid Materials. https://doi.org/10.1016/B978-0-12-822694-0.00012-0

for homogenous application of metallic nanoclusters over the surface of CNTs compared with untreated doping, which mostly led to cluster formation over the CNT surfaces. Moreover, metal oxides have shown high sensitivity for detecting a wide range of gases. In particular, tin oxides (SnO_2) and tungsten oxides (WO_3) exhibit high detection efficiency against toxic gases, especially when worked at high temperatures in a range of $200-500°C$. In addition, it was revealed that the efficiency of metal oxide-based nanosensors for gas detection can be increased by adding metallic nanoclusters or using dopants in bulk amounts over the surface of sensing material.

Metal oxide-based sensors work by sensing changes in the morphology of the gas-sensitive film of the nanosensor. The number and mobility of oxygen vacancies are enhanced at elevated temperatures, and conduction occurs through both mechanisms, i.e., ionic and electronic. The most commonly proposed conduction mechanism in metal oxide nanosensors involves the long-term diffusion of oxygen vacancies. Such structural changes may lead to permanent changes in the surface morphology of these sensors; however, these can be avoided by operating the sensors at temperatures low enough to regenerate the induced variations in surface films, but the gas-sensing reaction must occur at a considerable rate.[3]

CNTs with a nanoscale structure and increased surface area can be employed to functionalize a metal oxide structure by exposing many sites for gas reaction. The sensitivity of hybrid sensors is much increased because the enhanced electrical conductivity of the CNTs reduces the resistance of the sensing material. Hybrid gas sensors are more sensitive because metal oxide films have n-type conducting behavior and MWCNTs with p-type behavior that produce diverse depletion regions. Adsorbed gas molecules change the thickness of these depletion regions by creating nanochannels and making a heterojunction that causes this hybrid sensor to be more sensitive.

Modern research introduced the novel gas-sensing materials with enhanced sensing potential designed by decorating the surface with conjugated hybrid films of CNTs and SnO_2 or WO_3. This research revealed that the sensing potential of these gas sensors could be increased by uniformly dispersing an appropriate quantity of CNTs into the metal oxide matrix over the surface of such gas nanosensors.

17.2 Carbon-based nanomaterials

17.2.1 Carbon nanotubes

A CNT made from graphite sheets is a one-dimensional hollow, tubular, and cylinder-shaped nanomaterial. It shows excellent thermal, mechanical, chemical, and electrical properties and represents an essential nanosubstance group that has gained impressive considerations since its first isolation in 1991.[2]

CNTs are also called buckytubes, as they are obtained from fullerenes. Whereas a CNT has no bend produced in the axial direction of the nanotube, fullerenes are curved in all directions in space. In CNTs, sp^2 hybridization is present among carbon atoms. Nanotubes possess unique strength. As a result of sp^2 hybridization, 2-D sheets are formed. These sheets can roll and cause the formation of tubular structures. MWCNTs have π-π

interaction that is an essential character of graphite. These structures are well suited for electrochemistry, optics, nanotechnology, electronics, and other fields owing to their amazing superconductive properties like high thermal conductivity and extreme strength (100 times stronger than steel).[4]

Graphene sheets are made of carbon atoms resembling benzene rings in which atoms are curled, forming a cylindrical shape. MWCNTs form cylinders in which layers of graphite are present (diameter 2−100 nm). The diameter and chirality greatly affect the CNTs. Due to diameter and chirality, single-walled CNTs (SWCNTs) show semiconductor or metallic electronic properties. CNTs exhibit metallic electronic properties. Three methods are used to synthesize CNTs: discharge, catalyst-assisted vapor deposition, and laser ablation.[5]

17.2.2 Graphene

Graphene is a planar, layered structure with sp^2 hybridization among carbon atoms. It is called a supermaterial because it is a basic element of various carbon structures with a variety of functionalities. Graphene is always the basic building block of CNTs, whether it is engrossed into zero-dimensional fullerenes or rolling into one-dimensional CNTs or three-dimensional vertically fixed graphite. The charge carriers in the graphene are tuned between electrons and holes, and their concentration ranges from $1013 cm^2$ with mobilities above 15,000 cm^2/vs in ambient conditions.[6]

That is why graphene shows electrical sensitivity to electron releasing species and electron-withdrawing species. There is the formation of a covalent bond of one carbon atom with three closely placed carbon atoms that form other chemical bonds. Graphene becomes conductive because of free electrons. Graphene and its derivatives form single-, double-, and multilayered structures like graphene oxide and reduced graphene oxide (rGO).[7]

Van der Waals forces of attraction are present between layers of graphene with 0.35 A° and interlayer spacing. Two techniques are used to synthesize graphene: top-down and bottom-up. In the top-down method, physical and chemical exfoliation processes and electrochemical techniques are used. In the bottom-up technique, chemical vapor deposition is involved. Graphene possesses excellent strength, great elasticity, and high thermal conductivity. It also possesses mechanical properties for which it is used in electronics, medicine, thermal therapy, and optics. Graphene is used in various sensor arrays due to amazing properties such as a high specific area and easy dispersion. The synthesis of graphene oxide occurs by the oxidation of exfoliated graphite, which can be achieved with many oxidizing agents—for example, sulfuric acid, phosphoric acid, or nitric acid. Due to its higher oxygen content, graphene oxide exhibits interlayer distance and hydrophobic behavior. In addition, many functional moieties are present in graphene, like hydroxyl, epoxide, phenol, and carboxylic groups, that provide pathways for DNA/aptamers, antibodies, and proteins with biosensor implications.[8,9]

Graphene oxide reduces to rGO through various techniques: using the chemical agent hydrazine, applying reduction potential, or using elevated temperature. Reduction is used to remove structural and chemical defects in graphene oxide and hence restores the semiconductor properties for which it is used in electronics, specifically field effect transistor-based biosensing.[8]

17.2.3 Diamond

Diamond is a solid, pure form of carbon and the stiffest material in the world. It is highly stable electrically with a low current and is biocompatible. It is considered a good electrochemical transducer. The potential for oxygen and the release of hydrogen are responsible for its wide working range. For high quality of nanodiamond films on electrode 3.25 V potential window is normally used. Diamond nanostructure surfaces are important for determining the biocompatibility and utility of nanostructures in medical and biological applications like selective biosensing, targeted drug delivery, and effective therapy. Surface termination of electrolytes with oxygen, hydrogen, and OH groups can manage their electronic properties. The diamond surface is hydrophobic for the hydrogen termination, and it is hydrophilic for the oxygen termination. Diamond shows its strongest bond stability with DNA. Diamond can be made through high-temperature, high-pressure, and low-pressure methods.[10]

By irradiating energy particles like electrons, neutrons, protons, ions, and gamma particles, a vacancy defect is produced in lattice. Up to 800°C, the diamond lattice experiences strain due to the presence of nitrogen coupled with vacancies that form a nitrogen-vacancy color strain center. In the chemical vapor deposition method, diamond films contain a mixture of gas with 0.7% CH_4, 99.2% H_2, and 0%−0.1% N that forms the nitrogen-vacancy centers.

Nanodiamonds are small and have superlative properties. They are called cogent materials because they possess stable and low current, wide working potential range, high mechanical and thermal properties, and good corrosion resistance. Every particle in the nanodiamond possesses sp^3-bonded carbon atoms and is covered by sp^2 hybridization. They are small particles with a diameter of 4−6 nm and a 300 m^2/g surface area. Compared with other carbon materials like SWCNTs, double-walled carbon nanotubes, and carbon black, nanodiamonds possess high compatibility.[11]

Diamond nanowires are 4.58 nm^2 cross-sectional at room temperature (300 K). They have yield strength of 688 GPa young's modulus and 91 GPa fracture strength. When nanodiamond and bulk diamond are compared, they can be taken as single photon source possessing great flux and low power. Diamond nanowires are increasingly used in microelectromechanical systems, electrochemical gene-sensing platforms, and quantum information-processing applications.[12] Different types of carbon-based nanomaterials are shown in Fig. 17.1.

17.3 Types of carbon nanotubes

CNTs are actually graphene sheets covalently bonded with carbon molecules rolled to form hollow cylinders. CNTs are classified into two main types, single-walled and multiwalled. SWCNTs have one carbon layer with a diameter of 1−5 nm. MWCNTs possess many layers of carbon atoms concentrically placed together. Structures of different types of CNTs are shown below in Fig. 17.2. MWCNTs have an outer diameter of 5.5 nm and an inner diameter of 2.3 nm. Many methods which have led to the increased production of CNTs like laser ablation, arc discharge and chemical vapor

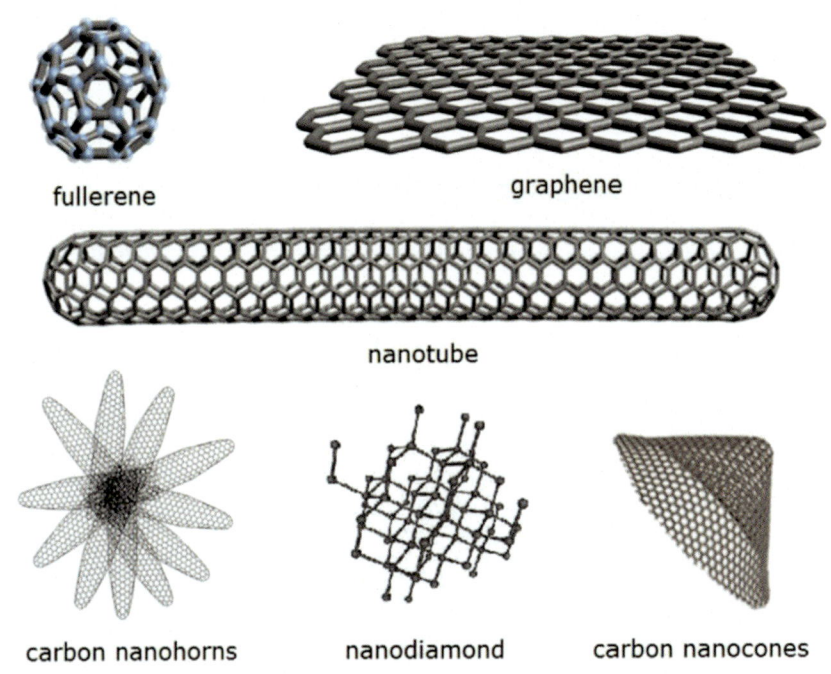

fullerene graphene

nanotube

carbon nanohorns nanodiamond carbon nanocones

Figure 17.1 The schematics of the representative carbon-based nanomaterials.
Reprinted from Ref. 12, copyright (2019), with permission from Elsevier.

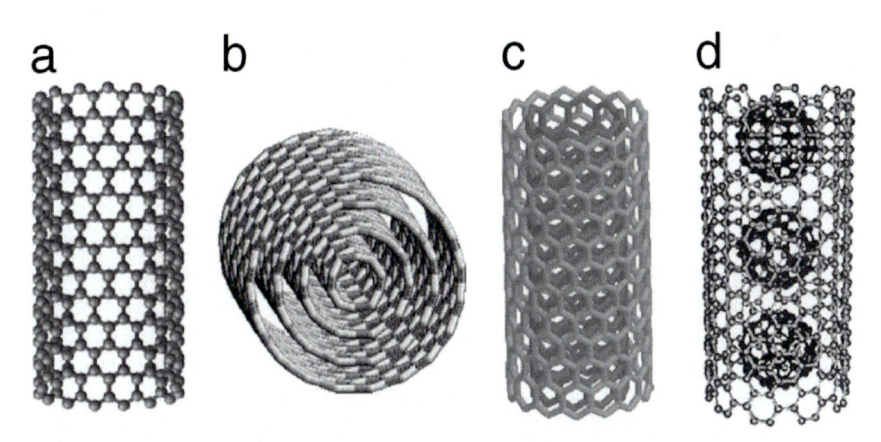

Figure 17.2 Schematic diagram of (A) a single-walled carbon nanotube (SWCNT),
(B) a multiwalled carbon nanotube, (C) a double-walled carbon nanotube, and (D) a peapod
nanotube consisting of an SWCNT filled with fullerenes.
Reprinted from Ref. 13, copyright (2003), with permission from Elsevier.

deposition are available. CNTs demonstrate unique thermal, electrical, mechanical, and optical properties. CNTs have potential applications in science and technology, sensors, displays, nanoelectronics, batteries, hydrogen storage, etc.

17.4   Metal oxide nanostructures

Gas sensors play a key role in monitoring the environment, safeguarding personal health, and controlling chemical processes because of their ability to detect toxic and volatile organic compounds. Gas sensors have promising characteristics for detecting toxic gases and volatile organic compounds compared with traditional detection techniques. The sensing capability and limit of detection of these gas sensors can be improved using a range of metal oxide nanostructures because of the increased surface-to-volume ratio and sites exposed to gases.

Metal oxide gas sensors are widely used in gas detection systems because of their easy production, low cost, and compact size. The working of these sensors is influenced by the structure of the sensing materials, and because of this, which can pose a significant obstacle to obtaining highly sensitive properties. Gas sensors based on nanomaterials are a good choice for making improvements in gas-sensing properties like selectivity, sensitivity, and response speed.[3]

17.5   Carbon nanotube—metal oxide hybrid structures and their features

There is increased demand for highly sensitive, cheap, and selective gas sensors for homeland security, safety, environmental monitoring, and quality control. The development of gas nanosensors from CNTs has led to additional research interest over the last few years because of these sensors' rapid detection of many gaseous species through novel nanostructures present in electronics that consume low power. The hole density of nanotubes changes with the interaction of electron withdrawing molecules (e.g., NO_2, O_2) and p-type semiconducting CNTs or the interaction of electron-donating molecules (e.g., NH_3) with p-type semiconducting CNTs. CNT conductance changes with changes in density. It forms the basis for applying CNTs in chemical gas sensors.[14]

The use of pristine CNT for sensors is not usually promising. MWCNTs have a high surface area to expose a great number of sites at which gas reacts and a nanoscale structure, so they are used for the functionalization of metal oxide gas-sensitive material. By preheating nanosensors at different temperatures and applying adaptive signal processing algorithms, various gases can be detected, thus providing possibilities to perform selective sensing. Compared with metal oxides, CNTs possess much higher electrical conductivity, which enhances the electron transfer, causing a speedier redox reaction and sensitivity for analyte. Metal oxide nanoparticles mostly control sensing properties. There are two depletion regions in metal oxide and CNTs. Metal oxide possesses mostly n-type conductivity, and MWCNTs have p-type conductivity. P-type semiconductors hold charges because of their large hole concentrations compared with n-types, for which electrons are the majority carriers. The first depletion region is present on the surface of metal oxide, and the second depletion region is located between nanoparticles and MWCNTs. Heterojunction formation and the presence of nanochannels enhance the sensitivity of hybridized gas sensors. Any change in conductivity or barrier height changes the depletion layer at the p—n junction.

When MWCNTs or SnO_2 composites are exposed to NH_3, its molecules react with the MWCNTs and replace the preadsorbed oxygen. As a result, NH_3 oxidation occurs, and oxygen is removed. In this way, the conductivity of the composite material is changed by formation of the heterojunction barrier by MWCNTs and SnO_2.[15] The adsorption of simple and polluted air modulate the surface charge which affect the electron transfer in between junction and change the resistance of sensing layer. For example, in the case of toxic gas ammonia, adsorption on the hybrid layer causes enhancement in negative charge by decreasing the number of positive holes in the p-type material and thus increases resistance. On the other hand, oxygen—gas adsorption on the hybrid layer causes a decrease in negative charge and thus decreases resistance. Gas sensors work on the principle of changes in conductance caused by oxidizing or reducing the sensed gas. To increase their selectivity, doping is done at optimum temperature to make them selective for specific gases.

As mentioned, MWCNT membranes prepared at the Szeged and Yerevan state universities were used to manufacture nanocomposite SnO_2 and MWCNT thin-film gas sensors by the sol—gel method. By modifying CNT or metal oxide surfaces, gas sensors using noble gases (Pt, Pd, Au, Ru, Rh) achieve improved sensitivity and selectivity. The sensitivity of MWCNT, SnO_2, and Pd sensors to gas is determined by the ratio of sensor resistance in the air to the sensor resistance in the presence of the pollutant after they have reached the steady state. In detecting reducing gases, semiconductor resistance is reduced by increasing the concentration of electrons in the n-type conduction band. Ru is a sensitizer for SnO_2 and increases the sensor response and selectivity to hydrocarbons. Ruthenium increases the oxidation rate and other surface reactions involving oxygen adsorption from air on the surface structure, which increases the depletion layer of the semiconductor near the surface region and causes increased sensor response. Isobutane vapors are dangerous for human health and are also used as fuel. We did not use sensors made of pure SnO_2 and CNTs to detect isobutane. For sensors made of the nanocomposite MWCNT/SnO_2/Pd sensitized with a 0.01 M (a) and 0.03 M (b) Ru $(OH)Cl_3$ solution, the sensor response for 5000 ppm isobutane was registered at $100-350°C$. Resistance of sensor was decreased 10 times in ~ 10 s after injection of isobutane and in ~ 30 s after injection of hydrogen. The response for MWCNT/SnO2/Pd sensors depends linearly on the isobutane concentration.[16]

17.6 Gas sensors and their uses

Chemical sensors and gas sensors are widely used in many fields like biomedicine, pharmaceutics, space exploration, and environmental monitoring. Gas sensors detect toxic and harmful gases, and this is done by sensors with high selectivity and sensitivity. Environmental monitoring is an important area of research in today's world of increasing pollution and global warming. Gas sensors detect various gases present in the atmosphere and are used as important instruments in experiments that involve imitation of atmospheric conditions of various planets. NASA scientists use gassed-up sensors to formulate constituents of the atmospheres of different planets. The following are the prominent parameters for an efficient gas-sensing system: (1) sensor portability (2) low detection limit (3) stable performance (4) high selectivity, response, and sensitivity (5) low operating temperature (6) the effect of temperature on fast sensor response and recovery.[17]

Ideal gas storage employs a material with a pitted structure and a high surface-to-volume ratio, so nanomaterial-based gas sensors like nanowires, nanofibers, CNTs, and nanoparticles are widely explored. The study of CNTs has increased to meet the demand for gas-detection technologies. The basic principles of gas sensors are adsorption and desorption of molecules of gas on the sensing materials. CNTs in fullerene structures are the best materials for gas sensing due to their high surface areas. CNTs have provided research pathways in many fields because of their amazing thermal, mechanical, and electrical properties. Electronic transport in CNTs takes place through quantum effects and ballistic transport along the nanotube axis without scattering.[18]

17.6.1 Zinc oxide-decorated multiwalled carbon nanotubes for NO$_2$ gas sensors

CNTs are basically one-dimensional material contrasting with other carbon allotropes including graphite, diamond, and fullerene and possessing a ratio of length to diameter of more than 1000. After the discovery of CNTs by Iijima in 1991, scientists started using them in a variety of technological advances because their enhanced surface-to-volume ratio and high chemical, mechanical, and thermal stability. MWCNTs possess tremendous potential in next-generation sensor technology. MWCNTs possess exclusive strength because of their sp^2 chemical bonding in contrast to diamond and alkane, which have sp^3 bonding. This strong bonding causes them to be less chemically reactive with certain molecules, which is why bare CNTs react with strong reducing and oxidizing gases. Bare CNTs possess weak sensing properties and are not selective, are irreversible, and require long recovery times. For example, an MWCNT-based gas sensor with a change in conductance value of SWCNTs of $5 \times 10^{-6-1}$ was reported earlier, but precise and accurate device for measuring conductance is required.

The sensitivity of CNT-based gas sensors can be improved by various adjustments, including coating and functionalization with polymers and decorating with metal and metal oxide nanostructures.[19] MWCNTs also have a tendency to agglomerate because of van der Waals forces. That is why decoration with other materials can increase the uniform scattering of MWCNTs and result in increased interactions of MWCNTs and gases.

Nitrogen dioxide is extremely harmful and toxic gas which damage ozone layer, cause acid rain and many health issues like acute respiratory illness, olfactory paralysis and many more. Occupational Safety and Health Administration has decided its threshold limit value up to 5 ppm but in current scenario, because of increased traffic on roads and number of buildings the concentration of NO_2 has crossed this limit. Therefore, gas sensors are highly required in this scenario. Gas sensors based on bare MWCNTs are not sensitive enough for NO_2, that is why, functionalizing them with other materials is best solution.[20] Hybrid gas sensors based on composite MWCNTs and metal oxides are of two types depending on how high the relative concentration of materials is. Metal oxide-decorated MWCNTs show more sensitivity to gas detection than MWCNT-doped metal oxide semiconductors. MWCNTs are combined with metal oxides by either the in situ or the ex situ method. The in situ approach involves growing MWCNTs within the original growth process, and the ex situ approach involves decorating MWCNTs with metal oxides after synthesis. The gas sensor uses an in situ approach in which MWCNTs serve as support and zinc oxide serves as a crystalline film with a controlled thickness or as separate nanostructures with increased sensitivity over bare MWCNTs. Microscopic and spectroscopic characterization results have validated the decoration of zinc oxide nanoparticles on MWCNT surfaces. A schematic of the various steps following the synthesis of zinc oxide and MWCNT nanocomposites is shown in Fig. 17.3. Gas-sensing studies have confirmed improved gas-sensing capability after the decoration of MWCNTs with zinc oxide. This research has opened a window to developing highly advanced gas sensors by decorating MWCNTs with other metal oxides.[19]

17.6.2 Ethanol-sensing features of carbon nanotube–ZnSnO₃ nanocomposite

The currently used metal oxide semiconductors include Co_3O_4, Zn_2SnO_4, SnO_2, $ZnSnO_3$ TiO_2, and $ZnFe_2O_4$. Among these, $ZnSnO_3$ has gained much attention for its multifunctionality and use in many applications, such as photocatalysts, energy harvesters, and lithium-ion batteries. $ZnSnO_3$ is thermally stable below 750°C; above this temperature, it decomposes into Zn_2SnO_4 and SnO_2. Hollow structured gas sensors with large surface areas are conducive to increased gas-sensing capability. Surfactants are used to create hollow structures, which makes synthesis somewhat complicated and adds impurities to our desired products. Metal–organic frameworks have gained attention for their porous structures and large surface areas without surfactants.[18]

Surface-functionalized CNTs are used considerably in gas sensor development because of their large surface areas and unique chemical, electrical, and mechanical properties. Gas sensors based on CNTs and SnO_2 nanoparticles have reportedly been developed via an ion-exchange method with enhanced sensing for ethanol by dropping the resistance from 17 to 4 MΩ. An ethanol sensor based on CNTs and $ZnSnO_3$, as shown in Fig. 17.4, exhibits an enhanced sensing response with maximum selectivity and sensitivity eight times greater than those of $ZnSnO_3$ nanocubes.[21]

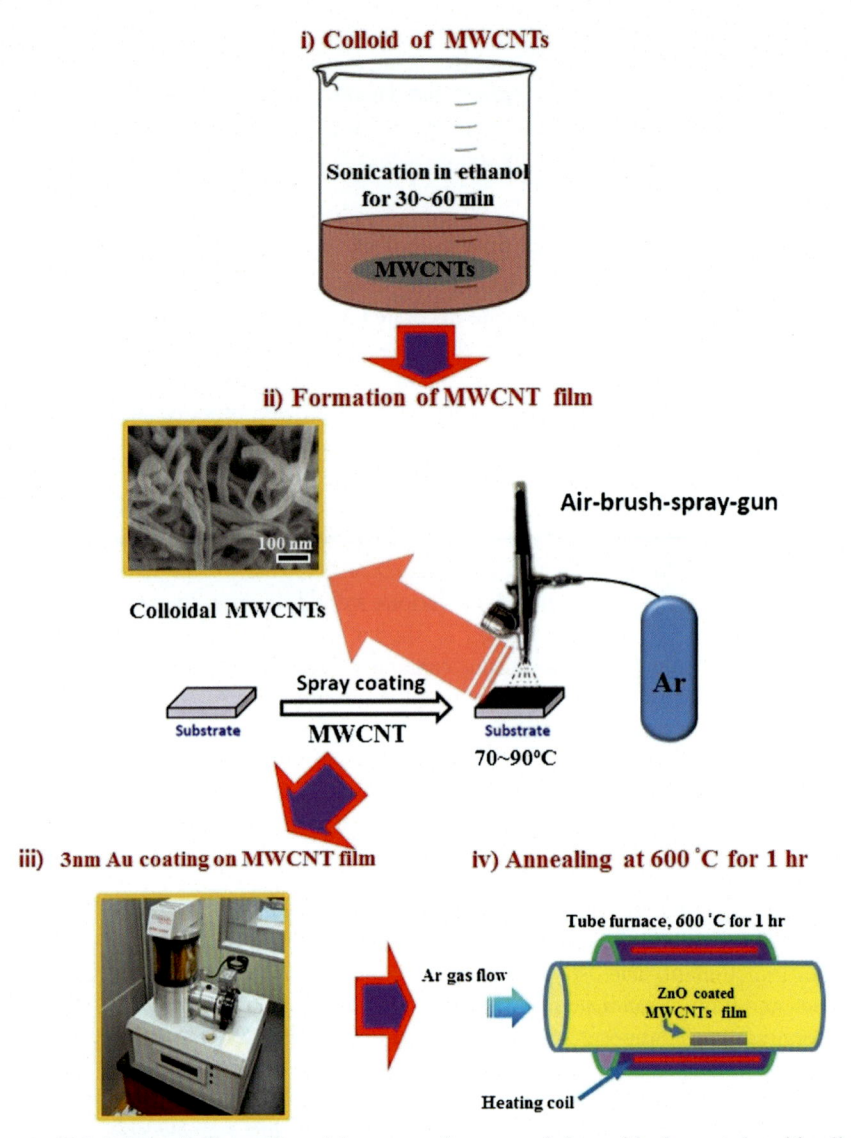

Figure 17.3 Schematic illustration of the preparation steps of zinc oxide-decorated multiwalled carbon nanotube nanocomposites.
Reprinted from Ref. 19, copyright (2017), with permission from Elsevier.

17.6.3 *Multiwalled carbon nanotube/tin oxide nanocomposite-based sensors for detection of various gases*

Modern research has shown that graphene and graphene-derived materials are quite effective for the fabrication of nanoscale gas sensors because they impart excellent properties, including enhanced surface area, high physical and chemical stability,

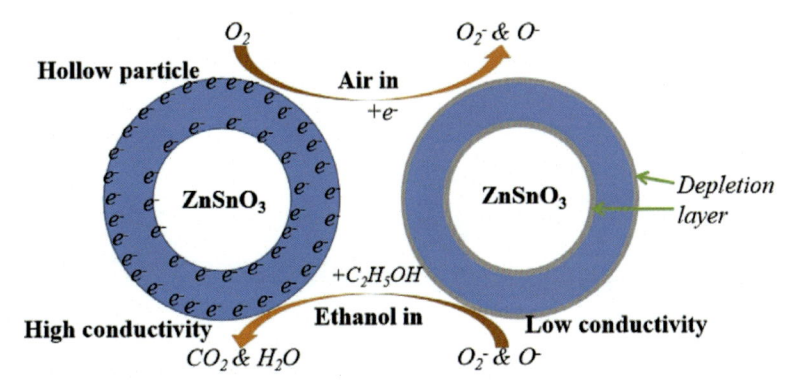

Figure 17.4 Ethanol-sensing mechanism of ZnSnO$_3$ hollow particles. Reprinted from Ref. 21, copyright (2020), with permission from Elsevier.

and measurable change in resistance upon reaction with the test gas and high mobility of carrier groups at ambient temperature. These graphene materials are generated by various physical and chemical methods, such as chemical vapor deposition and chemical or thermal reduction of graphene oxide.

Gas sensors based on rGO are highly reliable for advantages including low cost, greater control over semiconducting properties, and high productivity. However, some limitations have been reported, such as long recovery time, slow and delayed response, and low robustness. Thus, there is a requirement for high-performance gas sensors, with the performance of such sensors depending on the properties of the sensing material. High-performance rGO-based materials can be developed by controlling their semiconducting properties via surface modifications through covalent or noncovalent interactions.[6]

In modern research on nanosensors, the sensing efficiency of rGO-based gas sensors has been increased by modifying the semiconducting properties of rGO by coupling with a typical N-type semiconducting material, SnO$_2$. Graphene loaded with SnO$_2$ in variable morphologies—such as SnO$_2$ nanofibers coupled with rGO, SnO$_2$ nanorods attached to graphene sheets, and SnO$_2$ nanoparticles conjugated with rGO—can be efficiently applied for sensing acetone, H$_2$S, and NO$_2$, but sensing efficiency is maximized at a relatively high temperature.

Sensors were recently developed for working at ambient temperatures by hybridizing SnO$_2$ with rGO, but they exhibited reduced performance, such as longer recovery times, slow response, and less sensitivity. A graphical representation of the fabrication of a gas sensor with this composite is shown in Fig. 17.5. In one recent study, the sensing potential of NO$_2$ sensors was enhanced by fabricating them with SnO$_2$−rGO hybrids, but the maximum response was obtained by heating the sensor at about 50°C. Ternary hybrids (rGO−CNT−SnO$_2$) generated by incorporating CNTs into SnO$_2$−rGO hybrids were obtained by the hydrothermal procedure. The sensing potential of such ternary hybrids was seen to increase enormously, especially compared with the performance of SnO$_2$-rGO hybrids and pure rGO.[22]

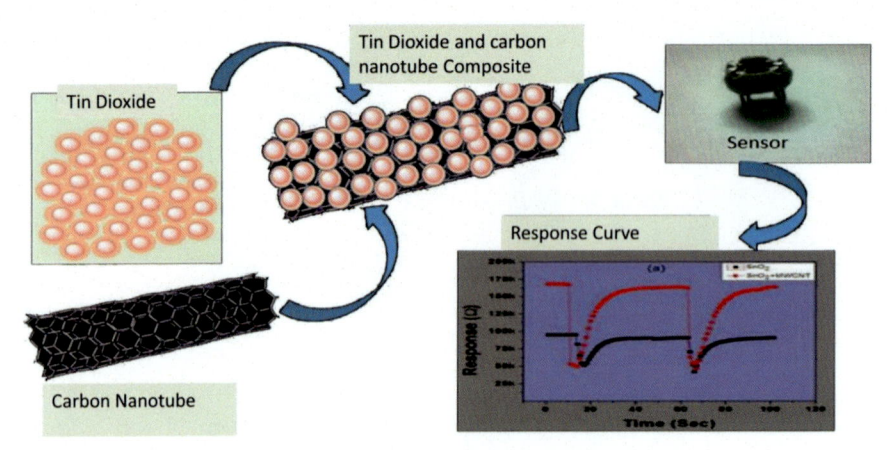

Figure 17.5 Multiwalled carbon nanotube/tin oxide nanocomposite-based sensors. Reprinted from Ref. 23, copyright (2017), with permission from Elsevier.

17.6.4 Promise of iron oxide/multiwalled carbon nanotube hybrid nanostructures for gas detection

CNTs present a wide range of gas-sensing applications, mainly due to their excellent structural properties, large surface areas, and wide diversity of nanoscale morphologies. The potential of a material to serve as a sensor depends on chemical resistance, a property related to MWCNTs rather than SWCNTs. Despite their common limitations, CNTs can be effectively used to design gas nanosensors by coupling with metals (Au, Pb, Pt) or metal oxides (SnO_2, WO_3); or by using CNT-based gas sensors at elevated temperatures. Researchers developed nanosensors for NH_3 and NO_2 gases using MWCNT nanocomposites loaded with SnO_2 nanoparticles and reported their superior sensing performance over unloaded MWCNTs. However, such nanocomposites suffer from the poor selectivity, which is overcome in modern research by using ferrites that exhibit higher selectivity and chemical inertness than metal oxide-based hybrid materials.

Modern research covers a wide range of ferrites, including spinel-type ferrites (MFe_2O_4, where M = Co, Mg, Mn, Ni, and Zn). Gas-sensing specificity can be controlled by monitoring the nature of metal ions at place M in the spinel-type ferrite. Scientists have observed that the gas-sensing potential of nickel ferrite is improved by conjugating it with Co and Mn. Moreover, higher efficiency and improved sensitivity have been observed for nanocomposites derived by coupling MWCNTs with ferrites rather than with discrete CNTs or ferrites. The schematics of iron oxide and MWCNT composite synthesis for gas sensors are shown in Fig. 17.6.

A quite low reaction temperature, in the range of 180°C, was required for nanocomposites assimilated by loading MWCNTs with Co and Ni-ferrites produced via the solvothermal process. Nanoparticles of Co and Ni-loaded ferrites were generated in different grades and directly dumped onto the surfaces of MWCNTs. Such nanocomposites were investigated for their sensitivity and selectivity and exhibited enhanced values for benzene, ethanol, formaldehyde, methanol, NH_3, and toluene in the gas phase.

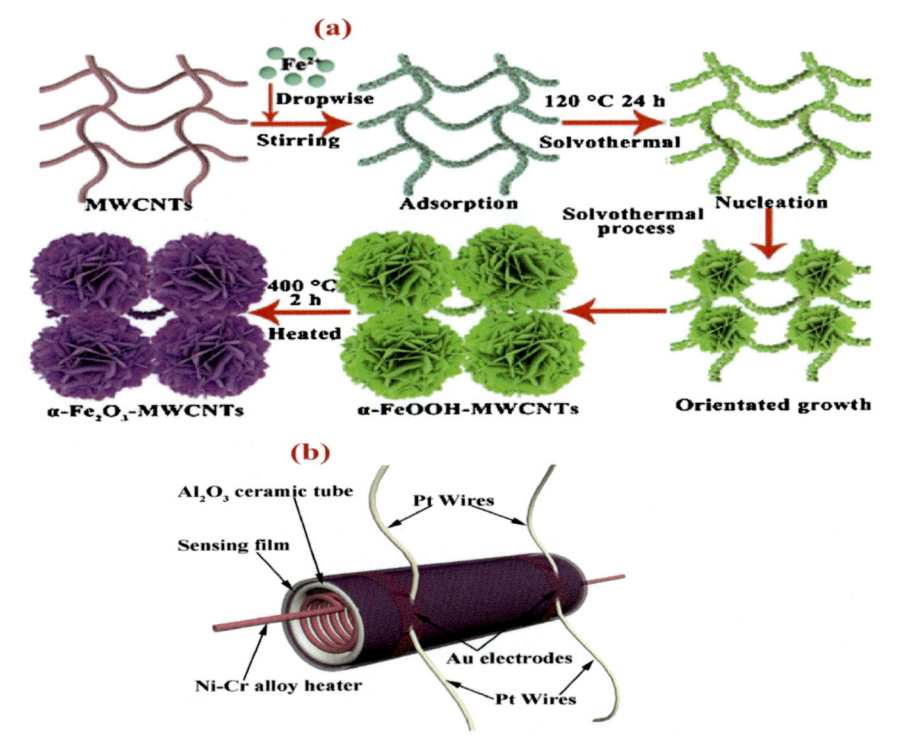

Figure 17.6 Schematics of iron oxide and multiwalled carbon nanotube composite synthesis for gas sensors.
Reprinted from Ref. 24, copyright (2017), with permission from Elsevier.

17.6.5 Multiwalled carbon nanotube–tungsten oxide-based devices for gas-sensing applications

One of the most versatile sensing materials is WO_3, which has been extensively analyzed because of its quick response and high sensitivity for many toxic gases, including NO_x, So_x, NH_3, CO, H_2S, ethanol, and ozone. When applied for H_2, it presented comparatively low sensitivity, as H_2 molecules showed the least reactivity with the smooth surface of WO_3 crystals. However, its sensitivity and selectivity can be enhanced by adding inert metals like Au, Pd, or Pt to WO_3. Modern procedures, such as the sol-gel method, screen printing, and sputtering, are applied for doping WO_3 and similar gas-sensing materials.

In a recent study on H_2 gas sensing, the sensors were equipped with thin films of MWCNTs doped with WO_3, which were further processed by an evaporation procedure carried out by an electron beam. Such electron beam-based evaporation processing methods help control the film's morphology, thickness, and desired characteristics, such as high thermal efficiency, high productivity, dense coating, low contamination, and high reliability. MWCNTs are the most common choice for doping purposes because of their large surface area, hollow geometry, and large number of active sites available to capture gas molecules. Moreover, MWCNTs were found to exhibit high sensitivity with reliable recovery times.

17.6.6 Gas-sensing features of copper oxide/zinc oxide/ multiwalled carbon nanotube composites

Gas sensors based only on CNTs exhibit poor sensing response, weak selectivity, and long recovery time. Decorating CNTs with metal oxide nanoparticles can enhance their performance because of the unique physical and chemical characteristics of CNTs and their semiconducting nature with various structural properties of metal oxides. Synthesis of this composite is a simple, quick, efficient, and cost-effective process. A gas sensor based on MWCNTs, copper oxide, and zinc oxide for detection of H_2S has been reported, as dual sensitization is much effective for sensing response.[25] The gas sensor based on dual sensitization was studied and compared with the gas sensor based on pristine MWCNTs, and increased selectivity and sensitivity were observed in the dual sensitized sensor. This gas sensor exhibits an enhanced response toward sensing H_2S and C_2H_5OH gases at varying temperatures. The selectivity of the sensor for H_2S and C_2H_5OH can be adjusted by selecting the respective sensing temperature. This designed gas sensor is sensitive enough for H_2S and C_2H_5OH and can be used for the early diagnosis of diseases using breath analysis.[26]

17.7 Conclusions

Process control has become integral to the functioning of modern industry and machinery through the development of technology. Sensors are essential to such progress, where high sensitivity, rapid response, recovery, and low energy consumption are imperative at acceptable costs. The application of nanomaterials has resulted in major improvements in the sensing efficiency of the system because of the wide applications area. The large surface-to-volume ratio, layered structure, and electric and thermal conductivity make them extremely suitable materials in applications for gas sensing. Significant developments in CNT gas sensors include improving the sensitivity and selectivity of nanotubes. CNTs are combined by special functionalization with a range of nanomaterials, such as metal oxides. Combining CNTs with certain metal oxide nanomaterials for sensor applications can increase CNT selectivity and sensitivity to several harmful gases. These CNT—metal oxide composites can be used for commercial applications with better efficiency and lower economic cost. This chapter summarized a variety of composites with their recent advances in gas-sensing applications. The preparation of nanocomposites, including functionalization with other metal oxides and their advantages and performance as gas sensors, has been described.

Acknowledgments

Prof. Dr. Waheed S. Khan acknowledges Pakistan Science Foundation (PSF) for financial support under Research Grants No. PSF-NSFC III/Med/P-NIBGE (14). We are also grateful to *National Institute for Biotechnology and Genetic Engineering (NIBGE)* Faisalabad Pakistan for providing conducive environment for such type of writing work.

References

1. Ionescu R, Espinosa E, Leghrib R, et al. Novel hybrid materials for gas sensing applications made of metal-decorated MWCNTs dispersed on nano-particle metal oxides. *Sensor Actuator B Chem* 2008;**131**(1):174−82.
2. Wang Y, Yeow JT. A review of carbon nanotubes-based gas sensors. *J Sensors* 2009;**2009**.
3. Dey A. Semiconductor metal oxide gas sensors: a review. *Mater Sci Eng, B* 2018;**229**: 206−17.
4. Bondavalli P, Legagneux P, Pribat D. Carbon nanotubes based transistors as gas sensors: state of the art and critical review. *Sensor Actuator B Chem* 2009;**140**(1):304−18.
5. Bondavalli P. Carbon nanotubes based transistors composed of single-walled carbon nanotubes mats as gas sensors: a review. *Compt Rendus Phys* 2010;**11**(5−6):389−96.
6. Yuan W, Shi G. Graphene-based gas sensors. *J Mater Chem* 2013;**1**(35):10078−91.
7. Varghese SS, Lonkar S, Singh K, Swaminathan S, Abdala A. Recent advances in graphene based gas sensors. *Sensor Actuator B Chem* 2015;**218**:160−83.
8. Basu S, Bhattacharyya P. Recent developments on graphene and graphene oxide based solid state gas sensors. *Sensor Actuator B Chem* 2012;**173**:1−21.
9. Singhal AV, Charaya H, Lahiri I. Noble metal decorated graphene-based gas sensors and their fabrication: a review. *Crit Rev Solid State Mater Sci* 2017;**42**(6):499−526.
10. Gurbuz Y, Kang W, Davidson J, Kinser D, Kerns D. Diamond microelectronic gas sensors. *Sensor Actuator B Chem* 1996;**33**(1−3):100−4.
11. Mochalin VN, Shenderova O, Ho D, Gogotsi Y. The properties and applications of nanodiamonds. *Nat Nanotechnol* 2012;**7**(1):11−23.
12. Rauti R, Musto M, Bosi S, Prato M, Ballerini L. Properties and behavior of carbon nano-materials when interfacing neuronal cells: how far have we come? *Carbon* 2019;**143**: 430−46.
13. Dresselhausa MS, Lin YM, Rabin O, Jorio A, Souza Filho AG, Pimenta MA, Saito R, Samsonidze GG, Dresselhaus G. *Mater Sci Eng C* 2003;**23**:129−40.
14. Mallakpour S, Khadem E. Carbon nanotube−metal oxide nanocomposites: fabrication, properties and applications. *Chem Eng J* 2016;**302**:344−67.
15. Navazani S, Hassanisadi M, Eskandari M, Talaei Z. Design and evaluation of SnO$_2$-Pt/MWCNTs hybrid system as room temperature-methane sensor. *Synth Met* 2020;**260**: 116267.
16. Aroutiounian VM, Adamyan AZ, Khachaturyan EA, et al. P1. 7.10 Methanol and ethanol vapor sensitivity of MWCNT/SnO$_2$/Ru nanocomposite structures. *Tagungsband* 2012: 1085−8.
17. Kumar S, Pavelyev V, Mishra P, Tripathi N. A review on chemiresistive gas sensors based on carbon nanotubes: device and technology transformation. *Sensor Actuator Phys* 2018; **283**:174−86.
18. Aroutiounian V. Semiconductor gas sensors made from metal oxides functionalized with carbon nanotubes. *Sensors Transducers* 2018;**228**(12):1−16.
19. Kwon YJ, Mirzaei A, Kang SY, et al. Synthesis, characterization and gas sensing properties of ZnO-decorated MWCNTs. *Appl Surf Sci* 2017;**413**:242−52.
20. Vyas R, Sharma S, Gupta P, et al. CNT-ZnO nanocomposite thin films: O$_2$ and NO$_2$ sensing. In: *Advanced materials research*, vol. 585. Trans Tech Publ; 2012. p. 235−9.
21. Guo R, Wang H, Tian R, et al. The enhanced ethanol sensing properties of CNT@ ZnSnO$_3$ hollow boxes derived from Zn-MOF (ZIF-8). *Ceram Int* 2020;**46**(6):7065−73.

22. Liu S, Wang Z, Zhang Y, Zhang C, Zhang T. High performance room temperature NO_2 sensors based on reduced graphene oxide-multiwalled carbon nanotubes-tin oxide nanoparticles hybrids. *Sensor Actuator B Chem* 2015;**211**:318−24.

23. Narjinary M, Rana P, Sen A, Pal M. Enhanced and selective acetone sensing properties of SnO_2-MWCNT nanocomposites: promising materials for diabetes sensor. *Mater Des* 2017; **115**:158−64.

24. Jia X, Cheng C, Yu S, Yang J, Li Y, Song H. Preparation and enhanced acetone sensing properties of flowerlike α-Fe_2O_3/multi-walled carbon nanotube nanocomposites. *Sensor Actuator B Chem* 2019;**300**:127012.

25. Mirzaei A, Kim SS, Kim HW. Resistance-based H_2S gas sensors using metal oxide nanostructures: a review of recent advances. *J Hazard Mater* 2018;**357**:314−31.

26. Choi MS, Bang JH, Mirzaei A, et al. Dual sensitization of MWCNTs by co-decoration with p-and n-type metal oxide nanoparticles. *Sensor Actuator B Chem* 2018;**264**:150−63.

Drug-detection performance of carbon nanotubes decorated with metal oxide nanoparticles

Anam Munawar[1], Rao F. Hussain Khan[2], M. Zubair Iqbal[3], Asma Rehman[1], Sadia Zafar Bajwa[1] and Waheed S. Khan[1]

[1]Nanobiotechnology Group, Industrial Biotechnology Division, National Institute for Biotechnology and Genetic Engineering (NIBGE), Faisalabad, Punjab, Pakistan; [2]Department of Radiation Oncology, Washington University School of Medicine, Missouri, WA, United States; [3]Department of Materials Engineering, College of Materials and Textiles, Zhejiang Sci-Tech University, Hangzhou, Zhejiang, PR China

18.1 Introduction

Over the last 20 years, extensive research has focused on the hottest topic in nanotechnology, the fabrication and functionalization of carbon nanotubes (CNTs). Based on the structure and extraordinary features of modified CNTs,[1] functionalized CNTs offer a series of applications, specifically in the field of chemosensors/biosensors and drug delivery. CNTs are acknowledged for their definitive carbon fiber's strength, thermal conductivity, and exceptional field-emission characteristics. Metallic CNTs are good conductors of electricity without dissipating heat. In different nanodevices, CNTs can act as active semiconductors because of their electronic properties. In this regard, CNTs functionalized by various metallic nanoparticles (NPs) have been explored, showing elevated activity and extraordinary features.[2] The most important challenge is how to uniformly disperse metal NPs on the available external surfaces of CNTs because of the integral inertness present in CNT walls. Earlier research described the inertness of CNTs in whole or its specific aspects (gold NPs deposited on CNTs or Pd and functionalization by bimetallic NPs and inner space filled with CNTs by metallic NPs).[3] The surface modification of CNTs can be accomplished by creating a connection of metal atoms with CNTs through covalent connection, hydrophobic linkage, hydrogen bonding, P-stacking, or electrostatic interfaces.

Scientists consider the methods listed here the most suitable for décorating CNTs, although many others could be included.[4] Indirect methods include covalent and non-covalent linkage (electrostatic, p-stacking, hydrogen-bond connections, or hydrophobic). Direct methods include physical and chemical approaches (in situ development of metallic nanomaterials with different morphology, chemical reduction, and different electrochemical deposition techniques). Physical methods include various techniques (evaporation, deposition, sputtering, and ion/electron beam irradiation).

Metal Oxide-Carbon Hybrid Materials. https://doi.org/10.1016/B978-0-12-822694-0.00001-6

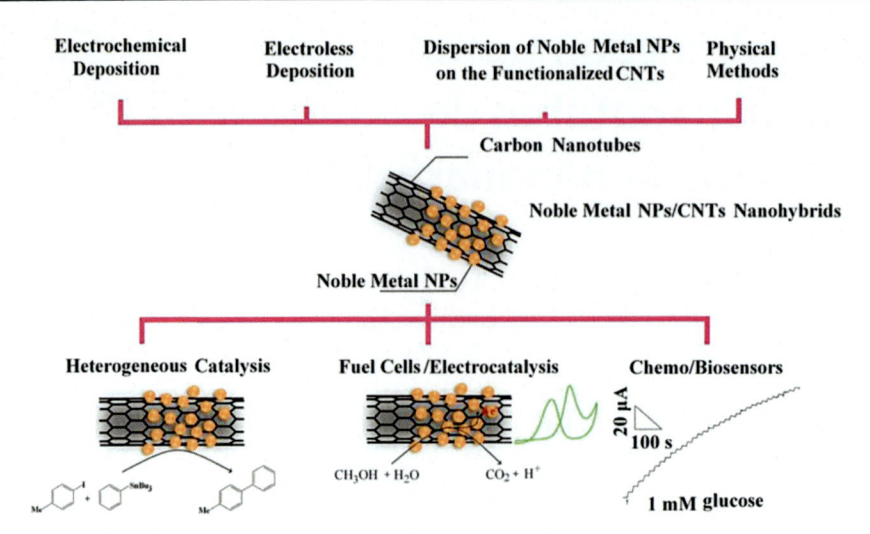

Figure 18.1 Summary fabrication of nanohybrids based on noble metal nanoparticles/carbon nanotubes.
Reprinted from Ref. 5, copyright (2011), with permission from Elsevier.

The most common CNTs and metal nanohybrids include noble metallic NPs (PtSn, PtRu, Pt, Au, Pd, Ag, and Ru; the size of these NP/CNT nanohybrids ranges from 1.3 to 7.0 nm), their synthesis methods and main applications are shown in Fig. 18.1.

The use of illicit drugs is increasing significantly worldwide, resulting in various global concerns, including national and international security and public health issues. The wide use of illicit drugs is converted into abused drugs. The traditional surveys to determine community drug use were extensive but less accurate. Recently, drug-detection measurement has become easier by quantifying drug metabolites or chemical residues in different biological samples.

This chapter presents advanced fabrication methods for modifying the external surface of CNTs using elemental metal NPs; the nanohybrid material is systematized, conferring to the nature of the precursor.

18.2   Carbon-based nanomaterials

Carbon-based nanomaterials (CBNs) possess distinctive chemical and physical properties (i.e., optical properties, great mechanical strength, and electrical and thermal conductivity). Wide-ranging research efforts are under way to investigate the unique advantageous features of CBNs for their application in a variety of biomedical, environmental, agricultural, and industrial fields. CBNs include fullerenes, graphene, nanodiamonds, CNTs (SWCNTs, MWCNTs), carbon nanocones, nanofibers, nanohorns, and nanodisks. The classifications of a few CBNs are depicted in Fig. 18.2.

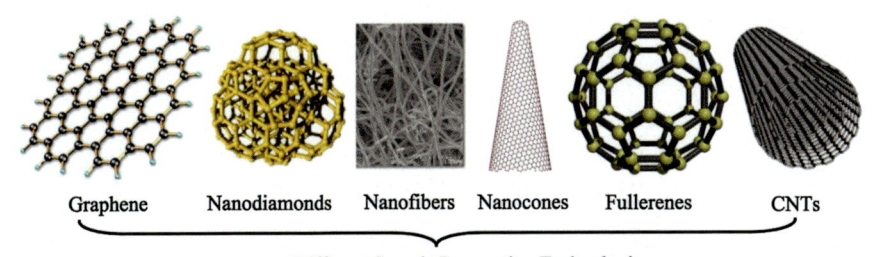

Graphene   Nanodiamonds   Nanofibers   Nanocones   Fullerenes   CNTs

Different Sample Preparation Technologies

Figure 18.2 Classification of carbon-based nanomaterials used in fabricating composite materials.
Reprinted from Ref. 6, copyright (2013), with permission from Elsevier.

18.3 Classification of carbon nanomaterials

CBNs include CNTs, fullerenes, and their derivatives, including graphite nanoparticles, carbon black, graphene nanoparticles, graphene oxide (GO), and nanodiamonds are measured as fundamentals of nanotechnology, as depicted in Fig. 18.3.[7] These materials possess various potential applications in different fields ranging from biomedical to mechanical engineering.

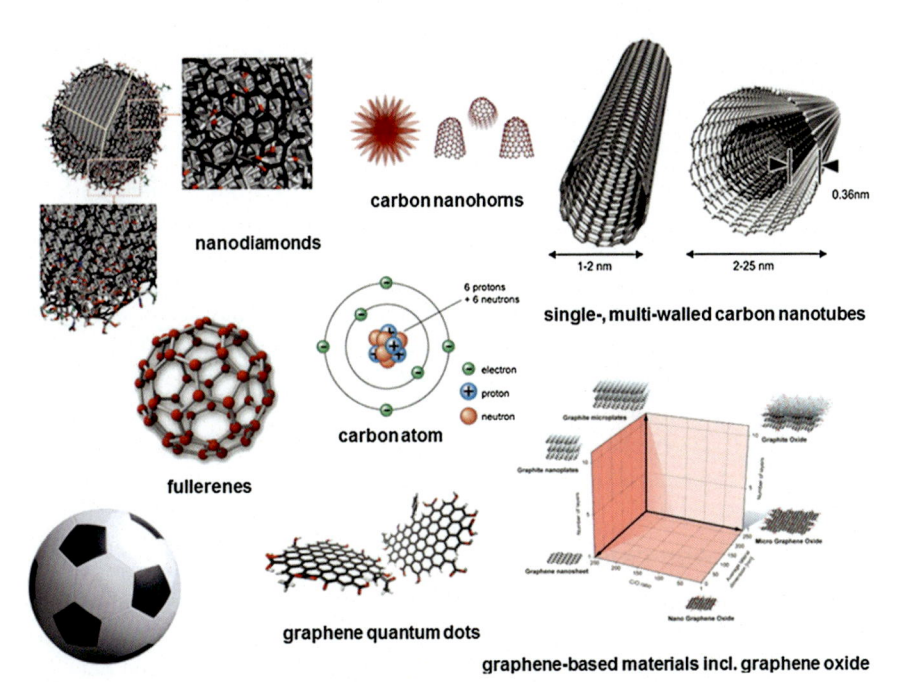

Figure 18.3 Carbon-based nanomaterial sizes, morphologies, and properties.
Reprinted from Ref. 7, copyright (2016), with permission from Elsevier.

The most commercial CBNs are CNTs, and theoretically, this material possesses a wide range of applications: health care, energy storage, sensing, and many more. Various potential industrial applications of CBNs are now in the market. Sheets of graphite material are rolled to form the cylinder-like shape of CNTs. Single-walled CNTs (SWCNTs) are made up of one sheet of a graphite material, while for multi-walled CNTs (MWCNTs), concentric tubes are made by rolling more than one graphite sheet. Currently, there is a gap in knowledge between the fabrication and application of nanomaterials, including CBNs. It is mandatory to investigate the possible health risks related to CBNs.[8]

18.3.1 Fullerenes

Using spectrometric measurements, fullerenes were discovered in 1985. An experiment was performed on interstellar dust. Fullerenes containing carbon molecules were arranged in closed-cage form, with carbon in penta- and 21 hexagonal rings. The formula of fullerenes is C20 + m, (m = integer number), while 21 represents the number of homologous and isomer series. Most fullerene-related studies include $C_{22,}$ C_{60}, and $C_{70,}$ leading toward the so-called advanced and complex fullerenes, for example, C_{540}, C_{240}, and C_{720}. Fullerenes possess a relatively hydrophobic surface, increased electron affinity, and a large surface area/volume ratio. Due to these exceptional properties, fullerenes can be used for the adsorption of organic molecules, and this feature makes them suitable candidates for extraction. Functionalized fullerenes can be used for sensing purposes. Impregnated chemical entities enhance selectivity and sensitivity.[9]

18.3.2 Carbon nanotubes

CNTs were first discovered by Iijima in 1991.[10] CNTs range from fractions to tens of nanometers in diameter with lengths up to a few micrometers. Graphene sheets can be considered a basic subunit of CNTs. CNTs present a cylinder-like shape with a cap of fullerene-like structures. One or more layers of graphene seamlessly rolled up to form Single-walled (SWCNTs) or multi-walled (MWCNTs) nanotubes, respectively. In the reported features, surface areas for CNTs range from 150 to 1500 nm. This wide range of surface area is basic to be used as good sorbents. Moreover, CNT surfaces can be functionalized by covalent or noncovalent attachment of various organic particles. Functionalized CNTs can be used for selective and sensitive detection of different analytes. Till now, the most abundantly used CBNs have included CNTs.[11]

18.3.3 Graphene oxide

During recent years, many research papers reported on the study of electronic features of graphene sheets. Graphene can be used to develop different graphitic forms consisting of graphite, GO, graphene quantum dots, fullerene, CNTs, and carbon dots. One layer of carbon atoms is arranged in a honeycomb-like structure to form GO. It presents in two-dimensional lattice form. The theoretical value for the GO surface area is 2630 m^2/g. GO is also considered a miracle material because of its unique planar geometry and a large surface area that provides more analyte loading. The low bandgap is suitable for electrical contact between the targeted substance and electrode surfaces with low signal-to-noise ratios. The planar sheetlike structure and its sides make it the best candidate for molecular adsorption. Additionally, in graphene, the arrangement of the delocalized π-electron system results in a resilient π-stacking linkage with a benzene ring, from which graphene becomes a suitable material for separating benzenoid from mixtures. Lastly, it is easier to modify graphene within functional groups, and the most important form is GO. Due to these extraordinary features of graphene, it is an important candidate as an adsorbent during the fabrication procedures of various materials.[12]

18.3.4 Graphene quantum dots

Graphene sheets break into tiny fragments to form graphene quantum dots (GQDs). The transport of GQDs is confined in all dimensions. GQDs possess different innate properties, such as semiconductors with no bandgap and infinite Bohr exciton diameter. Consequently, small broken fragments of graphene have confinement. GQDs have a size of less than 20 nm. GQDs can be synthesized in vitro by breaking graphene sheets. But the edge effects and quantum confinement of GQDs introduce photoluminescence (PL). There is zero bandgap, which is why it is necessary to excite GQDs to show PL. By changing the surface chemistry and size, one can tune the bandgap of GQD.

However, GQDs are considered an example of carbon dots (CDs). The distinct structure is an important characteristic of CDs. In general, crystalline, amorphous, and spherical NPs (quasi) are common features of CDs. According to many researchers, the crystallinity of quantum dots came from sp^2 carbon atoms; on the other side, GQDs have poorer crystallinity. In contrast, most GQDs are produced from graphene, GO, and molecules with specific structures such as benzene rings. Accordingly, graphene lattices are present inside GQDs, resembling the crystalline arrangement of one or more graphene sheets. Remarkably, however, there is different core morphology of GQDs and CDs. Multifarious surface groups can be used to modify both GQDs and CDs, most importantly functional groups related to oxygen, for example, hydroxyl and carboxyl. Such kinds of surface modifications increase the optical properties of GQDs and CDs. Functionalized GQDs and CDs are easy to disperse in a liquid. Other atoms, such as sulfur, nitrogen, copper, lithium, and many others, heighten the electrical conductivity and luminescence of GQDs and CDs significantly via fine-tuning of electronic arrangements.

18.3.5 Carbon quantum dots

Carbon quantum dots (CQDs) are unique zero-dimensional CBNs acknowledged for their moderately durable fluorescence properties and nano size. One of the fluorophores, CQDs are a center of global attention because of their easy fabrication and unique characteristics like excellent biocompatibility, nano size, upconversion, chemical stability, good photostability, and tunable PL.

The core material for small NPs of CQDs is carbon. The surface of CNTs can be modified or functionalized. The fabricated CQDs can possess amorphous or crystalline nature. Usually, the hybridization of carbon in CQDs is sp^2, while some reported research indicates that sp^3 hybridization is also possible. The size of CQDs is less than 10 nm, which is why these NPs are categorized as zero-dimensional. The crystalline CQD lattice parameter is about 0.34 nm, conforming interlayer spacing for graphite (002). A wide range of materials can be used to functionalize the surface of CQDs. Most of the reported research papers are based on the functionalization of CQDs using oxygen-consisting groups (carboxyl and hydroxyl). Such functionalized CQDs are easily soluble in water. The important properties of CQDs depend on the functional groups used for surface modification. The functionalization effect on solubility makes them suitable to form colloids in polar organic solvents or in aqueous. Fluorescence properties can be imparted on CQDs, and this depends on surface groups used for functionalization.

18.4 Nanosensors and their types

A promising advanced tool in the area of detection is the nanosensor. The nanosensor is a field instrument based on nanotechnology that can detect chemical and physical changes[13]; monitor biochemical and biomolecular changes in vivo[14]; and measure toxic material released by industry into the environment.[15] Principally, the main components of a sensor include the recognition and transduction elements. When an analyte is specific for its recognition element, it produces signals that amplify or transform measured signals—for example, current change, frequency change, or impedance variation.

Nanosensors can be classified according to the mechanism involved, the nanostructure, the energy source, and applications, as shown in Fig. 18.4. Sensors based on nanomaterials offer substantial improvements over traditional biological and chemical techniques in selectivity, diversity, stability, accuracy, sensitivity, and speed. Based on the nanostructure and entities used for functionalization, the designed device can be used for many purposes like determining contaminants, pollutants, and microbes. Chemical nanosensors can be used for

- **Industrial**: detecting release of harmful gases and chemicals
- **Environmental**: measuring the pollutants present in water and air
- **Military**: monitoring terrorism-related activities
- **Aerospace**: chemical analysis of atmospheric and soil constituents

Nanosensors can be categorized on different bases—for example, structure, use, and energy source:

- Classification based on structure includes (1) electromagnetic nanosensors, (2) mechanical nanosensors, and (3) optical nanosensors.
- Use-based classification includes (1) deployable nanosensors, (2) biosensors, (3) electrometers, and (4) chemical sensors.

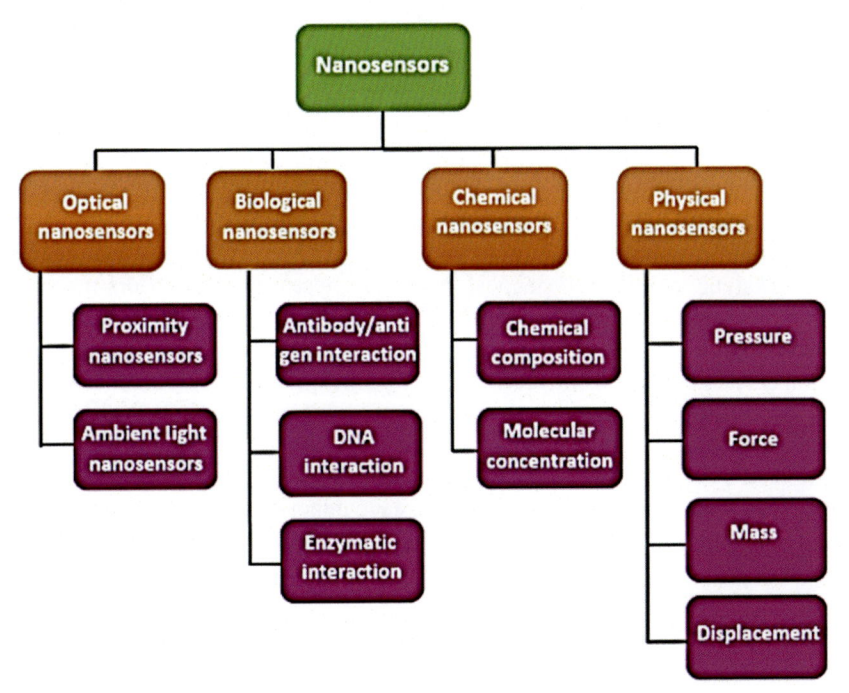

Figure 18.4 Nanosensor types and subcategories based on sensing elements. Reprinted from Ref. 16, copyright (2020), with permission from Elsevier.

- Classification based on energy sources includes (1) active nanosensors, for which energy is required (thermistors, for example), and (2) passive nanosensors, for which no energy is required (piezoelectric and thermocouple sensors, for example).

Nanosensors based on the associated knowledge of chemistry, biology, and nanotechnology are called nanobiosensors. The arrangement of nanosensors is just like that of ordinary sensors but with results produced at the nanoscale. Consequently, the definition of a nanosensor is a particularly small device selectively bound to an analyte that produces a certain signal. Such nanosensors can sense and respond in the form of signals, such as biological signals (detected by biosensors) and physicochemical signals (nanosensors). Compared with the shortcomings of traditional sensors, nanosensors possess various advantageous characteristics, for example, high sensitivity, stability, robust response, portability, real-time detection, and selectivity. Nanomaterials used in their nanosensors can enhance other essential attributes. Nanosensors can be classified based on fabrication techniques, for example:

- self-assembly of molecules
- bottom-up arrangement approaches
- top-down lithography

Nanosensors based on nanomaterials can be classified as

- porous materials like silicon
- metallic NP-based sensors
- nanowire nanosensors

- nanoelectromechanical systems
- nanoprobes
- nanosystems: atomic force microscope (cantilevers)

18.5 Nanosensor application

Special catalytic and physicochemical properties present in nanosensors make them suitable for application in various fields, including agriculture, environmental, and biomedical. The characteristics of nanosensors that make them suitable candidates for biomedical applications include their nano size, water solubility, and biocompatibility (minor toxicity). Characteristics of carbon composites used in drug detection are summarized in Table 18.1.[17]

These NPs can act as drug delivery vehicles; because of the PL property, they can be observed within the human body. CQDs have minor toxicity, high hydrophilicity, chemical stability, and water solubility. Based on nanosensor properties, their uses can be defined.

Table 18.1 Carbon-based composite materials reported for drug detection.

Sr. No	Nanocomposite	Analyte	Limit of detection(M)	Reference
1.	MWCNT/B-CD modified	Uric acid	6.70×10^{-6}	18
2.	SWCNT/Ppy,Surf.cov.bend	Ascorbic acid	5.00×10^{-6}	19
3.	NH_2/MWCNT and ZnO	Pimozide	1.02×10^{-11}	20
4.	CNTs and zirconium oxide NPs	Nebivolol	1.2×10^{-11}	21
5.	Graphene Nafion/GCE	Nebivolol	4.6×10^{-8}	22
6.	CNTs @ Cu	Chloramphenicol	10×10^{-6}	23
7.	CNTs@Bi_2WO_6	Rifampicin	0.16×10^{-6}	24

18.6 Drug molecules and their detection

Drug detection using nanosensors is a technology in which a chemical reaction happens between targeted analytes and nanomaterials. Different transduction approaches can be applied to convert this reaction into a sensory signal that may be changed in current, mass change, or impedance variation. The nanodetection approach has been used to detect many drugs—for example, cyclophosphamide, paclitaxel, gemcitabine, chloramphenicol, rifampicin, and dihydroartemisinin. In vivo sensing instruments should be capable enough to monitor signals produced within a living body. Some important features must be present in nanobiosensors, such as biocompatibility (nontoxicity) and no harm to the examined system.[25] The difficult part of nanobiosensor application is the approach of in vivo tissues and essential features required to be developed in nanobiosensors. Recently, however, a range of different, unique technologies have been improved and designed to resolve the possible challenges and permit in vivo biological detection for the first time in history. Biocompatibility, nano size, selectivity, and sensitivity make nanosensors suitable candidates for in vivo applications.

Other in vivo disease states, for example, cardiac arrest or apoptosis of cancerous cells, can be investigated with nanodevices. Different high potential in vivo sensing implantable devices are receiving more attention. A device for monitoring glucose levels in diabetic patients is clinically approved. These nanosensors are clinically validated to control glycemic levels in the interstitial liquid of patients by constantly monitoring glucose levels. According to scientists, the list of nanosensors and nanotechnologies will grow and revolutionize the healthcare system in the near future. This book chapter highlights the potential and breadth of major challenges for fabricating the carbon-based devices used for in vivo sensing from bench to bedside.

18.7 Role of zinc oxide–carbon nanotube nanocomposite in morphine detection

We all are aware of the unique mechanical, thermal, and electrical properties of CNTs. In recent years, CNTs have gained tremendous attention because they have desirable and unique properties and wide uses in analytical chemistry.[26] Different reviews and book chapters focus on the manifestation of the tendency of CNTs. The composites of CNTs with different nanomaterials have useful properties like sensing applications and catalysis. However, NPs of ZnO are very useful for decorating CNTs because of their extremely small size and desirable biocompatibility. Recently, a new technique has been widely used to fabricate ZnO/CNT nanocomposites wherein ionic liquids known as room temperature ionic liquids are deposited on the carbon surface. ZnO/CNT nanocomposites are commonly referred to as the presence of a compound completely made up of inorganic anions and various organic cations.[27] ZnO/CNT nanocomposite at room temperature exists in the liquefied state because ZnO/CNT nanocomposite has various special chemical and physical properties, such as increased ionic conductivity, a large electrochemical window, and good solubility. Zinc oxide-based CNT composites have been documented for numerous electrochemical practices because it provides a nonaqueous medium.

The alkaloid present in opium is morphine. Different levels of aches (severe to moderate) can be released using morphine.[28] This painkiller remains active for a short duration. Due to its repetitive use in our daily life, there must be some analytical methods to determine the concentration of morphine. Many previous techniques include gas chromatography–mass spectroscopy, electrochemical methods, high-performance liquid chromatography (HPLC), surface plasma resonance (SPR), chemiluminescence, and fluorimetry.[29] Recently, new research reported on the fabrication of unique

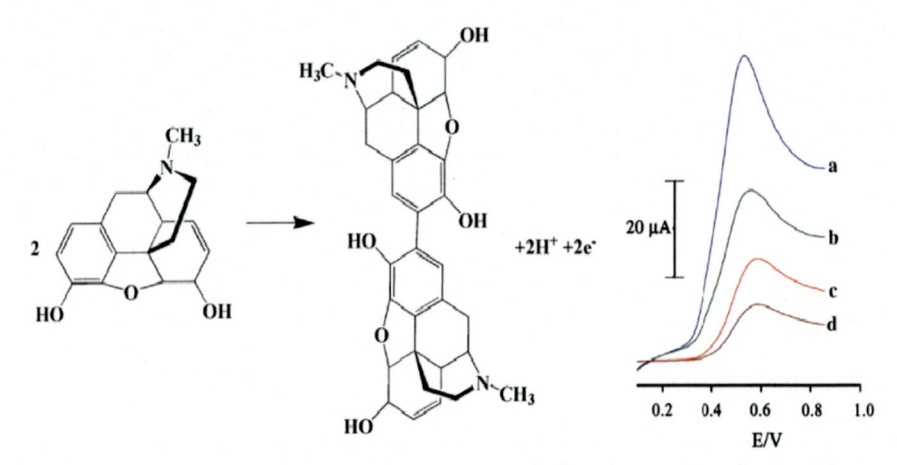

Figure 18.5 Zinc oxide—carbon nanotube nanocomposite-based nanosensor for the voltammetric determination of morphine.
Reprinted from Ref. 30, copyright (2013), with permission from Elsevier.

electrochemical sensors based on liquid ionic form ZnO/CNT paste electrodes. This novel sensitive and robust electrochemical sensor voltammetrically determines the concentration of morphine. Efforts continue toward improving the sensitivity, selectivity, speed, and cost-effectiveness of sensors for the analysis of morphine. A novel application of ZnO/CNT nanocomposite is reported where 1-methyl-3-butylimidazolium bromide enhances the selectivity for voltammetric detection of morphine; this extreme selectivity is due to the presence of binder, and sensitivity is due to the ionic liquid composite.[30] Fig. 18.5 shows a nanosensor-based on ZnO/CNT nanocomposite and 1-methyl-3-butylimidazolium bromide carbon paste electrode for voltammetric determination of morphine.

18.8 Cerium oxide nanoparticle-decorated carbon nanotubes as an effective platform for acetaminophen

Small biomolecules coexisting in biological matrices have an immensely vital and significant role in the physiological functions of organisms. In recent years, researchers have diverted their attention to developing sensitive and efficient detection systems with the capability to detect substances associated with physiological functionalities present in biological systems.[31] Numerous analytical methods have been employed for the detection of one or more than one element associated with physiological factors like electrochemical devices, chemiluminescence, HPLC, capillary electrophoresis, and spectrometry. Most of these reported techniques are laborious, demanding, time-consuming, and complex, excluding electrochemical detectors.[32] Electrochemical techniques are gaining in research importance for sensing various analytes owing to their various features like favorable stability, convenience, robust responsiveness, cost-effectiveness, higher selectivity, and greater sensitivity.

Acetaminophen is important for its contribution to physiological performance. AC, commonly known as paracetamol, is a broadly prescribed drug used as a painkiller and antipyretic throughout the globe, with rather minute anti-inflammatory characteristics. In general observation, AC does not exhibit any deleterious side effects at recommended dosages, but overdose or prolonged use may lead to acute liver and kidney failure. Thus, AC monitoring is essential for drug safety. Conventional techniques (mentioned above) used to quantify paracetamol (i.e., AC) are suitable as they are costly, laborious, and demanding with respect to experienced staff and expensive apparatuses. Recently, electrochemical methods have emerged as the most suitable routes for determining paracetamol based on its electrochemically active molecular structure. The electrochemical process involves the chemical modification of the electrode.[33]

Carbon nanotubes (CNTs) offer good options for their use as electrocatalysts in electrochemical sensors because of striking features like electrical conductivity and good electrochemical strength. Several researchers reported that decorating NPs on CNT electrode matrices enhances electrochemical devices' electrocatalytic activity and results in multifarious benefits. Cerium oxide (CeO2), also called ceria, is considered an active material for electrochemical devices due to its redox features. Sensing activities can be elevated by the interaction between nanoceria and well-selected support like CNTs. CeO2/CNT nanocomposite-based electrochemical sensor has been recently reported[31] that detects not only acetaminophen but also dopamine (DA), uric acid (UA), and ascorbic acid (AA), as depicted in Fig. 18.6. The electrode

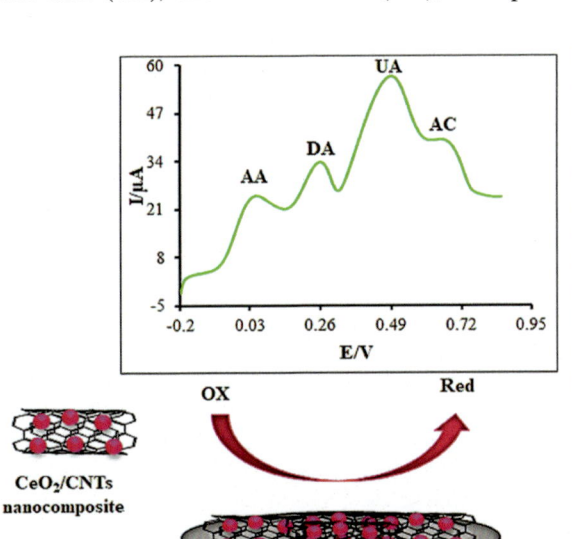

Figure 18.6 Electrochemical sensor based on cerium oxide (CeO$_2$) nanoparticle-decorated carbon nanotube (CNT) modified electrode for simultaneous detection of ascorbic acid (AA), dopamine (DA), uric acid (UA), and acetaminophen (AC). Reprinted from Ref. 31, copyright (2020), with permission from Elsevier.

modified with CeO2/CNT nanocomposite exhibits high sensitivity and electrocatalytic activities toward oxidizing AC, DA, UA, and AA. These features may be ascribed to higher surface areas, higher levels of conductivity, very good catalytic activities, and amplification impact of CeO2/CNT nanocomposite.

18.9 Efficient electrochemical detection of cetirizine antiinflammatory drug using titanium dioxide—carbon nanotube nanohybrid

Electrochemical detection is an analytical technique to sense the electric currents generated from redox reactions between the analyte and device surface. Nonsteroidal anti-inflammatory drugs (NSAIDs) have been widely used to treat aches and inflammation-associated difficulties. One of the most important NSAID drugs, cetirizine (CTZ), is in the second generation of antihistamines. Chemically, it is recognized as 2-[2-[4-[(4-chlorophenyl)-phenyl-methyl] piperazin-1-yl] eth-oxy] acetic acid. When hydroxyzine metabolizes, a carboxylated group with histamine receptors (H1) called CTZ is produced. The oral administration of CTZ in tablet or syrup form can be rapidly absorbed; nonetheless, its overdosage can cause various issues like headache, mild drowsiness, fatigue, and dry mouth. Other adverse effects include edema, diabetes, cardiac failure, and tachycardia. It is openly found in the medicine market in different forms: compounded capsules, syrup, and tablets. The residues from all forms can be excreted through urine in unmodified form. In previously reported research, many analytical methods have been used to analyze CTZ—for example, gas chromatography, HPLC, calorimetry, spectrophotometry, and potentiometry.[34] However, as discussed, such techniques have many drawbacks, like the tiresome steps required for sample preparations, time-consuming procedures, and high costs. On the other hand, the electrochemical approach has some benefits because it requires superior sensitivity, minimal cost, an anti-interference outcome, and high selectivity. Due to the extensive use of CTZ, it has become important to sense residues of CTZ in different biological samples.[35] Different chemically modified electrodes with a unique material nanohybrid of TiO$_2$/CNTs have been widely used to detect CTZ. In TiO$_2$/CNT nanomaterials, modified electrodes raise the charge density of TiO$_2$ and CNTs, providing increased surface area. The combined synergistic effect of TiO$_2$ and CNTs imparts increased edge density and electroactivity of the device to detect CTZ. In this regard, electrodes with carbon paste burdened with different NPs have become the center of attention because of their unique morphologies, physical and chemical features, and various benefits like low fouling effect, reduced background current, many applications, easily renewable surface, and most important, inertness.[36]

Carbon-based nanosensors and CNTs composited with other metals can be used to estimate and detect bioactive molecules. The surface modification of working electrodes using different nanomaterials, nanohybrids, and nanocomposites has interesting and unique features. Surface modification is related to the coating of material on the exposed surface of the electrode.

18.10 CuCo₂O₄/nitrogen-doped carbon nanotubes for electrochemical sensor for metronidazole detection

Different nanomaterials can be used to modify electrochemical sensors, for example, CNTs, CQDs, and metal NPs. CNTs are used extensively among these materials because of their unique properties, such as a large specific surface-area-to-volume ratio and increased electrical conductivity. Nitrogen doping on the surface is an active and effective way to improve performance. On the other side, doping with nitrogen empowers the properties of CNTs, increasing their conductivity and CNT stability.[37] Nitrogen doped on CNT surfaces acts as an electron-accepting site that increases electron transfer efficiency among the mediator (redox-active site) and the electrode surface. For example, CNTs doped with nitrogen materials were applied to modify electrode surfaces for increased surface area selectivity, lower detection limit, and enhanced conductivity, which are very useful for detecting metronidazole. Some metal oxides possess extraordinary electrochemical activity, such as transition metal.[38] Between them, CuCo₂O₄/N-doped CNT nanocomposite materials become the center of attention because of their various features like catalytic properties, excellent stability, and cost-effectiveness. The deposition of CuCo₂O₄ nanostructures on CNT surfaces has improved their conductivity, electrochemical activity, and large selective surface area. A new system designed for electrochemical detection of metronidazole (MNZ) based on nanohybrid of CuCo₂O₄/N-Doped CNTs. An enhanced response was recorded in the case of nanocomposite modified electrode; this is because of good conductivity, increased electron transfer rate, and large surface-area-to-volume ratio. Many have already been reported based on graphene nanosheets decorated with GQDs to increase selectivity molecularly imprinted polymer loaded on the modified electrochemical devices used for metronidazole sensing.[37] A nickel framework with a nonporous texture has been reported for an electrochemical detection approach to sensing MNZ. Molecularly imprinted polymers enhance electrochemical selectivity and sensitivity toward metronidazole.

18.11 Carbon nanotube–Fe₃O₄ magnetic composites for electrochemical detection of triclosan

The preservative and antimicrobial reagent "triclosan" has been used in various pharmaceuticals and personal products, such as hand soaps, toothpastes, plastics, antiseptic creams, surgical disinfectants, and foodstuffs for many years.[39] Nonetheless, the massive use of triclosan is chronically harmful to living organisms; even its residues present in wastewater can cause side effects on human health.[40] That's why it is important to develop some analytical methods that are reliable, robust, cost-effective, and sensitive for sensing of triclosan. Until now, several methods have been reported that can be applied for triclosan detection, for example, liquid chromatography ultraviolet sensing, gas chromatography-mass spectrometry, liquid chromatography-mass spectroscopy, capillary electrophoresis ultraviolet detector, and SPR.[41] No doubt the abovementioned analytical techniques are great achievements, but there are still some limitations like the complicated operation of instruments, the extreme cost for running the instruments, and presample preparation. Due to these limitations, further applications are hindered. But due to its wide use of this electrochemical detection analytical method, triclosan sensing has gained much attention and attraction from scientists because of its simple fabrication, high sensitivity and selectivity, and robust response. Modification of CNTs with magnetic NPs Fe_3O_4 has fascinated many scientists because of the special attributes of this composite material, such as high adsorption ability, easy preparation, low toxicity, magnetic properties, biocompatibility, and rapid electrical response, which enhances direct electrochemical triclosan detection. This is why CNT-based Fe_3O_4 nanocomposites have been extensively researched and used for electrochemical detection systems for chemical and biological systems.[42] However, naked magnetic NPs Fe_3O_4 can be easily oxidized and removed from CNT surfaces. Due to which these nanomaterials are poorly magnetized and dispersed and restrict their applications in other fields. Therefore, it has become a scientific challenge to fabricate CNTs and magnetic NP composite with increased electrochemical activity and high stability. The most effective reported method to solve this problem is to magnetize CNTs and cover their surfaces with inorganic or polymeric materials, for example, oxides or metals.[43] Such protective coverings prevent the clumping of Fe_3O_4 NPs but act as a suitable platform for additional modification and functionalization. Among these coating materials, the most important material is PPy because of its unique properties like irreversible redox property, discrete electrical properties, easy preparation, and extraordinary environmental stability. For the time being, this material proved a good supporting coating to load metallic NPs on the surface of CNTs.[44]

18.12 Nickel oxide/carbon nanotube/PEDOT composite for simultaneous detection of dopamine, serotonin, and tryptophan

Important neurotransmitters are serotonin (5-HT) and dopamine (DA). The 5-HT precursor is tryptophan (Trp), and in the human body, it belongs to the most important amino acids. Simultaneously, these described three species reside in human body fluids [158]. If there is an issue in their level, it can cause serious issues with fluctuating levels directly related to diseases. Quantitative analyses based on electrochemical sensing techniques that can be used to detect the level of these biomarkers in human blood are operational. These sensors can cause trouble for patients but are fast and cost-effective for diagnostic use.[45] Different electrochemical sensors based on unique nanomaterials have been reported, comprising detection associated with Trp, 5-HT, and DA. These analytes are basically electroactive substances. Different neutral conditions are suitable for cyclic voltammetry, square-wave voltammetry, or differential pulse voltammetry detection of these electroactive analytes.[46]

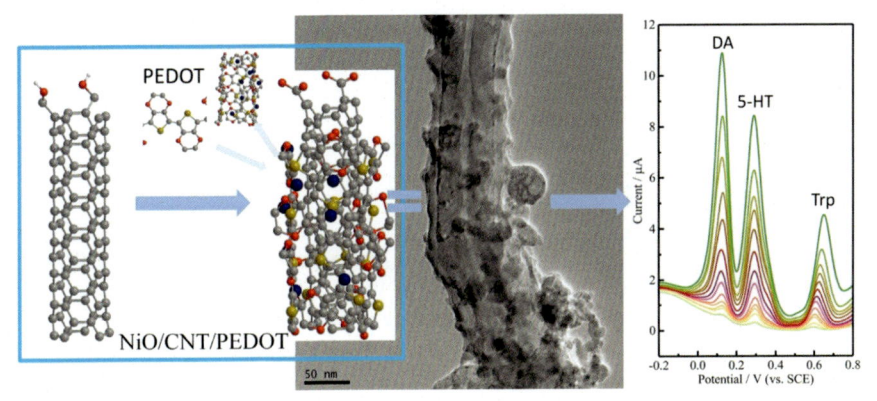

Figure 18.7 Nickel oxide (NiO)/carbon nanotube (CNT)/PEDOT nanocomposite sensors used to detect dopamine (DA), serotonin (5-HT), and tryptophan (Trp).
Reprinted from Ref. 47, copyright (2018), with permission from Elsevier.

The potentials for oxidation peaks related to Trp, 5-HT, and DA ranges from 500 to 800, 200−500, and 100−300 mV, respectively. As a result, the three neurotransmitters previously discussed can be identified using a nickel oxide/PEDOT/CNT nanocomposite-based electrochemical sensor with excellent sensitivity and selectivity, as depicted in Fig. 18.7. Such electrochemical methods can be used for the quantitative determination of analytes.[48] Currently, the main way to design and fabricate an electrochemical detection system involves synthesizing unique nanocomposite material. The layer of nanocomposite material provides a sensitive and selective platform for detecting analytes. Each nanomaterial component has its own catalytic effect. When one nanomaterial combines with another, synergistic effects are produced that show excellent sensitivity and improved sensing response toward analytes.[49] Additionally, combinations can lead to various electrochemical approaches such as physical adsorption, effective electrical performance, excellent chemical combinations for detecting chemical entities and different biomolecules, and different features like effectiveness, strength, selectivity, cost-effectiveness, and device sensitivity.[50]

18.13   Conclusion

For successful modification of external CNT surfaces, metal NPs can be prepared from bulk materials or already fabricated metal NPs. Metal salts (organic or inorganic acids) can be added during the formation of composite materials. In particular, different noble metal NPs like Pd, Pt, Ag, and Au (mainly), and transition metals like Ti, Ni or many others are applied for modification CNTs surface to obtain unique features and properties. Mostly used fabrication techniques include evaporation or thermal methods for synthesis of metallic NPs and combination with CNTs. A list of various other ways can be used to fabricate nanohybrids of CNTs and metal NPs formed using metal salts {metal acetates, $PdCl_2$, $HAuCl_4$, $RuCl_3.6H_2O$, $(NH_4)_2PdCl_4$, H_2PtCl_6 or K_2PtCl_4, etc.), and their complexes. Physical and physicochemical methods are either indirect or direct techniques—a few examples are sputtering, electron-beam evaporation, thermal CVD, thermal covalent coupling, photoreduction, and pyrolysis.

Acknowledgments

Authors acknowledge Pakistan Science Foundation (PSF) for financial support under Research Grants No. PSF-NSFC III/Med/P-NIBGE (14). Dr. Waheed S. Khan is also grateful to their parent department *National Institute for Biotechnology and Genetic Engineering (NIBGE)* Faisalabad Pakistan for providing conducive environment for such type of writing work.

References

1. Cha J, Jin S, HunShim J, Park CS, Ho JR, Soon HH. Functionalization of carbon nanotubes for fabrication of CNT/epoxy nanocomposites. *Mater Des* 2016;**95**:1−8.
2. Im JS, Bai BC, Lee Y-S. The effect of carbon nanotubes on drug delivery in an electrosensitive transdermal drug delivery system. *Biomaterials* 2010;**31**(6):1414−9.
3. Liu X, Marangon I, Melinte G, Wilhelm C, Ménard-Moyon C, Pichon BP, Ersen O, Kelly A, Baaziz W, Pham-Huu C, Bégin-Colin S, Bianco A, Gazeau F, Dominique B. Design of covalently functionalized carbon nanotubes filled with metal oxide nanoparticles for imaging, therapy, and magnetic manipulation. *ACS Nano* 2014;**8**(11):11290−304.
4. Li H, Wang W, Lv Q, Xi G, Bai H, Zhang Q. Disposable paper-based electrochemical sensor based on stacked gold nanoparticles supported carbon nanotubes for the determination of bisphenol A. *Electrochem Commun* 2016;**68**:104−7.
5. Wu B, Kuang Y, Zhang X, Chen J. Noble metal nanoparticles/carbon nanotubes nanohybrids: synthesis and applications. *Nano Today* 2011;**6**(1):75−90.
6. Zhang B-T, Zheng X, Li H-F, Lin J-M. Application of carbon-based nanomaterials in sample preparation: a review. *Anal Chim Acta* 2013;**784**:1−17.
7. Bhattacharya K, Mukherjee SP, Gallud A, Burkert SC, Bistarelli S, Bellucci S, Bottini M, Star A, Fadeel B. Biological interactions of carbon-based nanomaterials: from coronation to degradation. *Nanomed Nanotechnol Biol Med* 2016;**12**(2):333−51.
8. Xue M, Tu X, Gao F, Xie Y, Huang X, Fernandez C, FengliQu, Liu G, Lu L, Yu Y. Hierarchical porous MXene/amino carbon nanotubes-based molecular imprinting sensor for highly sensitive and selective sensing of fisetin. *Sensor Actuator B Chem* 2020;**309**:127815.
9. Zhan L, Li S, Lau T-K, Cui Y, Lu X, Shi M, Li C-Z, Li H, Hou J, Chen H. Over 17% efficiency ternary organic solar cells enabled by two non-fullerene acceptors working in an alloy-like model. *Energy Environ Sci* 2020;**13**(2):635−45.
10. Pan Y, Yang L, Fu Q, Luo Q, Wang J, Hao Y, Yang R, Lai X, Zhao X, Wang N, Gao Q, Thakur R, Carreo C, Qian S, Hu L, Guo Z. Anchoring carbon nanotubes and post-hydroxylation treatment enhanced Ni nanofiber catalysts towards efficient hydrous hydrazine decomposition for effective hydrogen generation. *Chem Commun* 2019;**55**(61): 9011−4.
11. Mu P, Zhang Z, Bai W, He J, Sun H, Zhu Z, Liang W, An L. Superwetting monolithic hollow-carbon-nanotubes aerogels with hierarchically nanoporous structure for efficient solar steam generation. *Adv Energy Mater* 2019;**9**(1):1802158.

12. Feng J, Dong L, Li X, Li D, Lu P, Hou F, Liang J, Shi XD. Hierarchically stacked reduced graphene oxide/carbon nanotubes for as high performance anode for sodium-ion batteries. *Electrochim Acta* 2019;**302**:65−70.
13. Vikesland PJ. Nanosensors for water quality monitoring. *Nat Nanotechnol* 2018;**13**(8): 651−60.
14. Gao G, Jiang Y-W, Jia H-R, Yang J, Wu F-G. On-off-on fluorescent nanosensor for Fe3+ detection and cancer/normal cell differentiation via silicon-doped carbon quantum dots. *Carbon* 2018;**134**:232−43.
15. Idrissi ME, Meyer CE, Zartner L, Meier W. Nanosensors based on polymer vesicles and planar membranes: a short review. *J Nanobiotechnol* 2018;**16**(1):1−14.
16. Nazari A. Nanosensors for smart cities: an introduction. In: *Nanosensors for smart cities.* Elsevier; 2020. p. 3−8.
17. Dou M, Sanjay ST, Dominguez DC, Zhan S, Li XJ. A paper/polymer hybrid CD-like microfluidic SpinChip integrated with DNA-functionalized graphene oxide nanosensors for multiplex qLAMP detection. *Chem Commun* 2017;**53**(79):10886−9.
18. Wayu MB, Schwarzmann MA, Gillespie SD, Leopold MC. Enzyme-free uric acid electrochemical sensors using β-cyclodextrin-modified carboxylic acid-functionalized carbon nanotubes. *J Mater Sci* 2017;**52**(10):6050−62.
19. Fayemi OE, Adekunle AS, Ebenso EE. Metal oxide nanoparticles/multi-walled carbon nanotube nanocomposite modified electrode for the detection of dopamine: comparative electrochemical study. *J Biosens Bioelectron* 2015;**6**:190.
20. Aftab S, Kurbanoglu S, Ozcelikay G, Shah A, Ozkan SA. NH2-Functionalized multi walled carbon nanotubes decorated with ZnO nanoparticles and graphene quantum dots for sensitive assay of pimozide. *Electroanalysis* 2019;**31**(6):1083−94.
21. Sadiković M, Nigović B. Development of electrochemical platform based on carbon nanotubes decorated with zirconium oxide nanoparticles for determination of nebivolol. *Int J Electrochem Sci* 2017;**12**:9675−88.
22. George JM, Antony A, Mathew B. Metal oxide nanoparticles in electrochemical sensing and biosensing: a review. *Microchim Acta* 2018;**185**(7):1−26.
23. Munawar A, Tahir MA, Shaheen A, Lieberzeit PA, Khan WS, Bajwa SZ. Investigating nanohybrid material based on 3D CNTs@ Cu nanoparticle composite and imprinted polymer for highly selective detection of chloramphenicol. *J Hazard Mater* 2018;**342**:96−106.
24. Munawar A, , et alSchirhagl R, Rehman A, Shaheen A, Taj A, Bano K, Bassous NJ, Webster TJ, Khan WS, Bajwa SZ. Facile in situ generation of bismuth tungstate nanosheet-multiwalled carbon nanotube composite as unconventional affinity material for quartz crystal microbalance detection of antibiotics. *J Hazard Mater* 2019;**373**:50−9.
25. Verma I, Sidiq S, Pal SK. Poly (l-lysine)-coated liquid crystal droplets for sensitive detection of DNA and their applications in controlled release of drug molecules. *ACS Omega* 2017;**2**(11):7936−45.
26. Barthwal S, Singh NB. ZnO-CNT nanocomposite: a device as electrochemical sensor. *Mater Today: Proceedings* 2017;**4**(4):5552−60.
27. Barthwal S, Singh B, Singh NB. A novel electrochemical sensor fabricated by embedding ZnO nano particles on MWCNT for morphine detection. *Mater Today Proc* 2018;**5**(3):9061−6.
28. Abraham P, Renjini S, Vijayan P, Nisha V, Krishna S, Anithakumary V. Review on the progress in electrochemical detection of morphine based on different modified electrodes. *J Electrochem Soc* 2020;**167**(3):037559.
29. Saravanakkumar D, Uma Devi S, Sivaranjani S, Gnanasaravanan S, Ayeshamariam A, Ravikumar B, Pandiarajan S. Structural investigation on synthesized Ag doped ZnO-MWCNT and its applications. *J Nanosci Nanoeng Appl* 2018;**8**(2):2321−5194. ISSN: 2231-1777.

30. Afsharmanesh E, Karimi-Maleh H, Ali P, Vahedi J. Electrochemical behavior of morphine at ZnO/CNT nanocomposite room temperature ionic liquid modified carbon paste electrode and its determination in real samples. *J Mol Liq* 2013;**181**:8−13.

31. Iranmanesh T, Foroughi MM, Jahani S, Shahidi Zandi M, Hassani Nadiki H. Green and facile microwave solvent-free synthesis of CeO_2 nanoparticle-decorated CNTs as a quadruplet electrochemical platform for ultrasensitive and simultaneous detection of ascorbic acid, dopamine, uric acid and acetaminophen. *Talanta* 2020;**207**:120318.

32. Yang H, Li B, Cui R, Xing R, Liu S. Electrochemical sensor for rutin detection based on Au nanoparticle-loaded helical carbon nanotubes. *J Nanoparticle Res* 2017;**19**(10):354.

33. Bouabi YEL, Farahi A, Labjar N, Hajjaji SE, Bakasse M, Mhammedi MAE. Square wave voltammetric determination of paracetamol at chitosan modified carbon paste electrode: application in natural water samples, commercial tablets and human urines. *Mater Sci Eng C* 2016;**58**:70−7.

34. Shetti N, Malode SJ, Nayak D, Reddy KR. Novel heterostructured Ru-doped TiO_2/CNTs hybrids with enhanced electrochemical sensing performance for cetirizine. *Mater Res Express* 2019;**6**(11):115085.

35. Shashanka R, Kumara Swamy B. Simultaneous electro-generation and electro-deposition of copper oxide nanoparticles on glassy carbon electrode and its sensor application. *SN Appl Sci* 2020;**2**:1−10.

36. Malode SJ, Keerthi PK, Shetti NP, Kulkarni RM. Electroanalysis of carbendazim using MWCNT/Ca-ZnO modified electrode. *Electroanalysis* 2020;**32**:1590−9.

37. Mohamed R, Attia AK, Mohamed HY, Elshahed MS, Rizk M. Validated voltammetric method for the simultaneous determination of anti-diabetic drugs, linagliptin and empagliflozin in bulk, pharmaceutical dosage forms and biological fluids. *Electroanalysis* 2020;**32**.

38. Wen L-N, Wang A-T, Guo N, Han Y, Yan C, Yan D. Bio-characteristic profiling related to clinic: a new technology platform for quality evaluation of Chinese materia medic. *TMR Modern Herbal Med* 2019;**2**(4):215−24.

39. Zheng J, Zhang M, Yang L, Xu J, Hu S, Hayat T, Alharbi NF, Fan Y. Fabrication of one dimensional CNTs/Fe_3O_4@ PPy/Pd magnetic composites for the accumulation and electrochemical detection of triclosan. *J Electroanal Chem* 2018;**818**:97−105.

40. Ndagijimana P, Liu X, Li Z, Yu G, Wang Y. The synthesis strategy to enhance the performance and cyclic utilization of granulated activated carbon-based sorbent for bisphenol A and triclosan removal. *Environ Sci Pollut Control Ser* 2020:1−14.

41. Santana ER, Spinelli A. Electrode modified with graphene quantum dots supported in chitosan for electrochemical methods and non-linear deconvolution of spectra for spectrometric methods: approaches for simultaneous determination of triclosan and methylparaben. *Microchim Acta* 2020;**187**:250.

42. Bozal-Palabiyik B, Erkmen C, Uslu B. Molecularly imprinted electrochemical sensors: analytical and pharmaceutical applications based on ortho-phenylenediamine polymerization. *Curr Pharmaceut Anal* 2020;**16**(4):350−66.

43. Yin Z, Cui C, Chen H, Duoni XY, Qian W. The application of carbon nanotube/graphene-based nanomaterials in wastewater treatment. *Small* 2020;**16**(15):1902301.

44. Rekos K, Kampouraki Z-C, Sarafidis C, Samanidou V, Deliyanni E. Graphene oxide based magnetic nanocomposites with polymers as effective bisphenol−A nanoadsorbents. *Materials* 2019;**12**(12):1987.

45. Ghasemi A, Amiri H, Zare H, Masroor M, Hasanzadeh A, Ali B, Aref AR, Karimi M, Hamblin MR. Carbon nanotubes in microfluidic lab-on-a-chip technology: current trends and future perspectives. *Microfluid Nanofluidics* 2017;**21**(9):151.

46. Mishra NS, Kuila A, Ahmad N, Saravanan Pichiah D, Kah Hon Leong D, Min Jang D. Engineered carbon nanotubes: review on the role of surface chemistry, mechanistic features, and toxicology in the adsorptive removal of aquatic pollutants. *Chemistry* 2018;**3**(4): 1040—55.
47. Sun D, Li H, Li M, Li C, Dai H, Sun D, Yang B. Electrodeposition synthesis of a NiO/CNT/ PEDOT composite for simultaneous detection of dopamine, serotonin, and tryptophan. *Sensor Actuator B Chem* 2018;**259**:433—42.
48. Gholivand MB, Shamsipur M, Paimard G, Feyzi M, Jafari F. Synthesis of $Fe-Cu/TiO_2$ nanostructure and its use in construction of a sensitive and selective sensor for metformin determination. *Mater Sci Eng C* 2014;**42**:791—8.
49. Tanaka S, Kaneti YV, Septiani NLW, Dou SX, Yoshio B, Hossain M, Shahriar A, Kim J, Yamauchi Y. A review on iron oxide-based nanoarchitectures for biomedical, energy storage, and environmental applications. *Small Methods* 2019;**3**(5):1800512.
50. Chakraborty P, Das S, Nandi AK. Conducting gels: a chronicle of technological advances. *Prog Polym Sci* 2019;**88**:189—219.

Role of functionalized metal oxide–carbon nanocomposites in biomolecule detection

19

Sumaira Younis[1,2], Rabisa Zia[1], Ayesha Taj[1], Amna Rafiq[1], Hunza Hayat[1], Nafeesa Nayab[1], Waheed S. Khan[1] and Sadia Zafar Bajwa[1]

[1]Nanobiotechnology Group, Industrial Biotechnology Division, National Institute for Biotechnology and Genetic Engineering (NIBGE), Faisalabad, Punjab, Pakistan; [2]Department of Chemistry, University of Agriculture, Faisalabad, Punjab, Pakistan

19.1 Introduction

Semiconductor metal oxides, carbon-based materials, and their composites have extensive applications in energy storage, high-performance electronics, environmental remediation, and energy conversion procedures. They have potential applications in solar cells, supercapacitors, photocatalysts, and biosensors due to their good catalytic and biocompatible properties. Nanomaterials play a significant role in designing innovative analytical probes that generate measurable analytical signals on exposure to the desired analyte molecules.[1]

The progress of recent diagnostic techniques having remarkable high sensitivity, simple operation, and low cost addresses fundamental needs of advanced medical diagnostics, environmental monitoring, and clinical investigations.[2] Many scientists have offered a variety of procedures such as chromatography, fluorescence, microdialysis, and surface-enhanced Raman scattering *to detect* biomolecules, disease-causing agents, viruses, and bacteria. All these techniques have their advantages but have some limitations to overcome.[3] Numerous researchers have conducted substantial research to develop highly selective, sensitive, and accessible detection techniques to overcome the limitations of previous conventional approaches. Detection based on sensing platforms with embedded nanostructures offers remarkable benefits over conventional examining measures due to sensitivity and practical feasibility. In such instances, sensitivity considerably depends on the properties of transducer biointerfaces, the sizes, shapes, and types of nanostructures, and their modification with bioreceptors such as aptamers, antibodies, and nucleic acids. Generally, nanomaterials smaller than 100 nm are used for biosensing platforms that exhibit a good stoichiometry with biomolecule size and comprise nucleic acids, proteins, bacteria, viruses, etc. These bimolecular stoichiometric interactions with nanomaterials contribute to increasing the sensitivity of the biosensing platform. They also reduce the number of bioreceptors to even a single molecule as they are anchored on the nanomaterials layered on the transducer surface,[2] which results in detection of minute quantities of targeted analytes with a minimum detection limit. Additionally, nanosized materials show unique chemical and physical characteristics different from those exhibited by bulk materials. These properties play fundamental roles in increasing the sensitivity of detection systems.[4]

The functionalization of nanostructures is of prime importance for fabricating sensitive and selective biosensors to detect the specified target biomolecule or infectious agent. Functionalization occurs through covalent or noncovalent interfaces by introducing various organic and inorganic functional groups to augment the binding of analyte molecules by creating electrostatic or bonding interactions. Such interactions between organic and inorganic matter lead to the stabilization of nanosensors by decreasing the agglomeration of pure inorganic and organic materials. For this purpose, various organic compounds, such as thiols, carboxylic acids, and amines, are also in practice to functionalize nanomaterials. Nanomaterials can also be modified by immobilizing recognition elements, including protein receptors, enzymes, antibodies, and probe molecules, to create bioselective receptor interfaces sensitive to capturing the specific analyte molecules.[1] Modification of electrodes with metallic, carbon-based nanomaterials and their composites gained much attraction in the electrochemical analysis due to their high surface area, extraordinary electrocatalytic properties and good biocompatibility.[3]

19.1.1   Functionalized metal oxide–carbon nanocomposites

The development of innovative materials is highly required in different scientific areas. Two or more elements are combined to obtain the desired material with remarkable properties, termed composite material. Such materials consist of two or more phases or materials on a scale greater than the size of an atom. Such as composites of carbon-based nanomaterials with metallic nanomaterials, i.e., carbon nanotube (CNT)–metallic nanomaterials, graphene metal/metal oxides, etc. have multiple applications in energy and health monitoring procedures.[5]

Recently, carbon-based nanoarchitectures have become more attractive choices due to their existence in various allotropic forms. Carbon is considered a basic nanomaterial due to its occurrence in zero-dimensional (nanospheres, fullerenes, quantum dots, and nanodiamonds, etc.), one-dimensional (carbon nanowires and carbon nanotubes), two-dimensional (graphene oxide), and their composites have distinctive chemical and physical properties.[4] Graphene oxide (GO) and carbon nanotubes (CNTs) are considered promising materials in science and technology. Their unique properties play a vital role in multiple biotic applications such as tissue engineering, cancer therapy, diagnosis, imaging, drug delivery, and biosensing procedures.[6] Due to their similar surface structures, comparable biofunctionalization strategies could be adopted, but the level of functionalization will be different according to their respective surface area. Biomolecules, such as proteins or nucleic acids or biorecognition elements with aromatic ring structures, are anchored through π-π interactions on the hydrophobic surfaces of CNTs and GO, offering a highly sensitive biosensing platform. Carbon-based nanocomposites are also functionalized through covalent interactions using photochemical, electrochemical, and thermally activated chemical reactions to generate active functional groups on GO or CNTs surfaces owing to the occurrence of oxidation. Then these functional groups such as hydroxyl, carboxyl, and epoxide behave as active sites for binding the bioreceptor element through chemical interactions.[4]

19.1.2 Carbon nanotubes

Carbon nanotubes, pseudo-1-D carbon allotropes, are considered a network of carbon atoms arranged in continuous cylinders consisting of one (single-walled CNTs) or multiple layers (double-walled CNTs) either with close or open ends. CNTs are used in medical and sensing devices due to their large surface area, outstanding mechanical strength, electrical and thermal conductivity, hydrophobic nature, and durable intermolecular π-π interactions. Functionalized CNTs are used to overcome toxicity and biocompatibility issues. Functionalization through chemical adsorption augments the stability and solubility of CNTs.[7] Typically, they form composites with metal oxides through their homogenous adornment around CNTs, and the resulting composite has remarkably high electrochemical properties. The high aspect ratio (ratio of width to height) of CNTs in a proceeding network assists the electron/charge transfer by reducing contact resistance toward adjacent nanomaterials, and they facilitate the charge transport pathway along the longitudinal direction. Because of this, CNT-based nanomaterials and their composites are significantly used as coelectrode materials to improve the rate aptitude of electrochemical energy devices/sensors.[8]

19.1.3 Graphene

Two-dimensional graphene nanosheets comprising tightly packed sp^2 hybridized carbon atoms result in a flat monolayer with high chemical stability, a large surface area, outstanding mechanical and electronic characteristics, and good biocompatibility. Due to its exceptional features, it is used for the adsorption of nanomaterials during the formation of metallic nanocomposites with graphene.[9] The aromatic structure of graphene plays an important role as a substrate for capturing target analyte molecules, whereas the intrinsic flexibility moderates the steric hindrance for elevated interactions between graphene, substrate, and analyte molecules. This contributes to effective adsorption compared with sphere-shaped affinity materials. Due to their distinctive properties, graphene and its derivatives have been explored as credible substrates to immobilize and separate biomolecules. For this, researchers modified GO by a combination of aptamers to bind and detect the targeted analytes from biological fluids. Graphene has been utilized extensively as an electrode/transducer modifying material to enhance the active surface area and accelerate smooth electron transfer, which facilitates the detection of biological and environmental analytes.[10]

Graphene composites with metallic nanomaterials gained much interest due to the synergy effect of both metal nanoparticles (NPs) and graphene, promising to augment electrocatalytic behavior. Graphene and its derivatives offer a versatile scaffold for NPs to develop hybrid nanocomposites with upgraded characteristics due to their oxygen functionalities and defects.[3]

19.1.4 Preparation of metal oxide–carbon nanocomposites

Generally, there are two ways to develop carbon-based inorganic nanocomposites, i.e., (1) ex situ hybridization and (2) in situ decoration.

In the case of ex situ hybridization, both inorganic and carbon-based nanomaterials are synthesized separately through their respective traditional procedures such as hydrothermal, coprecipitation, solvothermal, chemical vapor deposition, and Hummers' method. The inorganic (metallic) NPs are then adsorbed on the surface of carbon-based nanomaterials—GO, reduced graphene oxide (rGO), CNT— through either covalent or noncovalent interactions. This process results in metallic nanostructures with controlled sizes and well-defined shapes. Moreover, the surfaces of inorganic materials can be modified before hybridization with carbonaceous nanomaterials.

In situ methods offer a one-pot synthesis of hybrid materials in which metal precursors are put directly into carbon-based materials. During this process, inorganic nanostructures are adsorbed on the surface of graphene or any other carbonaceous nanomaterial and form novel nanocomposites having excellent physicochemical properties. The oxygenated functional groups in carbon nanomaterials give nucleation sites to facilitate the crystallization of metallic nanostructures during the synthetic reaction. Moreover, the same reducing agent simultaneously reduces the GO and metal precursors in the solution.[7,11,12]

19.2 Detection of biomarkers

Biochemical functions in living things are regulated by numerous biomolecules like metabolites, amino acids, vitamins, neurotransmitters, minerals, complex proteins, and nucleic acids that can influence metabolism and serve as biomarkers in case of diseases like cancer, metabolic disorders, and others. The biomarkers can be detected by stable, accurate, and highly sensitive systems.[13] Biomolecules are used as disease biomarkers for many point-of-care diagnostics applications.[14] Normal biological processes can be compared with altered biological processes, e.g., pathogenic processes or the responses of therapeutic pharmacological interventions using biomarkers. Biomarkers can be predictive, diagnostic, or prognostic. Diagnostic biomarkers help with appropriate diagnosis, as in the case of cancers and cardiovascular diseases, these markers can identify, for example, the type of cancer, such as the size of prostate identified by prostate-specific antigen potentially indicates the prostate cancer.[15] Antibody–antigen interaction-based immunoassays, enzyme-linked immunosorbent assays (ELISAs), and radioimmunoassays are techniques for detecting cardiac protein markers.[16,17]

Some diseases are linked to a specific biomarker—e.g., cancers such as prostate and ovarian cancer—while others, like cardiac disorders, are related to multiple biomarkers because there are various heart disorders and a large number of associated proteins not specific to just one disorder. Such diseases are detected using multiplex biosensing devices.[18]

Biomarkers can be detected by fabricating a suitable biosensor platform. The combination of analyte (biomarker) and bioreceptor generates signals transduced into a readable signal, e.g., an optical or electrochemical signal, or an affinity-based interaction may occur, e.g., antibody and antigen interaction or DNA and its complement hybridization. Advanced nanotechnology has enabled us to construct more sophisticated, sensitive, and high-throughput analytical devices. This has significantly increased the time window for analysis and reduced assay costs. Such devices come under the term of point-of-care diagnostic platforms. These interfacial biosensors comprise metal—biomolecular interfaces that detect disease-associated biomarkers by simply analyzing their altered interaction behavior that makes them an attractive technique for the development of advanced biomarker diagnostic and therapeutic platforms.[19]

19.2.1 Detection of cardiac biomarkers

Cardiovascular diseases are the leading cause of death globally, with an estimated 17.9 million deaths in 2016 alone accounting for 31% of all global deaths. Heart attack and stroke account for 85% of all deaths attributed to cardiovascular diseases, and 75% of these deaths are in underdeveloped and developing countries. People who suffer from cardiovascular diseases along with diabetes, hypertension, or hyperlipidemia have a higher mortality risk.[20]

19.2.2 Types of cardiac biomarkers

C-reaction proteins (CRPs), creatinine kinase-MB (CK-MB), cardiac troponins (troponin I and troponin T), myeloperoxidase, myoglobin, and fatty acid-binding protein, are the cardiac biomarkers used to detect acute myocardial infarction (AMI).[21] Oxidation of biomolecules at the cellular level with the formation of carbonylated reactive oxygen and nitrogen species is related to pathological conditions including metabolic disorders, cardiovascular diseases, etc. therefore, biomolecules carbonylation is considered a biomarker for pathologies related to oxidative stress.[22]

CK-MB and CRP are well-established, while D-dimer and PAPP-A are the emerging cardiac markers for cardiovascular diseases. CK-MB is a creatine kinase isoenzyme in heart muscle used in routine biochemical myocardial infarction. CRP is an early inflammatory diagnostic marker with increased expression in the blood vessel wall. D-dimer is the by-product of the degradation of fibrin, a biomarker for the early diagnosis of deep vein thrombosis and atherosclerosis. Dimeric pregnancy-associated plasma protein A, a recently discovered biomarker for AMI and acute coronary syndromes, is highly expressed in unstable coronary atherosclerotic plaques.[18]

Troponin-T and troponin-I are both markers used in the diagnosis of cardiovascular diseases, especially AMI. Biosensors based on the troponin biomarker are used for quick, selective, sensitive, and portable detection.[23] The cardiac troponin I (cTn-I) concentration in a patient's blood is between 0.1 and 1 ng/mL.[24]

19.2.3 Role of functionalized metal oxide–carbon nanocomposites in cardiac biomarker detection

Graphene is a unique planar structure with a honeycomb-like carbon arrangement with structural, optical, and electrochemical properties suitable for biosensing. Molybdenum tetraselenide, nMo3Se4 with an rGO-based electrochemical immunosensor, is one of the biosensors used to detect cardiac cTn-I. The rGO is functionalized with 3-aminopropyltriethoxysilane. The surface is immobilized with monoclonal anticardiac troponin I, and nonspecific binding is avoided through bovine serum albumin. The metal oxide, carbon nanocomposite-based immunosensor has higher sensitivity, stability, low detection limit, and better linear detection range for cardiac biomarker troponin-I.[23]

19.2.4 Comparison and prospects

Cardiac biomarkers must be predictive, specific, and sensitive because the damage caused by AMI to myocardial cells is abrupt, irreversible, and life-threatening for the patient.

There are certain precautions associated with it such that monoclonal antibodies must be attached in a way that they have Fab site exposed for cardiac biomarker (analyte) attachment to retain maximum functionality.[18] This is a highly specific but difficult and expensive method; thus, other methods are being explored for matching its specificity at a lower cost.

There are certain limitations of biosensors in AMI diagnosis with cardiac biomarkers such as cardiac troponin I (cTn-I), which has a very low concentration approximately between 0.1 and 1 ng/mL in the patient's blood. As an alternate, a sensitive label-free fiber-optic-based immunosensor can sense a cTn-I concentration as small as 10.8 pg/mL with high specificity and in a short period. 193 nm laser inscription can further improve the limit of detection (LOD) and spectral resolution, which makes it a suitable POC diagnostic device for AMI.[25] Nanoscale cost-effective biosensors are under study for early detection of disease; thus, smartphone-based visual analysis and monitoring are the potential future of nanobiosensing.[26]

19.2.5 Detection of cancer biomarkers

Cancer is the term used for a large group of diseases related to the abnormal growth of cells that metastasize to the whole body, leading to patient death. Globally, it is the second leading cause of death, with an estimated 9.6 million deaths in just 2018, which means every sixth death in the world is due to one or another type of cancer.[27,28] The death toll of cancer patients is highest in lung, colorectal, stomach, liver, and breast cancers. Genetic factors and physical, chemical, and biological carcinogens are the causative agents of cancer.[29] The diagnostic and treatment facilities in high-income countries are threefolds that of the low-income countries. Thus, there is room for cost-effective early diagnosis and treatment to reduce cancer mortality.[30] Basic screening to check abnormalities related to cancer development is the prerequisite that can identify the threat. Screening methods include cancer biomarker detection, visual inspection with acetic acid, cytology testing, and mammography screening, as in the case of breast cancer. Clinical evaluation at an early stage makes treatment easier and curative.[31]

19.2.6 Types of cancer biomarkers

There are different types of cancers, and some biomarkers are specific for each cancer type like carcinoembryonic antigen (CEA) is a biomarker for lung cancer, carbohydrate antigen CA for breast cancer, cytokeratin for oral cancer, and cell carcinoma antigen for ovarian cancer, while others like 8-hydroxy-2-deoxyguanosine (8-OHdG), glutathione, uric acid are biomarkers of stress and reactive oxygen species oxidative DNA damage.[32] Circulating tumor cells (CTCs), as the name indicates, are present in the body during various types of cancers. CTCs are not present in the bloodstream otherwise and can be detected by microfluidic platforms that can functionally characterize individual cancer cells at the molecular level.[33] MiRNAs are major tumor suppressor families, and MicroRNA-34 is a biomarker for lung cancer as it targets CDK4/6, a cell cycle protein; BCl2, an antiapoptosis protein; MYC and CD44, metastasis-related proteins.[34]

Colon cancer biomarkers like CEA, cancer antigen 19-9 (CA 19-9), cancer antigen 125, miRNA (miRNA-29a and miRNA-92a), ssDNA (colorectal cancer gene) have been used for diagnosis, and they have certain limitations like specificity, stability, etc. Endothelin-1 (ET-1) is a relatively new biomarker under study with elevated levels in colon cancer patients' blood plasma.[35]

Breast cancer is the highest diagnosed cancer among women worldwide. Diagnostic methods like Raman spectroscopic analysis of cells, tissues, and biofluids and metabolic end products of cancer cells are being studied to explore novel tumor biomarkers like serum-based tumor markers (CA 15-3, CA 27-29), tryptophan, tyrosine, phenylalanine, carotenoids, and lipids.[36]

19.2.7 Role of functionalized metal oxide—carbon nanocomposites in cancer biomarkers' detection

Graphene has physicochemical properties that favor its application in biosensing for reliable and sensitive detection of biomarkers like CEA, carbohydrate antigen CA and cell carcinoma antigen. Sulfur-doped reduced graphene oxide (SrGO) is used to fabricate a highly sensitive electrochemical biosensor to detect the 8-OHdG molecule, oxidative stress, a cancer biomarker (Fig. 19.1).

Fig. 19.2 shows that SrGO offers excellent sensitivity(~ 1 nM) and a very wide detection window (20—0.002 M) along with good selectivity, high stability and reproducibility, and excellent recoveries for the detection of an 8-OHdG biomarker in various spiked urine samples as suggested by the abovementioned voltammograms. The electrochemical sensitivity of SrGO sensor is because of sulfur's strong electron-donating ability and strong catalytic activity of the doping sites in S-doped

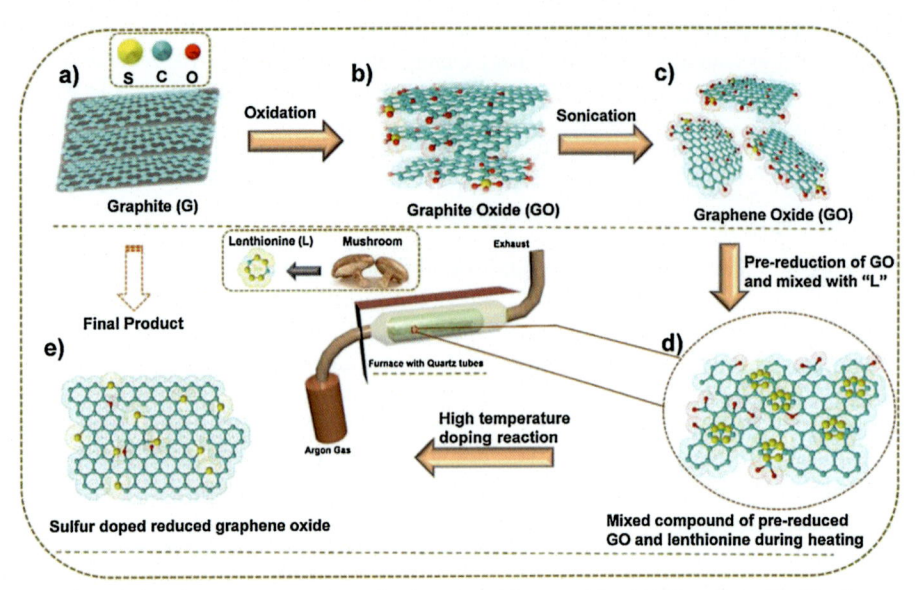

Figure 19.1 Synthesis of sulfur-doped reduced graphene oxide: (A) Graphite; (B) Graphene oxide (GO); (C) Exfoliated graphene oxide; (D) Mixing pre-rGO with lenthionine; (E) Sulfur-doped reduced graphene oxide (SrGO) is adapted from.[32]

graphene. In addition, the 8-OHdG oxidative biomarker has high conductivity, increased electrode surface area, and relatively high adsorption capacity. This work is among the new potential applications for functionalized metal oxide–carbon nano-composites for electrochemical detection of cancer biomarkers.[32]

19.2.8 *Comparison and prospects*

It is important to detect biomolecules or CTCs at the single-molecule or single-cellular level to understand cancer heterogeneity and its early diagnosis. The complexity of blood media and the low concentration of the biomarker makes it extremely difficult; however, label-free techniques are a promising option. Its integration with microfluidic technology is serving as a powerful tool in cancer research, especially when it comes to cancer biomarker diagnosis.[33]

Electrochemical-based sensors are cost-effective, easy to handle, signal sensitive, and portable in the form of POC devices to detect cancer biomarkers. On the other hand, aptasensors and DNA are difficult to handle and depend on multiple factors like specific DNA sequences, temperature, and other experimental measures.[32]

Paper-based analytical sensors designed for noninvasive biomarker detection constitute an advanced field that has opened up avenues for diagnostic research because of the sensors' low cost, stability, and efficiency. Paper-based biosensors have also been applied in the detection of exosomes released by cancer cells. Carbon nanotubes, graphene, and GO-based nanocomposites enhance biomarker sensitivity

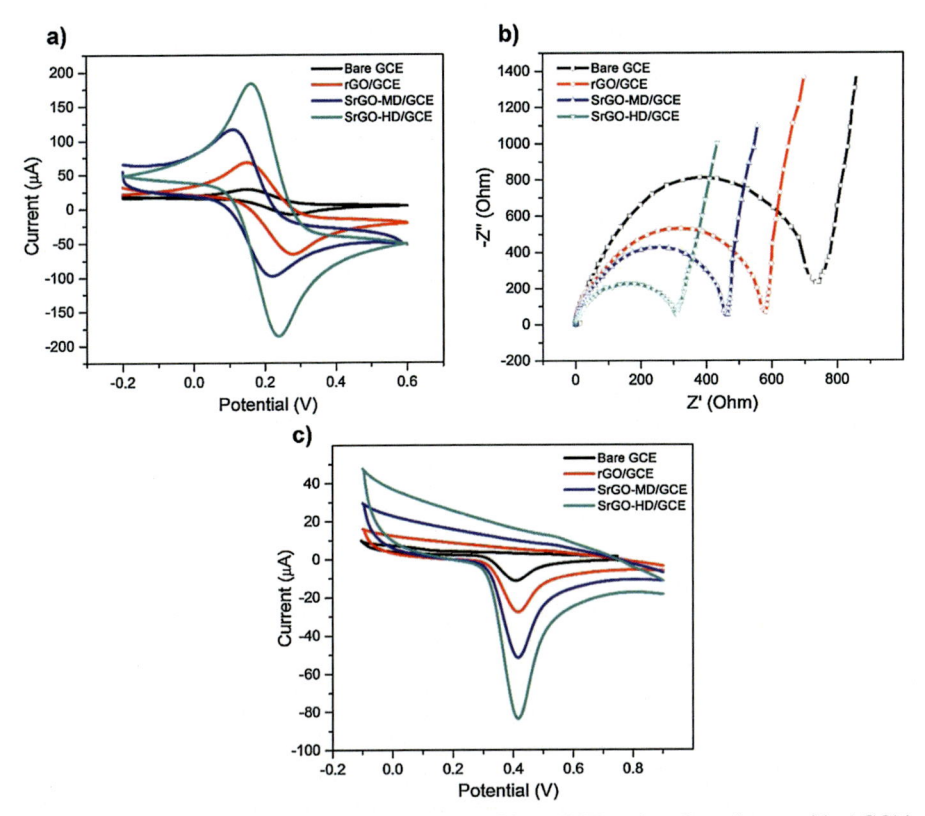

Figure 19.2 (A) Typical cyclic voltammograms of bare GCE, reduced graphene oxide (rGO)/ GCE, sulfur-doped reduced graphene oxide (SrGO)−MD/GCE and SrGO−HD/GCE (B) Typical EIS analysis of various modified GCE. (C) Cyclic voltammograms of bare GCE, rGO/ GCE, SrGO−MD/GCE, and SrGO−HD/GCE in the presence of 10 M 8-OHdG in 0.1 M PBS (pH 7.2) at a scan rate of 100 mV/s is an adaptation from.[32]

and specificity. Materials such as polylactic acid, polypropylene, Teflon, and glass fibers are also being explored for their mechanical strength, porosity, surface hydrophobicity, and interaction capabilities. Polydimethylsiloxane is a combination of polymer and cellulose with improved biosensing performance and durability.[37] The cancer biomarkers and biosensors are evolving with the understanding of the material and advancement in biotechnology.

19.3 Detection of biomolecules

19.3.1 Uric acid

UA is a waste product created when purines are normally metabolized in the human body. Purine is a double-ring compound present in DNA, and their normal level in urine and serum is about 250–520 µm and 1.49–4.46 mM, respectively.[38–40] The UA level is also related to several disorders such as gout, hypertension, obesity, diabetes, high cholesterol, and heart diseases. Some conventional techniques used to detect uric acid include high-performance liquid chromatography, enzyme testing, and capillary electrophoresis.[41,42]

These methods are time-consuming, cumbersome, costly, technically challenging, and require skilled personnel. Electrochemical methods are preferred because they are simple, cost-effective, rapid, and sensitive compared with traditional methods.[40,43] Nanomaterials modify the interfaces for innovative applications such as electrochemical detection of important biological molecules. Two or more nanomaterials are combined to fabricate a hybrid nanostructure where each contributes its characteristics in combination with others, leading to diverse and interesting applications, including device fabrication and disease diagnostics.[44]

An innovative but simple, in situ biobased approach allows the effective growth of gold nanostructures on graphene surfaces to synthesize a three-dimensional Au NS@ GO hybrid. The morphology of a hybrid is investigated by various microscopic techniques that show uniform dispersion of ultrasmall gold nanostructures with a diameter of about 2–8 nm anchored on graphene sheets. The hybrid is reported as an effective receptor material for electrochemical binding efficiency for uric acid, which constitutes an essential biomolecule for the metabolism of human beings. The designed hybrid materials can detect the uric acid level to 30 nm. Fig. 19.3 shows the cyclic voltammetry responses of the bare glassy carbon electrode (A), gold nanostructure (B), GO (C), and Au NS@GO hybrid (D) against 0.3 µM uric acid. These responses showed that the Au NS@GO hybrid provides a sharper electrochemical reaction to uric acid than the other methods listed. The response is dependent on uric acid concentration (0.001–0.8 µM) (E), and the scan rate proves that it is an efficient charge transfer kinetics process (F). This approach is appropriate for designing novel hybrid materials with an attractive morphology and outstanding functionality to distinguish significant therapeutic biomolecules.[45]

For sensitive uric acid (UA) determination, a novel and simple ratiometric electrochemical sensor is designed. The gold–silver bimetallic NPs (Au–Ag NPs) are fabricated by one-step coreduction of $HAuCl_4$ and $AgNO_3$ (as precursors) and used to deposit on a glassy carbon electrode surface. The self-assembly of GO and thionine (TH) formed a GO–TH complex by electrostatic interactions. To construct the Au–Ag/GO/TH@GCE sensing network, the complex is drop-coated on Au-Ag NPs. Square wave voltammetry is reported for the electrochemical signal responses against UA. The redox current peak intensity gradually increases with the increase in UA concentration (1–100 µM) with an LOD of 0.3 µM. An effective sensor is designed on the sensing platform to enable UA's responsive ratiometric sensing and possible interferents in urinary fluids and total human serum. The experimental findings demonstrated the strong reliability of the sensor and stability, suggesting its strong UA tracking capability in biological fluids. Different nanocomposites used in the detection of UA with their respective detection limit are compiled in Table 19.1.

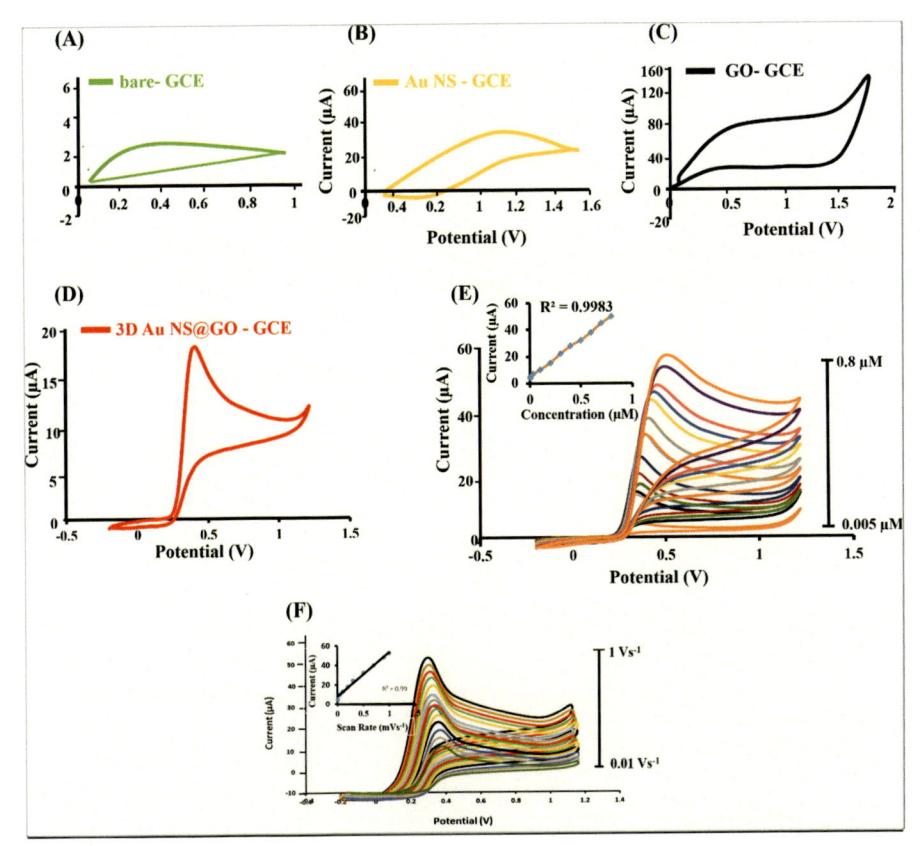

Figure 19.3 Cyclic voltammetry (CV) signals of bare, graphene oxide, gold nanostructure, and 3-D Au Ns@GO hybrid with a modified electrode at 0.3 μM uric acid background electrolyte NaH$_2$PO$_4$. H$_2$O (50 mM), a scan rate of 50 mV/s. (A) CV signal response of bare, (B) gold nanostructures, (C) graphene oxide (D) 3-D Au Ns@GO hybrid, (E) Response of 3-D Au Ns@ GO hybrid for different concentrations of uric acid (0.001−0.8 μM), a scan rate of 50 mV/s with background electrolyte NaH$_2$PO$_4$. H$_2$O (50 mM), Outset demonstrates linear regression plot of the anodic current against different uric acid concentrations, (F) CVs of 3-D Au Ns@ GO hybrid against 0.8 μM; inset shows a linear relation plot; experimental conditions are with background electrolyte NaH$_2$PO$_4$. H$_2$O (50 mM), scan range 0.1−1 V/s.[45]

19.3.2 Ascorbic acid

Ascorbic acid plays an important role in initiating a defense mechanism by triggering several critical enzymatic reactions. It is an effective drug to cure minor diseases such as the common flu with serious diseases such as cancer and AIDS. Ascorbic acid was used extensively for its antioxidant properties in the medicinal, dairy, cosmetics, and chemical industries.[48] The bulk industrial processes require strict quality control that requires a reliable, accurate, and robust method to track the concentration of ascorbic acid in routine examinations.[40] Materials with good stability and high electrical conductivity, such as metallic and nonmetallic NPs and carbon-based nanohybrids,

Table 19.1 Values of analytical parameters of various nanomaterials for the detection of uric acid.

Material	Linear range (μM)	Detection limit (μM)	Reference
(ERGO)/GCE	0.5−60	0.5	41
AuNPs−β-CD−Gra/GCE	0.5−60	0.2	42
Au−Ag NPs/GO/TH@GCE	1−100	0.3	41
Au/rGO/GCE	8.8−530	1.8	40
CTAB-GO/MWCNT/GCE	3.0−60	1.5	39
Chitosan-graphene modified GCE	2.0−45	2	46
rGO−PAMAM−MWCNT−AuNPs	20−1800	0.3	43
Fe_3O_4@Au−S−Fc/GS-chitosan modified GCE	1−300	0.2	47
Au Ns@GO biohybrid	0.001−0.8	0.1	45

have been reported as a transducer material for the electrochemical investigation of ascorbic acid.[49]

Composite materials with fascinating morphological and chemical configurations can be tailored by combining the intriguing properties of different nanostructures. Assembling nanomaterials with different shapes, morphology, and chemical complexities may result in high-performance composites.[50−52]

CeO_2 nanocubes are grown on carbon nanotubes to fabricate a synergistic nanocomposite. Microscopic analysis demonstrated that the extremely dense population of ceria cube on the surface of CNTs. A unique three-dimensional morphology offered a network of microcavities and pits for better transport and a wider region to promote exceptional electrical catalysis. The hybrid purity is verified by X-ray diffraction analysis, whereas Fourier transform infrared spectroscopy tests the existence and use of functional groups in hybrid formation. The hybrid exhibited a strong electrical potential to detect ascorbic acid and very low potential for rapid oxidation (0.1 V). It is noticed that the detection limit is to be 70 nM (S/N = 3). The 3-D morphology and versatile chemical composition with better catalytic properties can provide pathways to developing high-performance functional hybrids for the determination of other specimens or analytes.[53] Different metal and metal oxide nanocomposites with their analytical parameters for detecting ascorbic acid are listed in Table 19.2.

19.3.3 Glucose

Glucose is a six-carbon compound with aldohexose contributed by the aldehyde group and a monosaccharide known as dextrose. Glucose exists in both cyclic and acyclic conformations. It occupies a pivotal role in animal metabolism. It is the main product of photosynthesis and produces ATP and NADH in cellular respiration.[59] In contexts of human health, glucose homeostasis is of central importance; brain cells survive on external supplies of glucose because these cells are unable to synthesize glucose. Physiological concentrations of glucose in humans must be less than 5.5 mM.[60] Therefore,

Table 19.2 Values of analytical parameters of various nanocomposites for the detection of ascorbic acid.

Material	Linear range (mM)	Detection limit	Reference
Graphene oxide (GO)/multiwalled carbon nanotube (MWCNT) GCE	5.0—300	100	39
f-MWCNT—NF—PtAu composite GCE	1.2—4.8	8823	54
MWCNT/CPE	0.6—112	20,000	55
Pd/carbon nanofiber composite GCE	0.05—4	1500	56
Electrochemically reduced GO	0.5—0.2	30,000	39
(AuNCs)/(AGR)/MWCNT nanocomposite	10—150	270	57
Reduced GO—ZnO composite	0.05—2.35	371	58
CeO$_2$ NCs@CNTs hybrid	0.05—0.5	70	53

maintenance of an optimal glucose level is critical for the body. A high glucose level is a primary symptom of diabetes, and a low glucose level is an indication of hypoglycemia.[61]

Accuracy in glucose detection is critical for the clinical diagnosis of diabetes.[62] Conventionally, blood glucose levels are monitored by an amperometric detection system consisting of a highly selective glucose oxidase (GOx) immobilized electrode. Glucose oxidase oxidizes glucose, and electrons are generated; thus, glucose concentration is estimated from the number of electrons produced.[63] In terms of enzyme specificity, the GOx-based system is highly specific, but enzyme purification, immobilization, and protection from denaturants are drawbacks. Moreover, the redox center of the GOx enzyme is embedded deep into the electrode, demanding certain electron shuttles in the samples. For these reasons, the nonenzymatic detection of glucose has recently gained significant interest.[64] Carbon-based nanocomposites have been largely explored for electrochemical sensors fabrication due to their superior electrical conductivity, low cost, and corrosion resistance in the presence of various electrolytes.[65] Generally, metal oxide NPs offer high electrocatalytic activity owing to the larger surface area that offers more active sites for electrochemical reactions.[66] In a recent study, a screen-printed carbon electrode modified with copper oxide (CuO) NPs has been employed for glucose detection. The sensor offers a flexible, modifiable, nonenzymatic amperometric detection system. The sensor exhibited high sensitivity and selectivity for glucose as well as robust analysis. Moreover, the biosensor offered good recoveries for real samples. Fig. 19.4 outlines the synthesis of the nanocomposite and further detection of glucose by the synthesized composite.[67]

Xiao et al. reported a metal oxide/metal—carbon-based nanocomposite for the detection of glucose. The composite was synthesized by the direct carbonization of bimetallic metal—organic framework. The synthesized composite has a well-dispersed bimetallic material over the carbon matrix. The combined properties of both metals and carbon matrix exhibited exceptional sensitivity, selectivity, and reproducibility for glucose sensing. The LOD was calculated to be 0.06 μM. Additionally, results of glucose detection in real samples signified the potential application of glucose biosensors in diagnostics.[68]

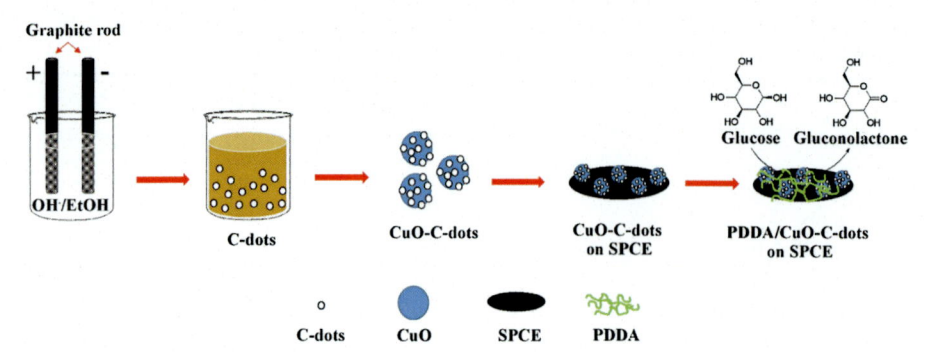

Figure 19.4 Schematic illustration for the synthesis of the CuO—carbon composite and detection of glucose by the assembly.
Reproduced with permission.[67]

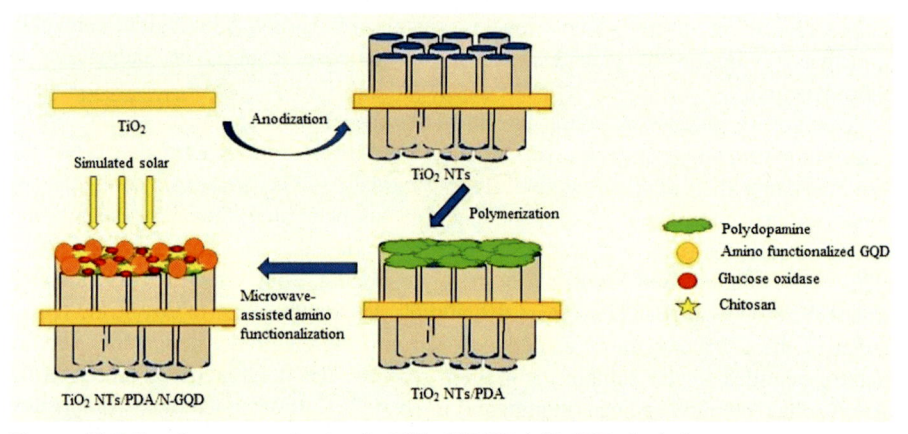

Figure 19.5 Step-by-step synthesis of a TiO$_2$ NTs/PDA/N-GQD dual-electron-acceptor biosensor.[69]

Another rapid and sensitive method for the detection of glucose was proposed by Yang et al. (2020); the photoelectrochemical biosensor consisted of titanium oxide nanotubes (TiO$_2$ NTs) modified with polydopamine (PDA), amino-functionalized graphene quantum dots (N-GQDs), and glucose oxidase (GOx) immobilized on the electrode. The dual electron acceptor structure was constructed by the synergistic effect of GQD-PDA on the surface of TiO$_2$ NTs, which offered enhanced electron transport. PDA and N-GQDs acted as active sites for the photogenerated holes, the enzyme reacted at these holes, and in this way, both PDA and N-GQD acted as electron acceptors, imparting a high-sensitivity sensor response. The LOD for this sensor was calculated as 0.015 mM, with a broader linear range of 0.11 mM. A dual electron acceptor provided enhanced sensitivity and linear range. Furthermore, excellent stability and uninterrupted activity in the presence of interfering agents signified the selectivity of the assembly. Fig. 19.5 outlines the preparation of the TiO$_2$ NTs/PDA/N-GQD dual-electron-acceptor biosensor.[69]

19.3.4 Dopamine

Neurotransmitters are the nuncio of the nervous system. These are tiny molecules that could transmit signals across a synapse from one nerve cell to another. These entities are crucial for controlling animal behaviors. In general terms, neurotransmitters are membrane-enclosed molecules near the nerve junctions, synapses, and the nerve cells housing these vesicles are presynaptic cells. Whenever these vesicles are stimulated, they release the neurotransmitter across the synapse, and released neurotransmitters bind with their respective receptors on the membranes of postsynaptic cells, resulting in depolarization of the membrane cascading the nerve-impulse.[70,71] Acetylcholine was the first-ever neurotransmitter to be described in 1914. So far, a plethora of neurotransmitters have been identified. Among these, dopamine, serotonin, noradrenaline, and acetylcholine are significant in the etiology of mental disorders.[72]

Dopamine (DA, 3, 4-dihydroxyphenylalanine) is a vital neurotransmitter in the brain and body.[73,74] It is critical for the transmission of signals to the brain. The physiological concentration of DA in the body is 10—1000 nM/L. Fluctuation in DA levels significantly increases the risk of neurological disorders such as schizophrenia, Parkinson's disease, and depression, etc.[75] Therefore, the significance of DA in clinical practices has attracted researchers to develop robust, highly sensitive, and selective techniques for DA detection.[76] The available methods for DA determination are spectrophotometry, flow injection, chemiluminescence, surface-enhanced Raman scattering spectroscopy, and electrophoresis, but these methods are expensive, laborious, and require highly expensive systems and are therefore undesirable for routine diagnostic practices.[77]

Due to the characteristic electrochemical behavior of DA, it can easily be detected by the electrochemical approach. Moreover, the electrochemical technique is highly sensitive, selective, cost-effective, and simple. However, some interfering agents like ascorbic acid and UA possessing oxidation potential similar to DA render DA detection unsuccessful by the bare electrode.[78] To overcome this issue, modified electrodes with hybrid materials have been developed, and recently functionalized multiwalled carbon nanotubes (f—MWCNTs) with metal oxide NPs have attracted much interest because they improve electrochemical properties.[79]

In a study, for the first time, multiwalled carbon nanotubes (MWCNTs) functionalized with molybdenum NP Mo NP@f—MWCNT core—shell hybrid nanocomposite was synthesized by acid condensation and employed in the electrochemical detection of DA. Negatively charged functional groups on the surface of MWCNTs served as an attachment site for positively charged Mo. The Mo NPs@f—MWCNT composites exhibited excellent electrocatalytic properties for the detection of DA. The LOD was calculated as 1.26 nM, and moreover, the sensor was highly selective, sensitive, and stable for DA.[78] Fig. 19.6 explains the preparation of the Mo NPs@f—MWCNT core—shell nanocomposite and detection of DA.

In another study, rGO functionalized with platinum—cobalt alloy NP composite (Pt-Co@rGO) was synthesized by microwave-assisted technique and employed the simultaneous electrochemical detection of DA, AA, and UA. Differential pulse voltammetry (DPV) and cyclic voltammetry (CV) developed well-defined oxidation peaks of

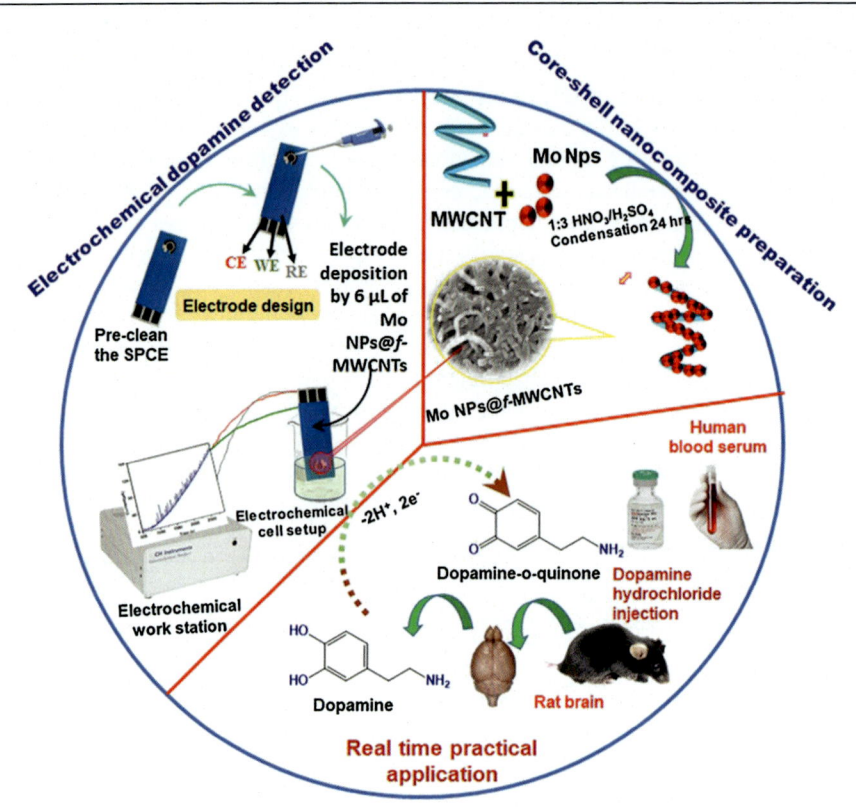

Figure 19.6 Schematic illustration of the preparation of the Mo NPs@f−MWCNT core−shell nanocomposite and electrochemical detection of dopamine.
Reproduced with permission.[78]

these analytes. The LOD for DA was 0.051 μM. The Pt-Co@rGO-based biosensor was of superior activity in terms of selectivity and sensitivity, even in real samples.[80]

19.3.5 L-tryptophan

L-tryptophan (L-Trp) is an essential amino acid in humans that is acquired exclusively from the diet. Tryptophan has an immense role in physiological processes, including cellular growth and repair and protein building blocks. Tryptophan metabolites are engaged in the coordination of sensory responses of an organism toward its environment, and L-Trp metabolites also serve as neurotransmitters and signaling modalities. Basically, L-Trp is a precursor of various neurotransmitters like 5-hydroxytryptamine, also known as serotonin, quinolinic acid, etc.[81,82] It has its role in the human circadian rhythm and also improves sleep and immunity. In addition, L-Trp is employed in the treatment of Parkinson's disease.[83,84] Overall, the role of L-Trp propounds that it evolved as a vital entity in cellular metabolism, organismal behaviors, and the dietary responses of organisms.[85]

Since humans and animals cannot synthesize their own L-Trp, its requirement must be fulfilled by food and supplements. The dietary intake of L-Trp recommended by the World Health Organization is 4 mg/kg/day. However, errors in the trip metabolism lead to the accumulation of toxic waste products in the brain, leading to hallucinations and illusions resulting in neurological disorders.[86,87] In this regard, a dire need is to establish a facile approach to quantify L-Trp in dietary products, pharmaceutical preparations, and living bodies.

So far, the available analytical techniques for the determination of L-Trp are high-performance liquid chromatography, spectroscopy, fluorescence, capillary electrophoresis, and chemiluminescence. These techniques are undoubtedly sensitive, but their high cost, complicated sample preparation, and requirement for human-sized instruments render these techniques less desirable.[88]

Recent advances in nanotechnology are coming up with a new paradigm in diagnostics. Reducing the size to nanoscale endows materials with versatile characteristics like enhanced surface area and quantum confinement. These novel physical features could be harnessed to develop robust sensors and biosensors to detect various biological analytes.[89]

In the last few decades, electrochemical detection of L-Trp has gained momentum due to the simplicity and excellent selectivity of the method. However, the electrochemical behavior of L-Trp is poor at the bare electrode. To circumvent this issue, researchers focused on the modification of glassy carbon electrodes.[90,91] In a study, electrochemical detection of L-Trp was performed using tin nanoflowers functionalized rGO (SnO_2@rGO). The modified electrode demonstrated fast direct electron transfer and significant electrocatalytic activity toward L-Trp owing to the excellent catalytic activity of SnO_2, enhanced specific surface area of rGO, and superior conductivity of the composite. The biosensor exhibited significantly high selectivity, sensitivity, and reproducibility of the reaction for L-Trp. The LOD was calculated to be 0.04 μM.[92]

Li et al. (2012) synthesized a composite of silver oxide NPs and GO by green synthesis approach with glucose simultaneously acting as reducing and stabilizing agent. DPV and CV were employed to explore the electrochemical properties of the composite toward L-Trp. The composite manifested distinctively high selectivity, stability, and repeatability. The detection limit was 2.0 nM. This method was free of interfering agents like tyrosine, and the biosensor showed good recoveries for real samples analysis.[93]

Another study conducted by Zhou and coworkers (2019) reported the detection of L-Trp through an electrochemical route using Tantalum pentoxide and rGO composite (Ta_2O_5—rGO). The composite was synthesized by hydrothermal and calcination methods. The composite was employed to detect L-Trp from human serum. An as-prepared Ta_2O_5—rGO composite showed better electrochemical activity for L-Trp than bare GCE or GCE modified with rGO alone. The lower detection range for the reaction setup was 0.84 μM. The prepared composite exhibited significantly higher recoveries of L-Trp from human serum. Fig. 19.7 explains the synthesis and electrochemical detection of L-Trp by the Ta_2O_5—rGO composite.[94]

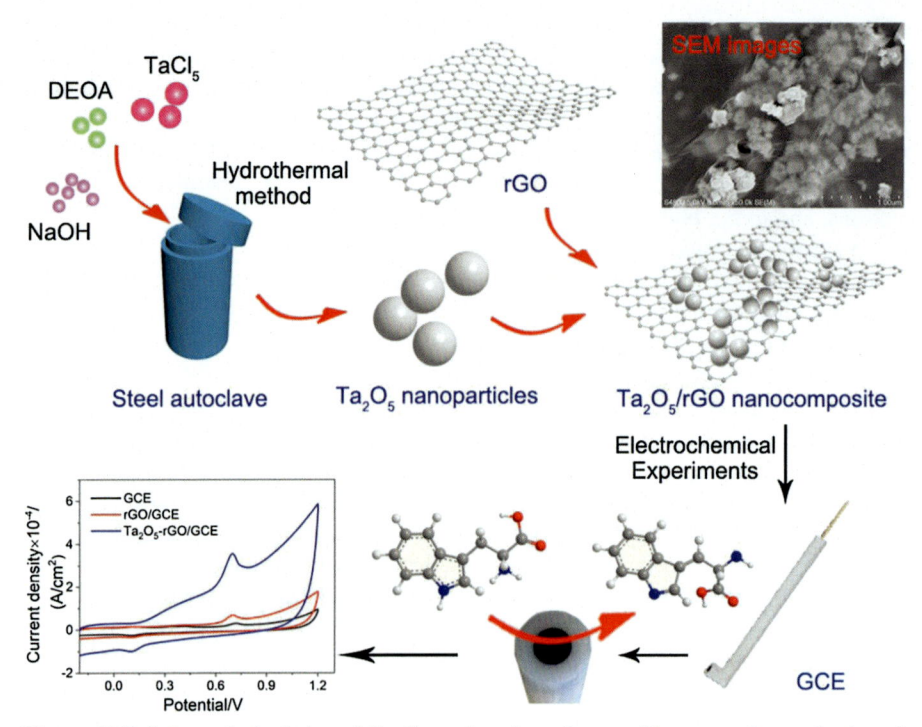

Figure 19.7 Schematic depiction of Ta_2O_5—reduced graphene oxide composite synthesis and further detection of L-Trp by the electrochemical route. Reproduced with permission.[94]

19.3.6 *Pathogens*

Pathogens are disease-causing microorganisms classified into different types, i.e., viruses, bacteria, parasites, and fungi. They can cause mild to life-threatening infections to the host cell.[95] The human body contains an immune system and helpful bacteria to protect the host against infectious agents. However, there are many different pathways whereby a pathogen can invade and disrupt the cell's mechanisms.[96] Numerous microorganisms are responsible for great casualties and have had severe effects on tormented groups. For example, *Neisseria meningitidis* causes inflammation of meninges, SARS coronavirus causes severe acute respiratory syndrome, different kinds of bacterial agents, e.g., *Staphylococcus, Salmonella, Shigella, Clostridium, Bacillus*, and *Campylobacter*, cause food poisoning. There are numerous such examples; from the beginning of humankind to forever, microorganisms have shaped human history, evolution, and development.[97] Today, while many clinical advances have been made to protect the host against such infectious agents through the use of fungicides, antibiotics, and vaccines, these pathogens keep on undermining human health.[98]

Early and efficient detection of these deadly microorganisms is what humankind needed most. For instance, the conventional detection methods for meningitis[99] include the tumbler test, polymerase chain reaction (PCR), real-time PCR, fluorescence in situ hybridization,[100] and immunological approaches. Though these approaches are reliable and give a sensitive response, they are expensive and time-consuming.

Sensitive, specific, and robust pathogen detection is still a key requirement in environmental monitoring, food safety, and clinical diagnosis. Advances in nanotechnology have enabled rapid and sensitive detection.[101] Because of their size, NPs possess unique properties such as surface plasmon resonance, enhanced surface reactivity, superparamagnetism, and quantum confinement, making their best fit as a base for making next-generation biosensors. The novel biosensor for *N. meningitidis* gives time and sensitivity of 20.20 μA/decade and 45 s of hybridization. A zinc oxide (ZnO)–MWCNT nanocomposite fabricated on indium tin oxide glass surface allows sensitive, specific, and on-time detection of *N. meningitidis* compared with PCR or other previous approaches. Fig. 19.8 shows another novel electrochemical genosensor for detecting

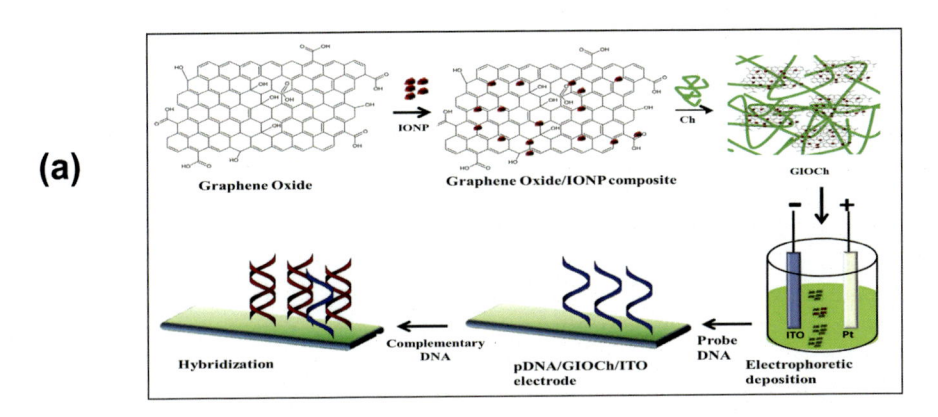

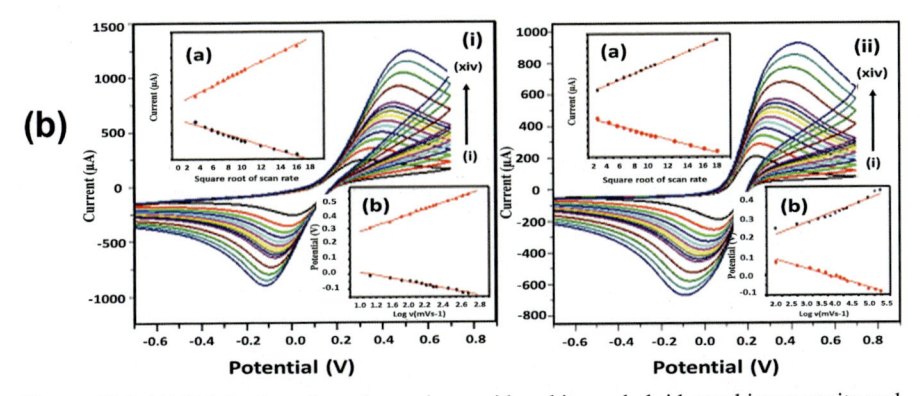

Figure 19.8 (A) Fabrication of graphene–iron oxide–chitosan hybrid nanobiocomposite and nucleic acid-functionalized GIOCh/ITO electrode (B) response of (i) GIOCh/ITO electrode and (ii) pDNA/GIOCh/ITO electrode with scan rate in PBS (100 mM, pH 7.4%, 0.9% NaCl) solution containing 5 mM [Fe(CN)6].[102]

pathogens; the sensor is fabricated using GO-modified iron oxide−chitosan hybrid nanocomposite (GIOCh) film. *Escherichia coli* O157:H7 (*E. coli*) specific probe oligonucleotide sequences, then immobilized onto the film for the detection purpose. The sensor exhibits a linear response to target DNA in the concentration range of 10^{-6} to 10^{-14} M with a detection limit of 1×10^{-14} M.[102]

Another impedance-based electrochemical biosensor was fabricated in which a sensing region (Man/MUA-MH/Au) was proposed with good stability (man: mannose; MUA: 11-mercapto 11 acids; MH: 6-mercapto hexanol). The designed sensor can capture antigens of *E. coli* JM109, *E. coli* DH5α, and *S. typhimurium* ATCC14028 by Electrochemical impedance spectroscopy.[100] The sensing surface was shown to have better affinity (2.16×10^6 CFU/mL) to bind with *S. typhimurium* ATCC 14028. The detection limit for *E. coli* JM109 and *E. coli* DH5α was 5.84×10^3 CFU/mL. The novel electrochemical impedance biosensor has been proved to be faster and more dependable in the analysis of interactions between pathogen and glycan, and a new diagnostic tool for assessing the pathogenicity has also been developed[103]

19.3.7 Detection of DNA

The detection of biological molecules such as DNA solves many problems in numerous fields like agriculture, forensic science, medicine, and more. DNA is the basic building block of life. The hereditary information stored in DNA is encoded by a chemical language that is then reproduced in many living organisms. Each DNA helix is made up of three compounds: Nitrogenous bases, phosphate, and deoxyribose sugar. Each unit of deoxyribose, phosphate, and nitrogenous base is a nucleotide about 0.3 nm long.[104] Thousands and millions of nucleotides join together to form readable sequences. And these sequences, in return, translate into particular proteins. These proteins then shape the whole organism—how it will work and what it will look like. DNA has a strict base-pairing pattern, making both the strands reverse complement each other.[105] In any DNA detection method, the key is to recognize these sequences. Because of its uniqueness, DNA serve as the core of many detection methods. It can serve as a biomolecule or even as a biomarker against many cancerous diseases.

Several technologies are available to detect the particular sequences or even a single base pair change at the DNA level, including enzymatic activity, surface plasmon resonance spectroscopy, Fluorescence In situ hybridization, PCR, DNA sequencing isothermal amplification, and many more.[106]

Although numerous efficient detection methods are available, the cost and major time consumption have made them unfavorable to use. In the last few years, electrochemical DNA biosensors have received attention for their high sensitivity and specificity, fast response rate, and most importantly, their low cost.[107] To date, many electrochemical DNA biosensors have used different nanoparticles to improve sensitivity and specificity, including gold NPs, metal oxides, and CNTs.[108]

Locating DNA sequences are of great importance to the virus, genetically modified organism, and detection of pathogens for early diagnosis. Chen et al. reported a novel

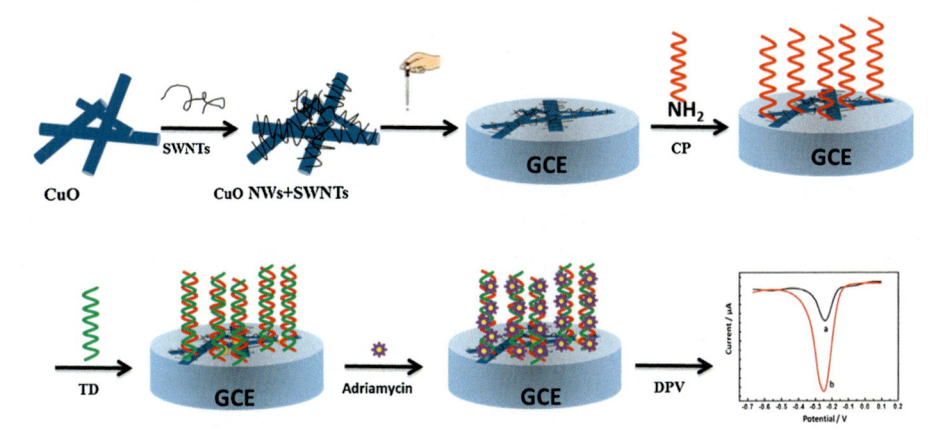

Figure 19.9 Fabrication of copper oxide nanowires on single-walled carbon nanotubes for detection of sequence-specific DNA.[109]

ultrasensitive biosensor that detects sequence-specific target DNA. The biosensor deployed hybrid nanocomposite consisting of carboxyl-functionalized single-walled carbon nanotubes (SWCNTs) and copper oxide nanowires (CuO NWs). The synergistic effect of this nanocomposite improved the immobilization of the probe DNA on the electrode surface. The biosensor exhibited an LOD of 3.5×10^{-15} M (signal/noise ratio of 3). The biosensor was also easy to implement with high sensitivity and stability compared with other available detection methods (Fig. 19.9). Furthermore, the biosensor also displayed high selectivity to single-base mismatched DNA.[109]

Another ultrasensitive novel DNA biosensor was designed with low detection limits of 1.18×10^{-13} M (at S/N = 3) to detect a biomarker of non-small-cell lung cancer. The electrochemical DNA biosensor was fabricated using CuO NWs and carboxyl-functionalized GO. The nanocomposite displayed higher stability, sensitivity, and linearity ($R^2 = 0.9750$) from 1.0×10^{-12} to 1.0×10^{-6} M. The biosensor even discriminates sequences with one to three base pair differences.[110]

19.4 Viruses

Viruses are small infectivity units that can surpass minute filters enough to refrain the passage of bacteria. For their existence, viruses need a living cellular system, either prokaryotic or eukaryotic cells, as they do not contain replication components and required machinery. Therefore, these are unable to replicate or amplify their genomic information in the absence of cellular scaffolding.[111]

Due to their non-cellular-like existence, vaccines are the only way to control virus spread and resultant chronic diseases. But there are some viral infections like chronic hepatitis C infection caused by HCV. No vaccine has been developed to date because of its extremely high genetic variability and introduced various immune escape strategies.[112] To combat such viral infections, early disease detection has utmost importance. Established detection methods include conventional molecular diagnostics enabling precise detection of nucleic acids, ELISA, etc.[113−115] With the routine development of scientific knowledge, efforts were made to develop new diagnostic platforms that are user-friendly, perform on-site detection, and do not require cost-intensive resources or high level of expertise. To fulfill desired goals, nanotechnology has offered various nanomaterials (only metallic nanomaterials are described in this chapter) as a step forward to developing smart and fully integrated diagnostic tools.[116] Here, a general idea has been discussed to design miniaturized setups using nanomaterials for the detection of a few very important human viruses, especially metal oxide NPs and their components with carbon-based nanomaterials like CNTs, graphene, etc. as former shows wide applications in diagnostic as most of them fluoresce or have magnetic property allowing visualization with a high level of precision leading to real-time quantification of disease.[117]

19.4.1 Human immunodeficiency virus

The human immunodeficiency virus is a devastating virus belonging to the family Reteroviridae, which transfers through the bloodstream and becomes highly infectious. This virus is almost spherical, with an estimated diameter of 120 nm. The HIV genome comprises positive-sense ssRNA and contains two copies, each encoding nine viral genes, including reverse transcriptase enzyme and protease and integrase. Structural analysis of this virus revealed a lipid bilayer composed envelope taken from the human cellular membrane when it buds from the host cell.

Like other viruses, at first, its genome was studied, opening various aspects for developing diagnostic systems. Later, due to its infectivity, scientists have tried to develop user-friendly detection platforms. For the miniaturization of detection setups, nanotechnology played a crucial role in introducing various nanomaterials as well as nanocomposites and nanoclusters containing more than one type of nanomaterials enhancing the properties of the resultant product. For instance, a study reported a detection system for HIV developed using biocompatible zinc oxide NPs employing their piezotronics effect when used as nanowires. This provides very selective in situ detection of HIV.[118] Another report presents biosensors having metal oxide NPs with magnetic character. In this nanosensor, iron oxide NPs were used to detect HIV with detection limits as low as 160 pmol. In this report, iron oxide NPs were first functionalized with streptavidin, and then these functionalized NPs were immobilized on a bare gold electrode. Later, biotinylated DNA probes designed for HIV detection were labeled and attached to these NPs through biotin−streptavidin linkage.[119]

Carbon nanotubes were wrapped using Au hollow nanospheres to immobilize horseradish peroxidase enzyme with a capture probe for carbon-based nanomaterials. This composite enhances the surface area for immobilization and capturing of the target molecule, thus enhancing the sensitivity of the designed sensor with a calculated LOD of 0.8863 fM/mL.[120,121] In another study, a nanocomposite of carbon nanotubes with graphene was synthesized and then encapsulated with horseradish peroxidase enzyme through an electrochemical detection system leading to higher sensitivity for HIV, i.e., calculated as 0.15 pg/mL.[120,122] Moreover, GO is extensively studied to design a label-free detection platform. In this case, DNA aptamers without labeling were immobilized directly on the GO surface. The designed DNA aptamer has two parts, i.e., immobilization side and probe part. When target DNA is present, hybridization occurs with the probe part; thus, an increase in negative charges was recorded by electrochemical impedance spectroscopy with an LOD of 1.1×10^{-13}.[120,123]

19.4.2 Dengue

Dengue is another threatening arbovirus pandemic globally, spreading through the female *Aedes* mosquito. Dengue virus belongs to the family of Flaviviridae consisting of positive-sense single-stranded RNA genome.[124]

Dengue is one of the fastest-growing viral infections, estimated at 400 million cases per annum. Therefore, detection of disease in the early infection period becomes clinically important for which different diagnostic strategies have been studied, including various biomarkers. Efforts for miniaturization of available diagnostic setups are done utilizing the properties of nanomaterials, especially composite of metal NPs and carbon-based nanomaterials.[125] Therefore, a two-electrode system was developed for the detection of dengue virus in which positive electrode consists of manganese oxide/polyaniline hybrid, while the other consists of rGO synthesized on porous cellulose paper, filter paper-based sensor gives a wide window for operational voltage along with specific capacitance makes it cost-effective. In addition, a foldable supercapacitor is used that provides the grounds for wearable sensors.[126] Similarly, antibody immobilized aluminum NPs were used to detect dengue virus, and change in impedance was studied on interaction with dengue virus. The designed sensor shows higher specificity for dengue virus even in the presence of closely related viruses with an LOD of 1–900 pfu/mL.[127]

19.4.3 Zika virus

Recent outbreaks of Zika virus infection, a vector-borne virus, occurred in 2015 near the United States of America. The disease has taken high attention due to the long-term impacts of Zika, including Guillain—Barre Syndrome, as well as other brain defects. Zika virus belongs to the family Flaviviridae, thus showing the structure and symptom similarities to dengue virus, causing difficulty to distinguish between the two. Earlier, infection detection methods show a sensitivity of less than 50% due to cross-reactivity by coviruses, e.g., dengue. Therefore, nanomaterials with more sensitivity as well as specificity for target molecules were in high demand, especially carbon-based nanocomposites, due to their promising properties.

Graphene, separately, was used as a biosensor when immobilized with specific antibodies against the Zika virus antigen. This biosensor gives real-time quantification of virus titer with an LOD as low as 450 pM, thus making it clinically significant. Moreover, the specificity of the designed system was validated using antigen of viruses belonging to the same family as that of Zika virus and shows structure homology.[128]

Molecularly imprinted GO immobilized miniaturized electronic devices were developed to detect viruses in blood serum. For metals, another proof of concept was presented by Kaushik and coworkers. A biosensor chip with interdigitated electrode fingers made of gold was immobilized with dithiobis (succinimidyl propionate) followed by antibody (that is, specifically against the envelop protein of the Zika virus) coating. The detection limit for this biosensor was noted as 10 pM. Later, this electrode system can be linked to a small potentiostat integrated into a smartphone for on-site detection of viruses.[129] In the case of metal oxides, zinc oxide (ZnO) star-shaped NPs were used to develop electrochemiluminescence biosensors for early Zika virus detection by immobilizing NS1 antibody after cystamine and glutaraldehyde layer used for surface functionalization instead of thiol assembly, etc. LOD for this setup was calculated as 1.00 pg/Ml.[130]

Together, combined metal oxide and carbon materials add properties to the resultant hybrid. A metal−organic framework was developed in which porous carbon material was nitrogen-doped, making a stable nanobiosensor for the Zika virus utilizing its own fluorescent property. This designed doped material can absorb TAMRA wavelength, which is a fluorescent tag, and thus TAMRA-tagged virus RNA were used as a capture probe that results in quenching of fluorescence. When the target molecule is present, binding results in hybrid release with recovery in fluorescence recorded by a battery-enabled imaging system. The whole sensor setup was validated for the Zika genome with an LOD of 0.23 nM.[131]

19.4.4 Human metapneumovirus

Respiratory viruses remained the problem of immediate concern since a long time ago as they tend to develop pandemic threats because newer virus types continuously emerge that were previously known to cause diseases only in birds and poultry. Human metapneumovirus (HMPV) is an example resulting from these zoonotic transmissions. It was first isolated from bronchiolitis specimens from children showing respiratory illnesses in 2001[132] before spreading worldwide and causing severe respiratory problems in older adults.[133] HMPV belongs to the family of Pneumoviridae, which was previously a subfamily of Paramyxoviridae. Viruses of this family contain a single-stranded RNA genome, i.e., negative-sense and nonsegmented. Based on genetic and antigenic diversity, HMPV is subdivided into two subgroups (A and B), which are further divided into two clades (A1, A2, B1, and B2). Scientific studies indicate that each year, one (and different) of these four HMPV subtypes circulates concurrently and remains predominant, causing conjunctivitis, mild pneumonia, and severe respiratory distress syndrome. Bhattacharya et al. 2011 designed an SWCNT-based sensor for detecting these HMPVs. The sensor system design is a perfect example of a layer-by-layer materials assembly, in which gold electrodes were immobilized on Si/SiO$_2$ substrate followed by poly(diallyldimethylammonium chloride) and then poly(styrene sulfonate) and SWCNTs, respectively. Afterward, viral antibodies were first treated with poly (L-lysine) to enhance their immobilization between electrodes. This sensor system efficiently captured immobilized antibody-specific antigens, resulting in a change in conductance that correlated to successfully capturing antigens to a virus titer of 102 TCID50/ml, that is, a 50% tissue culture infective dose.[134]

19.4.5 Influenza

Influenza is a transmittable respiratory disorder caused by the human influenza virus and transfers through respiratory droplets. This broad-spectrum virus shows four types (A, B, C, and D); A and B are specific to humans, while others are restricted to other species depending on receptors that define their host specie. There are various subtypes of influenza A virus, depending on the combination of its two surface proteins, i.e., the hemagglutinin (H) and neuraminidase (N) proteins that define the H and N influenza A viruses—H7N9, H5N1, and others. There are 18 H type (H1−18) an 11 N-type (N1−11) proteins. Influenza is worrisome, as it can be a health emergency potentially of international concern because most animal-originated influenza viruses change their antigenic characters and are thus capable of infecting humans. A sensitive point-of-care diagnostic setup is required to monitor the spread of viruses—especially influenza A, given its high mutation rate and ability to develop new antigenic characters.[135]

As discussed, NPs have unique, astonishing chemical and physical properties that can be changed by optimizing size, dimension, shape, and geometry. Moreover, scientists studied that these properties can be enhanced by making different composite materials of two NP types, and these can be hybridized to 2-D or 3-D nanomaterials to obtain synergistic characters. A study presented a biosensor for the human influenza virus with a binary NP system (bNP). This binary NP system comprised gold NPs and magnetic iron oxide NPs. This bNP composite was then hybridized to carbon nanotubes (bNP—CNTs or Au/MNP—CNT). Later, this hybrid was aligned on interdigitated electrode made of platinum. Au NPs were used in this assembly to provide an attachment site for the thiol-modified DNA capture probe. This study reports an LOD of 8.4 pM for this binary NP system for diagnosis of influenza virus with higher specificity.[136]

Another hybrid of metal oxide NP along with two-dimensional nanomaterials was presented by.[137] For the detection of influenza virus. In this study, zinc oxide NPs were synthesized, and their pure-graphene composite was made by the homogenous distribution of ZnO NPs onto graphene sheets. This nanohybrid was coated on the carbon electrode of the electrochemical detection system through the just-mentioned screen-printing method, and the modified electrode showed an active reduction of hydrogen peroxide. The LOD for this ZnO-NPs@GO sensor showed an analytical sensitivity of 7.4357 μM, significantly more sensitive than other available methods.

19.5 Conclusion

The fascinating characteristics of nanomaterials, including metal oxides, carbon-based materials, and their composites, directed the development of novel techniques in clinical diagnosis. Nanomaterials and their composites are considered multifunctional materials in various fields, such as health care, energy storing devices, and environmental monitoring and biosensing applications. They behave as electrode comaterials to enhance the effective active surface area and activate the signal production to increase the reaction rate in electrochemical energy storing and sensor devices. Because of their unique properties such as chemical and thermal stability, large surface area, biocompatibility due to their good stoichiometry with biomolecule size, and remarkable electrochemical properties, nanomaterials are considered receptor materials to fabricate sensing platforms. Nanosensors are widely used to detect multiple analytes such as detection of cancer biomarkers aids to the early diagnosis of cancer, detection of abnormal levels of biomolecules such as UA, ascorbic acid, dopamine, glucose, pathogens, and DNA provide information regarding health status, life-threatening infectious agents, in biomedical, forensic, and agricultural areas.

References

1. Shetti NP, Bukkitgar SD, Reddy KR, Reddy CV, Aminabhavi TM. ZnO-based nanostructured electrodes for electrochemical sensors and biosensors in biomedical applications. *Biosens Bioelectron* 2019;**141**:111417.
2. Balasubramanian K. Challenges in the use of 1D nanostructures for on-chip biosensing and diagnostics: a review. *Biosens Bioelectron* 2010;**26**:1195−204.
3. Lee CS, Yu SH, Kim TH. One-step electrochemical fabrication of reduced graphene oxide/gold nanoparticles nanocomposite-modified electrode for simultaneous detection of dopamine, ascorbic acid, and uric acid. *Nanomaterials* 2018;**8**:17.
4. Sabherwal P, Mutreja R, Suri CR. Biofunctionalized carbon nanocomposites: new-generation diagnostic tools. *Trac Trends Anal Chem* 2016;**82**:12−21.
5. Sengupta J. *Different synthesis routes of graphene-based metal nanocomposites*. 2019. arXiv preprint arXiv:1911.01720.
6. Maiti D, Tong X, Mou X, Yang K. Carbon-based nanomaterials for biomedical applications: a recent study. *Front Pharmacol* 2019;**9**:1401.
7. Sireesha M, Jagadeesh Babu V, Kranthi Kiran AS, Ramakrishna S. A review on carbon nanotubes in biosensor devices and their applications in medicine. *Nanocomposites* 2018;**4**:36−57.
8. Seok D, Jeong Y, Han K, Yoon DY, Sohn H. Recent progress of electrochemical energy devices: metal oxide-carbon nanocomposites as materials for next-generation chemical storage for renewable energy. *Sustainability* 2019;**11**:3694.
9. Zhang Y, Tian J, Li H, Wang L, Qin X, Asiri AM. Biomolecule-assisted, environmentally friendly, one-pot synthesis of CuS/reduced graphene oxide nanocomposites with enhanced photocatalytic performance. *Langmuir* 2012;**28**:12893−900.
10. Cheng G, Wang ZG, Denagamage S, Zheng SY. Graphene-templated synthesis of magnetic metal organic framework nanocomposites for selective enrichment of biomolecules. *ACS Appl Mater Interfaces* 2016;**8**:10234−42.
11. Wu X, Xing Y, Pierce D, Zhao JX. One-pot synthesis of reduced graphene oxide/metal (oxide) composites. *ACS Appl Mater Interfaces* 2017;**9**:37962−71.
12. Asal M, Özen Ö, Şahinler M, Baysal HT, Polatoğlu İ. An overview of biomolecules, immobilization methods and support materials of biosensors. *Sens Rev* 2019. https://doi.org/10.1108/SR-04-2018-0084.

13. Sinha A, Lu X, Wu L, Tan D, Li Y, Chen J, et al. Voltammetric sensing of biomolecules at carbon based electrode interfaces: a review. *Trac Trends Anal Chem* 2018;**98**:174–89.
14. Labib M, Sargent EH, Kelley SO. Electrochemical methods for the analysis of clinically relevant biomolecules. *Chem Rev* 2016;**116**:9001–90.
15. Nalejska E, Mączyńska E, Lewandowska MA. Prognostic and predictive biomarkers: tools in personalized oncology. *Mol Diagn Ther* 2014;**18**:273–84.
16. Stone MJ, Willerson JT, Gomez-Sanchez CE, Waterman M. Radioimmunoassay of myoglobin in human serum. Results in patients with acute myocardial infarction. *J Clin Invest* 1975;**56**:1334–9.
17. Katus HA, Remppis A, Looser S, Hallermeier K, Scheffold T, Kübler W. Enzyme linked immuno assay of cardiac troponin T for the detection of acute myocardial infarction in patients. *J Mol Cell Cardiol* 1989;**21**:1349–53.
18. Mitsakakis K, Gizeli E. Detection of multiple cardiac markers with an integrated acoustic platform for cardiovascular risk assessment. *Anal Chim Acta* 2011;**699**:1–5.
19. Ibn Sina A, Koo K, Ahmed M, Carrascosa L, Trau M. *Interfacial biosensing: direct biosensing of biomolecules at the bare metal interface.* 2018.
20. WHO. In: *who.int*; 2020.
21. Wang J, Wang X, Ren L, Wang Q, Li L, Liu W. Conjugation of biomolecules with magnetic protein microspheres for the assay of early biomarkers associated with acute myocardial infarction. *Anal Chem* 2009;**81**:6210–7.
22. Coliva G, Duarte S, Pérez-Sala D, Fedorova M. Impact of inhibition of the autophagy-lysosomal pathway on biomolecules carbonylation and proteome regulation in rat cardiac cells. *Redox Biol* 2019;**23**:101123.
23. Chauhan D, Nirbhaya V, Srivastava CM, Chandra R, Kumar S. Nanostructured transition metal chalcogenide embedded on reduced graphene oxide based highly efficient biosensor for cardiovascular disease detection. *Microchem J* 2020;**155**:104697.
24. Tuteja SK, Bhalla V, Deep A, Paul A, Suri CR. Graphene-gated biochip for the detection of cardiac marker Troponin I. *Anal Chim Acta* 2014;**809**:148–54.
25. Liu T, Liang LL, Xiao P, Sun LP, Huang YY, Ran Y. A label-free cardiac biomarker immunosensor based on phase-shifted microfiber Bragg grating. *Biosens Bioelectron* 2018;**100**:155–60.
26. El-Safty SA, Shenashen MA. Nanoscale dynamic chemical, biological sensor material designs for control monitoring and early detection of advanced diseases. *Mater Today Bio* 2020:100044.
27. Forouzanfar MH, Afshin A, Alexander LT, Anderson HR, Bhutta ZA, Biryukov S. Global, regional, and national comparative risk assessment of 79 behavioural, environmental and occupational, and metabolic risks or clusters of risks, 1990–2015: a systematic analysis for the Global Burden of Disease Study 2015. *Lancet* 2016;**388**:1659–724.
28. Plummer M, de Martel C, Vignat J, Ferlay J, Bray F, Franceschi S. Global burden of cancers attributable to infections in 2012: a synthetic analysis. *Lancet Glob Health* 2016;**4**: e609–616.
29. Ferlay J, Soerjomataram I, Ervik M, Dikshit R, Eser S, Mathers C, et al. *France: International Agency for Research on Cancer; 2013*, vol. 10; 2012. Cancer Incidence and Mortality Worldwide: IARC CancerBase.
30. Stewart B, Wild CP. *World cancer report 2014*. 2014.
31. Adjei AA. Lung cancer worldwide. *J Thorac Oncol* 2019;**14**:956.
32. Shahzad F, Zaidi SA, Koo CM. Highly sensitive electrochemical sensor based on environmentally friendly biomass-derived sulfur-doped graphene for cancer biomarker detection. *Sensor Actuator B Chem* 2017;**241**:716–24.

33. Leung CH, Wu KJ, Li G, Wu C, Ko CN, Ma DL. Application of label-free techniques in microfluidic for biomolecules detection and circulating tumor cells analysis. *Trends Anal Chem* 2019;**117**:78–83.

34. Rupaimoole R, Yoon B, Zhang WC, Adams BD, Slack FJ. A high-throughput small molecule screen identifies ouabain as synergistic with miR-34a in killing lung cancer cells. *Iscience* 2020;**23**:100878.

35. Narayan T, Kumar S, Kumar S, Augustine S, Yadav B, Malhotra BD. Protein functionalised self assembled monolayer based biosensor for colon cancer detection. *Talanta* 2019;**201**:465–73.

36. Krishnamoorthy C, Prakasarao A, Srinivasan V, GN SP, Singaravelu G. Monitoring of breast cancer patients under pre and post treated conditions using Raman spectroscopic analysis of blood plasma. *Vib Spectrosc* 2019;**105**:102982.

37. Ratajczak K, Stobiecka M. High-performance modified cellulose paper-based biosensors for medical diagnostics and early cancer screening: a concise review. *Carbohydr Polym* 2019:115463.

38. Fiorentino TV, Sesti F, Succurro E, Pedace E, Andreozzi F, Sciacqua A, et al. Higher serum levels of uric acid are associated with a reduced insulin clearance in non-diabetic individuals. *Acta Diabetol* 2018:1–8.

39. Yang YJ, Li W. CTAB functionalized graphene oxide/multiwalled carbon nanotube composite modified electrode for the simultaneous determination of ascorbic acid, dopamine, uric acid and nitrite. *Biosens Bioelectron* 2014;**56**:300–6.

40. Wang C, Du J, Wang H, Zou Ce, Jiang F, Yang P. A facile electrochemical sensor based on reduced graphene oxide and Au nanoplates modified glassy carbon electrode for simultaneous detection of ascorbic acid, dopamine and uric acid. *Sensor Actuator B Chem* 2014;**204**:302–9.

41. Gao X, Gui R, Xu KQ, Guo H, Jin H, Wang Z. A bimetallic nanoparticle/graphene oxide/ thionine composite-modified glassy carbon electrode used as a facile ratiometric electrochemical sensor for sensitive uric acid determination. *New J Chem* 2018;**42**:14796–804.

42. Tian X, Cheng C, Yuan H, Du J, Xiao D, Xie S, et al. Simultaneous determination of l-ascorbic acid, dopamine and uric acid with gold nanoparticles—β-cyclodextrin—graphene-modified electrode by square wave voltammetry. *Talanta* 2012;**93**:79–85.

43. Wang S, Zhang W, Zhong X, Chai Y, Yuan R. Simultaneous determination of dopamine, ascorbic acid and uric acid using a multi-walled carbon nanotube and reduced graphene oxide hybrid functionalized by PAMAM and Au nanoparticles. *Anal Methods* 2015;**7**: 1471–7.

44. He C, Liu D, Lin W. Nanomedicine applications of hybrid nanomaterials built from metal—ligand coordination bonds: nanoscale metal—organic frameworks and nanoscale coordination polymers. *Chem Rev* 2015;**115**:11079–108.

45. Taj A, Shaheen A, Xu J, Estrela P, Mujahid A, Asim T, et al. In-situ synthesis of 3D ultrasmall gold augmented graphene hybrid for highly sensitive electrochemical binding capability. *J Colloid Interface Sci* 2019;**553**:289–97.

46. Kang X, Wang J, Wu H, Aksay IA, Liu J, Lin Y. Glucose oxidase—graphene—chitosan modified electrode for direct electrochemistry and glucose sensing. *Biosens Bioelectron* 2009;**25**:901–5.

47. Liu M, Chen Q, Lai C, Zhang Y, Deng J, Li H, et al. A double signal amplification platform for ultrasensitive and simultaneous detection of ascorbic acid, dopamine, uric acid and acetaminophen based on a nanocomposite of ferrocene thiolate stabilized Fe3O4@ Au nanoparticles with graphene sheet. *Biosens Bioelectron* 2013;**48**:75–81.

48. Khan MYA, Zahoor M, Shaheen A, Jamil N, Arshad MI, Bajwa SZ, et al. Visible light photocatalytic degradation of crystal violet dye and electrochemical detection of ascorbic acid & glucose using BaWO4 nanorods. *Mater Res Bull* 2018;**104**:38–43.

49. Culver HR, Steichen SD, Herrera-Alonso M, Peppas NA. Versatile route to colloidal stability and surface functionalization of hydrophobic nanomaterials. *Langmuir* 2016;**32**: 5629–36.

50. Nsabimana A, Kitte SA, Wu F, Qi L, Liu Z, Zafar MN. Multifunctional magnetic Fe3O4/ nitrogen-doped porous carbon nanocomposites for removal of dyes and sensing applications. *Appl Surf Sci* 2019;**467**:89–97.

51. Punetha VD, Rana S, Yoo HJ, Chaurasia A, McLeskey Jr JT, Ramasamy MS. Functionalization of carbon nanomaterials for advanced polymer nanocomposites: a comparison study between CNT and graphene. *Prog Polym Sci* 2017;**67**:1–47.

52. Yue L, Pircheraghi G, Monemian SA, Manas-Zloczower I. Epoxy composites with carbon nanotubes and graphene nanoplatelets—dispersion and synergy effects. *Carbon* 2014;**78**: 268–78.

53. Arshad A, Taj A, Rehman A, Bajwa SZ, Mujahid A, Kaleem I, et al. In situ synthesis of highly populated CeO$_2$ nanocubes grown on carbon nanotubes as a synergy hybrid and its electrocatalytic potential. *J Mater Res Technol* 2019;**8**:5336–43.

54. Yogeswaran U, Thiagarajan S, Chen SM. Nanocomposite of functionalized multiwall carbon nanotubes with nafion, nano platinum, and nano gold biosensing film for simultaneous determination of ascorbic acid, epinephrine, and uric acid. *Anal Biochem* 2007; **365**:122–31.

55. Rafati AA, Afraz A, Hajian A, Assari P. Simultaneous determination of ascorbic acid, dopamine, and uric acid using a carbon paste electrode modified with multiwalled carbon nanotubes, ionic liquid, and palladium nanoparticles. *Microchim Acta* 2014;**181**: 1999–2008.

56. Huang J, Liu Y, Hou H, You T. Simultaneous electrochemical determination of dopamine, uric acid and ascorbic acid using palladium nanoparticle-loaded carbon nanofibers modified electrode. *Biosens Bioelectron* 2008;**24**:632–7.

57. Abdelwahab AA, Shim YB. Simultaneous determination of ascorbic acid, dopamine, uric acid and folic acid based on activated graphene/MWCNT nanocomposite loaded Au nanoclusters. *Sensor Actuator B Chem* 2015;**221**:659–65.

58. Zhang X, Zhang YC, Ma LX. One-pot facile fabrication of graphene-zinc oxide composite and its enhanced sensitivity for simultaneous electrochemical detection of ascorbic acid, dopamine and uric acid. *Sensor Actuator B Chem* 2016;**227**:488–96.

59. Dzoyem JP, Kuete V, Eloff JN. 23 - Biochemical parameters in toxicological studies in Africa: significance, principle of methods, data interpretation, and use in plant screenings. In: Kuete V, editor. *Toxicological survey of African medicinal plants*. Elsevier; 2014. p. 659–715.

60. N. I. f. C. Excellence. *Type 2 diabetes: prevention in people at high risk*. NICE guideline (PH38). 2012.

61. Hruby VJ. Chapter 16 - Glucagon:: molecular biology and structure-activity. In: Bittar EE, Bittar N, editors. *Principles of medical biology*, vol. 10. Elsevier; 1997. p. 387–401.

62. Lang XY, Fu HY, Hou C, Han GF, Yang P, Liu YB, et al. Nanoporous gold supported cobalt oxide microelectrodes as high-performance electrochemical biosensors. *Nat Commun* 2013;**4**:1–8.

63. Si P, Ding S, Yuan J, Lou XW, Kim DH. Hierarchically structured one-dimensional TiO$_2$ for protein immobilization, direct electrochemistry, and mediator-free glucose sensing. *ACS Nano* 2011;**5**:7617–26.

64. Guo C, Huo H, Han X, Xu C, Li H. Ni/CdS bifunctional Ti@ TiO$_2$ core—shell nanowire electrode for high-performance nonenzymatic glucose sensing. *Anal Chem* 2014;**86**: 876—83.

65. Yang N, Chen X, Ren T, Zhang P, Yang D. Carbon nanotube based biosensors. *Sensor Actuator B Chem* 2015;**207**:690—715.

66. Beitollahi H, Movahedifar F, Tajik S, Jahani S. A review on the effects of introducing CNTs in the modification process of electrochemical sensors. *Electroanalysis* 2019;**31**: 1195—203.

67. Sridara T, Upan J, Saianand G, Tuantranont A, Karuwan C, Jakmunee J. Non-enzymatic amperometric glucose sensor based on carbon nanodots and copper oxide nanocomposites electrode. *Sensors* 2020;**20**:808.

68. Xiao X, Peng S, Wang C, Cheng D, Li N, Dong Y. Metal/metal oxide@carbon composites derived from bimetallic Cu/Ni-based MOF and their electrocatalytic performance for glucose sensing. *J Electroanal Chem* 2019;**841**:94—100.

69. Yang W, Xu W, Zhang N, Lai X, Peng J, Cao Y, et al. TiO$_2$ nanotubes modified with polydopamine and graphene quantum dots as a photochemical biosensor for the ultra-sensitive detection of glucose. *J Mater Sci* 2020;**55**:6105—17.

70. Breed MD, Moore J. Chapter 2 - Neurobiology and endocrinology for animal behaviorists. In: Breed MD, Moore J, editors. *Animal behavior*. San Diego: Academic Press; 2012. p. 25—65.

71. Webb WG. 4 - Neuronal function in the nervous system. In: Webb WG, editor. *Neurology for the speech-language pathologist*. 6th ed. Mosby; 2017. p. 74—92.

72. Stevens L, Rodin I. Introduction to drug treatments. In: Stevens L, Rodin I, editors. *Psychiatry*. 2nd ed. Churchill Livingstone; 2011. p. 20—1.

73. Tukimin N, Abdullah J, Sulaiman Y. Review—electrochemical detection of uric acid, dopamine and ascorbic acid. *J Electrochem Soc* 2018;**165**:B258—67.

74. Lammel S, Steinberg EE, Földy C, Wall NR, Beier K, Luo L, et al. Diversity of transgenic mouse models for selective targeting of midbrain dopamine neurons. *Neuron* 2015;**85**: 429—38.

75. Robinson DL, Venton BJ, Heien ML, Wightman RM. Detecting subsecond dopamine release with fast-scan cyclic voltammetry in vivo. *Clin Chem* 2003;**49**:1763—73.

76. Patrice FT, Zhao LJ, Fodjo EK, Li DW, Qiu K, Long YT. Highly sensitive and selective electrochemical detection of dopamine using hybrid bilayer membranes. *ChemElectroChem* 2019;**6**:634—7.

77. Lu S, Hummel M, Kang S, Gu Z. Selective voltammetric determination of nitrite using cobalt phthalocyanine modified on multiwalled carbon nanotubes. *J Electrochem Soc* 2020;**167**:046515.

78. Keerthi M, Boopathy G, Chen SM, Chen TW, Lou BS. A core-shell molybdenum nanoparticles entrapped f-MWCNTs hybrid nanostructured material based non-enzymatic biosensor for electrochemical detection of dopamine neurotransmitter in biological samples. *Sci Rep* 2019;**9**:13075.

79. Wang H, Yan N, Li Y, Zhou X, Chen J, Yu B. Fe nanoparticle-functionalized multi-walled carbon nanotubes: one-pot synthesis and their applications in magnetic removal of heavy metal ions. *J Mater Chem* 2012;**22**:9230—6.

80. Demirkan B, Bozkurt S, Şavk A, Cellat K, Gülbağca F, Nas MS. Composites of bimetallic platinum-cobalt alloy nanoparticles and reduced graphene oxide for electrochemical determination of ascorbic acid, dopamine, and uric acid. *Sci Rep* 2019;**9**:12258.

81. Sainio EL, Pulkki K, Young S. L-tryptophan: biochemical, nutritional and pharmacological aspects. *Amino Acids* 1996;**10**:21—47.

82. Sa M, Ying L, Tang AG, Xiao LD, Ren YP. Simultaneous determination of tyrosine, tryptophan and 5-hydroxytryptamine in serum of MDD patients by high performance liquid chromatography with fluorescence detection. *Clin Chim Acta* 2012;**413**:973—7.

83. Zhou Z, He L, Mao Y, Chai W, Ren Z. Green preparation and selective permeation of d-Tryptophan imprinted composite membrane for racemic tryptophan. *Chem Eng J* 2017; **310**:63—71.

84. Lin C, Top D, Manahan CC, Young MW, Crane BR. Circadian clock activity of cryptochrome relies on tryptophan-mediated photoreduction. *Proc Natl Acad Sci U S A* 2018; **115**:3822—7.

85. Platten M, Nollen EAA, Röhrig UF, Fallarino F, Opitz CA. Tryptophan metabolism as a common therapeutic target in cancer, neurodegeneration and beyond. *Nat Rev Drug Discov* 2019;**18**:379—401.

86. Babaei A, Zendehdel M, Khalilzadeh B, Taheri A. Simultaneous determination of tryptophan, uric acid and ascorbic acid at iron (III) doped zeolite modified carbon paste electrode. *Colloids Surf B Biointerfaces* 2008;**66**:226—32.

87. He Q, Liu J, Liang J, Liu X, Li W, Liu Z. Towards improvements for penetrating the blood—brain barrier—recent progress from a material and pharmaceutical perspective. *Cells* 2018;**7**:24.

88. He Q, Tian Y, Wu Y, Liu J, Li G, Deng P. Electrochemical sensor for rapid and sensitive detection of tryptophan by a Cu_2O nanoparticles-coated reduced graphene oxide nanocomposite. *Biomolecules* 2019;**9**:176.

89. Viter R, Iatsunskyi I. Chapter 2 - Metal oxide nanostructures in sensing. In: Zenkina OV, editor. *Nanomaterials design for sensing applications*. Elsevier; 2019. p. 41—91.

90. Alaejos MS, Garcia Montelongo FJ. Application of amperometric biosensors to the determination of vitamins and α-amino acids. *Chem Rev* 2004;**104**:3239—66.

91. Ya Y, Luo D, Zhan G, Li C. Electrochemical investigation of tryptophan at a poly (p-aminobenzene sulfonic acid) film modified glassy carbon electrode. *Bull Kor Chem Soc* 2008;**29**:928.

92. Haldorai Y, Yeon SH, Huh YS, Han YK. Electrochemical determination of tryptophan using a glassy carbon electrode modified with flower-like structured nanocomposite consisting of reduced graphene oxide and SnO_2. *Sensor Actuator B Chem* 2017;**239**: 1221—30.

93. Li J, Kuang D, Feng Y, Zhang F, Xu Z, Liu M. Green synthesis of silver nanoparticles—graphene oxide nanocomposite and its application in electrochemical sensing of tryptophan. *Biosens Bioelectron* 2013;**42**:198—206.

94. Zhou S, Deng Z, Wu Z, Xie M, Tian Y, Wu Y. Ta_2O_5/rGO nanocomposite modified electrodes for detection of tryptophan through electrochemical route. *Nanomaterials* 2019; **9**:811.

95. Balloux F, van Dorp L. Q&A: what are pathogens, and what have they done to and for us? *BMC Biol* 2017;**15**:91.

96. Ribet D, Cossart P. How bacterial pathogens colonize their hosts and invade deeper tissues. *Microb Infect* 2015;**17**:173—83.

97. Douglas AE. Multiorganismal insects: diversity and function of resident microorganisms. *Annu Rev Entomol* 2015;**60**:17—34.

98. Łukasiewicz K, Fol M. Microorganisms in the treatment of cancer: advantages and limitations. *J Immunol Res* 2018;**2018**.

99. El Bashir H, Laundy M, Booy R. Diagnosis and treatment of bacterial meningitis. *Arch Dis Child* 2003;**88**:615—20.

100. Poppert S, Essig A, Stoehr B, Steingruber A, Wirths B, Juretschko S. Rapid diagnosis of bacterial meningitis by real-time PCR and fluorescence in situ hybridization. *J Clin Microbiol* 2005;**43**:3390−7.

101. Seeman NC, Sleiman HF. DNA nanotechnology. *Nat Rev Mater* 2017;**3**:1−23.

102. Tiwari I, Singh M, Pandey CM, Sumana G. Electrochemical genosensor based on graphene oxide modified iron oxide−chitosan hybrid nanocomposite for pathogen detection. *Sensor Actuator B Chem* 2015;**206**:276−83.

103. Cui F, Xu Y, Wang R, Liu H, Chen L, Zhang Q. Label-free impedimetric glycan biosensor for quantitative evaluation interactions between pathogenic bacteria and mannose. *Biosens Bioelectron* 2018;**103**:94−8.

104. Travers A, Muskhelishvili G. DNA structure and function. *FEBS J* 2015;**282**:2279−95.

105. Drury MD, Kmiec EB. DNA pairing is an important step in the process of targeted nucleotide exchange. *Nucleic Acids Res* 2003;**31**:899−910.

106. Saad R. Discovery, development, and current applications of DNA identity testing. *Proc (Bayl Univ Med Cent)* 2005;**18**:130−3.

107. Hammond JL, Formisano N, Estrela P, Carrara S, Tkac J. Electrochemical biosensors and nanobiosensors. *Essays Biochem* 2016;**60**:69−80.

108. Umemura K. Hybrids of nucleic acids and carbon nanotubes for nanobiotechnology. *Nanomaterials* 2015;**5**:321−50.

109. Chen M, Hou C, Huo D, Yang M, Fa H. An ultrasensitive electrochemical DNA biosensor based on a copper oxide nanowires/single-walled carbon nanotubes nanocomposite. *Appl Surf Sci* 2016;**364**:703−9.

110. Chen M, Hou C, Huo D, Fa H. A highly sensitive electrochemical DNA biosensor for rapid detection of CYFRA21-1, a marker of non-small cell lung cancer. *Anal Methds* 2015;**7**.

111. Diemer GS, Stedman KM. A novel virus genome discovered in an extreme environment suggests recombination between unrelated groups of RNA and DNA viruses. *Biol Direct* 2012;**7**:13.

112. Zingaretti C, De Francesco R, Abrignani S. Why is it so difficult to develop a hepatitis C virus preventive vaccine? *Clin Microbiol Infect* 2014;**20**:103−9.

113. Hamza M, Tahir MN, Mustafa R, Kamal H, Khan MZ, Mansoor S, et al. Identification of a dicot infecting mastrevirus along with alpha-and betasatellite associated with leaf curl disease of spinach (*Spinacia oleracea*) in Pakistan. *Virus Res* 2018;**256**:174−82.

114. Shafiq M, Iqbal Z, Ali I, Abbas Q, Mansoor S, Briddon RW. Real-time quantitative PCR assay for the quantification of virus and satellites causing leaf curl disease in cotton in Pakistan. *J Virol Methods* 2017;**248**:54−60.

115. Hohenlohe PA, Hand BK, Andrews KR, Luikart G. Population genomics provides key insights in ecology and evolution. In: *Population genomics*. Springer; 2018. p. 483−510.

116. Tahir MA, Hameed S, Munawar A, Amin I, Mansoor S, Khan WS. Investigating the potential of multiwalled carbon nanotubes based zinc nanocomposite as a recognition interface towards plant pathogen detection. *J Virol Methods* 2017;**249**:130−6.

117. Tahir MA, Bajwa SZ, Mansoor S, Briddon RW, Khan WS, Scheffler BE. Evaluation of carbon nanotube based copper nanoparticle composite for the efficient detection of agroviruses. *J Hazard Mater* 2018;**346**:27−35.

118. Cao X, Cao X, Guo H, Li T, Jie Y, Wang N. Piezotronic effect enhanced label-free detection of DNA using a Schottky-contacted ZnO nanowire biosensor. *ACS Nano* 2016;**10**:8038−44.

119. Hassen WM, Chaix C, Abdelghani A, Bessueille F, Leonard D, Jaffrezic-Renault N. An impedimetric DNA sensor based on functionalized magnetic nanoparticles for HIV and HBV detection. *Sensor Actuator B Chem* 2008;**134**:755−60.

120. Mokhtarzadeh A, Eivazzadeh-Keihan R, Pashazadeh P, Hejazi M, Gharaatifar N, Hasanzadeh M. Nanomaterial-based biosensors for detection of pathogenic virus. *Trends Anal Chem* 2017;**97**:445−57.

121. Liu F, Xiang G, Zhang L, Jiang D, Liu L, Li Y. A novel label free long non-coding RNA electrochemical biosensor based on green L-cysteine electrodeposition and Au—Rh hollow nanospheres as tags. *RSC Adv* 2015;**5**:51990−9.

122. Fang YS, Huang XJ, Wang LS, Wang JF. An enhanced sensitive electrochemical immunosensor based on efficient encapsulation of enzyme in silica matrix for the detection of human immunodeficiency virus p24. *Biosens Bioelectron* 2015;**64**:324−32.

123. Hu Y, Li F, Han D, Wu T, Zhang Q, Niu L. Simple and label-free electrochemical assay for signal-on DNA hybridization directly at undecorated graphene oxide. *Anal Chim Acta* 2012;**753**:82−9.

124. Brady OJ, Hay SI. The global expansion of dengue: how *Aedes aegypti* mosquitoes enabled the first pandemic arbovirus. *Annu Rev Entomol* 2020;**65**:191−208.

125. Eivazzadeh-Keihan R, Pashazadeh-Panahi P, Mahmoudi T, Chenab KK, Baradaran B, Hashemzaei M. Dengue virus: a review on advances in detection and trends—from conventional methods to novel biosensors. *Microchim Acta* 2019;**186**:329.

126. Hekmat F, Shahrokhian S, Taghavinia N. Ultralight flexible asymmetric supercapacitors based on manganese dioxide—polyaniline nanocomposite and reduced graphene oxide electrodes directly deposited on foldable cellulose papers. *J Phys Chem C* 2018;**122**: 27156−68.

127. Nguyen BTT, Peh AEK, Chee CYL, Fink K, Chow VT, Ng MM. Electrochemical impedance spectroscopy characterization of nanoporous alumina dengue virus biosensor. *Bioelectrochemistry* 2012;**88**:15−21.

128. Afsahi S, Lerner MB, Goldstein JM, Lee J, Tang X, Bagarozzi Jr DA. Novel graphene-based biosensor for early detection of Zika virus infection. *Biosens Bioelectron* 2018; **100**:85−8.

129. Kaushik A, Yndart A, Kumar S, Jayant RD, Vashist A, Brown AN. A sensitive electrochemical immunosensor for label-free detection of Zika-virus protein. *Sci Rep* 2018;**8**:1−5.

130. Faria AM, Mazon T. Early diagnosis of Zika infection using a ZnO nanostructures-based rapid electrochemical biosensor. *Talanta* 2019;**203**:153−60.

131. Li J, Yang K, Wu Z, Li X, Duan Q. Nitrogen-doped porous carbon-based fluorescence sensor for the detection of ZIKV RNA sequences: fluorescence image analysis. *Talanta* 2019;**205**:120091.

132. Casas I, Pozo F. SARS, avian influenza, and human metapneumovirus infection. *Enferm Infecc Microbiol Clín* 2005;**23**:438−47. quiz 448.

133. Fouchier RA, Rimmelzwaan GF, Kuiken T, Osterhaus AD. Newer respiratory virus infections: human metapneumovirus, avian influenza virus, and human coronaviruses. *Curr Opin Infect Dis* 2005;**18**:141−6.

134. Bhattacharya M, Hong S, Lee D, Cui T, Goyal SM. Carbon nanotube based sensors for the detection of viruses. *Sensor Actuator B Chem* 2011;**155**:67−74.

135. Kawahara T, Hiramatsu H, Ohmi Y, Hayashi T, Suzuki Y, Sriwilaijaroen N. Effect of clustering on fluctuations in binding activity of sugar chains to influenza viruses. In: *25th international conference on noise and fluctuations (ICNF 2019)*; 2019.

136. Lee J, Morita M, Takemura K, Park EY. A multi-functional gold/iron-oxide nanoparticle-CNT hybrid nanomaterial as virus DNA sensing platform. *Biosens Bioelectron* 2018;**102**:425−31.

137. Low SS, Tan MT, Loh H-S, Khiew PS, Chiu WS. Facile hydrothermal growth graphene/ZnO nanocomposite for development of enhanced biosensor. *Anal Chim Acta* 2016;**903**: 131−41.

Metal oxide/carbon nanotube hybrid nanomaterials as ultraviolet photodetectors

Gul Naz[1], Muhammad Ramzan[1], Muhammad Latif[2], Muhammad Bilal Tahir[3] and Muhammad Arshad[4]

[1]Institute of Physics, Faculty of Science, The Islamia University of Bahawalpur, Baghdad-ul-Jadid Campus, Bahawalpur, Pakistan; [2]Department of Physics, University of Balochistan, Quetta, Pakistan; [3]Department of Physics, Khawaja Fareed University of Engineering and Information Technology, Rahim Yar Khan, Pakistan; [4]Nanosciences and Nanotechnology Department, National Centre for Physics, Quaid-i-Azam University Islamabad, Islamabad, Pakistan

20.1 Introduction

Stratospheric ozone layer depletion caused by human economic activities has produced chlorofluorocarbon-based compounds that have adversely disrupted the ecosystem with increased ultraviolet (UV) doses through the atmosphere.[1,2] Enhanced UV exposure produces adverse impacts on human health, especially skin-related diseases.[3] Thus, UV sensors/photodetectors are imperative for monitoring UV radiation effectively to avoid damage from excessive UV exposure. These photodetectors are also of fundamental importance in other areas—for example, imaging, environmental and space monitoring, fiber optic communication systems, and security-related applications.

Due to their widespread applications in technology, photodetectors first appeared as a hot research subject in the early 1910s. Photodetectors are photosensors that usually convert light photon energy into an electrical signal. Indeed, they are essential components of various scientific implementations, such as optic fibers, process control, flame sensing, environmental safety, and security, and are also used in military applications. Higher photoresponsivity and lower noise levels are basic requirements for photodetectors in such applications. Nowadays, research and development in photodetector technology mostly focus on nanomaterials because of their large surface area-to-volume ratios and small absorption regions.[4] Such characteristics contribute to higher photoresponse speeds followed by short charge transport times and allow for superior structural compatibility with nanoscale technologies (see Fig. 20.1)[5,6].

Metal oxides (MOs) are well-known as materials for sensing applications.[8] Indeed, the MO surface is important for effective interaction with light photons.[9,10] Modification of MO surface properties may result in improved photosensing. Various physical and chemical routes-established synthetic approaches have been used to develop

Metal Oxide-Carbon Hybrid Materials. https://doi.org/10.1016/B978-0-12-822694-0.00011-9

Figure 20.1 Schematic representation of the working principle of a nanohybrid photodetector.
Reprinted with permission from Ref. 7, copyright ©2020 MDPI.

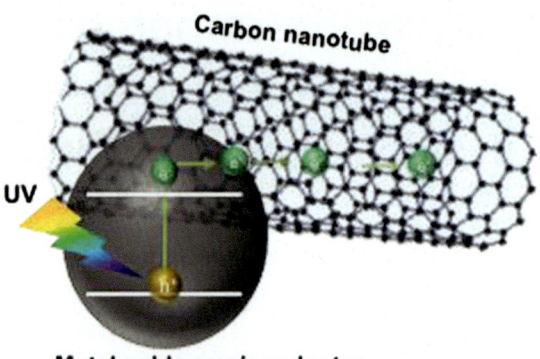

MO-based nanocomposites.[11–17] The approaches of such modifications have been reported many times. However, in recent years, hybrid photodetectors have revealed substantial photocurrent enhancements compared with the functionality of their component photodetectors. As one example, perovskite/Au nanorod hybrid photodetectors have achieved responsivity of 317 A/W at a very low bias voltage of −1 V, showing development of 60% with reference to the photoresponsivity of pure perovskite ($CH_3NH_3PbI_3$) film (≈ 200 A/W).[18] Thus, high-level responsivity and low-level voltage bias place hybrid photosensitive devices made of nanomaterials among the best-performing photodetectors. In addition, the incorporation of highly conductive carbon-based nanomaterials such as carbon nanotubes (CNTs) in MOs has established effective means to facilitate charge mobility while extending electron lifetimes, thus leading to enhanced efficiency of photovoltaic (PV) devices.[19,20]

20.2 Metal oxide materials

MOs, well known for good semiconducting devices, possess exceptional optical features as well as charge-transport properties.[21–23] Their properties can be optimized at the nanoscale through morphology-control, surface modification, ion-doping, etc. MOs have been extensively introduced in PV appliances as charge-carrying layers and transparent electrodes as well as in photoactive films. For instance, MoO_3, TiO_2, VO, and ZnO are used at a large scale as charge transferring and extraction layers in perovskite and organic PVs. Among many, ZnO and TiO_2 have been largely employed as quantum dot-sensitive solar cells and as photoelectrodes in dye-sensitive solar cells.[24–28]

Metals are acquired for their oxides by the simple chemical reduction method. Usually, organic compounds are used to reduce MOs. The most common and economical reducing agent may be coke carbon. Primarily, in an oxide, there is at least an oxygen atom along with one other element to develop its chemical formula. MOs typically hold an oxygen anion of −2 oxidation state. Solid oxides are abundant in most of the Earth's crust and are oxidized by the oxygen in the environment. Carbon dioxide and carbon monoxide (CO) are the two principal carbon oxides that emerge from hydrocarbon combustion processes in air. Even pure elements often grow an oxide coating on their surface, i.e., aluminum foil develops a thin layer of Al_2O_3 as a passivation layer that effectively prevents foil corrosion.

MOs are a class of ionic compounds made of metallic (M^+) and oxygen (O^-) ions. Usually, the d-shells of MOs are partially filled by electrons; however, the s-shells (only of metallic ions) are always completely occupied by the electrons. Therefore, semiconductor MOs are classified as p-type (holes as the majority charge carriers) and n-type (electrons as the majority charge carriers). Though most MOs are polymeric, some oxides appear as monomeric molecules such as CO. An example of polymeric oxides is phosphorus pentoxide as a rather complex molecular oxide with a deceptive name, P_4O_{10}. Both acids and bases can confront MOs, and those that react only with acids are called basic oxides, and those that react with bases are called acidic oxides. In general, metals tend to provide basic oxides, and in contrast to metals, nonmetals tend to furnish acidic oxides.

Various synthesis strategies like chemical and physical vapor deposition, surfactant-assisted, template-reduction, electrochemical, and exfoliation have been adopted for designing various MOs nanostructures showing different geometrical morphologies, which may be zero-dimensional, one-dimensional, two-dimensional, or three-dimensional.[29−32] In contrast to traditional three-dimensional nanostructures, one-dimensional MO nanostructures such as In_2O_3, ZnO, Cu_2O, Ga_2O_3, SnO_2, Fe_2O_3, CeO_2, etc. have earned fascinating achievements in sensing applications owing to several advantages, namely, a large length-to-diameter ratio with photons and charge carriers confinement in two dimensions, superior crystalline stability, and potential surface functionalization.[33]

The photoharvesting capability of MO nanostructures is mainly associated with the long lifetime of the photogenerated electrons, which results from the hole-trapping impact caused by O_2 desorption at their surfaces, which potentially decreases the electron-hole pairs recombination rate. For MOs nanostructures, it is well-believed that oxygen adsorption $[O_{2(g)} + e^- \rightarrow O_{2\,(ad)}^-]$ and desorption $[O_{2\,(ag)}^- + h^+ \rightarrow O_{2\,(g)}]$ processes on nanostructure surfaces dominate photocarrier transference and photoresponse activities. Although these MO nanostructure-based sensors exhibit unique properties such as size confined to the nanoscale, low costs, faster response, and easy recovery, they still suffer from the disadvantage of a high operational temperature that may result in high power consumption and trouble in device integration.

20.3 Photodetectors

Photodetectors are the key part of an optical receiver, converting incoming photon signals into electric current, which defines them as O/E convertors. Semiconductor-based photodetectors, typically called photodiodes and phototransistors, are the principal photodetectors used in optical communication systems owing to their dimensional confinement, faster detection speed, and pronounced detection efficiency. Resembling

LASER diodes, photodiode structures also encounter *pn* junctions. Though the *pn* junction of a LASER diode is forward-biased, a photodiode's *pn* junction is reverse-biased, with only a minute reversed saturation current flowing through the device deprived of any input optical signal. Various device modifications have been adopted to enhance the quantum efficiency of practical photodiodes. The popular pin structure holds a sandwiched-type intrinsic layer between the p- and n-type films, making it a semiconductor photodiode also recognized as a pin-diode. Another commonly used photodetector, the avalanche photodiode, shows significant photoamplification via avalanche gain under a sufficiently high bias voltage.

Because the photoinduced current is linearly proportional to the input optical power, a photodiode must directly transform each photon of the incoming optical signal into a free electron. However, practically a semiconductor material does not incorporate the creation of an electron for every incoming photon, which may be unavoidable due to nonefficient photon absorption and charge carrier collection. Thus, the photoresponsivity and the detection speed of a photodiode are the functions of many factors like the semiconductor bandgap structure, material's nature, and electrode design.[34]

Practically, the photodetectors also generate noises in the photosensing process, which is unfavorable to device performance due to the signal-to-noise ratio. Main noise sources may comprise thermal noise, shot noise, and dark-carrier noise. Thermal noise can be controlled by optimizing the load resistance. Dark-carrier noise can be reduced by reducing the reversed saturation current through material upgrading and optimizing the junction structure. However, reducing shot noise is very difficult because it is intrinsically related to the photodetection mechanism. A high-performance photodetector must be provided with very high photosensitivity and wide spectral selectivity. It must own a fast and linear response speed, accompanied by high light transmission and enhanced physical and chemical stability.[8]

20.3.1 Photodetector parameters

A photodetector's most important parameters include device sensitivity, photoresponsivity, and external quantum efficiency (EQE). Such parameters are generally used to label the UV detector performance. The photodetector sensitivity (S) is assumed as the ability to detect even weak signals and is typically limited by the detector noise, calculated using[35]

$$S\,(\%) = \frac{I_{Ph} - I_d}{I_d} \times 100 \tag{20.1}$$

where I_{ph} and I_d denote the photo- and dark-current, respectively. The photodetector responsivity (R), an important factor for photodetectors, is the ratio of photocurrent generated (as an output signal) and the incident light power and is calculated using the following relation:

$$R = \frac{I_{ph} - I_d}{P_{in}} \tag{20.2}$$

where P_{in} is the incident light source power. The EQE measures the number of electrons detected per incident photon and quantifies the probability of electron/hole pairs that stimulate the external photocurrent generated upon a single photon illuminated in the device. The EQE value can be calculated using the following equation[36]:

$$\text{External Quantum Efficiency (EQE)} = \frac{R}{\lambda_{inc}} \times hc \times 100\% \qquad (20.3)$$

where R represents the responsivity, h is the Planck constant, c is the speed of light photon, and λ_{inc} is the incident wavelength.

20.4 Metal oxide-based hybrid photodetectors

Perceptibly, a large photocurrent is critical for achieving high photoresponsivity of the detector. For this purpose, efforts have been made to accomplish high photocarriers and increased photoresponsivity for detectors. Photodetectors based on commonly available MO semiconductor nanostructures, such as ZnO and SnO_2, usually exhibit high absorption coefficient values and photoresponsivity resulting from the pronounced surface effects.[37−39] But these surface effects usually lead to the slow oxygen absorption/desorption procedures at the surface of MO nanostructures that may cause the long rise/decay times of these photodetectors, leading to the decay of photocurrent.[40] Also, the rapid recombination of photogenerated electron/hole pairs in terms of the short carriers lifetime and low absorption coefficient of semiconductor nanomaterials leads to a drop in quantum efficiency followed by poor photodetection performance. Compromises between enhanced photoresponsivity and reduced rise/decay times can seriously limit the hands-on applications of MO photodetectors. Several strategies have been proposed to meet these challenges and resolve the poor photoresponse and slow response speed of MO-based photodetectors, such as amendment of light absorption, defects-engineering induction, development of novel heterostructure-based photodetectors, tailoring nanostructure geometries, and modification of electrode configurations, to accomplish enhanced responsivity and photocurrent gain.

For MO-based photodetectors, the role of oxygen adsorption/desorption in sensing the light signal is of significant interest, whereas in hybrid photodetectors, developing the depletion region at the interface of two different materials, as well as the passivation effect, needs to be explored. For the fabrication of MO-based hybrid UV photodetectors, the following methods are usually adopted:

(i) RF sputtering is applied to deposit ITO thin film layers on a washed and cleaned glass substrate. Then a small chunk of the ITO film layer is covered using an appropriate mask, which behaves as a back-electrode of an as-fabricated device.

(ii) Using a spray pyrolysis technique, the optimal MO layer of high transmittance and low resistance values is deposited on the ITO-coated glass substrate.

(iii) A thin layer of other material with low electrical resistivity and high hole concentration is deposited onto the MO layer using the spin coating technique. Finally, thermal evaporation methods are used to deposit silver (Ag) as top electrodes.

As one example, the photoresponsivity of ITO/ZnO/PEDOT:PSS/Ag hybrid photodetectors, where poly (3,4 ethylene dioxythiophene):poly (styrene sulphonate) (PEDOT:PSS) used as the hole transporting layer, was significantly enhanced by the generation of a large number of photocarriers and speedy separation of the carriers due to integrated electric field and O_2 passivation during adsorption/desorption procedures.[41] Comparatively, the pristine ITO/ZnO/Ag photodetector photoresponse was slow because of the short-term oxygen adsorption/desorption procedures. Thus, it was concluded that the ZnO/PEDOT:PSS-based hybrid junctions possessed (i) fast separation of photoinduced charge carriers and (ii) O_2 molecules passivation during adsorption/desorption procedures.

20.5 Carbon nanotube structures and characteristics

Over the last 2 decades, CNTs as one type of carbonaceous nanofillers have established part of extensive research and challenges due to their superior characteristics and extended applications over other materials. Also, the exceptional characteristics of CNTs—for instance, 1-D structure, highly exposed surface area, chemically and thermally improved stability, and excellent mechanical strength—result in the development of CNT-based technologies in thin-film electronics, supercapacitors, energy storage, health care, biosensing, and even in terahertz devices.

CNTs generally exist within three different possible geometries—armchair, zigzag, and chiral—depending on how the graphite is wrapped during its formation process.[42] The chiral vector and its corresponding pairs of integers are responsible for the electrical, optical, and mechanical properties of CNTs.[43] A pair of indices, nn and mm, represents the chiral vector; the two integers correspond to the magnitude of unit vectors along the two directions in the honeycomb type crystal lattice of graphene. When mm $m\, m$, the nanotube (NT) is called a "zigzag"; when $nn\, m\, mm$, the NT is called an "armchair"; and all other configurations are "chiral." Fig. 20.2 represents the three types of CNTs—armchair, zigzag, and chiral.

The sp^2 carbon—carbon bonding is accountable for the unusual mechanical strength of CNTs, which may exhibit a Young's modulus value of 1000 GPa, indicating a mechanical strength five times greater than steel. Similarly, the tensile strength of CNTs can be about 50 times higher than steel, and at approximately 63 GPa.[45] CNT thermal conductivity is as high as $2000-6000$ W/m·K, which is remarkably greater than bulk metallic silver and copper.[46] Having a 1-D graphitic character, CNTs possess extraordinary electrical properties owing to their peculiar electronic structure. As resistance occurs during collisions of electrons with defects in the crystal structure through which they pass, CNTs are provided with extremely low electrical resistivity because electrons are not easily scattered in CNTs. The CNT characteristics mentioned above

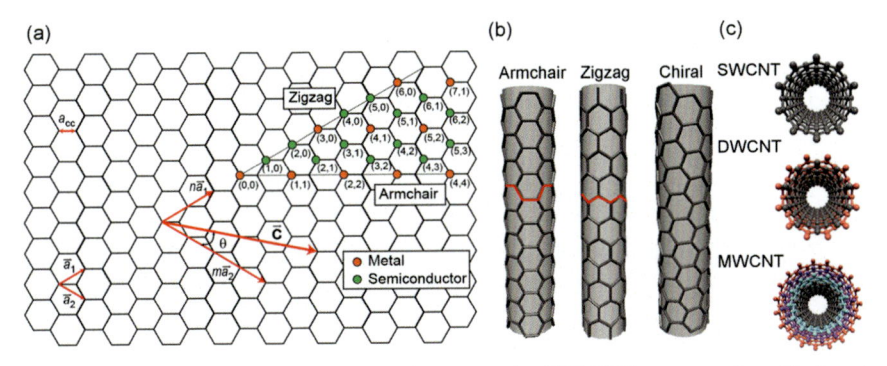

Figure 20.2 (A) Unrolled single-walled carbon nanotube (SW-CNT) representing chiral vector C and the effect of different values of the integers (*n* and *m*) on electronic properties of SW-CNTs; (B) Appearance of nanotube directed by chiral vector orientation. Examples of CNTs include (4,4) armchair shape, (6,0) zigzag shape, and (5,3) chiral shape; and (C) "Ball and Stick" representation of SW-CNT, double-walled CNT, and multiwalled CNT.
Reprinted with permission from Ref. 44, Copyright © 2016 American Chemical Society.

are synthetic procedure-dependent and can be tailored by monitoring the wall diameter and structure, chirality, and length of the NT.

Different CNT variants, such as single-walled carbon nanotubes (SW-CNTs), double-walled carbon nanotubes (DW-CNTs), and multiwalled carbon nanotubes (MW-CNTs), possess astonishing physical properties attributed to their intrinsic morphological configurations and structural characteristics. Containing entirely carbon, the structure of SW-CNT appears as a rolled-up cylindrical shell of graphene sheet representing crystalline graphite of single atomic-layered thickness and in the form of hexagonal benzene type rings of carbon atoms. The SW-CNT is 1 atom thick while containing about 10 atoms around its circumference and generally having a large aspect ratio (length-to-diameter ratio) of about 1000.[47] On the other hand, an MW-CNT consists of a rolled-up cylindrical stack of graphene sheets. Each NT is like a single molecule comprising millions of atoms. The dimension of this molecule usually exists in tens of micrometers along its length with a diameter around 10 nm. MW-CNTs are sufficiently large to include many SW-CNTs stacked one inside the other. MW-CNTs are nanostructures with an outer diameter of the value below 15 nm, above which these nanostructures are recognized as carbon nanofibers.

A variety of techniques has been established to produce CNTs of different structures and morphologies in excess amounts. The well-established three methods commonly used to synthesize CNTs include arc discharge, LASER ablation, and chemical vapor deposition (CVD),[48] while new approaches of their facile and large-scale synthesis need to be developed and optimized. Interestingly, lower-dimensional CNT structures such as SW-CNTs and DW-CNTs exhibit increased photoexcitation effects when combined with higher-dimensional MW-CNT structures or metal electrodes to develop heterodimensional ohmic contacts. In such cases, the photoexcited electrons are transported from the lower-dimensional structure to the higher-dimensional or metal electrodes.[49,50] In a typical nonselective synthesis, if all chiral species grow with the same probability, one-third of the CNTs are metallic, and two-thirds are semiconducting.

20.6 Metal oxide/carbon nanotube hybrid nanomaterials as ultraviolet photodetectors

Photodetectors in the UV light region possess numerous applications, including flame detection, space survey, and military-based uses. The photosensing features of MO-based photodetectors are principally directed by the oxygen molecules' adsorption/desorption behaviors on MO surfaces. Therefore, developing practical methods for the fabrication of photodetectors that can reduce the gap between surface chemical reactions and effective electrical signal production is imperative. Carbon-based nanocomposites are valued for their high thermal and electrical conductivity, high adhesion capability, large surface area, and high flexibility with zero bandgaps. With these properties, they may appear as special compounds to be called nanohybrids that can act as photocatalysts[51−53] and electrocatalyst.[54]

Recently, many combinations of CNTs and MO such as V_2O_5, TiO_2, and SnO_2 have expanded considerable attention owing to their greater applications in photocatalysis, energy storage, and gas sensor technologies.[16,55−58] As one more example, ZnO coupled with CNTs are promising hybrid nanomaterials for applications in optoelectronic-based devices.[59] Coaxial-type heterostructured NTs combining a *p*-channel CNT-core with an *n*-channel ZnO-shell may be integrated into logical inverters.[60] The photosensing applications of some MO/CNT-based hybrid nanomaterials are very interesting and therefore make them preferable materials for upgrading the heterojunction-based photodetectors in a wide-ranging light wavelength. This section highlights the development, the photoresponse mechanism, and recent progress concerning some kinds of MO/CNT heterostructure-based hybrid photodetectors, including TiO_2/CNT, ZnO/CNT, and tungsten oxide (WO_3)/CNT nanohybrid photodetectors.

20.6.1 Titanium dioxide/carbon nanotube nanohybrid photodetectors

Due to the exceptional properties of TiO_2 nanostructures, much interest has been focused on UV photodetectors founded by TiO_2 nanostructures, especially 1-D TiO_2 nanostructures such as nanorods, nanowires, and NTs.[61−64] These nanostructures possess enough surface-to-volume ratio, low-cost fabrication, long-lasting stability, and extensive functionality while making great interest in photodriven applications. Among these 1-D nanostructures, TiO_2 NTs are evident due to their vertically directed and extremely ordered structure. Such TiO_2 NTs arrays have been widely investigated for efficient UV photodetection applications as they exhibit high performance with lower detection limit, fast responsivity, increased stability, and noticeable reproducibility.[64,65] Such parameters are crucial for ozone layer monitoring, space communication, blaze sensing, etc.

In the past few years, the fabrication of TiO_2/CNT-based nanohybrids has been successfully reported, with considerable photodriven activities in the visible light range because of the electron-capturing capability and conductivity of the CNTs, making these nanohybrids ideal for reducing electron/hole recombination in the MO.[66] As one specialty, fabrication of TiO_2/CNT arrays facilitates their significance for various photodriven applications, ranging from photocatalysis[53] to photoelectrode.[19]

The dimensionality difference effect of the DW-CNT film/TiO_2 NTs array-based photodetector may exhibit a much higher photoresponse relative to other composite materials where the heterodimensional nonohmic contacts can also make very efficient charge generation and electron transfer from DW-CNTs into higher oriented TiO_2 NTs arrays under light illumination with modulating bias voltages.[67] An as-fabricated photodevice showed high photoconductivity not only under illumination at 340 nm UV photons, but also at 532 nm which is lower than the bandgap of TiO_2. Incorporation of Cu_2O nanoparticles film between DW-CNT film and TiO_2 NTs array further reduced dark current and enhanced visible photoresponse of the heterojunctions[68] (see Fig. 20.3).

In addition, some photodetector devices are preferred in the vacuum due to the photodesorption of oxygen molecules from CNTs causing an intense drop in the current.[69] This requires additional modification in the photodetector device like deposition of Au nanoparticles on the DW-CNT film, the Au nanoparticle/DW-CNT film/TiO_2 NTs arrays/Ti heterojunctions are constructed for use as broadband photodetectors that show incredible enhancement in the photoresponsivity, and the intrinsic mechanism behind this is associated with enhanced photoabsorption by the Au nanoparticles.[70]

On the other hand, the core/shell nanowire morphology has also been proved a new trail toward supersensitive, ultrafast, and broadband photosensing materials. Following the case of MW-CNT/TiO_2 core/shell architecture, a radial Schottky barrier was developed at the interface of core/shell geometry which not only regulated electron transporting but also facilitated the creation of photogenerated electron/hole pairs.[71] Consequently, a highest photogain of 1.4×10^4 was achieved along with 4.3/10.2 ms fastest response/recovery times. The radial Schottky junction and the

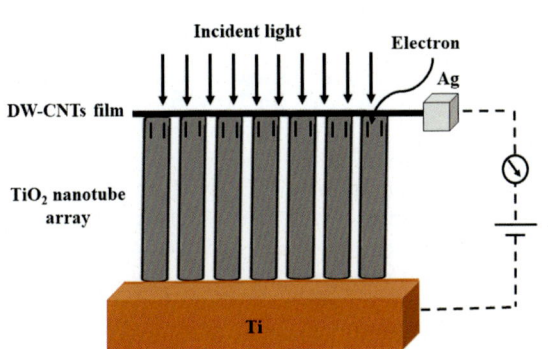

Figure 20.3 A general schematic representing the dimensionality difference effect of the double-walled carbon nanotube film/titanium dioxide nanotube array-based photodetector. Redrawn with permission from Ref. 67, Copyright © 2011 Springer.

defect band absorption also accomplished the broad-band photodetection probably through additional absorption from MW-CNTs. Indeed, the photoresponse in the regime higher than 400 nm wavelength is evidently attributed to two possible mechanisms linked with the structural effects: (i) the photocarriers are excited through metallic MW-CNTs over the Schottky barrier, owing to an exclusive light absorption by the device. (ii) Another possibility for the observed broad-band absorption is the presence of defects-induced sub-bandgap in TiO_2 and at the interface of the MW-CNT/TiO_2 heterojunction (see Fig. 20.4).

20.6.2 Zinc oxide/carbon nanotube nanohybrid photodetectors

ZnO, being an n-type II–IV semiconducting MO with a wide direct bandgap and a noticeable exciton binding energy (60 meV), has established extensive attention due to its higher sensitivity in the UV light range. In the 1940s, Mollow first observed the UV photoresponse in ZnO film.[72] However, the research on ZnO-based photodetectors grew steadily in the 1980s. Therefore, UV photodetectors based on various ZnO nanostructures such as nanoparticles, nanoflakes, nanorods, and nanowires, have been developed through various synthetic processes.[73–76] Most studies on ZnO-based photodetectors have essentially focused on tailoring various ZnO

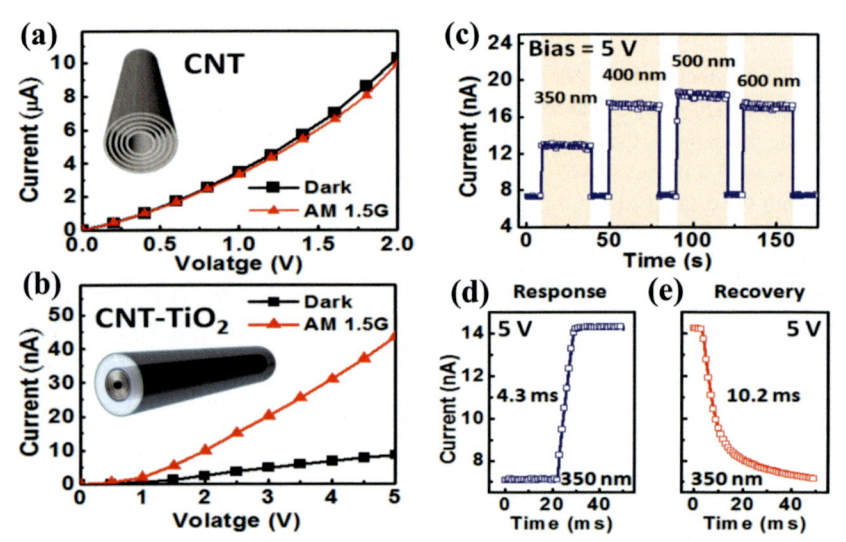

Figure 20.4 (A) IV characteristics of the MW-CNT device in the dark and upon air mass 1.5 global (AM 1.5G) illumination (1000 W/m^2). (B) IV characteristics of the TiO_2-coated MW-CNT photodetector in the dark and upon illumination by AM 1.5G. (C) Time-resolved photocurrents of the TiO_2-coated photodevice under various wavelengths illumination in the visible region at 5 V. Highlights of the (D) response speed and (E) recovery speed under 350 nm wavelength illumination at 5 V.
Reprinted with permission from Ref. 71, Copyright © 2012 American Chemical Society.

nanostructures to extend their surface area, which allows rich oxygen molecules adsorption onto the nanostructure's surface and higher charge carrier concentrations that finally increase the photoresponsivity of UV photodetectors.

The photocarrier generation mechanism is controlled by the O_2-mediated free electrons produced when ZnO is used alone as the active material in UV photodetector devices irrespective of morphology and the synthetic procedure for ZnO nanostructures. In this mechanism, the electronegative O_2 molecules already adsorbed onto the ZnO surface immediately desorbed upon UV light illumination. The generated holes cause the charge neutralization, resulting in free electrons generation in ZnO nanostructures. These free electrons then stream as photocurrent under external bias potential. However, such photodevices generally suffer from low photosensitivity, with a slow on/off mechanism and poor photoresponse performance due to long rise/decay time. The instability of free electrons is caused when the holes have not been completely combined with O_2 and relaxed O_2 adsorption/desorption procedures.[77] Furthermore, ZnO possesses intrinsic n-type characteristics, and therefore, the presence of p-type impurity in ZnO may cause complications in developing ZnO-based p-n homojunction devices.[41] Thus, an alternative approach including the fabrication of hybrid photodetectors is imperative.[78] So far, a large variety of heterojunctions of semiconductor nanostructures has been applied in photodetectors, and a greater enhancement in photoresponse has been observed because of heterojunction-based photodetectors, including ZnO nanostructures operated in the UV region.[79,80]

Modifications of ZnO nanostructures with CNTs for enhancing the photosensitive and electrical properties of hybrids demonstrate that hybrids' long-term operational stability and photoresponsivity can be significantly improved by introducing CNTs into the ZnO nanostructures rather than using solely ZnO nanostructures, which is attributed to the excellent electrical conductivity of CNTs. In ZnO/CNT nanohybrids, the holes generated under the UV illumination are partially trapped with negatively charged oxygen, while free electrons are transported mainly through the metallic CNTs under the applied voltage bias between the two electrodes. In this routine, ZnO/SW-CNT hybrid nanostructures are one of the technologically attractive nanocomposites investigated for their optoelectronic properties, and hence, ZnO nanowire-based transparent and flexible visible-blind UV photodetectors could be developed very efficient following a solution-based approach where SW-CNT thin film is used only as electrodes.[81] The improved carrier transport and photoresponse efficiency have also been encountered by incorporating MW-CNTs into ZnO nanowires photodetector.[82] The photoresponsivity of ZnO/CNT hybrid photodetectors is found to be in proportion to both ZnO nanowire concentration and CNT content.[81,83]

Thus far, various techniques have been developed to fabricate ZnO/CNT nanohybrids, including largely thermal CVD, electrochemical deposition, covalent coupling, pulsed-LASER deposition, and water-assisted growth. However, atomic layer deposition (ALD) has been considered as an efficient route to fabricate ZnO/CNT nanohybrids by the ALD of ZnO on MW-CNTs for use of the UV photodetectors.[59,84] ALD is a cyclic and self-limiting CVD technique, which accomplishes the conformal and uniform coating of thin films at angstrom level. This technique can be applied to deposit a range of materials including metals and MOs on various nonplanar

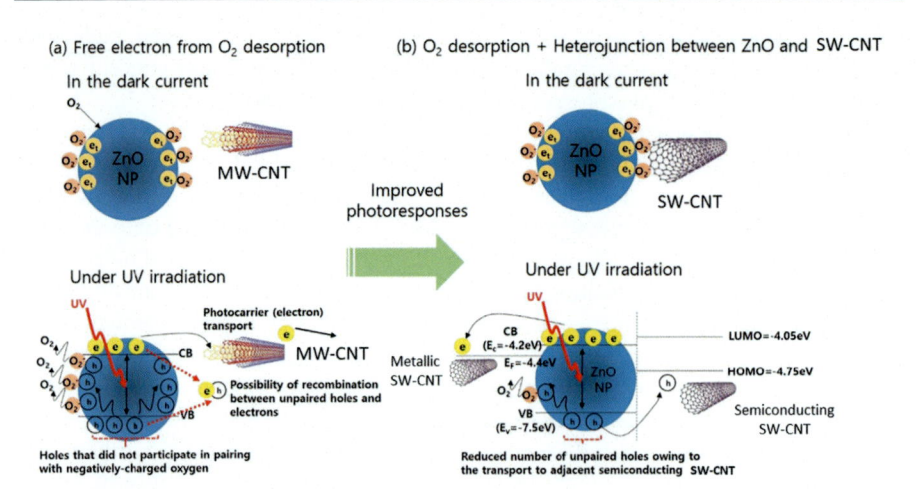

Figure 20.5 Schematic representation of the photocurrent mechanism for (A) ZnO/MW-CNTs, free electrons generation from O_2 desorption under UV irradiation, and (B) ZnO/SW-CNTs, combination of O_2 desorption and hole depletion resulting in electron enrichment at the p-n heterojunction under UV irradiation.
Reprinted with permission from Ref. 77, Copyright © 2020 MDPI.

substrates. It is an essential part of semiconductor industry to promote the development of high *k*-dielectric layers of the MOs field-effect transistors and film capacitors of dynamic random-access memory.[85]

Practically, the introduction of SW-CNTs (having both metallic and semiconducting characters) exhibits better performance than the MW-CNTs (only being metallic). SW-CNTs greatly promote the stability of free electrons released due to the O_2 desorption on to ZnO surface when exposed to an UV light source and develop a built-in potential between ZnO and SW-CNT heterojunctions. Hence, they permit efficient electron transport trails through high-aspect-ratio SW-CNTs with reduced defect density[77] (see Fig. 20.5).

20.6.3   Tungsten oxide/carbon nanotube nanohybrid photodetectors

Nanohybrid photodetectors have attracted extraordinary research interest because of their enhanced photoresponsivity and photoconductive response in optoelectronic fields.[86] In this instance, MO-based nanocomposites have already been use to fabricate UV photodetectors. According to some research reports, WO_3 nanostructures have shown a greater capability to be fabricated as a UV photodetector.[87] In this regard, $WO_3 \cdot H_2O$ (tungstite)- and WO_3-decorated MW-CNTs develop ultrafast UV photodetectors, wherein the faster charge transfer through the MW-CNTs greatly provides to the speedy UV photodetection of the tungstite and WO_3-decorated MW-CNT photodetectors.[88] The recovery and response times recorded were as low as 400 and 500 μs for a WO_3-decorated MW-CNT photodetector. Thus, developing efficient methods for the fabrication of WO_3-based nanohybrid photodetectors is demanding for those that can reduce the gap between surface-active reactions and effective electrical signal production through the device.

20.7 Conclusion

Modifications of MO nanostructures with CNTs are demonstrated to enhance the photosensitive and charge-transport properties of the hybrids. The metal oxide/CNT hybrid photodetector devices show a fast response compared to pristine devices. Basic strategies behind the photocurrent generation mechanism in nanohybrids require superior photosensitive capabilities for a broad range of UV light applications. Advances in photodetection efficiency have been ascribed to the incorporation of CNTs into MOs as well as concurrent improvements in charge transport and extraction. The distinct surface effects and dimensionality difference effects of nanohybrid-based UV photodetectors can manifest themselves as possessing much higher photo-responsivity than their counterparts. The very efficient charge separation and electron transfer from CNTs into well-structured MOs usually result from the Schottky barriers developed at the heterojunctions when exposed to UV light at modulating bias voltages. It is advisable to accept that both the selected materials and the heterojunction structures along dimensionality difference influence have to be taken in the advancement, strategy, and upgradation of high-performance nanohybrid photodevices. Fabricating novel MO/CNT nanohybrid photodetectors may facilitate their significance for other photodriven applications such as photoelectrode, photocatalysis, photoluminescence, etc.

References

1. Solomon KR. Effects of ozone depletion and UV-B radiation on humans and the environment. *Atmos-Ocean* 2008;**46**:185−202.
2. Norval M, Lucas RM, Cullen AP, de Gruijl FR, Longstreth J, Takizawa Y, van der Leun JC. The human health effects of ozone depletion and interactions with climate change. *Photochem Photobiol Sci* 2011;**10**:199−225.
3. Lucas RM, Yazar S, Young AR, Norval M, de Gruijl FR, Takizawa Y, Rhodes LE, Sinclair CA, Neale RE. Human health in relation to exposure to solar ultraviolet radiation under changing stratospheric ozone and climate. *Photochem Photobiol Sci* 2019;**18**: 641−80.
4. Filatzikioti A, Glezos N, Kantarelou V, Kyriakis A, Pilatos G, Romanos G, Speliotis T, Stathopoulou DJ. Carbon nanotube Schottky type photodetectors for UV applications. *Solid State Electron* 2019;**151**:27−35.
5. Lam K-T, Hsiao Y-J, Ji L-W, Fang T-H, Hsiao K-H, Chu T-T. High-sensitive ultraviolet photodetectors based on ZnO nanorods/CdS heterostructures. *Nanoscale Res Lett* 2017;**12**:31.

6. Wang Q, Zhang D, Wu Y, Li T, Zhang A, Miao M. Fabrication of supercapacitors from $NiCo_2O_4$ nanowire/carbon-nanotube yarn for ultraviolet photodetectors and portable electronics. *Energy Technol* 2017;**5**:1449—56.

7. Pathak P, Park S, Cho HJ. A carbon nanotube—metal oxide hybrid material for visible-blind flexible UV-sensor. *Micromachines* 2020;**11**:368.

8. Nunes D, Pimentel A, Gonçalves A, Pereira S, Branquinho R, Barquinha P, Fortunato E, Martins R. Metal oxide nanostructures for sensor applications. *Semicond Sci Technol* 2019; **34**:043001.

9. Akram M, Alshemary AZ, Butt FK, Goh Y-F, Ibrahim WAW, Hussain R. Continuous microwave flow synthesis and characterization of nanosized tin oxide. *Meter Lett* 2015;**160**: 146—9.

10. Naz G, Shamsuddin M, Butt FK, Bajwa SZ, Khan WS, Irfan M, Irfan M. Au/Cu_2O core/ shell nanostructures with efficient photoresponses. *Chin J Phys* 2019;**59**:307—16.

11. Fan Z, Chen J, Wang M, Cui K, Zhou H, Kuang Y. Preparation and characterization of manganese oxide/CNT composites as supercapacitive materials. *Diam Relat Mater* 2006; **15**:1478—83.

12. Richter K, Birkner A, Mudring A-V. Stabilizer-free metal nanoparticles and metal—metal oxide nanocomposites with long-term stability prepared by physical vapor deposition into ionic liquids. *Angew Chem, Int Ed* 2010;**49**:2431—5.

13. Hu C, Lu T, Chen F, Zhang R. A brief review of graphene—metal oxide composites synthesis and applications in photocatalysis. *J Chin Adv Mater Soc* 2013;**1**:21—39.

14. Sreeprasad TS, Maliyekkal SM, Lisha KP, Pradeep T. Reduced graphene oxide—metal/ metal oxide composites: facile synthesis and application in water purification. *J Hazard Mater* 2011;**186**:921—31.

15. Atchudan R, Jebakumar Immanuel Edison TN, Perumal S, RanjithKumar D, Lee YR. Direct growth of iron oxide nanoparticles filled multi-walled carbon nanotube via chemical vapour deposition method as high-performance supercapacitors. *Int J Hydrogen Energy* 2019;**44**:2349—60.

16. Seekaew Y, Wisitsoraat A, Phokharatkul D, Wongchoosuk C. Room temperature toluene gas sensor based on TiO_2 nanoparticles decorated 3D graphene-carbon nanotube nanostructures. *Sensor Actuator B Chem* 2019;**279**:69—78.

17. Visakh PM, Raneesh B. *Metal oxide nanocomposites: state-of-the-art and new challenges, metal oxide nanocomposites.* 2020. p. 1—26.

18. Wang H, Lim JW, Quan LN, Chung K, Jang YJ, Ma Y, Kim DH. Perovskite—gold nanorod hybrid photodetector with high responsivity and low driving voltage. *Adv Opt Mater* 2018; **6**:1701397.

19. Batmunkh M, Macdonald TJ, Shearer CJ, Bat-Erdene M, Wang Y, Biggs MJ, Parkin IP, Nann T, Shapter JG. Carbon nanotubes in TiO_2 nanofiber photoelectrodes for high-performance perovskite solar cells. *Adv Sci* 2017;**4**:1600504.

20. Macdonald TJ, Batmunkh M, Lin C-T, Kim J, Tune DD, Ambroz F, Li X, Xu S, Sol C, Papakonstantinou I, McLachlan MA, Parkin IP, Shapter JG, Durrant JR. Origin of performance enhancement in TiO_2-carbon nanotube composite perovskite solar cells. *Small Methods* 2019;**3**:1900164.

21. Akram M, Butt FK, Alshemary AZ, Goh Y-F, Ibrahim WAW, Hussain R. Continuous microwave flow synthesis (CMFS) of nanosized titania: structural, optical and photo-catalytic properties. *Mater Lett* 2015;**158**:95—8.

22. Akram M, Razali IR, Ghafoor F, Ur-Rahman S, Butt FK, Hussain R. Continuous facile synthesis of nano-sized zinc oxide and its optical properties. *Mater Res Express* 2018;**5**: 075901.

23. Shin SS, Lee SJ, Seok SI. Metal oxide charge transport layers for efficient and stable perovskite solar cells. *Adv Funct Mater* 2019;**29**:1900455.

24. Aboulouard A, Gultekin B, Can M, Erol M, Jouaiti A, Elhadadi B, Zafer C, Demic S. Dye sensitized solar cells based on titanium dioxide nanoparticles synthesized by flame spray pyrolysis and hydrothermal sol-gel methods: a comparative study on photovoltaic performances. *J Mater Res Technol* 2020;**9**:1569–77.

25. Zhang Y, Zhong X, Zhang D, Duan W, Li X, Zheng S, Wang J. TiO$_2$ nanorod arrays/ZnO nanosheets heterostructured photoanode for quantum-dot-sensitized solar cells. *Sol Energy* 2018;**166**:371–8.

26. Boro B, Gogoi B, Rajbongshi BM, Ramchiary A. Nano-structured TiO$_2$/ZnO nano-composite for dye-sensitized solar cells application: a review. *Renew Sustain Energy Rev* 2018;**81**:2264–70.

27. Idris MI, Abidin YZ, Abdullah H, Shafie S, Chachuli SAM, Rashid M, Xian KJ. Zinc oxide quantum dots as photoanode for dye-sensitized solar cell. In: *2020 IEEE international conference on semiconductor electronics (ICSE)*; 2020. p. 57–60.

28. Omar A, Ali MS, Abd Rahim N. Electron transport properties analysis of titanium dioxide dye-sensitized solar cells (TiO$_2$-DSSCs) based natural dyes using electrochemical impedance spectroscopy concept: a review. *Sol Energy* 2020;**207**:1088–121.

29. Sun Z, Liao T, Kou L. Strategies for designing metal oxide nanostructures. *Sci China Mater* 2017;**60**:1–24.

30. Peng Z, Liu Z, Chen J, Ren Y, Li W, Li C, Chen J. Influence of ZnO nano-array interlayer on the charge transfer performance of quantum dot sensitized solar cells. *Electrochim Acta* 2019;**299**:206–12.

31. Chavali MS, Nikolova MP. Metal oxide nanoparticles and their applications in nanotechnology. *SN Appl Sci* 2019;**1**:607.

32. Xiao X, Song H, Lin S, Zhou Y, Zhan X, Hu Z, Zhang Q, Sun J, Yang B, Li T, Jiao L, Zhou J, Tang J, Gogotsi Y. Scalable salt-templated synthesis of two-dimensional transition metal oxides. *Nat Commun* 2016;**7**:11296.

33. Zhai T, Fang X, Liao M, Xu X, Zeng H, Yoshio B, Golberg D. A comprehensive review of one-dimensional metal-oxide nanostructure photodetectors. *Sensors* 2009;**9**:6504–29.

34. Tian W, Wang Y, Chen L, Li L. Self-Powered nanoscale photodetectors. *Small* 2017;**13**: 1701848.

35. Selman AM, Hassan Z. Highly sensitive fast-response UV photodiode fabricated from rutile TiO$_2$ nanorod array on silicon substrate. *Sensor Actuator Phys* 2015;**221**:15–21.

36. Meng M, Wu X, Ji X, Gan Z, Liu L, Shen J, Chu PK. Ultrahigh quantum efficiency photodetector and ultrafast reversible surface wettability transition of square In$_2$O$_3$ nanowires. *Nano Res* 2017;**10**:2772–81.

37. Pan X, Zhang T, Lu Q, Wang W, Ye Z. High responsivity ultraviolet detector based on novel SnO$_2$ nanoarrays. *RSC Adv* 2019;**9**:37201–6.

38. Choi H, Seo S, Lee J-H, Hong S-H, Song J, Kim S, Yim S-Y, Lee K, Park S-J, Lee S. Solution-processed ZnO/SnO$_2$ bilayer ultraviolet phototransistor with high responsivity and fast photoresponse. *J Mater Chem C* 2018;**6**:6014–22.

39. Ma H, Liu K, Cheng Z, Zheng Z, Liu Y, Zhang P, Chen X, Liu D, Liu L, Shen D. Speed enhancement of ultraviolet photodetector base on ZnO quantum dots by oxygen adsorption on surface defects. *J Alloys Compd* 2021;**868**:159252.

40. Ouyang W, Teng F, He J-H, Fang X. Enhancing the photoelectric performance of photodetectors based on metal oxide semiconductors by charge-carrier engineering. *Adv Funct Mater* 2019;**29**:1807672.

41. Rasool A, Santhosh Kumar MC, Mamat MH, Gopalakrishnan C, Amiruddin R. Analysis on different detection mechanisms involved in ZnO-based photodetector and photodiodes. *J Mater Sci Mater Electron* 2020;**31**:7100−13.

42. Zhang F, Hou P-X, Liu C, Cheng H-M. Epitaxial growth of single-wall carbon nanotubes. *Carbon* 2016;**102**:181−97.

43. Rahman G, Najaf Z, Mehmood A, Bilal S, Shah AuHA, Mian SA, Ali G. An overview of the recent progress in the synthesis and applications of carbon nanotubes. *Chimia* 2019;**5**:3.

44. Yu L, Shearer C, Shapter J. Recent development of carbon nanotube transparent conductive films. *Chem Rev* 2016;**116**:13413−53.

45. Yu M-F, Files BS, Arepalli S, Ruoff RS. Tensile loading of ropes of single wall carbon nanotubes and their mechanical properties. *Phys Rev Lett* 2000;**84**:5552−5.

46. Ghalandari M, Maleki A, Haghighi A, Safdari Shadloo M, Alhuyi Nazari M, Tlili I. Applications of nanofluids containing carbon nanotubes in solar energy systems: a review. *J Mol Liq* 2020;**313**:113476.

47. Dresselhaus MS, Dresselhaus G, Jorio A. Unusual properties and structure of carbon nanotubes. *Annu Rev Mater Res* 2004;**34**:247−78.

48. Saifuddin N, Raziah AZ, Junizah AR. Carbon nanotubes: a review on structure and their interaction with proteins. *J Chem* 2013;**2013**:676815.

49. Sun J-L, Wei J, Zhu J-L, Xu D, Liu X, Sun H, Wu D-H, Wu N-L. Photoinduced currents in carbon nanotube/metal heterojunctions. *Appl Phys Lett* 2006;**88**:131107.

50. Xu J, Sun J-L, Wei J, Xu J. The wavelength dependent photovoltaic effects caused by two different mechanisms in carbon nanotube film/CuO nanowire array heterodimensional contacts. *Appl Phys Lett* 2012;**100**:251113.

51. Jain B, Hashmi A, Sanwaria S, Singh AK, Susan MABH, Singh A. Zinc oxide nanoparticle incorporated on graphene oxide: an efficient and stable photocatalyst for water treatment through the Fenton process. *Adv Compos Hybrid Mater* 2020;**3**:231−42.

52. Ong CB, Mohammad AW, Ng LY, Mahmoudi E, Azizkhani S, Hayati Hairom NH. Solar photocatalytic and surface enhancement of ZnO/rGO nanocomposite: degradation of perfluorooctanoic acid and dye. *Process Saf Environ Protect* 2017;**112**:298−307.

53. Dai K, Zhang X, Fan K, Zeng P, Peng T. Multiwalled carbon nanotube-TiO_2 nanocomposite for visible-light-induced photocatalytic hydrogen evolution. *J Nanomater* 2014;**2014**:694073.

54. Yu J, Huang T, Jiang Z, Sun M, Tang C. Synthesis and characterizations of zinc oxide on reduced graphene oxide for high performance electrocatalytic reduction of oxygen. *Molecules* 2018;**23**:3227.

55. Yin B, Zhang S, Ke K, Xiong T, Wang Y, Lim BKD, Lee WSV, Wang Z, Xue J. Binderfree V2O5/CNT paper electrode for high rate performance zinc ion battery. *Nanoscale* 2019;**11**:19723−8.

56. Olowoyo JO, Kumar M, Jain SL, Babalola JO, Vorontsov AV, Kumar U. Insights into reinforced photocatalytic activity of the CNT−TiO_2 nanocomposite for CO_2 reduction and water splitting. *J Phys Chem C* 2019;**123**:367−78.

57. Shaban M, Ashraf AM, Abukhadra MR. TiO_2 nanoribbons/carbon nanotubes composite with enhanced photocatalytic activity; fabrication, characterization, and application. *Sci Rep* 2018;**8**:781.

58. Zhao Y, Zhang J, Wang Y, Chen Z. A highly sensitive and room temperature CNTs/SnO2/CuO sensor for H_2S gas sensing applications. *Nanoscale Res Lett* 2020;**15**:40.

59. Lin Y-H, Lee P-S, Hsueh Y-C, Pan K-Y, Kei C-C, Chan M-H, Wu J-M, Perng T-P, Shih HC. Atomic layer deposition of zinc oxide on multiwalled carbon nanotubes for UV photodetector applications. *J Electrochem Soc* 2011;**158**:K24.

60. Kim DS, Lee S-M, Scholz R, Knez M, Gösele U, Fallert J, Kalt H, Zacharias M. Synthesis and optical properties of ZnO and carbon nanotube based coaxial heterostructures. *Appl Phys Lett* 2008;**93**:103108.

61. Xie Y, Wei L, Wei G, Li Q, Wang D, Chen Y, Yan S, Liu G, Mei L, Jiao J. A self-powered UV photodetector based on TiO$_2$ nanorod arrays. *Nanoscale Res Lett* 2013;**8**:188.

62. Zhou H, Song Z, Tao P, Lei H, Gui P, Mei J, Wang H, Fang G. Self-powered, ultraviolet-visible perovskite photodetector based on TiO$_2$ nanorods. *RSC Adv* 2016;**6**:6205−8.

63. Wang Y, Cheng J, Shahid M, Zhang M, Pan W. A high-performance TiO$_2$ nanowire UV detector assembled by electrospinning. *RSC Adv* 2017;**7**:26220−5.

64. Zou J, Zhang Q, Huang K, Marzari N. Ultraviolet photodetectors based on anodic TiO$_2$ nanotube Arrays. *J Phys Chem C* 2010;**114**:10725−9.

65. Wang L, Yang W, Chong H, Wang L, Gao F, Tian L, Yang Z. Efficient ultraviolet photodetectors based on TiO$_2$ nanotube arrays with tailored structures. *RSC Adv* 2015;**5**:52388−94.

66. Abdi Y, Khalilian M, Arzi E. Enhancement in photo-induced hydrophilicity of TiO$_2$/CNT nanostructures by applying voltage. *J Phys Appl Phys* 2011;**44**:255405.

67. Yang M, Zhu J-L, Liu W, Sun J-L. Novel photodetectors based on double-walled carbon nanotube film/TiO2 nanotube array heterodimensional contacts. *Nano Res* 2011;**4**:901−7.

68. Yang M, Xu J, Wei J, Sun J-L, Liu W, Zhu J-L. Fabrication of double-walled carbon nanotube film/Cu$_2$O nanoparticle film/TiO$_2$ nanotube array heterojunctions for photosensors. *Appl Phys Lett* 2012;**100**:253113.

69. Chen RJ, Franklin NR, Kong J, Cao J, Tombler TW, Zhang Y, Dai H. Molecular photodesorption from single-walled carbon nanotubes. *Appl Phys Lett* 2001;**79**:2258−60.

70. Chen Y, Zhang G, Dong Z, Wei J, Zhu J-L, Sun J-L. Fabrication of Au nanoparticle/double-walled carbon nanotube film/TiO$_2$ nanotube array/Ti heterojunctions with low resistance state for broadband photodetectors. *Phys B Condens Matter* 2017;**508**:1−6.

71. Hsu C-Y, Lien D-H, Lu S-Y, Chen C-Y, Kang C-F, Chueh Y-L, Hsu W-K, He J-H. Supersensitive, ultrafast, and broad-band light-harvesting scheme employing carbon nanotube/TiO$_2$ core−shell nanowire geometry. *ACS Nano* 2012;**6**:6687−92.

72. Mollow E. In: Breckenridge RG, editor. *Proceedings of the photoconductivity conference.* New York, NY, USA: Wiley; 1954. p. 509.

73. Chang S-P, Chen K-J. Zinc oxide nanoparticle photodetector. *J Nanomater* 2012;**2012**:602398.

74. Deka Boruah B, Misra A. Energy-efficient hydrogenated zinc oxide nanoflakes for high-performance self-powered ultraviolet photodetector. *ACS Appl Mater Interfaces* 2016;**8**:18182−8.

75. Saleh Al-Khazali SM, Al-Salman HS, Hmood A. Low cost flexible ultraviolet photodetector based on ZnO nanorods prepared using chemical bath deposition. *Mater Lett* 2020;**277**:128177.

76. Butanovs E, Piskunov S, Zolotarjovs A, Polyakov B. Growth and characterization of PbI$_2$-decorated ZnO nanowires for photodetection applications. *J Alloys Compd* 2020;**825**:154095.

77. Choi M-S, Park T, Kim W-J, Hur J. High-performance ultraviolet photodetector based on a zinc oxide Nanoparticle@Single-walled carbon nanotube heterojunction hybrid film. *Nanomaterials* 2020;**10**:395.

78. Li G, Suja M, Chen M, Bekyarova E, Haddon RC, Liu J, Itkis ME. Visible-blind UV photodetector based on single-walled carbon nanotube thin film/ZnO vertical heterostructures. *ACS Appl Mater Interfaces* 2017;**9**:37094−104.

79. Xiong D, Deng W, Tian G, Gao Y, Chu X, Yan C, Jin L, Su Y, Yan W, Yang W. A piezo-phototronic enhanced serrate-structured ZnO-based heterojunction photodetector for optical communication. *Nanoscale* 2019;**11**:3021—7.

80. Wang X, Xu K, Yan X, Xiao X, Aruta C, Foglietti V, Ning Z, Yang N. Amorphous ZnO/PbS quantum dots heterojunction for efficient responsivity broadband photodetectors. *ACS Appl Mater Interfaces* 2020;**12**:8403—10.

81. Ates ES, Kucukyildiz S, Unalan HE. Zinc oxide nanowire photodetectors with single-walled carbon nanotube thin-film electrodes. *ACS Appl Mater Interfaces* 2012;**4**:5142—6.

82. Shao D, Yu M, Lian J, Sawyer SML. Ultraviolet photodetector fabricated from multiwalled carbon nanotubes/zinc-oxide nanowires/p-GaN composite structure. *IEEE Electron Device Lett* 2013;**34**:1169—71.

83. Lupan O, Schütt F, Postica V, Smazna D, Mishra YK, Adelung R. Sensing performances of pure and hybridized carbon nanotubes-ZnO nanowire networks: a detailed study. *Sci Rep* 2017;**7**:14715.

84. Li XL, Li C, Zhang Y, Chu DP, Milne WI, Fan HJ. Atomic layer deposition of ZnO on multi-walled carbon nanotubes and its use for synthesis of CNT—ZnO heterostructures. *Nanoscale Res Lett* 2010;**5**:1836.

85. Johnson RW, Hultqvist A, Bent SF. A brief review of atomic layer deposition: from fundamentals to applications. *Mater Today* 2014;**17**:236—46.

86. Liu Q, Gong M, Cook B, Ewing D, Casper M, Stramel A, Wu J. Transfer-free and printable graphene/ZnO-nanoparticle nanohybrid photodetectors with high performance. *J Mater Chem C* 2017;**5**:6427—32.

87. Cheng W, Niederberger M. Evaporation-induced self-assembly of ultrathin tungsten oxide nanowires over a large scale for ultraviolet photodetector. *Langmuir* 2016;**32**:2474—81.

88. Majumder R, Kundu S, Ghosh R, Pradhan M, Ghosh D, Roy S, Roy S, Pal Chowdhury M. Expeditious UV detection of tungstite ($WO_3 \cdot H_2O$) and tungsten oxide (WO_3) decorated multiwall carbon nanotubes (MWCNT) based photodetector: ultrafast response and recovery time. *SN Appl Sci* 2019;**2**:81.

Index

Printed in the United States
by Baker & Taylor Publisher Services